D1309674

DATE		
MAY 7 1990		
Aug 27 90		
MAR 17 '95		

The Biochemistry of Plants:

A COMPREHENSIVE TREATISE

Volume 1

P. K. Stumpf and E. E. Conn

EDITORS-IN-CHIEF

*Department of Biochemistry
and Biophysics
University of California
Davis, California*

THE BIOCHEMISTRY OF PLANTS

A COMPREHENSIVE TREATISE

Volume 1
The Plant Cell

N. E. Tolbert, editor

Department of Biochemistry
Michigan State University
East Lansing, Michigan

1980

ACADEMIC PRESS

A Subsidiary of Harcourt Brace Jovanovich, Publishers
New York   London   Toronto   Sydney   San Francisco

ACADEMIC PRESS, INC.
111 Fifth Avenue, New York, New York 10003

United Kingdom Edition published by
ACADEMIC PRESS, INC. (LONDON) LTD.
24/28 Oval Road, London NW1   7DX

Library of Congress Cataloging in Publication Data
Main entry under title:

The Biochemistry of plants.

    Includes bibliographies and index.
    CONTENTS:
——v. 2. Metabolism and respiration.
    1. Botanical chemistry. I. Stumpf, Paul Karl,
Date. II. Conn, Eric E.
QK861.B48   581.19'2   80–13168
ISBN 0–12–675401–2 (v. 1)

PRINTED IN THE UNITED STATES OF AMERICA

80 81 82 83   9 8 7 6 5 4 3 2 1

Contents

3   The Primary Cell Walls of Flowering Plants
ALAN DARVILL, MICHAEL McNEIL, PETER
ALBERSHEIM, AND DEBORAH P. DELMER

4   The Plasma Membrane
ROBERT T. LEONARD AND THOMAS K. HODGES

List of Contributors

Numbers in parentheses indicate the pages on which the authors' contributions begin.

Peter Albersheim (91), Department of Chemistry, University of Colorado, Boulder, Colorado 80309

Daniel Branton (625), The Biological Laboratories, Harvard University, Cambridge, Massachusetts 02138

Peter S. Carlson (55), Department of Crop and Soil Sciences, Michigan State University, East Lansing, Michigan 48824

Maarten J. Chrispeels (389), Department of Biology, University of California at San Diego, La Jolla, California 92093

Alan Darvill (91), Department of Chemistry, University of Colorado, Boulder, Colorado 80309

Eric Davies (413), School of Life Sciences, University of Nebraska, Lincoln, Nebraska 68588

D. A. Day (315), Department of Botany, University of Illinois, Urbana, Illinois 61801

Deborah P. Delmer (91), MSU-DOE Plant Research Laboratory and Department of Biochemistry, Michigan State University, East Lansing, Michigan 48824

J. B. Hanson (315), Department of Botany, University of Illinois, Urbana, Illinois 61801

Thomas K. Hodges (163), Department of Botany and Plant Pathology, Purdue University, West Lafayette, Indiana 47907

Richard G. Jensen (273), Department of Biochemistry, University of Arizona, Tucson, Arizona 85721

E. G. Jordan (489), Biology Department, Queen Elizabeth College, Campden Hall, London W8 7AH, England

Grahame J. Kelly (183), Botanisches Institut der Universität, Schlossgarten 3, D-4400 Münster, West Germany

Brian A. Larkins (413), Department of Botany and Plant Pathology, Purdue University, West Lafayette, Indiana 47907

Erwin Latzko (183), Botanisches Institut der Universität, Schlossgarten 3, D-4400 Münster, West Germany

Roger A. Leigh* (625), Botany School, University of Cambridge, Cambridge CB2 3EA, England

Robert T. Leonard (163), Department of Botany and Plant Sciences, University of California, Riverside, California 92521

John N. A. Lott (589), Department of Biology, McMaster University, Hamilton, Ontario L8S 4K1, Canada

Paul Ludden (55), Department of Biochemistry, University of California, Riverside, California 92521

Michael McNeil (91), Department of Chemistry, University of Colorado, Boulder, Colorado 80309

Francis Marty (625), The Biological Laboratories, Harvard University, Cambridge, Massachusetts 02138

Hilton H. Mollenhauer (437), Veterinary Toxicology and Entomology Research Laboratory, United States Department of Agriculture, SEA/AR, College Station, Texas 77840

D. James Morré (437), Department of Medicinal Chemistry, Purdue University, West Lafayette, Indiana 47907

Eldon H. Newcomb (1), Botany Department, University of Wisconsin, Madison, Wisconsin 53706

Jerome A. Schiff (209), Institute for Photobiology of Cells and Organelles, Brandeis University, Waltham, Massachusetts 02154

J. N. Timmis (489), Department of Botany, University College, Dublin, Ireland

N. E. Tolbert (359), Department of Biochemistry, Michigan State University, East Lansing, Michigan 48824

A. J. Trewavas (489), Department of Botany, University of Edinburgh, Edinburgh EH 3JH, Scotland

C. Peter Wolk (659), MSU-DOE Plant Research Laboratory, Michigan State University, East Lansing, Michigan 48824

---

* Present address: Department of Soils and Plant Nutrition, Rothamsted Experimental Station, Harpenden, Hertfordshire, England.

General Preface

In 1950, James Bonner wrote the following prophetic comments in the Preface of the first edition of his "Plant Biochemistry" published by Academic Press:

> There is much work to be done in plant biochemistry. Our understanding of many basic metabolic pathways in the higher plant is lamentably fragmentary. While the emphasis in this book is on the higher plant, it will frequently be necessary to call attention to conclusions drawn from work with microorganisms or with higher animals. Numerous problems of plant biochemistry could undoubtedly be illuminated by the closer application of the information and the techniques which have been developed by those working with other organisms . . .
>
> Certain important aspects of biochemistry have been entirely omitted from the present volume simply because of the lack of pertinent information from the domain of higher plants.

The volume had 30 chapters and a total of 490 pages. Many of the biochemical examples cited in the text were derived from studies on bacterial, fungal, and animal systems. Despite these shortcomings, the book had a profound effect on a number of young biochemists since it challenged them to enter the field of plant biochemistry and to correct "the lack of pertinent information from the domain of higher plants."

Since 1950, an explosive expansion of knowledge in biochemistry has occurred. Unfortunately, the study of plants has had a mixed reception in the biochemical community. With the exception of photosynthesis, biochemists have avoided tackling for one reason or another the incredibly interesting problems associated with plant tissues. Leading biochemical journals have frequently rejected sound manuscripts for the trivial reason that the reaction had been well described in *E. coli* and liver tissue and thus was of little interest to again describe its presence in germinating pea seeds! Federal granting agencies, the National Science Foundation excepted, have also been reluctant to fund applications when it was indicated that the principal experimental tissue would be of plant origin despite the fact that the most prevalent illness in the world is starvation.

The second edition of "Plant Biochemistry" had a new format in 1965 when J. Bonner and J. Varner edited a multiauthored volume of 979 pages; in 1976, the third edition containing 908 pages made its appearance. A few textbooks of limited size in plant biochemistry have been published. In addition, two continuing series resulting from the annual meetings and symposia of phytochemical organizations in Europe and in North America provided the biological community with highly specialized articles on many topics of plant biochemistry. Plant biochemistry was obviously growing.

Although these publications serve a useful purpose, no multivolume series in plant biochemistry has been available to the biochemist trained and working in different fields who seeks an authoritative overview of major topics of plant biochemistry. It therefore seemed to us that the time was ripe to develop such a series. With encouragement and cooperation of Academic Press, we invited six colleagues to join us in organizing an eight volume series to be known as "The Biochemistry of Plants: A Comprehensive Treatise." Within a few months, we were able to invite over 160 authors to write authoritative chapters for these eight volumes.

Our hope is that this Treatise not only will serve as a source of current information to researchers working in plant biochemistry, but equally important will provide a mechanism for the molecular biologist who works with *E. coli* or the neurobiochemist to become better informed about the interesting and often unique problems which the plant cell provides. It is hoped, too, the senior graduate student will be inspired by one or more comments in chapters of this Treatise and will orient his future career to some aspect of this science.

Despite the fact that many subjects have been covered in this Treatise, we make no claim to have been complete in our coverage nor to have treated all subjects in equal depth. Notable is the absence of volumes on phytohormones and on mineral nutrition. These areas, which are more closely associated with the discipline of plant physiology, are treated in multivolume series in the physiology literature and/or have been the subject of specialized treatises. Other topics (e.g., alkaloids, nitrogen fixation, flavonoids, plant pigments) have been assigned single chapters even though entire volumes, sometimes appearing on an annual basis, are available.

Finally, we wish to thank all our colleagues for their enthusiastic cooperation in bringing these eight volumes so rapidly into fruition. We are grateful to Academic Press for their gentle persuasive pressures and we are indebted to Ms. Barbara Clover and Ms. Billie Gabriel for their talented assistance in this project.

P. K. Stumpf
E. E. Conn

Preface to Volume 1

The organization of this volume was based on the precept of devoting one chapter to each subcellular compartment of a plant cell. Each chapter therefore serves as introduction to the various parts of the cell and to the basic biochemistry carried out in the different subcellular components. Subsequent volumes will go into details about these biochemical processes, such as respiration involving the mitochondria, microbodies or cytosol, or photosynthesis in the chloroplasts. Chapter length is due in part to self restraint by the authors, as there was little attempt to set arbitrary importance among the various cellular compartments.

The term subcellular compartment has in general been used rather than subcellular organelle, because many parts of a cell can hardly be considered as an organelle. Thus the cytosol is not a specific organelle, yet the cytosol is a subcellular compartment bounded by the cell membrane and the other organelles. Perhaps when a second edition of these volumes appears, more will be known about the organization of macromolecular complexes within the cytosol. Knowledge of the vacuole, as a lysosomal compartment, has developed rapidly, but the question still remains as to whether it is a compartment or a particle in the plant cell. For the cell wall and membrane, the term "cell component" is more appropriate than organelle. The cell wall has been extensively investigated by plant biochemists and much that is known about it is presented in Chapter 4 of this volume and continued in Volume 3 on carbohydrates. One subcellular component, the microtubules, is not included in this volume.

Each portion of the continuous flow of carbon among many compounds in the cells has been called a specific metabolic pathway and is to be found in different locations of the cell. These spatial divisions necessitate regulated transport between compartments. Only to a limited extent has transport in and out of organelles been considered in Volume 1, but investigators are aware that this is an important and growing area of research. Included within this purview are translocases of the chloroplast and mitochondrial membranes and the role of ATPase in chemiosmotic reaction systems.

There are examples of duplicate metabolic pathways in the cell which are spatially separated by compartments. Within each compartment mass carbon flow must have a direction dictated by thermodynamic considerations and catalyzed by enzymes located in that compartment. There may be a pathway in one compartment for catabolism, such as glycolyis in the cytoplasm, and a similar pathway in another organelle for synthesis, such as the reductive photosynthetic carbon cycle in the chloroplasts. Another example of similar pathways is the tricarboxylic acid cycle in the mitochondria and the glyoxylate cycle in the microbody. Enzymes catalyzing the same reaction in the different locations may be isoenzymic or different from each other, such as multiple NAD:malate dehydrogenases isoenzymes and different aminotransferases. When common reactions are catalyzed by enzymes with a different cofactor specificity (i.e., NADP:malate dehydrogenase versus NAD:malate dehydrogenase), they are considered different enzymes. Exactly how the various isoenzymes are derived and directed to their specific compartment is not clear. The possibility of identical primary amino acid sequence at the active sites of these different enzymes and isoenzymes is one example of additional problems to occupy the biochemists.

Although this volume indicates that plant scientists are making progress at breaking the cell apart and examining its various components, successful regulation of the cell and plant growth remains a complex task for the future.

N. E. Tolbert

The General Cell | 1

ELDON H. NEWCOMB

The Biochemistry of Plants, Vol. 1

I. INTRODUCTION

A. Prokaryotes and Eukaryotes

It is now recognized that in their structural organization, all organisms belong to one of two distinct types—the prokaryotes, including bacteria and blue–green algae, and the eukaryotes, comprising all other plants and animals.

Prokaryotes (*pro:* before; *karyon:* nucleus) lack a membrane-bounded cell nucleus. The contents of the cell are dispersed into two recognizably distinct regions, the cytoplasm and the nucleoid, the latter consisting of one or more less-dense regions containing fine DNA fibrils. Only three cellular components occur in all prokaryotic cells: plasma membrane, ribosomes, and nucleoid (Avers, 1976). A cell wall is usually present but is absent in the mycoplasmas. Prokaryotes lack microtubules and do not divide by mitosis, rather by binary fission or other means. They are gen rally smaller than eukaryotic cells, and mostly fall in the size range from 1–10 μm.

In Eukaryotes (*eu:* true; *karyon:* nucleus) the nuclear material is enclosed within a pair of membranes, the nuclear envelope. The cells are generally larger than those of prokaryotes, usually ranging from 10–100 μm. Cell division is by a mitotic process involving microtubules. The cytoplasm is highly compartmentalized by membranous systems.

General recognition of the existence of these two fundamentally different kinds of cellular organization dates only from the early 1960s. However, according to Lwoff (1971), credit for drawing the distinction between prokaryotes and eukaryotes should go to the French microbiologist Edouard Chatton, who first used the two terms in 1928.

The blue–green algae (Cyanophyta) contain chlorophyll a, and in this respect resemble eukaryotes. However, blue–green algae contain neither chlorophyll b nor c, whereas either b or c is found in all photosynthetic eukaryote groups except the Rhodophyta (red algae). The ancillary pigments in blue–green algae are the water-soluble linear tetrapyrroles, phycocyanin and phycoerythrin. In association with specific proteins, these occur in small particles termed phycobilisomes. The phycobilisomes are loosely attached to the external surfaces of the internal membranes, or thylakoids, of the blue–green algal cell. They are usually small and closely packed, and cannot always be observed (Coombs and Greenwood, 1976).

In recent years an anomalous group of algae has been discovered (Lewin, 1975; Newcomb and Pugh, 1975) whose members possess prokaryotic cellular organization (Fig. 1) but resemble eukaryotes in their photosynthetic pigments. These algae not only contain chlorophyll b as well as a, but lack the phycobilin and phycobilisomes typical of blue–green algae (Lewin and

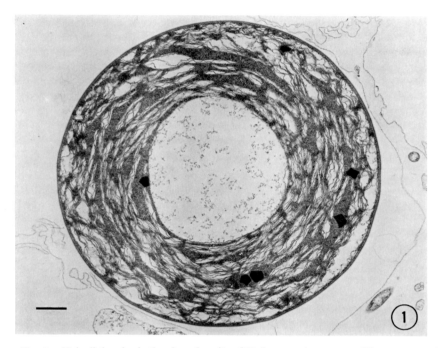

Fig. 1. Unicellular alga in the cloacal cavity of *Diplosoma virens,* an ascidian common along the Great Barrier Reef. Prokaryotic in organization, the alga lacks a nucleus and organelles. Scale bar = 1 μm ($\times$ 8000). Micrograph by T. D. Pugh.

Withers, 1975; Thorne *et al.,* 1977). The algae, which are spherical and unicellular, inhabit the cloacal cavities of a few species of colonial ascidians of the Great Barrier Reef and other coral reefs in the tropical waters of the Pacific and Indian Oceans. Photosynthetic assimilation in these algae appears to proceed via the initial carboxylation of ribulose-1,5-P_2, and the principal product of carbon fixation is an α1-4 glucan (Akazawa *et al.,* 1978). The algae have been placed in a new genus (*Prochloron*) in a new phylum (Prochlorophyta) (Lewin, 1977), but comparisons of base pair sequences will have to be made before conclusions can be drawn regarding their evolutionary affinities.

B. Cell Organelles

One of the major contributions of electron microscopy to biology has been the revelation that eukaryotic cells, both plant and animal, possess for the most part the same fundamental ultrastructural components. Diversity of cellular structure and function is achieved through alterations in the

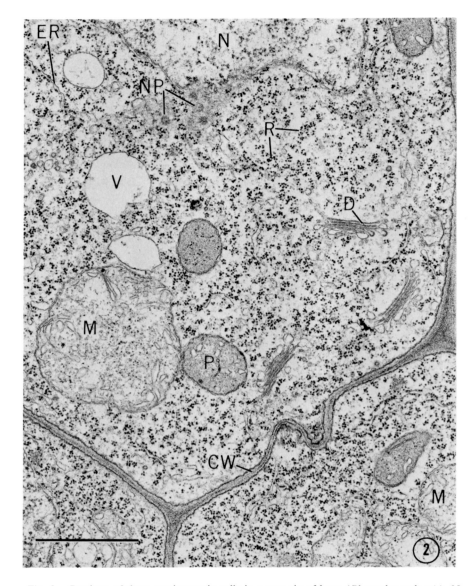

Fig. 2. Portions of three meristematic cells in a root tip of bean (*Phaseolus vulgaris*). N, nucleus; NP, nuclear pores; CW, primary cell wall; M, mitochondrion, P, proplastid; ER, rough endoplasmic reticulum; D, dictyosome; V, vacuole; R, ribosomes. Scale bar = 1 μm ($\times$ 27,000).

amounts, structural details, and biochemical activities of these few kinds of components.

A number of the components of the cell are considered to be *organelles,* diminutive intracellular organs of characteristic structure and function. Several of the organelles are membranous and, indeed, the extent to which the cell employs membrane systems to accomplish its ends is remarkable. Some of the organelles are not, however, constructed of membranes; microtubules and ribosomes are good examples.

The young, undifferentiated meristematic cells of plants contain a basic set of these organelles and associated ultrastructural components, some of which are seen in the portion of a meristematic cell shown in Fig. 2.

Meristematic cells are found in the embryonic tissue zones that give plants their unique, open type of growth. Cell division continues in these meristems while cells in other parts of the plant differentiate and reach maturity. Cells in the apical meristems at the tips of root and shoot are small, i.e., usually no more than 20–40 μm on a side. They tend to be isodiametric, and have thin walls which at this stage possess little mechanical rigidity. Lining the wall on the inside is the plasma membrane (or plasmalemma), the outer boundary of the living protoplast. The protoplast consists of nucleus and cytoplasm. The nucleus in meristematic cells is about 10 μm in diameter and usually occupies the center of the cell. Vacuoles are mostly quite minute at this stage.

A few estimates have been made of the numbers of organelles in meristematic cells. In *Epilobium* there are 50–120 mitochondria in a young cell of the shoot apex (Anton-Lamprecht, 1967). Central cells of the root cap of maize have about 200 mitochondria when young (Clowes and Juniper, 1968). There are 10–20 proplastids and 20–35 dictyosomes in meristematic cells of the shoot apex of *Epilobium* (Anton-Lamprecht, 1967).

II. THE PLASMA MEMBRANE

The plasma membrane is the differentially permeable barrier that separates the living cell from the external environment. When cells are fixed and stained by the usual methods for electron microscopic examination, cross sections of the plasma membrane commonly reveal a "unit membrane" structure (Robertson, 1967). This means that the membrane has a "railroad track" appearance, where two dark lines, each about 25 Å thick, are separated by a light space about 40 Å thick (Fig. 3). The interpretation according to the unit membrane hypothesis is that a layer of protein coats both surfaces of a bilayer of lipid molecules. Although the fluid mosaic model of membrane structure (Singer and Nicolson, 1972) has superseded the unit membrane hypothesis as a better interpretation of the facts, it is still convenient to refer to the "unit membrane structure" of a membrane that appears as described

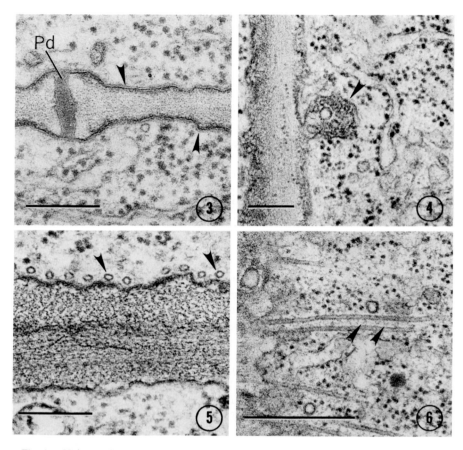

Fig. 3. Unit membrane structure visible in the plasma membranes (arrowheads) next to the primary wall separating two cells in a bean root. Pd, plasmodesma. Scale bar = 0.2 μm (× 98,000).

Fig. 4. Plasmalemmasome (arrowhead) blebbing from a primary wall in a bean root cell. Scale bar = 0.2 μm (× 60,000).

Fig. 5. Cortical microtubules (arrowheads) seen in transverse section near a growing primary side wall in a bean root tip. Scale bar = 0.2 μm (× 96,000). Micrograph by B. A. Palevitz.

Fig. 6. Cross bridges (arrowheads) between microtubules near a side wall in an enlarging cell in a root of tobacco (*Nicotiana tabacum*). The wall has been sectioned obliquely. Scale bar = 0.5 μm (× 58,000). Micrograph by E. L. Vigil.

above. As ordinarily fixed, however, not all membranes in the cell exhibit unit membrane structure.

Care must be used not to misinterpret the *pairs* of membranes (or envelope) surrounding the nucleus, mitochondria, and plastids as the unit membrane structure of a single membrane. It is helpful, when in doubt as to whether an organelle is surrounded by an envelope or only a single membrane, to note the magnification of the micrograph and also to examine closely the appearance of nearby membranes, since the envelope will be several times as thick as a unit membrane. As illustrated in Fig. 2, the two membranes of an envelope can be readily distinguished at a magnification too low to see the unit membrane structure of the membranes.

Various infoldings and inrollings of the plasma membrane into the cytoplasm, some containing vesiclelike inclusions (Fig. 4), have been described from time to time in plant cells from many species (Marchant and Robards, 1968; Cox and Juniper, 1973). These structures are known by several names, including *lomasomes, paramural bodies,* and *plasmalemmasomes;* at least some of them are probably artifactual.

III. MICROTUBULES

Microtubules are long, slender, unbranched, cylindrical structures of widespread occurrence in the cytoplasm of plant and animal cells (Newcomb, 1969; Hepler and Palevitz, 1974). They are about 24 nm thick, with an electron-lucent core about 10 nm across and an electron-opaque cortex or wall about 7 nm thick (Fig. 5). On the outside, arms or bridges can sometimes be seen that link them to each other or to other structures (Fig. 6). They appear to be rather stiff and rigid structures of variable lengths up to many micrometers, and they usually follow straight or gently curving paths through the cytoplasm. They are built up of subunits of a group of closely related proteins known as tubulins (Snyder and McIntosh, 1976). Colchicine acts as a potent inhibitor of microtubule polymerization by combining with the subunits. The Vinca alkaloids, vinblastine and vincristine, from periwinkle (*Vinca minor*) have effects somewhat similar to those of colchicine.

In many animal and protistan cells microtubules play important cytoskeletal roles in controlling cell asymmetries. In most plant cells, surrounded as they are by a rigid wall, microtubules do not have a directly comparable cytoskeletal function but do nevertheless engage in several important activities. In dividing nuclei, bundles of microtubules (as the spindle and chromosomal fibers) are instrumental in moving the chromosomes to the poles. During cell division following mitosis, they form the fibers of the phragmoplast that controls the orderly deposition of cell plate materials (Fig. 7). Then, in enlarging cells, microtubules appear aligned in the cortical cytoplasm just beneath the growing wall and presumably control the orientation of cellulose microfibrils during their deposition in the wall. In those cell types where secondary wall thickenings are deposited after primary wall growth ceases, microtubules are present in the cytoplasm next to the developing wall bands, and may play a similar orienting role.

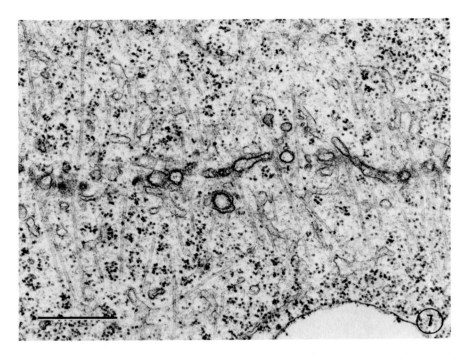

Fig. 7. Cell plate formation in a dividing cell of a bean root tip. The numerous microtubules run out at right angles from the plate region. Scale bar = 0.5 μm (× 42,000).

IV. PLASMODESMATA

Plasmodesmata are narrow, more or less cylindrical strands of cytoplasm that connect neighboring plant cells by penetrating through the intervening cell walls (Gunning, 1976) (Figs. 8–11). Their internal diameters fall generally between 30 and 60 nm. They are always bounded by the plasmalemma (Figs. 9 and 11), which is continuous from one cell to the other. Ordinarily a fine, tubulelike structure, termed the desmotubule, also runs through a plasmodesma from a nearby cisterna of endoplasmic reticulum in one cell to a similar cisterna in the other. The desmotubule, which is 16–20 nm thick, frequently contains an axial central rod (Figs. 9 and 11). It is considered to be a modification of the endoplasmic reticulum (Robards, 1976).

Plasmodesmata connect most of the living cells of a higher plant (Robards, 1976). In the apt expression of Gunning (1976), plasmodesmata "elevate a plant from a mere collection of individual cells to an interconnected commune of living protoplasts." By diffusion or bulk flow of materials through them, plasmodesmata may serve to nourish cells and tissues remote from direct sources of nutrient. They are also potential pathways for the passage

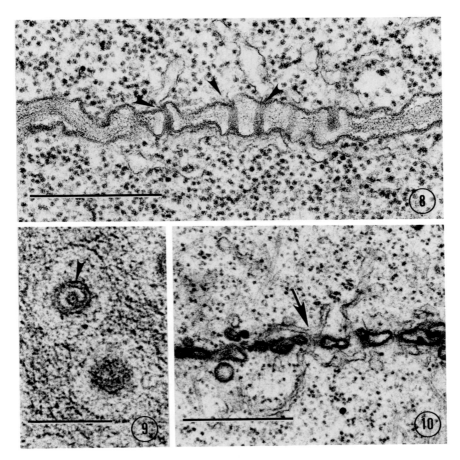

Fig. 8. Several plasmodesmata running through a primary side wall between young adjacent cells in a bean root. Entry of endoplasmic reticulum into the plasmodesmata is indicated by arrowheads. Scale bar = 0.5 μm ($\times$ 70,000). Micrograph by W. P. Wergin.

Fig. 9. Structure of a plasmodesma as seen in transverse view. Note unit membrane structure of the plasma membrane (arrowhead), and presence of an electron opaque central rod within the desmotubule. Scale bar = 0.1 μm ($\times$ 225,000).

Fig. 10. Early stage in plasmodesma formation in a dividing cell of a bean root. At arrow, a cisterna of endoplasmic reticulum runs from one daughter protoplast to the other through a gap between consolidating masses of plate material. Scale bar = 0.5 μm ($\times$ 58,000). Micrograph by P. K. Hepler.

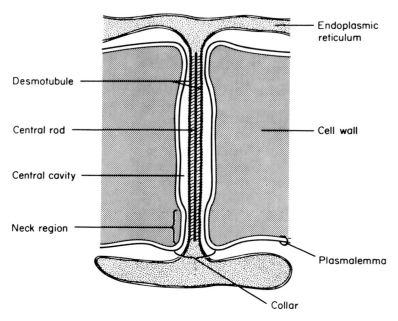

Fig. 11. Diagram of the components of a simple plasmodesma as seen in longitudinal section. Modified from Robards (1976).

of electrical or hormonal stimuli that might serve to regulate and coordinate the activities of different parts of the plant body. According to Gunning (1976), their frequency varies from less than one to more than 15 per square micrometer of cell wall. Thus they could occupy as much as 1% or more of the cell surface, and might constitute as much surface area of plasmalemma as does the remainder of the bounding membrane of the cell.

It is unclear, however, to what extent plasmodesmata represent open channels of transport between cells. They may possess two different transport channels—the cytoplasm between plasmalemma and desmotubule and an intracisternal channel represented by a lumen within the desmotubule itself. In the first channel, plasmodesmata are commonly restricted at the neck, or entrance from each cell by appression of the inner face of the plasmalemma to the desmotubule (Fig. 11); in the second channel, the exact nature of the desmotubule remains unclear, and whether it represents an opening between adjacent cisternae is questionable (Robards, 1976).

Plasmodesmata are normally formed as the cell plate is being laid down during cytokinesis. Elements of endoplasmic reticulum become oriented at right angles to the developing plate and run from one side of the plate to the other in plasmalemma-delimited isthmuses of cytoplasm isolated by the coalescing masses of plate material (Fig. 10). Continued fusion and growth of the plate material constrict the strands of cytoplasm and isolate the centrally located tubules of endoplasmic reticulum. Presumably microtubules of the phragmoplast are involved in aligning the endoplasmic reticulum, but how the size, frequency, and distribution of the plasmodesmata are controlled during cytokinesis remain obscure (Jones, 1976).

Assuming that secondary formation of plasmodesmata does not take place, their frequency must obviously fall as the meristematic cell enlarges. For example, in the transverse walls of the developing root cap of *Zea mays,* the frequency of plasmodesmata per unit area is estimated to be 4.5 per square micrometer in the meristematic cells, and only 0.81 per square micrometer in the peripheral cells of the cap apex (Clowes and Juniper, 1968).

V. CELL VACUOLES

The origin of the small vacuoles seen in the cytoplasm of young plant cells is unclear. As the cells grow, these vacuoles enlarge and coalesce (Fig. 12), resulting ultimately in a single large central vacuole. In mature storage parenchyma cells, for example, the vacuole may occupy more than 90% of the cell volume, while the cytoplasm constitutes only a thin peripheral layer next to the cell wall.

The bounding membrane of the vacuole, named the tonoplast, resembles the plasmalemma in having a tripartite dark-light-dark appearance, but reacts somewhat differently to the stains used in electron microscopy. Crystals and amorphous materials frequently occur as deposits in vacuoles (Fineran, 1971), while in the cells of flower petals, the anthocyanin pigments are present in vacuoles in solution in the "cell sap."

Maintenance of cell turgidity is one of the principal roles of the vacuole and tonoplast. Normally the high solute concentration in the vacuole (e.g., 0.4–$0.6\,M$) insures that there will be a tendency for water to diffuse into it, causing it to press the surrounding cytoplasm against the cell wall and maintain the cell in a turgid condition. In a young growing cell, the vacuoles tend to swell and coalesce owing to the continual uptake of solutes and accompanying inflow of water. The pressure thus exerted on the cell wall as it is relaxed during growth brings about cell enlargement. As explained in Section VII,B,2, the shape assumed as many cell types enlarge is believed to be controlled by cortical microtubules, since these apparently determine the alignment of new cellulose microfibrils as they are being deposited.

Plant cell vacuoles may also play a digestive role. Animal cells contain lysosomes, single membrane-bounded organelles that contain a large number of acid hydrolases and provide the cells with the capability of digesting all the biologically significant groups of macromolecules (Avers, 1976). These enzymes remain latent until the organelle membrane is damaged or some appropriate substrate enters the organelle. Considerable evidence indicates that lysosomes develop within terminal dilations of Golgi cisternae.

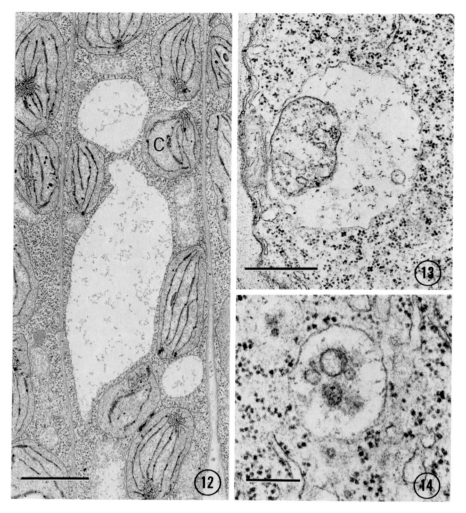

Fig. 12. Vacuoles expanding and coalescing in an enlarging palisade parenchyma cell in a young bean leaf. C, developing chloroplast. Scale bar = 1 μm ($\times$ 89,000). Micrograph by P. J. Gruber.

Fig. 13. Vacuole containing vesicle of cytoplasm in a differentiating protophloem sieve element in a bean root. Indistinctness of ribosomes in the vesicle may indicate digestion of contents. Scale bar = 0.5 μm ($\times$ 38,000).

Fig. 14. Small vacuole containing both amorphous and vesiculate material in a differentiating protophloem sieve element of a bean root tip. Scale bar = 0.2 μm ($\times$ 70,000).

Plant cells lack a comparable organelle, but considerable evidence suggests that a comparable lytic function is carried out in the cells by the vacuoles (Matile, 1974, 1978). For example, the invaginations into vacuoles sometimes observed in electron micrographs (Figs. 13 and 14) may represent cytoplasmic material being brought into contact with a wide variety of digestive enzymes. Also the rapid lysis of the cytoplasm that takes place late in the differentiation of xylem elements and in senescing cells may result from exposure of the cytoplasm to hydrolytic enzymes released when the tonoplast breaks down. Nishimura and Beevers (1978, 1979) have shown that vacuoles in the endosperm of young castor bean seedlings arise when the outer matrix of the storage protein bodies is dissolved. When these vacuoles containing protein crystalloids are isolated, hydrolysis of the endogenous protein continues in the isolated vacuoles, providing a direct demonstration of their proteolytic function.

VI. THE NUCLEUS

The nucleus is the largest and most prominent body in the cell, and is the cell's center of regulatory activity. It is bounded by a pair of membranes, the nuclear envelope, which bears numerous pores as described in Section IX,B,2. The nucleus contains the major part of the DNA, the genetic material of the cell, and is the site of gene replication and transcription. The DNA strands in association with protein complexes constitute the chromosomes. During interphase, the period between mitoses, the DNA–protein complexes are evident as an irregular network of chromatin; the chromosomes become most clearly visible during mitosis. The nucleus contains one to several nucleoli, densely staining spherical bodies involved in the formation of the subunits of cytoplasmic ribosomes, as discussed in Section IX,B. The nucleus is treated in detail in this volume in Chapter 13.

VII. THE CELL WALL

A. Primary and Secondary Walls

Plant cells differ from animal cells most obviously in possessing a wall surrounding the protoplast. The wall is an extra-cytoplasmic product, and is not considered to be a living part of the cell. In vascular plants, a distinction is made between primary and secondary walls. Each cell of the plant body is surrounded by a primary wall; it encloses the meristematic cell, and when that cell divides, each daughter protoplast deposits a new primary wall that separates it from its sister. The primary wall is present throughout the growth of the cell and it undergoes irreversible plastic extension and growth as the cell enlarges. In some cell types, e.g., parenchyma cells, the primary wall is the only wall present at maturity.

As some cell types mature, however, a secondary wall of considerably different properties is deposited on the inner surface of the primary wall, i.e., between primary wall and plasmalemma. This wall is less hydrated and more compact than the primary wall and differs substantially in chemical composition. It may be deposited in massive, uniformly thick layers, as in the mechanically supportive fibers of xylem and phloem, or in the form of annular, helical, or ladderlike thickenings, as in the water-conducting vessels and tracheids of xylem (Fig. 15) (Esau, 1977).

B. Cellulose Microfibril Deposition and Cell Enlargement

1. The Pattern of Deposition

Cellulose constitutes 20–30% of the dry weight of various primary walls that have been analyzed (Preston, 1974). It is a key wall component, but is not universally present. It is known to be absent in certain algae; for example, it is replaced by mannan in the green alga *Codium* (Preston, 1974). Also it is missing from many taxa of the fungi, where it is replaced by chitin (Burnett, 1976). In the wall the cellulose molecules are aggregated into microfibrils held to one another by extensive hydrogen bonding. With its great tensile strength, cellulose is the component that contributes strength to the wall, and whose bonds must be loosened if the wall is to expand and permit the cell to enlarge. The hemicellulosic polysaccharides are hydrogen bonded to the cellulose, and the pectic polysaccharides are covalently bound to the hemicellulosic polysaccharides (Albersheim, 1976; Keegstra et al., 1973).

Microfibrillar structure in the primary wall can readily be seen in favorable electron micrographs of thin sections stained with uranium and lead salts (Figs. 16 and 17). These microfibrils have dimensions (3.5–4.0 nm) appropriate for those of cellulose, but since cellulose would not be expected to react with heavy metal salts, it seems likely that the cellulose microfibrils are being delineated and visualized by a coating of more reactive polysaccharides, such as the pectic polysaccharides.

The deposition of new microfibrils at the inner surface of the wall can be either at random or in an oriented pattern. Various cell types can be recognized (Roelofsen, 1959) depending upon the manner in which the cellulose is deposited, since the mode of deposition strongly influences the direction of cell enlargement. In parenchymatous cells in storage tissues and in tissue cultures, for example, microfibril deposition is generally a random process, and consequently the wall yields multidirectionally and the cell enlarges isodiametrically. In enlarging cells of the root and shoot axis, microfibril deposition in the end walls is random, but deposition in the side walls is

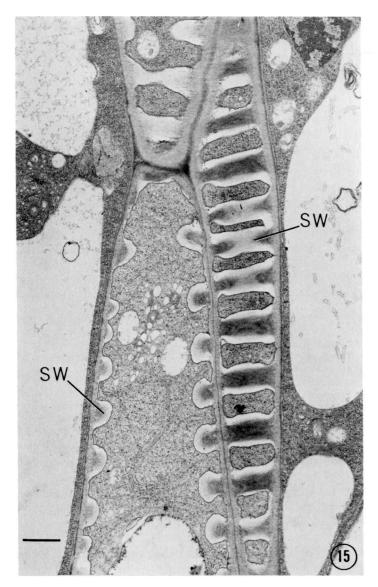

Fig. 15. Bands of secondary wall (SW) in differentiating tracheids in a root of lentel (*Lens culinaris*). Bands in profile view on left and in face view on right. Scale bar = 1 μm (× 10,000). Micrograph by W. P. Wergin.

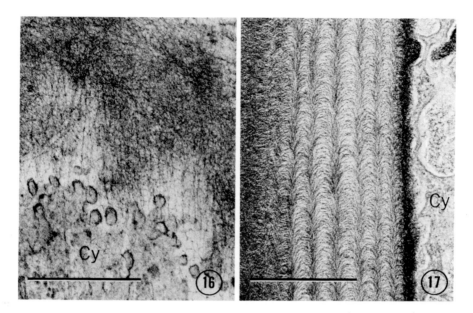

Fig. 16. Microfibrillar structure in the wall of a root hair of radish (*Raphanus sativus*) as seen in an oblique section. Cy, cytoplasm. Section poststained in uranyl acetate and lead citrate. Scale bar = 0.5 μm (× 59,000). Micrograph by H. T. Bonnett, Jr.

Fig. 17. Polylamellate wall showing helicoid structure in an epidermal cell of a leaf of a seagrass (*Cymodocea rotundata*). Cy, cytoplasm. Scale bar = 0.5 μm (× 56,000). Micrograph by M. E. Doohan.

transverse to the plant axis. Elongation of the cell will be in the direction of the axis since the microfibrils encircling the side walls are much more resistant to the turgor pressure of the cell circumferentially than to the pressure exerted at right angles. Progressively deeper in the wall, owing to continued elongation of the cell, the microfibrils depart more and more from the transverse orientation they possess when first deposited (Roelofsen, 1965).

Some cells or cell processes that elongate greatly do so through tip extension only, while others grow both at the tip and along the side walls. The root hair and pollen tube are both examples of growth confined to the tip (Roelofsen, 1965). Microfibril deposition at the tip is random. Since the side walls do not elongate, the microfibrils deposited at the tip remain fixed in random arrangement as they become part of the side wall structure owing to continued tip growth. On the other hand, in the rapidly extending cotton hair (a prolongation of an epidermal cell of the seed coat), elongation occurs at the tip as well as along the sides (Willison and Brown, 1977). As expected, the newly deposited microfibrils in the side walls lie transverse to the axis of elongation.

There are a number of instances in which successive multifibrillar lamellae occur in primary walls, resulting in either a crossed-fibrillar ("herringbone") or helicoidal architecture. An example of helicoidal structure is illustrated in Fig. 17. Recently several models of the architecture of these walls have been proposed to account for the fibrillar patterns observed (Neville *et al.*, 1976; Roland *et al.*, 1977; Sargent, 1978).

2. *Microtubules and the Oriented Deposition of Microfibrils*

Ever since the discovery of microtubules in plant cells (Ledbetter and Porter, 1963), attention has been focused on the cortical microtubules located close to the plasmalemma and their possible roles in wall growth. In cells in which the deposition of microfibrils is a random process, e.g., parenchymatous cells in tissue culture and the tips of root hairs, cortical microtubules are absent or rare. However, in cells in which the newly deposited microfibrils are oriented, a remarkable correlation has been repeatedly noted in direction of alignment between the cortical microtubules and the microfibrils (Newcomb, 1969). In elongating cells the microtubules lying just beneath the plasmalemma along the sides of the cell are always transversely aligned (Fig. 18), as are the most recently deposited microfibrils in the wall. The microtubules may occur in a single row along the plasmalemma (Fig. 5), or in clusters two or three deep, the outermost lying about 100 nm from the plasmalemma. The same correspondence between microtubules and microfibrils is found in secondary wall deposition (Fig. 19). A few exceptions to this generalization relating microtubules to wall microfibril orientation have been noted, but can be plausibly explained without seriously weakening the generalization (Hepler and Palevitz, 1974). Strong supportive evidence for a role of microtubules in controlling microfibril alignment has come from experiments with colchicine, which disrupts microtubules by binding to the tubulin subunits. Application of colchicine leads not only to disappearance of the cortical microtubules, but also to disorientation of the microfibrils being deposited (Hepler and Fosket, 1971).

The preceding correspondence has generated considerable speculation but no clear explanation of the role microtubules might play. Preston (1974) proposes that cellulose fibril directions are determined by an ordered lattice of cellulose synthetase granules on the plasma membrane. Arms or bridges between the microtubules and the plasma membrane have been reported and have encouraged the suggestion that through oscillatory or other movement the microtubules might bring about an alignment either of the cellulose synthetase granules or of the nascent cellulose microfibrils themselves on the outer surface of the plasma membrane (Hepler and Palevitz, 1974). Why the microtubules are formed in the cortex of cells that are beginning to enlarge, and how they themselves become aligned, remain unexplained phenomena of differentiation, although recently Gunning *et al.* (1978) have described

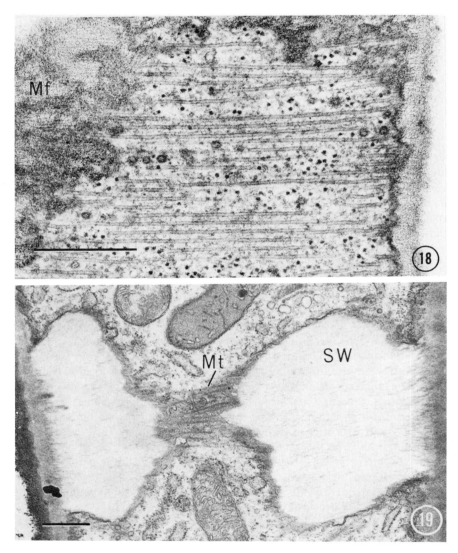

Fig. 18. Tangential section of a young cell in a bean root tip showing correlation between the alignment of side-wall microfibrils (Mf) and the cortical microtubules. Scale bar = 0.5 μm (× 54,000).

Fig. 19. Tangential section of a differentiating xylem tracheary element in a root of white sweet clover (*Melilotus alba*). The secondary wall (SW) microfibrils parallel the microtubules (Mt). The latter are revealed just beneath the band when the plane of section passes out of the wall into the cytoplasm. Scale bar 0.5 μm (× 25,000).

regions in the cortex that appear to be organizing centers of cortical microtubules.

C. Protein in the Primary Wall

Several other lines of inquiry of considerable importance in shaping our present concepts about the primary wall deserve mention. One relates to the presence of structural protein in the wall. From 3 to 10% of the dry weight of primary walls consists of protein (Albersheim, 1976). Some of this protein may arise from entrapped plasmodesmata and fragments of plasmalemma remaining in the wall fraction, and some may be due to enzymes in the wall proper. A number of enzymes have been identified as primary wall constituents, including peroxidases, pectin methylesterase, ascorbic acid oxidase, invertase, ATPase, UDPG pyrophosphorylase, and inorganic pyrophosphatase. A sizable proportion of the wall protein, however, appears to play a structural role.

The investigation of wall structural protein was triggered by the discovery by Steward and co-workers (1956) that a substantial amount of protein containing the unusual α-imino acid, hydroxy-L-proline, occurs in plant cells. Lamport and Northcote (1960) and Dougall and Shimbayashi (1960) demonstrated in early papers that the hydroxyproline occurs in actively growing primary walls. In a long series of subsequent investigations, Lamport and co-workers have explored the distribution and chemistry of the protein in plant cell walls that contains hydroxyproline, isolated a number of glycopeptides rich in the imino acid, established that side chains of tetra-arabinosides are attached to hydroxyproline through O-glycosidic links, and speculated on the possible roles of the protein in wall structure and growth (Lamport 1970, 1973; Lamport and Miller, 1971). The biosynthesis of hydroxyproline-rich glycoproteins in the cytoplasm and their secretion into the wall have been investigated in Chrispeels' laboratory (Chrispeels, 1976; Sadava and Chrispeels, 1973).

D. A Model for Primary Wall Structure

A milestone of progress in understanding wall structure was reached in 1973 with the publication from Albersheim's laboratory of three classic papers on the primary wall structure of suspension-cultured sycamore cells (Bauer *et al.,* 1973; Keegstra *et al.,* 1973; Talmadge *et al.,* 1973). By sequential application of wall-degrading hydrolytic enzymes of fungi and use of new chemical methods for the analysis of sugars, the group was able to identify the structural components and propose a model for the wall of sycamore cells based on the interconnections between the macromolecular components. In agreement with Lamport's findings, evidence was obtained for the

presence of a hydroxyproline-rich protein linked to other components of the
wall through an arabinogalactan. Attention was focused particularly on the
xyloglucan component of the wall, since it and the other hemicellulosic
polysaccharides have the ability to bind noncovalently through hydrogen
bonding to cellulose, and to bind covalently to the pectic polysaccharides.
Xyloglucan and the other hemicellulosic polysaccharides might therefore
serve to interconnect the cellulose fibrils and the pectic polysaccharides of
the wall. Considerable attention has subsequently been focused on the possible role of the xyloglucan component in cell enlargement mediated by auxin
(Labavitch and Ray, 1974a,b).

E. Site and Mode of Cellulose Microfibril Biosynthesis

The site and manner in which cellulose is biosynthesized, aggregated into
fibrils, and deposited have been continually of special interest to students of
the plant cell wall (Preston, 1964), but progress in this area has been slow. It
is known that polymerization of the hemicellulosic and pectic polysaccharides and glycosylation of the structural protein of the wall take place in
the dictyosomes, after which the products reach the plasma membrane via
vesicles arising from dictyosome cisternae (Morré and Mollenhauer, 1974;
Chrispeels, 1976). There is substantial evidence, however, that cellulose
biosynthesis takes place at the cell surface, either in the plasma membrane
or on its outer surface (Wooding, 1968; Northcote, 1972; Bowles and Northcote, 1974).

The work of Brown and co-workers (Brown et al., 1970, 1973; Brown and
Romanovicz, 1976; Herth et al., 1972; Romanovicz and Brown, 1976) on the
production of wall scales in the marine chrysophycean alga, *Pleurochrysis
scherffelii,* is pertinent because it establishes that cellulose similar to that of
higher plants is a major component of the scales, and that the scales—
cellulose and all—are fabricated in dictyosome cisternae. Subsequently the
cisternae migrate to the cell surface, fuse with the plasma membrane, and
release the scales to the surface. This work, like the classic earlier papers of
Manton on scale production in marine algae, is instructive in demonstrating
the capacity of the dictyosome cisternae to construct a highly organized
product containing several different types of macromolecular aggregates.
Also it is important because it demonstrates that dictyosome cisternae have
the capacity to synthesize and lay down oriented microfibrils of cellulose in a
highly organized pattern. The relevance for the problem of cellulose biosynthesis in higher plants is clear when it is recalled that the membrane of a
cisterna becomes part of the plasma membrane upon fusion, and that the
inner surface of the cisterna then becomes the outer surface of the plasma
membrane.

More recently, Brown and co-workers have investigated cellulose biosyn-

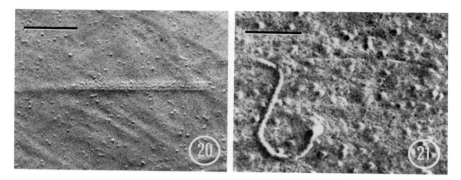

Fig. 20. Terminal complex on the fracture face of the outer leaflet of the plasma membrane of the unicellular green alga, *Oocystis apiculata*. A groove (on the left) terminates in a linear array of particles in three rows. Scale bar = 0.2 μm ($\times$ 70,000). Micrograph courtesy of D. Montezinos and R. M. Brown, Jr., University of North Carolina, Chapel Hill.

Fig. 21. Fracture face of the outer leaflet of the plasma membrane from a cortical parenchyma cell of a corn root. The looped structure is a microfibril that has been torn through the membrane leaflet (note tear mark in leaflet) and now lies on the face. The globular terminal knob is assumed to be a cellulose synthesizing enzyme complex. Scale bar = 0.1 μm ($\times$ 148,000). Micrograph courtesy of Susette Miller and R. M. Brown, Jr., University of North Carolina, Chapel Hill.

thesis using freeze-etch preparations of several different types of untreated living cells. In the green alga *Oocystis* (Brown and Montezinos, 1976; Montezinos and Brown, 1976), in cells of the stelar tissue of corn roots (Mueller *et al.*, 1976), and in the developing cotton fiber (Westafer and Brown, 1976; Willison and Brown, 1977), granular complexes associated with the outer leaflet of the plasmalemma have been observed (Figs. 20 and 21). In view of the location and pattern of distribution of these with respect to what appear to be nascent fibrils, the authors suggest that the granules are enzyme complexes engaged in the synthesis of new cellulose microfibrils.

VIII. THE ENDOMEMBRANE SYSTEM

A. The Endoplasmic Reticulum

Electron microscopy has revealed that the cell protoplast is highly compartmentalized by systems of membranes. Among these, the most pervasive is the endoplasmic reticulum, a system of flattened sacs and tubules, all interconnected and forming a continuous labyrinthine series of chambers and channels that ramifies through the cell and is compartmented off from the surrounding cytoplasm. The endoplasmic reticulum serves to anchor messenger RNA and attached polyribosomes engaged in protein synthesis and to sequester at least some of this protein for storage or transport, to serve as a system for channeling and distributing products, and to separate, localize, and organize the components of various metabolic sequences.

There are two forms of endoplasmic reticulum, one commonly grading into the other so that the two components belong to the same continuous system. The rough or granular endoplasmic reticulum (Fig. 22), which bears polyribosomes on the cytoplasmic face, generally comprises extensive saccules or cisternae interconnected three-dimensionally by numerous branches. The smooth or agranular reticulum can frequently be seen in continuity with the rough reticulum. It lacks polyribosomes and is generally differentiated into a maze of interconnected tubules. In most higher plant

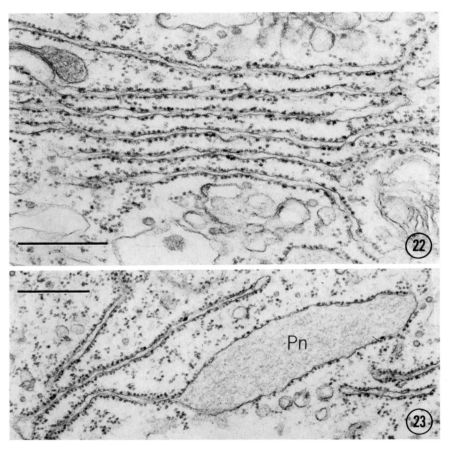

Fig. 22.   Parallel cisternae of granular endoplasmic reticulum in the cytoplasm of a radish root hair. Scale bar = 0.5 μm ($\times$ 47,000). Micrograph by H. T. Bonnett, Jr.

Fig. 23.   Dilated cisterna of granular endoplasmic reticulum containing protein (Pn) in a radish root hair. Scale bar = 0.5 μm ($\times$ 38,000). From Bonnett and Newcomb (1965).

cells, the smooth endoplasmic reticulum is not extensively developed, and may be hard to distinguish. It reaches its greatest degree of elaboration in certain specialized cells, including the suspensor of young embryos (Schnepf and Nagl, 1970), oil glands (Heinrich, 1973; Schnepf, 1969a,b,c,d), and the "farina" glands of members of the Primulaceae (Wollenweber and Schnepf, 1970).

The endoplasmic reticulum enfolds the nuclear material in a double membrane or envelope, ramifies through the cytoplasm, and connects with the endoplasmic reticulum of adjacent cells via the plasmodesmata. The endoplasmic reticulum is also closely related both developmentally and functionally to the Golgi apparatus, or system of dictyosomes, and the latter is in turn closely related to the plasmalemma. There is extensive evidence for the flow of membrane via vesicles from the rough endoplasmic reticulum to the forming face of the dictyosome cisternae, and from the maturing face of the dictyosome via vesicles to the plasmalemma at the cytoplasmic surface. This "endomembrane concept" has been developed and promulgated particularly by Morré, Mollenhauer, and Bracker, whose reviews should be consulted for details (Morré and Mollenhauer, 1974).

Admittedly a favorable orientation of the major components of the system and the flow of materials between them are much more clearcut and easily perceived in certain algae, fungi, and animal cells than in the cells of higher plants. A number of derivatives of the endoplasmic reticulum, including microbodies, which arise as protuberances (Frederick et al., 1968), storage protein bodies that accumulate in maturing seeds within sacs bounded by the endoplasmic reticulum (Larkins and Hurkman, 1978), and the vacuoles that result from the enlargement of vesicles of endoplasmic reticulum in plant cells generally, can also be considered components of the endomembrane system.

There is great variation among higher plant cells in the degree to which the endoplasmic reticulum is developed. The rough endoplasmic reticulum is much less commonly found in closely packed parallel arrays in plant cells than it is in the secretory cells of animals, but nevertheless substantial protein accumulation within cisternae of the endoplasmic reticulum is not unusual, especially in the Brassicaceae (Jørgensen et al., 1977). The example in Fig. 23 illustrates protein deposited within a cisterna of rough endoplasmic reticulum in a young root hair of radish (Raphanus sativus) (Bonnett and Newcomb, 1965). Another example is provided by beet root slices aerated for 2–3 days, whereupon crystalline protein begins to appear within the cisternae of endoplasmic reticulum of the parenchyma cells (van Steveninck and van Steveninck, 1971).

B. The Golgi Apparatus

The population of dictyosomes in a cell constitutes its Golgi apparatus. A dictyosome consists of a stack of several flattened sacs or cisternae (Fig. 24).

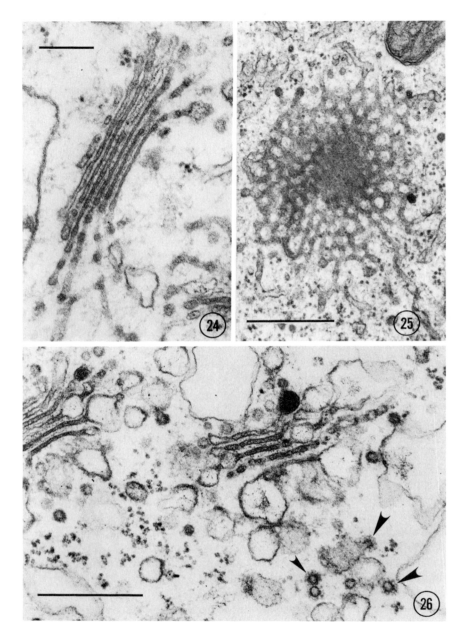

Fig. 24. Profiles of several cisternae of a dictyosome sectioned normally in bean root. Note evidence of electron opaque material between cisternae. Scale bar = 0.2 μm (× 70,000).

Fig. 25. Face view of a dictyosome in a young cell in a bean root tip. Note faint intercisternal rods running from upper left to lower right, and fenestrated periphery. Scale bar = 0.5 μm (× 46,000).

Each cisterna is commonly disc-shaped with fenestrated margins, and is 1–2 μm in diameter. The cisternal membrane lacks ribosomes on the outer surface. Although successive cisternae in the stack are separated from each other by a space of about 10 nm, they are held together by a bonding constituent (Franke *et al.*, 1972; Mollenhauer *et al.*, 1973). Parallel fibers spaced about 15 nm apart are often present in this intercisternal material between some of the cisternae (visible as fine striations in Fig. 25). Frequently the cisternae are curved, so that the organelle as a whole is concave on one side and convex on the other. At the periphery of the dictyosome the cisternae are usually either dilated or vesiculated (Fig. 26). Where dictyosomes are active, vesicles can usually be seen budding off the cisternae, while other vesicles are clustered at the margins.

Plant cells usually have 4–8 cisternae per dictyosome, but some algal cells have 20–30. Extensive evidence indicates that new cisternae are formed on one face of the dictyosome (the "proximal" or "forming" face) by fusion of transition vesicles arising from a locally smooth surface of rough endoplasmic reticulum (Morré and Mollenhauer, 1974). Progressive maturation across the stack leads to loss of cisternae by breakup into secretory vesicles on the "maturing" or "distal" face at the opposite surface. Changes in membrane composition, stainability, and particle number as seen in freeze-etched membranes have demonstrated that the cisternal membranes on the forming face resemble those of the endoplasmic reticulum, and that progressively across the stack they come to resemble the plasma membrane (Grove *et al.*, 1968; Morré *et al.*, 1971).

The number of dictyosomes in a cell varies widely, depending upon the plant, the stage of development of the cell, and the nature of its activities. Meristematic cells in the shoot apex have 20–35 dictyosomes (Anton-Lamprecht, 1967). Those of the root apex have a similar number but this increases to several hundred in the root cap cells as they mature (Clowes and Juniper, 1964). Extremes in the number of dictyosomes are found among the algae, where cells of *Chara* rhizoids have about 25,000, while numerous unicellular algae have but one (Gunning and Steer, 1975).

Major activities of dictyosomes include the glycosylation of proteins produced in the rough endoplasmic reticulum, the polymerization of hemicellulosic and pectic polysaccharides from sugar nucleotides, and the biosynthesis of new membrane (i.e., as cisternal membrane) destined for incorporation into the plasmalemma. The ultrastructural appearance of the cisternal contents frequently reflects the kinds of products being elaborated.

Fig. 26.   Dictyosomes in a young root hair of radish. The cisternae are highly vesiculate, especially at the margins. Several coated vesicles can be seen (arrowheads). These appear to arise secondarily from larger, uncoated vesicles. Scale bar = 0.5 μm ($\times$ 55,000). From Bonnett and Newcomb (1966).

C. Coated Vesicles

Coated vesicles are curious structures commonly observed in both plant and animal cells; most investigators believe they are involved in membrane transport. Although many questions about their functions remain unanswered, their similarity of structure and widespread occurrence suggest that their roles in cells are important ones. Accordingly, they have attracted considerable attention from animal cell biologists in recent years (Geisow, 1979; Ockleford and Whyte, 1979; Woods *et al.,* 1978).

In plant cells, coated vesicles arise from dictyosomes. They have been reported from algae to angiosperms and, in the latter, from a wide diversity of tissues and cell types (Newcomb, 1979). In most plants they are strikingly uniform in size and morphology. They are 85–90 nm in diameter, including coat. The latter is a reticulate or honeycomb-like layer outside the bounding membrane. In median sections this layer exhibits radiating spokes about 25 nm long (Fig. 27). In grazing sections the coat is seen to be constructed of hexagonally and pentagonally packed units (Fig. 28).

Pearse (1975) has devised a procedure for isolating coated vesicles from various animal tissues, and has shown that the latticelike coats are composed predominantly of one particular type of polypeptide. It has a molecular weight of 180,000 and has been named "clathrin." It is known from several studies that the coat lattice is formed from trimeric associations of clathrin to give hexagonal and pentagonal units (Crowther *et al.,* 1976; Woods *et al.,* 1978). Peptide maps prepared from different species of animals indicate that the clathrin protein is strongly conserved (Pearse, 1976).

In animals it has been established that coated vesicles participate in both endocytotic and exocytotic transport, i.e., transport both from and to the cell surface. In plants, coated vesicles are concentrated in the vicinity of the dictyosomes and also in the cortical cytoplasm just beneath the plasmalemma (Fig. 27). At least in the radish root hair (Bonnett and Newcomb, 1966) and in flagellated marine algae (Manton, 1967), the coated vesicles develop secondarily by budding off from larger uncoated vesicles arising from dictyosomes (Fig. 26).

Coated vesicles are especially numerous in the cytoplasm during cell plate formation and in ensuing stages of cell enlargement. They can be observed fusing with the cell plate (Hepler and Newcomb, 1967) (Fig. 29) during cytokinesis, and with the primary wall (Fig. 30) during cell enlargement. Numerous patches made up of polygonal units can often be seen lying in the cytoplasm near the plasmalemma (Fig. 28), suggesting that the coats have remained behind after the vesicles have fused with the cell surface.

Franke and Herth (1974) found that rapidly growing cultured cells of the composite *Haplopappus gracilis* were especially favorable material in which to observe coated vesicles fusing with the cell plate. As much as 60% of the

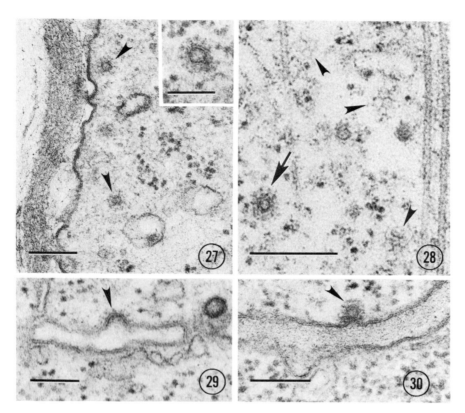

Fig. 27. Coated vesicles (arrowheads) near the plasma membrane in a young, rapidly growing radish root hair. Inset: coated vesicle at higher magnification showing spokelike reticulate surface outside the vesicle membrane. Scale bar = 0.2 μm ($\times$ 63,000). Inset: scale bar = 0.1 μm ($\times$ 124,000). From Bonnett and Newcomb (1966).

Fig. 28. Arrowheads point to patches of polygonal structures in the cytoplasm next to the primary cell wall in a young cell of a tobacco root tip. Arrow indicates a coated vesicle. Scale bar = 0.2 μm ($\times$ 115,000). Micrograph by E. L. Vigil.

Fig. 29. Coated vesicle (arrowhead) fusing with the cell plate in a dividing cell of a bean root tip. Scale bar = 0.2 μm ($\times$ 65,000). Micrograph by P. K. Hepler.

Fig. 30. Coated vesicle (arrowhead) fusing with the primary wall in a young cell of a tobacco root. Scale bar = 0.2 μm ($\times$ 80,000). Micrograph by E. L. Vigil.

new plasmalemma bore the coat pattern, strongly suggesting that in a rapidly growing plasma membrane where cell plate formation is completed within a few minutes, membrane material of the vesicles, identified by their coat markers, becomes an integral part of the plasmalemma.

It remains a possibility that in some cases coated vesicles serve as a mechanism for the conservation of excess plasma membrane material in plant cells by developing endocytotically from the plasma membrane and migrating inward to the dictyosomes. Roles that might be played by the cagelike coat structures of coated vesicles have been considered briefly by Geisow (1979).

IX. POLYRIBOSOMES AND PROTEIN SYNTHESIS

A. Free and Bound Polyribosomes in Protein Synthesis

The cytoplasmic ribosomes occur either "free" in the hyaloplasm, or bound as polyribosomes on the cytoplasmic face of the rough endoplasmic reticulum. Many of the free ribosomes appear to be present as polyribosomes, usually in the form of small clusters or rosettes (Fig. 31), and occasionally as helical arrays. The polyribosomes on the endoplasmic reticulum usually take the form of coils or loops (Fig. 32).

The relative proportions of free and bound ribosomes, and the amounts, form, and intracellular distribution of the endoplasmic reticulum, vary greatly in plant cells, depending upon the type of cell and stage of development. For example, in very young cells the hyaloplasm is populated with large numbers of free ribosomes, and there is relatively little rough endoplasmic reticulum (Fig. 31). However, generally as the cells enlarge and differentiate, rough endoplasmic reticulum covered with polyribosomes becomes more abundant and free polyribosomes less so (Fig. 22).

It is obvious that such changes in the distribution of polyribosomes must be intimately related to the types and fates of the proteins being synthesized, although these relationships are only in the early stages of clarification. In summarizing and extrapolating from the available information on polyribosome–membrane interactions in protein synthesis, Shore and Tata (1977) have emphasized that the synthesis and packaging of secretory proteins by the rough endoplasmic reticulum is only one expression of the more general role that this system plays in transferring proteins across cellular membranes. Presumably, proteins that come to lie "free" in the hyaloplasm, e.g., the tubulin of microtubules and the enzymes participating in the glycolytic sequence of respiration, are synthesized by free polysomes. Also the proteins that normally exist as peripheral proteins on the cytoplasmic surface of membranes, as well as some of the intrinsic proteins of various types of membranes, may be synthesized in the cytoplasm. But synthesis on ribosomes bound to the endoplasmic reticulum may represent the major route not only for secreted proteins, but also for proteins that form the structural elements of the endoplasmic reticulum, either as intrinsic proteins buried within the membrane lipid bilayer, or as proteins that are deposited asymmetrically in the membrane toward the endoplasmic space.

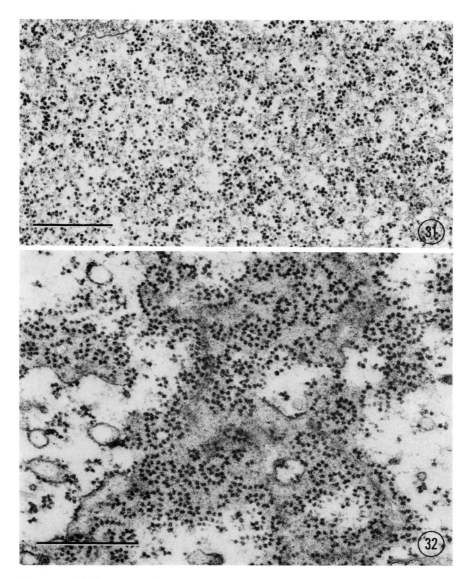

Fig. 31.   Polyribosomes lying free in the cytoplasm of a young meristematic cell in a bean root. Rosettes and helical forms can be identified. Scale bar = 0.5 μm ($\times$ 42,000).

   Fig. 32.   Face view of granular endoplasmic reticulum in young epidermal cell of a radish root, showing numerous polyribosome coils on the surface. Scale bar = 0.5 μm ($\times$ 54,000). From Bonnett and Newcomb (1965).

B. Ribosome Production and Transport

1. The Numbers of Cytoplasmic Ribosomes

Estimates place the numbers of cytoplasmic ribosomes in a young plant cell in the millions. In a meristematic cell undergoing division, growth, and then division again, the number of ribosomes must be doubled between successive divisions. For example, in a meristematic cell in a pea root tip, about 9 million ribosomes are made in a growth and division cycle time of 10 h. Ribosomes must therefore be produced in this case at an average rate of 15,000/min (Gunning and Steer, 1975). The proteins that go into the assembly of the ribosomes must be made by polyribosomes in the cytoplasm, then pass into the nucleus through the envelope, be assembled with rRNA in the nucleus, then pass back into the cytoplasm through the nuclear pores as components of the preribosomal particles.

2. Nuclear Pore Structure

The nuclear envelope is perforated by numerous pores. The pores are not simply openings, but are exceedingly complex structures that have a universality of size and architecture in eukaryotic cells (Franke, 1969). The structure of the nuclear pores is of great interest in considering how the large and small subunits of the preribosomal particles, partially assembled in the nucleolus, reach the cytoplasm. Studies on animal cells have given estimates of nuclear pore flow rates of 1–3 rRNA molecules per pore per minute (Franke, 1970; Scheer, 1973).

When the pores are viewed in profile (Fig. 33), it is apparent that the inner and outer membranes are continuous at the margins so that the nucleoplasm is in communication with the hyaloplasm of the cell rather than with the space between the envelope membranes. Surprisingly, in face view (Fig. 34) the pores are seen to be octagonal in outline (Gall, 1967), not circular as might be expected. The pore diameter from a face of the octagon to the one opposite is 65–75 nm. The borders of the pore bear the annulus, believed to be composed of two rings of eight granules each, one on the nucleoplasmic face and the other on the cytoplasmic. The annulus is not clearly seen in face view in Fig. 34, but its presence nevertheless obscures the vertices of the octagon in some of the pores, causing the outlines to appear circular.

The lumen of the pore contains a complex of small granules and delicate fibrils. In profile view these various structures appear as fuzzy fibrils surrounding and filling the pore (Fig. 33). Surprisingly, the pore complexes appear to be components completely separate from the nuclear envelope. When the two membranes of the envelope are completely removed by gentle treatment with detergents, the pore complexes remain in their characteristic locations on the nucleoplasm, and retain their attachments to it (Aaronson and Blobel, 1975). Frequently, an electron-dense granule can be seen in the

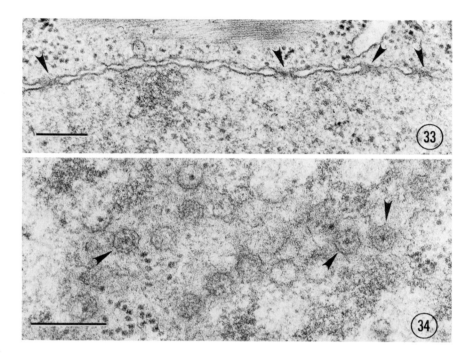

Fig. 33. Profile view of nuclear pores (arrowheads) in a differentiating phloem cell of a bean root. Note the two membranes of the nuclear envelope. Scale bar = 0.2 μm ($\times$ 66,000). Micrograph by W. P. Wergin.

Fig. 34. Face view of nuclear pores in a bean root cell. The octagonal shape and other features of pore substructure can be particularly well seen in the pores indicated by arrowheads. An electron-dense object can be seen in the center of several pores. Scale bar = 0.2 μm ($\times$ 100,000). Micrograph by W. P. Wergin.

pore center (Fig. 34). In view of the structural complexity of the pore, it seems likely that it plays roles in addition to transport.

There is considerable variation in pore frequency, so that pores can occupy anywhere from 8 to 20% of the envelope surface in plant cell nuclei. Correspondingly, the pore density can vary from 6 to 25 per micrometer (Lott *et al.,* 1972; Roberts and Northcote, 1971; Thair and Wardrop, 1971). The pattern of pore distribution over the envelope is also variable.

X. MICROBODIES

Plant microbodies fixed in glutaraldehyde possess a single, smooth membrane surrounding a matrix of moderate electron opacity (Fig. 35). Most commonly they are round to ovoid, with diameters ranging from 0.5 to 1.5

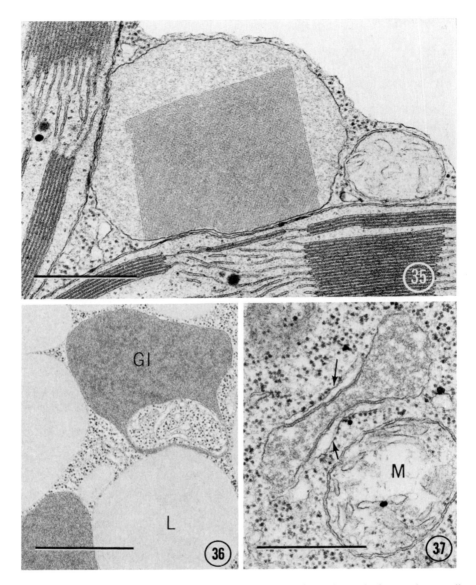

Fig. 35. Leaf peroxisome appressed to two chloroplasts in a tobacco leaf parenchyma cell. High catalase activity can be demonstrated in crystals similar to the one present in this peroxisome. A mitochondrion lies to the right of the peroxisome. Scale bar = 0.5 μm (× 54,000). From Frederick and Newcomb (1969b).

Fig. 36. Glyoxysomes (Gl) among lipid bodies (L) in a cotyledonary cell of a tomato seedling. Scale bar = 1 μm (× 26,000). From Frederick *et al.* (1975).

Fig. 37. Young microbody flanked on both sides by granular endoplasmic reticulum (arrows) in a meristematic cell of a bean root. M, mitochondrion. Scale bar = 0.5 μm (× 58,000). From Frederick *et al.* (1968).

μm, but may exhibit highly elongate or irregular shapes. The matrix is usually uniform and finely granular in consistency. Depending upon the species and cell type, crystalline or fibrillar inclusions or densely amorphous regions (nucleoids) may occur within the matrix, but ribosomes and internal membranes are never present (Frederick et al., 1968).

"Microbody" is a morphological term for a distinct class of organelle. Typically, microbodies contain catalase and one or more hydrogen peroxide-producing oxidases. To emphasize their potential role in various peroxidatic reactions, they were termed "peroxisomes" when isolated from animal cells and biochemically characterized (de Duve and Bauduin, 1966).

Microbodies apparently occur in all major eukaryotic groups in the plant kingdom, and are widely distributed throughout the tissues of higher plants. In particular tissues they are highly specialized to play key roles in certain metabolic processes. In other tissues, where they are less abundant, their functions are still obscure. They are especially abundant and well developed in the fat-storing cotyledons or endosperm of germinating fatty seeds, where they are called "glyoxysomes" (Breidenbach and Beevers, 1967; Breidenbach et al., 1968), and in green leaves, where they are termed "leaf peroxisomes" (Tolbert et al., 1968; Tolbert, 1971). In fatty seedlings they possess the glyoxylate cycle enzymes and participate in the conversion of fat to carbohydrate. Appropriately, they usually are observed closely appressed to lipid bodies (Gruber et al., 1970; Trelease et al., 1971; Vigil, 1973) (Fig. 36).

In green leaves, microbodies are associated with chloroplasts and mitochondria in photorespiration. In C_3 plants, whose photorespiratory activity is high, they are large and abundant, and tend to be appressed to the chloroplasts (Fig. 35) (Frederick and Newcomb, 1969a). They are less conspicuous in C_4 plants (Frederick and Newcomb, 1971).

Considerable ultrastructural evidence suggests that plant microbodies arise as evaginations from the endoplasmic reticulum (Frederick et al., 1968). In young cells a close association between the microbody and one or two cisternae of rough endoplasmic reticulum is consistently observed (Fig. 37), although an opening from the cisterna into the microbody is noted very rarely. Ribosomes are always absent on the face of the endoplasmic reticulum next to the microbody. Substantial biochemical evidence also supports the belief that the microbodies arise from the endoplasmic reticulum (Shore and Tata, 1977). The presence of catalase in microbodies can be demonstrated by employing 3,3'-diaminobenzidine at high pH in an extraordinarily effective electron cytochemical method (Frederick and Newcomb, 1969b; Sexton and Hall, 1978) (Fig. 38).

Since the ultrastructural studies on animal cells have shown that microbodies bud from the endoplasmic reticulum, it has been supposed that the peroxisomal enzymes are synthesized by ribosomes bound to the endoplasmic reticulum as in the case of secretory proteins, and that a subsequent

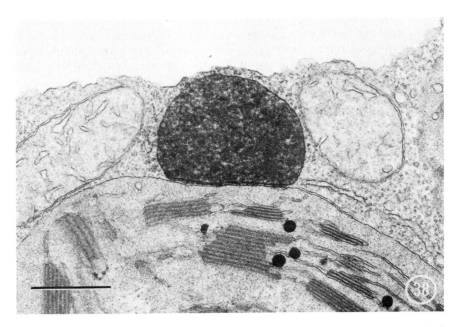

Fig. 38.   Leaf peroxisome heavily stained in the electron cytochemical method employing
3,3'-diaminobenzidine to detect catalase activity. The peroxisome is flanked by mitochondria in
a tobacco leaf parenchyma cell. Scale bar = 0.5 μm ($\times$ 42,000). From Frederick *et al.* (1975).

sorting mechanism operates within the lumen of the endoplasmic reticulum
to separate peroxisomal and secretory proteins. However, Goldman and
Blobel (1978) have shown that in rat liver the two peroxisomal enzymes,
catalase and uricase, are synthesized by free ribosomes, not by membrane-
bound ones. Thus, transfer of these enzymes across the membrane must
occur after protein synthesis and without mediation by a ribosome–
membrane junction. The connection between the endoplasmic reticulum and
the microbody may serve to allow the flow of certain integral proteins from
their site of synthesis and insertion on the rough endoplasmic reticulum to
positions in the microbody membrane.

XI. MITOCHONDRIA

As the respiratory centers in which reactive phosphate (as ATP) is generated through operation of the citric acid cycle and associated electron transport, mitochondria are essential organelles and occur in all eukaryotic cells.
Their architecture is basically the same in both plants and animals, and they
are immediately recognizable regardless of source. Plant mitochondria are

generally spherical to ellipsoid, and are commonly about 1 μm thick and 1–3 μm long (Fig. 39). They are stained blue in living cells by Janus green B. This "supravital" stain was the subject of a classic paper by Lazarow and Cooperstein (1953) establishing that the dye is reduced to a colorless form elsewhere in the cell but in mitochondria it is maintained in the blue (oxidized) form by activity of the cytochrome–cytochrome oxidase system.

Like plastids, mitochondria are bounded by a pair of membranes, i.e., an envelope. The inner membrane of the pair extends into the interior matrix in a number of flattened sacs or cristae (Figs. 39–41). Mitochondria are semi-autonomous organelles; that is, they possess a genome of DNA, ribosomes, and the ancillary components necessary for the synthesis of a number of their own proteins. The DNA is in the form of one or more nucleoids that reside in relatively clear regions of the organelle and are visible as a fine tangle of strands when poststained with heavy metal salts (Fig. 39). The DNA in the nucleoid, formerly believed to be linear in plant mitochondria, is now known to be a closed circular molecule as it is in animal cells (Birky, 1978). The DNA molecule has proved to be considerably longer in plant cells than in animal cells, an interesting fact since the number of organelle genes is limited by the size of the DNA molecule. In animal cells the mtDNA is about 5 μm long, corresponding to about 10×10^6 daltons or 15×10^3 base pairs, whereas in the pea plant, for example, it is about 30 μm long, equal to about 60×10^6 daltons and 90×10^3 base pairs (Birky, 1978). The ribosomes of mitochondria are visible as electron opaque particles about 15 nm in diameter lying in the matrix (Fig. 39). These do not ordinarily appear to be grouped into polyribosomes.

Both mitochondria and plastids resemble prokaryotes in numerous ways (DNA in a nucleoid region, similarity in properties of ribosomes, etc.), an observation that has spawned considerable literature in recent years, developing the view that both organelles represent the invasion of the eukaryotic cell by prokaryotes that were originally endosymbionts and in the course of evolution became essential cell components. The subject has been reviewed frequently (Bücher et al., 1976; Schwartz and Dayhoff, 1978).

It is now well established that, like prokaryotes, mitochondria reproduce by fission. An organelle possibly undergoing division is shown in Fig. 41. While the profile shown could have resulted simply from thin-sectioning a lobed mitochondrion, the narrow neck and presence of a nucleoid in both bodies strongly suggests that fission is taking place and is nearly complete.

Unit membrane structure (i.e., dark–light–dark) cannot ordinarily be seen in either membrane of the mitochondrial envelope. The outer membrane, highly permeable to small molecules, contains only a small percentage of the total mitochondrial mass and only relatively few of the enzymes known for the organelle (DePierre and Dallner, 1975; DePierre and Ernster, 1977). The inner membrane (including cristae) contains 80–95% of the total membrane-

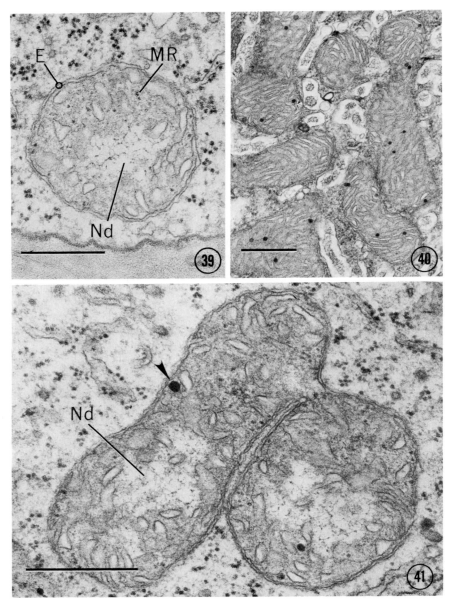

Fig. 39. Mitochondrion in a differentiating phloem cell of a bean root. E, envelope; MR, mitochondrial ribosome; Nd, nucleoid region containing presumptive DNA fibrils. Scale bar = 0.5 μm (× 44,000).

Fig. 40. Mitochondria containing dense matrix and numerous cristae in an epidermal cell of a sea grass (*Cymodocea rotundata*). The cell contains a large number of plasma membrane-bounded wall ingrowths characteristic of transfer cells. These ingrowths can be seen among the numerous mitochondria. Scale bar = 0.5 μm (× 29,000). Micrograph by M. E. Doohan.

based protein and over 90% of the total mitochondrial lipid. It is the site of the respiratory chain and related energy-transducing processes, and carrier-facilitated ion translocations. The inner membrane has very low permeability, even for relatively small molecules (DePierre and Dallner, 1975). The presence of the cytochrome–cytochrome oxidase system in the inner membrane and cristae can be demonstrated with the electron microscope by means of an electron–cytochemical method that employs 3,3′-diaminobenzidine at a lower pH (7.4) than is used to localize catalase (pH 9.0). The method has been thoroughly reviewed by Sexton and Hall (1978).

When isolated mitochondria are negatively stained with phosphotungstic acid, lollipop-like particles can be seen protruding from the surface of the inner membrane and cristae facing the matrix. The projections include a spherical head about 8.5 nm in diameter and a stalk about 5 nm in length. These structures, termed the F_1 coupling factor, have ATPase activity and are believed to participate in conversion of inorganic phosphate to the reactive phosphate of ATP during electron transport. In 1964 Nadakavukaren demonstrated the particles in mitochondria prepared from the endosperm of germinating castor bean seeds; in 1965 Parsons *et al.* did the same, using mitochondria from a variety of tissues representing six different plant species.

Plant mitochondria have relatively few cristae, and a low crista-to-matrix ratio when compared with mitochondria from animal tissues with high energy demands, such as muscle. Also plant mitochondria generally lack the diversity in cristal structure and types of inclusions in the matrix exhibited by animal mitochondria. In appearance, the mitochondrion in Fig. 39 is much more typical of plant mitochondria generally than are those of Fig. 40. The latter mitochondria are found in the leaf epidermal cells of a seagrass, a submerged aquatic angiosperm (Doohan and Newcomb, 1976). These cells have the wall ingrowths and greatly increased surface area of plasma membrane typical of transfer cells (Gunning and Pate, 1969), presumably correlated with their greater activity in solute absorption and possibly salt secretion. The large numbers, dense matrix, and numerous cristae of these mitochondria are consistent with a role in solute transport.

Electron-dense granules considerably larger than ribosomes are commonly observed in the matrix of plant mitochondria (Figs. 40 and 41). These granules are probably rich in calcium phosphate, as they are in animal mitochondria (Weinbach and von Brand, 1967). The accumulation of calcium has been studied in mitochondria isolated from a variety of higher plant

Fig. 41. Mitochondrion possible undergoing division in meristematic cell of a bean root tip. Note abundant fibrillar material (DNA?) in the nucleoid regions (Nd) in each body. An electron-dense granule (arrowhead) may contain calcium phosphate. Scale bar = 0.5 μm (× 59,000).

species (Chen and Lehninger, 1973). When mitochondria isolated from plant tissues are incubated in a medium containing Sr^{2+}, inorganic phosphate, and an energy source such as succinate, massive electron-dense deposits of strontium phosphate accumulate in the matrix (Ramirez-Mitchell et al., 1973).

In higher plants the number of mitochondria per cell is usually in the hundreds or thousands, depending on cell type and stage of development. In the stem apex of willow-herb (Epilobium), a cell in interphase just before division has about 120 mitochondria, and the daughter cells just after division have about half this number (Anton-Lamprecht, 1967). It has been estimated that the central cells of the maize root cap have about 200 mitochondria when young, and 2000–3000 when fully enlarged and mature (Clowes and Juniper, 1968). The number per unit volume of cytoplasm, however, is not believed to change greatly. In very active cells, including transfer cells, up to 20% of the cytoplasmic volume is attributable to mitochondria (Gunning and Steer, 1975).

XII. PLASTIDS

Plastids, like mitochondria, are semi-autonomous organelles enclosed within an envelope (i.e., two membranes). Unlike mitochondria, however, they are confined to the plant kingdom. There are several types, all more or less developmentally related to one another, and they occur as colorless or pigmented forms in virtually every cell of the plant body (Kirk and Tilney-Bassett, 1978). Because plastid terminology can be confusing, it will be helpful to define each of the principal kinds.

Proplastids (Figs. 42–44) are small precursors of other plastids found in young cells in the growing regions of the plant.

Amyloplasts (Fig. 45), proteinoplasts, and elaioplasts are plastids especially rich in deposits of starch, protein, and lipid, respectively. These are terms that refer simply to a principal storage product in unpigmented plastids that have developed beyond the proplastid stage.

Leucoplasts are colorless plastids. The term "leucoplast" is imprecise and is best not applied to proplastids. It was commonly used in the older literature as a synonym for amyloplasts.

Etioplasts (Fig. 46) are plastids whose development from proplastids into chloroplasts has been arrested by absence of light. The most prominent feature of the etioplast is the prolamellar body, a semi-crystalline lattice composed of membranes in tubular form. The prolamellar body comprises membranes in temporary storage.

Chloroplasts (Figs. 47 and 48) are the green, photosynthetic plastids. Typ-

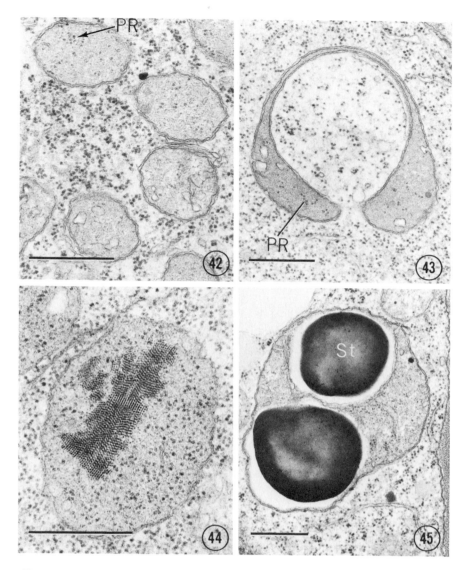

Fig. 42. Several proplastids in a young cell of a bean root tip. PR, plastid ribosomes. Scale bar = 0.5 μm (× 45,000).

Fig. 43. Amoeboid plastid in an enlarging cell in a bean root tip. In this example the plastid is umbonate (shield-shaped). PR, plastid ribosomes. Scale bar = 0.5 μm (× 35,000). From Newcomb (1967).

Fig. 44. Paracrystalline deposit of phytoferritin in a proplastid in meristematic cell in a root tip of soybean (*Glycine max*). Note numerous plastid ribosomes. Scale bar = 0.5 μm (× 55,000). Micrograph by W. P. Wergin.

Fig. 45. An amyloplast with two starch (St) deposits in cell of a bean root tip. The clear space around the starch bodies is characteristic of material fixed in glutaraldehyde. Scale bar = 0.5 μm (× 30,000). Micrograph by B. A. Palevitz.

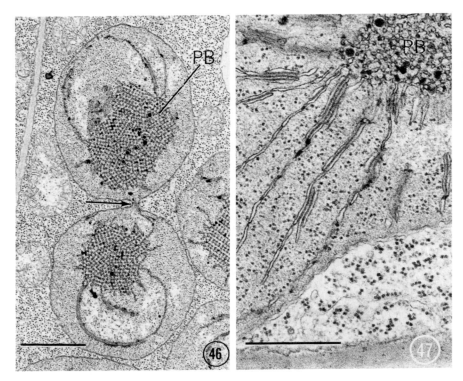

Fig. 46. Etioplast apparently undergoing division in a palisade parenchyma cell of an 11-day-old etiolated bean seedling. A narrow neck (arrow) still connects the daughter plastids. Note numerous ribosomes. PB, prolamellar body. Scale bar = 1 μm (× 17,000). Micrograph by P. J. Gruber.

Fig. 47. Portion of a young chloroplast in a leaf cell of timothy (*Phleum pratense*) illustrating the presence of numerous ribosomes at this stage of plastid development. Note lamellar outgrowths from the prolamellar body (PB). Scale bar = 0.5 μm (× 50,000).

ically they are about 5 μm long, 2 μm wide, and 1–2 μm thick. As shown in Fig. 48, their principle features include the following:

1. The *envelope*, consisting of a pair of concentric membranes separated by a gap 10–20 nm wide.

2. The *stroma*, a finely granular matrix in which the dark reactions of photosynthesis are localized.

3. The internal membrane system of the chloroplast, in which the photosynthetic pigments and light reactions of photosynthesis are localized. This includes the *grana* (singular, granum), cylindrical stacks of flattened discoid components. Each disclike component in the granum is a *granum thylakoid*. (Thylakoids are closed, flattened sacs into which the internal

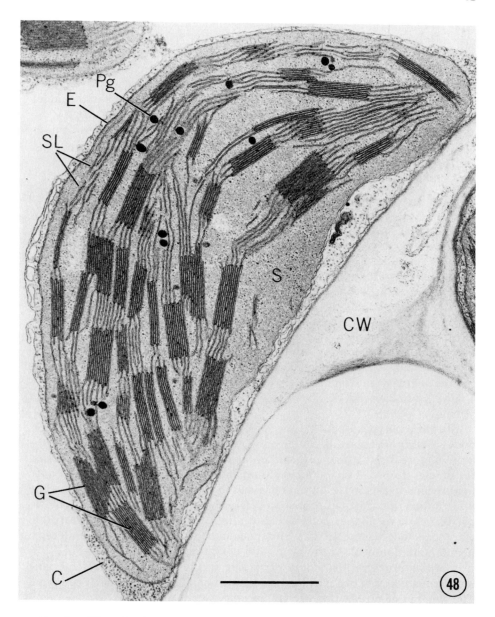

Fig. 48. Chloroplast in a leaf of corn (*Zea mays*). C, cytoplasm surrounding the chloroplast; CW, cell wall; E, chloroplast envelope; G, grana; Pg, plastoglobulus; S, stroma; SL, stroma lamellae. Scale bar = 1 μm (× 26,000). Micrograph by W. P. Wergin.

membrane system of the chloroplast is organized.) There are usually 40–60 grana in a mature chloroplast.

4. Also part of the internal membrane system are the *stroma lamellae,* the membranes that traverse the stroma and interconnect the grana. These lamellae, also known as *frets* and as *stroma thylakoids,* consist of two closely spaced membranes enclosing an interior channel. The channels of the lamellae open into the loculi of the granal thylakoids.

5. One or more *nucleoids,* relatively clear regions of the stroma containing fibrils of DNA.

6. *Plastid ribosomes,* electron-dense particles lying in the stroma. They are smaller than cytoplasmic ribosomes.

The structure and development of chloroplasts are considered in detail by Gunning and Steer (1975) and Kirk and Tilney-Bassett (1978).

Chromoplasts (Fig. 49) are yellow, orange, or red plastids containing carotenoid pigments. They can develop directly from proplastids or by modification of chloroplasts.

Plastids reproduce by fission (Fig. 46) and are remarkable in retaining the capacity to do so. Proplastids, etioplasts, and chloroplasts are all capable of division. In meristematic cells the division of proplastids appears to keep pace approximately with cell division. However, during the development of leaves from leaf primordia, the proplastids mature into chloroplasts, and the latter continue to multiply after cell division has ceased (Possingham and Saurer, 1969). A subpopulation of smaller chloroplasts may be responsible for the increase in numbers, since they can be seen to possess dumbbell shapes characteristic of division whereas the larger chloroplasts do not (Honda *et al.,* 1971).

The various types of plastids are also remarkable for their degree of interconvertibility. Proplastids can develop into any of the other types, while the more mature forms can undergo differentiation into one another, or reversion to simpler forms. For example, illumination of potato tubers can cause mature leucoplasts in the outer cell layers to develop into chloroplasts, while illumination of carrot tap roots can cause chromoplasts to revert to chloroplasts. Also, in mature cells that are stimulated to divide, or in arrested meristems that resume activity, the specialized plastids are able to undergo a dedifferentiation back to the proplastid condition.

It seems probable that for a particular species there is a characteristic approximate number of plastids per cell, which is reached in each cell type. The number of proplastids per cell in angiosperm meristems has been estimated to be perhaps 10–20 (Anton-Lamprecht, 1967), or about 20 (Clowes and Juniper, 1968), and the number of amyloplasts in the mature cells of the root cap of *Zea mays,* between 20 and 30 (Clowes and Juniper, 1964). A

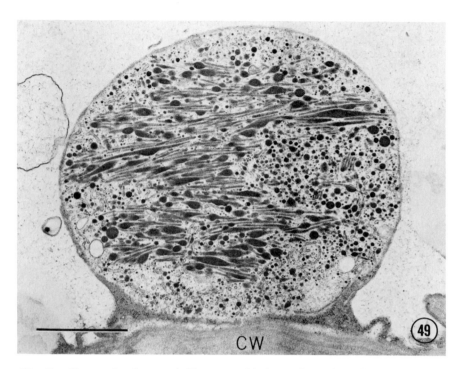

Fig. 49. Chromoplast (surrounded by a very thin layer of cytoplasm) in the wall of a ripe fruit of Jerusalem cherry (*Solanum pseudo-capsicum*). The carotenoid pigments are localized in the numerous electron-dense globules and spindle-shaped bodies in the plastid. CW, cell wall. Scale bar = 1 μm ($\times$ 24,000). Micrograph by W. P. Wergin.

palisade mesophyll cell of a leaf of castor bean (*Ricinus communis*) contains about 36 chloroplasts, while a spongy mesophyll cell contains only about 20 (Haberlandt, 1914). Per square millimeter of castor bean leaf area, there are a total of about 500,000 chloroplasts, of which 82% are in the palisade mesophyll cells (Haberlandt, 1914). It is estimated that the leaf parenchyma cells of beet (*Beta vulgaris*) contain 40–50 chloroplasts (Birky, 1978).

Although the smallest proplastids are even smaller than mitochondria (Fig. 42), many proplastids are somewhat larger. Though frequently spherical, they exhibit a wide variety of shapes and when somewhat older may be ameboid in appearance (Fig. 43). They have a rather uniform, dense matrix where a relatively small number of plastid ribosomes (compared to the number of ribosomes in developing etioplasts and chloroplasts) are found. One or more nucleoids are sometimes identifiable in the stroma as electron lucent regions containing a tangle of fine (3 nm) DNA fibrils. Like more complex plastids, proplastids have an internal membrane system; however, this is poorly developed and consists only of a few cisternal invaginations of the inner membrane of the envelope and one or two lamellae lying in the stroma. Occasionally one or more microtubule-like structures and a few spherical, amorphous lipid-like droplets are also present in the stroma. The latter, known as plastoglobuli (Lichtenthaler, 1968), contain plastid quinones and are especially prominent in etioplasts and chloroplasts (Lichtenthaler, 1973). Deposits of starch and, less commonly protein, may be present.

The proplastid stroma may also contain deposits of phytoferritin (Hyde *et al.,* 1963) (Fig. 44), a storage form of iron sometimes seen also in more mature plastids. Phytoferritin is similar to animal ferritin, and consists of a core of ferric oxide or hydroxide and phosphate surrounded by a shell of protein (Seckbach, 1972). In the electron microscope the protein is ordinarily not visible, and the phytoferritin molecule is apparent as four dense points, each about 15 Å in diameter, arranged at the corners of a square 5.5–7 nm on each side. Each of the four points contains several hundred iron atoms. When the phytoferritin deposits are extensive, the molecules often occur in paracrystalline array (Fig. 44).

Although in animal cells the ferritin deposits occur in the nucleus and mitochondria, and are also distributed through the cytoplasm, in plant cells the phytoferritin is confined to plastids and to an occasional association with lipid droplets in the cytoplasm. Phytoferritin is most abundant in storage organs, where it may serve as a temporary storage form of iron. The demands of chloroplasts for the iron of phytoferritin may be particularly heavy since they contain up to 80% of the iron content of leaves. Seckbach (1972) has shown that if extra iron is supplied to iron-starved cocklebur plants, increased deposits of phytoferritin appear in the chloroplasts.

Ribosomes are remarkably numerous in young, rapidly developing chloroplasts (Fig. 47). The genetic system of chloroplasts is based on a double-helical DNA circle, average length 45 μm, and a 70 S ribosomal protein-synthesizing machinery which can account for as much as 50% of the total ribosomal complement of photosynthetic cells (Highfield and Ellis, 1978). Since many of the chloroplast polypeptides are synthesized by cytoplasmic ribosomes, it is clear that the development of chloroplasts requires the integrated activities of both chloroplast and nuclear genomes. One of the major proteins of chloroplasts is ribulose bis-phosphate carboxylase/ oxygenase, the key enzyme in photosynthesis and photorespiration. While the small subunit of this enzyme is encoded in nuclear DNA and translated on cytoplasmic ribosomes, the large subunit is encoded in chloroplast DNA. Since the enzyme comprises up to 50% of the total soluble protein in leaves and may be the most abundant protein in nature, this may account for the large number of ribosomes found in chloroplasts (Highfield and Ellis, 1978).

XIII. THE CYTOPLASMIC GROUND SUBSTANCE

A. General Properties of the Ground Substance

The various organelles and other discrete components of the cytoplasm are suspended in the ground substance, also known as the hyaloplasm. Cytoplasm, or more specifically the ground substance of the cytoplasm, has a number of unexplained properties that have long challenged cytologists. Cytoplasm is viscous, and its viscosity varies spatially within the cell. There is commonly a peripheral region of high viscosity (the ectoplasm) and an inner region of lower viscosity (the endoplasm). Also, structural elements in the cytoplasm are dynamic in the sense that regions of high and low viscosity are interconvertible. Cytoplasm is a thixotropic gel, i.e., its viscosity can be decreased by gentle mechanical disturbance. A violent shock, however, can induce cytoplasm to set to a firm jelly in which all Brownian movement ceases. High pressure liquifies cytoplasm, showing that gel formation involves a change in volume (Pollard, 1976).

Closely related to the foregoing properties is the remarkable phenomenon known as cytoplasmic streaming, in which particles as large as plastids are swept along by a flow of the ground substance. Many cells in both lower and higher plants possess this ability. Periodic reversals of flow and simultaneous movement in opposite directions of two streams within a strand of cytoplasm 1 μm or less in diameter are intriguing processes for which explanations have been sought for nearly two centuries (Kamiya, 1959, 1962; Noland, 1957). Many algal and higher plant cells possess a type of streaming known as cyclosis, in which the cytoplasm circles continuously around the central vacuole. The flow occurs in a fluid, sol-like region of the cytoplasm lying within a cortical, stationary gel.

B. Preservation of the Ground Substance for Ultrastructural Study

Until recently, electron microscopy provided little encouragement to those seeking a structural basis for such phenomena as changes in cytoplasmic viscosity, streaming, and cell migration. In the 1950s and early 1960s, when osmium tetroxide was the fixative employed with animal tissues, electron micrographs failed to reveal the network of proteinaceous fibers in gelled regions of cytoplasm that might have been expected from indirect evidence. Fixation of plant cells by osmium gave even poorer results, as evidenced by badly disrupted organelles. Potassium permanganate became a popular fixative for plant cells for several years owing to its sharp delineation of membranous systems, but its use in routine fixation has been largely abandoned because its strong oxidizing action completely destroys several kinds of organelles and causes the ground substance to appear as a homogeneous, bland continuum devoid of useful information.

The introduction in 1963 by Sabatini *et al.* of glutaraldehyde followed by osmium tetroxide in a two-step fixative procedure for electron microscopy greatly improved preservation of the contents of both plant and animal cells, and led to the recognition of microtubules as widespread components of eukaryotic cells. It was realized, however, that microtubules could not be responsible for the sol–gel properties of cytoplasm, nor for streaming. Ledbetter and Porter (1963), in their classic paper on microtubules in plant cells, did in fact suggest that these organelles might provide the motive force for streaming; however, it was soon pointed out that microtubules, when seen in nondividing cells, are generally located in the stationary cortex immediately beneath the plasmalemma, and furthermore are not usually aligned in the direction of streaming.

C. Microfilaments

In the last few years a combination of biochemical and ultrastructural work has revealed that nonmuscle animal cells contain cytoplasmic microfilaments composed of contractile proteins that generate the forces for cellular movements and act as cytoskeletal elements through their ability to form a solid gel in the cytoplasmic matrix (Wessells *et al.*, 1971). These contractile proteins are composed of actin similar to that of muscle, and are in the form of long filaments 50–70 Å in diameter. Myosin and other proteins have been implicated as components of the system in some cells.

Recently it has become possible to study the assembly and disassembly of actin microfilaments *in vitro*. Actin filaments in extracts of sea urchin eggs and other nonmuscle cells, including the cellular slime mold *Dictyostelium discoideum* (Condeelis and Taylor, 1977), form a gel when the extracts are warmed in the presence of ATP. In *Dictyostelium,* gelation involves a specific Ca^{2+}-sensitive interaction between actin and several other components. Myosin, which interacts with actin in muscle contraction, is an absolute requirement for contraction of the extract.

D. Streaming

By far the greatest amount of work on streaming in green plants has been carried out on the giant internodal cells of the Characean algae *Nitella* and *Chara*. The internodal cells of *Nitella* are cylinders 2–5 cm long with a large central vacuole and a thin peripheral layer of cytoplasm lining the walls. The cytoplasm consists of an outer, stationary, gel-like ectoplasm or cortex, and an inner, fluid endoplasm. Chloroplasts one layer deep are embedded in the cortex in helical files just beneath the interface with the endoplasm. The entire endoplasm streams around, virtually as a belt, in a helically wound pathway at a velocity of about 0.1 mm s^{-1} (R. D. Allen, 1974). The direction of streaming is parallel to the files of stationary chloroplasts (Kamitsubo, 1972b).

Many of the classical studies on streaming in *Nitella* and *Chara* were conducted by Kamiya (1959, 1960, 1962; Kamiya and Kuroda, 1956) and Jarosch (1956, 1958, 1964). Kamiya and Kuroda (1956) proposed in their "active shearing theory" that the cytoplasm is driven by a shearing or sliding force developed at the interface between the cortical gel and the streaming endoplasm. Kamitsubo (1966, 1972a) later detected subcortical fibrils apparently attached to the undersurface of the chloroplasts; these fibrils were assumed to provide the structural basis for development of the motive force for streaming. An electron microscopic investigation by Nagai and Rebhun (1966) revealed that the fibrils seen by Kamitsubo in living *Nitella* cells were in fact bundles of microfilaments oriented in the direction of streaming. Microtubules were present, but were located in the ectoplasm just beneath the plasma membrane, and were 1 μm or more distant from the moving endoplasm.

More recently it has been shown that bundles of microfilaments are present in cytoplasm expressed from *Nitella* (Fig. 50), and that the microfilaments resemble muscle actin in their capacity to bind heavy meromyosin in characteristic arrowhead arrays removable by treatment with ATP (Palevitz *et al.*, 1974) (Fig. 51). Palevitz and Hepler (1975) have since shown that the bundles of microfilaments present *in situ* at the ectoplasm–endoplasm interface of *Nitella* also bind heavy meromyosin. As indicated by the arrowheads, all microfilaments in a bundle have a uniform polarity, with arrowheads pointing opposite to the direction of flow, analogous to the pattern of polarity and movement in a striated muscle sarcomere (Kersey *et al.*, 1976).

Palevitz (1976) has considered the significance of the organization of microfilaments in bundles on the pattern and mechanism of streaming. Although it has been claimed in a recent report that myosin occurs in *Nitella* (Kato and Tonomura, 1978), it is conceivable that a mechanism not utilizing actin–myosin interactions functions in *Nitella*. Also, it has not yet been unequivocally established that the bundles of microfilaments are responsible for the motive force in streaming. By techniques utilizing light microscopy and motion analysis, N. S. Allen (1974) has detected the presence of an additional component in the form of "active" filaments which branch from the bundles into the endoplasm. Her evidence suggests that waves are propagated along these filaments, driving the endoplasm forward.

It should be noted that although the ectoplasm in some way fixes both chloroplasts and microfilament bundles in position, providing further evidence that this layer differs structurally from the endoplasm, no single characteristic differentiating the gel-like ectoplasm from the mobile endo-

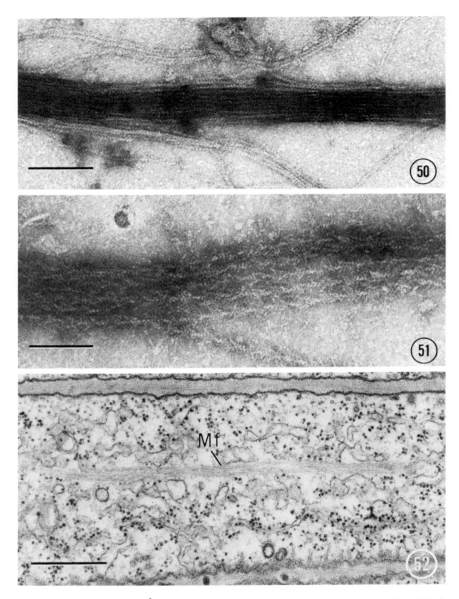

Fig. 50. A bundle of 50-Å beaded microfilaments in cytoplasm expressed from *Nitella*. Scale bar = 0.1 μm ($\times$ 170,000). From Palevitz *et al.* (1974), courtesy of B. A. Palevitz.

Fig. 51, A bundle of microfilaments expressed from *Nitella* and reacted with heavy meromyosin (HMM). The HMM has been bound in characteristic arrowhead arrays. Scale bar = 0.1 μm ($\times$ 170,000). From Palevitz *et al.* (1974), courtesy of B. A. Palevitz.

Fig. 52. Bundle of microfilaments (Mf) in an elongating cell of a bean root. Scale bar = 0.5 μm ($\times$ 40,000). From Newcomb (1969).

plasm has been detected in the electron microscope (Palevitz, 1976). In light of considerable evidence that actin microfilaments are involved in gelation in a number of other systems, one possibility is that in addition to the interface bundles, the entire ectoplasm may contain a large number of individual microfilaments not evident in material fixed for electron microscopy because they are not stabilized, as they may be in the bundles, e.g., by cross-linking. In recent work that may be pertinent, it has been found that during the fixation of isolated actin, exposure to osmium tetroxide converts the long, straight, unbranched actin filaments into branching meshworks of fibrous material (Maupin-Szamier and Pollard, 1978). A similar process can occur when cells are fixed in glutaraldehyde/osmium tetroxide under the conditions ordinarily employed. A degradation of the microfilaments in this way might make it impossible to detect structural differences between different regions of the cytoplasm in fixed material.

Bundles of microfilaments, each 50–60 Å thick, have been observed in many elongating cells of higher plants (Parthasarathy and Mühlethaler, 1972) (Fig. 52). The bundles are about 0.1–0.3 μm wide and up to 12 μm long, and are usually oriented parallel to the longitudinal axis of the cell. They have been observed in a variety of cells known to exhibit vigorous rotational streaming, but not nearly so consistently as expected if they do indeed underlie the streaming process. It seems likely that they are labile and are not well preserved during processing for electron microscopy.

The observations of O'Brien and McCully (1970) on living cells are particularly suggestive of a causative role of microfilaments in streaming. Using phase contrast, they have observed the high speed movement of cytoplasm close to cytoplasmic fibers in petiolar hairs of cow-parsnip (*Heracleum mantegazzianum*). When organelles happened into the zone of influence of a fiber, they were swept away instantly in the stream. The authors suggest that the fibers correspond to the bundles of microfilaments seen in various plant tissues by electron microscopy.

The actinlike nature of microfilaments in higher plants has been demonstrated by means of the reaction with heavy meromyosin using material from the African blood lily (*Haemanthus katherinae*) by Forer and Jackson (1975) and from *Amaryllis belladonna* by Condeelis (1974).

E. Structure in the Ground Substance

In recent years, several investigators working with cultured animal cells have accumulated evidence that the cytoplasmic ground substance has a three-dimensional mesh or lattice constructed of strands 3–6 nm thick (Buckley, 1975; Buckley and Porter, 1975; Lenk *et al.*, 1977; Wolosewick and Porter, 1976, 1979). The strands, referred to as "microtrabeculae," are different from the previously considered microfilaments that contain actin, and also from a class of filaments 6–11 nm thick which have been observed in a variety of animal cells and are collectively termed "intermediate-sized filaments."

The lattice is best seen when entire cells are fixed in glutaraldehyde/osmium, dried by the critical point method, and examined stereoscopically in the high voltage electron microscope (Buckley and Porter, 1975; Wolosewick and Porter, 1976). It is continuous with the surfaces of microtubules, and also with microfilaments, the plasma and nuclear membranes, the endoplasmic reticulum, and free ribosomes.

The significance of the lattice structure remains in doubt. Gray (1975) considers the "stereo framework" and other examples of a three-dimensional cytonet to be an "artifact, the result of denatured protein complexes," but Wolosewick and Porter (1976, 1979) suggest that it may provide a structured frame for the nonrandom distribution of organelles and membranous systems, and may be an important structural feature in eukaryotic cells generally.

In plant cells fixed in glutaraldehyde/osmium the ground substance has a fine fibrogranular appearance. The possibility that if 50–70 Å microfilaments are present in the cytoplasm, they might be destroyed by glutaraldehyde/osmium (Maupin-Szamier and Pollard, 1978) has already been mentioned. Whether other components of the ground substance of plant cells might be organized as a lattice or cytonet appears not to have been determined using critical point drying and high voltage electron microscopy. Until now cultured animal cells, growing as single individuals over a substratum as a very thin layer of protoplasm, have provided the material for these techniques.

REFERENCES

Aaronson, R. P., and Blobel, G. (1975). *Proc. Natl. Acad. Sci. U.S.A.* **72**, 1007–1011.

Akazawa, T., Newcomb, E. H., and Osmond, C. B. (1978). *Mar. Biol.* **47**, 325–330.

Albersheim, P. (1976). *In* "Plant Biochemistry" (J. Bonner and J. E. Varner, eds.), 3rd ed., pp. 225–274. Academic Press, New York.

Allen, N. S. (1974). *J. Cell Biol.* **63**, 270–287.

Allen, R. D. (1974). *Symp. Soc. Exp. Biol.* **28**, 15–26.

Anton-Lamprecht, I. (1967). *Ber. Deutsch Bot. Ges.* **80**, 747–754.

Avers, C. J. (1976). "Cell Biology." Van Nostrand-Reinhold, New York.

Bauer, W. D., Talmadge, K. W., Keegstra, K., and Albersheim, P. (1973). *Plant Physiol.* **51**, 174–187.

Birky, C. W., Jr (1978). *Annu. Rev. Genet.* **12**, 471–512.

Bonnett, H. T., Jr., and Newcomb, E. H. (1965). *J. Cell Biol.* **27**, 423–432.

Bonnett, H. T., Jr., and Newcomb, E. H. (1966). *Protoplasma* **62**, 59–75.

Bowles, D. J., and Northcote, D. H. (1974). *Biochem. J.* **142**, 139–144.

Breidenbach, R. W., and Beevers, H. (1967). *Biochem. Biophys. Res. Commun.* **27**, 462–469.

Breidenbach, R. W., Kahn, A., and Beevers, H. (1968). *Plant Physiol.* **43**, 705–713.

Brown, R. M., Jr., and Montezinos, D. (1976). *Proc. Natl. Acad. Sci. U.S.A.* **73**, 143–147.

Brown, R. M., Jr., and Romanovicz, D. K. (1976), *Appl. Polym. Symp.* **28**, 537–585.

Brown, R. M., Jr., Franke, W. W., Kleinig, H., Falk, H., and Sitte, P. (1970). *J. Cell. Biol.* **45**, 246–271.

Brown, R. M., Jr. Herth, W., Frank, W. W., and Romanowicz, D. (1973). *In* "Biogenesis of Plant Cell Wall Polysaccharides" (F. Loewus, ed.), pp. 207–257. Academic Press, New York.

Bücher, T., Neupert, W., Sebald, W., and Werner, S., eds. (1976). "Genetics and Biogenesis of Chloroplasts and Mitochondria." North-Holland Publ., Amsterdam.

Buckley, I. K. (1975). *Tissue Cell* **7**, 51–72.

Buckley, I. K., and Porter, K. R. (1975). *J. Microsc.* **104**, 107–120.

Burnett, J. H. (1976). "Fundamentals of Mycology." Arnold, London.

Chen, C. H., and Lehninger, A. L. (1973). *Arch. Biochem. Biophys.* **157**, 183–196.

Chrispeels, M. J. (1976). *Annu. Rev. Plant Physiol.* **26**, 13–29.

Clowes, F. A. L., and Juniper, B. E. (1964). *J. Exp. Bot.* **15**, 622–630.

Clowes, F. A. L., and Juniper, B. E. (1968). "Plant Cells." Blackwell, Oxford.

Condeelis, J. S. (1974). *Exp. Cell. Res.* **88**, 435–439.

Condeelis, J. S., and Taylor, D. L. (1977). *J. Cell Biol.* **74**, 901–927.

Coombs, J., and Greenwood, A. D. (1976). *In* "The Intact Chloroplast" (J. Barber, ed.), pp. 1–51. Elsevier, Amsterdam.

Cox, G. C., and Juniper, B. E. (1973). *Nature (London) New Biol.* **243**, 116–117.

Crowther, R. A., Finch, J. T., and Pearse, B. M. F. (1976). *J. Mol. Biol.* **103**, 785–798.

de Duve, C., and Bauduin, P. (1966). *Physiol. Rev.* **46**, 323–357.

DePierre, J. W., and Dallner, G. (1975). *Biochim Biophys. Acta* **415**, 411–472.

DePierre, J. W., and Ernster, L. (1977). *Annu. Rev. Biochem.* **46**, 201–262.

Doohan, M. E., and Newcomb, E. H. (1976). *Aust. J. Plant Physiol.* **3**, 9–23.

Dougall, D. K., and Shimbayashi, K. (1960). *Plant Physiol.* **35**, 396–404.

Esau, K. (1977). "Anatamy of Seed Plants." Wiley, New York.

Fineran, B. A. (1971). *Protoplasma* **72**, 1–18.

Forer, A., and Jackson, W. T. (1975). *Cytobiology* **10**, 217–226.

Franke, W. W. (1969). *Z Zellforsch.* **105**, 405–429.

Franke, W. W. (1970). *Naturwissenschaften* **57**, 44–45.

Franke, W. W., and Herth, W. (1974). *Exp. Cell. Res.* **89**, 447–451.

Franke, W. W., Kartenbeck, J., Krien, S., van der Woude, W. G., Scheer, U., and Morré, D. J. (1972) *Z. Zellforsch.* **132**, 365–380.

Frederick, S. E., and Newcomb, E. H. (1969a). *Science* **163**, 1353–1355.

Frederick, S. E., and Newcomb, E. H. (1969b). *J. Cell Biol.* **43**, 343–353.

Frederick, S. E., and Newcomb, E. H. (1971). *Planta* **96**, 152–174.

Frederick, S. E., Newcomb, E. H., Vigil, E. L., and Wergin, W. P. (1968). *Planta* **81**, 229–252.

Frederick, S. E., Gruber, P. J., and Newcomb, E. H. (1975). *Protoplasma* **84**, 1–29.

Gall, J. G. (1967). *J. Cell Biol.* **32**, 391–399.

Geisow, M. (1979). *Nature (London)* **277**, 90–91.

Goldman, B. M., and Blobel, G. (1978). *Proc. Natl. Acad. Sci. U.S.A.* **75**, 5066–5070.

Gray, E. G. (1975), *J. Neurocytol.* **4**, 315–339.

Grove, S. N., Bracker, C. E., and Morré, D. J. (1968). *Science* **161**, 171–173.

Gruber, P. J., Trelease, R. N., Becker, W. M., and Newcomb, E. H. (1970). *Planta* **93**, 269–288.

Gunning, B. E. S. (1976). *In* "Intercellular Communication in Plants. Studies on Plasmodesmata" (B. E. S. Gunning and A. W. Robards, eds.), pp. 1–13. Springer-Verlag, Berlin and New York.

Gunning, B. E. S., and Pate, J. S. (1969). *Protoplasma* **68**, 107–133.

Gunning, B. E. S., and M. W. Steer. (1975). "Ultrastructure and the Biology of Plant Cells." Arnold, London.

Gunning, B. E. S., Hardham, A. R., and Hughes, J. E. (1978). *Planta* **143**, 161–179.

Haberlandt, G. F. J. (1914). Macmillan, New York.

Heinrich, G. (1973). *Planta Med.* **23**, 154–166.

Hepler, P. K., and Fosket, D. E. (1971). *Protoplasma* **72**, 213–236.

Hepler, P. K., and Newcomb, E. H. (1967). *J. Ultrastruct. Res.* **19**, 498–513.

Hepler, P. K., and Palevitz, B. A. (1974). *Annu. Rev. Plant Physiol.* **25**, 309–362.

Herth, W., Franke, W. W., Stadler, J., Bittiger, H., Keilich, G., and Brown, R. M., Jr. (1972). *Planta* **105**, 79–92.

Highfield, P. E., and Ellis, R. J. (1978). *Nature (London)* **271**, 420–424.

Honda, S. I., Hongladarom-Honda, T., and Kwanyuen, P. (1971). *Planta* **97**, 1–15.

Hyde, B. B., Hodge, A. J., Kahn, A., and Birnstiel, M. L. (1963). *J. Ultrastruct. Res.* **9**, 248–258.

Jarosch, R. (1956). *Phyton (Buenos Aires)* **6**, 87–108.

Jarosch, R. (1958). *Protoplasma* **50**, 93–108.

Jarosch, R. (1964). *In* "Primitive Motile Systems in Cell Biology" (R. D. Allen and N. Kamiya, eds.), pp. 599–622. Academic Press, New York.

Jones, M. G. K. (1976). *In* "Intercellular Communication in Plants: Studies on Plasmodesmata" (B. E. S. Gunning and A. W. Robards, eds.), p. 81. Springer-Verlag, Berlin and New York.

Jørgensen, L. B., Behnke, H. D. and Mabry, T. J. (1977). *Planta* **137**, 215–224.

Kamitsubo, E. (1966). *Proc. Jpn. Acad. Sci.* **42**, 640.

Kamitsubo, E. (1972a). *Protoplasma* **74**, 53–70.

Kamitsubo, E. (1972b). *Exp. Cell Res.* **74**, 613–616.

Kamiya, N. (1959). *Protoplasmatologia* **8** No. 3a, 1–199.

Kamiya, N. (1960). *Annu. Rev. Plant Physiol.* **11**, 323–340.

Kamiya, N. (1962). *Handb. Pflanzenphysiol.* **XVII** (2), 979–1035.

Kamiya, N., and Kuroda, K. (1956). *Bot. Mag.* **69**, 544–554.

Kato, T., and Tonomura, Y. (1978). *J. Biochem.* **82**, 777–782.

Keegstra, K., Talmadge, K. W., Bauer, W. D., and Albersheim, P. (1973). *Plant Physiol.* **51**, 188–196.

Kersey, Y. M., Hepler, P. K., Palevitz, B. A., and Wessells, N. K. (1976), *Proc. Natl. Acad. Sci. U.S.A.* **73**, 165–167.

Kirk, J. T. O., and Tilney-Bassett, R. A. E. (1978). "The Plastids: Their Chemistry, Structure, Growth and Inheritance," 2nd ed. Elsevier/North-Holland Biomed. Press, Amsterdam.

Labavitch, J. M., and Ray, P. M. (1974a). *Plant Physiol.* **53**, 669–673.

Labavitch, J. M., and Ray, P. M. (1974b). *Plant Physiol.* **54**, 499–502.

Lamport, D. T. A. (1970). *Annu. Rev. Plant Physiol.* **21**, 235–270.

Lamport, D. T. A. (1973). *In* "Biogenesis of Plant Cell Wall Polysaccharides" (F. Loewus, ed.), pp. 149–164. Academic Press, New York.

Lamport, D.T. A., and Miller, D. H. (1971). *Plant Physiol.* **48**, 454–456.

Lamport, D. T. A., and Northcote, D. H. (1960). *Nature (London)* **188**, 665–666.

Larkins, B. A., and Hurkman, W. J. (1978). *Plant Physiol.* **62**, 256–268.

Lazarow, A., and Cooperstein, S. J. (1953). *J. Histochem. Cytochem.* **1**, 234–241.

Ledbetter, M., and Porter, K. R. (1963). *J. Cell Biol.* **19**, 239–250.

Lenk, R., Ransom, L., Kaufmann, Y., and Penman, S. (1977). *Cell* **10**, 67–78.

Lewin, R. A. (1975). *Phycologia* **14**, 153–160.

Lewin, R. A. (1977). *Phycologia* **16**, 217.

Lewin, R. A., and Withers, N. W. (1975). *Nature (London)* **256**, 735–737.

Lichtenthaler, H. K. (1968). *Endeavor* **27**, 144–149.

Lichtenthaler, H. K. (1973). *Ber. Deutsch Bot. Ges.* **86**, 313–329.

Lott, J. N. A., Larsen, P. J., and Whittington, C. M. (1972). *Can. J. Biol.* **50**, 1785–1787.

Lwoff, A. (1971). *Annu. Rev. Microbiol.* **25**, 1–26.

Manton, I. (1967). *J. Cell Sci.* **2**, 411–418.

Marchant, R., and Robards, A. W. (1968). *Ann. Bot. (London)* **32**, 457–471.

Matile, Ph. (1974). *In* "Dynamic Aspects of Plant Ultrastructure" (A. W. Robards, ed.), pp. 178–218. McGraw-Hill, New York.

Matile, Ph. (1978). *Annu. Rev. Plant Physiol.* **29**, 193–213.

Maupin-Szamier, P., and Pollard, T. D. (1978). *J. Cell Biol.* **77**, 837–852.

Mollenhauer, H. H., Morré, D. J., and Totten, C. (1973). *Protoplasma* **78**, 443–459.

Montezinos, D., and Brown, R. M., Jr. (1976). *J. Supramolec. Struct.* **5**, 277–290.

Morré, D. J., and Mollenhauer, H. H. (1974). *In* "Dynamic Aspects of Plant Ultrastructure" (A. W. Robards, ed.), pp. 84–137. McGraw-Hill, New York.

Morré, D. J., Mollenhauer, H. H., and Bracker, C. E. (1971). *In* "Origin and Continuity of Cell Organelles" (J. Reinert and H. Ursprung, eds.), pp. 82–126. Springer-Verlag, Berlin and New York.

Mueller, S. C., Brown, R. M., Jr., and Scott, T. K. (1976). *Science* **194**, 949–951.

Nadakavakaren, M. J. (1964). *J. Cell Biol.* **23**, 193–195.

Nagai, R., and Rebhun, L. I. (1966). *J. Ultrastruct. Res.* **14**, 571–589.

Neville, H. C., Gubb, D. C., Crawford, R. M. (1976). *Protoplasma* **90**, 307–317.

Newcomb, E. H. (1969). *Annu. Rev. Plant Physiol.* **20**, 253–288.

Newcomb, E. H. (1967). *J. Cell Biol.* **33**, 143–163.

Newcomb, E. H. (1979). *In* "Coated Vesicles" (C. Ockleford and A. Whyte, eds.). Cambridge Univ. Press, London and New York.

Newcomb, E. H., and Pugh, T. D. (1975). *Nature (London)* **253**, 533–534.

Nishimura, M., and Beevers, H. (1978). *Plant Physiol.* **62**, 44–48.

Nishimura, M., and Beevers, H. (1979). *Nature (London)* **277**, 412–413.

Noland, L. E. (1957). *J. Protozool.* **4**, 1–6.

Northcote, D. H. (1972). *Annu. Rev. Plant Physiol.* **23**, 113–132.

O'Brien, T. P., and McCully, M. E. (1970). *Planta* **94**, 91–94.

Ockleford, C., and Whyte, A., eds. (1979). "Coated Vesicles." Cambridge Univ. Press, London and New York.

Palevitz, B. A. (1976). *Cell Motility, Cold Spring Harbor Conf. Cell Prolif.,* pp. 601–611.

Palevitz, B. A., and Hepler, P. K. (1975). *J. Cell Biol.* **65**, 29–38.

Palevitz, B. A., Ash, J. F., and Hepler, P. K. (1974). *Proc. Natl. Acad. Sci. U.S.A.* **71**, 363–366.

Parsons, D. F., Bonner, W. D. and Verboon, J. G. (1965). *Can. J. Bot.* **43**, 647–655.

Parthasarathy, M. V., and Mühlethaler, K. (1972). *J. Ultrastruct. Res.* **38**, 46–62.

Pearse, B. M. F. (1975). *J. Mol. Biol.* **97**, 93–98.

Pearse, B. M. F. (1976). *Proc. Natl. Acad. Sci. U.S.A.* **73**, 1255–1259.

Pollard, T. D. (1976). *J. Supramolec. Struct.* **5**, 317–334.

Possingham, J. W., and Saurer, W. (1969). *Planta* **86**, 186–194.

Preston, R. D. (1964). *In* "The Formation of Wood in Forest Trees" (M. A. Zimmerman, ed.), p. 169. Academic Press, New York.

Preston, R. D. (1974). "The Physical Biology of Plant Cell Walls." Chapman & Hall, London.

Ramirez-Mitchell, R., Johnson, H. M., and Wilson, R. H. (1973). *Exp. Cell Res.* **76**, 449–451.

Robards, A. W. (1976). *In* "Intercellular Communication in Plants: Studies on Plasmodesmata" (B. E. S. Gunning and A. W. Robards, eds.), pp. 15–57. Springer-Verlag, Berlin and New York.

Roberts, K., and Northcote, D. H. (1971). *Microsc. Acta* **71**, 102–120.

Robertson, J. D. (1967). *Protoplasma* **63**, 218–245.

Roelofsen, P. A. (1959). *In* "Encyclopaedia of Plant Anatomy" (W. Zimmerman and P. G. Ozenda, eds.), Vol. IV, Part 3. Borntraeger, Berlin.

Roelofsen, P. A. (1965). *Adv. Bot. Res.* **2,** 69–149.

Roland, J.-C., Vian, B., and Reis D. (1977). *Protoplasma* **91,** 125–141.

Romanovicz, D. K., and Brown, R. M., Jr. (1976). *Appl. Polym. Symp.* **28,** 587–610.

Sabatini, D. D., Bensch, K., and Barrnett, R. J. (1963). *J. Cell Biol.* **17,** 19–58.

Sadava, D., and Chrispeels, M. J. (1973). *In* "Biogenesis of Plant Cell Wall Polysaccharides" (F. Loewus, ed.), pp. 165–174. Academic Press, New York.

Sargent, C. (1978). *Protoplasma* **95,** 309–320.

Scheer, U. (1973). *Dev. Biol.* **30,** 13–28.

Schnepf, E. (1969a). *Protoplasma* **67,** 185–194.

Schnepf, E. (1969b). *Protoplasma* **67,** 195–203.

Schnepf, E. (1969c). *Protoplasma* **67,** 205–212.

Schnepf, E. (1969d). *Protoplasma* **67,** 375–390.

Schnepf, E., and Nagl, W. (1970). *Protoplasma* **69,** 133–143.

Schwartz, R. M., and Dayhoff, M. O. (1978). *Science* **199,** 395–403.

Seckbach, J. (1972). *J. Ultrastruct. Res.* **39,** 65–76.

Sexton, R., and Hall, J. L. (1978). *In* "Electron Microscopy and Cytochemistry of Plant Cells" (J. L. Hall, ed.), pp. 63–147. Elsevier/North-Holland Biomed. Press, Amsterdam.

Shore, G. C., and Tata, J. R. (1977). *Biochim. Biophys. Acta* **472,** 197–236.

Singer, S. J., and Nicolson, G. L. (1972). *Science* **175,** 720–731.

Snyder, J. A., and McIntosh, J. R. (1976). *Annu. Rev. Biochem.* **45,** 699–720.

Steward, F. C., Bidwell, R. G. S., and Yemm, E. W. (1956). *Nature (London)* **178,** 734–738.

Talmadge, K. W., Keegstra, K., Bauer, W. D., and Albersheim, P. (1973). *Plant Physiol.* **51,** 158–173.

Thair, B. W., and Wardrop, A. B. (1971). *Planta* **100,** 1–17.

Thorne, S. W., Newcomb, E. H., and Osmond, C. B. (1977). *Proc. Natl. Acad. Sci. U.S.A.* **74,** 575–578.

Tolbert, N. E. (1971). *Annu. Rev. Plant Physiol.* **22,** 45–74.

Tolbert, N. E., Oeser, A., Kisaki, T., Hageman, R. H., and Yamazaki, R. K. (1968). *J. Biol. Chem.* **243,** 5179–5184.

Trelease, R. N., Becker, W. M., Gruber, P. J., and Newcomb, E. H. (1971). *Plant Physiol.* **48,** 399–426.

van Steveninck, M. E., and van Steveninck, R. F. M. (1971). *Protoplasma* **73,** 107–119.

Vigil, E. L. (1973). *Sub-Cell. Biochem.* **2,** 237–285.

Weinbach, E. C., and von Brand, T. (1967). *Biochim. Biophys. Acta* **148,** 256–266.

Wessells, N. K., Spooner, B. S., Ash, J. F., Bradley, M. O. L., Luduena, M. A., Taylor, E. L., Wrenn, J. T., and Yamada, K. M. (1971). *Science* **171,** 135–143.

Westafer, J. M., and Brown, R. M., Jr. (1976). *Cytobios* **15,** 111–138.

Willison, J. H. M., and Brown, R. M., Jr. (1977). *Protoplasma* **92,** 21–41.

Wollenweber, E., and E. Schnepf. (1970). *Z. Pflanzenphysiol.* **62,** 216–227.

Wolosewick, J. J., and Porter, K. R. (1976). *Am. J. Anat.* **147,** 303–323.

Wolosewick, J. J., and Porter, K. R. (1979). *J. Cell Biol.* **82,** 114–134.

Wooding, F. B. P. (1968). *J. Cell Sci.* **3,** 71–80.

Woods, J. W., Woodward, M. P., and Roth, T. F. (1978). *J. Cell Sci.* **30,** 87–97.

Use of Plant Cell Cultures in Biochemistry

2

PAUL LUDDEN
PETER S. CARLSON

I. INTRODUCTION

To the nonpractitioners of plant cell culture, the mention of the technique often evokes the picture of single plant cells, in high concentrations, growing

The Biochemistry of Plants, Vol. 1

55

with uniform morphology and developmental state. In short, a bacterial-like culture is envisioned. This vision has been shared by many workers in the field and is surely the ideal but, in reality, never the case. Plant cell cultures are derived from complex eukaryotic organisms and these cells are capable of many developmental states. Plant cells are large, roughly 1000 times the volume of a bacterial cell, and thus rapidly settle out of an unstirred or unshaken medium. They are also, of course, incapable of self locomotion. As a result of this physical property of the cells, the simple measurement of cell density that is performed on bacterial cultures is difficult and error-prone with plant cell cultures. Furthermore, the cells do not grow as single cells but, rather, proliferate as small clumps of cells. Even in the most finely divided cultures, those of sycamore and scarlet rose, only about 30% of the cells exist in clumps of less than 10 cells. In contrast to bacteria, where 10^7 cells per milliliter results in just barely turbid media, plant cells have reached stationary phase of growth at a density of 10^6–10^7 cells per milliliter and the majority of the volume of the culture medium is taken up by plant cells. Furthermore, plant cells do not grow when inoculation densities fall below 10^3 cells per milliliter and, therefore, a cell growth experiment will often involve only four to five cell doublings. The rate of growth of plant cells is slow when compared to bacteria or even transformed lines of mammalian cells; the fastest growing plant cell suspension cultures have doubling times of 18–20 h and a more common doubling time would be 40–50 h. This slowness of growth rate increases the time needed for experiments, and the extended duration of experiments (during which the culture is being sampled and manipulated) increases the possibility of bacterial or fungal contamination. These growth rates also need to be considered when determining how long of a wait will be required when enriching for cells expressing a mutant phenotype (Parke and Carlson, 1978).

Plant cells in culture do not necessarily exhibit any of the patterns of development observed for cells in intact whole plants; plant cells in culture are in a unique developmental state. Furthermore, there are variations among cells in a single batch culture. For example, different isozymes of aspartate kinase are produced by carrot cells during different phases of culture growth (Davies and Miflin, 1978) and variations in the expression of peroxidase isozymes have been noted for cells in different clump sizes (Verma and Van Huystee, 1970). Thus, it should not be assumed that cells have identical metabolic activities during all phases of growth or that a completely homogeneous population of cells is being used. In addition to variations in the expressed phenotype of cells, the karyotypes of cultured cells show great variation, generally with increases rather than decreases in chromosome number. Nevertheless, plant cells in culture offer the potential of a far more homogeneous population of cells than do whole plants which have numerous tissue complexities.

Despite the warnings above that plant cells are not bacteria, the application of microbiological techniques to plant cells offers some distinct advantages to the plant researcher. Plant cells can be grown in uniform, well-defined sterile medium on a large scale. Large populations of plant cells can therefore be obtained rapidly in a relatively uniform state. Large numbers of cells can also be screened for mutant or variant phenotypes; for the species that can be regenerated to whole plants, each cell potentially represents a whole plant. For comparison, corn plants are planted at approximately 10^5 per acre in the field; as cells, they can be plated at 10^4–10^5 per petri plate (a petri plate is approximately 10^{-7} acres). To the extent that the phenotype to be selected for can be observed in cells on plates, the advantages of cultured plant cells are obvious. To the biochemist, it is not how many cells but how much DNA, RNA, or protein can be obtained that is important. Plant cells vary in their content of these macromolecules, but it can be calculated from King's (1973) data that in 10^6 sycamore cells, representing 35 mg of fresh weight or about 4 mg of dry weight, 2–8 μg of DNA, 30–100 μg of RNA, and 170–660 μg of protein would be found. Another useful example is the purification of phenylalanine ammonia lyase (PAL) (Zimmerman and Hahlbrock, 1975). Starting with 3 kg of parsley cells, they were able to purify about 5 mg of PAL to homogeneity. For comparison, 3 kg of sliced and illuminated potato tuber tissue (a particularly rich source of the enzyme) yields about 40 mg of PAL. Thus, the application of microbiological techniques to plant cells, although not as easy as might be imagined, offers an adequate and attractive system to both the biochemist and the geneticist. This chapter seeks to describe areas of research in which plant cell culture has been applied to questions of plant biochemistry that could not have been as easily approached using whole plants or their tissues.

II. CHOICE OF EXPERIMENTAL ORGANISM

Biochemical and molecular biological investigations with cultured plant cells and protoplasts have been performed with a wide variety of cell developmental types spanning essentially the entire spectrum of the plant kingdom. Techniques for the proliferation of cells from plant tissues and for the preparation of protoplasts from both cell cultures and from plant organs have advanced so that, technically, no tissue is thought to be inaccessible. Several plants have been used extensively in the development of tissue culture techniques and have come to be used more widely than others in biochemical studies. These are carrot (*Daucus carrota*), tobacco (*Nicotiana* sp.), sycamore (*Acer pseudoplantanus*), soybean (*Glycine max*), and pea (*Pisum sativum*).

Carrot tissue was used in early studies by Gautharet and by Steward's group. It was the first species utilized that demonstrated plant regeneration and totipotency of cultured cells. Plant regeneration occurs via embryogenesis rather than via shoot formation followed by root formation as seen in tobacco. Carrot tissues require only an auxin for continued growth in culture. Thus, it could be grown on defined medium before many other cell types from different species.

The use of *Nicotiana* sp. derives largely from White's finding that the product of the hybrid cross between *N. langsdorfii* and *N. glauca* produced a plant which produced tumorous tissue that would continue to proliferate in simple nutrient medium, and since these tumors were autotrophic for plant hormones, they did not require the addition of plant hormones. *N. tabaccum* was a species widely used by plant physiologists for developmental studies, and Skoog and Miller (1957) showed the interaction of auxin and cytokinin for regeneration. In addition, tobacco cells grow rapidly *in vitro* and readily form suspension cultures. *Nicotiana sylvestra,* a presumed true diploid, has become more popular in recent years.

Sycamore (*Acer pseudoplantanus*) cells were first established in suspension culture by Lamport (1964). The advantage of sycamore cells over other cell types is that the sycamore cells in suspension formed a fine suspension of cells that was more "pipettable" than any other cell line used. Since plant cells grow in small clumps in suspension culture rather than as single cells, cell lines which grow as relatively fine suspensions are much easier to reproducibly sample; any biochemical measurement involves sampling of the experimental material (in this case, the cell suspension) so that a "pipettable" cell line becomes the line of choice. Street's group adopted sycamore as their standard cell line, and the sycamore cell line is undoubtedly the best biochemically characterized of all the available plant cell lines (King and Street, 1977).

Soybean cells have been used extensively by Gamborg's group for studies on metabolism and for the development of media and techniques for producing protoplasts. Consequently, a large body of knowledge about soybean cells in culture (in particular, the SB-1 line) has accumulated. One attraction to the use of soybean cells is the capacity of soybean plants to fix molecular nitrogen in association with *Rhizobium* sp. However, a true nodule-type association has not yet been seen *in vitro*. Cells of pea have been used in Torrey's laboratory and is a favorite of investigators studying the development of vascular elements in cells in culture (e.g., Roberts, 1976). Pea, of course, shares with soybean the capacity to fix nitrogen.

Cell culture work has by no means proceeded to the point that a single organism is preeminent to the extent that *Escherichia coli* is in bacterial genetics. Nor has the mass of knowledge with any single cell line accumulated so that choice of a cell line other than tobacco, carrot, soybean, or sycamore is unwise. In fact, it is well worth considering the properties of other plant cells that have advantages to the plant biochemist or molecular biologist. Table I lists the primary properties of a number of plants with respect to potential for biochemical and molecular biological investigations.

Regeneration is of importance to some but not all studies. Experiments which depend on regeneration of mutant cells into whole plants to determine the effect of a mutation on whole plant physiology or development require the ability of the experimental material to pass easily from plant to *in vitro* cell culture and back to whole plant. The ability to prove by Mendelian analysis that a mutation in a cell culture has occurred also requires regeneration of whole plant from the presumed mutant. Regeneration parameters for tobacco, petunia, and carrot from suspension cultures and from single cells are well established; carrot regeneration occurs via embryogenesis while tobacco and petunia produce shoots and roots on separate media containing different ratios of phytohormones. Tomato and corn can be regenerated from callus tissue and potato plants have recently been regenerated from single cells. Sugar cane is an economically important C_4 plant for which regeneration from callus tissue can readily be achieved. The grain legumes—soybean, pea, and others—have resisted efforts to regenerate whole plants from cultured cells; rice, barley, and corn have shown the most promise for regeneration among the grasses. Sycamore, the best biochemically characterized plant cell culture system, has shown no inclination to revert to whole plant status. Lack of a regeneration system is not an insurmountable problem, and more species are sure to be placed in the category of routine regeneration in the future.

To the extent that the molecular biology of eukaryotes is a fusion of biochemistry and classical genetics, the following parameters become extremely important in choosing plant material: chromosome number, ploidy, available genetic markers, and characterized linkage groups as well as general cytology. Susceptibility to potential genetic vectors which then allow genetic modifications must also be considered. Tomato, pea, barley, and corn are economically important diploids with relatively low chromosome numbers. As a result of breeding programs, these four plants have large numbers of genetic markers that have been mapped in linkage groups corresponding to the individual chromosomes. Carrot and petunia also are diploids without excessive numbers of chromosomes but lack the extensive characterization of chromosomal or cytological markers available in the first four species. Tobacco, sycamore, soybean, and potato have high chromosome numbers (48, 52, 40, and 48, respectively), are not true diploids, and have relatively few genetic markers.

The ability to recover either haploid cells or haploid plants is an advantage

TABLE I

Biochemical and Molecular Properties of Various Plants

Plant	Regeneration	2N	Cytology	Genetic markers	A. tumefaciens	Virus[a]	Monoploids	True diploid	Other
Tomato	+	24	Good	Many	Yes	Yes	Yes	Yes	
Barley	+	10	Excellent	Many	No	Yes	Yes	Yes	
Pea	−	14	Good	Many	Yes	No	No	Yes	N_2 fixation
Corn	+	20	Good	Many	No	Yes	Yes	Yes	C_4 plant
N. tabaccum	+++	48	Poor	Few	Yes	Yes	Yes	No	
Carrot	+++	18	Poor	Few	Yes	No	Yes	Yes	
Petunia	++	14	Poor	Few	Yes	Yes	Rare	Yes	
Soybean	−	40	Poor	Some	Yes	No	Rare	No	N_2 fixation
Sycamore	−	52	Poor	Few	Yes	No	No	No	
Haplopapus	−	4	Excellent	Few	Yes	No	No	Yes	
Potato	+	48	Poor	Few	Yes	No	Yes	No	
Wheat	−	42	Poor	Many	No	No	Yes	No	
N. sylvestris	++	18	Poor	Few	Yes	No	Yes	Yes	
Arabidopsis thaliana	+	16	Poor	Many	Yes	No	Rare	Yes	

[a] The ability of protoplasts to be infected by a plant virus (see review by Takebe, 1978).

in genetic manipulation; mutants can be selected without the problem of dominant vs. recessive mutations. Haploid cells or plants have been obtained from most of the plants listed in Table I; however, the generation of haploids from tobacco via anther culture has been more successful than from other species.

The karyotype of the cells being used is also important to the extent that chromosomes need to be distinguishable from each other. Barley, which has seven pairs of large distinguishable chromosomes, and *Haplopapus*, which has only four large chromosomes ($2N = 4$), are clearly the organisms of choice for this property. Tomato, pea, and corn have cytologically analyzable karyotypes; however, several steps in mitosis in pea are difficult to observe. Tobacco has 48 ($2N = 48$) small chromosomes and little beyond chromosome counts can be performed as a routine analysis. There is a great need for improved cytological techniques for analysis of plant chromosomes; development of chromosome banding techniques is of particular importance.

Susceptibility to potential genetic vectors may be of great future importance. In this respect, monocots are at a disadvantage to the dicots, because part of the Ti plasmid from *Agrobacterium tumefaciens* (the organism responsible for crown gall in plants) is transmitted to dicots but not to monocots. The potential of the Ti plasmid from *Agrobacterium* as vector lies in the fact that part of the large plasmid is incorporated into the plant genome during tumor formation and its genetic information is expressed (Schell and Van Montagu, 1977; Nester *et al.*, 1977). *Agrobacterium* has recently been reported to bind to tobacco and carrot cells *in vitro* (Matthysse and Wyman, 1978); however, no evidence for Ti plasmid transfer was observed. Other potential vectors of genetic information are some plant viruses. DNA viruses seem most desirable, but only a few double-stranded DNA viruses which infect plants are known, cauliflower mosaic virus and dahlia mosaic virus being the best examples. The viruses have fairly limited host specificities (but the dahlia virus will infect a number of solanaceous species) and complex population requirements for infectivity (Meagher, 1977). The apparent constitutive presence of reverse transcriptase in plant cells (Ikegami and Fraenkel-Conrat, 1978) may lessen the need for DNA viruses as vectors and some RNA viruses may be useful as vectors. A large number of systems for the infection of plant cell protoplasts by viruses are known (see review by Takebe, 1978).

A final consideration to be made is the possession of specialized genetic information in the plant to be studied. The possession of the C_4 photosynthetic pathway is one character that might be considered, as is the ability to form nodules in association with nitrogen-fixing bacteria. The importance of these features to the plant and the fact that the biochemistry of the systems is well characterized makes them more attractive for studies in cell culture.

III. REGULATION OF DEVELOPMENT AND METABOLISM

While plant cells in culture do not accurately reflect any one developmental state found in the whole plant, the potential for regulation of metabolism can be studied. The metabolism of plant cells in culture can be studied without the constraints of autotrophy, diffusion barriers, diverse cell types, and differential response to plant hormones. A number of systems have been investigated.

A. Phenylalanine Ammonia Lyase (PAL)

Since the discovery of this enzyme in plants by Koukol and Conn (1961) and correlation of increases in its activity on illumination by Zucker (1965), the enzyme has been extensively studied both in whole plants and in tissue culture (Zucker, 1972). PAL is the enzyme at the committed step to biosynthetic pathways resulting in polyphenols, lignin, and flavone glycosides. Figure 1 shows the pathway and enzymes involved in flavone glycoside synthesis (Hahlbrock, 1976). In studies on the induction of this pathway in parsley cells, Hahlbrock and his associates found that the first three enzymes, which are also involved in lignin and polyphenol biosynthesis and are termed the group 1 enzymes, can be induced by both light and the dilution of the cells into fresh medium (Hahlbrock and Schroeder, 1975). In contrast the remaining enzymes unique to flavone glycoside biosynthesis can be induced by light but not by dilution of the cells. The induction of PAL activity by dilution was not seen if cells were diluted into conditioned medium and induction was also not seen if more than 5 g fresh weight of cells were diluted into 40 ml of fresh medium. It is not known which compound (or compounds), when lowered in concentration by dilution, causes the large increase (twentyfold) in PAL activity. It was also shown that, when light was used to induce both groups of enzymes, the lag time before increases in activities of the two sets of enzymes observed were different; the group 1 enzymes showed a lag of 2.5 h before increases in activity were seen, while the group 2 enzymes showed a lag of 4 h (Hahlbrock, 1976). Thus, there is evidence both from differential induction of activity (by light and dilution) and from differential lag periods for enzyme synthesis (or activation) that there are two sets of enzymes that are not coordinately regulated. PAL from light-induced parsley cells has been purified (Zimmerman and Hahlbrock, 1975) and shown to be immunologically identical to the enzyme produced by cells that are induced to make high levels of PAL by dilution (Schroeder, *et al.*, 1977). That the increase in PAL activity was due to *de novo* synthesis of the enzyme rather than to the activation of enzyme already present was shown by demonstrating that increases in [^{35}S]methionine (^{35}S-met) incorporation into PAL correlated with increases in enzyme activity. Increases in

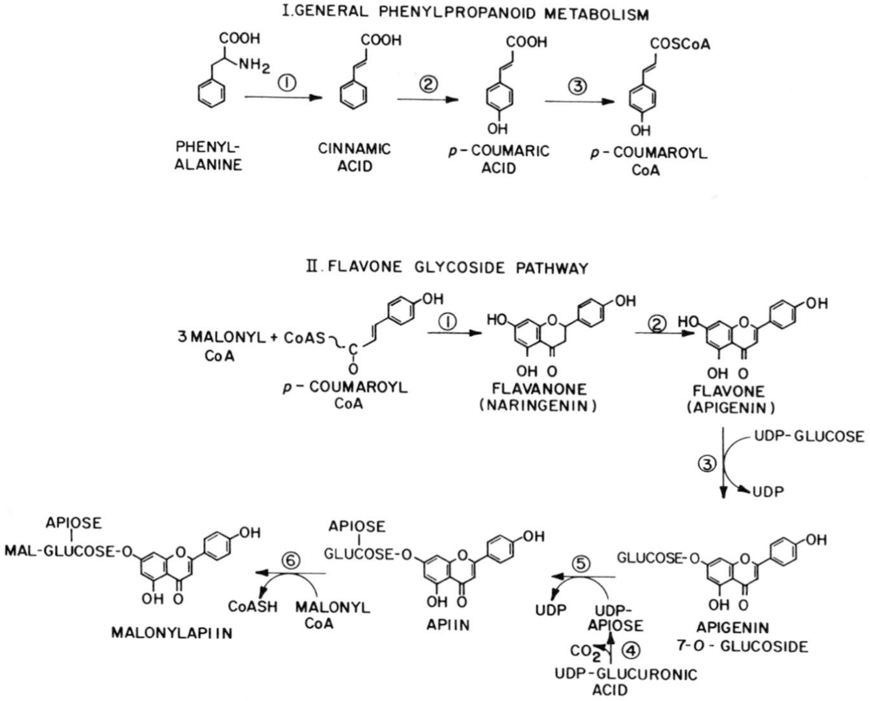

Fig. 2. Sequence of reactions involved in the light-modulated formation of the flavone glycoside "malonyl apiin" in cell suspension cultures. Group I comprises the enzymes of the general phenylpropanoid metabolism (1) phenylalanine ammonia lyase, (2) cinnamate 4-hydroxylase, and (3) p-coumarate:CoA ligase. The enzymes of the flavone glycoside pathway are: (1) flavone synthase, (2) flavone oxidase, (3) UDP-glucose:flavone flavonol-7-O-glucosyl transferase, (4) UDP-apiose synthase, (5) UDP-apiose flavone-7-O-glucoside 2"-O-apiosyl-transferase, and (6) malonyl-CoA:flavonoid glycoside malonyl transferase. From Hahlbrock (1976). Reprinted by permission of *Eur. J. Biochem.*

[35]S-met label in PAL during induction of the enzyme could be seen both in enzyme extracted from cells and in enzyme synthesized *in vitro* on mRNA isolated from cells being induced for PAL. The induction of PAL and the other enzymes did not noticeably affect the pattern of total ^{35}S-labeled peptides from parsley cells since the enzymes make up a relatively small proportion of the total proteins; however, ^{35}S-labeled PAL could be detected on acrylamide gels after precipitation of PAL by antibody to purified enzyme (Fig. 2).

The increases in activity of enzymes in both groups were specific to the extent that increases in the level of activity of other enzymes tested (GDH, glucose-6-P DH) were not seen. It is interesting, and fortunate for this study,

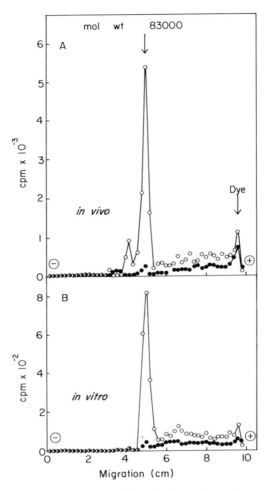

Fig. 2. Gel electrophoretic analysis of PAL subunits synthesized either (A) *in vivo* or (B) *in vitro*. In experiment A, dark-grown cell suspensions, which had been diluted into water, were labeled *in vivo* for 20 min with [^{35}S]methionine immediately prior to harvest either 1.5 h (0) or 4.5 h (0) after dilution. Crude extracts were treated with the antiserum, and the immunoprecipitates were analyzed by electrophoresis on 7.5% polyacrylamide gels in the presence of 0.1% SDS. The total amounts of radioactivity incorporated into extractable protein were 1.5 × 10⁶ and 1.8 × 10⁶ cpm/mg, respectively. The arrow indicates the position and the mol wt of the subunits of highly purified enzyme from light-induced cells. In experiment B, polyribosomal RNA from undiluted cultures (0) or from cultures which had been diluted into water for 5 h (0) was incubated *in vitro* in the rabbit reticulocyte lysate. The immunoprecipitates obtained from the incubations were analyzed by SDS-gel electrophoresis on 7.5% polyacrylamide gels. The cell-free incubations contained 59 μg (0) and 65 μg (0) of parsley polyribosomal RNA. From Schröder *et al.* (1977). Reprinted by permission of *Plant Physiology.*

that a single PAL enzyme type was found; Scandalios (1974) does not list PAL as an enzyme for which isoenzymes have been reported.

PAL has also been studied in varient cell lines of tobacco and carrot cells (Berlin and Widholm, 1978). One interesting varient line of tobacco (resistant to p-fluorophenylalanine) showed elevated levels of PAL (10 times normal) at all times in the growth cycle of a cell culture. The varient line did not show induction by light, while the normal, p-fluorophenylalanine sensitive line showed 3- to 5-fold induction by light, analogous to the parsley cell system. Hence, the cell line with high PAL is a presumptive constituative mutant for induction of PAL and may be of use in determining the conditions that cause the induction of PAL.

The use of cell culture as a tool to study the induction of the enzymes of the flavone glucoside pathway is a model for the type of studies which can be done and the answers that can be obtained. Many basic biochemical and microbiological techniques were applied to the problems and answers were obtained that might not have been easily acquired using whole plants as the experimental system.

B. Cytokinin Control of Polyribosome Formation and the Cell Cycle

The cell cycle is particularly amenable to study with cell culture because large numbers of relatively uniform cells can be easily assayed for their phase of cell division and because the cell cycle can be more easily controlled by chemical or environmental conditions. The cytokinins, named for their ability to promote cell division, have been proposed to play a role in a number of cell division events.

One point of cytokinin control of plant cell metabolism has been shown to be at the level of translation (Fosket et al., 1977; Muren and Fosket, 1977; Fosket and Tepfer, 1978). Using cotyleden-derived soybean cells which require an auxin and a cytokinin for growth, the role of cytokinin in cell division was investigated. It was found that cells deprived of cytokinin stop the cell division process either during the G-2 stage of cell division or just after G-2, before mitosis. However, the significance of the cells ceasing cell division should not be overinterpreted; many types of stress result in interruption of cell division just before mitosis. The addition of cytokinin allowed resumption of cell division, with the first cell division occurring within 36 h. By 15 h after the addition of hormone, increases in RNA and protein synthesis can be observed. The initial event appears to be the increased formation of polyribosomes which occurs within 3 h, but polysome formation is not, in itself, sufficient for the induction of cell division, since dilution of cells into fresh medium also caused increases in polysomes but not in cell division. Dilution of stationary phase cells into fresh medium containing cytokinin, however, causes the formation of polysomes as well as the appearance of several new protein bands on SDS gels run on extracts of cytokinin-treated cells. The RNA synthesis inhibitors, actinomycin D and 5-fluorouridine, did not completely overcome the cytokinin-induced polysome formation. From double label experiments, pulse chase experiments, and inhibitor studies, it was concluded that cytokinin did not influence the number of ribosomes per polysome, the rate of initiation, or the rate of peptide chain elongation. Fosket and Tepfer (1978) have postulated that a population of mRNAs are made but not translated in the absence of cytokinin. The addition of cytokinin is thought to permit translation, and the protein products of translation of these mRNAs are thought to be involved in regulating the cell cycle. Fox's finding (Pratt and Fox, 1978) of a cytokinin binding site in plant ribosomes is compatible with Fosket and Tepfer's model of cytokinin control of translation. The fitting together of these results into a model where a ribosomal binding site for cytokinin regulates at least the translation of some classes of plant mRNAs must be approached with great caution; a binding site for cytokinin-like molecules could be predicted from the knowledge of bases with cytokinin activity found in tRNAs (Skoog et al., 1966). Ascribing a regulatory role to such a binding site would require overwhelming evidence, none of which is available at the present time.

C. Nitrate Assimilation

Nitrate reductase is one of the few enzymes of plant cells which is known to be inducible. Both nitrate reductase and nitrite reductase are induced by nitrate; nitrogen starvation or growth on a poor nitrogen source (urea) is not sufficient to cause the induction of nitrate reductase. Molybdenum (nitrate reductase is a molybdenum-containing enzyme) or tungstate is required for the synthesis of nitrate reductase. In the presence of Mo, the enzyme is active; when tungstate is used, the nitrate reductase that is synthesized is inactive but can be activated by the addition of Mo to the cells. The activation does not require protein synthesis, as cycloheximide does not inhibit the activation. Presumably the activation involves conversion of Mo to the molybdenum cofactor (Ketchum et al., 1970; Shah and Brill, 1977) and the insertion of the cofactor into the enzyme. The nitrite reductase as well as the transport system for nitrate is synthesized and active when tungstate replaces Mo in the medium. It should be noted that nitrate reductase can be induced in embryos of *Agrostemma githago* by benzyladenine in the absence of nitrate (Dilworth and Kende, 1974; Hirsehberg et al., 1972).

Recently, several chlorate resistant lines of tobacco cells have been selected after mutagenesis with N-ethyl-N-nitrosourea (Müller and Grafe, 1978; Mendel and Miller, 1978). The cells used in these experiments were allodihaploid cell lines ($n = 24$) and it was possible to recover chlorate-resistant colonies after selection. Two different phenotypes were recovered in the selected colonies, one of which lacked both nitrate reductase activity and xanthine oxidase activity. Since both of these enzymes are molybdoenzymes, it is possible that a cell line deficient in the synthesis of molybdenum cofactor (MoCo) has been recovered. In agreement with this is the finding that low levels of nitrate reductase activity can be reconstituted when extracts of the putative MoCo$^-$ cell line are mixed with extracts of a cell line that is lacking nitrate reductase but which has xanthine oxidase activity. In contrast to the XD line of tobacco cells used by Filner, the cell lines used by Mendel, Muller, and Grafe had a measureable level of nitrate reductase in the absence of nitrate. In light of the work by Dilworth and Kende (1974), it seems possible that the difference in the level of nitrate reductase in the two cell lines in the absence of nitrate may be a consequence of the hormone concentrations in the media used for the growth of the XD line of tobacco cells and the lines used by Muller *et al.* Filner's medium contains only the auxin, 2,4-dichlorophenoxyacetic acid (2,4-D) at a level of 0.5 mg/liter; Muller *et al.*'s medium contains 1 mg/liter 2,4-D, 0.2 mg/liter kinetin, and 0.1 mg/liter gibberellin.

D. ATP Sulfurylase

The regulation and induction of ATP sulfurylase, the first enzyme in the sulfate assimilation pathway, has been studied in tobacco cells (Reuveny and Filner, 1977; Reuveny, 1977). Although this enzyme was always found, it was synthesized in much greater quantities when cells were starved for sulfur. Sulfate, cysteine, and methionine, but not L-djenkolate, repress the synthesis of the enzyme. The inhibitors of ATP sulfurylase, selenate, and molybdate both derepress the synthesis of the enzyme, presumably because the cells become starved for sulfur-containing amino acids in the presence of the inhibitors. The addition of cysteine or sulfate to cell cultures with high levels of ATP sulfurylase causes the loss of enzymatic activity; the loss of activity was inhibited by protein synthesis inhibitors. The assimilatory systems for nitrate and sulfate can be contrasted as to their modes of regulation. Nitrate reductase is induced by the presence of nitrate while the synthesis of ATP sulfurylase does not require the presence of sulfate. More work is needed to determine if these two assimilatory systems are examples of positive and negative control of protein synthesis in plant cells and the extent of their coordinate control.

E. Urease

A number of types of plant cells will grow with urea as the sole nitrogen source, and the assimilation of urea in soybean cells has been studied

(Polacco, 1976, 1977). Urease was present in cells grown with either nitrate plus ammonia or urea as the nitrogen source but was present in 10- to 20-fold higher levels in cells grown on urea. Stationary phase cells that had been grown on nitrate plus ammonia showed only traces of urease activity. Citrate was found to inhibit the growth of soybean cells on urea but not when ammonia and nitrate served as the nitrogen source; the interpretation offered was that urease, being a nickel enzyme, was deprived of the required cofactor due to chelation with citrate. Both ammonia and methylammonia inhibited the formation of increased levels of urease. Nickel at low levels ($5 \times 10^{-5} M$) has also been shown to stimulate the growth of one whole plant (Lemna) where urea is the nitrogen source (Gordon et al., 1978).

In contrast to the rapid rise in urease activities observed in soybean cells, a very slow adaptation of tobacco cells to growth on urea was found (Skokut, 1978). Urease levels increased 4- to 5-fold when cells were provided with urea as the sole source, but the increase began only after 10 or more cell generations (Fig. 3). The selection of spontaneous high-urease mutants and the presence of a high-urease containing subpopulation were ruled out as explanations. Mixtures of urea-adapted and nitrate grown cells, when grown together, had urease activities that would have been predicted if the cells were grown separately, then mixed; thus, the urea-adapted cells did not rapidly induce high urease in unadapted cells. When urease-adapted cells are transferred to ammonia medium, the urease level drops slowly over several generations. The author notes the similarity of the slow rise and decline of urease activity to the behavior of dihydrofolate reductase in sarcoma cells challenged by the inhibitor methotrexate (Alt et al., 1978). In these cells, there is amplification (200 times) of the gene coding for dihydrofolate reduc-

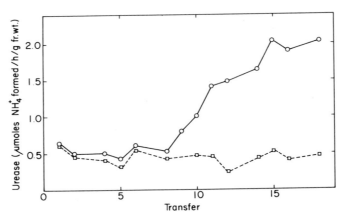

Fig. 3.   Increase in urease activity in tobacco cells grown with urea as the sole nitrogen source. From Skokut (1978). Reprinted by permission of the author.

tase. Experimental evidence is lacking for amplification of the urease gene, or any gene in plants, but the similarities of the urease system and the methotrexate-resistant sarcoma cells are sufficient to warrant further investigation.

F. Phosphatase

The induction of phosphatase activity by phosphate starvation has been studied in cultured tobacco cells (Ueki and Sato, 1977). Three phosphatases were observed and could be easily separated by gel filtration. One phosphatase increased dramatically when cells were starved for phosphorus. Other enzymes which were measured under phosphate stress conditions did not increase in specific activity (α-amylase, galactosidase, succinic dehydrogenase, and catalase).

G. Infection of Plant Cells with Viruses

The development of techniques for producing protoplasts of plant cells has been of prime importance to several aspects of plant biochemistry. Among these is the ability to synchronously infect plant protoplasts with viruses and to thereby study the time course of virus replication. Cocking first observed the uptake (Cocking, 1960) and infection (Cocking and Pojnar, 1969) of tomato protoplasts with tobacco mosaic virus (TMV). However, the efficiency was quite low. Takebe, using protoplasts isolated by enzyme treatment, was able to show rapid uptake of TMV by tobacco mesophyll protoplasts followed by replication of the virus (Takebe and Otsuki, 1969). The addition of a polycation, poly-L-ornithine, was essential for high efficiency infection. The mechanism by which poly-L-ornithine acts to enhance infection is not clear, but Takebe notes that TMV has a net negative charge and that poly-L-ornithine might function to neutralize the charge; poly-L-ornithine also enhances the frequency of endocytosis in protoplasts.

The extent and efficiency of infection of tobacco mesophyll cells by TMV is extremely good for biochemical investigations. Eighty to 90% of the protoplasts present are infected after only a 10-min incubation with virus particles. (This is using protoplasts of mesophyll cells; the percentage of infection is less when protoplasts from suspension culture are used.) Takebe has calculated that there are 8000 virus particles present per protoplast when the infection is most efficient. In contrast, 50,000–1,000,000 particles are present for each infected center on a tobacco leaf. The yield of TMV is 1×10^6 to 9×10^6 particle per infected protoplast. Tobacco protoplasts have also been infected by isolated TMV RNA.

The time course of infection has been studied by Aoki and Takebe (1975) (Fig. 4) by taking advantage of the synchronous infection that can be

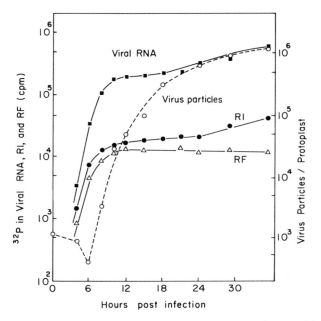

Fig. 4. Time course of synthesis of virus-specific RNAs and virus particle formation in protoplasts. From Aoki and Takebe (1975). Reprinted by permission of *Virology*.

achieved with protoplasts. Three virus-dependent RNAs were found in protoplasts; these corresponded to the TMV RNA, the replicative form (RF), and the replicative intermediate (RI). No virus-dependent, small (2.5 × 10⁶ dalton) RNA was observed, although an RNA of this size has been observed in infected leaves and shown to direct the synthesis of TMV coat protein (Hunter *et al.*, 1976). From the time course in Fig. 4, it can be seen that TMV RNA is synthesized at an exponential rate at 4–8 h postinfection, while very little synthesis of virus particle was observed until 8 h postinfection. Since the timing and rates of appearance of coat protein and virus particle were essentially identical, it would appear that most of the viral RNA is synthesized before packaging into virus particles begins.

Three virus specific proteins have also been identified in the tobacco protoplast-TMV system. The major protein is, of course, the 21,000-dalton TMV coat protein which can be identified antigenically, electrophoetically, and by its lack of methionine and histidine. Two other proteins are seen on SDS gels: a 140,000-MW protein which is thought to be the RNA replicase and a 180,000-MW protein of unknown function. Since there is not sufficient information on the TMV RNA to code for all three proteins, several models have been invoked to explain the three proteins. These include (1) overlapping cistrons within the RNA; (2) the coding of one of the proteins by the minus strand of RNA produced by the replicase; (3) the 180,000-dalton protein being coded for by the host genome but which is synthesized only when induced by TMV infection; and (4) the large protein is a read through of replicase plus coat protein. Hunter *et al.* (1976) have argued against the last possibility from the physical properties of the proteins. TMV RNA also directs the synthesis of two large proteins in *Xenopus* oocytes and in a reticulocyte lysate (Hunter *et al.*, 1976). If these are the same proteins that are found in infected leaves and protoplasts (and this has not been directly demonstrated), it would argue against the minus strand directing synthesis of one of the peptides. Since Aoki and Takebe (1975) did not see an additional message that might code for the large protein, model 3 would also seem to be unlikely. At the present time there is no evidence for or against the overlapping cistron model, but this could be tested with the appropriate mutants.

The study of virus-directed protein synthesis in protoplasts is complicated by the fact that host protein synthesis is scarcely affected by infection (Sakai and Takebe, 1970). Although actinomycin D does not inhibit virus replication, it is of little use in suppression of host protein synthesis because of the relatively long average half-life of tobacco mRNAs. However, uv treatment was helpful in the inhibition of tobacco cell protein synthesis; presumably, it causes degradation of endogenous host mRNA populations. Chloramphenicol does not inhibit virus replication, suggesting that the protein synthesis system of organelles are not involved in virus replication.

A number of other systems have been used in the study of plant–virus interaction. One particularly interesting system discussed by Takebe (1978) is the turnip yellow mosaic virus–Chinese cabbage protoplast system. On infection with TYMV, the chloroplasts of Chinese cabbage protoplasts agglutinate (Ushiyama and Matthews, 1970). Although chloramphenicol did not inhibit replication, the agglutination suggests a role for the chloroplast in the viral replication process; furthermore, no RNA replication occurred in protoplasts incubated in the dark (Renaudin *et al.*, 1976). In his review of plant cell–virus interactions, Takebe (1978) clearly outlines the major problems remaining to be solved. These include the mechanism of RNA uncoating upon infection, the site of virus RNA replication in the plant cell, the role of the minus strand in encoding for a viral protein synthesis, and the possible regulatory properties of the virus-specific proteins.

Cultured plant cells have also been used to study possible inhibitors of viral infection. Takayama *et al.* (1977) selected *Agrostema githago* (corncockle) cells as the producer of the most or the most potent inhibitor of TMV infection of phaseolus. Sucrose concentration, pH, and other media conditions were varied to optimize the production of inhibitor. The inhibitor has been characterized as a protein, but additional information about the properties of the protein and the mode of action are needed.

IV. AUTOTROPHY BY PLANT CELLS IN CULTURE

A. Carbon

Cell cultures able to fix their own carbon have been sought for many years (Gautheret, 1955) and, until recently, there has been a general lack of success. A number of workers have reported photosynthetic activity in cell cultures (Hansen and Edelman, 1972; Davey *et al.*, 1971) and increases in total carbon in the absence of other carbon sources for short periods of time have been reported (Corduan, 1970; Hansen and Edelman, 1972). The first report of sustained photosynthetic growth of plant cells in liquid culture in the absence of any other carbon source at substrate level was by Chandler *et al.* (1972) using tobacco cells. In the absence of sucrose, the total chlorophyll content of cultures increased at least sixfold and the photosynthetic activity (as measured by O_2 evolution) increased about eightfold over cultures grown in sucrose medium. Haploid tobacco cells have been grown continuously in shallow liquid medium in the absence of sucrose (Berlyn and Zelitch, 1975). These cells showed a threefold increase in dry weight in 21 days. The photosynthetic capacity of these cultures has been further characterized (Berlyn *et al.*, 1978) and shown to have rates of about 125 μmoles CO_2 fixed per milligram chlorophyll per hour. In addition, a functioning photorespiratory system was present. Cultures of *Chenopodium rubrum* (Huseman and Barz, 1977) and *Ruta graveolens* (Polevaya *et al.*, 1975) have been coerced to grow autotrophically.

Since sucrose is the primary transported carbon source in whole plants, it is interesting that sucrose inhibits the synthesis of chlorophyll in some carrot cell lines (Edelman and Hansen, 1971a). Sucrose, but not glucose, fructose, or a combination of glucose plus fructose, specifically inhibited chlorophyll synthesis and lamellar development in proplastids. It has been suggested that the site of the sucrose inhibition was at the committed step to chlorophyll and heme synthesis, aminolevulinic acid synthetase (Pamplin and Chapman, 1975). However, these studies used labeled glycine and succinyl-CoA as the precursors to aminolevulinic acid (ALA). Subsequent to these studies, it has been found that chlorophyll synthesis in higher plants, unlike chlorophyll synthesis in photosynthetic bacteria and heme synthesis in bacteria and animals, proceeds through glutamate (or α-ketoglutarate) rather than through succinate and glycine (Beale, 1978). The site of action of sucrose in suppressing chlorophyll synthesis should be reinvestigated in light of this result.

B. Nitrogen

Plants are not capable of converting N_2 to ammonia alone, but a number of plant species provide favorable environments (root nodules) for specific species of nitrogen-fixing bacteria. A number of attempts have been made to achieve symbiosis between nitrogen-fixing bacteria and plant cells in culture. None of these attempts has been entirely successful. The first attempt was not so much an effort to achieve symbiosis but, rather, to use alder tissue in culture to maintain the growth of the actinomycete endophyte of alder nodules (Becking, 1965) because alder endophyte has proved resistant to free-living growth until recently. While the endophyte did grow with the alder tissue, it was generally not possible to produce active nodules on alder plants by inoculation with the tissue culture propagated endophyte. Since the experiments were done before the simple acetylene reduction technique for measurement of nitrogenase activity became available, no tests were done to see if the tissue culture association provided a suitable environment for nitrogen fixation.

Mixtures of Rhizobia and various types of legume cells in culture have resulted in enhanced acetylene reduction (Holsten *et al.*, 1971; Phillips, 1974) and subsequently free-living Rhizobia have been shown to be capable of expressing nitrogenase activity [Parkhurst and Craig (1978) and references therein]. None of the bacteria–plant cell associations have been shown to produce any leghemoglobin—a protein that is found in large amounts in competent root nodules formed by Rhizobium on legume roots. Verma *et al.* (1978) have called attention to the parasitic or saprophytic nature of mixtures of Rhizobium and plant cells which do not have leghemoglobin and do not have high rates of nitrogen fixation; these cautions are well worth considering in evaluating experiments involving legume cells and Rhizobium. Plant cells have the capacity to elicit nitrogenase activity in Rhizobium species that do not show activity on defined media (Reporter, 1976; Hess and Schetter, 1978; Reporter and Hermina, 1975). Reporter has separated two inducing factors from conditioned plant medium. However, the induction of nitrogenase in Rhizobium by plant cells does not require that the plant cells be derived from a plant capable of forming nitrogen-fixing nodules; both non-nodulating mutants of legumes as well as nonlegumes that do not form root nodules under any circumstances have been shown to enhance nitrogenase activity (Hess and Schetter, 1978). The effects of plant hormones on Rhizobium nitrogenase activity have been studied very little; Child and LaRue (1976) reported that the hormone composition of plant growth medium used for experiments in which plant cells and Rhizobium cells were mixed could cause the nitrogenase activity to vary twentyfold.

An association between an adenine requiring strain of *Azotobacter vinelandii* and carrot cells in culture has been described (Chaleff and Carlson, 1974). The adenine auxotrophy of the bacterial member of the association prevented overgrowth of the carrot callus. Increases in fresh weight were recorded for the association on both low nitrogen and no nitrogen medium, while the uninfected controls showed no increase on either medium. Bacterial cells were found in the intercellular spaces but not inside the carrot cells.

V. UPTAKE AND TRANSPORT STUDIES

One area of plant biochemistry research that is amenable to, in fact almost demands, the use of cultured cells or protoplasts is the area of transport of small molecules into the plant cell. Single cells or small clumps of cells (as mentioned previously, very few plant cell lines even approach single cell suspensions) allow easy, uniform application of radioactively labeled compounds to plant cells. The cells can be washed on a filtration apparatus to facilitate the rapid removal of exogenous label as time points are taken. All of the problems associated with comparing plant cells in culture to plant cells *in vivo* apply to studies on uptake by cultured plant cells; however, as a system for investigating the potential of plant cells to transport nutrients, it is unparalleled. Despite the obvious advantages, relatively little work has been done; only a few major nutrients have been studied. The systems for uptake of glucose, sulfate, nitrate, and several amino acids are described below and in Table II.

The general methodology for uptake studies is quite simple, and no deviation from the methods that have been established for uptake studies of metabolites by bacteria are required (see Passow and Stampfli, 1969). The compound to be transported is added at a range of concentrations, and the cells plus compound are shaken to facilitate mixing and to prevent cell settling. At time intervals, samples are taken and applied to a filtration device. The cells are retained by the filter while the metabolite that remains outside of the cell passes through the membrane filter. The cells are washed, either with water or buffer to remove traces of material bound to the outside of the cells. Usually when a radioactive label is being used, the cells are washed with a solution of $10\times$ concentrated unlabeled metabolite to ensure the exchange of labeled material off of the cell wall. The cells on the filter are then dissolved and the amount of labeled material in the cell is quantitated by scintillation counting. Note that some compounds, e.g., nitrate, do not have radioactive isotopes that are easy to work with. Heimer and Filner (1971) measured the nitrate that was taken up by tobacco cells using an enzymatic assay.

In addition to knowing the rate of uptake, it is important to know the rate of efflux of a compound from cells, since net uptake reflects both rates. For efflux measurements, cells are loaded with high concentrations of labeled compound for relatively long periods of time and the release of the label from the loaded cell is measured. The major problem in studying uptake in plant cells is sampling. The cell clumps are not as reproducibly pipetted as bacteria are, and care must be taken to ensure that a large enough sample is taken so that the data obtained will not be hopelessly scattered.

The uptake of glucose by sugar cane cells was measured by Maretzki and Thom (1972a,b). Two systems of glucose uptake were observed, a low K_m

TABLE II

Transport Systems in Plant Cells

Compound	Cell type	K_m (mM)	Energy dependent	Inducible	Inhibitor	Reference
Lysine	Sugarcane	2.45	Yes	?	Arginine	Maretzki and Thom, 1970
Arginine	Sugarcane	0.1	Yes	?	Lysine	Maretzki and Thom, 1970
Cysteine (2)	Tobacco	0.017	Yes	?	Met, Ala	Harrington and Smith, 1977
		0.350	Yes	?	Met, Ala	Harrington and Smith, 1977
Alanine (2)	Soybean	0.046	Yes	?	Tyr, Phe, Met	King, 1976
Arginine	Soybean	?	?	?	Lys	King and Hirji, 1975
Glutamate	Soybean	?	?	?	Ala, Asp	King and Hirji, 1975
SO_4	Tobacco	0.015–0.020	Yes	?	Cys, Met	Hart and Filner, 1969
Urea	Soybean	?	?	?		Smith, 1975
NO_3	Tobacco	0.4	Yes	Yes	Trp	Polacco, 1976
						Heimer and Filner, 1971
Glucose (2)	Sugarcane	0.02	Yes	?	3-O-Me glucose	Maretzki and Thom, 1972a
		1.40	Yes	Yes	3-O-Me glucose	Maretzki and Thom, 1972a

and a high K_m system. The low K_m system had kinetic parameters of 20 μM (K_m) and 0.91 nmoles mg cells^{-1} × min^{-1} (V_{max}). The high K_m system exhibited a much lower affinity for glucose ($K_m = 1.40$ mM) as well as a lower V_{max} (0.45 nmoles × min^{-1} × mg cells^{-1}). 3-O-Methyl glucose, fructose, and galactose were also transported by the low K_m system and were competitive vs glucose. Sucrose was a very poor inhibitor and pentoses did not inhibit at all. Both systems were energy dependent as determined by loss of uptake ability in the presence of $10^{-4} M$ dinitrophenol or $10^{-3} M$ azide. In addition, the high K_m system was sensitive to floride ion. When sugar cane cells are grown with 2% sucrose as the carbon source, only the low K_m system is seen; if glucose is added to the medium for several days prior to the uptake measurements, the high K_m system is also observed. This implies the inducibility of the second system, but does not rule out the possibility that it is only activated in the presence of high glucose concentrations. Another possibility is the conversion of the low K_m form to the high K_m system. The fate of transported glucose was also investigated in these studies. Sugar phosphates were observed after only 5 s, whereas labeled fructose was seen in 15 s and labeled sucrose in 30 s. A large pool of nonphosphorylated glucose was also seen, indicating that the sugar was not phosphorylated during transport. 3-O-Methyl glucose was readily transported but did not become phosphorylated.

Heimer and Filner (1970, 1971) have done extensive work on the nitrate uptake and assimilation system in cultured tobacco cells (XD line). This system was not produced when the cells were starved for nitrogen in the absence of nitrate (either by removal of all nitrogen from the medium or by supplying urea, a poor nitrogen source). The presence of nitrate was required for the induction of the uptake system. Tungstate, which inhibits formation of active nitrate reductase, did not inhibit the formation of the uptake system. The uptake system concentrated nitrate eightyfold over the medium and was determined to be energy dependent by showing inhibition of uptake by KCN and dinitrophenol. The inhibition by KCN, however, was not apparent until 2 h after the addition of the inhibitor, while inhibition by dinitrophenol began immediately. The K_m for the nitrate uptake system was found to be 0.4 mM and the V_{max} was 2-5 μmoles of nitrate taken up per hour per gram (fresh wt.) of cells. There was no evidence for more than a single nitrate uptake system. Ammonia and nitrile at levels below the concentration at which they become toxic to cell growth and casein hydrolyzate all inhibited nitrate accumulation from the medium. The inhibitory effect of casein hydrolyzate was independent of the presence of tungstate. The effect of casein led to the finding that certain amino acids inhibited growth of cells on nitrate but not when urea or γ-aminobutyric acid served as the nitrogen source. Threonine at 100 μM was a potent inhibitor of growth on nitrate. A varient line of tobacco cells (XDR^{thr}) which was stably resistant to threonine inhibition of growth on nitrate was selected after treatment of cells with nitrosoguanidine. Amino acids were still able to inhibit the formation of active nitrate reductase in XDR^{thr} cells, but nitrate accumulation was not affected. Thus, it would appear that the transport system and the primary assimilatory enzyme are not coordinately regulated. It was not determined to what degree the transport system was specific for nitrate, but it was pointed out that sulfate and phosphate in the medium did not induce the nitrate transport system, so it does not seem likely that there is a general transport system for inorganic nutrients.

The uptake of another nitrogen source, urea, has been investigated by Polacco (1976) in soybean cells. In these studies, the effect of ammonium ion and methylammonia on the uptake of urea (1 mM) was determined. Little effect was seen if cells that had been grown on ammonia and nitrate were used, but when urea grown cells were used, both compounds stimulated urea uptake. Note that these experiments were done with a single urea concentration and a single concentration of ammonia and methylammonia so that kinetic parameters could not be determined.

The energy dependent uptake of sulfate has also been studied in tobacco cells. Both Hart and Filner (1969) and Smith (1975) have found a single uptake system with a K_m for sulfate of 15–20 μM. Thus the K_m for sulfate uptake is tenfold lower than for nitrate in tobacco cells; the cells transport nitrate into the cell 10 times more rapidly than sulfate at their respective V_{max}. Hart and Filner reported that cysteine inhibited sulfate uptake after a 2-h lag. Smith, however, did not report a lag in the inhibition; he also notes that reducing agents in general inhibit the sulfate uptake system. Selenate and chromate also competitively inhibited sulfate uptake. Selenite inhibited with a K_i that is as low as the K_m for sulfate. Nitrate inhibited very little.

It was not determined if sulfate is an inducer of the sulfate uptake system or whether it is a constitutive component of the cell membrane. Efflux studies showed that very little sulfate was lost from preloaded cells and the conclusion was reached that rates of [^{35}S]sulfate uptake reflect net uptake into the cells. The fate of transported sulfate was investigated by Smith (1975). He found 75% of the sulfate was incorporated into cysteine, glutathione, and methionine (free pools).

Several studies on the uptake of amino acids by soybean cells *in vitro* have been done. King and Hirji (1975) studied the uptake of ^{14}C-labeled arginine, glutamate, and alanine, representing the basic, acidic, and neutral amino acid groups. Three systems were claimed to be present since there was relatively little interaction among the three types of amino acid, although alanine did show inhibition of both arginine and glutamate. Arginine and glutamate did not significantly inhibit each other's uptake, nor did they inhibit the uptake of alanine. Alanine transport was studied in more detail by King (1976). It was found that kinetically distinguishable systems were present for alanine uptake: a high affinity, energy-dependent system and a low affinity, high capacity system that was unaffected by metabolic inhibitors or by shifts in temperature. The K_m for the high affinity system was determined to be 46 μM. King also pointed out the possibility that some transport of alanine at high concentrations may be mediated by transport systems that are much more efficient with other amino acids (for example, glutamate or arginine as noted above).

The transport of cysteine into cultured tobacco cells was investigated by Harrington and Smith (1977). Once again, two systems were found with K_m values of 17 and 350 μM. Both systems were energy dependent as determined by inhibition by metabolic inhibitors. A number of amino acids inhibited cysteine uptake, in particular, methionine, alanine, and D-cysteine.

The transport of the basic amino acids, lysine and arginine, has been studied in sugarcane cells (Maretzki and Thom, 1970). On the basis of differential inhibition of uptake of the two amino acids by cycloheximide treatment and on the noncompetitive inhibition pattern seen for lysine inhibition of arginine uptake, it was concluded that there are at least two transport systems of basic amino acid uptake in sugarcane cells. Since arginine did inhibit lysine competitively, the suggestion was made that arginine could be taken up by either system, lysine by only one of the two systems. Both systems were energy dependent. The K_m values for lysine and arginine are relatively high, 2.45 and 0.1 mM, respectively, compared to the K_m values for transport of other amino acids.

Pea leaf protoplasts were used in the measurement of uptake of 3-O-methyl glucose and α-aminoisobutryic acid (Guy *et al.*, 1978). While the uptake by protoplasts was not well characterized, this work showed the potential of using protoplasts prepared from selected plant organs. The argument was put forth by the authors that the use of protoplasts eliminated concern for nonspecific binding of metabolites to cell wall; however, the use of protoplasts necessitates the presence of high osmotic strength medium and cells are plasmolyzed. It seems likely that quantitative data on uptake of small molecules by plant protoplasts would reflect the plasmolyzed state of the cells and might not reflect the true kinetic parameters for the same cells *in vivo*. Since protoplasts can be prepared from specific plant organs and since specific cell types can often be isolated, the true value of uptake studies using protoplasts lies in the ability to determine which uptake systems are operating *in vivo* in a specific cell type. The advantage to doing uptake studies using plant cells in culture is that the total uptake potential of the plant cell can be determined without regard for developmental state of the cells being used.

It is surprising that more studies on uptake systems have not been performed, especially uptake systems that might be involved in the transport of plant hormones, vitamins, and herbicides. Much work is also needed to determine if the systems that have been observed are inducible, repressible, or constitutive and if, as in the case of the nitrate and sulfate systems, there is any feedback inhibition of uptake.

VI. ISOLATION AND CHARACTERIZATION OF ORGANELLES FROM PLANT CELLS IN CULTURE

Relatively few studies on cell organelles from plant cells *in vitro* have been attempted; however, the major organelles have at least been shown to be present in several cell types. The use of plant cells *in vitro* allows the development of cell organelles to be studied under conditions that whole plants or plant organs would not tolerate (i.e., copper deficiency as described below). Although it has not been reported, it would be interesting to know if cells could be induced to produce more of an organelle or alter the enzyme composition of an organelle by the introduction into the growth medium of compounds metabolized in specific organelles (D-amino acids, H_2O_2, glycolate, etc.). Sucrose has been shown to inhibit the development of chloroplasts in carrot cells (Edelman and Hanson, 1971a,b), but no changes in the enzyme composition of chloroplasts from sucrose grown or glucose grown cells were noted. Since cell cultures can by synchronized (Wilson *et al.,* 1971), it should be possible to investigate the presence and importance of different organelles at different stages of the cell cycle as well as during the growth phases of a culture of cells as has been done for microbodies of soybean cells *in vitro* (Moore and Beevers, 1974).

Organelles from soybean suspension cultures were prepared by Moore and Beevers (1974). They found grinding cells with mortar and pestle to be the method of cell breakage which gave the best yield of organelles. Sucrose gradients were used to separate organelles that were identified on the basis of marker enzymes that have been established from whole plant studies. Only the mitochondria and microbodies gave symmetric peaks on graphs of fractions from the sucrose gradients; other organelles were not easily separated from other membrane and organelle fractions. It was noted in these studies that both catalase and glycolate oxidase activities were lost rapidly as the cells entered stationary phase. Since both catalase and glycolate oxidase were lost at the same rate and neither enzyme increased in concentration in the soluble fraction, it was suggested that microbodies and their contents were destroyed as the cells ceased to divide. The lack of a proplastid fraction in these studies was attributed to the fact that proplastids are more fragile than other cell organelles.

A. Proplastids and Chloroplasts

Proplastids from tobacco cells have been isolated and extensively studied (Washitani and Sato, 1977a,b,c,d, 1978). The purification technique used involved a gel filtration method devised by Wellburn and Wellburn (1971) instead of density centrifugation. While the original paper of Wellburn and Wellburn showed fairly well resolved peaks of proplastids and mitochondria, the studies of proplastids from tobacco cells did not show two peaks, so it would seem likely that the proplastids and mitochondria are present together. Electron micrograph and marker enzyme (succinic dehydrogenase) analysis, however, indicated minimal contamination of the proplastid fraction with mitochondria. On the basis of increase in specific activity of enzymes in the proplastid fraction as compared to crude extracts of tobacco cells, a number of enzymes were proposed to be located in the proplastids (Table III). From these studies and from whole plant studies (Miflin and Lee, 1977), it is clear that proplastids and chloroplasts play a major role in nitrogen metabolism of plant cells as well as their carbon metabolism.

Few studies have been done on chloroplasts isolated from cultured plant cells because of the low photosynthetic activity of green callus tissue and from the availability of techniques for the isolation of highly active chloroplasts from the mesophyll cells of leaves. Photosynthetically active chloroplasts have been isolated from *Screptanthus tortuosos* cells grown in culture (Smith and Sjoland, 1975). The cells were not autotrophic for carbon and the

TABLE III

Proplastid Enzymes

Enzymes found	Enzymes not found
Nitrite reductase	Hexokinase
NADP-dependent glutamic dehydrogenase	NADP-dependent glyceraldehyde dehydrogenase
Glucose-6-P dehydrogenase	NAD-dependent glyceraldehyde dehydrogenase
NADP-dependent malic enzyme	
Glutamate–OAA transaminase	Nitrate reductase
Glutamate–pyruvate transaminase	
Glutamine synthetase	
Glutamate synthase	
RuDP carboxylase (trace levels)	
Fructose-1,6-diphosphatase	

CO_2 assimilation activity of the isolated chloroplasts was low compared to the activity of chloroplasts isolated from spinach leaves.

B. Mitochondria

A careful characterization of mitochondria from cultured sycamore cells has been done by Wilson (1971). It is not surprising that no major differences between mitochondria of cultured cells and those from whole plants were found, although mitochondria from cultured plant cells were somewhat less tightly coupled than preparations of mitochondria from other plant sources. An interesting pattern of cyanide insensitive respiration by mitochondria was observed during the 20-day growth cycle of a culture of sycamore cells. The highest level of cyanide insensitive respiration was seen during the lag phase of culture growth. Such results would not be consistent with the theory often proffered that cyanide insensitive respiration precedes cell death. Variations in cyanide insensitive respiration among types of plant cells should be further investigated since it could provide a valuable selection system for the somatic cell geneticist (Carlson and Polacco, 1975). Sycamore cell mitochondria have also been isolated by Bligny and Douce (1977) from cells which were grown in medium with very low copper (< 0.5 μg/liter). It was found that cytochrome oxidase, but none of the other electron carriers investigated, was made in greatly reduced amounts. An important point, however, was that mitochondria with 5% of the normal amount of cytochrome oxidase had full activity of O_2 uptake. Thus, plant cells apparently make cytochrome oxidase in great excess; on a mole for mole basis, cytochrome oxidase is made in an amount roughly equivalent to the amount of cytochrome c when excess copper is available to the cells.

C. Plasma Membrane

The presence of a sturdy cell wall makes the preparation of plasma membrane from plant material difficult. The application of force sufficient to disrupt the cell wall results in disruption of the cell and its organelles and a well-mixed membrane population is obtained. If protoplasts devoid of cell wall are produced using the methods of Cocking (1960) and Takebe (1975), then the cells can be gently lysed without disruption of cell organelles. Galbraith and Northcote (1977) have used these procedures to purify plasma membrane from protoplasts of cultured soybean cells. They also used diazotized [^{35}S]-sulfanilic acid, which does not penetrate the plasma membrane, to specifically label the outer membrane of the cell. Thus, both radioactivity and marker enzymes could be followed during purification of plasma membrane by density centrifugation. MgATPase, acid phosphatase, and inosine diphosphate phosphatase were found to be associated with a radioactively labeled membrane band with a density of 1.14 g/ml. Specific labeling of plasma membranes of plant cells (pea) has also been accomplished using radioactively labeled UDP-glucose (Anderson and Ray, 1978).

D. Nuclei

A method for preparation of nuclei from soybean protoplasts was described by Ohyma et al. (1977). Nuclei are separated from cytoplasmic organelles by centrifugation through 0.4 M sorbitol at low speed. The isolated nuclei are capable of protein and RNA synthesis, but DNA synthesis could not be detected. Analysis of the DNA in the isolated nuclei by density centrifugation showed a single band with a density of 1.693 g/ml. No contaminating organelle DNA was detected. As methods for the uptake of organelles by plant protoplasts improve, the ability to isolate undamaged nuclei will be of tremendous importance in the transfer of genetic information.

E. Vacuoles

Vacuoles have been isolated from tobacco protoplasts using the methods that have been used for vacuole isolation from flower petals. A cAMP and bis-p-nitrophenylphosphate phosphodiesterase was found to be associated with vacuolar membranes (Boller and Kende, 1978).

F. Spherosomes

Carrot cells in culture have been found to be a good source of spherosomes (oleosomes) (Kleinig et al., 1978). These triglyceride-containing organelles were found to be lacking a unit membrane and no enzymatic activity was found to be specifically and unambiguously associated with the spherosomes.

VII. PLANT CELLS AND PROTOPLASTS AS GENETIC SYSTEMS

Plant cells and protoplasts have been used both to introduce variability into the plant genome and to obtain high degrees of homogeneity in plant populations. The latter goal has been achieved with remarkable success, especially with horticultural plants (Murashige, 1974). The approach to obtaining homogeneous populations of plants involves producing shoot tip on meristem tissue, propagating this tissue, and, finally, regenerating large numbers of plants derived from a single explant. Another advantage to producing plants through tissue culture propagation is that plant cells can often be rid of viruses via meristem culture; thus, more vigorous, virus-free plants can be obtained.

The introduction of new genetic information and the selection of altered genotypes has met with less success, but definite progress has been made. Attempts to introduce new genetic information into the plant genome have included studies with the uptake of exogenous DNA from bacterial, plant, animal, and viral sources which has been applied to plant cells and proto- plasts. Doy (1975) has proposed the term "transgenosis" to apply to the uptake, translation, and transcription of foreign DNA. The term is intended also to imply that the genetic information has in some way been stably integrated into the plant genome. However, this term appears superfluous at present. All observed genetic changes can be adequately described by more classical and accepted terminology. Kleinhoffs and Behki (1977) have criti- cally examined the numerous claims of DNA uptake by plant cells. They have concluded that claims of uptake of DNA followed by integration of the genetic information into the plant genome have yet to be rigorously demon- strated.

Exogenous genetic information has been taken up and expressed by plant cells (usually protoplasts). Most notably, isolated RNA from tobacco mosaic virus can be taken up and expressed, since replication of infective virus particles is observed. The DNA of several nonplant DNA viruses has been expressed (i.e., transcribed and translated) in plant cells (Doy et al., 1973; Carlson, 1973b; Johnson et al., 1973). In some of these experiments it was claimed that bacterial genes that had been incorporated into the viral genome were expressed in the plant cells. Integration of new genetic information into the plant genome was apparently not achieved.

Another approach to introducing new genetic information into plant cells is to fuse somatic cells from different plant types to obtain a fusion product with new combinations of genetic information. In order for somatic cell fusions of plant cells to be recovered, several technical barriers had to be overcome. The first and most obvious is the presence of a cell wall that would prevent intimate contact of cell membranes. This problem was over- come by the application of cellulase and pectinase to plant cells to recover viable plant cell protoplasts (Takebe and Otsuki, 1969). The second problem was the absence of agents that would promote fusion of plant protoplasts, since the agents used for the fusion of mammalian cells (viruses and plant lectins) did not work. For a discussion of mammalian cell fusion and other aspects of mammalian cell culture, see Puck (1972). Fusion of plant proto- plasts was initially achieved with sodium nitrate (Power et al., 1970) and later with a much more effective fusing agent, polyethylene glycol (PEG) (Kao and Michayluk, 1974).

Unlike mammalian systems, where extensive chromosome banding tech- niques exist, the identification of fusion products is a problem. Chromosome counts cannot always rule out self fusions and, since plant cells in culture often have aberrant chromosome numbers, chromosome counts are sometimes of little value in determining the somatic parents of the fused product. In several instances, the fusion of a chloroplast-containing cell with a non-chlorophyllous cell (usually derived from cells in suspension culture) has been used to monitor cell fusion. This technique, however, loses its value after several cell divisions for it does not allow the use of chemical selective agents. At present, there is a great need for a fusion system for plant cells analogous to the man–mouse hybrid system. The fusion products of the man–mouse hybrids lose human chromosomes until a stable chromosome number is obtained. Often only a single human chromosome remains (Ruddle and Creagan, 1975). The value of such a system for assigning various functions to different chromosomes is obvious. The requirements for such a system are chemical selection, the ability to recognize the individual chromosomes of a species, and the ability to generate true mutants so that a given chromosome can be shown to repair the defect resulting from the mutant.

If the products of somatic cell fusions between divergent cell types are to be recognized, screening and selection techniques will have to be developed. Visible screening techniques have the advantage that the desired colony can be grown in a lawn of cells that, while still growing, do not show the desired phenotype (i.e., green vs. white). The advantage of a screening system lies in the fact that plant cells do not have the high plating efficiency of bacteria or mammalian cells. Plant protoplasts have a lower population limit for plating at about 10^3 protoplasts per milliliter. Melchers and Labib (1974) have used a screening system which involves the use of two nuclear encoded mutants of tobacco which are incapable of greening. Fusion products of the somatic cells from each mutant complement each other and dark green fusion products are easily picked out. A selection can be imposed on this screening system since many of the chlorophyll deficient lines of cells are light sensitive. Carlson et al. (1972) used a selection based on growth of the fusion products of N. glauca × N. langsdorfii on a medium that did not permit the growth of either parental type. Further selection was achieved based on the knowledge that the sexual cross of the parents produces plants prone to genetic tumors; cells of these tumors will grow on medium lacking plant hormones. This selection system has subsequently been used to study the fate of chloroplast genome markers (the large subunit of RuDP carboxylase) in somatic cell fusions (Chen et al., 1977). Power et al. (1977) and Cocking et al. (1977) have described selection techniques for use with Petunia species based on natural differences in drug sensitivity and albino complementation. Another potential selection system might take advantage of the variation in activities of the cyanide-resistant respiration found in plant mitochondria. Mok et al. (1978) have recently investigated structure–activity relationships of cytokinins for growth of Phaseolus tissue. The information on differential growth obtained in these studies could presumably be used for selection systems similar to those used by Power *et al.* (1977). However, care would have to be taken when using differential hormone requirements as a selection system since habituated strains are often found in plant cell cultures (Einset and Skoog, 1973).

A wisely used selection system in animal cells, the HAT system, is based on knowledge of nucleic acid metabolism in animal cells (Littlefield, 1964). There is a vast amount of information available concerning purine and pyrimide metabolism by animal cells, primarily as a result of the efforts of cancer researchers over the last 25 years. While much of this information may be applicable to plant cells, it should not be assumed that it can be. Thymidine phosphorylation in plants, for example, has been shown to use the phosphate of AMP as a donor to thymidine; thymidine kinases from other organisms use the γ phosphate of ATP as the donor. A detailed knowledge of the enzymology of purine and pyrimidine metabolism in plant cells would be of great value at this time.

Varient lines of plant cells (the XDR[thr] line of tobacco by Heimer and Filner, 1970) have been selected in culture which show decreased sensitivity to amino acid inhibition of growth on nitrate, resistance to amino acid analogues (Widholm, 1977), and resistance to chlorate (Müller and Grafe, 1978). Temperature sensitive cell lines have been selected from haploid tobacco tissue after BuDR treatment at high temperature (Malmberg, 1978) and the BuDR selection technique has been used to obtain auxotrophs of tobacco after EMS mutagenesis (Carlson, 1970). It should be pointed out that tobacco is not an ideal plant from which to select auxotrophs since it is thought to be an allodihaploid rather than a true diploid. All of the auxotrophs of tobacco selected by Carlson (1970) were leaky, presumably because a functional copy of the gene remained after mutation of the other copy. Another class of mutants are those which show increased disease tolerance. Colonies of tobacco cells resistant to methionine sulfoximine, an analogue of wildfire disease toxin, were selected by Carlson (1973a) and plants regenerated from these colonies were shown to be tolerant to *Pseudomonas tabacci,* the causative agent of the disease. It is of interest that the glutamine levels of these plants were not elevated, as might have been expected, since methionine sulfoximine is a potent inhibitor of glutamine synthetase. Rather, the methionine pools in leaves of these plants were elevated severalfold. In another study, Gengenbach *et al.* (1977) have selected maize cells resistant to *Helminthosporium maydis* toxin and regenerated plants which show stable resistance to the pathogen.

It is clear that care must be taken to demonstrate that putative mutants are, in fact, mutants rather than epigenetic events. Epigenetic events are stable changes in the expressed phenotype of cells in the absence of mutation. In order for a varient cell line to be declared a mutant, either direct biochemical or genetic evidence must be obtained. Direct genetic evidence is easy to obtain only if whole plants can be regenerated from the cells in culture and the trait is sexually transmitted; the maintenance of the trait through repeated transfer of the cell line is not definitive evidence for a genetic lesion, nor is the maintenance of the trait after regeneration followed by reestablishing the cell line in culture. Of 33 methionine sulfoximine-resistant cell lines obtained by Carlson, only three retained the character after passing through meiosis, although a greater number retained the trait when the regenerated tissue was reestablished as callus. Direct biochemical evidence is not easy to obtain; a change in the primary sequence of a protein or RNA is the most acceptable type of direct evidence. Indirect biochemical evidence that would imply (but not prove) that a cell line differed by a mutation would be changes observed in positions of proteins on gels or clear changes in the physical properties of an enzyme.

VIII. CONCLUDING REMARKS

It should be clear from the preceding sections that research on plant cells in culture is very much a patchwork, both in terms of the research areas that have been attacked and in the degrees of success of the various investigations in answering basic questions about the workings of plant cells. This patchwork reflects the refractory nature of plant cells in culture; as much as we would like them to, plant cells in culture do not behave as bacteria. Not only do plant cells taunt us with a slow growth rate, but they undergo extensive fluctuations in developmental state during growth. These fluctuations affect the cell cycle, cell morphology, enzyme activities, and metabolic pathways; it is as if the uncertainty principle holds special influence over plant cells and that any attempt to know their metabolic state results in unpredictable changes. But today's situation with plant cells is not that different from the situation that existed in the 1940s for bacterial cells when genetics and physiology merged to provide an explosion on information about the metabolism of prokaryotes. As our understanding of eukaryotic cells in culture becomes more complete, the fluctuations we observe may become more predictable (and hence controllable) and less frustrating.

The plant geneticist and the plant physiologist/biochemist have used quite different tolls in past scientific generations. The lab bench of the plant geneticist has been, of necessity, the field plot while the plant biochemist has often proceeded with "material obtained from the local market" with little other regard for the source, physiological state, or genetic background of the material. Both approaches have yielded valuable information, but rarely have genetic and biochemical techniques been brought to bear together on the problems facing plant scientists. It is hoped that cell culture as a tool for both the geneticist and the biochemist will allow the synergism of the two approaches to take place. The methods of the biochemist are readily applicable to work with plant cells; the detection and quantitation of plant DNA, RNA, proteins, and metabolites is performed as it is in other systems. What is needed is the development of techniques for manipulation of the plant genome. Specific and general vectors, either viral or plasmid, are necessary; the ability to select and demonstrate mutants of plant cells is also required. Considering the intensity with which such systems are being sought, it would be surprising if methods for reproducible plant genome modification did not arrive soon.

One question to which there is not a clear answer is the question of which organism, if any, should become the standard cell line for plant molecular genetics. Many workers are tied to a specific plant type by their research goals or to a specific cell line by tradition. Tremendous variation exists among the labs doing cell culture work in the type of cells used, the tissue from which the cells are derived, and the medium used to grow the cells (especially in the hormone composition of the medium). In reality, it seems unlikely that a standard cell line (analogous to *E. coli* or Chinese hamster ovary cells) will be adopted. The basis for this is the dilemma of the plant cell researcher; that is, the longer a cell line exists *in vitro* and thus becomes more stable and better characterized, the less easy it becomes to regenerate whole plants. Regeneration is the ultimate technique in plant cell research and affords the capability of obtaining direct genetic evidence that modification (if not improvement) of the plant genome has occurred. Unless regeneration of whole plants becomes possible from established cell lines, it seems likely that many workers will opt for the ability to regenerate from less well-characterized cell lines over the advantages of using established cell lines.

The attractiveness of plant cell culture lies in its ability to "reduce" a complex organism to single cells insofar as growth requirements are concerned. The cell cycle and the total metabolic capability of plant cells are best studied in this reduced state. The study of whole plants derived from mutant cell lines obtained and characterized *in vitro* will aid greatly in the understanding of the growth and development of whole plants. Research with somatic cells has passed through its "gee-whiz" stage, where all positive results are exciting in and of themselves, and now must be applied as a tool in answering the complex questions of plant biology.

ACKNOWLEDGMENTS

The authors wish to thank R. L. Malmberg, T. Orton, M. L. Christianson, and J. Hunsperger for helpful discussions during the preparation of this chapter.

REFERENCES

Alt, F. W., Kellems, R. E., Bertino, J. R., and Schimke, R. T. (1978). *J. Biol. Chem.* **253**, 1357–1370.

Anderson, R. L., and Ray, P. M. (1978). *Plant Physiol.* **61**, 723–730.

Aoki, S., and Takabe, I. (1975). *Virology* **65**, 343–354.

Beale, S. I. (1978). *Annu. Rev. Plant Physiol.* **29**, 95–120.

Becking, J. H. (1965). *Nature (London)* **207**, 885–887.

Berlin, J., and Widholm, J. M. (1978). *Phytochemistry* **17**, 65–68.

Berlyn, M. B., and Zelitch, I. (1975). *Plant Physiol.* **56**, 752–756.

Berlyn, M. B., Zelitch, I., and Beaudette, P. D. (1978). *Plant Physiol.* **61**, 606–610.

Bligny, R., and Douce, R. (1977). *Plant Physiol.* **60**, 675–679.

Boller, T., and Kende, H. (1978). *Plant Physiol.* **61**, Abstr. 533.

Carlson, P. S. (1970). *Science* **168**, 487–489.

Carlson, P. S. (1973a). *Science* **180**, 1366–1368.

Carlson, P. S. (1973b). *Proc. Natl. Acad. Sci. U.S.A.* **70**, 598–602.

Carlson, P. S., and Polacco, J. C. (1975). *Science* **188**, 622–625.

Carlson, P. S., Smith, H. H., and Dearing, R. D. (1972). *Proc. Natl. Acad. Sci. U.S.A.* **69**, 2292–2294.

Chaleff, R. S., and Carlson, P. S. (1974). *Nature (London)* **252**, 393–395.

Chandler, M. T., Tandeau-de-Marsac, N., and Kouchkovsky, Y. (1972). *Can. J. Bot.* **50**, 2265–2270.

Chen, K., Wildman, S. G., and Smith, H. H. (1977). *Proc. Natl. Acad. Sci. U.S.A.* **74**, 5109–5112.

Child, J. J., and LaRue, T. A. (1976). *Proc. Int. Symp. Nitrogen Fixation 1st,* pp. 447–455.

Cocking, E. C. (1960). *Nature (London)* **187**, 927–929.

Cocking, E. C., and Pojnar, E. (1969). *J. Gen. Virol.* **4**, 305.

Cocking, E. C., George, D., Price-Jones, M. J., and Power, J. B. (1977). *Plant Sci. Lett.* **10**, 7–12.

Corduan, G. (1970). *Planta* **91**, 291–301.

Davey, M. R., Fowler, M. W., and Street, H. E. (1971). *Phytochemistry* **10**, 2559–2575.

Davies, H. M., and Miflin, B. J. (1978). *Plant Physiol.* **61**, Abstr. 526.

Dilworth, M. F., and Kende, H. (1974). *Plant Physiol.* **54**, 826–828.

Doy, C. H. (1975). *In* "The Eukaryote Chromosome" (W. J. Peacock and R. D. Brock, eds.), pp. 447–457. Australian Natl. Univ. Press, Canberra, Australia.

Doy, C. H., Gresshoff, P. M., and Rolfe, B. G. (1973). *Nature (London) New Biol.* **244**, 90–91.

Edelman, J., and Hanson, A. D. (1971a). *Planta* **98**, 97–108.

Edelman, J., and Hanson, A. D. (1971b). *Planta* **98**, 150–156.

Einset, J. W., and Skoog, F. (1973). *Proc. Natl. Acad. Sci. U.S.A.* **70**, 658–660.

Fosket, D. E., and Tepfer, D. A. (1978). *In Vitro.* **14**, 63–75.

Fosket, D. E., Volk, M., and Goldsmith, M. (1977). *Plant Physiol.* **60**, 554–562.

Galbraith, D. W., and Northcote, D. H. (1977). *J. Cell. Sci.* **24**, 295–310.

Gautheret, R. J. (1955). *Annu. Rev. Plant Physiol.* **6**, 433–484.

Gengenbach, B. G., Green, C. E., and Donovan, C. M. (1977). *Proc. Natl. Acad. Sci. U.S.A.* **74**, 5113–5117.

Gordon, W. R., Schwemmer, S. S., and Hillman, W. S. (1978). *Planta* **140**, 265–268.

Guy, M., Reinhold, L., and Laties, G. (1978). *Plant Physiol.* **61**, 593–596.

Hahlbrock, K. (1976). *Eur. J. Biochem.* **63**, 137–145.

Hahlbrock, K., and Schroder, J. (1975). *Arch. Biochem. Biophys.* **171**, 500–506.

Hanson, A. D., and Edelman, J. (1972). *Planta* **102**, 11–25.

Harrington, H. M., and Smith, I. K. (1977). *Plant Physiol.* **60**, 807–811.

Hart, J. W., and Filner, P. (1969). *Plant Physiol.* **44**, 1253–1259.
Heimer, Y. M., and Filner, P. (1970). *Biochem. Biophys. Acta.* **215**, 152–165.
Heimer, Y. M., and Filner, P. (1971). *Biochem. Biophys. Acta.* **230**, 362–372.
Hess, D., and Schetter, C. (1978). *Z. Pflanzenphysiol.* **86**, 177–185.
Hirschberg, K., Hübner, G., and Borries, H. (1972). *Planta* **108**, 333–337.
Holsten, R. D., Burns, R. C., Hardy, R. W. F., and Herbert, R. R. (1971). *Nature (London)* **232**, 173.
Hunter, T. R., Hunt, T., Knowland, J., and Zimmern, D. (1976). *Nature (London)* **260**, 759.
Husemann, W., and Barz, W. (1977). *Physiol. Plant.* **40** 77–81.
Ikegami, M., and Fraenkel-Conrat, H. (1978). *Proc. Natl. Acad. Sci. U.S.A.* **75**, 2122–2124.
Johnson, C. B., Griedson, D., and Smith, H. (1973). *Nature (London) New Biol.* **244**, 105–106.
Kao, K. N., and Michayluk, M. R. (1974). *Planta* **115**, 355–367.
Kleinhoffs, A., and Behki, R. (1977). *Annu. Rev. Genet.* **11**, 79–102.
Kleinig, H., Steinki, C., Kopp, C., and Zarr, K. (1978). *Planta* **140**, 233–237.
Ketchum, P. A., Cambier, H. Y., Frazer, W. A., Madansky, C. H., and Nason, A. (1970). *Proc. Natl. Acad. Sci. U.S.A.* **66**, 1016–1023.
King, J. (1976). *Can. J. Bot.* **54**, 1316–1321.
King, J., and Hirji, R. (1975). *Can. J. Bot.* **53**, 2088–2091.
King, P. J. (1973). "The Continuous Culture of Plant Cells." Ph.D. Thesis, Univ. of Leicester, Leicester, England.
King, P. J., and Street, H. E. (1977). *In* "Plant Tissue and Cell Culture" (H. E. Street, ed.), 2nd ed., pp. 307–389. Univ. of California Press, Berkeley.
Koukol, J., and Conn, E. E. (1961). *J. Biol. Chem.* **236**, 2692–2698.
Lamport, D. T. A. (1964). *Exp. Cell Res.* **33**, 195–206.
Littlefield, J. W. (1964). *Science* **145**, 709–710.
Malmberg, R. L. (1978). *Genetics* **88**, S61–S62.
Maretzki, A., and Thom, M. (1970). *Biochemistry* **9**, 2731–2736.
Maretzki, A., and Thom, M. (1972a). *Biochem. Biophys. Res. Commun.* **47**, 44–50.
Maretzki, A., and Thom, M. (1972b). *Plant Physiol.* **49**, 177–182.
Matthysse, A. G., and Wyman, P. M. (1978). *Plant Physiol.* **61**, Abstr. 396.
Meagher, R. B. (1977). *In* "Genetic Engineering for Nitrogen Fixation" (A. Hollaender, ed.), pp. 129–158. Plenum, New York.
Melchers, G., and Labib, G. (1974). *Mol. Gen. Genet.* **135**, 277–294.
Mendel, R. R., and Müller, A. J. (1978). *Mol. Gen. Genet.* **161**, 77–80.
Miflin, J., and Lee, P. J. (1977). *Annu. Rev. Plant Physiol.* **28**, 299–329.
Mok, M. C., Mok, D. W. S., and Armstrong, D. J. (1978). *Plant Physiol.* **61**, 72–75.
Moore, T. S., and Beevers, H. (1974). *Plant Physiol.* **53**, 261–265.
Müller, A. J., and Grafe, R. (1978). *Mol. Gen. Genet.* **161**, 67–76.
Murashige, T. (1974). *Annu. Rev. Plant Physiol.* **25**, 135–166.
Muren, R. C., and Fosket, D. E. (1977). *J. Exp. Bot.* **28**, 775–784.
Nester, E. W., Merlo, D. J., Drummond, M. H., Sciaky, D., Montoya, A. L., and Chilton, M. D. (1977). *In* "Genetic Engineering for Nitrogen Fixation" (A. Hollaender, ed.), pp. 181–196. Plenum, New York.
Ohyama, K., Pelcher, L. E., and Horn, D. (1977). *Plant Physiol.* **60**, 179–181.
Pamplin, E. J., and Chapman, J. M. (1975). *J. Exp. Bot.* **26**, 212–220.
Parke, D., and Carlson, P. S. (1978).
Parkhurst, C. E., and Craig, A. S. (1978). *J. Gen. Microbiol.* **106**, 207–219.
Passow, H., and Stampfli, R. (1969). "Laboratory Techniques in Membrane Biophysics." Springer-Verlag, Berlin and New York.
Phillips, D. A. (1974). *Plant Physiol.* **53**, 67.
Polacco, J. C. (1976). *Plant Physiol.* **58**, 350–357.
Polacco, J. C. (1977). *Plant Physiol.* **59**, 827–830.

Polevaya, V. S., Smolov, A. R., and Ignate'v, A. R. (1975). *Dokl Akad. Nauk. SSSR* **225**, 230–231.

Power, J. B., Cumming, S. E., and Cocking, E. C. (1970). *Nature (London)* **225**, 1016–1018.

Power, J. B., Berry, S. F., Frearson, E. M., and Cocking, E. C. (1977). *Plant Sci. Lett.* **10**, 1–6.

Pratt, H. M., and Fox, E. J. (1978). *Plant Physiol.* **61**, Abstr. 49.

Puck, T. T. (1972). "The Mammalian Cell as a Microorganism." Holden-Day, San Francisco.

Renaudin, J., Gandar, J., and Bove, J. M. (1976). *Ann. Microbiol.* **127**, A:61.

Reporter, M., and Hermina, N. (1975). *Biochem. Biophys. Res. Commun.* **64**, 1126–1133.

Reporter, M. (1976). *Plant Physiol.* **57**, 651–655.

Reporter, M. (1978). *Plant Physiol.* **61**, 753–756.

Reuveny, Z. (1977). *Proc. Natl. Acad. Sci. U.S.A.* **74**, 619–622.

Reuveny, Z., and Filner, P. (1977). *J. Biol. Chem.* **252**, 1858–1864.

Roberts, L. W. (1976). "Cytodifferentiation in Plants. Xylogenesis as a Model System." Cambridge Univ. Press, London and New York.

Ruddle, F. H., and Creagan, R. P. (1975). *Annu. Rev. Genet.* **9**, 407–486.

Sakai, F., and Takebe, I. (1970). *Biochem. Biophys. Acta* **224**, 531–540.

Scandalios, J. G. (1974). *Annu. Rev. Plant Physiol.* **25**, 225–258.

Schell, J., and Van Montagu, M. (1977). *In* "Genetic Engineering for Nitrogen Fixation" (A. Hollaender, ed.), pp. 159–179. Plenum, New York.

Schröder, J., Betz, B., and Hahlbrock, K. (1977). *Plant Physiol.* **60**, 440–445.

Shah, V. K., and Brill, W. J. (1977). *Proc. Natl. Acad. Sci. U.S.A.* **74**, 3249–3253.

Skokut, T. (1978). Ph.D. Thesis, Michigan State University, East Lansing, Michigan.

Skoog, F., and Miller, C. O. (1957). *Symp. Soc. Exp. Biol.* **11**, 118–130.

Skoog, F., Armstrong, D. J. Cherayll, J. D., Hampel, A. E. and Bock, R. M. (1966). *Science* **165**, 1354–1356.

Smith, D. D., and Sjolund, R. D. (1975). *Plant Physiol.* **55**, 520–525.

Smith, I. K. (1975). *Plant Physiol.* **55**, 303–307.

Takayama, S., Misawa, M., Ko, K., and Miseto, T. (1977). *Physiol. Plant* **41**, 313–320.

Takebe, I. (1975). *Annu. Rev. Phytopathol.* **13**, 105–125.

Takebe, I. (1978). *Compr. Virol.* **11**, 237–283.

Takebe, I., and Otsuki, Y. (1969). *Proc. Natl. Acad. Sci. U.S.A.* **64**, 843–848.

Takebe, I., Otsuki, Y., and Aoki, S. (1968). *Plant Cell Physiol.* **9**, 115.

Ueki, K., and Sato, S. (1977). *Plant Cell Physiol.* **18**, 1253–1263.

Ushiyama, R., and Matthews, R. E. F. (1970). *Virology* **42**, 293.

Verma, D. P. S., and Van Huystee, R. (1970). *Can. J. Biochem.* **48**, 444–449.

Verma, D. P. S., Hunter, N., and Bal, A. K. (1978). *Planta* **138**, 107–110.

Washitani, J., and Sato, S. (1977a). *Plant Cell Physiol.* **18**, 117–125.

Washitani, J., and Sato, S. (1977b). *Plant Cell Physiol.* **18**, 505–512.

Washitani, J., and Sato, S. (1977c). *Plant Cell Physiol.* **18**, 1235–1241.

Washitani, J., and Sato, S. (1977d). *Plant Cell Physiol.* **18**, 1243–1251.

Washitani, J., and Sato, S. (1978). *Plant Cell Physiol.* **19**, 43–50.

Wellburn, A. R., and Wellburn, F. A. M. (1971). *J. Exp. Bot.* **22**, 972–979.

Widholm, J. (1977). *Crop Sci.* **17**, 597–600.

Wilson, S. B. (1971). *J. Exp. Bot.* **22**, 725–734.

Wilson, S. B., King, P. J., and Street, H. E. (1971). *J. Exp. Bot.* **21**, 177–207.

Zimmerman, A., and Hahlbrock, K. (1975). *Arch. Biochem. Biophys.* **166**, 54–62.

Zucker, M. (1965). *Plant Physiol.* **40**, 779–84.

Zucker, M. (1972). *Annu. Rev. Plant Physiol.* **23**, 133–156.

The Primary Cell Walls of Flowering Plants

3

ALAN DARVILL
MICHAEL McNEIL
PETER ALBERSHEIM
DEBORAH P. DELMER

The Biochemistry of Plants, Vol. 1
Copyright © 1980 by Academic Press, Inc.
All rights reserved.
ISBN 0-12-675401-2

91

INTRODUCTION

A. Why Study the Structures of Cell Walls?

Cell walls are responsible for the shape of plants, for the walls of plant cells are analogous to the skeletons of animals. The walls control the growth rate of plant cells and thus of plants. The walls are a structural barrier to some molecules and to invading pests. Cell walls are also a source of food, fiber, and energy. Thus, knowledge of the structure of cell walls, of the mode of synthesis of cell walls, and of the function of cell walls is of great importance.

Plant cell walls are of two general types: primary and secondary. Primary cell walls are laid down by undifferentiated cells that are still growing. The primary walls control the rate of cell growth and form the basic structural backbone of growing plant cells and tissues. Secondary walls are derived from primary cell walls by cells that have stopped or are stopping growth and that are differentiating into cells with specialized functions.

This chapter is concerned with the structure and biosynthesis of the primary cell walls of monocotyledonous and dicotyledonous plants, the two subclasses of flowering plants (angiosperms). These two subclasses of plants differ in several aspects (Marsland, 1964) and, in particular, in the number of cotyledons present in their seeds. Of particular interest to this chapter is the degree to which the primary cell walls of monocots and dicots are similar.

B. The Goals of Cell Wall Structural Research

The general structure of the primary cell walls of both monocots and dicots has been envisioned for many years to be composed of cellulose fibers embedded in an amorphous mixture of polysaccharides and glycoproteins. Although this picture of primary walls appears to be accurate, it obviously lacks considerable detail. A more detailed description of the primary cell wall will eventually include the following: (1) isolation and identification of each of the individual macromolecular components of the cell wall; (2) determination of the primary structure of each of these macromolecules; (3) determination of the three-dimensional structures of these macromolecules; (4) determination of how and where these macromolecules are biosynthesized within the plant cells; (5) determination of how these macromolecules are attached to one another or how they are interrelated; (6) determination of how the interrelated macromolecules are distributed throughout the thickness and the length of the wall; and (7) determination of how newly synthesized macromolecules are inserted into the wall and how the wall grows.

Plant cell wall biochemistry is still at the stage of identifying and elucidating the covalent structures of the macromolecular components of the walls (points 1 and 2 above). Hence, this chapter will primarily consider the identity and structure of the major macromolecular components of the primary cell walls, as well as summarize what is known about the biosynthesis of cell wall macromolecules (point 4). The known structural features of each of the known wall components will be described and used to compare the structural components of monocot and dicot primary cell walls. A part of this chapter will discuss the available information about the chemical bonding (covalent or otherwise) which exists between the structural components of primary cell walls (point 5). Although the three-dimensional structures of polysaccharides have received attention recently (Dea *et al.,* 1977; Grant *et al.,* 1973; Rees, 1972; Rees and Welsh, 1977), little is known about the secondary, tertiary, and quaternary structures of plant cell wall polymers (point 3). Further, although the distribution of the polysaccharides throughout the wall has been studied at the ultrastructural level (point 6), no generally accepted description of this important phenomenon is available, and it will not be considered in this chapter. Essentially nothing is known about the integration of newly synthesized macromolecules into walls and little is known about the biochemistry of wall growth (point 7).

Most of the plant polysaccharides that have been studied have not been obtained from isolated primary cell walls, but rather from other plant organelles and differentiated tissues. We believe that in order to describe the structure of primary cell walls, it will eventually be necessary to study polysaccharides that have been isolated directly from primary cell walls. This chapter will comprehensively consider all those polysaccharides that have been isolated from dicot and monocot primary cell walls. We have also considered a selection of those studies of plant polysaccharides which have been isolated from sources other than primary cell walls, but which are thought to resemble polysaccharides of the primary cell wall.

C. Experimental Problems Associated with Cell Wall Structural Research

There are significant technical problems facing those who wish to study the structure of primary cell walls. An important and sometimes overlooked problem is the purification of the cell walls. Generally, cell walls are purified by insolubility in buffered salt solutions and in organic solvents. Such purification procedures undoubtedly remove some of the molecules present in the walls of intact plant tissues, and some of the discarded molecules may have a structural function within the wall. Any molecule solubilized by the purification procedures employed is not considered, in this chapter, as a structural component of the wall.

The purity of the wall preparations that have been studied can also be questioned. Even though elaborate washing of the wall preparations is customary, the walls may be contaminated by some of the cytoplasmic components that attach to or sediment with the walls following tissue homogenation. Starch grains, for example, are difficult to remove from cell walls.

The insolubility of the cell wall structural components is the cause of another technical problem. In order to study the structure of the individual wall components, these components must be solubilized and purified. It is impossible to do this without alteration of the structures of the wall components. All of the methods currently used for solubilization of the structural components have some disadvantages. Most chemical solubilization techniques are suspected or known to break covalent bonds and to solubilize a size-heterogeneous mixture of cell wall components. Enzymes that solubilize wall polymers do so by hydrolyzing covalent bonds and, thereby, alter the polymers that they are solubilizing. Purified enzymes do have the advantage of extracting the wall polymers in a predictable manner. However, the possibility always exists that even highly purified enzymes are contaminated with undetected degradative activities. This problem is minimized by careful examination of the ability of the enzymes to degrade model substrates. Enzymes often have the same deficiency as most chemical extraction procedures in failing to extract all of their substrate from the walls. It is usually not apparent why an extraction procedure fails to completely solubilize a particular wall component. The difficulties associated with the solubilization of cell wall polymers remain, perhaps, as the major impediments to progress in cell wall analysis.

It is no easy task to purify a wall component to homogeneity even after the component has been successfully extracted from the wall. The available methods for the purification of polysaccharides are being improved dramatically but, even so, the heterogeneity and complexity of the interconnected cell wall polymers makes this a difficult problem.

Once the wall polysaccharides and proteins are purified, there remains the problem of determining the primary structures of these molecules. The evolution of powerful and sensitive methods for determining the sequence of glycosyl residues in a polysaccharide and amino acid residues in a protein makes structural analysis one of the most tractable problems facing those who study cell wall structure. Nevertheless, the complexity of the wall and the problems associated with its study suggest that complete elucidation of the structure of primary cell walls will not come for some time. However, those of us working in the area of cell wall structural research are excited by the knowledge that the methods available today and those which are being developed make this very necessary research feasible and attractive.

D. The Types of Cell Wall Polysaccharides

Early workers considered the wall to be composed of three polysaccharide fractions, i.e., cellulose, the hemicelluloses, and the pectic polysaccharides. Grouped in the pectic fraction are all of the polysaccharides extracted from cell walls by hot water, ammonium oxalate, weak acid, or chelating agents. Hemicelluloses are not extracted by weak acids but by relatively strong alkali. The wall residue remaining after alkali extraction is mostly composed of cellulose. These extraction techniques have led to some confusion and contradictions in the literature. This has mainly been due to incomplete and overlapping extraction of the wall polymers by the chemical procedures employed. Nevertheless, recent work has shown that the classification of the wall polysaccharides into cellulose, hemicelluloses, and pectic polysaccharides is reasonably accurate.

Presently we classify the pectic polysaccharides as those polymers found in covalent association with galacturonosyl-containing polysaccharides. The hemicelluloses are those polysaccharides noncovalently associated with cellulose. It has been proposed that the hemicelluloses are capable of hydrogen bonding strongly to cellulose (Bauer *et al.*, 1973). The hemicelluloses of primary cell walls are the xyloglucans and various xylans and heteroxylans (Bauer *et al.*, 1973; Darvill, 1976, J. Darvill, A. Darvill, M. McNeil, and P. Albersheim, unpublished results; Keegstra *et al.*, 1973; McNeil *et al.*, 1975; Talmadge *et al.*, 1973).

The original classification of wall polysaccharides can be related to the modern terminology; mild acid preferentially extracts the pectic polysaccharides, while subsequent extraction with alkali preferentially solubilizes the hemicelluloses. Therefore, in this chapter discussion of the noncellulosic cell wall polysaccharides will appear under the general headings of the pectic polysaccharides and hemicelluloses. Cellulose as well as nonpolysaccharide components of primary cell walls will also be considered.

II. THE SUGAR NOMENCLATURE AND ABBREVIATIONS USED

A. Glycosyl Residues

A sugar residue glycosidically linked through its reducing carbon (C-1) is called a glycosyl residue, e.g., 4-linked glucosyl residues are glucosyl residues glycosidically linked at C-1 and which also have another glycosyl residue attached to them at C-4. Sugars with their reducing carbons free, whether or not the sugars have other glycosyl residues attached to them, are called glycoses, e.g., 4-linked glucose indicates a glucose that is located at the reducing end of an oligo- or polysaccharide and which has another glycosyl residue attached to it at C-4.

Abbreviations used in this chapter include Glc = glucose; Gal = galactose; Man = mannose; Xyl = xylose; Api = apiose; Ara = arabinose; Rha = rhamnose; Fuc = fucose; GlcA = glucuronic acid; GalA = galacturonic acid; p = pyranose ring form; f = furanose ring form.

B. Absolute Configuration

All of the sugars which compose the plant cell wall polymers, except for arabinose, rhamnose, and fucose, are invariably found in the D configuration. However, galactose has been found in the L configuration in plant tissue (Roberts and Harrer, 1973) although most of the galactosyl residues of plant cell wall polysaccharides are in the D configuration. Arabinose, rhamnose, and fucose have consistently been found in the L configuration. The D or L configuration is omitted from the nomenclature used in this chapter. Those instances where the absolute configuration has been specifically determined are generally noted in the text.

C. Anomeric Configuration

The anomeric configuration, α or β, of the glycosidic linkages is designated when known.

D. Ring Size

Except for most of the arabinosyl residues, all of the primary cell wall glycosyl residues of dicots have been found to be in the pyranose ring form. It is possible that some of the glycosyl residues that have been determined to be 4-linked pyranosyl residues are, in fact, 5-linked furanosyl residues, since methylation analysis does not distinguish between the two possibilities. The ring forms for all of the glycosyl residues, except arabinose, are not designated. The ring form of arabinosyl residues is designated when known, as Ara_f for arabinofuranosyl, and Ara_p for arabinopyranosyl.

E. Linkage Analysis

Methylation data are expressed using a simplified "linkage" notation. The linkages of the glycosyl residues are determined from the position of the O-methyl groups introduced during methylation analysis. Methyl groups are never attached to either C-1 (protected from O-methylation by its participation in glycosidic linkage) or C-5 (protected from O-methylation by its participation in the pyranose ring). Arabinofuranosyl residues are an exception and may have methyl groups on C-5 but never on C-4. In the notation adopted, all carbons designated as "linked" do not have O-methyl groups attached (protected from methylation by glycosidic linkage with another sugar), whereas all of the remaining carbons except C-1 and C-5 (or C-4 in the case of arabinofuranosyl residues) do have O-methyl groups attached. For example, a glycosyl residue designated as "terminal" (T) is glycosidically linked to another glycosyl or glycose residue only through C-1 and contains no glycosyl residues linked to other carbons. A glycosyl residue designated as 2-linked is glycosidically linked to another glycosyl or sugar residue through C-1 and has another glycosyl residue linked to it at C-2. A glycosyl residue designated as 3-6-linked is glycosidically linked to another sugar through C-1 and has glycosyl residues linked to it at C-3 and C-6; such a residue therefore represents a branch point in a chain molecule.

F. Polymer Names

Cell wall polymers are often referred to by the quantitatively dominant glycosyl residues of which they are composed (e.g., xyloglucan). These names do not mean the polymers are composed solely of the glycosyl residues referred to in the polymer's name (e.g., xyloglucan also contains arabinosyl, fucosyl, and galactosyl residues).

III. METHODS USED IN THE STRUCTURAL ANALYSIS OF CELL WALL POLYSACCHARIDES

A. Introduction

Many structural studies have relied upon the use of a few well-defined experimental techniques, which cannot be described in detail in this chapter but are referenced.

B. Solubilization and Fractionation of Cell Wall Polysaccharides

A successful first step in fractionating primary sycamore cell walls (a dicotyledon) has been the treatment of the walls with a purified endo-α1-4-galacturonase (A. Darvill, M. McNeil, and P. Albersheim, unpublished results; Talmadge *et al.*, 1973). This enzyme hydrolyzes α4-linked galacturonosyl linkages resulting in the solubilization of approximately 18% of the mass of the wall. The cell wall residue remaining after endopolygalacturonase treatment can be extracted with alkali to yield additional pectic polysaccharides and the hemicelluloses. However, the endopolygalacturonase-treated cell walls can be extracted with a second enzyme; an endo-β1-4-glucanase has been of value in this regard. The endoglucanase specifically fragments xyloglucan, a primary wall hemicellulose (Bauer *et al.*, 1973).

The endopolygalacturonase and alkali-solubilized cell wall polysaccharides have each been fractionated by ion exchange and gel filtration chromatography (J. Darvill, A. Darvill, M. McNeil, and P. Albersheim, unpublished results). Ion exchange chromatography is particularly valuable in separating the acidic polysaccharides from the neutral polysaccharides and is also useful in separating those polysaccharides that contain differing amounts of acidic residues. For example, anion exchange chromatography can separate the acidic uronosyl-containing xylan and the neutral xyloglucan solubilized by alkali extraction of endopolygalacturonase-treated walls. Five noncellulosic primary cell wall polysaccharides of dicots have been highly purified by a combination of anion exchange and gel filtration chromatography. These five polysaccharides are xyloglucan (Bauer *et al.,* 1973), glucuronoarabinoxylan (J. Darvill, A. Darvill, M. McNeil, and P. Albersheim, unpublished results), homogalacturonan and rhamnogalacturonan I (A. Darvill, M. McNeil, and P. Albersheim, unpublished results), and rhamnogalacturonan II (Darvill *et al.,* 1978).

It is important to be able to determine which chromatography column fractions contain polysaccharides and, specifically, which fractions contain hexosyl, pentosyl, or uronosyl residues. It is also important to detect in the fractions the presence of proteins and the presence of the specific amino acid characteristic of wall proteins, hydroxyproline. The detection of these substances is carried out by facile and sensitive colorimetric procedures. The most frequently used colorimetric assays in our laboratory are the anthrone assay for detection of hexosyl residues (Dische, 1962), the orcinol assay for detecting pentosyl residues (Dische, 1962), the *m*-hydroxyl-diphenyl assay for detection of uronosyl residues (Blumenkrantz and Asboe-Hansen, 1973), the Lowry assay for detection of proteins (Lowry *et al.,* 1951), and the Kivirikko and Liesma (1959) assay for the detection of hydroxyprolyl residues.

C. Quantitative Analysis of the Glycosyl Residues of Oligo- or Polysaccharides

The most frequent procedure used by those studying polysaccharide structures is the determination of the glycosyl composition of the sample being investigated. This assay is used both for determining the purity of polysaccharides (Darvill *et al.,* 1978) and for identifying which polysaccharides are present in a particular chromatographic fraction. The most commonly used and most accurate method for the quantitative analysis of the glycosyl residues involves the conversion of the glycosyl residues into their corresponding volatile alditol acetates. The alditol acetates are conveniently separated and quantitated by flame ionization gas chromatography (Albersheim *et al.,* 1967).

D. Uronic Acid Quantitation

Polysaccharides that contain uronosyl residues, such as the acidic pectic polysaccharides and the glucuronoarabinoxylans, pose a particular problem when trying to quantitatively determine glycosyl compositions. Uronosidic bonds are resistant to acid hydrolysis. Conditions sufficiently harsh to hydrolyze the uronosidic bonds often result in significant degradation of the uronic acids as well as degradation of some of the neutral sugars. Uronic acids, even when converted into their monomeric form, do not yield stable alditol acetates as the reduced form of an uronic acid is an aldonic acid which cannot be acetylated by the standard procedures.

The problems associated with the presence of uronosyl residues can be bypassed by converting the uronosyl residues to the corresponding dideutero-labeled hexosyl residues (Taylor and Conrad, 1972). Acidic polysaccharides are converted by this procedure into neutral polysaccharides and the neutral polysaccharides can then be analyzed by formation of the corresponding alditol acetates. The alditol acetates derived from deuterium-labeled hexosyl residues are quantitatively distinguished from the unlabeled hexosyl residues by combined gas chromatography–mass spectrometry.

E. Glycosyl-Linkage Composition Analysis

Once the glycosyl composition of a polysaccharide is known, the next step is to determine the glycosyl linkage compositions. This analysis allows one to quantitatively determine the amounts of the differently linked glycosyl residues, such as the amount of a polysaccharide composed of 3-linked glucosyl residues, 4-linked glucosyl residues, and 3-4 linked glucosyl residues. The method of choice for glycosyl linkage analysis is by formation of partially methylated acetylated alditols. The partially methylated aldoses are reduced to the corresponding alditols and then acetylated. The partially methylated alditol acetates are volatile and are quantitated and tentatively identified by flame ionization gas chromatography (Sandford and Conrad, 1966; Talmadge *et al.*, 1973). The identity of the partially methylated alditol acetates is confirmed by combined gas chromatography–mass spectrometry (Björndal *et al.*, 1970). Chemical ionization mass spectrometry (McNeil and Albersheim, 1977) has proven to be of value in augmenting electron ionization mass spectrometry (Björndal *et al.*, 1970) in identifying these derivatives.

Uronosyl residues of the polysaccharides offer the same problems for glycosyl linkage analysis as they do for glycosyl composition analysis. Before the linkages to the uronosyl residues are ascertained, the uronosyl residues are usually reduced by the carbodiimide method (Taylor and Conrad, 1972) to their corresponding deuterium-labeled alditols.

F. Sequencing the Glycosyl Residues in Polysaccharides

New methods for sequencing the glycosyl residues of oligo- and polysaccharides are now being developed. The new methods are sophisticated elaborations of the rather commonplace procedure of converting polysaccharides by partial hydrolysis into structurally analyzable oligosaccharides. The conversion of polysaccharides into oligosaccharides is generally achieved by partial acid hydrolysis, acetolysis (Danishefsky *et al.*, 1972), or formolysis

(Darvill et al., 1978b). These methods may differ in the rate at which they catalyze the hydrolysis of different glycosidic linkages and, therefore, the methods yield different sets of oligosaccharides from the same polysaccharide.

The goal of converting polysaccharides into manageable oligosaccharides can also be achieved with the assistance of highly purified endoglycanases. It is frequently rather laborious to obtain these enzymes, but their value cannot be questioned. One example of the value of such enzymes is the formation of a set of identifiable oligosaccharides from xyloglucan with the aid of an endo-β1-4-glucanase (Bauer et al., 1973). Other useful enzymes which have been purified with a goal of studying the structures of the primary cell wall polysaccharides include the widely used endo-α1-4-galacturonase (A. Darvill, M. McNeil, and P. Albersheim, unpublished results; Talmadge et al., 1973), an endo-β1-4-galactanasae (Labavitch et al., 1976) and an endo-α1-5-arabanase (Kaji and Saheki, 1975; Weinstein and Albersheim, 1979).

Polysaccharides may also be specifically cleaved into sets of analyzable oligosaccharides by periodate oxidation and, in particular, by using the procedure known as Smith degradation (Danishefsy et al.; 1970; Sharon, 1975). This procedure involves periodate oxidation of those glycosyl residues of an oligo- or polysaccharide which possess vicinal hydroxyls. The aldehyde groups of the glycosyl fragments resulting from periodate oxidation are reduced with sodium borohydride and then the glycosidic bonds of these fragments are preferentially cleaved by acid hydrolysis. Analysis of the Smith degradation products gives information on the sequence of glycosyl residues in the original polysaccharide.

IV. THE PECTIC POLYSACCHARIDES

A. The Pectic Polysaccharides of Dicots

1. Introduction

The primary cell walls of dicots are characterized by a relatively high content ($\approx 35\%$) of pectic polysaccharides. The most characteristic component of the pectic polysaccharides are galacturonosyl residues (Worth, 1967). The most characteristic physical property of the pectic polysaccharides is an ability to form gels (Grant et al., 1973; Rees and Welsh, 1977). The area between primary cell walls of adjoining cells, known as the middle lamella, is thought to be particularly rich in pectic polysaccharides (Hall, 1976; B. Viand and J. C. Roland, unpublished results). In addition to galacturonosyl residues, the pectic polysaccharides are characterized by the presence of rhamnosyl, arabinosyl, and galactosyl residues. The rhamnosyl residues are closely associated with galacturonosyl residues in that both are integral components of the same polysaccharide chain. A considerable portion of the arabinosyl and galactosyl residues appear to be components of araban and of galactan side chains which are covalently attached to the rhamnogalacturonan backbone. A portion of the cell wall arabinosyl and galactosyl residues are likely to be constituents of arabinogalactan chains, but these heteropolysaccharides may not be covalently attached to the galacturonosyl-containing polymers. Discussion of the araban, galactan, and arabinogalactan as well as the homogalacturonan and rhamno-galacturonan are presented below. The evidence that the araban, galactan, and rhamnogalacturonan polymers are covalently linked to one another is discussed in Section IX.

2. *Rhamnogalacturonan I*

Polysaccharides containing only rhamnose and galacturonic acid have never been isolated; such polysaccharides always have other sugars covalently attached to them. However, rhamnogalacturonans are thought to be the backbone chains of the pectic polymers. The rhamnogalacturonan described in this section is called rhamnogalacturonan I in order to distinguish it from an entirely different type of pectic polysaccharide, rhamnogalacturonan II, which is discussed in Section IV,A,7.

Rhamnose has long been known to be associated with the galacturonosyl residues of the pectic polysaccharides (Worth, 1967). More recently, rhamnosyl residues have been found to be glycosidically linked to galacturonosyl residues. The present structural knowledge of rhamnogalacturonan I has largely been obtained by isolation of oligosaccharide fragments of this polymer (Aspinall and Jiang, 1974; Aspinall and Molloy, 1968; Aspinall *et al.*, 1967b, 1968a,c; Siddiqui and Wood, 1976; Talmadge *et al.*, 1973) and by methylation analysis of the intact polymer (Aspinall and Jiang, 1974; Aspinall and Molloy, 1968; Aspinall *et al.*, 1968a; Talmadge *et al.*, 1973). The oligosaccharides have been obtained by either partial acid hydrolysis or by acetolysis. Partial acid hydrolysis of rhamnogalacturonan I has yielded the disaccharide GalA-(α1-2)-Rha as well as the tetrasaccharide GalA-(1-2)-Rha-(1-4)-GalA-(1-2)-Rha. Acetolysis has yielded the trisaccharides GalA-(1-4)-GalA-(1-2)-Rha and GalA-(1-2)-Rha-(1-2)-Rha as well as the tetrasaccharide GalA-(1-4)-GalA-(1-2)-Rha-(1-2)-Rha (Aspinall *et al.*, 1967b, 1968c).

Methylation analysis was used to demonstrate that the rhamnosyl residues of the above oligosaccharides are 2-linked. Methylation analysis of the intact polysaccharide has demonstrated that about 50% of the rhamnosyl residues are 2-linked, but the other 50% are 2-4 linked (Aspinall and Jiang, 1974; Aspinall and Molloy, 1968; Talmadge *et al.*, 1973). No aldobiuronic acid with a galacturonosyl residue attached to C-4 of a rhamnose has been isolated. It has generally been assumed that the C-4 of rhamnose is a point of attachment of other neutral glycosyl residues (Aspinall *et al.*, 1967b; Talmadge *et al.*, 1973). Recently, it has been established that 5-linked arabinosyl and several differently linked galactosyl residues are attached to the C-4 residues of rhamnogalacturonan I (M. McNeil, A. Darvill, and P. Albersheim, unpublished results). The manner in which the oligosaccharides that have been characterized are arranged in the intact rhamnogalacturonan has not yet been established (Aspinall *et al.*, 1963a; Talmadge *et al.*, 1973).

Further information about the structure of rhamnogalacturonan I has recently been obtained by M. McNeil, A. Darvill, and P. Albersheim (unpublished results). The walls of suspension-cultured sycamore cells were exhaustively treated with endopolygalacturonase. This enzyme solubilizes rhamnogalacturonan I and II as well as homogalacturonans. These acidic polysaccharides were separated from one another by ion exchange chromatography on DEAE Sephadex. Rhamnogalacturonan I contains, in addition to rhamnosyl and galacturonosyl residues, substantial amounts of arabinosyl and galactosyl residues (as noted above, rhamnogalacturonans have never been isolated free of other neutral glycosyl residues). The ratio in rhamnogalacturonan I of rhamnosyl to galacturonosyl to arabinosyl to galactosyl residues is $1:2:1.5:1.5$.

Rhamnogalacturonan I is very large as it only partially includes in an agarose 5-m column (exclusion limit of 5×10^{-6} MW for globular proteins). Comparison of the elution volume of rhamnogalacturonan I with the elution volume of dextrans of known molecular weights suggests rhamnogalacturonan I has a degree of polymerization of about 2000. The apparent size of rhamnogalacturonan I is not altered when it is chromatographed in a solvent of $0.5\,M$ NaCl containing 5 mM EDTA. This suggests that the apparent size of rhamnogalacturonan I is not due to noncovalent aggregation. If the backbone of rhamnogalacturonan I is a single linear chain, then this chain contains about 300 rhamnosyl residues and 600 galacturonosyl residues uninterrupted by regions of homogalacturonans. This is a major upward revision of the size of rhamnogalacturonan I compared to that envisioned earlier by this laboratory (Talmadge *et al.*, 1973).

Linkage analysis of the rhamnogalacturonan I isolated from the walls of suspension-cultured sycamore cells shows the polysaccharide contains 2- and 2-4 linked rhamnosyl residues and 4-linked galacturonosyl residues in a ratio of $1:1:4$. The manner in which the galacturonosyl and rhamnosyl residues are arranged is assumed to be as in the oligosaccharides discussed above; indeed, the disaccharide GalA-(1-2)-Rha has been isolated from sycamore cell walls (Talmadge *et al.*, 1973).

The presumed structural features of rhamnogalacturonan I are illustrated in Fig. 1. Note that this is not an exact structure. Rather, the structure presented is a pictorial summation of the data presented above. No informa-

$[(1\text{-}4)\text{-GalA-}(\alpha1\text{-}4)\text{-GalA-}(\alpha1\text{-}2)\text{-Rha-}(1\text{-}4)\text{-GalA-}(\alpha1\text{-}2)\text{-Rha-}(1\text{-}2)\text{-Rha-}(1\text{-}]_{100\text{-}200}$

Fig. 1. One possible sequence of rhamnosyl and galacturonosyl residues of rhamnogalacturonan I. Approximately half the rhamnosyl residues have an unidentified glycosyl residue attached to C-4 as well as having a galacturonosyl or another rhamnosyl residue attached to C-2. Approximately 5% of the galacturonosyl residues have an unidentified glycosyl residue attached to C-3. It is known that arabans and galactans are covalently linked to the rhamnogalacturonan backbone. It is not known how the arabans and galactans are attached. It is possible that one of these polysaccharides is attached to the C-4 of rhamnosyl residues and the other polysaccharide is attached to the C-3 of galacturonosyl residues.

tion is available on whether the rhamnosidic bonds are in the α or β configuration. The attachment of araban and galactan to this polymer is discussed in Section IX.

3. Homogalacturonan

The acidic pectic polysaccharides are characterized not only by regions of the large molecular weight rhamnogalacturonan I, but also by regions of unbranched $\alpha4$ linked galacturonosyl residues. Like the rhamnogalacturonan regions, the homogalacturonan regions are larger than was predicted (Talmadge *et al.*, 1973). The existence of the homogalacturonan regions was established by examining the polysaccharides released by purified endopolygalacturonase from the walls of suspension-cultured sycamore cells. Approximately 5% of the wall is converted by endopolygalacturonase into mono- , di- , and trigalacturonic acid. These are the expected products of the action of the endopolygalacturonase on an $\alpha4$ linked galacturonan. The endopolygalacturonase also releases an $\alpha4$ linked polygalacturonan from sycamore walls which accounts for 1–2% of the starting wall material. This galacturonan is stable to further enzyme degradation due to esterification of its uronosyl carboxyl groups (English *et al.*, 1972; Talmadge *et al.*, 1973). Similar $\alpha4$ linked galacturonans have been isolated from sunflower seeds (Zitko and Bishop, 1966) as well as from apple pectin (Barrett and Northcote, 1965). The sycamore homogalacturonan has an apparent degree of polymerization greater than 25 as deduced by gel filtration chromatography. It seems probable that the homogalacturonan regions of the pectic polysaccharides are considerably longer than 25, since these polysaccharides have been exposed to the action of the endopolygalacturonase which would hydrolyze any region of the homogalacturonans possessing sufficient deesterification to be susceptible to the enzyme. Certainly, four consecutive unesterified galacturonosyl residues are susceptible to the action of the enzyme (English *et al.*, 1972).

The fact that the galacturonosyl residues of pectic polymers are 4-linked has been established by converting the galacturonan to the corresponding galactan by reduction of the uronosyl carboxyl groups. The resulting galactan was then subjected to methylation analysis. In addition, Aspinall and

Jiang (1974) have methylated unreduced pectic polymers and have isolated from these polymers 2,3-dimethyl-galacturonic acid. These workers also have methylated carboxyl-reduced pectins and have isolated 2,3,6-trimethyl galactose, the product expected from 4-linked galacturonosyl residues. Further evidence of the 4-linked nature of the galacturonosyl residues is provided by the successful hydrolysis of sycamore cell wall galacturonans with the endopolygalacturonase specific for α4 linked galacturonosyl residues (Talmadge *et al.*, 1973).

The fact that the 4-linked galacturonosyl residues of the cell wall galacturonans are in the α-anomeric configuration has been demonstrated by characterization of the galacturonosyl-containing oligosaccharides derived from wall polymers (Aspinall *et al.*, 1967c), and by the fact that both the galacturonan and the derived oligomers have highly positive optical rotations (Aspinall and Jiang, 1974).

The carboxyl groups of the galacturonosyl residues of the cell wall pectic polysaccharides are known to be highly methyl esterified (Aspinall and Fanshave, 1961; Aspinall and Jiang, 1974; Aspinall and Molloy, 1968; Aspinall *et al.*, 1968a,c; Siddiqui and Wood, 1976). The degree of esterification of the carboxyl groups varies depending on the source of the pectic polymers (Aspinall *et al.*, 1968a). It is not known how the methyl esters are distributed along the polygalacturonan backbone. However, it is clear that there are regions which are highly methyl esterified and, therefore, are not susceptible to the endopolygalacturonase which requires free carboxyl groups (A. Darvill, M. McNeil, and P. Albersheim, unpublished results), as well as regions which are relatively free of methyl esters and, therefore, are susceptible to the endopolygalacturonase.

4. Araban

Arabans have been isolated from the cell walls of many dicotyledonous plants. Until recently, no homo-araban has been isolated specifically from primary cell walls. However, methylation analysis of the primary walls of suspension-cultured sycamore (Talmadge *et al.*, 1973) and pea (Gilkes and Hall, 1977) cells strongly suggested that these primary cell walls possess arabans which are structurally similar to arabans obtained from other tissues or organelles. An araban, essentially free of other polysaccharides, has now been isolated from a methylated primary cell wall polysaccharide fraction of suspension-cultured sycamore cells (A. Darvill, M. McNeil, and P. Albersheim, unpublished results).

The structures of plant arabans have been investigated by methylation analysis (Aspinall and Cottrell, 1971; Hirst and Jones, 1947; Joseleau *et al.*, 1977; Karacsonyi *et al.*, 1975; Rees and Richardson, 1966; Siddiqui and Wood, 1974), by Smith degradation, and by ^{13}C-nmr spectroscopy (Joseleau *et al.*, 1977). All of the arabans that have been investigated have similar structures. The arabans are highly branched polymers; the arabinosyl residues are largely in the furanose ring form; and the glycosidic linkages are uniformly in the α-anomeric configuration. In addition, arabinose is universally the L rather than the D isomer.

The glycosyl linkage compositions of the arabans from a number of dicots are compared in Table I. It can be concluded, by the presence of an *O*-methyl group on carbon 5, that the terminal- and 3-linked arabinosyl residues are in the furanose ring form. The fact that all the arabinosyl linkages are susceptible to hydrolysis by relatively mild acidic conditions is evidence that all the arabinosyl residues, including the 5-linked, 3-5 linked, and 2-5 linked residues, are in the furanose ring form (Karacsonyi *et al.*, 1975; Talmadge *et al.*, 1974).

Nuclear magnetic resonance analysis of *Rosa glauca* araban has provided additional evidence that the arabinosyl residues are α5 linked and in the furanose configuration. In ^{13}C-nmr analysis only the C-1 resonance expected of α-arabinofuranosyl residues was detected (Joseleau *et al.*, 1977). The proton nmr spectra is consistent with α- or β-furanosyl residues as well as with β-pyranosyl residues, but not with α-pyranosyl residues. The α-anomeric nature of these linkages is confirmed by the negative optical rotations, from

TABLE I

Glycosyl Linkage Compositions (Mole %) of the Arabans of Dicots

Arabinosyl linkage	Soy-bean meal[a]	Lemon peel[b]	Mus-tard cotyle-dons[c]	Mus-tard seed[d]	Rape-seed[e]	Rose bark[f]	Syc-amore[g]	White willow bark[h]	Aspen bark[i]
T-furanosyl	39.2	30.0	36.0	39.6	34.0	44.8	31.7	40.0	31.0
T-pyranosyl	0	0	0	0	0	0	0	2.4	2.7
5-	30.0	38.4	38.0	25.4	25.7	21.0	22.5	24.0	39.0
3-	0	0	0	0	0	8.0	0	4.4	0
3-5-	14.2	15.0	21.0	28.6	31.5	13.9	18.5	8.9	11.4
2-5-	6.0	4.6	0	tr	tr	3.5	8.0	5.7	6.5
2-3-5-	10.5	12.0	4.2	6.3	8.7	8.5	7.2	13.8	9.5

[a] *Glycine max* (Aspinall and Cottrell, 1971).
[b] *Citrus limon L.* pectin (Aspinall and Cottrell, 1971).
[c] *Sinapis alba L.* (Rees and Richardson, 1966).
[d] *Sinapis alba L.* (Aspinall and Cottrell, 1971).
[e] *Brassica campestris* (Siddiqui and Wood, 1974).
[f] *Rosa glauca* (Joseleau *et al.*, 1977).
[g] A mixture of methylated polysaccharides from sycamore, *Acer pseudoplatanus*, cell walls (A. Darvill, M. McNeil, and P. Albersheim, unpublished).
[h] *Salix alba L.* (Karacsonyi *et al.*, 1975).
[i] *Populus tremuloides* (Jiang and Timell, 1972).

−181 to −108, exhibited by such arabans (Aspinall and Cottrell, 1971; Joseleau et al., 1977).

The degree of polymerization of arabans has been estimated by converting the reducing end of the polymers to arabitol with sodium borohydride, hydrolyzing the reduced polymers, and determining the ratio of arabitol to arabinose. This method has provided evidence that two different arabans isolated from the bark of *Rosa glauca* have degrees of polymerization of 34 and 100 (Joseleau et al., 1977), while an araban from willow has a degree of polymerization of 90 (Karacsonyi et al., 1975).

There is not much information about the arrangement of the differently linked arabinosyl residues in arabans. The best work to date is that of Rees and Richardson (1966) who have studied an araban from mustard cotyledons using the Smith degradation. Their results ruled out the possibility of regions of long, unbranched 5-linked arabinosyl residues. The evidence obtained suggested that branched and unbranched arabinosyl residues occur near each other in the chain.

A number of complex pectic polysaccharides have been demonstrated to contain arabinosyl residues (Aspinall and Molloy, 1968; Rees and Wight, 1969; Siddiqui and Wood, 1966; Stoddart et al., 1967; Talmadge et al., 1973). These studies have generally not been carried to the point of determining whether the arabinosyl residues of the pectic polysaccharides exist as relatively long araban chains or whether the arabinosyl residues exist as mono- , di- , or trisaccharide side chains attached to the other pectic polysaccharides. One investigation using methylation analysis has provided evidence that the arabinosyl residues of rapeseed pectic polysaccharides are present as mono- or disaccharide side chains (Aspinall and Jiang, 1974). On the other hand, glycosyl linkage analyses of the pectic polysaccharides of primary cell walls (Talmadge et al., 1973) and studies of these polysaccharides, using mild acid hydrolysis for selective cleavage of the furanosyl linkages, suggest the presence of longer homo-arabans.

It is difficult to draw even a tentative structure of the primary cell wall araban. Clearly, branched arabans are important primary cell wall components. Efforts are currently under way in one of our laboratories (Department of Chemistry, University of Colorado) to isolate and structurally analyze the araban of sycamore cell walls. This study is augmented by the availability of two recently purified enzymes, an endo-α1-5-arabanase and an exo-α-arabinosidase (Kaji and Saheki, 1975; Weinstein and Albersheim, 1979).

5. Galactan

Galactans have been isolated from citrus pectin (Labavitch et al., 1976), white willow (Toman et al., 1972), and beech (Meier, 1962). As with the arabans, no homogalactan has ever been isolated directly from primary cell walls, although the glycosyl linkages which comprise those homogalactans that have been studied are also present, in similar ratios, in primary cell walls (Talmadge *et al.*, 1973).

The pectic galactans are primarily $\beta4$ linked polymers. The 4-linkage has been established by methylation analysis (M. McNeil and P. Albersheim, unpublished results; Toman *et al.*, 1972). The galactosidic linkages were shown to be in the β-anomeric configuration by the fact that these linkages are susceptible to hydrolysis by an endo-1-4-β-galactanase and by their low positive optical rotation (Labavavitch *et al.*, 1976). In addition, oligosaccharides produced from the intact galactan by partial acid hydrolysis (Toman *et al.*, 1972) are susceptible to further hydrolysis by a β-galactosidase. Finally, the β-configuration of some of the galactosidic linkages of oligosaccharides derived from a galactan by partial acid hydrolysis (Meier, 1962) has been established by chromatographic comparison to known standards.

Those galactans which have been studied have degrees of polymerization ranging from 33 (Toman *et al.*, 1972) to 50 (M. McNeil and P. Albersheim, unpublished results). These values were obtained by vapor pressure osmosis and by comparing the ratio of terminal to internal sugars as obtained by methylation analysis.

Galactans have been obtained which contain 6-linked galactosyl residues in addition to 4-linked residues. In two of the cases studied, the 6-linked residues accounted for approximately 4% of the polymer and are, therefore, quantitatively minor components of the polysaccharides. On the other hand, beech galactan is a polysaccharide with a major content of 6-linked galactosyl residues although the amount of the polysaccharide accounted for by the 6-linked residues has not been determined. The fact that 6-linked and 4-linked galactosyl residues are present in a single polymer has been established by the isolation of the trisaccharide: Gal-(β1-6)-Gal(β1-4)-Gal (Meier, 1962).

Homogalactans have not been isolated from primary cell walls, but the presence of such galactans in the walls is inferred by the detection of large amounts of 4-linked galactosyl residues upon methylation analysis of total cell walls and of pectic fractions of cell walls (Talmadge *et al.*, 1973). In addition, small oligomers of $\beta4$ linked galactosyl residues have been isolated in relatively large amounts from sycamore cell walls after treating the walls with an endo-β1-4-galactanase which can only hydrolyze galactans which contain four contiguous $\beta4$ linked galactosyl residues (Labavitch *et al.*, 1976).

Although the pectic polysaccharides probably do contain $\beta4$ linked homogalactans, many of the galactosyl residues of the pectic polysaccharides are probably not part of homogalactans (Aspinall and Cottrell, 1970; Aspinall and Jiang; 1974; Aspinall and Molloy, 1968; Aspinall *et al.*, 1967b; Siddiqui and Wood, 1976; Talmadge *et al.*, 1973; Toman *et al.*, 1975). The galactosyl residues on one pectic polymer have been shown to occur as $\beta 4$ linked dimers rather than as longer oligosaccharides or polymers (Aspinall and Jiang, 1974). There are several pectic polysaccharides that have been demonstrated to contain 3- and 6-linked galactosyl residues (Aspinall and Cottrell, 1970; Aspinall and Molloy, 1968; Talmadge *et al.*, 1973). The syca-more cell walls contain appreciable amounts of terminal and 3-, 6-, 3-6-, and 2-6 linked galactosyl residues (A. Darvill, M. McNeil, and P. Albersheim, unpublished results; Talmadge *et al.*, 1973). It is likely that these galactosyl residues are part of an arabinogalactan (Talmadge *et al.*, 1973); arabinogalac-tans are discussed in Section IV,A,6.

Several galactose-containing oligosaccharides have been isolated from plant polysaccharides. These include Gal-(1-2)-Xyl, GlcA-(1-6)-Gal, and GlcA-(1-4)-Gal (Aspinall *et al.*, 1967c) and GalA-(1-4)-Gal (Toman *et al.*, 1975). It is not known whether these oligosaccharides are constituents of the polymers of primary cell walls.

There are not sufficient data at this time to write a preliminary structure of the galactans of primary cell walls. Such galactans almost certainly exist, but whether they contain glycosyl constituents other than $\beta 4$ linked galactosyl residues has not been determined.

6. Arabinogalactan

Arabinogalactans have been isolated from the tissues of a variety of dicots. However, no arabinogalactan has been isolated from a source known to contain only primary cell walls. The glycosyl compositions of the arabinogalactans isolated from different sources are summarized in Table II. Unlike the arabans and galactans discussed earlier, there is considerable variation in the glycosyl compositions of the arabinogalactans. The arabinogalactan isolated from rapeseed (*Brassica campestris*) flour (Larm *et al.*, 1976) contains 90% arabinosyl residues while the arabinogalactan iso-lated from larch (*Larix leptolepis*) (Aspinall *et al.*, 1968b) contains 88% galac-tosyl residues. Three of the arabinogalactans that have been studied have been shown to contain rhamnosyl residues, whereas the other four arabinogalactans do not.

The glycosyl linkage analyses of these arabinogalactans are summarized in Table III. These data show that the arabinogalactan of soybean cotyledons is very different from the other arabinogalactans. The soybean arabinogalactan has a $\beta 4$ linked galactosyl backbone with arabinosyl dimers glycosidically linked to C-3 of some of the galactosyl residues (Aspinall *et al.*, 1967a; Morita, 1965a,b). The arabinosyl dimers have the structure Ara_f-(1-5)-Ara_f. The other arabinogalactans summarized in Table III are more similar to each other but still vary a great deal in the ratios of the arabinosyl and galactosyl residues. The differences in the glycosyl linkage compositions (Table III) reflect the differences in the glycosyl compositions (Table II). Except for the

TABLE II

The Glycosyl Compositions (Mole %) of a Variety of Arabinogalactans

Glycosyl residue	Rapeseed cotyledon[a]	Rapeseed flour[b]	Larch[c]	Maple sap[d]	Extra-cellular tobacco[e]	Extra-cellular Sycamore I[f]	Extra-cellular Sycamore II[g]	Soybean cotyledon[h]
Arabinose	48	90	14	45	46	31	34	30
Galactose	46	10	84	51	44	69	31	71
Uronic acid	6[i]	0	0	0	3[j]	0	12[j]	0
Rhamnose	0	0	Trace	5	8	0	4	0

[a] *Brassica campestris* (Siddiqui and Wood, 1972).
[b] *Brassica campestris* (Larm et al., 1976).
[c] *Larix leptolepis* (Aspinall et al., 1978b).
[d] *Acer saccharum* (Adams and Bishop, 1960).
[e] Isolated from the medium of suspension-cultured tobacco (*Nicotiana tabacum*) cells (Kato et al., 1977).
[f] Isolated from the medium of suspension-cultured sycamore (*Acer pseudoplatanus*) cells (Aspinall et al., 1969).
[g] Also contains 14% xylosyl residues which are thought to arise from a contaminating xylan (Keegstra et al., 1973).
[h] *Glycine max* (Aspinall et al., 1967; Morita, 1965a,b).
[i] Shown to be glucuronosyl residues.
[j] The type of uronic acid was not determined.

TABLE III

Glycosyl Linkage Compositions (Mole %) of the Arabinogalactans Listed in Table II[a]

Glycosyl linkage	Rapeseed cotyledon[b]	Rapeseed flour	Larch	Maple sap	Extra-cellular tobacco	Extra-cellular Sycamore I	Extra-cellular Sycamore II[d]	Soybean cotyledon
T-Ara$_f$	32	48	+[c]	22	23	+++	38	14
T-Ara$_p$	0	1	+	0	15	0	0	0
2-Ara$_f$	0	0	0	11	0	0	1	0
3-Ara$_f$	0	0	+	0	0	0	1	0
5-Ara$_f$	11	3	0	0	15	+	2	14
2-5-Ara$_f$	2	40	0	0	0	0	5	0
T-Gal	2	0	++	0	0	+	4	0
3-Gal	9	0	+	39	8	++	6	0
4-Gal	0	0	0	0	0	0	0	0
6-Gal	5	0	++	0	2	++	6	57
3-4-Gal	0	0	0	0	0	0	0	0
3-6-Gal	29	8	++++	22	33	+++	33	14
3-4-6-Gal	5	0	0	0	0	0	0	0
T-Rha	0	0	0	6	6	0	3	0

[a] The sources and references are the same as in Table II.
[b] Contains 7% T-GlcA.
[c] Approximate linkage quantitation is represented by + through ++++.
[d] The xylosyl residues have been left out of this calculation (see Table II).

soybean arabinogalactan, all of the arabinogalactans are characterized by the presence of significant amounts of 3-6 linked galactosyl residues and terminal arabinofuranosyl residues.

The arabinogalactans are also related to one another in that they all appear to have a galactan backbone with arabinosyl side chains. The structures of these polysaccharides have been further investigated by partial acid hydrolysis and by periodate oxidation (Aspinall *et al.*, 1968b; Kato *et al.*, 1977; Larm *et al.*, 1976; Siddiqui and Wood, 1972). The arabinosyl residues are hydrolyzed preferentially by mild acidic conditions, suggesting that the arabinosyl residues are attached to the galactan by furanosidic bonds.

The disaccharide L-Ara$_p$-(β1-3)-L-Ara$_f$ has been isolated from larch arabinogalactan (Aspinall *et al.*, 1968b), while the disaccharide L-Ara$_p$-(β1-5)-L-Ara$_f$ has been isolated from the arabinogalactan of the extracellular medium of suspension-cultured tobacco (*Nicotiana tabacum*) cells (Kato *et al.*, 1977). More extensive hydrolysis of the arabinogalactans isolated from larch, from the extracellular medium of suspension-cultured tobacco cells, and from the extracellular medium of suspension-cultured sycamore cells (Aspinall *et al.*, 1968b, 1969; Kato *et al.*, 1977) have led to the isolation from each arabinogalactan of Gal-(β1-3)-Gal and Gal-(β1-6)-Gal. Clearly, these arabinogalactans have galactosyl residues attached to one another by both 1-6- and 1-3 linkages. Smith degradation of the arabinogalactans isolated from larch (Aspinall *et al.*, 1968b), from rapeseed (Larm *et al.*, 1976; Siddiqui and Wood, 1972), and from the extracellular medium of suspension-cultured tobacco cells (Kato *et al.*, 1977) supports the results obtained by partial acid hydrolysis in that the periodate oxidation studies also demonstrate that the backbones of the polysaccharides are galactans and that the galactosyl residues of the galactans are glycosidically linked to one another through either or both C-3 and C-6.

The presence of arabinogalactans in the primary cell walls is supported primarily by the results of a single study of an endopolygalacturonase-released pectic fraction obtained from the walls of suspension-cultured sycamore cells (Talmadge *et al.*, 1973). The main arabinosyl and galactosyl containing components of these pectic polysaccharides appeared to originate from a β4 linked galactan and a highly branched araban. However, glycosyl linkages were also detected that are characteristic of arabinogalactans. The endopolygalacturonase-released pectic polysaccharides contained terminal and 3-, 6-, and 3-6 linked galactosyl residues. These residues were detected in amounts totaling approximately 5% of the pectic fraction. In addition, terminal, 3-, 5-, and 2-5 linked arabinofuranosyl linkages were also detected in substantial amounts in the pectic polysaccharides, but these residues could have originated from a branched araban.

Partial acid hydrolysis of the sycamore primary cell wall polysaccharides released by the endopolygalacturonase did not significantly alter the amounts of the branched galactosyl residues that were detected. This result suggests that the primary cell walls may contain an arabinogalactan which is similar to the one isolated from larch (Aspinall *et al.*, 1968b); the primary wall arabinogalactan may possess a low percentage of arabinosyl side chains. These results are also consistent with the presence in primary walls of a branched galactan that lacks arabinosyl side chains.

It is difficult to suggest a structural model for the arabinogalactan of primary cell walls. Indeed, since no arabinogalactan has been purified from the primary cell walls of dicots, it is not certain that arabinogalactans exist in these walls. In addition, the great variability among the arabinogalactans isolated from dicots would make it difficult to make any generalized model of arabinogalactans.

7. Rhamnogalacturonan II

A previously unknown pectic polysaccharide, rhamnogalacturonan II, has been isolated from the walls of suspension-cultured sycamore cells (Darvill *et al.*, 1978b). Hydrolysis of rhamnogalacturonan II yields the rarely observed cell wall sugars 2-*O*-methyl fucose, 2-*O*-methyl xylose, and apiose. The methylated sugars, 2-*O*-methyl xylose and 2-*O*-methyl fucose, have long been recognized as trace components of pectic polymers (Aspinall and Canas-Rodriguez, 1958; Aspinall and Fanshawe, 1961; Aspinall *et al.*, 1967c; Barrett and Northcote, 1965). Apiose has also been recognized as a component of the pectic polysaccharides of *Lemna* species (see Section IV,A,8). However, these three sugars have never previously been recognized to be associated in a single cell wall polysaccharide, although all three sugars have been isolated from leaves of deciduous trees (Bacon and Cheshire, 1971). The previously isolated apoise-containing pectic polysaccharide is not structurally related to rhamnogalacturonan II.

Rhamnogalacturonan II is solubilized from the walls of suspension-cultured sycamore cells by the action of endo-α1-4-galacturonase. Rhamnogalacturonan II is separated from the other pectic polysaccharides solubilized by the enzyme by anion exchange and gel permeation chromatography. As isolated, rhamnogalacturonan II is size homogeneous, containing between 25 and 50 glycosyl residues (Table IV). Methylation analysis has provided information about the glycosyl linkages in rhamnogalacturonan II. The polysaccharide is characterized by a wide variety of terminal glycosyl residues including T-galacturonosyl, T-galactosyl, T-arabinosyl, T-2-*O*-methyl xylosyl, T-2-*O*-methyl fucosyl, and T-rhamnosyl residues. Rhamnogalacturonan-II also contains 2-linked glucuronosyl, 3'-linked apiosyl, 3-linked rhamnosyl, 2-4 linked galactosyl, 3-4 linked rhamnosyl, and 3-4 linked fucosyl residues. The large amount of terminal glycosyl residues

Alan Darvill *et al.*

TABLE IV

Glycosyl Composition of Rhamnogalacturonan II

Glycosyl residue	Wt. % of recovered carbonhydrate	Number of residues in a polymer 39 residues long
Galacturonic acid	28	10
Rhamnose	18	7
Galactose	12	5
Arabinose	13	5
Apiose	7	3
2-*O*-methyl fucose	5	2
Glucuronic acid	3	2
2-*O*-methyl xylose	3	2
Fucose	4	2
Glucose	2	1

suggests a highly branched molecule. Indeed, rhamnogalacturonan II appears to be the most structurally complex plant polysaccharide known. It is not yet possible to draw even a partial structure for rhamnogalacturonan II.

8. Apiogalacturonan

A component of the cell walls of duckweed (*Lemna minor*) has been isolated and identified as an apiogalacturonan, with apiosyl and galacturonosyl residues as the only components (Beck, 1967; Duff, 1965; Hart and Kindel, 1970a,b). Apiose-containing galacturonans are also reported to be present in other plant tissues (Bacon and Cheshire, 1971), although these polysaccharides may contain other glycosyl components. Apiogalacturonans have never been isolated from a source containing only primary cell walls; therefore, their presence in such walls is still uncertain.

A partial structure of the apiogalacturonan from *Lemna minor* has been established by partial enzymic hydrolysis, by partial acid hydrolysis, and by periodate oxidation (Hart and Kindel, 1970a,b). A dimer of apiose (apiobiose) has been isolated following partial acid hydrolysis of the apiogalacturonan. Apiobiose has the structure D-Api$_f$-(1-3')-D-Api$_f$ (Hart and Kindel, 1970b). Most if not all of the apiose of the polysaccharide can be accounted for by this dimer. The homogalacturonan recovered following partial acid hydrolysis of the apiogalacturonan was degraded by a crude pectinase-containing preparation providing some evidence that the galacturonosyl residues are α4 linked (Hart and Kindel, 1970a). The apiogalacturonan possesses a very low content of methyl-esterified galacturonosyl residues. It has not been ascertained whether the apiobiosyl sidechains are attached to the galacturonosyl residues through C-2 or C-3.

B. The Pectic Polysaccharides of Monocots

The pectic polysaccharides of the primary cell walls of monocots have not been studied extensively. The reason for this appears to be that monocot primary cell walls contain only minor amounts of pectic polysaccharides. Indeed, monocot cell walls appear to contain less than 10% of the pectic polysaccharide content of dicot primary cell walls. The total amount of galacturonic acid in monocots has been estimated as being 3% of the cell walls of oat coleoptiles (Ray and Rottenberg, 1974), 6% of the cell walls of maize coleoptiles (Darvill, 1976), and only 1.3% of the cell walls of maize root meristem (Dever et al., 1968). These values may be low due to incomplete hydrolysis and recovery of the uronic acids. However, even if these values are incorrect by a factor of two, it is clear that monocot cell walls possess relatively small amounts of galacturonic acid.

Galacturonans have never been isolated from monocotyledons. The total pectic substances of oat (Avena sativa) coleoptile cell walls have been extracted by ammonium oxalate (Wada and Ray, 1978). Most of the polysaccharide solubilized by this procedure was not pectic in nature. However, electrophoresis of this ammonium oxalate extracted material did yield a component rich in galacturonosyl residues. This component did not yield by hydrolysis significant amounts of either galactose, arabinose, or rhamnose, sugars which are part of the pectic polymers of dicot cell walls. The electrophoretically purified component did yield by hydrolysis large amounts of glucose, but no evidence was presented as to whether the glucosyl residues were covalently linked in the polymer to the galacturonosyl residues. The structure of the galacturonosyl-containing polymers of oat coleoptile cell walls was not further investigated, although in an earlier report (Ray and Rottenberg, 1964), a disaccharide of galacturonic acid was isolated from oat coleoptile cell walls. The linkage of the disaccharide was not investigated. In a separate study, methylation analyses of maize coleoptile cell walls have demonstrated the presence of 4-linked galacturonosyl residues (Darvill, 1976; Darvill et al., 1977). This suggests that monocot cell walls may contain small amounts of 4-linked galacturonans.

Arabinogalactans have never been isolated from the primary cell walls of monocots. However, 3-, 6-, and 3-6 linked galactosyl residues are present in the walls of several suspension-cultured monocots (Burke et al., 1974). These glycosyl residues are characteristic of the arabinogalactans of dicots. Since no arabinogalactan has actually been isolated from monocot cell walls, the existence of arabinogalactans in monocot primary cell walls must be considered an open question.

Arabinogalactan and arabinogalactan-glycoproteins have been isolated from many monocotyledon tissues (Anderson et al., 1977; Fincher and Stone, 1974; Maekawa and Kitao, 1974; Neukom and Marwalker, 1975). These glycoproteins are not likely to be structural components of walls (Anderson *et al.*, 1977). However, their existence is mentioned because they contain similar linkages to the dicot arabinogalactans (Table III) and because the glycoproteins have been shown to bind carbohydrates. Their binding properties are discussed in Section VIII,B and C.

There is no evidence for the presence in monocot primary cell walls of pectic arabans or galactans. No rhamnogalacturonan has been isolated from monocot cell walls, although the disaccharide GalA-(1-2)-Rha has been obtained from oat coleoptile cell walls (Ray and Rottenberg, 1964). Trace amounts of 2-*O*-methyl xylose and 2-*O*-methyl fucose have been detected in cell walls of oat coleoptiles (Darvill *et al.*, 1978b). These methylated sugars are characteristic of the newly discovered polymer of dicot cell walls called rhamnogalacturonan II. If rhamnogalacturonan II is a constituent of monocot primary cell walls, it is present in these walls at a concentration at least tenfold less than the amount of this polymer in dicot cell walls.

In conclusion, the pectic polysaccharides of monocot primary cell walls have not been as thoroughly studied as those of dicots. The pecticlike constituents that are present in primary cell walls of monocots when compared to dicot cell walls account for a comparatively small amount of the walls; therefore, the pectic polysaccharides of monocots do not appear to play as important a role in primary cell wall structure of monocots as they do in primary cell walls of dicots.

V. THE HEMICELLULOSES

A. The Hemicelluloses of Dicots

1. *Xyloglucan*

Xyloglucan is perhaps the most thoroughly understood of the noncellulosic polysaccharides of primary cell walls (Bauer *et al.*, 1973). Xyloglucans were first characterized as an amyloid component of seeds (Aspinall *et al.*, 1977; Gould *et al.*, 1971; Hsu and Reeves, 1967; Kooiman, 1961; Siddiqui and Wood, 1971, 1977a). Considerably later, xyloglucans were isolated from the medium of suspension-cultured sycamore (Aspinall *et al.*, 1969) cells and then from the primary cell walls of suspension-cultured sycamore cells (Bauer *et al.*, 1973). The basic structure of xyloglucans was elucidated by Kooiman (1961) studying the amyloid of *Tamarindus indica* seeds. The widespread occurrence of the amyloids (Aspinall *et al.*, 1977; Hsu and Reeves, 1967; Kooiman, 1961; Siddiqui and Wood, 1971, 1977a) and xyloglucans (Aspinall *et al.*, 1969; Barnoud *et al.*, 1977; Bauer *et al.*, 1973) shows that

TABLE V

Glycosyl Compositions (Mole %) of a Variety of Xyloglucans and Amyloids

Glycosyl residue	Tama-rindus[a]	Nastur-tium[b,c]	Rape-seed I[d]	Rape-seed II[e]	Extracellular polysaccharides			Syc-amore cell wall[i,j]
					Bean[f]	Rose[g]	Syc-amore[c,h]	
Glucose	48	55	64	48	46	51	46	31
Xylose	36	27	24	34	36	30	37	36
Galactose	16	18	12	10	10	10	37	14
Fucose	0	0	0	7	8	7	7	7

[a] *Tamarindus indica* seeds (Courtois and LeDizet, 1974; Kooiman, 1961).

[b] *Tropeoleum majus* seeds (Courtois and LeDizet, 1974; Hsu and Reeves, 1967).

[c] For these sources of xyloglucan or amyloids, the glucosyl, galactosyl, and xylosyl residues were shown to be in the D-configuration.

[d] *Brassica campestris* seeds (Siddiqui and Wood, 1972).

[e] *Brassica campestris* seeds (Aspinall *et al.,* 1977; Siddiqui and Wood, 1977a).

[f] From the medium of suspension-cultured true bean (*Phaseolus vulgaris*) cells (calculated from data presented in Wilder and Albersheim, 1973).

[g] From the medium of suspension-cultured Rose (*Rosa glauca*) cells (Barnoud *et al.,* 1977).

[h] From the medium of suspension-cultured (*Acer pseudoplatanus*) cells (Aspinall *et al.,* 1969; Bauer *et al.,* 1973).

[i] From the cell walls of suspension-cultured sycamore (*Acer pseudoplatanus*) cells (calculated from the data presented in Bauer *et al.,* 1973).

[j] This preparation is known to be contaminated with pectic polysaccharides.

polysaccharides isolated from tissues other than primary cell walls can, at times, serve as excellent models for cell wall polysaccharides.

The composition of the xyloglucans isolated from a variety of sources as presented in Table V, indicate differences between the xyloglucans. For example, all of the cells possessing primary cell walls produce xyloglucans containing fucosyl residues. It is possible that all xyloglucans contain fucosyl residues when they are synthesized, but during differentiation (to secondary walls) the fucosyl residues may be removed. It is also possible that xyloglucans without fucosyl residues are present in primary cell walls but have not yet been detected.

The structure of the xyloglucans has been determined by a combination of methylation analysis and chromatographic separation of the oligosaccharides produced by partial enzymic digests. The only major differences in the methylation analyses (Table VI) of the various xyloglucans result from the presence or absence of the terminal fucosyl residues. The fucosyl-containing xyloglucans yield the same partially methylated partially acety-

TABLE VI

Glycosyl Linkage Compositions (Mole %) of the Xyloglucans and Amyloids Isolated from a Variety of Sources[a]

Glycosyl residue	Tamarindus indica[c]	Nasturtium	Rapeseed I[d]	Rapeseed II	BEPS	REPS	SEPS[e]	Sycamore cell walls[e]
4-Glc	16	+[f]	20	13	11	17	13	13
4,6-Glc	33	+	32	39	33	30	32	29
T-Xyl	16	+	20	25	24	28	28	29
2-Xyl[b]	16	+	4	6	7	21	8	7
T-Gal	16	+	12	3	2	2	2	3
2-Gal	0	0	0	6	7	+	6	7
T-Fuc	0	0	0	6	8	+	5	7

[a] Xyloglucan and amyloid sources and references are the same as in Table V.
[b] This figure for 2-linked Xyl may also contain some 4-linked Xyl in all the preparations except *Tamarindus indica* and nasturtium.
[c] Calculated from Siddiqui and Wood 1977a; original data from White and Rao, 1953.
[d] Also contained 4% T-Glc and 8% 6-Glc.
[e] Also contained between 1 and 2% of both T-Ara$_f$ and 2-4-6-Glc.
[f] + = detected but not quantitated.

lated alditols as the nonfucosyl-containing xyloglucans except for the addition of terminal fucosyl residues to C-2 of the majority of the otherwise terminal galactosyl residues.

The terminal xylosyl residues were shown by Kooiman (1967) to be linked via α-glycosidic bonds to the C-6 of glucosyl residues. Kooiman hydrolyzed *Tamarindus indica* xyloglucan with a crude commercial mixture of enzymes called "Luizym" and recovered from the enzymic digest almost all of the xylosyl residues as Xyl-(α1-6)-Glc. He methylated the disaccharide to establish that the xylosyl residues are attached to C-6 of the glucose. Kooiman also showed that the glycosidic linkage was most likely the α anomer by demonstrating that the disaccharide possesses a highly positive optical rotation. Kooiman (1961) also established that the α-linked disaccharide possessed a different melting point from that of the naturally occurring β-linked disaccharide, Xyl-(β1-6)-Glc. The disaccharide Xyl-(α1-6)-Glc has been isolated following acetolysis of rapeseed hull xyloglucan. The disaccharide was characterized by methylation analysis and by optical rotation (Aspinall *et al.*, 1977).

It is assumed, because of Kooiman's results with the amyloids, that all of the xylosyl residues of the cell wall xyloglucans are linked to C-6 of glucosyl residues. This linkage has been confirmed for at least a portion of the xylosyl residues of the sycamore wall and extracellular xyloglucans (Bauer *et al.*, 1973). This was established by analysis of enzyme-produced hepta- and nonasaccharides (see below). The anomeric configuration of the xylosidic bonds of the primary cell wall xyloglucans have not been determined but, by analogy to the amyloids, these bonds may be assumed to be in the α configuration.

The galactosyl residues of xyloglucans are linked through a β-galactosidic bond to the C-2 of xylosyl residues. This was established by the isolation of Gal-(β1→2)-Xyl following partial acid hydrolysis or acetolysis of the xyloglucans isolated from *Tamarindus indica* seeds (Kooiman, 1961), from nasturtium (*Tropeoleum majus*) seed (Hsu and Reeves, 1967), and from rapeseed (*Brassica campestris*) hulls (Aspinall *et al.*, 1977). The existence of the Gal-(1-2)-Xyl linkage in sycamore extracellular xyloglucan has been established by isolation and characterization (see below) of pentasaccharide "d" in Fig. 2 (B. S. Valent, M. McNeil, and P. Albersheim, unpublished).

At least a portion of the glucosyl residues of rapeseed hull xyloglucan has been shown to be β-linked by the isolation of cellobiose after partial acid hydrolysis (Aspinall *et al.*, 1973). Similarly, at least a portion of the glucosyl residues of the xyloglucans of soybean and true bean cell walls and extracellular polysaccharides have been established to be β4 linked by their susceptibility to hydrolysis by an endo-β1-4-glucanase (Wilder and Albersheim, 1973). The fact that xyloglucans hydrogen-bond to cellulose is evidence that the most of the backbone of the xyloglucans is a β4 linked glucan.

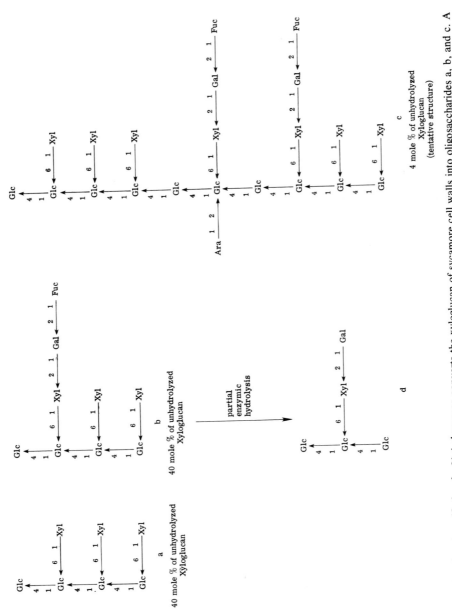

Fig. 2. A purified endo-β1-4-glucanase converts the xyloglucan of sycamore cell walls into oligosaccharides a, b, and c. A mixture of glycosidases converts oligosaccharide b into oligosaccharide d and thereby establishes the point of attachment of the fucosyl-galactosyl disaccharide present in oligosaccharides b and c. Experimental details are found in Bauer et al. (1973) and B. S. Valent, M. McNeil, and P. Albersheim, unpublished.

The terminal fucosyl linkages of sycamore extracellular xyloglucan were shown to be linked to the C-2 of the galactosyl residues by taking advantage of the acid lability of the fucosidic bonds. The fucosidic linkages are hydrolyzed when the xyloglucan is treated at pH 2 and at 110°C for 1 h. Following this treatment, most of the terminal fucosyl and 2-linked galactosyl residues disappear, while an equivalent amount of terminal galactosyl residues is formed (Bauer *et al.*, 1973). The anomeric configuration of this fucosidic linkage is not known, although the linkage is hydrolyzed by a mixture of enzymes known to contain an α1-2-fucosidase (Bahl, 1970; B. S. Valent, M. McNeil, and P. Albersheim, unpublished).

The structures of the xyloglucans isolated from the walls and culture medium of suspension-cultured sycamore cells have been further studied by digestion of the xyloglucans with a purified endo-β1-4-glucanase (Bauer *et al.*, 1973). The endoglucanase-produced oligosaccharides have been fractionated by Bio-Gel P-2 chromatography into four quantitatively major components: a void peak, and oligomers composed of 22, 9, and 7 glycosyl residues. The details of the chemical characterization of these oligosaccharides may be found in Bauer *et al.*, (1973). The structures and mole percentage of the total xyloglucan accounted for by each of these peaks are summarized in Fig. 2. The location of the Fuc-(1-2)-Gal disaccharide was not originally determined (Bauer *et al.*, 1973). It has since been determined by partial enzymic hydrolysis of the nonasaccharide and isolation of the resulting definitive pentasaccharide labeled "d" in Fig. 2.

The molecular weight of sycamore cell culture medium xyloglucan has been estimated as 7600, which represents about 50 glycosyl residues. This number is consistent with that found by Kooiman for *Tamarindus indica* seed xyloglucan (Kooiman, 1961).

The relative mole percentage of the xyloglucan oligosaccharides a, b, and c (Fig. 2) suggests that these occur in the xyloglucan chains in a ratio of 10 : 10 : 1, respectively. The smallest possible xyloglucan that can be constructed from 10 repeats each of oligosaccharides "a" and "b" and a single copy of oligosaccharide "c" would contain 182 glycosyl residues. Since this value is much larger than the experimentally determined 50 glycosyl residues, it suggests that there exists at least two different xyloglucan species. For example, there could be one species made of a dimer of oligosaccharide c and another species made of three molecules each of oligosaccharide a and b. If this were true, the ratio of the two species would be one molecule of the former to approximately seven molecules of the latter. Clearly, the exact nature of the xyloglucan is not known. Indeed, the following additional uncertainties concerning the structure of xyloglucans remain to be elucidated.

1. The Ara$_f$-(1-2)-Glu linkage shown in oligosaccharide "c" (Fig. 2) is based solely and insubstantially on the finding of equal molar amounts of terminal arabinofuranosyl and 2-4-6 linked glucosyl residues.

2. Some glucosyl residues could be attached to C-6 of other glucosyl residues with an equivalent amount of xylosyl residues attached to C-4 of glucosyl residues. This possibility has been more or less ruled out in the case of the seed xyloglucans, but has not been ruled out for the cell wall xyloglucans.

3. The anomeric linkages have not been carefully determined for cell wall xyloglucans. They are assumed to be consistent with the glycosidic linkages of the seed xyloglucans, i.e., all the glucosyl and galactosyl residues are β-linked, while the xylosyl residues are α-linked. The fucosyl linkage is also thought to be the α-anomer.

2. Xylan

Xylans, with a $\beta 4$ linked backbone, have generally been considered to be quantitatively major constituents of the *secondary* cell walls of dicots (Whistler and Richards, 1970). The secondary cell wall xylans of different plants often differ in the nature of the side chains glycosidically-linked to the xylan backbone.

The side chain reported most frequently is a terminal 4-*O*-methylglucuronosyl residue (Aspinall and McGrath, 1966; Bardalaye and Hay, 1974; Comtat *et al.*, 1974; Eda *et al.*, 1977; Siddiqui and Wood, 1977b; Toman, 1973). Xylans characterized by the presence of these monomethylated mono-uronosyl side chains have been isolated from rapeseed cotyledon meal (*Brassica campestris*), from midrib of tobacco (*Nicotiana tabacum*), from aspen wood (*Populus tremuloides*), and from the bark of white willow (*Salix alba L.*). The uronosyl residues are always linked to the C-2 of xylosyl residues. One uronosyl residue is linked to one in every seven to ten xylosyl residues (Aspinall and McGrath, 1966; Siddiqui and Wood, 1977b; Toman, 1973). The xylan isolated from soybean (*Glycine max*) hull cell wall contains an unmethylated glucuronosyl residue attached to only about one in every thirty xylosyl residues (Aspinall *et al.*, 1976).

Some purified xylans contain sugars other than xylose and glucuronic acid. For example, 4-linked glucosyl residues have been shown to be interspersed with the 4-linked xylosyl residues in the xylan backbone (Bardalaye and Hay, 1974; Henderson and Hay, 1972). Two such (4-*O*-methyl-D-glucuronosyl) glucoxylans have been isolated from the sapwood of the lateral roots of the sugar maple (*Acer saccharum Marsh*). These glucoxylans have molar ratios of D-glucosyl to D-xylosyl to 4-*O*-methyl-D-glucuronosyl residues of 3 : 36 : 1 and 0.5 : 25 : 1 (Bardalaye and Hay, 1974). Three neutral glucoxylans have been isolated from barberry (*Berberis vulgaris*) leaves; two of these glucoxylans have been reported to contain small proportions ($\approx 7\%$) of terminal, nonreducing galactosyl and arabinopyranosyl residues (Hender-

TABLE VII

The Glycosyl Composition (Mole %) of Sycamore Cell Wall Glucuronoarabinoxylan

Glycosyl residue	Mole %
Xylose	69
Arabinose	17
Glucuronic acid	10
4-O-methyl glucuronic acid	2

son and Hay, 1972). Galacturonosyl and rhamnosyl residues have also been reported in a xylan purified from birch (*Betula verrucosa*) meal (Shimizu and Samuelson, 1973). Three-linked rhamnosyl residues ($\approx 1\%$) have been reported to be interspersed with 4-linked xylosyl residues in the backbone of a xylan purified from lucerne (*Medicago sativa*) stems (Aspinall and McGrath, 1966).

A xylan has been isolated from the stalks of tobacco (*Nicotiana tabacum*) by alkaline extraction (Eda *et al.*, 1976). Methylation analysis, acid hydrolysis, and hydrolysis with an endo-β1-4-xylanase confirm that the tobacco xylan is a linear, unbranched chain of β4 linked xylosyl residues.

The first xylan, or, more accurately, a glucuronoarabinoxylan, to be clearly established as a constituent of the primary cell walls of a dicot has recently been characterized by J. Darvill, A. Darvill, M. McNeil, and P. Albersheim (unpublished). The glucuronoarabinoxylan, which constitutes 5% of the walls of suspension-cultured sycamore cells, was extracted with alkali from walls that had been preextracted with endopolygalacturonase to remove most of the pectic polysaccharides. The alkali soluble polysaccharides were fractionated by anion exchange chromatography. A xylose-rich fraction was further purified by gel filtration on Bio-Gel P-100 and then on Agarose 1.5 m. The first gel filtration column removed contaminating pectic polysaccharides and the second column removed a glucan. The xylose-containing fractions from the Agarose column were shown to possess a constant sugar composition across the polysaccharide peak. The glycosyl composition of the glucuronoarabinoxylan is presented in Table VII. The glucuronoarabinoxylan has been shown, by methylation analysis, to contain terminal and 4-, 2- and 3-4 linked xylosyl residues. The polysaccharide also contains terminal and 2-linked arabinofuranosyl residues as well as terminal 4-O-methyl glucuronosyl and terminal glucuronosyl residues. Future experiments, including β-elimination of the uronosyl residues, should yield a definitive covalent structure of this cell wall polysaccharide.

B. The Hemicelluloses of Monocots

1. Xyloglucan

The primary cell walls of monocots may contain xyloglucans similar to those found in dicots. The evidence which suggests that xyloglucans may be present in monocot primary cell walls comes from a combination of methylation analysis and the isolation and characterization of diagnostic oligosaccharides. Methylation analysis of six different suspension-cultured monocot cell walls has shown that these walls contain from 0.3% to 3.8% 4-6 linked glucosyl residues (Burke *et al.*, 1974). Such 4-6 linked glucosyl residues account for 1.9% of the cell walls of maize coleoptiles (Darvill *et al.*, 1977). The presence of 4-6 linked glucosyl residues is not proof that xyloglucans are present in the walls of the suspension-cultured cells but since this particularly-linked glucosyl residue has not been detected in any wall polymer other than xyloglucans, the presence of these residues is suggestive of the presence of xyloglucans in monocot cell walls.

A fraction has been isolated from oat coleoptile walls which is rich in 4-6 linked glucosyl residues (Labavitch and Ray, 1978). This polymer was subjected to the action of a crude enzyme preparation containing an endo-β1-4-glucanase. A trisaccharide and pentasaccharide were isolated from the products of the enzymic hydrolysis. Methylation analyses of the oligosaccharides indicated that the trisaccharide has a structure consistent with Xyl-(1-6)-Glc-(1-4)-Glc and the pentasaccharide has a structure consistent with:

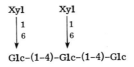

Such oligosaccharides could theoretically be isolated from the xyloglucans of dicots (Fig. 2). Thus, their isolation is further evidence that dicotlike xyloglucans are present in monocot primary cell walls. However, if the xyloglucans are present in the primary walls of monocots, they account for only about 2% of the walls while xyloglucans account for about 19% of dicot primary cell walls (see Section V,A,1).

2. Xylan

The quantitatively dominant component of monocot primary cell walls is the hemicellulosic xylan. Primary cell wall xylan has not been as well studied as secondary cell wall xylan. Nevertheless, the xylans of all of the monocots that have been studied are characterized by possessing a backbone of β4 linked D-xylopyranosyl residues (Aspinall and Greenwood, 1962; Hirst, 1962;

Whistler and Richards, 1970; Wilkie and Woo, 1977). The xylans from different sources possess a wide variety of side chains attached to the main xylan backbone. These side chains vary from species to species and also from tissue to tissue within a single species. The side chains of secondary cell wall xylans have, on the whole, been more extensively characterized than those of primary cell wall xylans although there appear to be similar side chains on xylans from both sources. The most common side chains are single L-arabinofuranosyl residues attached to C-3 of some of the xylosyl residues of the xylan backbone. The second most common side chains are single D-glucuronosyl or 4-O-methyl-D-glucuronosyl residues attached to C-2 of xylosyl residues in the xylan backbone. However, more complex side chains have also been characterized. The effect of the structural diversity of xylans on the function of the xylans is not known. A nonbranched homoxylan containing no side chains has been extracted from esparto grass (Chanda *et al.*, 1950; Ehrenthal *et al.*, 1954). The following sections will discuss some of the more common side chains of the xylans.

a. Terminal Arabinofuranosyl. Most xylans contain some single terminal arabinofuranosyl residue side chains. Usually these are found attached to C-3 of the xylosyl residues of the xylan backbone (Aspinall and Cairncross, 1960a; Aspinall and Sturgeon, 1957; Aspinall *et al.*, 1963; Barnoud *et al.*, 1973; Buchala, 1973, 1974; Buchala and Meier, 1972; Buchala and Wilkie, 1974; Buchala *et al.*, 1972; Joseleau and Barnoud, 1974; Maekawa and Kitao, 1973; McNeil *et al.*, 1975; Wilkie and Woo, 1977; Woolard *et al.*, 1976). On the other hand, single arabinofuranosyl side chains have been found attached to the C-2 on xylosyl residues in the xylans extracted from the husks of sorghum grain (Woolard *et al.*, 1976, 1977) and from barley aleurone cell walls (McNeil *et al.*, 1975).

The points of attachment of the arabinofuranosyl residues have been established in several ways. One method has taken advantage of the fact that the arabinofuranosyl linkages are more labile to acid hydrolysis than the pyranosyl linkages of the xylan backbone. Because of this, it has been possible to selectively hydrolyze the arabinofuranosyl residues and then to look for the positions on the backbone that have been vacated due to this release of the arabinose (Aspinall and Ferrier, 1960; Aspinall and Sturgeon, 1957). Another method involved the isolation and characterization of the trisaccharide Ara$_f$-(β1-3)-Xyl-(β1-4)-Xyl which was obtained as a partial hydrolysis product of the xylans of wheat straw and rye flour (Aspinall and Ferrier, 1960; Fincher, 1976). An ingenious method for identifying the point of attachment of arabinofuranosyl residues involved the oxidation of the primary alcohol of the arabinofuranosyl residues (there are no primary alcohols on the xylosyl residues of the xylan backbone). Conversion of the primary alcohols to carboxyl groups causes the arabinofuranosyl linkage to become relatively resistant to acid hydrolysis. This permitted the isolation of the aldobiuronic acid, arabinofuranouronosyl-(β1-3)-Xyl, which identified the C-3 of the xylosyl residues as the point of attachment, in this polysaccharide, of the arabinofuranosyl residues (Aspinall and Cairncross, 1960b).

b. Terminal 4-*O*-methyl Glucuronosyl and Terminal Glucuronosyl. Single uronosyl residue side chains are almost as commonly found in the monocot xylans as the single arabinofuranosyl side chains. The resistance of uronosyl bonds to acid hydrolysis has led to the isolation of aldobiuronic acids identifying the C-2 position of the xylosyl residues as the point of attachment of these side chains on the xylan backbone.

The terminal glucuronosyl residues of some xylans are all unsubstituted while other xylans contain terminal 4-*O*-methyl glucuronosyl residues and still other xylans contain a mixture of both of these side chains. Xylans isolated from *Cyperus papyrus* stalk contain only terminal glucuronosyl residues (Buchala and Meier, 1972), whereas xylans from bamboo culm (Maekawa and Kitao, 1973), reed stalk (Barnoud *et al.*, 1973), reed internode (Joseleau and Barnoud, 1974), barley leaves (Buchala, 1973), cocksfoot grass (Aspinall and Cairncross, 1960a), and rye grass root (Aspinall *et al.*, 1963) contain only terminal 4-*O*-methyl glucuronosyl residues. The xylans of oat stem (Buchala *et al.*, 1972), oat straw leaf and stem (Aspinall and Wilkie, 1956), bamboo leaves (Wilkie and Woo, 1977), guinea grass aerial tissue (Buchala, 1974), and sorghum grain husk (Woolard *et al.*, 1976, 1977) have both glucuronosyl and 4-*O*-methyl glucuronosyl side chains.

c. Gal-(β1-4)-Xyl-(β1-2)-Ara$_f$-. This trisaccharide side chain has been demonstrated to be attached to xylans of barley leaves (Buchala, 1973), bamboo leaves (Wilkie and Woo, 1977), guinea grass aerial tissue (Buchala, 1974), rye grass roots (Aspinall *et al.*, 1963), and oat stem (Buchala *et al.*, 1972). In each of these cases, the trisaccharide is attached to the C-3 of xylosyl residues.

d. Gal-(β1-5)-Ara$_f$-. This disaccharide side chain has been isolated from the xylans of bamboo leaves (Wilkie and Woo, 1977), guinea grass aerial tissues (Buchala, 1974), and oat stem (Buchala *et al.*, 1972) and, when present, is also attached to the C-3 of xylosyl residues.

e. GluA-(β1-4)-Xyl-(β1-4)-Gal-. This trisaccharide has been shown to be attached to the xylans of oat stems (Buchala *et al.*, 1962) and bamboo leaves (Wilkie and Woo, 1977). It has not been determined whether the glucuronosyl residues have endogenous methyl groups or not. It has also not been determined to which point on the xylan this trisaccharide is attached.

f. Xyl-(β1-2)-Ara$_f$-. This disaccharide is attached to the xylans of bamboo leaves (Wilkie and Woo, 1977), guinea grass aerial tissues (Buchala, 1974), rye grass root (Aspinall et al., 1963), and oat stem (Buchala et al., 1962). In each case, this disaccharide is attached to C-3 of xylosyl residues. It has been suggested that a trisaccharide Xyl-(1-2)-Ara-(1-2)-Ara- may be attached to bamboo xylan (Wilkie and Woo, 1977), although this trisaccharide has not been isolated.

Monocot xylans may possess some O-acetyl esters. Most of the extraction procedures that have been utilized to isolate xylans involved a base extraction which would result in hydrolysis of these esters. Xylans extracted under more mild conditions, for example, with dimethyl sulphoxide, have been shown to contain O-acetyl groups (Hirst, 1962).

There has been little effort to isolate and characterize the xylans from homogeneous preparations of monocot primary cell walls. Methylation analysis of the intact walls indicates that xylans are a major component of these walls. Methylation analysis of the primary cell walls of suspension-cultured monocots (wheat, oat, rye, sugarcane, brome grass, and rye grass) (Burke et al., 1974) and of maize coleoptiles (Darvill et al., 1977), and oat internodes (M. McNeil, A. Darvill, and P. Albersheim, unpublished) indicate that xylans are major components of these walls. This is inferred from the high content in these walls of 4- and 3-4 linked xylosyl residues. The walls also contain some 2-4 linked xylosyl residues. In addition, the walls contain a high content of terminal arabinofuranosyl residues. The detection in these walls of 2-linked xylosyl, 4-linked galactosyl, terminal galactosyl, and 5-linked arabinosyl residues also is suggestive of the side chains found on the secondary cell wall xylans. The primary cell walls also contain glucuronosyl and 4-O-methyl glucuronosyl residues (Darvill et al., 1977; Ray and Rottenberg, 1974).

A glucuronoarabinoxylan has been isolated from oat coleoptile cell walls, a tissue that contains a very small proportion of secondary cell walls (Labavitch and Ray, 1978; Wada and Ray, 1978). Partial acid hydrolysis of the oat coleoptile glucuronoarabinoxylan released most of the uronosyl residues as glucuronosyl-xylose and 4-O-methyl glucuronosyl-xylose. A glucuronoarabinoxylan has been partially purified from maize coleoptiles (Darvill et al., 1978a). Methylation analysis of this preparation indicated the presence of the side chains typically attached to xylan backbones (Darvill et al., 1977, 1978a).

A xylan has been isolated from the young internodes of the reed *Arundo donax* (Joseleau and Barnoud, 1974). Most of the cells of this tissue possess primary cell walls. The xylan isolated from the young internodes has a structure very similar to that of the xylan isolated from the older tissues of the same source (Barnoud *et al.*, 1973). The reed xylan has a 4-linked backbone with 4-O-methyl glucuronosyl residues attached to C-2 and terminal arabinofuranosyl residues attached to C-3 of some of the xylosyl residues in the backbone.

The xylans constitute a far higher percentage of the primary cell walls of monocots than they do of dicots. Structurally, the xylans from both sources appear similar; they both possess a β4 linked xylan backbone, and both can have neutral and/or acidic side chains with the acidic attached to C-2 or C-3 of xylosyl residues of the backbone.

VI. NONCELLULOSIC GLUCAN

A. Noncellulosic Glucans of Dicots

Noncellulosic β-linked glucans have frequently been obtained from the tissues of monocots (Buchala and Wilkie, 1970, 1971; Eda *et al.*, 1976; Nevins *et al.*, 1977; Wilkie and Woo, 1976; see also Section VI,B). There is only one report of the isolation of a β-glucan from the cell walls of a dicot, i.e., mung bean (*Phaseolus aureus*) hypocotyls (Buchala and Franz, 1974). The glucan was extracted by hot water from cell walls prepared from 3-day-old hypocotyls. Extracts of cell walls of older hypocotyls were deficient in this glucan. Methylation analysis and periodate oxidation studies (Buchala and Franz, 1974) of the hypocotyl glucan indicated that the glucan contains 3-linked and 4-linked glucopyranosyl residues in the molar ratio of 1.0 : 1.7. Partial hydrolysis of the glucan released oligosaccharides containing both 3- and 4-linked glucosyl residues as well as other oligosaccharides containing only 3- or 4-linked glucosyl residues. The hypocotyl tissue from which the β-glucan was obtained contains both primary and secondary cell walls. Thus, these results do not establish whether such glucans exist in the primary cell walls of dicots.

B. Noncellulosic Glucans of Monocots

Glucans consisting of a mixture of β3 linked and β4 linked glucosyl residues (mixed β-glucans) are among the most studied polysaccharides of monocot primary cell walls. These mixed β-glucans have been found in nearly all of the cell wall preparations obtained from monocots. The generalized occurrence of the mixed glucans and the inability to easily remove them from cell wall preparations indicates that these mixed β-glucans may be structural polymers of cell walls. However, the mixed β-glucans exhibit properties that question their role as structural polymers of primary cell walls (Nevins and Loescher, 1974; Nevins *et al.*, 1977). For example, mixed

β-glucans of oat coleoptiles are catabolized and disappear when this tissue is grown in the dark in the absence of an energy source, suggesting that the mixed β-glucan may be an energy reserve rather than a structural polymer (Loescher and Nevins, 1972; Nevins and Loescher, 1974; Nevins et al., 1977). More work is necessary before the function of the mixed β-glucans is ascertained. A description of the structure of the mixed β-glucan is given below.

Mixed β-glucans are widely distributed among the monocots and have been isolated from rye (Smith and Stone, 1973b), oat (Morris, 1942; Parrish et al., 1960; Peat et al., 1957), and barley endosperm tissues (Costello and Stone, 1968). The mixed β-glucans have also been isolated from maize (Buchala and Meier, 1973), barley (Buchala and Wilkie, 1970), and wheat stems (Buchala and Wilkie, 1970, 1973) as well as from wheat and oat leaves and oat hulls. In addition, the mixed β-glucans have been isolated from cell wall preparations of corn (Kivirikko and Liesmaa, 1959) and oat (Labavitch and Ray, 1978; Nevins et al., 1977; Wada and Ray, 1978) coleoptiles. These mixed β-glucans were not detected in five of six species of suspension-cultured monocot cell walls (Burke et al., 1974). However, these results are open to question because the walls were treated with a five times crystallized Bacillus subtilis α-amylase preparation to remove starch; the amylase preparation was subsequently shown to be contaminated with a β-glucanase capable of solubilizing mixed β-glucans (Huber and Nevins, 1977). On the other hand, the cell walls of wheat endosperm appear to contain little if any of the mixed glucans. The wheat endosperm walls of this study were not treated with the contaminated B. subtilis amylase preparation (Mares and Stone, 1973).

A wide variety of methods have been used to study the structures of the mixed β-glucans. Optical rotation measurements indicated that all or most of the glucosyl linkages of the polysaccharides are the β-anomers (Buchala and Meier, 1973; Fraser and Wilkie, 1971; Labavitch and Ray, 1978; Smith and Stone, 1973b; Wilkie and Woo, 1976). In addition, di- and trisaccharides have been isolated from a number of the mixed β-glucans and the linkages of the oligosaccharides isolated have been shown, by comparing their chromatographic mobilities to the mobilities of known standards, to be in the β-configuration (Fraser and Wilkie, 1971; Kivilaan et al., 1971; Labavitch and Ray, 1978; Smith and Stone, 1973b; Wilkie and Woo, 1976).

Methylation analysis of the mixed β-glucans has demonstrated that the glucans are linear polymers of glucose and that the polymers contain a mixture of 4-linked and 3-linked glucosyl residues (Buchala and Meier, 1973; Fraser and Wilkie, 1971; Kivilaan et al., 1971; Wilkie and Woo, 1976). Periodate oxidation of the mixed β-glucans results in the isolation, following reduction and mild acid hydrolysis, of erythritol and glucose (Smith and Stone, 1973b; Buchala and Meier, 1973; Fraser and Wilkie, 1971; Wilkie and

Woo, 1976). These are the products expected of a polymer containing both 3-linked and 4-linked glucosyl residues.

The ratio of 3-linked to 4-linked glucosyl residues has been determined by both quantitative methylation and periodate oxidation analyses. These methods give similar results (Buchala and Meier, 1973; Buchala and Wilkie, 1971). There are more 4-linked glucosyl residues in the polymers than 3-linked with the ratio of 4-linked to 3-linked residues varying from 1.7 to 4.0 (Buchala and Meier, 1973; Fraser and Wilkie, 1971; Labavitch and Ray, 1978; Smith and Stone, 1973b; Wilkie and Woo, 1976). The ratio of 4-linked to 3-linked glucosyl residues has been shown to vary with the age of the tissue from which the mixed β-glucans were isolated. The amount of 4-linked glucosyl residues increases in comparison to the amount of 3-linked glucosyl residues as the tissues age. (Buchala and Wilkie, 1970, 1971).

It has been established that both 3-linked and 4-linked glucosyl residues occur within a single polymer chain. The best evidence for this has come from the isolation of oligosaccharides containing both 3-linked and 4-linked glucosyl residues (Fraser and Wilkie, 1971). The oligosaccharides Glc-(β1-4)-Glc-(β1-3)-Glc (Buchala and Meier, 1973; Kivilaan *et al.*, 1971; Wilkie and Woo, 1976) and Glc-(β1-3)-Glc-(β1-4)-Glc (Buchala and Meier, 1973) have been characterized from partial acid hydrolyses of the mixed glucans. Similar oligosaccharides have been isolated following partial enzymolysis (Aspinall and Greenwood, 1962; Labavitch and Ray, 1978; Parrish *et al.*, 1960). A trisaccharide, tentatively identified as Glc-(β1-4)-Glc-(β1-3)-Glc, and a tetrasaccharide, tentatively identified as Glc-(β1-4)-Glc-(β1-4)-Glc(β1-3)-Glc, have been isolated following hydrolysis of an oat glucan with an enzyme from *Bacillus subtilis* (Nevins *et al.*, 1977).

Mixed oligosaccharides have also been isolated from the β-glucans following periodate oxidation, reduction with sodium borohydride, and mild acid hydrolysis. These oligosaccharides which include glucosyl-erythritol, laminarobiosyl-erythritol, and the higher monologs of this series through laminaropentosyl-erythritol have been isolated from rye and barley endosperm glucans (Smith and Stone, 1973b). The same oligomer series, through laminarotriosyl-erythritol, has been isolated from oats (Aspinall and Greenwood, 1962). The amounts of the polysaccharides accounted for by these oligosaccharides has not been determined and, therefore, contiguous β3 linked glucosyl residues may be minor constituents of the polymers. Enzymatic hydrolysis of primary cell wall oat β-glucans suggests that contiguous 3-linked glucosyl residues are relatively uncommon (Labavitch and Ray, 1978; Nevins *et al.*, 1977).

A recent study has shown that β-glucans isolated from barley endosperm contain from 1 to 3% peptide (Forrest and Wainwright, 1977). The molecular weight of the β-glucans is changed from approximately 4×10^7 to approximately 10^6 by cleavage of the peptide bonds either by hydrazinolysis or with a proteolytic enzyme. The authors suggest that the peptides are an integral part of the β-glucans.

VII. CELLULOSE

Cellulose is the most abundant polysaccharide in nature and is probably the most studied cell wall polymer. Most of the structural studies of cellulose have been carried out with material from secondary cell walls; little data are available for primary cell wall cellulose. The chemistry of cellulose has been reviewed numerous times (Gardner and Blackwell, 1974a; Kolpak and Blackwell, 1976; Preston, 1974). The present review considers only the major aspects of cellulose structure and assumes that the cellulose of primary walls is similar to that of secondary walls, and that monocot and dicot cell wall cellulose is similar.

Cellulose is composed of long, linear chains of β4 linked glucosyl residues (Gardner and Blackwell, 1974a,b; Kolpak and Blackwell, 1975, 1976; Preston, 1974). The most recent estimate of the cross-sectional dimensions of these fibers (by electron microscopy) is 4.5×8.5 nm (Preston, 1974). A single such fiber has been estimated to consist of 60–70 glucan chains. However, such measurements are difficult to make accurately. The inaccuracy is partly caused by the adherence of other cell wall polysaccharides to cellulose fibers.

The degree of polymerization of the glucan chains within the cellulose fibers has been measured, but the resulting estimates may not reflect the degree of polymerization *in vivo*. The degree of polymerization of the cellulosic glucan chains can only be estimated following solubilization of the glucans. The solubilization procedures are likely to break the glucan chains. The best available estimate of the degree of polymerization of the glucan chains of primary cell wall cellulose comes from studies of cotton fibers by Marx-Figini (1966) and Marx-Figini and Schulz (1966). These workers used viscometric methods to study derivatized glucan chains and determined a degree of polymerization of 6000–7000. However, the literature (Preston, 1974, and references cited therein) contains a variety of chain length values for the cellulosic glucans of a variety of tissues.

It may be that the glucan chains of cellulose have no natural ends, that is, once a chain is initiated, it never ends except when a fiber is physically separated from its synthetic enzymes. This idea is supported by the electron microscopic observation that the cellulose fibers do not appear to have natural termination points. It may also be that the fibers have unlimited length but that the individual glucan chains within the fibers have a finite length; the ends of the glucan chains may overlap to result in fibers of indeterminate length.

The aggregated glucans within a fiber are so ordered that they are, in fact, crystalline (Gardner and Blackwell, 1974a,b; Kolpak and Blackwell, 1975, 1976; Preston, 1974; Sarko and Mugli, 1974). x-Ray diffraction studies indicate that all of the glucan chains within cellulose fibers have a parallel orientation, that is, the reducing ends of the glucan chains face in the same direction. The x-ray diffraction studies were performed on the highly crystalline cellulose of the cell walls of the alga *Valonia ventricosa*. It seems likely, but it has not yet been established, that the glucan chains of primary cell wall cellulose also have a parallel orientation.

Purified cellulose invariably contains, in addition to a preponderance of glucosyl residues, minor amounts of other glycosyl residues (Mühlethaler, 1967; Preston, 1964). One must consider the possibility that the nonglucosyl residues are normal constituents of the glucan chains, perhaps representing glucan chain termination points. Alternatively, the nonglucosyl residues may originate from noncovalently although tightly bound hemicelluloses. The latter possibility is supported by the evidence which indicates that the cellulose fibers of dicot primary cell walls are completely covered by hemicelluloses hydrogen-bonded to the fiber surface (Bauer *et al.*, 1973; Valent and Albersheim, 1974). As the cellulose fibers of secondary cell walls have a considerably greater cross section than the cellulose fibers of primary cell walls (Mühlethaler, 1967; Preston, 1964), it may be that the primary cell wall fibers aggregate to form secondary cell wall fibers. The nonglucosyl residues of cellulose may, in fact, originate from hemicelluloses trapped between the aggregating cellulose fibers.

VIII. CELL WALL PROTEIN

A. Hydroxyproline-Rich Proteins of Dicots

The primary cell walls of dicots contain between 5 and 10% protein (Lamport, 1970; Preston, 1964; Talmadge *et al.*, 1973). The cell wall protein is exceptionally rich in hydroxyproline (20%). The wall protein also has a relatively high content of alanine, serine, and threonine. A high content of these amino acids is characteristic of the structural proteins of animals (Lamport, 1970). This characteristic amino acid composition and the inability to extract much of the protein from cell walls under nondegradative conditions (Lamport, 1973) indicates that the protein is a structural component of primary cell walls (Lamport, 1970; Lamport *et al.*, 1973).

Fragments of the hydroxyproline-rich protein obtained from the primary cell walls of dicots invariably contain arabinosyl and galactosyl residues. The hydroxyproline-rich protein fragments used for these studies have generally been isolated from the walls of suspension-cultured sycamore (*Acer pseudoplatanus*) and suspension-cultured tomato (*Lycopersicon esculentum*) cells. A series of hydroxyproline-arabinosides have been isolated from such wall preparations. The hydroxyproline-arabinosides are obtained by 0.2 M barium hydroxide hydrolysis of the peptide linkages of cell walls or of glycopeptides obtained by digestion of the walls with a crude mixture of polysaccharide- and protein-degrading enzymes. The hydroxyproline arabinosides are a mixture of mono-, di-, tri- and tetra-arabinosides glycosidically linked to the hydroxyl group of hydroxyproline. The mixture of hydroxyproline arabinosides has been separated chromatographically on Chromobeads B (Lamport and Miller, 1971). The hydroxyproline tetra-arabinoside is the predominant species of the dicot primary cell wall protein. In most analyses, no unglycosylated hydroxyproline is detected (Lamport, 1967, 1961; Lamport and Miller, 1971).

Methylation analysis of the tetra-arabinosides isolated from primary cell walls of tomato and sycamore indicate that the arabinosyl residues are terminal and 2- and 3-linked (Karr, 1972; Talmadge *et al.*, 1973). Periodate oxidation (Smith degradation), methylation analyses, nmr, and optical rotation have been used to show that the structure of the hydroxyproline tetra-arabinoside isolated from suspension-cultured cells of tobacco (*Nicotiana tabacum*) is: Ara_f-(β1-3)-Ara_f-(β1-2)-Ara_f-(β1-2)-Ara_f-(β1-4)-hydroxyproline (Akiyama and Kato, 1977).

Single galactosyl residues are glycosidically attached to the serine hydroxyls of the hydroxyproline-rich cell wall proteins of suspension-cultured tomato cells (Lamport, 1970, 1973). This was shown by removing the arabinosides from the intact cell walls by acid hydrolysis (pH 1 for 1 h at 100°C). The hydroxyproline-rich wall protein, with arabinosyl residues removed, is susceptible to proteolysis with trypsin. The resulting solubilized tryptides have been separated by cation exchange and gel filtration chromatography. The composition of the tryptides has been determined by amino acid analysis. Some of the tryptides have been sequenced by subtractive N-terminus identification and further partial acid hydrolysis. Each of the hydroxyproline-rich wall protein tryptides contains a pentapeptide of serine-(hydroxyproline)$_4$, while most of the tryptides contain one or more galactosyl residues. One tryptide, which was found to contain two residues each of galactose and serine, was subjected to β-elimination by several methods. The elimination procedures converted serine to either alanine or cysteic acid with a concomitant release of free galactose (Lamport *et al.*, 1973). These results demonstrated the covalent attachment of a single galactosyl residue to each seryl residue in the tryptide. Similar evidence has also been obtained for the existence of galactosyl-serine linkages in the hydroxyproline-rich glycoprotein of carrot cell walls (Cho and Chrispeels, 1976).

The covalent attachment of arabinose and galactose to the hydroxyproline-rich proteins of primary cell walls is a generally accepted fact (Lamport, 1970; Lamport *et al.*, 1973). However, the available evidence suggests that the hydroxyproline-rich glycoprotein is *not* covalently attached to any of the other cell wall polymers. The evidence does not rule out the possible existence of strong, noncovalent bonding between the hydroxyproline-rich glycoprotein and the other wall polymers.

A hydroxyproline-rich glycoprotein is secreted by suspension-cultured sycamore cells into their culture medium (Keegstra *et al.*, 1973). The carbohydrate component of the glycoprotein is an arabinogalactan. The structure of the arabinogalactan portion of this glycoprotein has been studied by methylation analysis and found to be structurally similar to a protein-free arabinogalactan which is also present in the culture medium of suspension-cultured sycamore cells (see Section IV,A,6). The hydroxyproline-rich proteins of the cell walls and extracellular culture medium of suspension-cultured sycamore cells both contain arabinosyl and galactosyl residues. In spite of these compositional similarities, it has found that the arabinogalactan-containing hydroxyproline-rich protein of the culture medium is structurally dissimilar from the hydroxyproline-rich protein of the cell wall (Pope, 1977). This dissimilarity was shown by comparing the hydroxyproline arabinosides obtained by barium hydroxide hydrolysis of these two glycoproteins. The tetra-arabinoside accounts for 80 mole % of the hydroxyproline arabinosides isolated from the cell wall glycoprotein while the tetra-arabinoside accounts for only 4 mole % of the hydroxyproline arabinosides obtained from the culture medium glycoprotein. Perhaps more importantly, no arabinogalactan is detectably associated with the cell wall hydroxyproline-rich protein while arabinogalactan accounts for fully 50% of the glycoprotein of the extracellular macromolecules.

It appears that the primary cell walls of suspension-cultured sycamore cells contain two hydroxyproline-rich proteins. One of these appears to be a structural protein, found only in the wall, while the second is a glycoprotein which, in culture, is present in the wall in only small amounts and is predominantly found in the culture medium. Of course, even the culture medium arabinogalactan hydroxyproline-rich protein is likely to be found in the cell wall in the intact tissues, as there is no culture medium for it to be dispersed in.

A composite model of a portion of the hydroxyproline-rich structural glycoprotein of dicot primary cell walls is depicted in Fig. 3.

B. Hydroxyproline-Rich Proteins of Monocots

The structural proteins of the primary cell walls of monocots have not been studied as extensively as the structural proteins of the primary cell walls of dicots. The cell walls isolated from suspension-cultured monocot cells have been reported to contain between 0.13 and 0.16% hydroxyproline. This

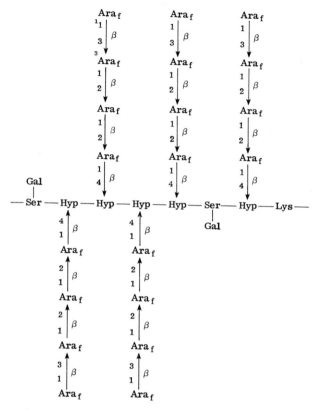

Fig. 3. A tentative model for a portion of the hydroxyproline-rich structural glycoprotein of dicot primary cell walls. This model is adapted from results described in Akiyama and Kato (1977) and Lamport *et al.* (1973).

may be compared to the dicot walls of suspension-cultured sycamore cells which have been shown to contain 2% hydroxyproline (Burke *et al.*, 1974). Maize coleoptiles have been shown to contain 2–3% hydroxyproline (Darvill, 1976). The total amount of protein in the walls of suspension-cultured monocots and in the walls of maize coleoptiles is equal to or larger than the total amount of protein in the walls of suspension-cultured dicots (Burke *et al.*, 1974).

Hydroxyproline arabinosides have been isolated from the walls of four different monocot species (Lamport and Miller, 1971). The majority of the hydroxyproline of monocot cell walls is unglycosylated as 65–75% of the hydroxyproline is isolated without arabinosyl residues attached. The hydroxyproline arabinosides which are detected are predominantly the tri-arabinoside. Smaller amounts of the tetra-arabinoside, the mono-arabinoside, and the di-arabinoside are also detected. This is in contrast to the hydroxyproline residues isolated from dicot cell walls in which all the hydroxyproline residues appear to be glycosylated and the tetra-arabinoside is by far the predominant species (Lamport and Miller, 1971).

It can be concluded that the structural proteins of monocots are different from those of dicots. Although there are arabinosyl hydroxyproline residues in monocot cell walls, the degree of polymerization of these arabinosides is not the same as the dicot cell walls and the percentage of glycosylated hydroxyproline residues is much less than in dicot cell walls.

C. Hydroxyproline-Rich Glycoproteins with Lectin-like Properties

Several of the hydroxyproline-rich glycoproteins extracted from plant tissues have carbohydrate-binding activity; these glycoproteins have the characteristics of lectins. These lectins or lectin-like glycoproteins have compositions which are similar to those of the cell wall hydroxyproline-rich glycoproteins.

A true lectin, which was isolated from potato tubers and shown to bind N-acetyl glucosamine residues, was the first lectin demonstrated to contain hydroxyproline. Allen and Neuberger (1963) showed that the potato tuber lectin is composed of 50% protein and 50% carbohydrate. Hydroxyproline accounts for 16% of the amino acids of the lectin while arabinosyl residues account for 92% of the carbohydrate.

Recently, the seeds and/or tissues of a wide variety of dicots and monocots have been found to possess hydroxyproline-containing glycoproteins with lectin-like properties (Jermyn and Yeow, 1975). It is not known whether these hydroxyproline-containing lectin-like glycoproteins are present in cell walls. Nevertheless, it is interesting that these glycoproteins are about 90% carbohydrate and only 10% protein; arabinosyl and galactosyl residues account for most of the carbohydrate of the lectins. These glycoproteins have been called "all-β" lectins as they bind to a variety of β-linked hexopyranosyl residues. The arabinogalactan portion of the molecules has been discussed in Section IV,A,6.

Two lectin-like protein fractions have been extracted from the cell walls of mung bean (*Phaseolus aureus*) seedlings (Kraus and Bowles, 1976; Kraus and Glasser, 1974). These lectin-containing fractions bind specifically to galactosyl residues. It has not yet been ascertained whether these lectin-like proteins contain hydroxyproline. The existence of these galactosyl-binding proteins in the wall has led to the suggestion that the lectins may be involved in establishing a noncovalent protein–glycan network.

IX. INTERCONNECTIONS BETWEEN THE PRIMARY CELL WALL POLYMERS

A. Interconnections between the Polymers of the Primary Cell Walls of Dicots

1. Introduction

Chemical studies of primary cell walls are still largely concerned with purifying and elucidating the identity and structures of the wall polymers. Sufficient progress in characterizing the wall components of dicots has been made to allow some efforts to determine the manner in which the wall polymers interact with each other. The major effort in this direction has been the work of Bauer et al. (1973), Keegstra et al. (1973), and Talmadge et al. (1973), which culminated in a preliminary model of the primary cell walls of dicots. This section will review our knowledge of the interconnections between primary wall polymers of dicots and will propose some changes in the preliminary model.

2. The Pectic Polysaccharides Covalently Interconnected

It has been established by many lines of evidence that the neutral pectic polysaccharides, the araban and galactan, are covalently attached to the acidic pectic polysaccharide, rhamnogalacturonan I (A. Darvill, M. McNeil, and P. Albersheim, unpublished; Talmadge et al., 1973). Araban, galactan, homogalacturonan, and both rhamnogalacturonan I and rhamnogalacturonan II are solubilized from isolated cell walls by a highly purified endopolygalacturonase. The endopolygalacturonase has been shown to be free of arabanase and galactanase activities. A solubilized wall fraction, containing the araban, galactan, and rhamnogalacturonan I, co-chromatograph as a single acidic polymer on DEAE-Sephadex. The araban and galactan would not be bound by the DEAE-Sephadex unless these neutral polysaccharides were strongly attached to the acidic rhamnogalacturonan. Further evidence for the interconnections of these polymers is provided by their co-chromatography on Agarose 5 m (A. Darvill, M. McNeil, and P. Albersheim, unpublished). Particularly strong evidence for the covalent connection of these polymers comes from studies in which the glycosidic linkages to C-4 of the uronosyl residues of rhamnogalacturonan I have been chemically cleaved. Beta-elimination of the uronosyl residues, under a variety of conditions, results in a drastic reduction in the apparent molecular size of both the araban and galactan. It is noteworthy that, in agreement with the proposed covalent bonding between these polysaccharides, no araban or galactan has ever been extracted from a primary cell wall free of rhamnogalacturonan I.

Homogalacturonans have always been assumed to be attached to rhamnnogalacturonans. One line of evidence for this attachment has been the isolation from sycamore cell walls of oligogalacturonides containing ten or more galacturonosyl residues in which the reducing ends of these oligogalacturonides are covalently attached to single rhamnose residues (Talmadge *et al.*, 1973). Another line of evidence for the interconnection of these polymers comes from the fact that both the homogalacturonan and the rhamnogalacturons, with associated galactans and arabans, are released from the cell walls by the same enzyme, the purified endopolygalacturonase. In other words, uronosyl bonds susceptible to the action of this enzyme result in the solubilization of all of these polymers. Evidence for the interconnection of these polymers has also been obtained in studies of a pectic fraction isolated from apple fruit (Barrett and Northcote, 1965). The neutral glycosyl residues and acidic galacturonosyl residues of this pectic fraction migrate as a single acidic component upon electrophoresis. In addition, a portion of these pectic polymers rich in neutral glycosyl residues can be separated by electrophoresis from a portion rich in galacturonosyl residues after the glycosidic bonds to some of the uronosyl residues have been cleaved by β-elimination.

The only evidence available at this time that rhamnogalacturonan II is covalently connected to the other pectic polysaccharides is that rhamnogalacturonan II is solubilized from the cell walls by the endopolygalacturonase that solubilizes the other pectic polymers. The ability of the endopolygalacturonase to solubilize rhamnogalacturon II suggests very strongly that this polymer is connected to the rest of the wall through a series of 4-linked galacturonosyl residues.

The pectic polysaccharides probably interact through noncovalent chemical bonding as well as through covalent bonding. Indeed, noncovalent interactions of the following type may well provide the most important interconnections between the pectic polysaccharides and the other polymers of the cell wall.

Calcium has long been known to confer rigidity to cell walls (Stoddart *et al.*, 1967). The "egg-box model" of Rees and his colleagues (Grant *et al.*, 1973; Rees, 1972; Rees and Richardson, 1972) is an attractive model for the manner in which calcium strengthens cell walls. Calcium ions can fit between two or more chains of unesterified polygalacturonosyl residues in such a fashion that the calcium ions chelate to the oxygen atoms of four galacturonosyl residues distributed between two galacturonan chains. The result would be that the calcium ions are packed somewhat like eggs in an egg-shaped box composed of polygalacturonans. Such chelation would result not only in increased rigidity of the galacturonans, but also in cross-linking of the galacturonan chains. The degree and strength of interchain cross-linking due to calcium ions is sensitive to the degree of methyl esterification of the galacturonans. Methyl esterification would interfere with the formation of bonds between uronosyl carboxylate ions and calcium ions.

Efforts have been made to correlate the degree of calcium cross-linking of galacturonans to the rate of cell wall elongation, but no such correlation has been established (Stoddart et al., 1967). The possibility of other types of noncovalent interactions between the pectic polysaccharides and other cell wall polymers must be considered (Dea et al., 1977; Gould et al., 1965), but there is little evidence available to indicate that such noncovalent interactions do exist in the cell wall. Efforts to determine whether such interactions are structurally important in cell walls are clearly called for.

3. Hemicelluloses of Dicot Primary Cell Walls Bond Strongly to Cellulose Fibers

Xyloglucan is the quantitatively predominant hemicellulose of the primary cell walls of dicots. It has been proposed that xyloglucan bonds strongly, through multiple hydrogen-bonds, to the surface of cellulose fibers. The following evidence supports this hypothesis (Bauer et al., 1973). (1) The primary cell walls of dicots contain sufficient xyloglucan to form a monolayer-coating of the cellulose fibers of the walls (Bauer et al., 1973; Keegstra et al., 1973). (2) Space filling molecular models of xyloglucan show that xyloglucan is capable of forming multiple hydrogen-bonds to cellulose. (3) Xyloglucan may be extracted from a cellulose-xyloglucan complex by either alkali or 8 M urea, solvents which are known to break hydrogen bonds, whether the complex has been formed in vitro or in the cell wall. (4) The binding of xyloglucan to the cell wall and to isolated cellulose is reversible, as would be expected for noncovalent hydrogen-bonding. (5) Xyloglucan reacts quickly with and bonds strongly to isolated cellulose fibers in the absence of enzymes or chemical catalysis (Aspinall et al., 1969; Bauer et al., 1973). (6) Xyloglucan can be extracted from the cell wall and from cellulose fibers, in cell free systems, by the action of an enzyme that hydrolyzes the xyloglucan chains into small fragments, fragments that are not long enough to form stable hydrogen-bond complexes with cellulose. (7) Xyloglucan fragments that are too short to form stable hydrogen-bond complexes with celluloses can be induced to form such complexes by reducing the water activity of the solvent and, thereby, reduce the opportunity for the fragments to hydrogen-bond with the solvent (Valent and Albersheim, 1974).

The bonding of xyloglucan to cellulose is one of the major interconnections of the cell wall polymers. This bonding may prevent the cellulose fibers from adhering to each other to form the large aggregates characteristic of secondary walls. Xyloglucan chains also offer the possibility of connecting the cellulose fibers to other polymers of the primary cell walls.

Xyloglucan is not the only hemicellulose in the primary cell walls of dicots. A glucuronoarabinoxylan, which has recently been characterized in the primary walls of suspension-cultured sycamore cells, is structurally related to the arabinoxylans and xylans, polysaccharides known to be capable of hydrogen-bonding to cellulose (Bauer *et al.*, 1973; McNeil and Albersheim, 1977; Northcote, 1972). In addition, examination of the data by Bauer *et al.* (1973) suggests that arabinoxylans (or glucuronoarabinoxylans) contaminated the xyloglucans extracted from primary cell walls by urea. It seems likely that glucuronoarabinoxylans, and perhaps xyloglucans, not only bind to cellulose in the cell wall, but also bind to themselves and to each other to form aggregates. This possibility is increased by the observation that plant arabinoxylans form aggregates in solution (Blake and Richards, 1971; Dea *et al.*, 1977). Evidence has been presented that, in solution, arabinoxylan exists as a mixture of random coils and aggregated linear chains (Dea *et al.*, 1977). Such structures can lead to gel formation and may, by this method, be involved in cross-linking of the primary cell wall polymers.

4. Is the Hydroxyproline-Rich Glycoprotein of the Cell Wall Connected to the Other Cell Wall Polymers?

There is no evidence which demonstrates a covalent linkage between the hydroxyproline-rich glycoprotein and the polysaccharides of cell walls. The hydroxyproline-rich glycoprotein which is present in the culture medium of sycamore cells grown in suspension culture has been shown to be covalently connected to an arabinogalactan (Keegstra *et al.*, 1973), and this interconnection was suggested as a possible model for an interconnection between the protein and polysaccharides of the primary cell walls. However, it has now been established (Pope, 1977) that the hydroxyproline-rich protein of the extracellular fluid is structurally different from the hydroxyproline-rich protein of the cell wall and, thus, the extracellular polysaccharide cannot be used as a model for the cell wall protein. There does not appear to be a covalent linkage between the cell wall hydroxyproline-rich protein and cell wall polysaccharide (Munro *et al.*, 1976). On the other hand, there is a very real, albeit undemonstrated, possibility that the hydroxyproline-rich structural glycoprotein is bonded through noncovalent interactions to the polysaccharides of the cell wall.

5. Are the Xyloglucan Chains Covalently Linked to the Pectic Polysaccharides?

Evidence has been obtained for a covalent linkage in primary cell walls between xyloglucan chains and pectic polysaccharides. The authors demonstrated that only a portion of the cell wall xyloglucan chains were covalently connected to the pectic polysaccharides (Bauer *et al.*, 1973; Keegstra *et al.*, 1973; Talmadge *et al.*, 1973). Later attempts in Albersheim's laboratory to isolate large amounts of solubilized xyloglucan attached covalently to the pectic polysaccharides have been unsuccessful (M. McNeil and P. Albersheim, unpublished), although small amounts of xyloglucan, in apparent covalent linkage with the pectic polysaccharides, have been isolated (Keegstra *et al.*, 1973; Munro *et al.*, 1976). The extent of the attachment of xyloglucans to the pectic polysaccharides and the structural importance of this interconnection remains unanswered.

6. The Current Cell Wall Model

There are four major types of polymers in the primary cell walls of dicots: hydroxyproline-rich glycoprotein, cellulose, hemicellulose, and pectic polysaccharides. The mole percentages of these polymers in suspension-cultured sycamore cell walls are presented in Table VIII. The pectic polysaccharides are covalently interconnected. The hemicelluloses are non-covalently bonded to the cellulose fibers. There is no known attachment between the hydroxyproline-rich glycoprotein and the other cell wall polymers. The extent of covalent attachment between the hemicelluloses and the pectic polysaccharides is unknown. The fact that removal of the xyloglucans from the cell walls is enhanced by solubilization of the pectic polysaccharides (Bauer *et al.*, 1973; Keegstra *et al.*, 1973; Talmadge *et al.*, 1973) suggests that the hemicelluloses and the pectic polysaccharides do interact, whether it is by covalent or noncovalent bonding. Our present picture of the dicot cell wall remains one in which the cellulose fibers are covered by a layer of hemicellulose, and these fibers are interconnected by the pectic polysaccharides. It may be that the hydroxyproline-rich glycoprotein or other proteins within the wall, acting possibly as lectins, participate in the

TABLE VIII

Polymer Composition of the Walls of Suspension-Cultured Sycamore Cells[a]

Wall component	Wt. % of cell wall	
A. Pectic Polysaccharides	34	
Rhamnogalacturonan I		7
Homogalacturonan		6
Arabinan		9
Galactan and possible arabinogalactan		9
Rhamnogalacturonan II		3
B. Hemicelluloses	24	
Xyloglucan		19
Glucuronoarabinoxylan		5
C. Cellulose	23	
D. Hydroxyproline-rich glycoprotein	19	

[a] Data from A. Darvill, M. McNeil, and P. Albersheim, unpublished; Keegstra *et al.*, 1973.

cross-linking of the cell wall polymers (Kauss and Bowles, 1976; Kauss and Glasser, 1974).

B. Interconnections between the Polymers in the Primary Cell Walls of Monocots

There have been few studies of the interconnections of the polymers in the primary cell walls of monocots. An arabinoxylan isolated from barley aleurone cell walls (McNeil *et al.,* 1975) and a glucuronoarabinoxylan isolated from maize coleoptile primary cell walls (Darvill *et al.,* 1978a) have been shown *in vitro* to reversibly bind to cellulose. This interconnection is important since xylans are quantitatively the major component of monocotyledon primary cell walls (see Section V,B,2). The binding between the heteroxylans and cellulose is very likely to occur via multiple hydrogen bonds, in a manner analogous to the way xyloglucan, isolated from dicots, is thought to bind to cellulose. It is possible that the heteroxylans participate in binding other cell wall components to cellulose (Darvill *et al.,* 1977).

A glucuronoarabinoxylan has been isolated from oat coleoptiles that does not bind to cellulose *in vitro* using reaction conditions similar to those that permitted other heteroxylans to bind (Wada and Ray, 1978). The oat coleoptile glucuronoarabinoxylan has such a high percentage of arabinosyl side chains that steric hindrance by the side chains may be responsible for the inability of this polysaccharide to bind to cellulose. A similar inability to bind to cellulose *in vitro* was exhibited by an arabinose-rich arabinoxylan isolated from barley aleurone cell walls (McNeil *et al.,* 1975).

It is likely that the xylans of monocots bind to themselves as well as to cellulose. This possibility is enhanced by the observation that monocot xylans form aggregates in solution (Blake and Richards, 1971; Dea *et al.,* 1977). Such structures can lead to gel formation and may, in fact, be involved in cross linking of the primary cell wall polymers.

Some monocot primary cell wall polysaccharides may be cross-linked by ferulic acid esters. It has been proposed that diferulic acid isomers may, via ester linkages, cross-link cell wall carbohydrates in Italia-rye grass (Hartley and Jones, 1976). It has also been suggested that diferulic acid cross-links the arabinoxylan component of the water insoluble pentosans of wheat endosperms (Markwaldes and Neukom, 1976). Ferulic acid is present in barley cell wall (Fincher, 1975), and has been reported to be released from the cell walls of Graminae by treatment with base, which supports the idea that ferulic acid is bound to the wall as an ester (Hartley and Jones, 1976). However, ferulic acid has not been reported to be present in the *primary* cell walls of either monocots or dicots and, therefore, consideration of the existence of such interconnections must await additional evidence.

X. ARE THE PRIMARY CELL WALLS OF DIFFERENT PLANTS STRUCTURALLY RELATED?

A. Introduction

The chemical nature of the structural polysaccharides and glycoproteins of the primary cell walls of monocots and dicots has been described. The available data were obtained not only from studies of homogeneous preparations of primary cell walls but also from studies of wall preparations obtained from differentiated plant tissues. In the latter studies, the information presented was considered to be representative of primary cell walls because of the vast majority of primary cell walls in the tissues studied. This assumption may not always be valid.

The detailed structure of any primary cell wall is not known. Nevertheless, the available information about primary cell wall polysaccharides, obtained from a variety of sources, does allow a comparison of the primary walls from which these polymers were obtained. This section compares the similarities exhibited by a variety of primary cell walls of dicots, by a variety of primary cell walls of monocots, and also compares monocot primary walls with dicot primary walls.

B. Similarity of Primary Cell Walls of Suspension-Cultured Cells and Intact Plant Tissues

1. Primary Cell Walls of Dicots Obtained from Suspension-Cultured Cells and Intact Tissues

The available evidence supports the view that suspension-cultured cells provide a source of primary cell walls that are extremely similar to those obtained from plant tissues. This is a very important consideration as cultured cells have been widely studied as these cells offer an easily obtained source of homogeneous primary cell walls which, unlike intact tissues, do not have the problems associated with the presence of secondary cell walls.

Comparison of the cell wall glycosyl and glycosyl-linkage compositions of young hypocotyl tissues of pea (Gilkes and Hall, 1977) and red kidney bean (B. Nusbaum and P. Albersheim, unpublished) with the same data obtained for walls isolated from suspension-cultured sycamore (Talmadge *et al.*, 1973) and red kidney bean cells (B. Nusbaum and P. Albersheim, unpublished) indicate that these walls are extremely similar. The similarity of the primary cell walls from these very different sources is a strong indication that the walls of suspension-cultured cells are very similar to the walls of intact plant tissues. In addition, it has been demonstrated that the primary walls of cambial cells, prepared from the branches of sycamore trees, are very similar in composition to the walls of suspension-cultured sycamore cells (D. H. Northcote, personal communication).

The hypothesis that the walls of suspension-cultured cells reflect the walls of intact plants is further substantiated by the fact that xyloglucan, a hemicellulosic component of the primary cell walls of bean plants, is structurally extremely similar to the xyloglucan of suspension-cultured sycamore cells (Wilder and Albersheim, 1973). Additional evidence for the above hypothesis is shown in Tables I, V, and VI where the arabans and xyloglucan of sycamore cell walls obtained from suspension cultures are shown to be similar to those obtained from a variety of plant tissues. The hypothesis that the walls of suspension-cultured cells reflect the walls of intact plants is still further supported by identification of the structurally complex polysaccharide rhamnogalacturonan II (see Section IV,A,7) in the walls of suspension-cultured sycamore cells as well as in the cell walls of intact tissues of pea, true bean, and tomato.

2. The Primary Cell Walls of Monocots Obtained from Suspension-Cultured Cells and Intact Tissues

The limited data that are available indicate that the primary cell walls isolated from suspension-cultured monocot cells are very similar to the primary cell walls isolated from intact monocot tissues. The compositions of the cell walls present in maize coleoptiles and oat internodes are very similar to the primary cells walls of six different suspension-cultured monocots (Table IX). Methylation analyses also indicate a high degree of similarity between the walls isolated from cultured cells and the walls isolated from intact plant tissues (Burke *et al.*, 1974; Darvill *et al.*, 1978; M. McNeil, A. Darvill, P. Albersheim, and P. Kaufman, unpublished). Unfortunately, no polysaccharides have been isolated from suspension-cultured monocot cell walls so that a detailed comparison of polysaccharide structures is not possible.

C. Similarity of Primary Cell Walls of a Variety of Dicots

All of the primary cell walls of the various dicots that have been studied are similar to each other. This similarity has been demonstrated by comparing the glycosyl and glycosyl linkage compositions of the walls of suspension-cultured sycamore, red kidney bean, tomato, and soybean cells (Albersheim, 1976). This similarity is further strengthened by the observation that the arabans and xyloglucans isolated from a wide spectrum of dicot species (both from cultured cells and intact tissues) show remarkable similarity in composition (Tables I, V, and VI) suggesting similarities in the cell walls from which they were isolated.

TABLE IX

Comparison of the Compositions of Monocot Primary Cell Walls Obtained from Plant Tissues and Cell Cultures

Component	Maize[a]	Oat[b]	Oat[c]	Wheat[c]	Rice[c]	Sugar-cane[c]	Brome grass[c]	Rye grass[c]
Glucose								
Noncellulosic	8.6	13.0	4.9[d]	3.9[d]	5.3[d]	4.1[d]	4.2[d]	27.0[d]
Cellulose	23.9	27.0	10.0	14.0	11.0	9.0	9.0	13.0
Total	30.5	40.0	14.9	17.9	16.3	13.1	13.2	40.0
Rhamnose	Tr	0.4	1.0	0.6	2.0	0.7	1.5	0.6
Fucose	Tr	Tr	1.5	0.1	1.0	0.3	Tr	0
Arabinose	20.0	12.0	19.4	20.7	27.1	24.7	29.3	20.3
Xylose	17.6	18.0	21.5	24.8	17.1	26.6	27.5	19.8
Mannose	1.7	3.0	0.8	1.0	0.3	0.5	0.1	0.1
Galactose	5.4	2.0	12.5	12.0	7.0	9.2	9.6	5.9
Uronic acids	8.9	17.0	11.0	12.0	18.0	10.0	7.0	7.0
Protein	12.9	6.5	16.0	11.0	17.0	15.0	14.0	7.0
Hydroxyproline	3.0	—	—	0.14	0.13	0.14	0.16	<0.05

[a] From coleoptile tissue (Darvill et al., 1978).

[b] From internode tissue (M. McNeil, A. Darvill, P. Albersheim, and P. Kaufman, unpublished).

[c] From suspension cultured cells (Burke et al., 1974).

[d] These noncellulosic glucose values may be underestimated because these tissues were treated with α-amylase known to be contaminated with a β-glucanase (see text).

Rhamnogalacturonan I (Section IV,A,2) and galactan (Section IV,A,5) are also found in a wide spectrum of dicot primary cell walls. Rhamnogalacturonan II has been seen in dicot primary cell wall tissues of suspension-cultured sycamore cells and evidence has been obtained for its presence in the walls of intact tissues of tomato, pea, and true bean. These are the only dicot tissues that have been examined for rhamnogalacturonan II. Cellulose is, of course, an important structural component of all higher plant cell walls. Hydroxyproline-rich glycoproteins are also present in all of the dicot primary cell walls examined for this component. In addition, tetra-arabinosides are the most common oligomer attached to the hydroxyproline residues of the glycoproteins of all of the dicot walls studied (Section VIII,A). Not only are the polymeric components of the primary cell walls of all of the dicots studied very similar, but these components are bound in the walls by the same linkages. This was shown by demonstrating the similarity of the fractions obtained from dicot primary cell walls by a variety of chemical and enzymatic extraction procedures (Albersheim, 1976; B. Nusbaum and P. Albersheim, unpublished).

D. Similarity of Primary Cell Walls of a Variety of Monocots

The primary cell walls of a variety of monocots appear to be similar to each other. This statement has to be qualified because of the limited data available pertaining to this point. The composition of the primary cell walls of six suspension-cultured monocots and the tissues of two intact monocots indicate the similar nature of the structures of the primary cell walls of a variety of monocots (see Table IX). Methylation analyses of the primary cell walls of the six suspension-cultured monocots further substantiates the close similarity of these walls (Burke *et al.,* 1974).

A substantial component of all monocot primary cell walls that have been studied is the hemicellulosic xylan (Section V,B,2). All of the xylans that have been studied possess a β4 linked xylan backbone and show similarities in the nature of the side chains attached to this backbone, although the detailed structure of the side chains does vary. All monocot primary cell walls are also very similar in the relatively small proportion of the walls that is composed of pectic polysaccharides (Section IV,B). The monocot primary walls are also similar to one another in that the majority of the hydroxy-proline residues of monocot walls are unglycosylated while the most prevalent arabinoside that is present is a trisaccharide.

E. Dissimilarity of Primary Cell Walls of Monocots and Dicots

The evidence given above indicates that all of the species of the subclass dicotyledon have very similar primary cell walls. Similarly, all of the species of the subclass monocotyledon appear to have very similar cell walls. The extent of the similarity of the primary cell walls of the various species of these subclasses must remain somewhat tentative until more data are available and, indeed, until a detailed structure of the walls is obtained. However, the available data make it quite clear that the primary cell walls of monocots are easily distinguished from the primary cell walls of dicots. Both monocots and dicots are composed of four principle components: cellulose, hemicellulose, pectic polysaccharides, and glycoproteins. The main differences between the walls of these two subclasses are in the quantities and nature of the hemicelluloses, pectic materials, and glycoproteins.

Dicot primary cell walls are composed of approximately 35% pectic polysaccharides while monocots contain only about 8 or 9% pectic polysaccharides (see Section IV). Data on the monocot pectin polysaccharides are very limited and do not allow a comparison of the structures of the pectic polysaccharides present in the walls of these two subclasses.

Monocot and dicot primary cell walls both are rich in hemicellulose although the proportion of the two predominant hemicelluloses, xylans and xyloglucans, present in the primary cell walls of the two subclasses is dramatically different. The primary cell walls of dicots contain at least four times as much xyloglucan as xylan; indeed, only recently has a xylan been shown to be present in the primary walls of dicots (J. Darvill, A. Darvill, M. McNeil, and P. Albersheim, unpublished). On the other hand, the primary cell walls of monocots are exceptionally rich in xylans and only recently have trace amounts of xyloglucan been suggested to be present in these walls (Labavitch and Ray, 1978). Regardless of the chemical nature of the hemicellulose, the ability of the hemicellulose to hydrogen-bond to cellulose appears to be a very important structural feature of both monocot and dicot primary cell walls (see Section IX).

Another major difference between the primary cell walls of these two subclasses is in the percentage of glycosylation and length of the side chains attached to the hydroxyproline residues of the wall glycoprotein (see Section VIII). The primary walls of monocots and dicots both contain approximately 10% total protein.

The available data lead us to speculate that, although compositionally very dissimilar, the primary cell walls of dicots and monocots may be arranged on similar architectural plans. The structure of primary walls of both subclasses appear to depend on cellulose fibers interconnected by hemicelluloses, glycoproteins, and pectic polysaccharides. More work is needed before we learn whether this model is structurally valid.

XI. BIOSYNTHESIS OF PRIMARY CELL WALL POLYMERS

A. Introduction

Much less is known about the processes involved in the biosynthesis of primary cell polymers than about the structure of these polymers. The reader is referred to several recent general reviews of cell wall biosynthesis (Delmer, 1977; Robinson, 1977; Kauss, 1974; see also Ericson and Elbein, this series, Vol. 3, Chapter 16).

It is difficult to study the biosynthesis of polymers whose structures and interconnections are not yet completely elucidated. But other factors have also contributed to the lack of progress. The structural studies described in this chapter have demonstrated that most of the polymers in primary cell walls contain a variety of monosaccharide units joined together in a variety of linkages. The catalytic systems responsible for synthesis of such complex polymers must, of necessity, involve multiple enzyme-catalyzed reactions which could require delicate coordination. Individual reactions can vary in requirements for substrate type and concentration, cofactors, and inherent stability of the enzymes involved. It is likely that lack of activity *in vitro* for even one of the multiple steps required for synthesis of a complex polymer could render the entire sequence inoperable. In this regard, it is surprising that so little work has been done to study the effects of combinations of substrates on polymer synthesis. Most of those biosynthetic studies which have been carried out have been concerned with the incorporation of but a single type of radioactive monomer into a polymer, a condition which for many primary wall components almost certainly cannot mimic the *in vivo* situation. The structural analysis of wall polysaccharides provide us with strong clues as to which combinations of substrates are likely to be required for successful *in vitro* synthesis.

Another serious criticism of many published reports has been the lack of adequate characterization of products synthesized *in vitro*. The usual enzymological approach to cell wall biosynthesis in plants has been to grind tissues, to isolate (usually) membrane preparations, to add a radioactive nucleotide sugar, and to look for incorporation of radioactivity into an insoluble product. The products of these reactions have seldom been carefully characterized with respect to glycosyl linkage compositions and only rarely have the products been even partially purified. Further, the degree of polymerization and molecular size of radioactive products have usually remained undetermined. It should be evident that methods for structural analysis of carbohydrates are applicable for structural characterization of the products of *in vitro* synthesized polysaccharides.

Except for the intriguing reports suggesting a role for glucosylinositol as a primer for glucan synthesis (Kemp and Loughman, 1973, 1974), essentially nothing is known about the endogenous acceptors present in the impure enzyme preparations used in these studies. Information on this rather crucial point may only come when the products of *in vitro* reactions are successfully isolated and characterized. One should also be aware of the possibility that the impure enzyme preparations that have been used are likely to catalyze the interconversion of substrates or to catalyze competing reactions which use up substrates. Degradation of the reaction products by the presence of hydrolases represents still another potential problem.

Another question as yet unresolved by *in vitro* studies is whether lipid intermediates participate in the synthesis of cell wall polymers. There is now clear evidence in bacterial and mammalian systems that phosphorylated polyprenols are required participants in the synthesis of polysaccharides or glycoproteins from nucleoside diphosphate sugars (Lennarz and Sher, 1973). Because the subject of plant glycolipids is covered elsewhere in this series (see Elbein, Vol. 3, Chapter 15), this topic will not be discussed in detail in this chapter. However, where appropriate, a few examples of possible involvement of such lipid intermediates in the biosynthesis of polymeric carbohydrates will be cited.

We have previously stated that the use of a single homogeneous cell type offers many advantages for structural studies; the same argument applies for biosynthetic studies. Clearly, cell wall synthesis is a developmentally regulated event. Extracts prepared from whole seedlings, even organs or tissues, reflect a variety of cell types, each of which was engaged at the time of harvest in synthesis of its own pattern of cell wall polymers. Cells grown in tissue culture and certain tissues rich in a single cell type, such as elongating coleoptiles and cotton fibers, would seem preferable for biosynthetic studies.

It should be pointed out that almost nothing is known about the regulation of cell wall biosynthesis. Are the types of polymers synthesized by various cell types controlled at the level of transcription or translation of the enzymes, by the levels of substrates, by small molecule activators or inhibitors, or by hormones? These questions will remain largely unanswered until the biosynthetic processes are understood, at which time regulation of cell wall biosynthesis will become an exciting area for study of developmental regulation in plants. Despite all the problems, some progress has been made in the area of biosynthesis and the subsections below summarize our current knowledge.

B. Biosynthesis of the Pectic and Hemicellulosic Polysaccharides

Most workers now believe that the noncellulosic polysaccharides of the cell wall, i.e., the pectins and hemicelluloses, are synthesized in cytoplasmic membrane systems and secreted into the cell wall as high molecular weight polymers. These conclusions have been drawn primarily from cytological and *in vivo* labeling studies in two systems: the elongating pea epicotyl and the root cap cells of corn. Two recent reviews (Northcote, 1974; Robinson, 1977) provide a detailed summary of the work with these systems. It is also pertinent to read the general discussion of plant organelles presented in other chapters of this volume. Therefore, only a brief summary is provided here.

Ray *et al.* (1976), Robinson (1977), and Robinson *et al.* (1976) have concluded that Golgi dictyosomes are the primary site of synthesis of the noncellulosic cell wall polysaccharides in the pea epicotyl. High molecular weight radioactive polysaccharides, with composition at least crudely resembling the noncellulosic polysaccharides of the cell wall, can be isolated from a dicotyosome fraction after feeding [^{14}C]glucose to pea stem sections. Pulse-chase experiments suggest the movement of these polysaccharides through the dictyosomes, to a vesicle fraction, and finally to the cell wall.

High molecular weight polysaccharides have been associated with a variety of membrane fractions in wheat seedling root cells (Jilka *et al.*, 1972). Polysaccharide material has also been associated with intracellular organelles in root cap cells of *Zea* (Northcote, 1974; Paull and Jones, 1976), although the endoplasmic reticulum (ER) rather than the Golgi dictyosomes has been proposed as the major site of synthesis of these polysaccharides (Northcote, 1974). Robinson (1977) has pointed out that it can be difficult to separate

Golgi-derived vesicles from ER, and he questions any role for ER in cell wall or slime synthesis.

Pulse-chase radioautography with root cap cells supports the concept of the Golgi as the principle site of slime polysaccharide synthesis (Northcote and Pickett-Heaps, 1966; Pickett-Heaps, 1968). Interesting preliminary evidence has been obtained (Green and Northcote, 1978) suggesting that the slime precursors are transported on protein carriers within the membranes. Autoradiographic studies using developing vascular tissue in sycamore have provided evidence that the Golgi are involved in the synthesis and secretion of secondary wall hemicelluloses (predominantly xylan) (Wooding, 1968).

It is generally accepted that nucleotide sugars are the most likely form of activated monomer substrate for polysaccharide synthesis. The synthesis and interconversions of nucleotide sugars is covered elsewhere in this series (see Feingold and Avigad, Vol. 3, Chapter 4) and will not be further discussed here, although the role of nucleotide sugars in cell wall polysaccharide synthesis is considered in the following paragraphs.

Early studies (Liu *et al.*, 1966; Villemez *et al.*, 1965), using particulate fractions isolated from *Phaseolus aureus,* demonstrated that UDP-galacturonic acid and, to a much lesser extent TDP-galacturonic acid, could serve as substrate for the synthesis of a polymeric product which could be degraded by polygalacturonase. Other similar early studies using particulate fractions from a variety of plant sources showed incorporation of radioactivity from UDP-arabinose (Odzuck and Kauss, 1972), UDP-xylose (Ben-Arie *et al.*, 1973; Odzuck and Kauss, 1972), and UDP-galactose (Panayotatos and Villemez, 1973) into insoluble products. In the first two cases, a mixture of arabinosyl and/or xylosyl-containing products was synthesized; in the case of UDP-galactose, some evidence was presented suggesting that the product was a β4-galactan. The methyl esterification of the uronosyl residues of a pectic polysaccharide has been studied (Kauss, 1974). The data suggest that methyl esterification occurs intracellularly after polymerization of uronosyl residues and involves S-adenosylmethione as methyl donor.

Three studies can be cited where knowledge of polysaccharide structures has been used in the design of biosynthetic experiments. The first study involved the biosynthesis of apiogalacturonan in *Lemna* (Mascara and Kindel, 1977; Pan and Kindel, 1977). Even though hampered by extreme instability of the enzymes, the results established an interaction between UDP-apiose and UDP-galacturonic acid in the synthesis *in vitro* of an apiogalacturonan which resembles the *Lemna* cell wall polymer. The second study involved attempts to synthesize xyloglucans *in vitro*. In two preliminary reports using preparations from peas, it has been shown that UDP-glucose and UDP-xylose can interact as substrates for the synthesis of a polymer which resembles in some respects cell wall xyloglucan (Ray, 1975; Villemez and Hinman, 1975). The third study involved the coordinate use of GDP-glucose and GDP-mannose for the synthesis of glucomannan in *Phaseolus*

aureus (Elbein, 1969; Heller and Villemez, 1972). The amounts of the polymers that were synthesized in all of these reactions were small, and only
limited characterization of the products were carried out.

There is now considerable evidence to support the concept of a role for a
mannolipid intermediate in the glycosylation of protein in plants (Elbein, this
series, Vol. 3, Chapter 15). The lipid portion of the mannolipid has been
shown by several criteria, including mass spectrometry, to be a dolichol
(Delmer *et al.*, 1978). In the developing cotyledons of *Phaseolus aureus,* both
glycoprotein and mannan are produced from GDP-mannose. Pulse-chase
studies do not rule out the possibility that the mannan as well as the glycoprotein are synthesized via the mannolipid intermediate (Ericson and Delmer,
1977, 1978), although much work remains to be done to prove this possibility
as another study with cultured sycamore cells implicates mannolipid involvement only in glycoprotein and not in mannan biosynthesis (Smith *et al.*,
1976). The studies of the synthesis of the mannose-containing polymer offer
the best evidence in a plant system for the involvement of a lipid intermediate in the polymerization reactions.

C. Biosynthesis of Cell Wall Glycoproteins

The synthesis of the peptide chain of the hydroxyproline-rich proteins
occurs on polysomes while the hydroxylation of the proline residues occur
intracellularly as a posttranslational modification of the polypeptide chain
(Sadava and Chrispeels, 1971a). The plant prolylhydroxylase has properties
closely resembling the animal enzyme (Sadava and Chrispeels, 1971b).
Pulse-chase experiments using [^{14}C]proline suggest that the hydroxyproline-
rich proteins pass through smooth membrane vesicles, probably Golgi (Gardiner and Chrispeels, 1975), and from these are transferred either to the cell
wall or are secreted as glycoproteins (Pope, 1977). The synthesis of hydroxyproline arabinosides from UDP-arabinose has been demonstrated by
Karr (1972), but the precise localization of this activity has not been determined. The attachment of galactosyl residues to seryl residues has not been
studied *in vitro*.

D. Biosynthesis of Noncellulosic Glucans and Cellulose

The biosynthesis of these two classes of glucans is considered together
because, unfortunately, the literature contains numerous studies which have
confused the synthesis of noncellulosic glucans with that of cellulose. A
recent review discusses this situation in some detail (Delmer, 1977). Colvin
(this series, Vol. 3, Chapter 14) also discusses cellulose biosynthesis and
includes studies on bacteria which are not discussed here. Basically, the
confusion in the literature has resulted from a lack of adequate characterization of the linkages between glucosyl residues in the glucans synthesized *in
vitro*. Two criteria have often been used by workers as evidence that an *in*

vitro product is cellulose: (1) that the product is insoluble in hot alkali, and (2) that the product can be degraded by cellulases. Unfortunately, use of these criteria alone is inadequate proof that a substance is cellulose. Several recent papers have shown that β3 linked glucans can also be insoluble in hot alkali (Anderson and Ray, 1978; Heiniger and Delmer, 1977; Raymond *et al.*, 1978) and the cellulase preparations that have been used have generally been contaminated with β-glucanases with specificities for other glucosyl linkages, most commonly β3 linked glucosyl linkages. Thus, the most common error found in the literature is a claim that β4 linked glucan was first synthesized *in vitro* (Brummond and Gibbons, 1965; Franz and Meier, 1965; Shore and Machlachlan, 1975; Shore *et al.*, 1975) when, in fact, the product was later shown to be predominantly β3 linked glucan (Anderson and Ray, 1978; Chambers and Elbein, 1970; Delmer *et al.*, 1977; Flowers *et al.*, 1969; Heiniger and Delmer, 1977; Raymond *et al.*, 1978). Since UDP-glucose is considered at present to be the most likely precursor of cellulose, and since many plants contain a highly active UDP-glucose: β3 linked glucan synthetase (see Delmer, 1977 and references cited therein), the necessity for adequate linkage analysis of putative cellulosic products is obvious. It should also be obvious that the purity of the products must be established.

The synthesis of β3 linked glucans, commonly referred to as callose, is of interest in itself. There are some reports which show that β3 linked glucans can be found in significant quantities in walls of plant cells such as pollen tubes (Herth *et al.*, 1974) and the secondary walls of cotton fibers (Meinert and Delmer, 1977; D. P. Delmer and D. Maltby, unpublished results; Maltby *et al.*, 1979). The question of whether this polymer exists as a significant structural component of primary walls remains unanswered. Callose is more commonly found associated with wound responses (Eschrich, 1965).

Beta-3 linked glucan synthetases have been reported in a variety of plants (Delmer, 1977), and are often activated *in vitro* by their substrate, UDP-glucose, and by β-linked disaccharides. A recent report indicates that this enzyme is localized on the plasma membrane of cells of pea epicotyls (Anderson and Ray, 1978), and preliminary data from our laboratory (MSU/DOE Plant Research Laboratory) indicates that the enzyme is similarly localized in cotton fibers and cultured cells of soybean. However, little is known about the *in vivo* regulation whereby callose is rapidly deposited upon wounding or, in the case of cotton fibers, where β3 linked glucan synthesis is initiated coincident with the onset of secondary wall cellulose synthesis.

Some attempts have been made to synthesize the mixed β3 linked and β4 linked glucans of the type found in cell walls of monocotyledonous plants (see Section VI,B). Several reports have demonstrated that the ratio of 3-linked to 4-linked glucosyl residues in glucan synthesized *in vitro* can be a function of the concentration of UDP-glucose used in the reaction mixture, with lower concentrations favoring synthesis of the β4 linked glucosyl residues (Ordin and Hall, 1968; Peaud-Lenoel and Axelos, 1970; Raymond *et al.,* 1978; Smith and Stone, 1973a). In one case (Smith and Stone, 1973a), it has been shown that at least some of the product synthesized *in vitro* at high UDP-glucose concentrations contains both $\beta3$ linked and $\beta4$ linked glucosyl residues in the same glucan molecule.

The only facet of cellulose synthesis in higher plants with which most workers in the field would agree is that polymerization of the glucan chains occurs at the cell surface. This conclusion is based partly on the fact that, in higher plants, intracellular cellulose fibrils have not been observed cytologically nor demonstrated chemically. Many algae with cellulosic walls, such as *Oocystis,* also apparently synthesize cellulose at the cell surface, although there are some notable exceptions in other algae (Allen and Northcote, 1965; Brown *et al.,* 1969). In *Oocystis,* freeze-fracture studies (Brown and Montezinos, 1976; Montezinos and Brown, 1976) of the plasma membrane show what appear to be multi-subunit particles on the interior face of the plasma membrane. These particles appear to be associated with the ends of cellulose fibrils. It has been suggested without any biochemical evidence, that these ultrastructurally observed particles are cellulose-synthesizing complexes. Similar, but smaller, globular particles have been observed on the plasma membrane of corn root cells and have been associated with cellulose fibrils (Mueller *et al.,* 1976). Since each cellulose fibril consists of a large number of glucan chains, a multi-subunit enzyme complex would be expected to participate in cellulose synthesis.

The manner by which the cellulose fibrils are arranged within the cell wall is poorly understood. The fact that colchicine interferes with orientation of newly synthesized cellulose fibrils has been interpreted to indicate that microtubules play some role in the orientation process (Pickett-Heaps, 1967; Robinson, 1977). This possibility is further supported by the observation that microtubule orientation itself is often (but not always) correlated with the orientation of the cellulose fibrils of the wall (Robinson, 1977). On the other hand, ordered granules present on the plasma membranes of algae, have been suggested by some to play a role in orientation (Brown and Montezinos, 1976; Kiermayer and Dobberstein, 1973; Montezinos and Brown, 1976; Preston, 1964).

We believe that there has as yet been *no* convincing demonstration of synthesis *in vitro* of cellulose. Such lack of success could be due to a variety of factors such as possible inherent instability of the cellulose synthesizing complex, disruption of the complexes when separating them from the cellulose fibrils during tissue homogenization, or use of inappropriate substrates and/or cofactors.

Most workers now agree that UDP-glucose is the most likely substrate for cellulose synthesis. The capacity for synthesis of UDP-glucose and the endogenous concentration of UDP-glucose are relatively high in most plant tissues (Delmer, 1977). This is in marked contrast to the lack of ability of plant tissue to synthesize and accumulate GDP-glucose, another proposed substrate for cellulose synthesis. For example, in the cotton fiber (Delmer, 1977; N. C. Carpita and D. P. Delmer, unpublished), GDP-glucose pyrophosphorylase activity is undetectable, but UDP-glucose pyrophosphorylase activity is found in extremely high levels throughout fiber development. Similarly, no GDP-glucose can be detected in extracts of cotton fibers, whereas UDP-glucose is easily detectable and the content is maximal (at least 5 mM) at the time of the maximum rate of cellulose deposition *in vivo*. Further support for the role of UDP-glucose as a precursor of cellulose comes from data which indicate that it can serve as substrate for cellulose synthesis in the bacterium *Acetobacter xylinum* (Colvin, this series, Vol. 3, Chapter 14).

Some workers have suggested that GDP-glucose is a substrate of cellulose synthesis in plants because this nucleoside diphosphate glucose can serve as a substrate *in vitro* in extracts from a variety of plants for the synthesis of insoluble material containing β4 linked glucosyl residues (see references cited in Delmer, 1977). However, the rate of incorporation is stimulated by the addition of GDP-mannose; therefore, it appears that this synthetic system may function in glucomannan rather than cellulose synthesis (Heller and Villemez, 1972; Villemez and Heller, 1970). Furthermore, this synthetic system is detectable in cotton fibers only during primary cell wall growth and disappears with the onset of secondary wall cellulose synthesis (Delmer *et al.,* 1974). However, a recent study (Hopp *et al.,* 1978a) of the nonphotosynthetic alga, *Prototheca,* suggests a role for *both* UDP-glucose and GDP-glucose as well as glucan-dolichol and glucan-protein intermediates in cellulose synthesis. Although these results are interesting, the evidence that has been presented must be considered preliminary.

Some support for the role of a protein intermediate in cellulose synthesis comes from a report that coumarin, an inhibitor of cellulose synthesis in some plants, inhibits the synthesis of a water-soluble glucoprotein in *Prototheca* (Hopp *et al.,* 1978b). Preliminary evidence for the glucosylation of protein from UDP-glucose has also recently been reported in pea (Pont-Lezica *et al.,* 1978). However, extensive efforts to implicate intermediates in cellulose synthesis in cotton have proved unsuccessful (D. P. Delmer, unpublished). Furthermore, combining UDP-glucose and GDP-glucose as substrates has not resulted in cellulose synthesis *in vitro* in cotton.

Cotton is typical of many higher plant systems in that UDP-glucose serves as an efficient substrate *in vitro* for β3 linked glucan synthesis, but not for cellulose synthesis (Delmer *et al.,* 1977; Heiniger and Delmer, 1977). It is possible that exposure of the outer face of the plasma membrane of plants to UDP-glucose results in preferential synthesis of β3 linked glucan, whereas UDP-glucose fed from inside the cell is used for cellulose synthesis. Recent results (A. Klein and D. P. Delmer, unpublished) on cell wall regeneration from protoplasts of cultured soybean cells could support this concept. When intact protoplasts are fed UDP-[^{14}C]glucose, significant synthesis of ^{14}C-labeled β3 linked glucan, but not cellulose, is observed. However, when [^{14}C]glucose is fed to such protoplasts, cellulose is one of the major polysaccharide products synthesized and far less β3 linked glucan is produced, compared to the situation with UDP-[^{14}C]glucose. A similar result has been observed using cut sections of pea epicotyls (Anderson and Ray, 1978). Of course, this discussion assumes that UDP-glucose is in the pathway of cellulose synthesis, but this is not a proven point. The inability to achieve synthesis of cellulose *in vitro* may be because some as yet unidentified activated form of glucose is the true precursor. What this is saying is that we are faced with a lack of fundamental biochemical knowledge about the properties of the cellulose synthesizing complex in higher plants. Indeed, we are faced with a lack of knowledge about the synthesis of all cell wall polysaccharides.

XII. THE FUTURE OF PRIMARY CELL WALL RESEARCH

This chapter has stressed the need to isolate and chemically characterize the structures of the polymers of primary cell walls. In the future, emphasis must be placed on working with cell wall polymers rather than with polymers obtained from other tissues and organelles. It is also important to obtain polymers from homogeneous preparations of primary cell walls rather than from walls isolated from tissues containing a mixture of wall types. Much of the early work on characterizing cell wall polymers was done with heterogeneous wall preparations. The availability of easily grown suspension-cultured cells offers the possibility of working with homogeneous preparations of primary cell walls. Hopefully, the structural work to be done in the future will be done with undifferentiated suspension-cultured cells or with other sources of homogeneous primary cell walls.

There is a continued need for new techniques to solubilize and fractionate the cell wall polymers. The power of purified cell wall-degrading enzymes in solubilizing cell wall polymers has been convincingly demonstrated (Talmadge *et al.*, 1973). Researchers in this field have often avoided using purified polysaccharide-degrading enzymes because of the large amount of labor necessary to obtain the purified enzymes. In almost every case, it has not been a simple task to obtain an enzyme which attacks only a single type of glycosidic linkage. The increasing use of affinity chromatography should help to ease the task of those attempting to purify cell wall-degrading enzymes. We believe that the availability of an array of pure cell wall-degrading enzymes is almost essential for furthering the structural knowledge of primary cell walls.

New chemical methods for cell wall structural analysis are needed and some are being developed. Methods have just become available which permit quantitative β-elimination of methylated polysaccharides containing methylated uronosyl residues (Aspinall and Rosell, 1977; Lindberg *et al.,* 1973, 1975). One of the newly characterized β-elimination methods has been extended for use with polysaccharides substituted with easily removed acetal groups instead of with methyl ether groups (Curvall *et al.,* 1975). Another β-elimination method is available which preserves the identity of the reducing sugar eliminated from the uronosyl residues (Aspinall and Chaudhari, 1975; Aspinall *et al.,* 1975). These reactions have been successfully used in the study of the pectic polysaccharides of suspension-cultured sycamore cells (A. Darvill, M. McNeil, and P. Albersheim, unpublished).

A powerful new technique for assisting in the study of the structure of plant cell wall polysaccharides is high pressure liquid chromatography. Chromatographic columns and solvents are now available which allow the rapid and quantitative separation of oligosaccharides differing in degree of polymerization. We have had success, too, in separating a variety of structurally similar equal-length oligosaccharides. In addition, reverse phase high pressure liquid chromatographic columns are available which allow the highly efficient separation of relatively large permethylated oligosaccharides. Although in its infancy, the high pressure liquid chromatographic separation of methylated oligosaccharides, when combined with subsequent gas chromatographic–mass spectrometric analysis of the derived partially methylated partially acetylated alditols, is opening the way for rapid and efficient sequencing of the sugars that compose oligo- and polysaccharides. These methods offer the possibility of sequencing oligosaccharides much faster than has been previously possible. Most impressively, the analysis can be carried out on amounts of oligomers as low as 1 mg.

Still another approach to the sequencing of the sugars of oligo- and polysaccharides is being pioneered by Sigfrid Svensson (personal communication) and his colleagues in the Department of Clinical Chemistry at the University Hospital in Lund, Sweden. These workers are developing methods using capillary column gas chromatography to separate methylated oligosaccharides and are combining these separation techniques with direct mass spectrometric identification of the oligosaccharides present in the capillary column effluent. These methods, although technically challenging, offer the possibility of still easier sequence analysis on submilligram quantities of oligo- and polysaccharides. The methodology available for structural analysis of polysaccharides, including the polysaccharides of primary cell walls, is undergoing a revolution. This technological revolution will permit dramatic advances in our knowledge of cell wall structure.

Increased knowledge of the structures of cell wall polymers, increased application of the new and powerful methods for structural analysis of polysaccharides, and increased use of homogeneous cell preparations should help promote the much needed studies of the biosynthesis of primary cell walls. We believe that only when the biochemistry of primary cell wall structure and synthesis is understood, will a fundamental understanding of plant growth be possible.

ACKNOWLEDGMENT

This research was supported by Department of Energy contract AY-76-S-02-1426.

REFERENCES

Adams, G. A., and Bishop, C. T. (1960). *Can. J. Chem.* **38**, 2380–2386.
Akiyama, Y., and Kato, K. (1977). *Agric. Biol. Chem.* **41**, 79–81.
Albersheim, P. (1976). *In* "Plant Biochemistry" (J. Varner and J. Bonner, eds.), pp. 225–274, Academic Press, New York.
Albersheim, P., Neukom, H., and Deuel, H. (1960). *Arch. Biochem. Biophys.* **90**, 46–51.
Albersheim, P., Nevins, D. J., English, P. D., and Karr, A. (1967). *Carbohydr. Res.* **5**, 340–345.
Allen, A. K., and Neuberger, A. (1973). *Biochem. J.* **135**, 307–314.
Allen, D. M., and Northcote, D. H. (1965). *Protoplasma* **83**, 389–412.
Anderson, R. L., and Ray, P. M. (1978). *Plant Physiol.* **61**, 723–730.
Anderson, R. L., Clarke, A. E., Jermyn, M. A., Knox, R. B., and Stone, B. A. (1977). *Aust. J. Plant Physiol.* **4**, 143–158.
Aspinall, G. O., and Cairncross, I. M. (1960a). *J. Chem. Soc.* **769**, 3877–3881.
Aspinall, G. O., and Cairncross, I. M. (1960b). *J. Chem. Soc.* **769**, 3998–4008.
Aspinall, G. O., and Canas-Rodriguez, A. (1958). *J. Chem. Soc.,* 4020–4026.
Aspinall, G. O., and Chaudhari, A. S. (1975). *Can. J. Chem.* **53**, 2189–2193.
Aspinall, G. O., and Cottrell, I. W. (1970). *Can. J. Chem.* **48**, 1283–1287.
Aspinall, G. O., and Cottrell, I. W. (1971). *Can. J. Chem.* **49**, 1019–1022.
Aspinall, G. O., and Fanshawe, R. S. (1961). *J. Chem. Soc. C,* 4215–4221.
Aspinall, G. O., and Ferrier, R. T. (1960). *J. Chem. Soc.* **769**, 3881–3990.
Aspinall, G. O., and Greenwood, C. T. (1962). *J. Inst. Brew. London* **68**, 167–178.
Aspinall, G. O., and Jiang, K. S. (1974). *Carbohyd. Res.* **38**, 247–255.
Aspinall, G. O., and McGrath, D. (1966). *J. Chem. Soc. C,* 2133–2139.
Aspinall, G. O., and Molloy, J. A. (1968). *J. Chem. Soc. C,* 2994–2999.
Aspinall, G. O., and Rosell, K. G. (1977). *Carbohyd. Res.* **57**, C23–C26.
Aspinall, G. O., and Sturgeon, R. J. (1957). *J. Chem. Soc.,* 4469–4471.
Aspinall, G. O., and Wilkie, K. C. B. (1956). *J. Chem. Soc.* 1072–1079.
Aspinall, G. O., Cairncross, I. M., and Ross, K. M. (1963). *J. Chem. Soc.,* 1721–1727.
Aspinall, G. O., Hunt, K., and Morrison, I. M. (1966). *J. Chem. Soc. C,* 1945–1949.
Aspinall, G. O., Begbie, R., Hamilton, A., and Whyte, J. N. C. (1967a). *J. Chem. Soc. C,* 1065–1070.
Aspinall, G. O., Cottrell, I. W., Egan, S. V., Morrison, I. M., and Whyte, J. N. C. (1967b). *J. Chem. Soc. C,* 1071–1080.
Aspinall, G. O., Hunt, K., and Morrison, I. M. (1967c). *J. Chem. Soc. C,* 1080–1086.
Aspinall, G. O., Craig, J. W. T., and Whyte, J. L. (1968a). *Carbohydr. Res.* **7**, 442–452.
Aspinall, G. O., Fairweather, R. M., and Wood, T. M. (1968b). *J. Chem. Soc. C,* 2174–2179.
Aspinall, G. O., Gestetner, B., Molloy, J. A., and Uddin, M. (1968c). *J. Chem. Soc. C,* 2554–2559.
Aspinall, G. O., Molloy, J. A. and Craig, J. W. T. (1969). *Can. J. Biochem.* **47**, 1063–1070.
Aspinall, G. O., Krishnamurthy, T. N., Mitura, W., and Funabashi, M. (1975). *Can. J. Chem.* **53**, 2182–2188.

Aspinall, G. O., Krishnamurthy, T. N., and Rosell, K. G. (1977). *Carbohydr. Res.* **55**, 11–19.
Bacon, J. S. D., and Cheshire, M. V. (1971). *Biochem. J.* **124**, 555–562.
Bahl, O. P. (1970). *J. Biol. Chem.* **245**, 299–304.
Bardalaye, P. C., and Hay, G. W. (1974). *Carbohydr. Res.* **37**, 339–350.
Barnoud, F., Dutton, G. G. S., and Joseleau, J. P. (1973). *Carbohydr. Res.* **27**, 215–223.
Barnoud, F., Mollard, A., and Dutton, G. G. S. (1977). *Physiol. Veg.* **15**, 153–161.
Barrett, A. J., and Northcote, D. H. (1965). *Biochem. J.* **94**, 617–627.
Bauer, W. D., Talmadge, K. W., Keegstra, K., and Albersheim, P. (1973). *Plant Physiol.* **51**, 174–187.
Beck, E. (1967). *Z. Pflanzenphysiol.* **57**, 444–450.
Ben-Arie, R., Ordin, L., and Kindinger, J. I. (1973). *Plant Cell Physiol.* **14**, 427–434.
Björndal, H., Hellerquist, C. G., Lindberg, B., and Svensson, S. (1970). *Angew. Chem. Int. Ed. Engl.* **9**, 610–616.
Björndal, H., Erbind, C., Lindberg, B., Fahraeus, G., and Ljunggren, H. (1971). *Acta Chem. Scand.* **25**, 1281–1286.
Blake, J. D., and Richards, G. N. (1971). *Carbohydr. Res.* **18**, 11–20.
Blumenkrantz, N., and Asboe-Hansen, G. (1973). *Anal. Biochem* **54**, 484–489.
Brown, R. M., Jr., Franke, W. W., Kleinig, H., Falk, H., and Sitte, P. (1969). *Science* **166**, 894–896.
Brown, R. M., Jr., and Montezinos, D. (1976). *Proc. Natl. Acad. Sci. U.S.A.* **73**, 143–147.
Brummond, D. O., and Gibbons, A. P. (1965). *Biochem. Z.* **342**, 308–318.
Buchala, A. J. (1973). *Phytochemistry* **12**, 1373–1376.
Buchala, A. J. (1974). *Phytochemistry* **13**, 2185–2188.
Buchala, A. J., and Franz, G. (1974). *Phytochemistry* **13**, 1887–1889.
Buchala, A. J., and Meier, H. (1972). *Phytochemistry* **11**, 3275–3278.
Buchala, A. J., and Meier, H. (1973). *Carbohydr. Res.* **26**, 421–425.
Buchala, A. J., and Wilkie, K. C. B. (1970). *Naturwissenschaften* **10**, 496–497.
Buchala, A. J., and Wilkie, K. C. B. (1971). *Phytochemistry* **10**, 2287–2291.
Buchala, A. J., and Wilkie, K. C. B. (1973). *Phytochemistry* **12**, 499–505.
Buchala, A. J., and Wilkie, K. C. B. (1974). *Phytochemistry* **13**, 1347–1351.
Buchala, A. J., Fraser, C. G., and Wilkie, K. C. B. (1972). *Phytochemistry* **11**, 2803–2814.
Burke, D., Kaufman, P., McNeil, M., and Albersheim, P. (1974). *Plant Physiol.* **54**, 109–115.
Chambers, J., and Elbein, A. D. (1970). *Arch. Biochem. Biophys.* **138**, 620–631.
Chanda, S. H., Hirst, E. L., Jones, J. K. N., and Percival, E. G. V. (1950). *J. Chem. Soc.*, 1289–1295.
Cho, Y. P., and Chrispeels, M. J. (1976). *Phytochemistry* **15**, 165–169.
Comtat, J., Joseleau, J. P., Bosso, C., and Barnoud, F. (1974). *Carbohydr. Res.* **38**, 217–224.
Costello, P. R., and Stone, B. A. (1968). *Proc. Austr. Biochem. Soc.*, p. 43.
Courtois, J. E., and Le Dizet, P. (1974). *An. Quim.* **70**, 1067–1072.
Curvall, M., Lindberg, B., and Lönngren, J. (1975). *Carbohydr. Res.* **41**, 235–239.
Danishefsky, I., Whistler, R. L., and Bettelheim, F. A. (1970). *In* "The Carbohydrates" (W. Pigman and D. Horton, eds.), Vol. IIA, pp. 375–410. Academic Press, New York.
Darvill, A. (1976). Doctoral thesis. University of Wales, Aberystwyth, Wales.
Darvill, A. G., Smith, C. J., and Hall, M. A. (1977). *In* "Regulation of Cell Membrane Activities in Plants" (E. Marré and O. Ciferri, eds.), pp. 275–282. North-Holland Publ., Amsterdam.
Darvill, A., Smith, C. T., and Hall, M. A. (1978a). *New Phytol.* **80**, 503–516.
Darvill, A., McNeil, M., and Albersheim, P. (1978b). *Plant Physiol.,* **62,** 418–422.
Dea, I. C. M., Rees, D. A., Beveridge, R. J., and Richards, G. N. (1973). *Carbohydr. Res.* **29**, 363–372.
Dea, I. C. M., Morris, E. R., Rees, D. A., Welsh, E. J., Barnes, H. A., and Price, J. (1977). *Carbohydr. Res.* **57**, 249–272.

Delmer, D. P. (1977). *Rec. Adv. Phytochem.* **11**, 45–77.

Delmer, D. P., Beasley, C. A., and Ordin, L. (1974). *Plant Physiol.* **53**, 149–153.

Delmer, D. P., Heiniger, U., and Kulow, C. (1977). *Plant Physiol.* **59**, 713–718.

Delmer, D. P., Kulow, C., and Ericson, M. C. (1978). *Plant Physiol.* **61**, 25–29.

Dever, J. E., Jr., Bandurski, R. S., and Kivilaan, A. (1968). *Plant Physiol.* **43**, 50–56.

Dische, Z. (1962). *In* "Methods in Carbohydrate Chemistry" (R. L. Whistler and M. L. Wolfrom, eds.), Vol. I, pp. 478–512. Academic Press, New York.

Duff, R. B. (1965). *Biochem. J.* **94**, 768–772.

Eda, S., Ohnishi, A., and Kato, K. (1976). *Agric. Biol. Chem.* **40**, 359–364.

Eda, S., Watanabe, F., and Kato, K. (1977). *Agric. Biol. Chem.* **41**, 429–434.

Ehrenthal, I., Montgomery, R., and Smith, F. (1954). *J. Am. Chem. Soc.,* p. 5509–5510.

Elbein, A. D. (1969). *J. Biol. Chem.* **244**, 1608–1616.

English, P. D., Maglothin, A., Keegstra, K., and Albersheim, P. (1972). *Plant Physiol.* **49**, 293–297.

Ericson, M. C., and Delmer, D. P. (1977). *Plant Physiol.* **59**, 341–348.

Ericson, M. C., and Delmer, D. P. (1978). *Plant Physiol.* **61**, 293–297.

Eschrich, W. (1965). *Planta* **65**, 280–285.

Fincher, G. B. (1975). *J. Inst. Brew.* **81**, 116–122.

Fincher, G. B. (1976). *J. Inst. Brew.* **82**, 347–349.

Fincher, G. B., and Stone, B. A. (1974). *Austr. J. Biol. Sci.* **27**, 117–132.

Flowers, H. M., Batra, K., Kemp, J., and Hassid, W. Z. (1969). *J. Biol. Chem.* **244**, 4969–4974.

Forrest, I. S., and Wainwright, T. (1977). *J. Inst. Brew. London* **83**, 279–286.

Franz, G., and Meier, H. (1969). *Phytochemistry* **8**, 579–583.

Fraser, C. G., and Wilkie, K. C. B. (1971). *Phytochemistry* **10**, 199–204.

Gardner, K. H., and Blackwell, J. (1974a). *Biopolymers* **13**, 1975–2001.

Gardner, K. H., and Blackwell, J. (1974b). *Biochim. Biophys. Acta* **343**, 232–237.

Gardiner, M., and Chrispeels, M. J. (1975). *Plant Physiol.* **55**, 536–541.

Gilkes, N. R., and Hall, M. A. (1977). *New Phytol.* **78**, 1–12.

Gould, S. E. B., Rees, D. A., Richardson, N. G., and Steele, I. W. (1965). *Nature (London)* **208**, 876–878.

Gould, S. E. B., Rees, D. A., and Wight, N. J. (1971). *Biochem. J.* **124**, 47–53.

Grant, G. T., Morris, E. R., Rees, D. A., Smith, P., and Thom, D. (1973). *FEBS Lett.* **32**, 195–198.

Green, J. R., and Northcote, D. H. (1978). *Biochem. J.* **170**, 599–605.

Hall, M. A., ed. (1976). "Plant Structure, Functions and Adaptation." MacMillan, New York.

Hart, D. A., and Kindel, P. K. (1970a). *Biochem. J.* **116**, 569–579.

Hart, D. A., and Kindel, P. K. (1970b). *Biochemistry* **9**, 2190–2196.

Hartley, R. D., and Jones, E. C. (1976). *Phytochemistry* **15**, 1157–1160.

Heiniger, U., and Delmer, D. P. (1977). *Plant Physiol.* **59**, 719–723.

Heller, J. S., and Villemez, C. L. (1972). *Biochem. J.* **129**, 645–655.

Henderson, G. A., and Hay, G. W. (1972). *Carbohydr. Res.* **23**, 379–398.

Herth, W., Franke, W. W., Bittiger, H., Kuppel, A., and Keilich, G. (1974). *Cytobiologie* **9**, 344–367.

Hirst, E. L. (1962). *Pure Appl. Chem.* **5**, 53–66.

Hirst, E. L., and Jones, J. K. N. (1947). *J. Chem. Soc.,* 1221–1225.

Hopp, H. E., Romero, P. A., Daleo, G. R., and Pont Lezica, R. (1978a). *Eur. J. Biochem.* **84**, 561–571.

Hopp, H. E., Romero, P. A., and Pont Lezica, R. (1978b). *FEBS Lett.* **86**, 259–262.

Hsu, D. S., and Reeves, R. E. (1967). *Carbohydr. Res.* **5**, 202–209.

Huber, D. J., and Nevins, D. J. (1977). *Plant Physiol.* **60**, 300–304.

Jermyn, M. A., and Yeow, Y. M. (1975). *Aust. J. Plant Physiol.* **2**, 501–508.

Jiang, K. S., and Timell, T. E. (1972). *Cellul. Chem. Technol.* **6**, 499–502.

Jilka, R., Brown, O., and Nordin, P. (1972). *Arch. Biochem. Biophys.* **152,** 702–711.
Joseleau, J. P., and Barnoud, F. (1974). *Phytochemistry* **13,** 1155–1158.
Joseleau, J. P., Chambat, G., Vignon, M., and Barnoud, F. (1977). *Carbohydr. Res.* **58,** 165–175.
Kaji, A., and Saheki, T. (1975). *Biochim. Biophys. Acta* **410,** 354–360.
Karacsonyi, S., Toman, R., Janecek, F., and Kubackova, M. (1975). *Carbohydr. Res.* **44,** 285–290.
Karr, A. L. (1972). *Plant Physiol.* **50,** 275–282.
Kato, K., Watanabe, F., and Eda, S. (1977). *Agric. Biol. Chem.* **41,** 533–538.
Kauss, H. (1974). *In* "Plant Carbohydrate Chemistry" (J. B. Pridham, ed.), pp. 191–205. Academic Press, New York.
Kauss, H., and Bowles, D. J. (1976). *Planta* **130,** 169–174.
Kauss, H., and Glaser, C. (1974). *FEBS Lett.* **45,** 304–307.
Keegstra, K., Talmadge, K. W., Bauer, W. D., and Albersheim, P. (1973). *Plant Physiol.* **51,** 188–196.
Kemp, J., and Loughman, B. C. (1973). *Biochem. Soc. Trans.* **1,** 446–448.
Kemp, J., and Loughman, B. C. (1974). *Biochem. J.* **142,** 153–159.
Kiermayer, O., and Dobberstein, B. (1973). *Protoplasma* **77,** 437–451.
Kivilaan, A., Bandurski, R. S., and Schulze, A. (1971). *Plant Physiol.* **48,** 389–393.
Kivirikko, K. I., and Liesmaa, M. (1959). *Scand. J. Clin. Lab. Invest.* **11,** 128–133.
Kolpak, F. J., and Blackwell, J. (1975). *Textile Res. J.* **45,** 568–572.
Kolpak, F. J., and Blackwell, J. (1976). *Macromolecules* **9,** 273–278.
Kooiman, P. (1961). *Recueil* **80,** 849–865.
Labavitch, J. M., and Ray, P. M. (1978). *Phytochemistry* **17,** 933–938.
Labavitch, J. M., Freeman, L. E., and Albersheim, P. (1976). *J. Biol. Chem.* **251,** 5904–5910.
Lamport, D. T. A. (1967). *Nature (London)* **216,** 1322–1324.
Lamport, D. T. A. (1969). *Biochemistry* **8,** 1155–1163.
Lamport, D. T. A. (1970). *Annu. Rev. Plant Physiol.* **21,** 235–270.
Lamport, D. T. A. (1973). *Colloq. Int. C N R S* **212,** 27–31.
Lamport, C. T. A., and Miller, D. H. (1971). *Plant Physiol.* **48,** 454–456.
Lamport, D. T. A., Katona, L., and Roerig, S. (1973). *Biochem. J.* **133,** 125–132.
Larm, O., Theander, O., and Aman, P. (1976). *Acta Chem. Scand. Ser. B.* **30,** 627–630.
Lennarz, W. J., and Sher, M. G. (1973). *In* "Membrane Structure and Mechanisms of Biological Energy Transduction" (J. Avery, ed.), pp. 441–453. Plenum, New York.
Lindberg, B., and Thompson, J. L. (1973). *Carbohydr. Res.* **28,** 351–357.
Lindberg, B., Lönngren, J., and Rudén, U. (1975). *Carbohydr. Res.* **42,** 83–93.
Loescher, W., and Nevins, D. J. (1972). *Plant Physiol.* **50,** 556–563.
Lowry, O. H., Rosebrough, N. J., Farr, L., and Randall, R. J. (1951). *J. Biol. Chem.* **193,** 265–275.
Liu, T. Y., Elbein, A. D., and Su, J. C. (1966). *Biochem. Biophys. Res. Commun.* **22,** 650–654.
Maekawa, E., and Kitao, K. (1973). *Agric. Biol. Chem.* **37,** 2073–2081.
Maekawa, E., and Kitao, K. (1974). *Agric. Biol. Chem.* **38,** 227–229.
Maltby, D., Carpita, N. C., Montezinos, D., Kulow, C., and Delmer, D. C. (1979). *Plant Physiol.* **63,** 1158–1164.
Mares, D. J., and Stone, B. A. (1973). *Aust. J. Biol. Sci.* **26,** 793–812.
Markwalder, H. U., and Neukom, H. (1976). *Phytochemistry* **15,** 836–837.
Marsland, D. (1964). "Principles of Modern Biology," 4th Ed. Holt, New York.
Marx-Figini, M. (1966). *Nature (London)* **210,** 754.
Marx-Figini, M., and Schulz, G. (1966). *Biochim. Biophys. Acta* **112,** 74–85.
Mascaro, L. J., and Kindel, P. K. (1977). *Arch. Biochem. Biophys.* **183,** 139–148.
McNeil, M., and Albersheim, P. (1977). *Carbohydr. Res.* **56,** 239–248.
McNeil, M., Albersheim, P., Taiz, L., and Jones, R. L. (1975). *Plant Physiol.* **55,** 64–68.

Meier, H. (1962). *Acta. Chem. Scand.* **16**, 2275–2283.
Meinert, M., and Delmer, D. P. (1977). *Plant Physiol.* **59**, 1088–1097.
Monro, J. A., Penny, D., and Bailey, R. W. (1976). *Phytochemistry* **15**, 1193–1197.
Montezinos, D., and Brown, Jr., R. M. (1976). *J. Supramol. Struct.* **5**, 277–290.
Morita, M. (1965a). *Agric. Biol. Chem.* **29**, 564–573.
Morita, M. (1965b). *Agric. Biol. Chem.* **29**, 626–630.
Morris, D. L. (1942). *J. Biol. Chem.* **142**, 881–891.
Mueller, S. C., Brown, Jr., R. M., and Scott, T. K. (1976). *Science* **194**, 949–951.
Mühlethaler, K. (1967). *Annu. Rev. Plant Physiol.* **18**, 1–24.
Neukom, H., and Deuel, H. (1958). *Chem. Industry* (*London*) 683–684.
Neukom, H., and Markwalker, H. (1975). *Carbohydr. Res.* **39**, 387–389.
Nevins, D., and Loescher, W. (1974). Plant Growth Subst. *Int. Conf. 8th,* pp. 828–837.
Nevins, D. J., Huber, D. J., Yamamoto, R., and Loescher, W. H. (1977). *Plant Physiol.* **60**, 617–621.
Northcote, D. H. (1972). *Annu. Rev. Plant Physiol.* **23**, 113–133.
Northcote, D. H. (1974). *In* "Plant Carbohydrate Chemistry" (J. B. Pridham, ed.), pp. 165–181. Academic Press, New York.
Northcote, D. H., and Pickett-Heaps, J. D. (1966). *Biochem. J.* **98**, 159–167.
Odzuck, W., and Kauss, H. (1972). *Phytochemistry* **11**, 2489–2494.
Ordin, L., and Hall, M. A. (1968). *Plant Physiol.* **43**, 473–476.
Pan, Y., and Kindel, P. K. (1977). *Arch. Biochem. Biophys.* **183**, 131–138.
Panayotatos, N., and Villemez, C. L. (1973). *Biochem. J.* **133**, 263–271.
Parrish, F. W., Perlin, A. S., and Reese, E. T. (1960). *Can J. Chem.* **38**, 2094–2104.
Paull, R. E., and Jones, R. L. (1976). *Plant Physiol.* **57**, 249–256.
Peat, S., Whelan, W. J., and Roberts, J. G. (1957). *J. Chem. Soc.,* 3916–3924.
Peaud-Lenoel, C., and Axelos, M. (1970). *FEBS Lett.* **8**, 224–228.
Pickett-Heaps, J. D. (1967). *Dev. Biol.* **15**, 206–236.
Pickett-Heaps, J. D. (1968). *J. Cell Sci.* **3**, 55–64.
Pont-Lezica, R., Romero, P. A., and Hopp, H. E. (1978). *Planta* **140**, 177–183.
Pope, D. G. (1977). *Plant Physiol.* **59**, 894–900.
Preston, R. D. (1964). *In* "The Formation of Wood in Forest Trees" (M. A. Zimmerman, ed.), p. 169. Academic Press, New York.
Preston, R. D. (1974). "The Physical Biology of Plant Cell Walls." Chapman and Hall, London.
Ray, P. M. (1963). *Biochem. J.* **89**, 144–150.
Ray, P. M. (1975). *Plant Physiol.* **56**, Suppl. 84.
Ray, P. M., and Rottenberg, D. A. (1964). *Biochem. J.* **90**, 646–655.
Ray, P. M., Eisinger, W. R., and Robinson, D. G. (1976). *Ber. Deutsch. Bot. Ges.* **89**, 121–146.
Raymond, Y., Fincher, G. B., and Machlachlan, G. A. (1978). *Plant Physiol.* **61**, 938–942.
Rees, D. A. (1972). *Biochem. J.* **126**, 257–273.
Rees, D. A., and Richardson, N. G. (1966). *Biochemistry* **5**, 3099–3107.
Rees, D. A., and Welsh, E. J. (1977). *Angew. Chem. Int. Ed. Engl.* **16**, 214–234.
Rees, D. A., and Wight, N. J. (1969). *Biochem. J.* **115**, 431–439.
Roberts, R. M., and Harrer, E. (1973). *Phytochemistry* **12**, 2679–2682.
Robinson, D. G. (1977). *Adv. Bot. Res.* **5**, 89–151.
Robinson, D. G., Eisinger, W. R., and Ray, P. M. (1976). *Ber. Deutsch. Bot. Ges.* **89**, 147–162.
Sadava, D., and Chrispeels, M. J. (1971a). *Biochemistry* **10**, 4290–4294.
Sadava, D., and Chrispeels, M. J. (1971b). *Biochim. Biophys. Acta* **227**, 278–287.
Sanford, P. A., and Conrad, H. E. (1966). *Biochemistry* **5**, 1508–1513.
Sarko, A., and Muggli, R. (1974). *Macromolecules* **7**, 486–494.
Sharon, N. (1975). "Complex Carbohydrates. Their Chemistry, Biosynthesis and Functions." Addison-Wesley Publ., Reading, Massachusetts.

Shimizu, K., and Samuelson, O. (1973). *Svensk Papperstidn.* **76,** 150–155.

Shore, G., and Machlachlan, G. (1975). *J. Cell. Biol.* **64,** 557–571.

Shore, G., Raymond, Y., and Machlachlan, G. A. (1975). *Plant Physiol.* **56,** 34–38.

Siddiqui, I. R., and Wood, P. J. (1971). *Carbohydr. Res.* **17,** 97–108.

Siddiqui, I. R., and Wood, P. J. (1972). *Carbohydr. Res.* **24,** 1–9.

Siddiqui, I. R., and Wood, P. J. (1974). *Carbohydr. Res.* **36,** 35–44.

Siddiqui, I. R., and Wood, P. J. (1976). *Carbohydr. Res.* **50,** 97–107.

Siddiqui, I. R., and Wood, P. J. (1977a). *Carbohydr. Res.* **53,** 85–94.

Siddiqui, I. R., and Wood, P. J. (1977b). *Carbohydr. Res.* **54,** 231–236.

Smith, M. M., and Stone, B. A. (1973a). *Biochim. Biophys. Acta* **313,** 72–94.

Smith, M. M., and Stone, B. A. (1973b). *Phytochemistry* **12,** 1361–1367.

Smith, M. M., Axelos, A., and Peaud-Lenöel, C. (1976). *Biochimie* **58,** 1195–1211.

Stoddart, R. W., Barrett, A. J., and Northcote, D. H. (1967). *Biochem. J.* **102,** 194–204.

Talmadge, K. W., Keegstra, K., Bauer, W. D., and Albersheim, P. (1973). *Plant Physiol.* **51,** 158–173.

Taylor, R. L., and Conrad, H. E. (1972). *Biochem.* **11,** 1383–1388.

Toman, R. (1973). *Cellul. Chem. Tech.* **7,** 351–357.

Toman, R., Karácsonyi, S., and Kovácik, V. (1972). *Carbohydr. Res.* **25,** 371–378.

Toman, R., Karácsonyi, S., and Kubacková, M. (1975). *Carbohydr. Res.* **43,** 111–116.

Valent, B. S., and Albersheim, P. (1974). *Plant Physiol.* **54,** 105–108.

Villemez, C. L., and Heller, J. S. (1970). *Nature (London)* **227,** 80–81.

Villemez, C. L., and Hinman, M. (1975). *Plant Physiol. Suppl.* 79.

Villemez, C. L., Liu, T. Y., and Hassid, W. Z. (1965). *Proc. Natl. Acad. Sci. U.S.A.* **54,** 1626–1632.

Wada, S., and Ray, P. M. (1978). *Phytochemistry* **17,** 923–932.

Weinstein, L., and Albersheim, P. (1979). *Plant Physiol.,* **63,** 425–432.

Whistler, R. L., and Richards, E. L. (1970). *In* "The Carbohydrates" (W. Pigman and D. Horton, eds.), Vol. IIA, Chapter 37. Academic Press, New York.

White, E. V., and Rao, P. S. (1953). *J. Am. Chem. Soc.* **75,** 2617–2619.

Wilder, B. M., and Albersheim, P. (1973). *Plant Physiol.* **51,** 889–893.

Wilkie, K. C. B., and Woo, S.-L. (1976). *Carbohydr. Res.* **49,** 399–409.

Wilkie, K. C. B., and Woo, S.-L. (1977). *Carbohydr. Res.* **57,** 145–162.

Wooding, F. B. P. (1968). *J. Cell Sci.* **3,** 71–93.

Woolard, G. R., Rathbone, E. B., and Novellie, L. (1976). *Carbohydr. Res.* **51,** 239–247.

Woolard, G. R., Rathbone, E. B., and Novellie, L. (1977). *Phytochemistry* **16,** 957–959.

Worth, H. G. J. (1967). *Chem. Rev.* **67,** 465–473.

Zitko, V., and Bishop, C. T. (1966). *Can. J. Chem.* **44,** 1275–1282.

The Plasma Membrane | 4

ROBERT T. LEONARD
THOMAS K. HODGES

I. INTRODUCTION

The protoplast of plant cells, as with cells of other organisms, is separated from its environment by the plasma membrane or plasmalemma. The term

The Biochemistry of Plants, Vol. 1

plasmalemma ("lemma," the shell of a fruit) is often used to refer to the surface membrane of the plant cell, but here we will refer to this membrane as the plasma membrane. Early attempts to verify the existence of the plasma membrane in plants were confounded by the presence of the thick and resistant cell wall. It was necessary to separate the plasma membrane from the cell wall by plasmolysis to conduct experiments designed to demonstrate the presence of an outer cell membrane. Even today our progress in understanding the structure and function of the plant plasma membrane is retarded by experimental difficulties associated with the presence of the cell wall.

While we presume that the structure and function of the plasma membrane in plants is fundamentally similar to that in animals, fungi, and bacteria, specific knowledge of many aspects of the plant plasma membrane is lacking. For example, the plasma membrane of plant cells is composed primarily of lipids and proteins, but little information is available on the nature of the lipids and proteins. It is recognized that the plasma membrane of plant cells may have a number of important functions, such as to mediate the transport of solutes into and out of the protoplast, to coordinate the synthesis and assembly of cell wall microfibrils, and to translate hormonal and environmental signals involved in the control of growth and differentiation; however, our understanding of the role of the plasma membrane in these biochemical and physiological processes is incomplete. The knowledge that has been gathered is in many instances extensively biased, and perhaps rightly so, by information from investigations on cells of other organisms. More detailed knowledge of the plasma membrane is essential to the complete understanding of plant biochemistry.

In this chapter we will discuss the structure and function of the plasma membrane of higher plant cells in general terms, but it should serve as a reasonably complete introduction to our present understanding of the plant plasma membrane. An attempt will be made to distinguish between what we know from experimental evidence available for plants and what has been extrapolated from results obtained with other organisms. In most instances review articles are cited rather than original papers to give the reader the broadest possible introduction to the relevant literature. As a result, many original contributions are not cited, but they can be found in the reviews.

II. MORPHOLOGY AS DISTINGUISHED WITH THE ELECTRON MICROSCOPE

A. Transmission Electron Microscopy

The plasma membrane, as revealed by conventional transmission electron microscopy (Fig. 1), appears as a thin undulating line around the periphery

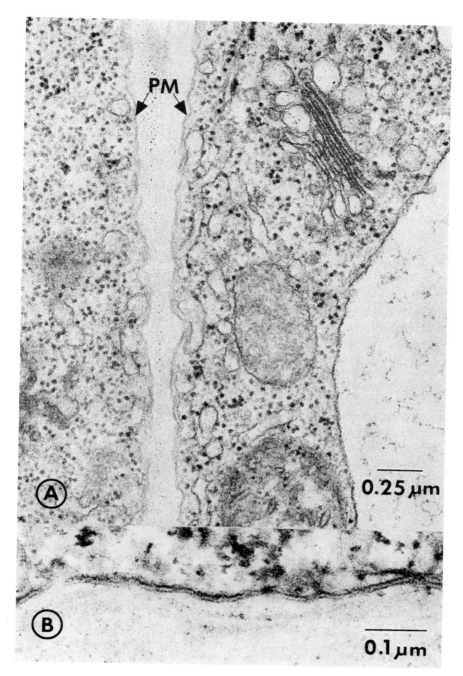

Fig. 1. The plasma membrane as viewed by transmission electron microscopy. (A) Portions of bean leaf cells showing typical undulating appearance of the plasma membrane (PM). (B) Higher magnification view of the plasma membrane of an oat root cell showing the dark–light–dark tripartite structure characteristic of all cell membranes. (Micrographs provided by C. E. Bracker.)

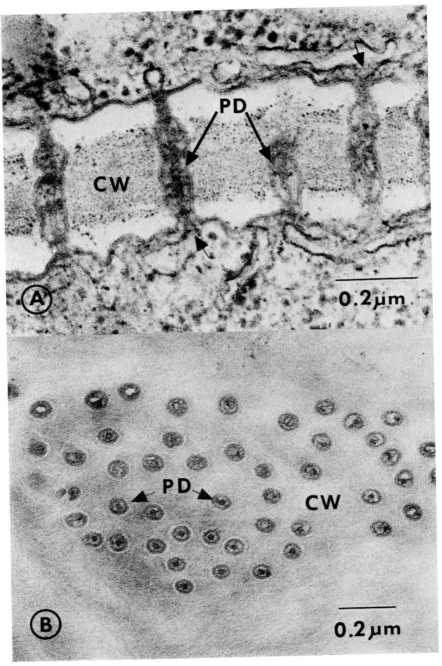

Fig. 2. Plasmodesmata connecting plant cells as viewed with the electron microscope. (A) Four plasmodesmata (PD) in the cell wall (CW) between two oat root cells showing the con-

of the protoplast separating the cytoplasmic contents from the cell wall and the external space. At higher magnification (Fig. 1), the dark–light–dark tripartite structure characteristic of all cell membranes can also be discerned for the plasma membrane. The electron-dense image of the plasma membrane is about 100 Å wide, but variations between 80 and 120 Å have been reported (Mooré and Bracker, 1976). The lighter staining region of the tripartite structure is about 35 Å thick and is bounded by two dark staining regions which usually range between 25 and 35 Å each. The outer dark region of the membrane sometimes stains more prominently than the region adjacent to the cytoplasm (Ledbetter and Porter, 1970) suggesting an asymmetry in membrane composition.

Several other structures characteristic of the plant plasma membrane are revealed by transmission electron microscopy. Complex invaginations of plasma membrane termed multivesicular bodies, lomasomes, or plasmalemmasomes are commonly observed in electron micrographs of plant cells (Gunning and Steer, 1975). The function (if any) of these structures is unknown, and the possibility that such structures are artifacts of fixation must be given careful consideration.

Numerous small (0.3 μm or less) vesicles are also associated with the plasma membrane of plant cells and are presumed to represent membrane fusion events involved in secretion of cell wall materials and other extracellular polysaccharides, plasma membrane turnover, or even the uptake of extracellular materials by reverse pinocytosis. All three possibilities seem likely and have been experimentally described for other organisms. Substantial experimental evidence (Gunning and Steer, 1975) is also available for higher plants to support the view that these vesicles do in fact represent membrane fusion events involved in changes in the cell surface.

A unique feature of the plant plasma membrane is that it is virtually continuous throughout the living cells of plant tissues, and possibly the whole plant (Gunning and Steer, 1975). The continuity is formed by plasmodesmata (Fig. 2), which are cytoplasmic strands bounded by plasma membrane extending from cell to cell through holes in the cell wall (Robards, 1975; Gunning and Robards, 1976; Evert et al., 1977). The average diameter of plasmodesmata is about 60 nm and typically the frequency ranges from 0.5 to 15 plasmodesma per μm^2 of wall surface (Gunning and Robards, 1976). Hence, two cells with 100 μm^2 of wall surface in common would share 50 to

---

tinuity of the plasma membrane between the protoplasts. Also, note the strands of membrane (presumably endoplasmic reticulum) extending through the center of the plasmodesma (arrows). (B) Numerous plasmodesmata (PD) grouped in a primary pit field viewed in cross section in the wall (CW) joining mesophyll and bundle sheath cells of the bermudagrass leaf. Note that the plasma membrane is intact around the circumference of each plasmodesma and that the lumen contains electron-dense material. (Micrographs A and B provided by C. E. Bracker and W. W. Thomson, respectively.)

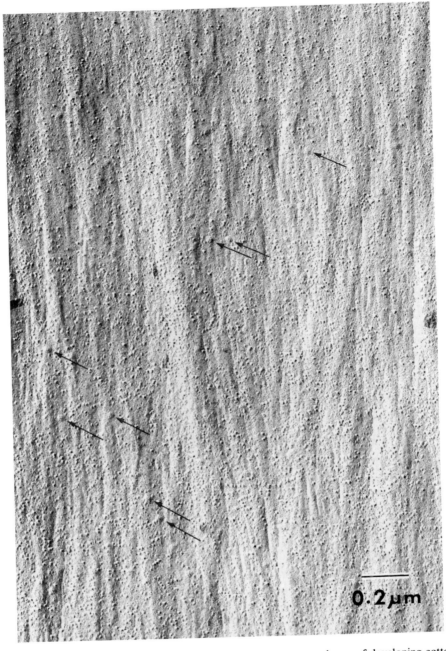

Fig. 3. The external fracture face (E face) of the plasma membrane of developing cotton fibers as viewed with freeze–fracture electron microscopy. This is how the membrane would appear if viewed from inside of the cell after stripping away the inner leaflet of the plasma membrane to expose the outer leaflet closely appressed to the cell wall. Note the impressions of cellulose microfibrils and of particle complexes (arrows) which are presumed to function in cellulose synthesis. From Fig. 14, Willison and Brown (1977).

1500 cytoplasmic connections. If the wall between the cells was 0.5 μm thick, the surface area of plasma membrane within the plasmodesmata (0.1 μm^2 for a plasmodesma 60 nm in diameter) would range between 5 and 150 μm^2 or from 2.5 to 75% of the plasma membrane surface area on the two sides of the 100 μm^2 wall area. Therefore, depending on the tissue or cell type, the amount of plasma membrane contained within the cell wall in plasmodesmata can approach that covering the remaining wall surface (Gunning and Robards, 1976). While the involvement of plasmodesmata in transport from cell to cell has been given careful consideration, the possibility that plasmodesmata may make a significant contribution to membrane functions such as transport from the apoplast (cell wall continuum) to the symplast (cytoplasmic continuum) needs to be investigated.

B. Freeze–Fracture Electron Microscopy

Freeze–fracture electron microscopy shows that the plasma membrane, like most cellular membranes, consists of particles embedded in an amorphous matrix (Fig. 3). The particles generally range from 80 to 150 Å in diameter and they are presumed to represent various functional proteins. The particle density varies depending on which fracture face (P or E, see Branton *et al.*, 1975) is being observed and on the stage of differentiation of the cells in question (Willison and Brown, 1977 and references therein). The amorphous matrix is believed to represent membrane lipids and hence, the plant plasma membrane, like all biological membranes, appears to be composed of various functional proteins floating in a fluid lipid bilayer. Some evidence is available for plant cells to support the idea that plasma membrane proteins are mobile within the plane of the membrane (Burgess and Linstead, 1977 and references therein).

Freeze–fracture studies also reveal an intimate association between the plasma membrane and the cell wall (Willison and Brown, 1977; Mueller *et al.*, 1976). The hypothesis has been proposed that cellulose biosynthesis and microfibril assembly are mediated by the plasma membrane in concert with microtubules (see Chapter 14). This hypothesis will be considered further in Section VI,B.

III. PURIFICATION OF THE PLASMA MEMBRANE

A. Generalized Procedure

Biochemical characterization of the plasma membrane depends on the availability of procedures for the separation of this membrane from other cellular components. Procedures have been developed for several plant species and tissues which result in partial purification of the plasma membrane (Nagahashi *et al.,* 1978 and references therein), but complete purification of the plant plasma membrane has not been achieved. The procedures available combine differential and density gradient centrifugation to yield membrane fractions which range between 60 and 80% pure. Such fractions are useful for studying plasma membrane associated activities such as transport ATPase, polysaccharide synthesis, and hormone or toxin binding, but because they are contaminated by other membranes they provide for only a first approximation of chemical composition. The development of a procedure for isolating large quantities of pure plasma membranes is essential for the ultimate elucidation of the biochemistry of this cell component.

The procedures that have been developed for purification of the plasma membrane vary in detail but the general approaches are similar (Falk and Stocking, 1976; Leech, 1977; Nagahashi *et al.,* 1978; Quail, 1979 and references therein). Usually the plant tissue is homogenized in a buffered osmoticum (e.g., 0.25–0.4 M sucrose in 25 mM tris-MES, pH 7.2–7.8) containing various additions to preserve the activities of marker enzymes usually by retarding the activities of hydrolytic or oxidative enzymes. The chelating agent, EDTA, is added to the homogenization medium to sequestor divalent cations. This reduces the activity of phospholipases leading to lower levels of free fatty acids and a preservation of membrane structure. Also the chelation of Mg^{2+} transforms rough endoplasmic reticulum to smooth endoplasmic reticulum reducing contamination by this cell component in the plasma membrane fraction (Table I). Homogenization techniques range from very gentle (chopping with razor blades) to mild (mortar and pestle) to fairly harsh (polytron). Upon cell disruption, the plasma membrane forms vesicles ranging in diameter from less than 0.1 μm to as large as 2 μm. The filtered homogenate is usually subjected to differential centrifugation to remove large cell components, mainly mitochondria and broken plastids, which equilibrate near plasma membrane vesicles in sucrose density gradients (Table I). The force necessary to sediment most of the mitochondria is about 13,000 g for 15 min. At this force, substantial amounts (30 to 50% depending on the method of cell disruption) of the plasma membrane vesicles are also sedimented, and in effect, discarded. Plasma membrane vesicles remaining in the supernatant are separated from other microsomal membranes (e.g., Golgi apparatus membranes, vesicles of the endoplasmic reticulum and of the tonoplast) by sucrose density gradient centrifugation of either the supernatant itself or, more commonly, a membrane fraction derived by differential centrifugation of the supernatant. The peak equilibrium density for plasma membrane vesicles in sucrose ranges between 1.14 and 1.17 g/ml (Nagahashi *et al.,* 1978 and references therein). Preparative sucrose gradients have been designed to collect microsomal membranes in this density range. These "plasma membrane" fractions consist primarily of smooth vesicles ranging in diameter

TABLE I

Generally Accepted Approximate Peak Equilibrium
Densities of Various Plant Cell Components in
Sucrose Gradients

Component	Densities (g/ml)
Tonoplast	Unknown[a]
Smooth endoplasmic reticulum	1.09–1.12
Golgi apparatus vesicles	1.12–1.15
Rough endoplasmic reticulum	1.15–1.17
Plasma membrane	1.14–1.17
Broken plastids	1.16–1.18
Mitochondria	1.18–1.20
Nuclear membrane	1.21–1.23
Intact plastids	1.21–1.24
Microbodies	1.19–1.27

[a] Some data available to indicate that the density is less than 1.12 g/ml.

from 0.1 to 0.5 μm with an average diameter of about 0.2 μm. Very little is known about the sidedness or permeability properties of these vesicles.

B. Identification of Plasma Membrane Vesicles in Subcellular Fractions

Since no unequivocal enzymatic markers are available, identification of smooth vesicles in subcellular fractions as plasma membrane vesicles has depended primarily on electron microscopy in combination with the periodate–phosphotungstate–chromate staining procedure (Nagahashi et al., 1978 for references). With this procedure, it is possible to preferentially stain the plasma membrane of plant cells after preparation of tissues for electron microscopy (Fig. 4). The procedure is also conducted on pelleted membrane fractions and the assumption is made that the plasma membrane retains its specific staining properties even after tissue homogenization and membrane purification. With this procedure, it has been possible to detect the presence of and estimate the amount of plasma membrane vesicles in subcellular fractions.

The chemical basis for the specificity of phosphotungstic acid–chromic acid for the plasma membrane is not known. In fact, the specificity of the staining procedure for the plasma membrane is not absolute and maintenance of specificity depends on rigorous adherence to the experimental protocol. Hence, the utility of the staining procedure as an unequivocal marker for the

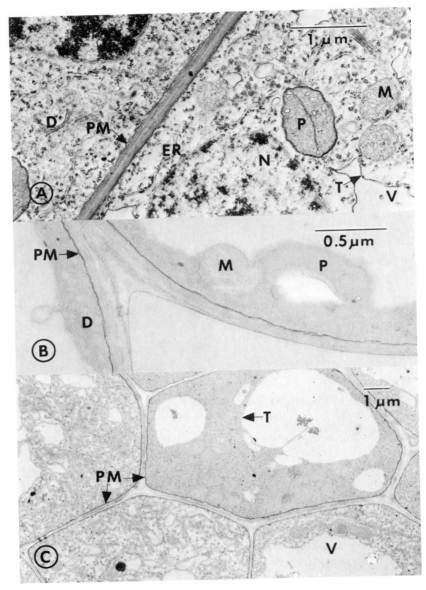

Fig. 4. Electron micrographs of cells of barley roots illustrating the specificity of phosphotungstic acid–chromic acid for the plasma membrane. (A) Portion of two stelar parenchyma cells showing cell components such as nucleus (N), mitochondria (M), plastids (P), endoplasmic reticulum (ER), vacuole (V), tonoplast (T), and plasma membrane (PM) as resolved with conventional uranylacetate–lead citrate staining. (B) Specific staining by phosphotungstic acid–chromic acid of cortical cell plasma membrane. (C) Lower magnification view of stelar parenchyma cells stained with phosphotungstic acid–chromic acid to resolve the plasma membrane (with permission from Nagahashi *et al.,* 1978).

plasma membrane has been questioned. Further research is needed to completely establish the validity of phosphotungstate–chromate staining as a marker for the plasma membrane. For the present, it can be concluded that the procedure, when used with care, has proven to be useful for identifying plasma membrane vesicles in subcellular fractions.

Plasma membrane fractions, as identified by phosphotungstate–chromate staining, are rich in Mg^{2+} requiring K^+-stimulated ATPase activity, glycosyl transferase activity at high UDP-glucose, latent cellulase activity, N-1-naphthylphthalamic acid- (antagonist of indole-3-acetic acid transport) binding activity, *Helminthosporium* toxin-binding activity, far-red absorbing form of phytochrome, and have a high sterol to phospholipid molar ratio. However, no one plasma membrane fraction has been shown to contain all these characteristics and none of these parameters has proven to be both convenient and/or unequivocal markers for the plant plasma membrane. ATPase activity is among the few markers which has been associated with the plasma membrane as identified by techniques other than phosphotungstate–chromate staining such as surface labeling of plant cells (Galbraith and Northcote, 1977; Anderson and Ray, 1978) or cytochemical localization (Gilder and Cronshaw, 1974; Sluiter *et al.*, 1977). Hence, ATPase comes the closest to a useful marker enzyme for the plasma membrane, but the existence of a variety of ATP hydrolyzing activities in subcellular fractions restricts the use of this enzyme activity to locate the plant plasma membrane. The discovery of a simple and convenient marker enzyme for the plant plasma membrane is needed for future research on the purification of this cell component.

C. Contaminants in the Plasma Membrane Fraction

Just as the unavailability of convenient markers for the plasma membrane has limited progress in purifying this cell component, the paucity of unequivocal markers for other cellular membranes has made it difficult to assess contamination in the plasma membrane fraction. Any cell component that generates vesicles during cell disruption which are similar in size and density to plasma membrane vesicles is a potential contaminant in the plasma membrane fraction. Hence, vesicles from the tonoplast, Golgi apparatus, and endoplasmic reticulum, and membranes of mitochondria and plastids are all potential contaminants during plasma membrane purification. Of these, vesicles of the Golgi apparatus and the tonoplast are the most difficult to recognize in subcellular fractions. Latent IDPase activity is a marker for Golgi apparatus membranes, at least in certain tissues. The distribution of this enzyme activity in sucrose gradients shows that membranes of the Golgi apparatus can represent a significant contaminant in the plasma membrane fraction. At present, there is no confirmed marker for the vacuole membrane (tonoplast), although research in progress should soon identify such an enzyme. Preliminary research indicates that tonoplast vesicles are not dense enough to significantly contaminate plasma membrane fractions from sucrose gradients. Thus, depending on the tissue, whether or not homogenization conditions favor conversion of rough endoplasmic reticulum to smooth endoplasmic reticulum, and the extent to which mitochondria and plastids are removed by differential centrifugation, the plasma membrane fraction is likely to be significantly contaminated by Golgi apparatus membranes and to contain measurable amounts of rough endoplasmic reticulum, mitochondria, and broken plastids.

IV. CHEMICAL COMPOSITION OF THE PLASMA MEMBRANE

A. Lipids

There are only a few reports describing the lipid composition of the plasma fraction from plant tissues (Mazliak, 1977). These reports have not been confirmed, and their validity can be questioned on grounds such as whether or not the fractions are of sufficient purity for lipid analysis and/or whether or not the action of phospholipases during membrane purification has significantly affected the results. Research is needed on the lipid composition of highly purified plasma membrane fractions from various plant tissues and species.

Because of the above, it is not possible to give a detailed description of plasma membrane lipids, but some generalities can be drawn from existing data. About 40–50% of the weight of the plasma membrane is accounted for by lipids and representatives of four major lipid classes, phospholipids, glycolipids, sterols, and neutral lipids, are present. The plant plasma membrane, like the outer membrane of animal cells, has a high sterol to phospholipid molar ratio. Cholesterol and various sterol esters are among the sterols in the plant plasma membrane.

B. Proteins

There are no confirmed reports on the protein composition of highly purified preparations of plant plasma membrane. Therefore, it is not possible to give a detailed description of the size or nature of the polypeptides associated with the plasma membrane. Proteins account for about 35–40% of the weight of the plasma membrane (Keenan *et al.,* 1973 and our other published results), and various surface labeling techniques confirm the expectation that plasma membrane proteins are exposed to the external solution (e.g., Galbraith and Northcote, 1977). Clearly, research is needed to characterize plasma membrane proteins.

C. Carbohydrates

As indicated above, lipids and proteins account for only about 80% of the weight of the plasma membrane. The third major component of the plant plasma membrane is probably carbohydrates in the form of glycolipids, glycoproteins, and various "cell wall" polysaccharides. This expectation has not been confirmed by rigorous chemical analysis; however, much indirect evidence such as the ability of lectins to bind to and agglutinate plant protoplasts (Burgess and Linstead, 1977) and the ability of UDP-glucose to surface label the plasma membrane (Anderson and Ray, 1978) supports the idea that carbohydrates of various kinds are commonly associated with this membrane. It would be most interesting to determine the carbohydrate content of highly purified plasma membrane vesicles prepared from both intact cells and isolated protoplasts.

V. ENZYMATIC COMPOSITION OF THE PLASMA MEMBRANE

A. K$^+$-ATPase

Both cytochemical (Gilder and Cronshaw, 1974; Bentwood and Cronshaw, 1976; Sluiter et al., 1977) and biochemical (Hodges, 1976) evidence supports the view that ion-stimulated ATPase activity is associated with the plasma membrane of all plant cells. The exact function of the enzyme is not known for sure, but the present evidence suggests that the ATPase is a primary energy transducer for the coupling of metabolic energy to ion transport across the plasma membrane. Presently, there is debate as to whether or not the ATPase is a primary cation carrier analogous to the (Na$^+$ + K$^+$)-ATPase of animal cells or a primary H$^+$ carrier similar to that of mitochondrial, chloroplast, and bacterial membranes (Poole, 1978). In the former instance, the ATPase would directly mediate the transport of cations through the membrane, while in the latter case, cation transport would be indirectly linked to ATPase function by energy conserved in an ATPase-generated charge and pH gradient (protonmotive force) across the plasma membrane. Much research is needed to distinguish between these alternatives. It is conceivable that future research will show that the action of the ATPase as a primary cation pump may also lead to the development of a protonmotive force across the plasma membrane (Hodges, 1973) or that in plants, like in some bacteria (Harold, 1977), the plasma membrane contains separate and distinct cation- and proton-translocating ATPases.

The ATPase activity of the plasma membrane fraction has been characterized for only a few plant species and tissues (Hodges, 1976; Nagahashi *et al.*, 1978 for references). The preferred substrate for the enzyme is Mg-ATP, and Ca^{2+} strongly inhibits enzyme activity probably by interfering with the substrate. Monovalent cations, particularly K^+, stimulate enzyme activity, but there is no confirmed synergistic effect of K^+ and Na^+ on the plasma membrane ATPase. The enzyme shows maximum activity at about pH 6.5 and 40°C, and is not inhibited by oligomycin or cardiac glycosides (e.g., ouabain). A specific inhibitor of the enzyme has not been reported; however, the ATPase is sensitive to N, N'-dicyclohexylcarbodiimide, diethylstilbesterol, octylguanidine and various sulfhydryl inhibitors.

The enzyme has unique characteristics with respect to K^+ stimulation. The activity of the ATPase as a function of K^+ concentration does not follow simple Michaelis-Menton enzyme kinetics. The kinetic data for the ATPase are similar to those observed for a negative cooperative enzyme, and identical in character to the kinetic data for K^+ transport into various plant tissues and species. This characteristic of the ATPase provides a major link between the enzyme and ion transport.

It should be noted that homogenates of plant tissues contain a number of enzymes which will hydrolyze phosphate from ATP and thus act like an ATPase. For example, ATP is an excellent substrate for nonspecific acid and alkaline phosphatases, and the enzymes of mitochondria and chloroplasts which function in oxidative phosphorylation and photophosphorylation, respectively, are also ATPases. Hence, ATPase activity is not an unequivocal marker for the plasma membrane. In addition, the reader should be aware that presumed studies on "transport ATPase" are in many cases studies on "nonspecific ATP hydrolyzing activity," and as such may have little relevance to transport ATPase or the plasma membrane.

B. Glycosyl Transferase

There is much evidence to indicate that the plasma membrane of at least certain plant tissues contains enzymes which catalyze the transfer of the sugar moiety of a nucleotide sugar to various components associated with the membrane. For example, plasma membrane fractions are rich in UDP-glucose polysaccharide synthetase activity (Chrispeels, 1976; Anderson and Ray, 1978 for references), and also appear to contain sterol glycosyl synthetase activity. At present, these enzymes are not extensively characterized and their function in cell wall synthesis or glycolipid metabolism is poorly understood. It is also clear that other cellular membrane systems such as

Golgi apparatus and endoplasmic reticulum contain the same or very similar enzyme activities. Like ATPase, glycosyl transferase activity is not an unequivocal marker for plasma membrane vesicles in homogenates of plant tissues.

C. Cellulase

There is limited, although convincing evidence that latent cellulase activity is associated with the plasma membrane of cells in tissues undergoing abscission (Koehler et al., 1976), and of cells of certain dicotyledonous species (Pierce and Hendrix, 1979). The enzyme appears to be intimately associated with the membrane because treatment with the nonionic detergent, Triton X-100, is required for maximum activity. The reason for this latency is not known, but it is tempting to speculate that this plasma membrane-associated cellulase may represent an enzyme being transported from the cytoplasm to the cell wall where it functions in the dissolution of cellulosic wall materials. Research is needed to determine whether or not cellulase is associated with the plasma membrane of cells of other tissues.

D. Other Functional Proteins

The plasma membrane contains specific binding sites for the auxin transport inhibitor, N-1-naphthylphthalamic acid (Kende and Gardner, 1976; Anderson and Ray, 1978). This binding activity appears to be exclusively associated with the plasma membrane and thus provides a marker for plasma membrane vesicles in subcellular fractions. The existence of a binding site for the transport inhibitor implies that the plasma membrane contains a protein which functions in auxin transport.

It also appears that the plasma membrane of at least certain plant species and tissues has a protein that functions in the transport of alpha-galactosides such as melibiose and raffinose. This can be inferred from research which shows that the plasma membrane of susceptible sugarcane plants contains a protein that binds the toxin, helminthosporoside, produced by the fungus *Helminthosporium sacchari* (Strobel, 1975). The toxin is a complex sugar similar in structure to alpha-galactosides. The toxin-binding protein has been purified and shown to bind the sugars more strongly than the toxin. It is not known how the binding of toxin to the sugar transport protein leads to disease symptoms, but available evidence suggests that toxin binding to the membrane-associated binding protein may alter other transport systems leading to a disruption of membrane function and cell death.

The toxin-binding protein and the presumed alpha-galactoside transport protein consists of four subunits, each with a molecular weight of about

12,000, giving an intact protein with a molecular weight of 48,000. This is small compared to transmembrane transport proteins such as the $(Na^+ + K^+)$-ATPase of animal cells which has a molecular weight in excess of 250,000. It is likely that the alpha-galactoside binding protein is only part of the sugar transport system and it may be analogous in function to amino acid binding proteins of bacterial cells.

It also appears that the plasma membrane of plants may be a major site for certain photoreceptors involved in the translation of environmental signals into physiological and developmental responses. For example, phytochrome is a major photoreceptive pigment involved in many plant responses. The red-light absorbing form of the pigment appears to be localized throughout the cytoplasm, but at least some of the physiologically active far-red form is associated with plasma membrane-rich fractions (Marmé, 1977). The nature of the association is not known. In addition, the blue-light photoreceptor, involved in such responses as phototropism, may be localized on the plasma membrane (Jesaitis *et al.,* 1977). In both cases the photoreceptors would presumably initiate the various physiological responses by changing plasma membrane properties or activating plasma membrane associated proteins.

VI. FUNCTIONS

It is clear from the preceding discussion that the knowledge of the plant plasma membrane is very incomplete, but the information available does suggest at least three major functions for this cell membrane: (1) the plasma membrane is highly specialized for the transport of a variety of substances into and out of the cell; (2) this membrane may have an important role in the synthesis and assembly of cell wall microfibrils; (3) the plasma membrane is charged with the reception of a variety of external signals and the translation of these signals into various biochemical and physiological responses. Certain aspects of these functions will be considered below, and for a detailed discussion the reader is directed to the list of references.

A. Transport

For cells to live, the plasma membrane must constantly struggle against the forces of equilibrium. That is, the plasma membrane must maintain and/or establish a gradient in cellular materials between the inside and outside. To do this, the membrane must provide an effective barrier to the passive leakage of metabolites from the cell, and at the same time regulate the transport of a wide variety of substances into and out of the cell. These conflicting objectives can be accomplished by a hydrophobic lipid layer which retards the passive permeability of water-soluble molecules and by various "carrier" proteins embedded in the lipid layer which facilitate the movement of certain solutes through the permeability barrier. Our knowledge of the structure of the plasma membrane suggests that it is ideally suited to perform its function.

The variety of substances transported through the plasma membrane is impressive. These substances range from the 16 essential mineral nutrients to various organic ions such as amino acids, to uncharged molecules such as sugars and low molecular weight polysaccharides. Despite the diversity of transport capabilities displayed by most plant cells, detailed biochemical knowledge of any of the presumed transport systems is lacking. The biochemistry of membrane transport in plants should provide a rewarding research area for years to come.

Of the various transport functions of the plasma membrane, the active pumping of inorganic ions is the best understood and perhaps the one which plays the broadest role in plant cell physiology (MacRobbie, 1977). Ion transport has two major functions which are fundamental to cellular activity. One, the transport of ions is necessary to establish and maintain the proper ionic environment required for cellular metabolism. This includes not only the transport of essential mineral nutrients, but also H^+ transport involved in the regulation of cytoplasmic pH and the generation of the electrical potential difference across the plasma membrane. Two, the accumulation of ions in the cell at high concentrations (compared to the surrounding medium) also provides the high osmotic pressure required for turgor-driven cell growth and for other turgor-related plant responses such as stomatal movement and leaf movements. Hence, by regulating ion transport it is possible to alter a variety of cellular characteristics including the ionic environment, pH, electrical properties, and growth rates. All of these properties in turn have far-reaching effects on cell form and function.

Ion movements through the plasma membrane are either directly or indirectly dependent on metabolic energy. The way in which cellular energy is coupled to ion transport is a major unresolved aspect of plasma membrane function. The evidence available suggests that even though many different kinds of molecules are transported through the plasma membrane, there may be only one energy-conserving process fundamental to the transport of this diverse group of molecules. This process may be an ATP-driven proton pump which functions in pH regulation and in establishing a protonmotive force for the transport of other solutes. The proton pump is very likely an ATPase and it may or may not be the K^+-ATPase which has been shown to be associated with the plasma membrane (see Section V,A). Coupling of solute movement to energy conserved in the charge and pH gradient is accomplished by various strategies depending on the nature of the solute in question (Poole, 1978), but the various strategies have a common theme. Solute movement through the membrane is facilitated by transmembrane proteins that possess specific binding sites for the various solutes. Techniques are available for solubilization, purification, and reconstitution of such transport proteins (Korenbrot, 1977). The application of these techniques to the study of membrane transport in plants promises to be an exciting area for future biochemical research.

B. Synthesis and Assembly of Cell Wall Microfibrils

The cell wall of higher plants is composed of cellulose microfibrils embedded in an amorphous matrix of noncellulosic substances consisting mainly of pectins, hemicelluloses, and proteins. Cell fractionation studies have shown that secretory vesicles derived from the Golgi apparatus are involved in the synthesis and deposition of the noncellulosic components of the cell wall (Crispeels, 1976). There is considerable evidence which suggests that cellulose biosynthesis and microfibril assembly occurs at the cell surface, probably in association with multienzyme complexes on the plasma membrane (Willison and Brown, 1977; Mueller *et al.*, 1976 for references). Freeze–fracture electron microscopy of the plasma membrane showing a close association of certain particles with cell wall microfibrils is among the strongest lines of evidence in support of the above hypothesis. The fracture face of the outer leaflet of the plasma membrane shows both randomly scattered particles and larger granules of 250–350 Å in diameter (Fig. 3) which are correlated with the impressions of microfibrils in the membrane. In addition, ultrastructural evidence indicates that microtubules near the inner surface of the plasma membrane are arranged parallel to the orientation of wall microfibrils. Hence, the view has developed that granular cellulose synthesis complexes associated with the outer leaflet of the plasma membrane are "spinning" cellulose microfibrils and being guided on "tracks" composed of microtubules. This idea is still quite speculative and much research is needed to verify this otherwise attractive hypothesis.

C. Receptor for Hormonal and Environmental Signals

Several lines of research suggest that the plasma membrane has a major function in the reception of environmental and perhaps certain hormonal signals involved in the initiation of a variety of plant growth responses. For example, developing evidence suggests that many phytochrome mediated responses may be initiated as a result of the conversion of the red-light absorbing form to the far-red absorbing form of the protein and its binding to the plasma membrane. The sequence of events which might occur after phytochrome binding is not known, but changes in permeability and/or electrical properties of the plasma membrane are implicated (Marmé, 1977). It seems possible that future research will show that phytochrome regulates various transport properties of the plasma membrane, and that changes in plasma membrane function translate the environmental signal, light quality, into a physiological response.

There is also substantial evidence which indicates that the action of auxin on cell elongation may be through increased activity of a plasma membrane-associated H^+ pump leading to acidification and "loosening" of the cell wall. This suggests that, at least for certain responses, auxin binds to the plasma membrane to initiate hormone action. However, available evidence (see Anderson and Ray, 1978 for references) indicates that the endoplasmic reticulum and not the plasma membrane is the primary binding site for auxin. While plasma membrane function may be a part of the action of auxin in inducing cell elongation, it may not be the receptor site for the response.

VII. CONCLUSIONS

Elucidation of the role of the plasma membrane in regulating solute transport, cellulose biosynthesis, and various developmental responses initiated by hormonal or environmental signals depends, at least to some extent, on detailed knowledge of the biochemistry of this cell component. Yet, the plasma membrane of plant cells is virtually an unknown biochemical entity. Six years have elapsed since this membrane was first purified to an extent which allowed for biochemical characterization. While some progress has been made in our understanding of the biochemistry of plasma membrane functions, the answers to many questions await the efforts of future researchers.

REFERENCES

Anderson, R. L., and Ray, P. M. (1978). *Plant Physiol.* **61**, 723–730.

Bentwood, B. J., and Cronshaw, J. (1976). *Planta* **130**, 97–104.

Branton, D., Bullivant, S., Gilula, N. G., Karnovsky, M. J., Moor, H., Muhlethaler, K., Northcote, D. H., Packer, L., Satir, B., Satir, P., Speth, V., Staehlin, L. A., Steere, R. L., and Weinstein, R. S. (1975). *Science* **190**, 54–56.

Burgess, J., and Linstead, P. J. (1977). *Planta* **136**, 253–259.

Crispeels, M. J. (1976). *Annu. Rev. Plant Physiol.* **27**, 19–38.

Evert, R. F., Eschrich, W., and Heyser, W. (1977). *Planta* **136**, 77–89.

Falk, R. H., and Stocking, C. R. (1976). *In* "Encyclopedia of Plant Physiology, Transport in Plants III" (C. R. Stocking and U. Heber, eds.), Vol. 3, pp. 3–50. Springer-Verlag, Berlin and New York.

Galbraith, D. W., and Northcote, D. H. (1977). *J. Cell Sci.* **24**, 295–310.

Gilder, J., and Cronshaw, J. (1974). *J. Cell Biol.* **60**, 221–235.

Gunning, B. E. S., and Steer, M. W. (1975). "Ultrastructure and the Biology of Plant Cells." Arnold, London.

Gunning, B. E. S., and Robards, A. W. (1976). "Intercellular Communication in Plants: Studies on Plasmodesmata." Springer-Verlag, Berlin and New York.

Harold, F. M. (1977). *Annu. Rev. Microbiol.* **31**, 181–203.

Hodges, T. K. (1973). *Adv. Agron.* **25**, 163–207.

Hodges, T. K. (1976). *In* "Encyclopedia of Plant Physiology, Transport in Plants II" (U. Lüttge and M. G. Pitman, eds.), Vol. IIA, pp. 260–283. Springer-Verlag, Berlin and New York.

Jesaites, A. J., Hevers, P. R., Hertel, R., and Briggs, W. (1977). *Plant Physiol.* **59**, 941–947.

Keenan, T. W., Leonard, R. T., and Hodges, T. K. (1973). *Cytobios* **7**, 103–112.

Kende, H., and Gardner, G. (1976). *Annu. Rev. Plant Physiol.* **27**, 267–290.

Koehler, D. E., Leonard, R. T., Vanderwoude, W. J., Linkins, A. E., and Lewis, L. N. (1976). *Plant Physiol.* **58**, 324–330.

Korenbrot, J. I. (1977). *Annu. Rev. Physiol.* **39**, 19–49.

Ledbetter, M. C., and Porter, K. R. (1970). "Introduction to the Fine Structure of Plant Cells." Springer-Verlag, Berlin and New York.

Leech, R. M. (1977). *In* "Regulation of Enzyme Synthesis and Activation in Higher Plants" (H. Smith, ed.), pp. 289–327. Academic Press, New York.

MacRobbie, E. A. C. (1977). *In* "International Review of Biochemistry, Plant Biochemistry II" (D. H. Northcote, ed.), Vol. 13, pp. 211–247. Univ. Park Press, Baltimore, Maryland.

Marmé, D. (1977). *Annu. Rev. Plant Physiol.* **28**, 173–198.

Mazliak, P. (1977). *In* "Lipids and Lipid Polymers in Higher Plants" (M. Tevini and H. K. Lichtenthaler, eds.), pp. 48–74. Springer-Verlag, Berlin and New York.

Mooré, D. J., and Bracker, C. E. (1976). *Plant Physiol.* **58**, 544–547.

Mueller, S. C., Brown, Jr., R. M., and Scott, T. K. (1976). *Science* **194**, 949–951.

Nagahashi, G., Leonard, R. T., and Thomson, W. W. (1978). *Plant Physiol.* **61**, 993–999.

Pierce, W. S., and Hendrix, D. L. (1979). *Planta* **146**, 161–169.

Poole, R. J. (1978). *Annu. Rev. Plant Physiol.* **29**, 437–460.

Quail, P. H. (1979). *Annu. Rev. Plant Physiol.* **30**, 425–484.

Robards, A. W. (1975). *Annu. Rev. Plant Physiol.* **26**, 13–29.

Sluiter, E., Lauchli, A., and Kramer, D. (1977). *Plant Physiol.* **60**, 923–927.

Strobel, G. A. (1975). *Sci. Am.* **232**, 80–88.

Willison, J. H. M., and Brown, Jr., R. M. (1977). *Protoplasma* **92**, 21–41.

The Cytosol* | 5

GRAHAME J. KELLY
ERWIN LATZKO

I. INTRODUCTION

The discovery of each cellular compartment, and elucidation of its structural and functional properties, has been almost invariably welcomed as a

* Unless otherwise stated, all sugars mentioned in this chapter are of the D configuration, and all amino acids are of the L configuration.

The Biochemistry of Plants, Vol. 1
Copyright © 1980 by Academic Press, Inc.
All rights of reproduction in any form reserved.
ISBN 0-12-675401-2

step forward in understanding how cells organize their biological activity. One exception is the cytosol. The cytosol can be viewed as one of the first recognized protoplasmic compartments, but for each organelle and membrane system subsequently found in the protoplasm, the cytosol has had to sacrifice a little more of what was previously attributed to it. Fortunately, this "metabolic juggling" now seems to have ceased since modern electron microscopy has introduced visualization of the cell structure in such detail that even the larger protein molecules can be seen, and hence the boundaries of the cytosol are presumably finally established. The business of defining the cytosol's biochemical function will be consequently that much easier. This chapter is an attempt to describe the major metabolic processes which are presently believed to occur in the cytosol, and how they connect with those of the membrane systems and organelles dispersed therein.

II. DEFINITION AND STRUCTURE

The cytosol is the protoplasmic material contained within the cell membrane (plasmalemma) but exterior to all other cell membranes. It is thus the "cellular soup" through which the endoplasmic reticulum extends, and in which the various cellular organelles are suspended. It is historically analagous to the so-called ground substance of the cytoplasm.

The cytosol is sometimes thought of as being more or less structureless. About 38 years ago Guilliermond (1941) adopted the description made over 100 years earlier by F. Dujardin, including mention that "one can distinguish in it (the cytoplasm) absolutely no trace of organization: neither fibre nor membrane nor an appearance of cellular form." Yet during the 1930s and 1940s studies of the physical properties of the ground substance raised suspicions that, despite the optical homogeneity seen through the light microscope, some form of cytoskeleton existed. These suspicions were confirmed when microfilaments and microtubules were observed by electron microscopists in the 1960s. Together with ribosomes (Chapter 11, this volume), they constitute the more obvious macromolecular entities of the cytosol. The remainder of the cytosol is, as far as we know, a colloidal solution of organic substances, mainly many different kinds of proteins, most of which are enzymes, together with small molecules and inorganic ions in true solution. A difficult challenge for the future will be to ascertain whether the many enzyme molecules are in any way organized. A high degree of organization is difficult to envisage when the flowing movement of the cytosol, usually termed cytoplasmic streaming, is viewed through the light microscope (e.g. with *Nitella*), and might not seem necessary when some processes, e.g., synthesis of amino acids, are so widely distributed between the cytosol and the various organelles. However, some metabolic pathways, glycolysis a classic example, are thought to be largely localized in the cytosol, and indeed some evidence for an aggregate of glycolytic enzymes in a bacterium has been obtained (Mowbray and Moses, 1976) and proposals for such an aggregate in plants are not lacking (Barker *et al.*, 1967). In fact, the considerations of Atkinson (1969) and Sols and Marco (1970), in which the concentration of "soluble" molecules in the cell is pointed out to be notably high, lead one to expect protein aggregation and, in addition, a high degree of binding of cellular metabolites by the proteins.

Whatever the true state of the enzyme molecules in the cytosol, they appear to be the major component (other than water) of this compartment, and their properties are therefore the central theme upon which this chapter is based.

III. CARBON METABOLISM

A. Sucrose Synthesis during Photosynthesis

It is appropriate to begin a consideration of cytosolic carbon metabolism by outlining the pathway from triose-P to sucrose, since triose-P is believed to be the major form in which photosynthetically reduced CO_2 is exported out of the chloroplast (Lilley *et al.*, 1977) and sucrose is the predominant form in which the reduced carbon is further transported to other parts of the plant. The synthesis of one molecule of sucrose from four molecules of triose-P is outlined in Fig. 1. The triose-phosphates are first condensed to two molecules of fructose-1,6-P_2 which are hydrolyzed to fructose-6-P; one fructose-6-P is converted in three reactions to UDP-glucose which then reacts with the second fructose-6-P to form sucrose-P. The sucrose-P is finally hydrolyzed to sucrose. Recognition that leaves can form UDP-glucose, as evidenced by the presence of UDP-glucose pyrophosphorylase (Burma and Mortimer, 1956), was a significant step in elucidating this pathway. The location of sucrose synthesis in the cytosol was largely confirmed by Bird *et al.* (1974) who showed that UDP-glucose pyrophosphorylase and sucrose-P synthetase (enzymes 5 and 8, Fig. 1) are predominantly or exclusively located in the cytosol.

Despite the double membrane separating cytosolic sucrose synthesis from the photosynthetic production of the triose-P precursor in the chloroplast, the relative rates of the two processes are closely linked in that the export of a triose-P from chloroplasts is dependent on the simultaneous import of a P_i: this counter-exchange of triose-P and P_i is facilitated by a specific phosphate translocator located on the inner membrane of the chloroplast envelope (Heldt, 1976). As can be seen from Fig. 1, all four P_i (released from the four triose-phosphates required to synthesize one molecule of sucrose) must be

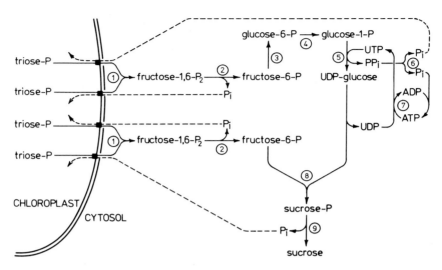

Fig. 1. The synthesis of sucrose from photosynthetically produced triose-P. The dashed lines indicate the return of liberated P_i to the chloroplast via phosphate translocators (represented by black squares on the chloroplast inner membrane), which is a transfer essential for sucrose synthesis to continue uninterrupted (see text). The sum of the reactions is: 4 triose-P $\rightarrow$ sucrose + 4 P_i. Circled numbers refer to the participating enzymes: 1, fructose-1,6-P_2 aldolase; 2, fructose bisphosphatase; 3, phosphohexose isomerase; 4, phosphoglucomutase; 5, UDP-glucose pyrophosphorylase; 6, pyrophosphatase; 7, nucleoside-P_2 kinase; 8, sucrose-P synthetase; 9, sucrose-P phosphatase. Reactions are shown in the direction of sucrose synthesis, but in fact all enzymes other than 2, 6, and 9 catalyze physiologically reversible reactions.

returned to the chloroplast envelope and exchanged for four more triose-phosphates before another molecule of sucrose can be synthesized. This dependence of chloroplast photosynthesis on a continuous supply of P_i from the cytosol during sucrose synthesis has been elegantly elaborated by Walker and Herold (1977), and emphasizes how easily the effects of any regulatory control on sucrose synthesis will be transmitted to the chloroplast.

However, there is at present little information on the regulation of sucrose synthesis. Of the nine enzymes involved, fructose biphosphatase and sucrose-P synthetase (enzymes 2 and 8, Fig. 1) are the two most likely to exert a regulatory influence since their activities are not much greater than observed rates of sucrose synthesis. The cytosolic fructose bisphosphatase from spinach leaves is quite unlike its chloroplast counterpart, but rather similar to the enzyme from gluconeogenic tissues. The enzyme activity is Mg^{2+}-dependent and inhibited by millimolar concentrations of AMP (Zimmermann et al., 1978). It may be relevant that low levels of Mg^{2+} not only reduce the fructose bisphosphatase activity, but also reduce the UDP-glucose pyrophosphorylase reaction in the direction of UDP-glucose formation (Gander, 1976). Sucrose-P synthetase, first detected by Leloir and Cardini (1955), has the characteristics of a regulatory enzyme (Turner and Turner, 1975) but has not received the attention it deserves; the enzyme from photosynthetic tissues has been notably neglected. Giaquinta (1978) has made the interesting observation that sucrose-P synthetase does not appear in sugar beet leaves until at least 40% leaf expansion, at which stage they are capable of exporting sucrose.

B. Sucrose Utilization

Most of the sucrose formed in mature leaves is transported to other parts of the plant, principally to growing and storage tissues. In storage organs such as sugar beet roots a reserve pool of sucrose, which can be later mobilized, is formed in the vacuole (Leigh *et al.*, 1979) but in other storage tissues the sucrose is converted to reserve compounds such as starch and fats. Sucrose moved to growing tissues is quickly utilized there. In each of these cases two possible reactions for cleaving the sucrose into its constituent monosaccharides must be considered: It may be hydrolyzed to glucose and fructose by acid invertase [soluble invertase is located in the vacuole (Leigh *et al.*, 1979)] or cleaved by sucrose synthase [which, despite its name, is understood to be involved predominantly in sucrose breakdown rather than synthesis *in vivo* (Turner and Turner, 1975; Gander, 1976)] to yield UDP-glucose and fructose:

$$\text{sucrose} + \text{UDP} \rightleftharpoons \text{UDP-glucose} + \text{fructose}$$

Sucrose synthase is a cytosolic enzyme (Nishimura and Beevers, 1979). First demonstrated in wheat germ by Leloir and Cardini (1953), it is most active in tissues expected to be utilizing sucrose. The products of the reaction can be metabolized to glucose-1-P and fructose-6-P by UDP-glucose pyrophosphorylase and fructokinase, respectively (Turner and Turner, 1975; Turner *et al.*, 1977a), and thereby directed into other metabolic activities of the cytosol. One advantage of utilizing sucrose synthase, rather than invertase, for sucrose cleavage is that at least part of the energy of the sucrose glycosidic bond is conserved (Gander, 1976).

Reserve starch or fat accumulated in seeds is degraded during germination to reform sucrose. Conversion of fat to sucrose is considered in Section III,H. Seeds that store starch degrade it to either glucose-1-P or glucose, depending on whether phosphorylase or amylases plus maltase are used. Glucose is phosphorylated to glucose-6-P by a hexokinase, possibly glucokinase, which has been confirmed in pea seeds (Turner *et al.*, 1977b). The glucose–phosphates may then be used to synthesize sucrose by reactions similar to those in photosynthetic cells (enzymes 3 to 9, Fig. 1). This sucrose is transported to the young shoot where it is rapidly utilized as described above.

C. Other Monosaccharide and Oligosaccharide Interconversions

Since sucrose metabolism is located in the cytosol, it is possible that reactions involving other sugars closely related to sucrose and UDP-glucose also take place in the cytosol. Some possible candidates are shown in Fig. 2, in which *myo*-inositol (1-L-*myo*-inositol) and glucuronate occupy central positions. The pathway from glucose-6-P to glucuronate via *myo*-inositol was recognized by the time Loewus (1971) reviewed the subject. The enzyme

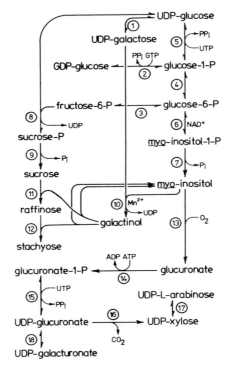

Fig. 2. Some carbohydrate interconversions involving UDP-glucose, sucrose, *myo*-inositol, and related compounds. Based on Gander (1976). Circled numbers refer to enzymes: 1, UDP-galactose 4-epimerase; 2, GDP-glucose pyrophosphorylase; 3, phosphohexose isomerase; 4, phosphoglucomutase; 5, UDP-glucose pyrophosphorylase; 6, glucose-6-P cycloaldolase; 7, *myo*-inositol-1-P phosphatase; 8, sucrose-P synthetase; 9, sucrose-P phosphatase; 10, UDP-galactose:inositol galactosyl-transferase; 11, galactinol:sucrose 6-galactosyltransferase; 12, galactinol:raffinose 6-galactosyltransferase; 13, *myo*-inositol oxygenase; 14, glucuronate kinase; 15, UDP-glucuronate pyrophosphorylase; 16, UDP-glucuronate decarboxylase; 17, UDP-xylose 4-epimerase; 18, UDP-glucuronate 4-epimerase.

catalyzing the cyclization of the carbon chain of glucose-6-P, an NAD^+-dependent cycloaldolase, has been purified from *Acer,* and separated from a phosphatase which hydrolyzes the product, *myo*-inositol-1-P, to free *myo*-inositol (Loewus and Loewus, 1971). An oxygenase-catalyzed cleavage of the *myo*-inositol ring to form glucuronic acid has also been identified in plant extracts. It is noteworthy that young wheat and bean seedlings do not seem to obtain *myo*-inositol from glucose, but rather from seed reserves in the form of the hexaphosphate ester of *myo*-inositol, phytic acid (Matheson and St. Clair, 1971).

Glucuronate is a precursor of the uronate and pentose residues of the cell wall [most probably via the pathway involving UDP-glucuronate pyrophosphorylase, as depicted in Fig. 2 (Roberts, 1971)] and thus, together with UDP-glucose and GDP-glucose, represents much of the starting material for cell wall biosynthesis. Enzyme activities catalyzing the necessary reactions have been identified in various plants (Gander, 1976), and since all were in the soluble portion of plant extracts, their location in the cytosol, at least in part, is not inconceivable.

Myo-inositol also plays an important role in the biosynthesis of raffinose and stachyose since it acts as a carrier of the galactose which is attached to sucrose to form raffinose, and to raffinose to form stachyose (Fig. 2). The three enzymes involved have been identified in plants (Gander, 1976).

A further discussion concerning *myo*-inositol metabolism is found in Chapter 2, Vol. 3 of this series.

The location of the terminal steps in the biosynthesis of plant C-, S- and O-glycosides might also be in the cytosol since it is during these steps that the sugar moiety is incorporated, and UDP-glucose is widely utilized as the sugar donor in these reactions (Gander, 1976).

D. Glycolysis

Some 22 years ago, Axelrod and Beevers (1956) wrote in their review that the Embden-Meyerhof-Parnas pathway of glycolysis "is so well known even outside the discipline of biochemistry that one hesitates to dwell on it in any detail." Since the existence of glycolysis in plant cells is, if anything, more established today than at that time, it might be wise not to ponder the even greater relevance of their statement to the present time. Rather, at the risk of repetition, and for the sake of completeness (glycolysis is traditionally accepted as a metabolic component of the cytosol), the pathway of glycolysis by which glucose (as the free sugar, or derived from starch) is degraded to pyruvate is presented in Fig. 3. Nevertheless, although the reactions comprising glycolysis are almost identical in all living cells, a number of the enzymes catalyzing these reactions vary quite remarkably in their properties, and these differences are emphasized below in an attempt to convey

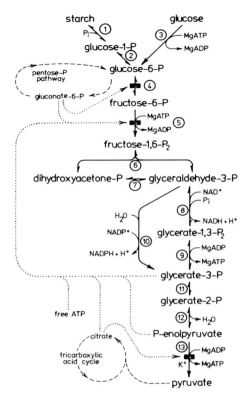

Fig. 3. Glycolysis in plants. The sequence begins with glucose or starch and ends with pyruvate. Dashed lines indicate continuations to other pathways; dotted arrows point to regulation by enzyme inhibition. Enzymes are indicated by circled numbers: 1, starch phosphorylase; 2, phosphoglucomutase; 3, hexokinase; 4, phosphohexose isomerase; 5, phosphofructokinase; 6, fructose-1,6-P_2 aldolase; 7, triose-P isomerase; 8, glyceraldehyde-3-P dehydrogenase (reversible); 9, glycerate-3-P kinase; 10, nonreversible glyceraldehyde-3-P dehydrogenase; 11, phosphoglyceromutase; 12, enolase; 13, pyruvate kinase.

some measure of uniqueness to glycolysis in plants. Only three of the glycolytic enzymes (phosphorylase, phosphofructokinase, and aldolase) had been purified to any extent at the time Stumpf (1952) reviewed their occurrence in plants, but almost all were obtained in the following 7 years (Gibbs, 1959).

Phosphorylase, the first enzyme of the sequence, was one of the first identified. There is now good evidence for more than one form of the enzyme in plants (Richardson and Matheson, 1977; Steup and Latzko, 1979), but the complex control mechanisms involving phosphorylation of the protein and activation by AMP, typical of muscle glycogen phosphorylase, have not been reported for the plant starch phosphorylase. The second glycolytic enzyme, phosphoglucomutase, is a relatively small protein (Takamiya and Fukui, 1978) which has not been extensively studied. Phosphoglucomutase is not required for the glycolysis of free glucose since hexokinase (reaction 3, Fig. 3) forms glucose-6-P directly. Plant tissues contain both soluble and particulate hexokinases, although the former tend to be rather specific for a particular hexose and are better designated as glucokinase, fructokinase, etc. (Medina and Sols, 1956; Turner *et al.*, 1977a,b).

Glucose-6-P produced by hexokinase or by phosphoglucomutase is next isomerized to fructose-6-P by phosphohexose isomerase which has been crystallized from pea seeds by Takeda *et al.* (1967); these authors showed that gluconate-6-P was a strong competitive inhibitor of the enzyme activity. Despite reports to the contrary, only one form of phosphohexose isomerase appears to be present in the cytosol (Phillips *et al.*, 1975). However, it is worth noting at this point that chloroplasts contain the majority (if not all) of the glycolytic enzymes in forms sufficiently different than those of the cytosol that separation by isoelectric focusing is often possible (Anderson and Advani, 1970).

The phosphorylation of fructose-6-P to fructose-1,6-P_2 has received special attention since the enzyme involved, phosphofructokinase, almost certainly plays a major role in regulating the overall flux through the glycolytic sequence (Latzko and Kotzé, 1965). The most notable control feature is the strong inhibition of enzyme activity by micromolar concentrations of glycerate-3-P, glycerate-2-P, and P-enolpyruvate; millimolar levels of gluconate-6-P and citrate also inhibit the enzyme (Kelly and Turner, 1970; Fig. 3). Free ATP (in contrast to $MgATP^{2-}$, which is the true substrate) also strongly reduces enzyme activity. Several of these effects are collected together in Fig. 3 to illustrate how accumulation of 3C-acids in glycolysis, citrate in the tricarboxylic acid cycle, or gluconate-6-P during operation of the oxidative pentose-P pathway may lead to feedback inhibition of phosphofructokinase.

The fructose-1,6-P_2 produced by phosphofructokinase is cleaved by fructose-1,6-P_2 aldolase into two triose-phosphates: dihydroxyacetone-P and glyceraldehyde-3-P. This aldolase is classified as being type I in that it forms a Schiff base-intermediate with dihydroxyacetone-P. Type II aldolases, which are metalloproteins and do not form this intermediate, are widespread among algae (Willard and Gibbs, 1968). The two triose-phosphates are interconverted by triose-P isomerase, a particularly active enzyme that has been demonstrated in a number of plant tissues, so that both halves of the cleaved fructose-1,6-P_2 may be ultimately oxidized by glyceraldehyde-3-P dehydrogenase. Plants are unique in that they possess three forms of this enzyme (Axelrod and Beevers, 1956). One form found in all plant cells catalyzes an NAD$^+$-specific, reversible oxidation (phosphorylating) of glyceraldehyde-

3-P to glycerate-1,3-P_2 and is quite similar to the enzyme from nonplant sources; highly purified preparations from the pea plant have been obtained (McGowan and Gibbs, 1974). It is this form which is generally believed to participate in cytosolic glycolysis. A second form catalyzes the same reaction, but with either $NADP^+$ or NAD^+; this enzyme is located in the chloroplast where it participates in photosynthesis. The third form oxidizes glyceraldehyde-3-P irreversibly to glycerate-3-P in a reaction which is $NADP^+$-specific. This nonreversible glyceraldehyde-3-P dehydrogenase is located in the cytosol; an odd characteristic is that it has been so far reported only in photosynthetic and gluconeogenic tissues of higher plants (Kelly and Gibbs, 1973a). Although it is quite conceivable that the enzyme could participate in the glycolysis of these tissues (Fig. 3), its irreversibility and high affinity for $NADP^+$ suggested an alternative role: that of participating in a triose-P/glycerate-3-P shuttle for transferring photosynthetically generated reducing power to the cytosol in the form of NADPH (Kelly and Gibbs, 1973b; Fig. 4). NADPH itself cannot cross the chloroplast envelope (Robinson and Stocking, 1968). An analagous shuttle that utilizes the cytosolic NAD^+-specific glyceraldehyde-3-P dehydrogenase was demonstrated earlier by Stocking and Larson (1969). This shuttle indirectly transfers reducing power (but as NADH rather than NADPH) and ATP from chloroplasts to the cytosol (Fig. 4).

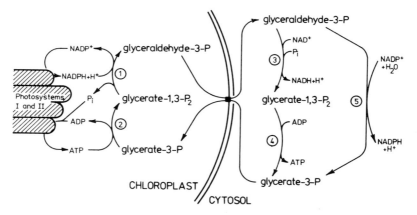

Fig. 4. Mechanisms for the indirect transfer of photosynthetically generated reducing power and ATP from chloroplasts to the cytosol. The black square represents a phosphate translocator (see Fig. 1). Circled numbers refer to enzymes: 1, chloroplast glyceraldehyde-3-P dehydrogenase; 2, chloroplast glycerate-3-P kinase; 3, NAD^+-linked glyceraldehyde-3-P dehydrogenase (phosphorylating); 4, cytosolic glycerate-3-P kinase; 5, nonreversible glyceraldehyde-3-P dehydrogenase. The inner membrane of the chloroplast envelope is largely impermeable to NADP(H) (Robinson and Stocking, 1968). Based on Stocking and Larson (1969), and Kelly and Gibbs (1973b).

Glycerate-1,3-P_2 produced by the conventional glyceraldehyde-3-P dehydrogenase is converted to glycerate-3-P by a kinase which has been little studied since the original early investigations (Axelrod and Beevers, 1956). The conversion of glycerate-3-P to P-enolpyruvate via glycerate-2-P involves the enzymes phosphoglyceromutase and enolase; potato tuber enolase has been purified, and like the enzyme from other sources, was found sensitive to inhibition by fluoride ions (Boser, 1959). The final enzyme of glycolysis, pyruvate kinase, is also thought to have a regulatory role. The activity of the enzyme from cotton seed, pea seed, and carrot root was inhibited by millimolar concentrations of ATP and citrate; unlike yeast and animal pyruvate kinases, these plant enzymes were not activated by fructose-1,6-P_2, although the *Euglena gracilis* enzyme was notably influenced by this sugar bisphosphate. Plant pyruvate kinases require either Mg^{2+} or Mn^{2+} and, in addition, a monovalent cation for activity; K^+ can satisfy the latter requirement (Tomlinson and Turner, 1973; Turner and Turner, 1975).

Stumpf (1952) noted six possible fates of pyruvate in plant tissues. Two of these (conversion to lactate, or to acetaldehyde and thence ethanol) are historically linked to glycolysis and can achieve some significance when conditions become relatively anaerobic, such as during seed germination. Lactate dehydrogenase has been well documented in plants (Gibbs, 1959) and was purified from potato tubers by Davies and Davies (1972) who demonstrated strong inhibition of the enzyme activity by ATP. Accumulation of lactate could lower the cytosol pH, a condition which Kenworthy and Davies (1976) have shown favorable for pyruvate decarboxylase which, together with alcohol dehydrogenase, leads to ethanol production:

$$\text{pyruvate} \rightleftharpoons \text{acetaldehyde} + CO_2$$
$$\text{acetaldehyde} + NADH + H^+ \rightleftharpoons \text{ethanol} + NAD^+$$

Alcohol dehydrogenase has been purified from maize seedlings by Leblová and Ehlichová (1972), who reported greater enzyme levels in plants held under anaerobiosis.

E. Oxidative Pentose Phosphate Pathway

An alternative to glycolysis for the breakdown of carbohydrates in the cytosol is the oxidative pentose-P pathway (Fig. 5). In contrast to the discovery of glycolysis, which principally involved studies with yeast and muscle extracts, elucidation of the oxidative pentose-P pathway in the early 1950s was an undertaking in which investigations with plant tissues made major contributions. The pioneering efforts of B. L. Horecker, E. Racker, B. Axelrod, H. Beevers, and M. Gibbs, to name a few, were concisely summarized by three of these workers shortly afterward (Axelrod and Beevers, 1956; Gibbs, 1959).

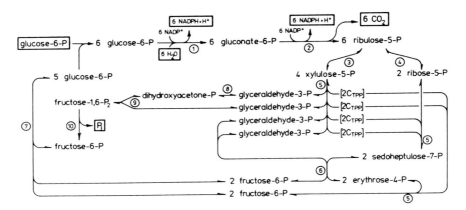

Fig. 5. The oxidative pentose-P pathway. The oxidation of one molecule of glucose-6-P to six CO_2 and one P_i is shown. Circled numbers indicate the participating enzymes: 1, glucose-6-P dehydrogenase; 2, gluconate-6-P dehydrogenase; 3, phosphopentose epimerase; 4, phosphopentose isomerase; 5, transketolase; 6, transaldolase; 7, phosphohexose isomerase; 8, triose-P isomerase; 9, fructose-1,6-P_2 aldolase; 10, fructose bis phosphatase. Abbreviation: $2C_{TPP}$, glycolaldehyde-thiamin-P_2 complex (attached to transketolase).

As depicted in Fig. 5, the oxidative pentose-P pathway is capable of completely degrading glucose-6-P according to the equation:

$$\text{glucose-6-P} + 6H_2O + 12\ NADP^+ \rightarrow 6\ CO_2 + 12\ NADPH + 12\ H^+ + P_i$$

Reactions 8, 9, and 10 in Fig. 5 are not normally included as part of the pathway in modern textbooks, although they were incorporated in the scheme presented by Axelrod and Beevers (1956), and as pointed out by these authors, are necessary if the cycle is to be completed; the three enzymes involved are believed to be present in the cytosol (Sections III,A and III,D), together with the other seven shown in Fig. 5. Nevertheless, it should be pointed out that there is no evidence as yet for this or any other cytosolic "pathway" to be operating independently of any other; thus it is equally realistic to consider that two of the four glyceraldehyde-3-phosphates released by transketolase (reaction 5, Fig. 5) may be fed into glycolysis and their carbon eventually released as CO_2 during mitochondrial respiration. In fact, early experiments of Gibbs and Beevers (1955) provided evidence for the simultaneous operation of both glycolysis and the oxidative pentose-P pathway, presumably both in the cytosol in a variety of plant tissues; only in some cases, notably meristematic tissues, was the almost exclusive operation of one pathway (glycolysis) detected. A second consideration is that the oxidative pentose-P pathway is not only a means of transferring chemical energy from sugars to NADPH, which (in contrast to NADH) is widely used in biosynthetic reactions (e.g., fatty acid synthesis; Section III,G), but it is also a principal source of ribose-5-P which is attached to purine and pyrimidine bases to form the respective nucleosides, and of erythrose-4-P which is utilized in the biosynthesis of tryptophan, tyrosine, phenylalanine (Section IV,B) and many secondary aromatic compounds.

There is evidence that all of the enzymes of the oxidative pentose-P pathway are in the cytosol (Simcox et al., 1977; Emes and Fowler, 1979). However, they have been much less investigated than those of glycolysis. Highly purified preparations of phosphopentose isomerase and transketolase were obtained from green leaves during the 1950s. Partially purified preparations of six of the enzymes (1, 2, and 4 to 7, Fig. 5) were recently obtained from *Phaseolus mungo* by Ashihara and Komamine (1974) who reported inhibition of both glucose-6-P dehydrogenase and gluconate-6-P dehydrogenase by NADPH, and of phosphohexose isomerase by erythrose-4-P. However, an oxidative pentose-P pathway is understood to operate also in chloroplasts, and in fact six enzymes (3 to 5 and 8 to 10, Fig. 5) are common to both this pathway and the reductive pentose-P cycle (Calvin cycle) of photosynthetic CO_2 fixation. Thus the number of preparations so far obtained of enzymes of the oxidative pentose-P pathway containing only cytosolic enzyme is probably quite small.

F. Phosphoenolpyruvate Carboxylase, Malate Dehydrogenase, and Malic Enzyme

One enzyme which has had a major influence on modern plant biochemistry is phosphoenolpyruvate carboxylase. This enzyme, which incorporates CO_2 into P-enolpyruvate to produce oxaloacetate (Fig. 6), was discovered in plants by R. S. Bandurski in the early 1950s (see Gibbs, 1959). Walker (1966)

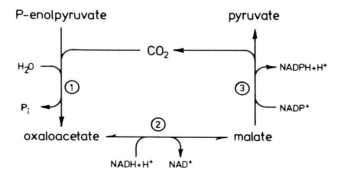

Fig. 6. Conversion of P-enolpyruvate to pyruvate by a sequence of reactions not requiring pyruvate kinase (enzyme 13, Fig. 3), and able to generate NADPH. Enzymes are designated by circled numbers: 1, phosphoenolpyruvate carboxylase; 2, cytosolic malate dehydrogenase; 3, malic enzyme.

summed up its metabolic role as replenishment of tricarboxylic acid cycle intermediates; a close association with photosynthesis was not particularly envisaged. The status of the enzyme seemed settled. But in the same year, Hatch and Slack (1966) reported the first of a series of investigations (see Hatch and Osmond, 1976) which established that phosphoenolpyruvate carboxylase is intimately involved in the photosynthetic assimilation of CO_2 in the leaves of tropical grasses and other plants which today are designated as C_4 plants [since the product of phosphoenolpyruvate carboxylase is a 4C compound; other plants are called C_3 plants after the 3C compound (glycerate-3-P) which is the product of carboxylation in the reductive pentose-P cycle]. In the leaves of these C_4 plants, phosphoenolpyruvate carboxylase is restricted to the cytosol of mesophyll cells (Hatch and Osmond, 1976). Some writers tend to give phosphoenolpyruvate carboxylase a status similar to that of the carboxylating enzyme in the pentose-P cycle (ribulose-1,5-P_2 carboxylase) when considering photosynthetic CO_2 fixation by C_4 plants, but as pointed out by Walker (1974) and emphasized by Hatch (1976) and by Kelly *et al.* (1976), only the latter is associated with a series of reactions permitting a net incorporation of CO_2 into sugars, thus making it possible for the plant to grow.

Nevertheless, phosphoenolpyruvate carboxylase is a particularly valuable enzyme in all groups of plants. In C_4 plants it acts as a "photosynthetic CO_2 antenna," collecting CO_2 for the reductive pentose-P cycle in bundle sheath cells, and most probably also efficiently capturing respiratory CO_2, including that lost by photorespiration which is restricted to the bundle sheath cells in these plants. In C_3 plants it is similarly active in capturing respired CO_2 (Hedley *et al.*, 1975), and in plants with crassulacean acid metabolism (cacti and other succulents are the best examples) the enzyme is indispensible for survival in arid environments; in conjunction with malate dehydrogenase it fixes sufficient atmospheric CO_2 into malate during the night so that photosynthesis the next day can proceed with endogenous CO_2 (that released by malic enzyme from the accumulated malate) (Walker, 1957). Stomata may therefore remain closed and loss of precious water by transpiration prevented during the hot days.

Several forms of phosphoenolpyruvate carboxylase exist in plants. The enzyme from the leaves of C_3, C_4, and crassulacean acid metabolism plants, and from a nonphotosynthetic plant tissue (roots), differed markedly with respect to affinity for P-enolpyruvate and calculated *in vivo* level of activity; leaves of C_4 plants contained over 15 times as much activity as the leaves of C_3 plants (Ting and Osmond, 1973). In fact, calculations by Uedan and Sugiyama (1976), who recently purified maize leaf phosphoenolpyruvate carboxylase, indicate that the enzyme may constitute over 10% of the cytosol protein in the cells of these leaves.

Although in C_4 plants the product of cytosolic P-enolpyruvate carboxylation (oxaloacetate) is reduced to malate in mesophyll cell chloroplasts, and this malate (or a 4C acid derived from it) later decarboxylated in bundle sheath chloroplasts or mitochondria (Hatch and Osmond, 1976), there is good evidence that these three reactions (Fig. 6) may all occur in the cytosol in some plant tissues. Danner and Ting (1967) reported all three enzymes to be present in the nonparticulate part of maize roots, and since then soluble malate dehydrogenase has been found in the green stem of the cactus *Opuntia,* spinach leaves, and maize seeds; in the latter case crystallized enzyme was obtained (Curry and Ting, 1973). These authors have proposed that the enzymes are involved in nonautotrophic CO_2 fixation, a process that may have possible benefit for NADPH generation (Peak *et al.,* 1973), especially in very young tissue with low oxidative pentose-P pathway activity (Fowler, 1974). However, these proposals are at present most relevant to roots, since it is less clear whether tissues other than roots contain appreciable levels of malic enzyme in the cytosol. Indeed, an alternative role for malate dehydrogenase in the cytosol of leaves has been proposed: this enzyme, in conjunction with a chloroplast malate dehydrogenase, may operate in a shuttle of malate and oxaloacetate between chloroplast and cytosol for the transfer of photosynthetically generated reducing power (in the form of NADH) to the cytosol (Heber, 1974).

G. Fatty Acid Synthesis

Three of the most predominant fatty acids in plants are palmitic, oleic, and linolenic acids; the latter two are derived from palmitic acid via stearoyl ACP (Stumpf, 1976 and Chapter 7, Vol. 4 of this series). The biosynthesis of one molecule of palmitic acid, which is a 16C acid, from eight molecules of acetate is outlined in Fig. 7. Centrally involved in this biosynthesis is a small protein, termed the acyl carrier protein (ACP, Fig. 7), which carries the acyl moiety during the reactions that lead to chain elongation; these reactions are catalyzed by enzymes collectively termed the fatty acid synthetase. The location of these enzymes in the cytosol of mammalian cells is well established; current research shows that fatty acid synthesis to stearoyl ACP is localized in chloroplasts in leaf cells (Ohlrogge *et al.,* 1979) and in proplastids in nonphotosynthetic tissue (Weaire and Kekwick, 1975; Simcox *et al.,* 1977).

In most tissues synthesizing fatty acids, the necessary acetate is derived from pyruvate supplied by glycolysis. The pyruvate enters mitochondria where it is converted to acetyl-coenzyme A (acetyl-CoA) which then condenses with oxaloacetate (obtained by dehydrogenation of malate) to form citrate. This citrate moves out of the mitochondria and is cleaved [possibly in the cytosol in some tissues (Nelson and Rinne, 1977)] by citrate cleavage enzyme to form oxaloacetate (which is reduced to malate and then returned

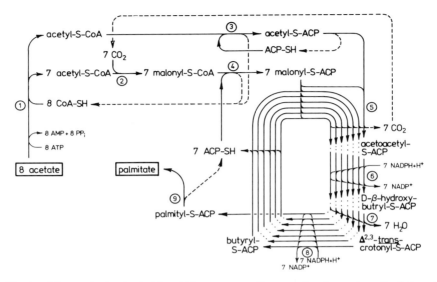

Fig. 7. The synthesis of palmitic acid from acetate. Enzymes 3 to 8 constitute the fatty acid synthetase complex. The intermediates formed by enzymes 5 to 8 are shown only for the first of the seven reactions which they catalyze during chain elongation. Dashed lines indicate return of components for which there is no net consumption during the fatty acid synthesis. Circled numbers refer to the enzymes: 1, acetate thiokinase; 2, acetyl-CoA carboxylase; 3, acetyl transacylase; 4, malonyl transacylase; 5, β-ketoacyl-ACP synthetase; 6, β-ketoacyl-ACP reductase; 7, β-hydroxyacyl-ACP dehydrase; 8, enoyl-ACP reductase; 9, palmityl-ACP deacylase. Abbreviation: ACP, acyl carrier protein.

to the mitochondria) and acetyl-CoA. Hence malate in effect carries acetate carbons out of the mitochondria and this circumvents the impermeability of the mitochondrial membrane to acetyl-CoA. See also Chapter 7, Volume 4 of this series for a further discussion of this aspect.

Eight molecules of acetyl-CoA are required to synthesize one molecule of palmitic acid. The first acetyl-CoA is attached to an acyl carrier protein and used as a primer for the eventual attachment of the seven other acetyl units in a system whereby each of these seven are first carboxylated to malonyl-coenzyme A, then connected to an acyl carrier protein and, finally, while attached to the elongating chain, lose the CO_2 originally incorporated in the carboxylating step (Fig. 7). The existence of this pathway in plants was demonstrated largely from investigations by P. K. Stumpf and co-workers with preparations from avocado mesocarp, safflower seeds, and potato tuber (see Stumpf, 1976).

Evidence indicates that there are two types of acetyl-CoA carboxylase (reaction 2, Fig. 7) in plant cells; see Chapter 7, Vol. 4, this series for a detailed discussion of this enzyme. Acetyl-CoA carboxylase has a principal regulatory function during fatty acid synthesis in mammalian systems, but indications that this might also be the case in plants are less convincing. In addition, Stumpf (1976) has pointed out that the plant may use an alternative malonyl-coenzyme A synthesizing system, i.e., decarboxylation of oxaloacetate to malonate from which malonyl-coenzyme A may be generated by malonate thiokinase.

Most of the other qualitatively important fatty acids are synthesized from palmitic acid by reactions not known to be located in the cytosol; these fatty acids are finally assembled into lipids by esterification with L-α-glycerol-P in the microsomal fraction of the cell (Stumpf, 1976). The origin of the L-α-glycerol-P in plant cells has been something of an enigma since numerous investigators have sought to demonstrate that the cytosolic enzyme L-α-glycerol-P dehydrogenase, which readily reduces dihydroxyacetone-P (available from glycolysis) to L-α-glycerol-P, is present in plants, but none were successful. Nevertheless, very recent evidence indicates that this enzyme does, after all, occur in plant leaves (Santora *et al.*, 1979). It is also worth noting that an alternative pathway to L-α-glycerol-P, i.e., glyceraldehyde-3-P $\rightarrow$ glyceraldehyde $\rightarrow$ glycerol $\rightarrow$ L-α-glycerol-P, has been proposed from studies with pea leaves (Hippmann and Heinz, 1976).

H. Gluconeogenesis

The synthesis of sugar from carboxylic acids is termed gluconeogenesis, and involves two phases: the formation of P-enolpyruvate from carboxylic acids, and the synthesis of hexose-P from P-enolpyruvate. The first phase usually involves mitochondrial enzymes which generate malate; this malate moves into the cytosol where it is oxidized to oxaloacetate by the cytosolic malate dehydrogenase (Section III,F) and then simultaneously decarboxylated and phosphorylated to produce P-enolpyruvate by the enzyme phosphoenolpyruvate carboxykinase:

$$\text{oxaloacetate} + \text{ATP} \leftrightarrows \text{P-enolpyruvate} + CO_2 + \text{ADP}.$$

The second phase involves a reversal of the glycolytic sequence (Fig. 3) and utilizes all the glycolytic enzymes between P-enolpyruvate and glucose-6-P, except for phosphofructokinase. This enzyme catalyzes a physiologically irreversible reaction and is replaced in gluconeogenesis by another, fructose bisphosphatase, which specifically hydrolyzes the phosphate from the C-1 of fructose-1,6-P_2:

$$\text{fructose-1,6-}P_2 \rightarrow \text{fructose-6-P} + \text{Pi}.$$

Gluconeogenesis is especially active during the germination of fat-storing seeds such as those of castor bean, peanut, and pumpkin. In these seeds, fats are hydrolyzed to fatty acids and glycerol; the fatty acids are oxidized to succinate by β-oxidation and the glyoxylate cycle in microbodies (see Chapter 9, this volume) while the glycerol is converted to dihydroxyacetone-P by a cytosolic glycerol kinase and mitochondrial L-α-glycerol-P oxidoreductase (Huang, 1975). The succinate produced in the microbodies moves to the mitochondria where it is converted by enzymes of the tricarboxylic acid cycle to malate; this malate is the substrate for gluconeogenesis as described above. The hexose-P produced is finally converted to sucrose (Section III,B) for transport to the growing shoot.

Enzyme localization studies with marrow (*Cucurbita pepo*) cotyledons (ap Rees *et al.*, 1975) and germinating castor bean (*Ricinus communis*) endosperm (Nishimura and Beevers, 1979) have confirmed that gluconeogenesis from oxaloacetate is located in the cytosol. Two key enzymes of the sequence, i.e., phosphoenolpyruvate carboxykinase and sucrose-P synthetase, are found only in the cytosol (Leegood and ap Rees, 1978; Nishimura and Beevers, 1979).

Since the latter phase of gluconeogenesis is located in the cytosol and is, in essence, a reversal of glycolysis, and since glycolysis itself is located in the cytosol, some means of metabolic regulation must be necessary: fructose bisphosphatase and phosphofructokinase together catalyze a "futile cycle" which achieves nothing other than ATP hydrolysis. In mammalian systems this cycle is prevented by the opposing effects of allosteric modifiers on the two enzymes, e.g., inhibition of fructose bisphosphatase by AMP which, however, stimulates phosphofructokinase. A similar system may operate in plants: the fructose bisphosphatase from castor bean endosperm is susceptible to inhibition by AMP (Youle and Huang, 1976), but it is not known what affect AMP has on the phosphofructokinase of this tissue. An alternative possibility is that, in gluconeogenic tissues, the activity of phosphofructokinase alone is controlled [perhaps inhibited by increased levels of ATP developed during gluconeogenesis (ap Rees *et al.*, 1975)] thus allowing gluconeogenesis to proceed as rapidly as the activity of fructose bisphosphatase will permit (Thomas and ap Rees, 1972).

The two characteristic enzymes of gluconeogenesis, i.e., phosphoenolpyruvate carboxykinase and fructose bisphosphatase, have been isolated and studied from gluconeogenic tissues of higher plants. The castor bean endosperm and spinach leaf cytosol fructose bisphosphatases are similar to each other (high affinity for fructose-1,6-P_2, and susceptibility to inhibition by AMP) but quite different from chloroplast fructose bisphosphatase which participates in photosynthesis (Youle and Huang, 1976; Zimmermann *et al.*, 1978). The properties of phosphoenolpyruvate carboxykinase from marrow cotyledons were similar to those of the enzyme from the leaves of C_4 plants (Leegood and ap Rees, 1978) and consistent with its function of generating P-enolpyruvate.

IV. NITROGEN METABOLISM

A. Nitrate Reduction

Nitrate absorbed by plant roots is either reduced by a cytosolic nitrate reductase to nitrite (which is further reduced to ammonium by a proplastid-located nitrite reductase) (Dalling *et al.*, 1972a; Miflin, 1974), or transported unaltered to the shoot (Wallace and Pate, 1967). Ammonium supplied to roots (by absorption from the soil, reduction of nitrate and, in the case of legumes, fixation of atmospheric nitrogen by bacteroids in root nodules) is incorporated into amino acids, some of which are transported to the shoot along with the nitrate (Wallace and Pate, 1967). In the shoot, nitrate is reduced by nitrate reductase in the cytosol (Dalling *et al.*, 1972b; Miflin, 1974), and the resultant nitrite moves into chloroplasts where it is reduced to ammonium which is then used in amino acid biosynthesis (Fig. 8).

Nitrate reductase is a complex enzyme which has been extensively investigated (Hewitt *et al.*, 1976). The plant leaf enzyme catalyzes a virtually irreversible reaction (reaction 1, Fig. 8). Most higher plant nitrate reductases use NADH as the effective or near-specific electron donor. This contrasts with fungal nitrate reductases which are more or less specific for NADPH, and the nitrate reductase of the blue–green alga *Anabaena cylindrica* which is particle-bound and accepts reduced ferredoxin. Purified nitrate reductases have been found to contain flavin adenine dinucleotide, cytochrome b_{557}, and molybdenum; these three components are understood to mediate electron transfer between NAD(P)H and nitrate (Hewitt *et al.*, 1976).

The regulation of nitrate reductase activity involves induction and repression of protein synthesis, reversible inactivation, and possibly allosteric regulation (Hewitt *et al.*, 1976). Many investigations have demonstrated induction of enzyme activity in higher plants, both by nitrate and a light-mediated process. The nitrate-induced increase is understood to reflect the synthesis of new enzyme, but this explanation probably does not apply to the effects of light. Rather, recent investigations emphasize that regulation by inhibitors and activators comprises the mechanism whereby enzyme activity responds to illumination. A protein with protease activity and able to inactivate nitrate reductase was first detected in maize roots by Wallace (1973). Jolly and Tolbert (1978) have now purified a second nitrate reductase inhibitor which occurs in an active form (i.e., capable of inhibiting the reductase) in the dark, and in an inactive form in the light. This inhibitor is also a protein but does not seem to act as a protease since its effect is immediate and reversible. Another characteristic of the dark inactivation of the enzyme *in vivo* is that oxygen is required; Sawhney *et al.* (1978) found that nitrite accumulated in leaves under anaerobic conditions in darkness. However, oxygen at a level

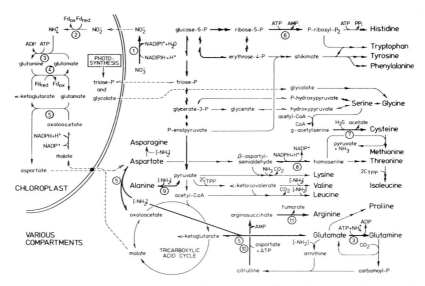

Fig. 8. The biosynthesis of amino acids in plant leaves. Reactions reported to occur in the cytosol are emphasized by thicker lines. Dashed lines represent movement of metabolites between chloroplast and cytosol; a dicarboxylate translocator (Heldt, 1976) is indicated by the black circle on the chloroplast inner membrane. Double arrows indicate that two or more reactions are involved. Not all of the transamination reactions are indicated. Enzymes are indicated by circled numbers: 1, nitrate reductase; 2, nitrite reductase; 3, glutamine synthetase; 4, glutamine:α-ketoglutarate aminotransferase; 5, aspartate transaminase; 6, phosphoribosyl pyrophosphate synthetase; 7, o-acetylserine sulfhydrylase; 8, homoserine dehydrogenase; 9, alanine transaminase; 10, arginosuccinate synthetase; 11, arginosuccinate lyase. Abbreviations: Fd_{ox} and Fd_{red}, ferredoxin (oxidized and reduced, respectively); $2C_{TPP}$, acetaldehyde-thiamin-P_2 complex.

of less than 1% of that in air prevented this nitrite formation. It was concluded that operation of the mitochondrial electron transport chain in darkness somehow contributed to the lack of nitrate reductase activity. Canvin and Woo (1979) have pointed out that the cytosolic supply of NADH for the enzyme may be a critical factor in this regard in that it is more likely to be limiting under dark aerobic conditions.

B. Amino Acid Biosynthesis

The nitrite produced by nitrate reductase is reduced to ammonium and incorporated into glutamate by the plastid enzymes nitrite reductase, glutamine synthetase, and glutamine:α-ketoglutarate aminotransferase (Lea and Miflin, 1974; Fig. 8). Glutamate is utilized by chloroplast transaminases for the formation of other amino acids in the chloroplast, or for the synthesis of certain amino acids such as aspartate which can cross the chloroplast envelope (Heldt, 1976) and initiate amino acid biosynthesis in the cytosol (Fig. 8). It is possible that in nonphotosynthetic tissues a similar system operates between plastids and the cytosol. At least 85 different enzymes participate in the synthesis of the 20 common protein amino acids in plants (Bryan, 1976), but few of these have been characterized to any extent from higher plants, and even fewer have been investigated with respect to their intracellular location. In this section only several enzymes known to exist in the cytosol are mentioned; for a complete list of all enzymes of amino acid biosynthesis the article by Bryan (1976) and the review by Miflin and Lea (1977) should be consulted.

Aspartate which moves out of plastids may donate its amino group to α-ketoglutarate to form glutamate in a reaction catalyzed by aspartate transaminase (reaction 5, Fig. 8). In some plant tissues over 90% of the activity of this enzyme was found in the cytosol (Wightman and Forest, 1978). Glutamate, the product of the reaction, is thought to be the principal amino-donor in the synthesis of practically all the other amino acids. A considerable amount of this synthesis also probably occurs in the cytosol since, in plant tissues, much of the total transaminase activity is found in the cytosol (Wightman and Forest, 1978).

In leaves at least some of the total o-acetylserine sulfhydrylase (Fankhauser et al., 1976; Ng and Anderson, 1978) and over 95% of the total phosphoribosyl pyrophosphate synthetase (Ashihara, 1977) are cytosolic. Two enzymes involved in arginine biosynthesis (enzymes 10 and 11, Fig. 8) were reported to be in the cytosol, but the preceding two (converting glutamine to citrulline, Fig. 8) were in plastids of soybean cells (Shargool et al., 1978). There is some evidence that portions of the cellular glutamine synthetase and homoserine dehydrogenase are in the cytosol (Miflin and Lea, 1977). On the other hand, certain enzymes such as acetolactate synthetase (Miflin, 1974), glutamine:α-ketoglutarate aminotransferase (Miflin and Lea, 1977) and those interconverting glycerate and hydroxypyruvate (Tolbert et al., 1970), and glycine and serine (Kisaki et al., 1971) appear to be largely, if not entirely, absent from the cytosol.

Although the intracellular distribution of the majority of the enzymes involved in the synthesis of amino acids is still unclear, it is rather well established that the amino acids, once produced, can be used for protein synthesis in the cytosol since the appropriate species of tRNA and aminoacyl-tRNA synthetases are present there (Lea and Norris, 1972).

C. Glutathione

The tripeptide glutathione (γ-glutamylcysteinylglycine) is widely distributed in plant tissues and is thought to have a number of stabilizing functions, including the prevention of oxidation of enzyme sulfhydryl groups (Jocelyn,

1972), the arrest of lipid peroxidation in membranes (Flohé and Menzel, 1971), and the maintenance of ascorbic acid in the reduced form (Foyer and Halliwell, 1976). Glutathione itself is maintained in the reduced form by an NADPH-dependent glutathione reductase which has been known to be present in both the photosynthetic and nonphotosynthetic parts of plants for over 25 years (Anderson *et al.*, 1952). However, the presence of this enzyme in the cytosol has not yet been unequivocally confirmed, although dehydroascorbate reductase activity (which utilizes reduced glutathione as the reductant) appears to be in the cytosol (Foyer and Halliwell, 1976), and there are certainly sulfhydryl-containing enzymes in the cytosol.

D. Purines and Pyrimidines

A number of enzymes which participate in the biosynthesis of purines and pyrimidines have been identified in higher plants, but detailed investigations of the intracellular locations are seldom undertaken. Exceptions are phosphoribosyl pyrophosphate synthetase, orotate P-ribosyltransferase, and orotidine-P decarboxylase, which Ashihara (1977, 1978) has demonstrated to be predominantly in the cytosol.

V. PHOSPHATE METABOLISM

Cellular phosphate exists in a metabolic pool (P_i, DNA and RNA, P-lipid, sugar-P, nucleoside-P) and in a storage pool [P_i in the vacuole, plus either phytate in the aleurone grains (higher plants) or polyphosphate as volutin granules in the cytosol (lower plants)] (Bieleski, 1973). Some of the metabolic routes followed by phosphate in the form of P_i, sugar-P, and nucleoside-P in the cytosol have been outlined in other sections.

Although P_i occurs in both the metabolic and storage pools, most of it is normally in the latter (Bieleski, 1973). In the leaves of spinach and *Elodea,* roughly one-sixth of the cellular P_i was present in the cytosol at a concentration of about 15 mM (Ullrich *et al.*, 1965). Experiments with *Spirodela* demonstrated that plants respond to phosphate deficiency by mobilizing their vacuolar P_i reserve and moving it into the cytosol where the P_i level is apparently maintained as constant as possible (Bieleski, 1973). Several advantages of buffering the cytosolic P_i concentration have been suggested by Bieleski (1973), and to these may be added the requirement (in chloroplast-containing cells) for exchanging just so much cytosolic P_i with chloroplast triose-P (Fig. 1) as is necessary to export from the chloroplast that proportion of its triose-P which represents net CO_2 fixation; excessive cytosolic P_i could conceivably exchange with so much chloroplast triose-P that no triose-P would be left to regenerate the CO_2-acceptor (ribulose-1,5-P_2) and photosynthesis would stop (Walker and Herold, 1977).

VI. CONCLUDING REMARKS

Modern plant biochemistry has tended to neglect the cytosol. Contained in an odd-shaped compartment, and surrounded by many membranes (plasmalemma, plastid and mitochondrial membranes, and others), the cytosol has been almost invariably sacrificed during efforts to isolate the contents of other compartments, usually in the form of intact organelles. The isolation of an intact cytosol has been seldom contemplated. This is unfortunate because of the wealth of metabolic activity which it is thought to support, and because it is not inconceivable that the isolation of relatively pure cytosol can be regularly achieved. The first step in this direction was made by Cocking (1960) who used enzymes to digest the cell walls of tomato root tips in order to obtain intact protoplasts. Theoretically, gentle breakage of these protoplasts (e.g., passage through a fine sieve) followed by centrifugations at appropriate speeds should produce a supernatant of "isolated cytosol," free of nuclei, organelles, and other cytoplasmic inclusions, and contaminated only by the soluble contents of the endoplasmic reticulum. Such an approach has been already elaborated by Edwards and Huber (1978) to a considerable degree of success; in those experiments emphasis was placed on the acquisition of chloroplasts, over 95% of which were intact, from tough grass leaves. It is likely that the emphasis change required to isolate cytosol will be greater than any necessary procedural change. The outlook for the cytosol in future plant biochemistry appears bright.

REFERENCES

Anderson, D. G., Stafford, H. A., Conn, E. E., and Vennesland, B. (1952). *Plant Physiol.* **27,** 675–684.

Anderson, L. E., and Advani, V. R. (1970). *Plant Physiol.* **45,** 583–585.

ap Rees, T., Thomas, S. M., Fuller, W. A., and Chapman, B. (1975). *Biochim. Biophys. Acta* **385,** 145–156.

Ashihara, H. (1977). *Z. Pflanzenphysiol.* **83,** 379–392.

Ashihara, H. (1978). *Z. Pflanzenphysiol.* **87,** 225–241.

Ashihara, H., and Komamine, A. (1974). *Z. Pflanzenphysiol.* **74,** 130–142.

Atkinson, D. E. (1969). *Curr. Top. Cell. Regul.* **1,** 29–43.

Axelrod, B., and Beevers, H. (1956). *Annu. Rev. Plant Physiol.* **7,** 267–298.

Barker, J., Khan, M. A. A., and Solomos, T. (1967). *New Phytol.* **66,** 577–596.

Bieleski, R. L. (1973). *Annu. Rev. Plant Physiol.* **24,** 225–252.

Bird, I. F., Cornelius, M. J., Keys, A. J., and Whittingham, C. P. (1974). *Phytochemistry* **13,** 59–64.

Boser, H. (1959). *Hoppe-Seyler's Z. Physiol. Chem.* **315**, 163–170.

Bryan, J. K. (1976). *In* "Plant Biochemistry" (J. Bonner and J. E. Varner, eds.), 3rd ed., pp. 525–560. Academic Press, New York.

Burma, D. P., and Mortimer, D. C. (1956). *Arch. Biochem. Biophys.* **62**, 16–28.

Canvin, D. T., and Woo, K. C. (1979). *Can. J. Bot.* **57**, 1155–1160.

Cocking, E. C. (1960). *Nature (London)* **187**, 962–963.

Curry, R. A., and Ting, I. P. (1973). *Arch. Biochem. Biophys.* **158**, 213–224.

Dalling, M. J., Tolbert, N. E., and Hageman, R. H. (1972a). *Biochim. Biophys. Acta* **283**, 513–519.

Dalling, M. J., Tolbert, N. E., and Hageman, R. H. (1972b). *Biochim. Biophys. Acta* **283**, 505–512.

Danner, J., and Ting, I. P. (1967). *Plant Physiol.* **42**, 719–724.

Davies, D. D., and Davies, S. (1972). *Biochem. J.* **129**, 831–839.

Edwards, G. E., and Huber, S. C. (1978). *Photosynthesis, Proc. Int. Congr. Photosyn. 4th,* pp. 95–106.

Emes, M. J., and Fowler, M. W. (1979). *Planta* **145**, 287–292.

Fankhauser, H., Brunold, C., and Erismann, K. H. (1976). *Experentia* **32**, 1494–1497.

Flohé, L., and Menzel, H. (1971). *Plant Cell Physiol.* **12**, 325–333.

Fowler, M. W. (1974). *Biochim. Biophys. Acta* **372**, 245–254.

Foyer, C. H., and Halliwell, B. (1976). *Planta* **133**, 21–25.

Gander, J. E. (1976). *In* "Plant Biochemistry" (J. Bonner and J. E. Varner, eds.), 3rd. ed., pp. 337–380. Academic Press, New York.

Giaquinta, R. (1978). *Plant Physiol.* **61**, 380–385.

Gibbs, M. (1959). *Annu. Rev. Plant Physiol.* **10**, 329–378.

Gibbs, M., and Beevers, H. (1955). *Plant Physiol.* **30**, 343–347.

Guilliermond, A. (1941). "The Cytoplasm of the Plant Cell." Chronica Botanica Co., Waltham, Massachusetts.

Hatch, M. D. (1976). *In* "Plant Biochemistry" (J. Bonner and J. E. Varner, eds.), 3rd. ed., pp. 797–844. Academic Press, New York.

Hatch, M. D., and Osmond, C. B. (1976). *In* "Encyclopedia of Plant Physiology, Transport in Plants" (C. R. Stocking and U. Heber, eds.), Vol. 3, pp. 144–184. Springer-Verlag, Berlin and New York.

Hatch, M. D., and Slack, C. R. (1966). *Biochem. J.* **101**, 103–111.

Heber, U. (1974). *Annu. Rev. Plant Physiol.* **25**, 393–421.

Hedley, C. L., Harvey, D. M., and Keely, R. J. (1975). *Nature (London)* **258**, 352–354.

Heldt, H. W. (1976). *In* "Encyclopedia of Plant Physiology, Transport in Plants" (C. R. Stocking and U. Heber, eds.), Vol. 3, pp. 137–143. Springer-Verlag, Berlin and New York.

Hewitt, E. J., Hucklesby, D. P. and Notton, B. A. (1976). *In* "Plant Biochemistry" (J. Bonner and J. E. Varner, eds.), 3rd. ed., pp. 633–681. Academic Press, New York.

Hippmann, H., and Heinz, E. (1976). *Z. Pflanzenphysiol.* **79**, 408–418.

Huang, A. H. C. (1975). *Plant Physiol.* **55**, 555–558.

Jocelyn, P. C. (1972). "Biochemistry of the SH Group." Academic Press, New York.

Jolly, S. O., and Tolbert, N. E. (1978). *Plant Physiol.* **62**, 197–203.

Kelly, G. J., and Gibbs, M. (1973a). *Plant Physiol.* **52**, 111–118.

Kelly, G. J., and Gibbs, M. (1973b). *Plant Physiol.* **52**, 674–676.

Kelly, G. J., and Turner, J. F. (1970). *Biochim. Biophys. Acta* **208**, 360–367.

Kelly, G. J., Latzko, E., and Gibbs, M. (1976). *Annu. Rev. Plant Physiol.* **27**, 181–205.

Kenworthy, P., and Davies, D. D. (1976). *Phytochemistry* **15**, 279–282.

Kisaki, T., Imai, A., and Tolbert, N. E. (1971). *Plant Cell Physiol.* **12**, 267–273.

Latzko, E., and Kotzé, J. P. (1965). *Z. Pflanzenphysiol.* **53**, 377–387.

Lea, P. J., and Miflin, B. J. (1974). *Nature (London)* **251**, 614–616.

Lea, P. J., and Norris, R. D. (1972). *Phytochemistry* **11**, 2897–2920.

Leblová, S., and Ehlichová, D. (1972). *Phytochemistry* **11**, 1345–1346.

Leegood, R. C., and ap Rees, T. (1978). *Biochim. Biophys. Acta* **524**, 207–218.

Leigh, R. A., ap Rees, T., Fuller, W. A., and Banfield, J. (1979). *Biochem. J.* **178**, 539–547.

Leloir, L. F., and Cardini, C. E. (1953). *J. Am. Chem. Soc.* **75**, 6084.

Leloir, L. F., and Cardini, C. E. (1955). *J. Biol. Chem.* **214**, 157–165.

Lilley, R. McC., Chon, C. J., Mosbach, A., and Heldt, H. W. (1977). *Biochim. Biophys. Acta* **460**, 259–272.

Loewus, F. (1971). *Annu. Rev. Plant Physiol.* **22**, 337–364.

Loewus, M. W., and Loewus, F. (1971). *Plant Physiol.* **48**, 255–260.

Matheson, N. K., and St. Clair, M. (1971). *Phytochemistry* **10**, 1299–1302.

McGowan, R. E., and Gibbs, M. (1974). *Plant Physiol.* **54**, 312–319.

Medina, A., and Sols, A. (1956). *Biochim. Biophys. Acta* **19**, 378–379.

Miflin, B. J. (1974). *Plant Physiol.* **54**, 550–555.

Miflin, B. J., and Lea, P. J. (1977). *Annu. Rev. Plant Physiol.* **28**, 299–329.

Mowbray, J., and Moses, V. (1976). *Eur. J. Biochem.* **66**, 25–36.

Nelson, D. R., and Rinne, R. W. (1977). *Plant Cell Physiol.* **18**, 1021–1027.

Ng, B. H., and Anderson, J. W. (1978). *Phytochemistry* **17**, 879–885.

Nishimura, M., and Beevers, H. (1979). *Plant Physiol.* **64**, 31–37.

Ohlrogge, J. B., Kuhn, D. N., and Stumpf, P. K. (1979) *Proc. Natl. Acad. Sci. U.S.A.* **76**, 1194–1198.

Peak, M. J., Peak, J. G., and Ting, I. P. (1973). *Biochim. Biophys. Acta* **293**, 312–321.

Phillips, T. L., Porter, D. W., and Gracy, R. W. (1975). *Biochem. J.* **147**, 381–384.

Richardson, R. H., and Matheson, N. K. (1977). *Phytochemistry* **16**, 1875–1879.

Roberts, R. M. (1971). *J. Biol. Chem.* **246**, 4995–5002.

Robinson, J. M., and Stocking, C. R. (1968). *Plant Physiol.* **43**, 1597–1604.

Santora, G. T., Gee, R., and Tolbert, N. E. (1979). *Arch. Biochem. Biophys.* **196**, 403–411.

Sawhney, S. K., Naik, M. S., and Nicholas, D. J. D. (1978). *Nature (London)* **272**, 647–648.

Shargool, P. D., Steeves, T., Weaver, M., and Russell, M. (1978). *Can. J. Biochem.* **56**, 273–279.

Simcox, P. D., Reid, E. E., Canvin, D. T., and Dennis, D. T. (1977). *Plant Physiol.* **59**, 1128–1132.

Sols, A., and Marco, R. (1970). *Curr. Top. Cell. Regul.* **2**, 227–273.

Steup, M., and Latzko, E. (1979). *Planta* **145**, 69–75.

Stocking, C. R., and Larson, S. (1969). *Biochem. Biophys. Res. Commun.* **37**, 278–282.

Stumpf, P. K. (1952). *Annu. Rev. Plant Physiol.* **3**, 17–34.

Stumpf, P. K. (1976). *In* "Plant Biochemistry" (J. Bonner and J. E. Varner, eds.), 3rd ed., pp. 427–461. Academic Press, New York.

Takamiya, S., and Fukui, T. (1978). *J. Biochem.* **84**, 569–574.

Takeda, Y., Hizukuri, S., and Nikuni, Z. (1967). *Biochim. Biophys. Acta* **146**, 568–575.

Thomas, S. M., and ap Rees, T. (1972). *Phytochemistry* **11**, 2177–2185.

Ting, I. P., and Osmond, C. B. (1973). *Plant Physiol.* **51**, 448–453.

Tolbert, N. E., Yamazaki, R. K., and Oeser, A. (1970). *J. Biol. Chem.* **245**, 5129–5136.

Tomlinson, J. D., and Turner, J. F. (1973). *Biochim. Biophys. Acta* **329**, 128–139.

Turner, J. F., and Turner, D. H. (1975). *Annu. Rev. Plant Physiol.* **26**, 159–186.

Turner, J. F., Harrison, D. D., and Copeland, L. (1977a). *Plant Physiol.* **60**, 666–669.

Turner, J. F., Chensee, Q. J., and Harrison, D. D. (1977b). *Biochim. Biophys. Acta* **480**, 367–375.

Uedan, K., and Sugiyama, T. (1976). *Plant Physiol.* **57**, 906–910.

Ullrich, W., Urbach, W., Santarius, K. A., and Heber, U. (1965). *Z. Naturforsch.* **20b**, 905–910.

Walker, D. A. (1957). *Biochem. J.* **67**, 73–79.

Walker, D. A. (1966). *Endeavour* **25**, 21–26.

Walker, D. A. (1974). *Plant Carbohydr. Biochem. Proc. Symp. Phytochem. Soc., 10th,* pp. 7–26.

Walker, D. A., and Herold, A. (1977). *In* "Photosynthetic Organelles" (S. Miyachi, S. Katoh, Y. Fujita, and K. Shibata, eds.), pp. 295–310. Japanese Soc. of Plant Physiol. Japan.

Wallace, W. (1973). *Plant Physiol.* **52,** 197–201.

Wallace, W., and Pate, J. S. (1967). *Ann. Bot. (London)* **31,** 213–228.

Weaire, P. J., and Kekwick, R. G. O. (1975). *Biochem. J.* **146,** 425–437.

Wightman, F., and Forest, J. C. (1978). *Phytochemistry* **17,** 1455–1471.

Willard, J. M., and Gibbs, M. (1968). *Plant Physiol.* **43,** 793–798.

Youle, R. J., and Huang, A. H. C. (1976). *Biochem. J.* **154,** 647–652.

Zimmermann, G., Kelly, G. J., and Latzko, E. (1978). *J. Biol. Chem.* **253,** 5952–5956.

Development, Inheritance, and Evolution of Plastids and Mitochondria

6

JEROME A. SCHIFF

I. INTRODUCTION

This chapter will attempt to present an overview of the two organelles possessing double membranes and DNA other than the nucleus (Schiff, 1973). Although the emphasis will be on plastids, particularly on their developmental interrelationships, it would be impossible to present a coherent picture without reference to the development and inheritance of mitochondria, which has been found to be extremely similar.

Particular reference will be made to the biochemistry of the developmental and genetic processes involved but the development of function and the biochemistry and metabolism of the mature organelles will be left largely to others.

II. ORGANELLE DEVELOPMENT AND INHERITANCE

A. Indications of Organelle Autonomy

As large, pigmented bodies the plastids received early attention by light microscopists. Especially among the algae with large chloroplasts, plastids could be seen to divide and the division products were apportioned to the daughter cells. Observations of this sort led early microscopists to suggest that organelles such as the chloroplasts might have some degree of autonomy and resemble cells within cells. It was but a step from here to suggest that such organelles might have arisen from the endosymbiotic invasion of one cell by another; for example, the plastid might have originated from the establishment of a blue–green algal (cyanobacterial) cell within a eukaryotic cell lacking plastids (see Margulies, 1970, for review). Similar conclusions about the mitochondrion had to await the development of the electron microscope since this organelle is close to the limit of resolution of the light microscope. We now know that mitochondria actively divide and fragment and, in the spectacular movies of tobacco leaf cells taken by Wildman (Wildman *et al.,* 1962), can also be seen to associate with and disassociate themselves from the plastids in the light microscope. In synchronous cells of algae, plastids and mitochondria can be seen to associate into larger rings and then disassociate at various phases of the cell cycle (Lefort-Tran, 1975). Thus, these organelles are in a dynamic state, dividing, fusing, and moving about. The amount of mitochondrial material can be changed in response to environmental needs; for example, in *Euglena* cells where photosynthesis has been eliminated by dark growth and the cells are dependent on respiration, a hypertrophy of mitochondria is observed (Lefort, 1964).

Further evidence came from genetic studies. Mendel's work was done in the nineteenth century at the same time that the early cytological studies of organelles were undertaken. Although Mendel's discoveries were a key to the understanding of both evolution and the development and inheritance of cellular organelles, his work did not enter the mainstream of scientific thought until it was rediscovered in the early twentieth century. Among the mutations found by early geneticists were plastid abnormalities, chiefly yellow or white plastids. These abnormalities were inherited in a normal Mendelian fashion and indicated that there are nuclear genes that affect the phenotypes of the plastids.

Like most scientific explanations of wide and general application the Mendelian view of inheritance, once accepted, became pervasive and was widely supported. As frequently happens, in the rush to support a new theory, seeming exceptions are swept under the rug, although they usually turn out not to be exceptions, but extensions of the theory. Such exceptions to the regularity of Mendelian inheritance were extensively investigated by Correns, for example, in maize (Rhoades, 1946). It was a tenet of Mendelian inheritance that reciprocal crosses should give the same results, it should not matter whether a given marker is carried by the male or female parent; the same ratios should be found in the progeny. (Of course we exclude such obvious exceptions as sex-linked or sex-limited inheritance.) Correns found certain mutations that produce chloroplast abnormalities (e.g., yellow or white plastids, rather than green) which did not behave the same way in reciprocal crosses. The phenotype of the offspring always resembled that of the female parent, not the male parent. He explained this matroclinal or maternal inheritance by assuming that the plastid markers were localized in the cytoplasm, perhaps in the plastids themselves. Since the pollen was thought to contribute only a nucleus to the cross while the egg contributed both nucleus and cytoplasm, it seemed reasonable that inheritance of plastid-localized mutations should be through the female only.

Strangely enough, the explanation for this form of non-Mendelian inheritance is still a vexing problem. One problem is the paucity of detailed information concerning the behavior of the plastids throughout the sexual cycle in eggs, pollen, apical meristems, etc. before they are established in the leaves. Cases are known of inheritance of plastid markers through the mother only, through the father only, and through both. Perhaps the most intensively studied system is that in *Oenothera* (Cleland, 1962) where incompatability groups exist which determine which plastid and nuclear genomes can coexist in the same cell. Thus the segregation of plastid markers cannot be completely explained on the basis of cytoplasm from the egg and its lack from the pollen. It has been suggested, on the basis of electron microscopical observations, that the inheritance of plastids in higher plants depends on events at the first mitosis in the formation of the pollen to give vegetative and generative cells (Hagemann, 1976). This is an unequal division and, in species that show maternal inheritance, no plastids segregate to the generative cell; however, in species showing biparental inheritance of plastids, plastids segregate to the generative cell.

However, things cannot be this simple, particularly in microorganisms. In *Neurospora,* for example, mitochondrial markers such as Pokey are found, which are inherited in a non-Mendelian manner. In *Chlamydomonas* there are markers that are inherited in a manner which deviates from the one observed for genes carried in the nuclear chromosomal linkage groups. This is called uniparental inheritance—where only the markers from one parent are transmitted to the progeny (Sager and Ramanis, 1967; Boynton *et al.*, 1976).

Evidence has been presented for the inheritance of mitochondrial markers in yeast (Dujon and Slonimski, 1976), which is consistent with their localization in the mitochondrion and with recombination among mitochondria (perhaps the fusion and fragmentation of mitochondria seen by cytological methods in other systems is a physical manifestation of this recombination).

As in higher plants, perfectly good Mendelian markers are found in these organisms which determine organelle phenotypes. Thus nuclear chromosomal genes are involved in the formation of mitochondria and chloroplasts.

One should not leave a discussion of the genetics of plastid inheritance in higher plants without mention of the curious case of iojap in maize (Rhoades, 1946). The evidence here is consistent with a situation where plastid abnormalities which are determined by a pair of homozygous recessive nuclear chromosomal markers behave in a normal Mendelian manner. However, the results indicate that once the abnormality in the plastid is established by the presence of the homozygous recessive iojap genes in the nucleus, it is inherited in a maternal manner even when the plastids are crossed back into cells lacking the homozygous iojap genes in the nucleus. One interpretation is that the presence of the homozygous recessive iojap genes in the nucleus leads to the production of an agent that mutates the plastid genome. Thus when the nuclear constitution is returned to wild type, the plastid genome continues to transmit the plastid abnormality.

Therefore, the genetic and cytological work which has been done is consistent with a cellular situation in which the nucleus, chloroplast, and mitochondrion each contain a genome. The mitochondrial and plastid genomes each provide genetic information for the construction of the respective organelle; the nuclear genome also contributes genetic information for the formation of each of the organelles. What emerges is a picture of semi-autonomous organelles, capable of division, but subject to genetic control both by their own genomes and by the nuclear genome.

B. Plastids and Mitochondria as Prokaryotic Residents in Eukaryotic Cells

The most primitive types of cells characteristic of the kingdom Monera (Fig. 1) are prokaryotic (pro = before, karyos = nucleus), representing the situation before the evolution of a true nucleus or other organelles visible in the light microscope and set off from the cytoplasm by limiting membranes. The known contemporary prokaryotic groups include the bacteria and the cyanobacteria or blue–green algae and, perhaps, a recently discovered green prokaryote, *Prochloron* (Lewin, 1975, 1976, 1977; Lewin and Withers, 1975;

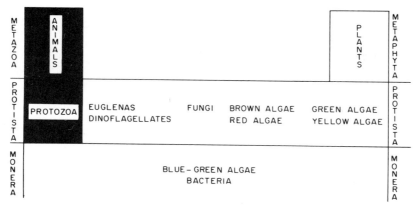

Fig. 1. Summary of phyllogenetic relationships of prokaryotes (Monera) and eukaryotes (Protista, Metazoa, Metaphyta). From Schiff and Hodson (1970).

Newcomb and Pugh, 1975; Whatley, 1977b; Thorne *et al.,* 1977; Perry *et al.,* 1978). There is little question among microbiologists that the blue–green algae represent a specialized group whose general characteristics indicate a close relationship to the gram negative bacteria (Fogg *et al.,* 1973; Wolk, 1973; Stanier *et al.,* 1971). The presence of oxygen-evolving (oxygenic) photosynthesis in this group distinguishes them from the other photosynthetic bacteria. However, the finding that cyanobacteria can carry out anaerobic (anoxygenic) photosynthesis facultatively after adaptation to anaerobic conditions (Padan, 1979), makes even this distinguishing characteristic less important.

All other cells of the biological world (the Protista, Metazoa, and Metaphyta, Fig. 1) are eukaryotic (eu = true, karyos = nucleus), i.e., they contain true nuclei and other organelles visible in the light microscope and set off from the cytoplasm by limiting membranes. It is these organisms, of course, which contain mitochondria and, in many groups, plastids.

Recent work indicates that there may be a close relationship between free-living prokaryotic cells and the organelles of eukaryotic cells, substantiating earlier speculations. Various characteristics of prokaryotic and eukaryotic cells, including their mitochondria and chloroplasts, are listed in Table I. There is an astonishing similarity among free-living prokaryotic cells, and the mitochondria and chloroplasts of eukaryotic cells and marked differences between the prokaryotic and eukaryotic cells themselves. Any discussion of the evolution, development, and inheritance of mitochondria and chloroplasts must take cognizance of the prokaryotic nature of these organelles and their relationship with the eukaryotic cell in which they are residents.

TABLE I

Some Properties of Prokaryotes Compared with Mitochondria, Plastids, and the Eukaryotic Cells Containing Them

Property	Prokaryotes (bacteria, blue–green algae, or cyanobacteria)	Mitochondria	Plastids	Rest of eukaryotic cell
Size (μm)	~1–10	~1–2	~1–10	~10–100
Membrane-bounded organelles within	None	None	None	Yes
Cell wall	Usually present	Absent	Absent	Present or absent
Endoplasmic reticulum and golgi	Absent	Absent	Absent	Present
Ribosomes	70 S type	70 S type (but variable in size)	70 S type	80 S type
Ribosomal proteins	~55	18–107	~55	~75
tRNAs	Complete set	Unique complete set?	Unique complete set	Unique complete set

214

Inhibition of protein synthesis by choramphenicol, streptomycin, etc.	Yes	Yes	Yes	No
Inhibition of protein synthesis by cycloheximide, etc.	No	No	No	Yes
DNA genome	Single molecule, circular	Single molecule, often circular	Single molecule, often circular	Many molecules in chromosomes
Spindle or comparable mitotic mechanism	No	No	No	Present
Reproduction	Fission, fragmentation, directional parasexual recombination	Fission, fragmentation	Fission	Mitosis, meiosis, sexual reproduction
Phosphorylative cell respiration (when present)	In cell membrane	In inner membrane	—	No, in mitochondria
Photosynthesis (when present)	In cell membrane, extensions, or free thylakoids	—	In thylakoids	No, in chloroplasts

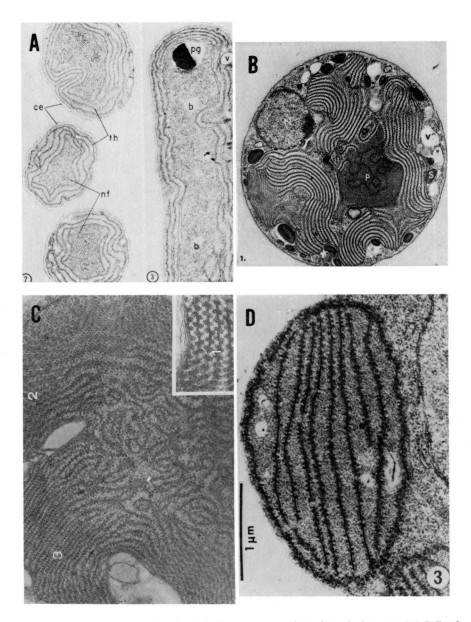

Fig. 2. Electron micrographs of thylakoid arrangements in various algal groups. (A) Cells of the blue–green alga (cyanobacterium) *Synechococcus* showing (in cross-section and longitudinal section) the unstacked thylakoids. b, Lighter bodies; ce, cell envelope; nf, nuclear fibrils; pg, polyphosphate granule; th, thylakoids; v, vesicle. From Edwards *et al.* (1968). (B) Cell of the unicellular red alga *Porphyridium* showing the large multilobed chloroplast containing unstacked thylakoids bearing phycobilisomes, and a pyrenoid. P, pyrenoid; N, nucleus; S, starch grain; V, vacuole. From Gannt *et al.* (1968). (C) Same as (B) but showing details of thylakoids and

C. Phyllogenetic Variation among Organelles

While mitochondria are fairly standardized throughout the eukaryotic world, there is a surprising variety of plastid morphologies. A variety of plastids from various groups is shown in Fig. 2. The most primitive situation occurs in the prokaryotes where the single cells of the blue–green algae or cyanobacteria contain unstacked single thylakoids which are free in the cytoplasm and, of course, are not surrounded by a limiting membrane. The plastid pigments found in each group may be seen in Table II. The cyano-bacteria contain chlorophyll a and varying amounts of the phycobilins, blue–green phycocyanin, and red phycoerythrin. The phycobilins are contained in bodies along the edges of the thylakoids called phycobilisomes (Gannt and Conti, 1966; Grabowski and Gannt, 1978). The pigments within the phycobilisomes are arranged in such a manner as to facilitate the transfer of energy from one to the next and eventually to the reaction centers given certain constraints to be detailed later.

This pattern of organization is retained in the chloroplasts of the red algae where unstacked thylakoids bearing phycobilisomes are' the rule (Fig. 2, Table II). The similarity of placement of the phycobilins in the cyanobacteria and red algal chloroplasts and the structural and immunological homology among the various phycocyanins and phycoerythrins of the two groups suggest a close evolutionary relationship between the cyanobacteria and the plastids of the red algae (Glazer et al., 1976; Troxler, 1977; see Ragan and Chapman, 1978). It might be noted that the rather heterogeneous flagellate group called the cryptomonads also contain phycocyanin and phycoerythrin, but here phycobilisomes are not present. It has been suggested that the phycobilins are contained within the thylakoids (Gannt et al., 1971), perhaps in the lumen, and structural and immunological studies suggest that the phycobilins of this group may be more distant from those of the cyanobacteria and red algae (MacColl and Berns, 1976; see Ragan and Chapman, 1978). This may be a case of convergent evolution where the same kinds of pigments and structures evolved separately in distantly related groups in response to similar environmental selection.

In the brown algae some stacking of thylakoids occurs and pyrenoids appear in many cases. These thylakoids contain chlorophyll a and fucoxan-thin. The pyrenoid in various groups of algae may be an aggregate of ribulose bisphosphate carboxylase and other enzymes of carbon dioxide fixation (Holdsworth, 1971; Schiff, 1973; Griffiths, 1970; Kirby and Evans, 1978; Salisbury and Floyd, 1978); in most groups the end products of photosyn-

---

phycobilisomes (inset). Phycobilisomes viewed in (1) cross-section; (2) tangential (grazing) section; and (3) longitudinal section. From Gannt et al. (1968). (D) Chloroplast of the red algal seaweed *Griffithsia* showing the unstacked thylakoids bearing phycobilisomes. From Waaland et al. (1974).

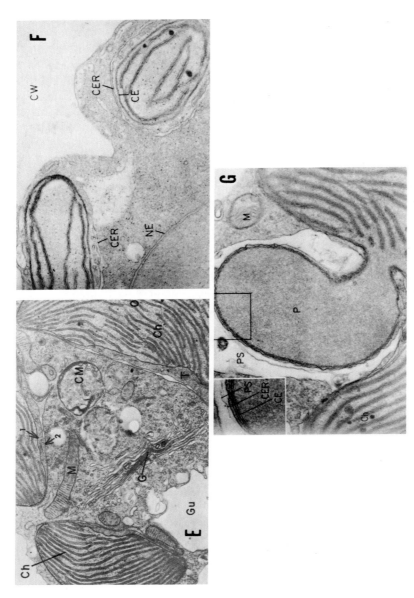

Fig. 2 (Continued). (E) Cell of the cryptophyte *Cryptomonas* showing chloroplasts containing stacked thylakoids. Ch, chloroplast; CM, corps de Maupas; G, golgi; Gu, gullet; M, mitochondrion. Symbols: (1) inner and (2) outer pairs of membranes enclosing plastid. From Lucas (1970). (F) Chloroplasts of the brown alga *Fucus* showing two-layered stacked thylakoids. CE, chloroplast envelope; CER, chloroplast endoplasmic reticulum; CW, cell wall; NE, nuclear envelope. From Bouck (1965). (G) Section through the protruding pyrenoid region of the chloroplast (Ch) of the brown alga *Chorda*. Ch, chloroplast; M, mitochondrion; P, pyrenoid; PS, pyrenoid sac. From Bouck (1965).

TABLE II

Pigments of Photosynthetic Organisms

Group	Chlorophylls	Carotenoids	Biliproteins
Eukaryotic organisms:			
Chlorophyta	Chlorophyll a	Carotenes	—
(green algae)	Chlorophyll b	Xanthophylls	
Euglenophyta	Chlorophyll a —	Carotenes	—
(*Euglenoids*)	Chlorophyll b	Xanthophylls	
Xanthophyta	Chlorophyll a	Carotenes	—
(yellow algae)		Xanthophylls	
Chrysophyta	Chlorophyll a	Carotenes	—
(yellow–brown or	Chlorophyll c	Xanthophylls	
golden algae)		*Fucoxanthin*, etc.	
Bacillariophyta	Chlorophyll a	Carotenes	
(diatoms)	Chlorophyll c	Xanthophylls	—
		Fucoxanthin, etc.	
Pyrrophyta	Chlorophyll a	Carotenes	—
(dinoflagellates)	Chlorophyll c	Xanthophylls	
		Fucoxanthin, etc.	
Pheophyta	Chlorophyll a	Carotenes	—
(brown algae)	Chlorophyll c	Xanthophylls	
		Fucoxanthin, etc.	
Rhodophyta	Chlorophyll a	Carotenes	Phycocyanin
(red algae)	Chlorophyll d	Xanthophylls	Phycoerythrin
Cryptophyta	Chlorophyll a	Carotenes	Phycocyanin
(cryptophytes)	Chlorophyll c	Xanthophylls	Phycoerythrin
Prokaryotic organisms:			
Cyanophyta	Chlorophyll a	Carotenes	Phycocyanin
(cyanobacteria)		Xanthophylls	Phycoerythrin
Prochlorophyta	Chlorophyll a	Carotenes	—
(*Prochloron*)	Chlorophyll b	Xanthophylls	
Photosynthetic	Bacteriochlorophylls a–e	Carotenes	—
bacteria		Xanthophylls	

thesis seem to be closely associated with this structure. The dinoflagellates show a higher degree of stacking and more membranes in their chloroplasts.

In the groups containing chlorophylls a and b such as the *Euglenas,* green algae, and multicellular higher plants, large numbers of thylakoid membranes and elaborate stacking are the rules. The function of stacking is not completely clear. Mutants of stacked organisms are known which lack normal stacking and show no significant impairment of photosynthesis—in some cases the degree of stacking can be varied by changing conditions, particularly cation concentrations (Miller, 1978; Surzycki *et al.,* 1970; Staehlin 1976). Stacking may be necessary for some subtle requirements of distance for energy transfer among pigments (Thornber *et al.,* 1979); if so, the

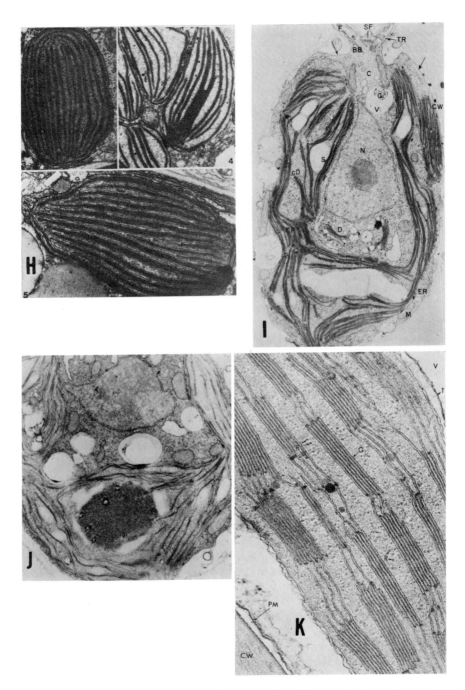

Fig. 2 (Continued). (H) Sections through the plastids of the dinoflagellates *Prorocentrum* (3), *Dissodinium* (4), and a symbiotic *Gymnodinium* (5). From Dodge (1975). (I) A section

cyanobacteria and red algae seem to have solved the energy transfer problem another way—through phycobilisomes that allow the accessory pigments to lie close to the reaction centers, but the phycobilisomes would actually thwart normal stacking by preventing the thylakoids from lying smoothly against each other (Gannt and Conti, 1966; Grabowski and Gannt, 1978). On the other hand it is possible that stacking is also a way to contain more thylakoids in a smaller space by pressing them closely together. In the higher plants, stacking organization reaches its most highly evolved form with separation of stroma and grana thylakoids, each with its own molecular organization and photosynthetic specialization (Arntzen and Briantais, 1975).

Many cases of symbiosis exist to provide adoptive photosynthetic structures for otherwise nonphotosynthetic organisms (Jennings and Lee, 1975). There are numerous examples of invertebrate animals and protozoa which harbor algae in their cells. The cases of *Glaucocystis, Glaucosphaera,* and *Cyanophora* are more interesting from the standpoint of plastid evolution (see Fogg *et al.,* 1973; Trench, 1979), where a blue–green algal (or cyanobacterial) cell appears to have become habituated in a flagellate leading to the establishment of a stable mutual relationship with the rest of the cell. Other views of this relationship have been offered. Perhaps the most self-serving of these relationships occurs among certain mollusks that eat algae, remove the chloroplasts, and retain them in the cells of their digestive pouches as active photosynthetic organelles (Trench, 1979). Certain amoebae lacking mitochondria harbor symbiotic bacteria (Whatley, 1976).

D. Arrested Development of Organelles: Proplastids and Promitochondria

1. A Comparative Overview

In most cells, mitochondria are constituitive; they are always present regardless of environmental conditions. This should not surprise us since most eukaryotes are obligate aerobes and the absence of mitochondrial respiration would be lethal. Certain eukaryotes among the fungi, however, have become facultatively anaerobic—they can obtain the energy necessary for growth

---

through a cell of the green alga *Chlamydomonas* showing the cup-shaped chloroplast containing highly stacked thylakoids. BB, basal body; C, contractile vacuole; CD, chloroplast DNA; CP, chloroplast; CW, cell wall; D, dictyosome; ER, endoplasmic reticulum; M, mitochondria; N, nucleus; S, starch grain; SF, striated fiber; TR, transition region; V, vacuole. From Goodenough (1969). (J) Same as (I) showing pyrenoid at base of chloroplast. From Goodenough (1969). (K) Section through a chloroplast of the grass *Phleum* (a flowering plant) showing grana thylakoid stacks linked by stroma thylakoids. CW, cell wall; G, grana thylakoid stack; PM, plasmalemma; R, ribosomes; T, tonoplast; V, vacuole. From Ledbetter and Porter (1970). (For a view of the pyrenoid and stacked thylakoids of the *Euglena* chloroplast, see Fig. 4.)

from either respiration or fermentation enabling them to exist as aerobes when oxygen is available and as anaerobes when it is not. Certain organisms of this type such as yeast have normal functional mitochondria when grown in the presence of oxygen but only rudimentary nonrespiratory promitochondria when grown anaerobically (Slonimski, 1953; Ephrussi and Slonimski, 1950; Watson et al., 1971). On exposure of anaerobically-grown yeast cells to oxygen, promitochondria are induced to become mitochondria (Perlman and Mahler, 1974). Although both mitochondria and promitochondria contain mitochondrial DNA (mtDNA), only a few of the normally occurring mitochondrial proteins are made under anaerobic conditions and when respiration is absent (Gunsalus-Cadavid, 1974) the cells must obtain energy through fermentation. On exposure to oxygen the respiratory system is induced to form and the cells become capable of utilizing molecular oxygen for energy production (Vary et al., 1970; Linnane and Crowfoot, 1975; Perlman and Mahler, 1974). The induction of development of a promitochondrion to form a mitochondrion is a substrate induction in that a utilizable substrate (in this case oxygen) induces the enzymes necessary for its utilization. Unlike simple cases of enzyme induction where one or a few proteins are involved, the formation of a mitochondrion on induction by its substrate, oxygen, requires the induction of many proteins and the synthesis of many lipids and other membrane constituents all properly coordinated and programmed in time—the process we call development.

A similar situation exists with respect to photosynthesis. In photosynthetic bacteria carrying out anoxygenic photosynthesis, photosynthesis occurs only under anaerobic conditions (Cohen-Bazire et al., 1957). In those organisms capable of growth in the presence of oxygen, the formation of photosynthetic structures is repressed by oxygen. On return to anaerobic conditions the photosynthetic structures are induced to form again. This situation apparently results from several causes. One is the necessity to avoid deleterious photooxidations in the presence of oxygen and light. Also, certain enzymes and intermediates appear to be involved in both photosynthesis and respiration, making it difficult for the organism to carry out both processes at the same time. Cyanobacteria are better adapted to aerobic conditions since they usually carry out oxygenic photosynthesis and respiration under normal conditions of growth. Under anaerobic conditions, however, and in the presence of a suitable electron donor such as H_2S, cyanobacteria can adaptively switch from oxygenic to anoxygenic photosynthesis (Padan, 1979). Perhaps the photometabolism of hydrogen (Kessler, 1974) seen during anaerobic adaptation of modern green algae is an evolutionary remnant of these types of reactions.

In most photosynthetic organisms chloroplasts and photosynthesis are probably constitutive. Many of the green algae (Kirk and Tilney-Bassett, 1967; Sager, 1958; Granick, 1949, 1960, 1971), for example, and the cotyledons of gymnosperms (Bogorad, 1950) form chlorophyll and chloroplasts in darkness as do two species of diatoms (White, 1974). However, in order to determine whether plastids and chlorophyll can be made in darkness, the organism in question must be capable of growth on a reduced substrate in the dark. Those that cannot be grown on a reduced carbon source cannot be tested, and are classed as obligate phototrophs. Here and there among photosynthetic organisms are situations in which the chloroplast is induced to form by light, another example of the elaborate developmental consequences of substrate induction; in this case the inducing substrate is light, the substrate for photosynthesis. In darkness, these organisms possess small undeveloped proplastids incapable of photosynthesis but containing plastid DNA (pDNA). On exposure to light these proplastids develop into chloroplasts and, concomitantly, photosynthetic activity appears. Examples of organisms showing induction of plastid development by light include all of the angiosperms or flowering plants (Kirk and Tilney-Bassett, 1967), a few species of *Euglena* (Leedale, 1967) (most *Euglenas* are obligate phototrophs), *Ochromonas,* a chrysomonad (Gibbs, 1962), and *Cyanidium caldarium* (Troxler and Bogorad, 1966) regarded as a red alga. Certain organisms having constitutive chlorophyll and chloroplasts can be mutated to become light-dependent for chloroplast development, e.g., *Chlamydomonas reinhardi* Y_1 (Hudock and Rosen, 1976; Sager, 1958), *Scenedesmus obliquus* C-2A1 (Senger and Bishop, 1972), and certain mutants of *Chlorella* (Granick, 1949, 1960, 1971). This suggests that both a dark system and a light-inducible system exists in constitutive organisms and that mutation eliminates the dark system rendering the organism light-dependent for plastid development.

Light is not the only inducer of plastid development in eukaryotes. In certain organisms such as *Chlorella protothecoides* (Hase, 1971), plastid formation is controlled by the carbon–nitrogen ratio in the medium. In *Euglena* the length of the lag period in plastid development is similarly controlled by the carbon–nitrogen ratio of the medium in which the cells are grown (Freyssinet, 1976). The higher the carbon, the more paramylum is stored and made available for plastid development, leading to a shortening of the lag period.

Not all arrested states of plastid development are the same (Fig. 3). In angiosperms, the apical cells and meristems contain tiny proplastids lacking internal membranes (Whatley, 1974). On further development in the dark, the young leaves contain small proplastids about 2 μm in diameter having a few internal membranes with poorly organized or no prolamellar bodies (Klein and Schiff, 1972). *Euglena* proplastids are of this type and remain this way throughout growth in darkness (Klein *et al.,* 1972). As angiosperm leaves age in darkness, the proplastids develop into etioplasts which are larger and contain extensive membranes and large crystalline prolamellar bodies (Klein and Schiff, 1972). More will be said of this when we discuss the formation of thylakoid pigments. In *Ochromonas,* a chrysomonad, there is a

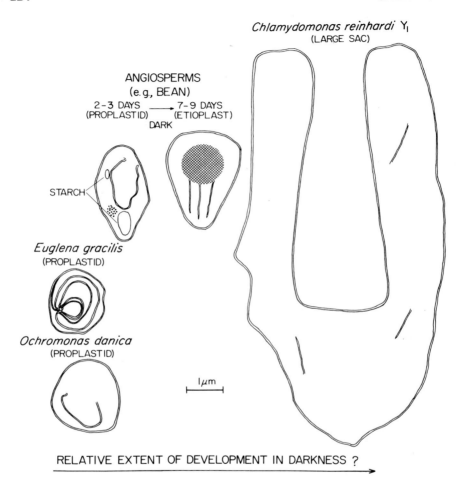

Fig. 3. Arrested stages of plastid development in various dark-grown organisms sketched roughly to scale. The proplastids of *Euglena* and *Ochromonas* are quite small, those of 2–3-day-old beans slightly larger but similar to *Euglena* in internal structure; 7–9-day-old beans have larger, more complex etioplasts with highly crystalline prolamellar bodies and *Chlamydomonas* Y_1 has a large sac about the same size as the mature chloroplast but little visible internal structure. After Klein *et al.,* 1972; Klein and Schiff, 1972; Gibbs, 1962; Sager, 1958.

small proplastid in darkness (Gibbs, 1962). In light-inducible mutants of normally constitutive organisms, e.g., *Chlamydomonas reinhardi* Y_1, the arrested condition is a sac, often as large as the plastid that will form (Sager and Palade, 1954; Sager, 1958). In such cases, there is frequently an accumulation of plastid constituents waiting to be organized on light exposure (Ohad, 1975). This is somewhat different from the proplastid–etioplast situation where many more constituents must be synthesized *de novo* on light induction. In any case all of these arrested structures are capable of forming fully functional chloroplasts on exposure to light.

It is apparent that the ability to make chloroplasts and chlorophyll in darkness is widespread and that the arrested condition inducible by light or other substrates is scattered widely among the various groups of protists and metaphytes. It is likely that the arrested condition showing substrate induction was selected during evolution in response to the environmental needs of particular organisms rather than predominating in one group or another. Perhaps dark arrest of plastid development is widespread among angiosperms because these plants all have similar developmental requirements brought about by having embryos in seeds that usually germinate in darkness; the seedlings then emerge into the light.

2. Chloroplast Development in Dark-Grown Euglena Cells

The *Euglena* system is an excellent one for the study of plastid development because the organism grows well on a defined medium and the development and inheritance of the plastid is under control by the experimenter. When grown in the light, *Euglena gracilis* Klebs var. *bacillaris* Cori or strain Z Pringsheim contains about 10 chloroplasts which divide at a rate sufficient to keep the plastid number constant from one generation to the next (Klein *et al.*, 1972) (see Fig. 1 in Schiff, 1973). When placed in darkness on a medium containing a carbon source, the plastid becomes reduced to a proplastid as the cells divide and the numbers of proplastids per cell are kept constant through proplastid division (Ben-Shaul *et al.*, 1965) Dark-grown cells, either dividing or placed in a medium lacking nitrogen and carbon sources to prevent division, are induced to form chloroplasts from their proplastids by exposure to light (Stern *et al.*, 1964a,b). This process in dividing or nondividing cells is called chloroplast development to distinguish it from proplastid or chloroplast replication, which occurs only in dividing cells.

The proplastid of the dark-grown cells is an irregularly-shaped organelle about 1-2 μm in diameter containing a few primary thylakoids and a disorganized prolamellar body, which expands about sixtyfold in volume during chloroplast development (Klein *et al.*, 1972) (Fig. 4). Along with the increase in volume, many new membranes (Bingham and Schiff, 1976; 1979a,b), proteins (Bovarnick *et al.*, 1974), lipids (Erwin, 1968), and other molecules are formed (Schiff, 1973, 1978). Concomitant with the formation of structure (see Fig. 3 in Klein *et al.*, 1972), various physiological parameters increase as well (see Fig. 6 in Schiff, 1973; Stern *et al.*, 1964a,b). There is a lag of 12 h under these conditions during which there is little increase in plastid size or in the number or extent of membranes (Klein *et al.*, 1972). During this period chlorophyll and carotenoid synthesis is slow and photosynthetic CO_2 fixation and O_2 evolution appear at about 3-4 h of development (Stern *et al.*, 1964a,b).

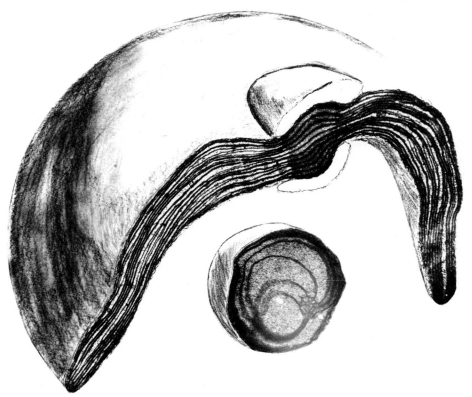

Fig. 4. Three-dimensional visualization of the proplastid and chloroplast from thin sections of dark-grown and light-grown *Euglena*. From Schiff (1973).

After 12 h the plastid enters a period of rapid expansion and membrane synthesis, the pyrenoid differentiates at about 48 h of development, and development is complete by 72 to 96 h.

Since the organism plates well, it is possible to isolate mutants which generally appear as cells or colonies of unusual color (Schiff *et al.,* 1971). The nuclear genome of *Euglena* appears to be more than haploid (perhaps diploid or octaploid) (Hill *et al.,* 1966; Rawson, 1975) and, therefore, expressed nuclear mutants are rare or nonexistent (unless dominant). Since the organism is an obligate aerobe, mitochondrial mutants affecting function would be lethal. Experience indicates that most mutants obtained in *Euglena* are plastid mutations (Schiff *et al.,* 1971) which must be rigorously cloned to obtain nonreverting strains since only mitosis occurs, no meiosis or segregation as far as is known. To date no sexual recombination, transformation, or transduction has been observed in *Euglena*.

3. Chloroplast Development in Etiolated Higher Plants

Another system that has been used extensively for studies of plastid development is the etioplast of older etiolated leaves of monocots and dicots (Boardman, 1966; Von Wettstein, 1958; Kirk and Tilney-Bassett, 1967; Bogorad, 1976). One of the most popular has been the bean *Phaseolus vulgaris* L. Etiolated leaves of dark-grown plants are much smaller than their light-grown counterparts and for this reason most workers have preferred to use older material which has been kept in darkness for 7–15 days. At this time the primary leaves of the bean contain enlarged proplastids called etioplasts which have an impressive crystalline prolamellar body that has been the delight of electron microscopists (Fig. 5A) although its function in chloroplast development is not clear (Lütz and Klein, 1979). Both the structure of the etioplast and the mode of light-induced development into a chloroplast are much different in this older leaf material than in dark-grown *Euglena*.

If one looks at younger leaf material, however, a much different picture is seen (Klein and Schiff, 1972; Whatley, 1977a). Starting with the young stem apex in the seed, very small proplastids lacking much internal structure are found. By 2-3 days after germination in darkness (Fig. 5B) the primary leaves contain proplastids very similar to those of *Euglena*, being small and containing only a few primary thylakoids. Unlike *Euglena*, starch is present within the plastids, a characteristic of higher plants and their precursors, the green algae. Either a small prolamellar body lacking the high crystallinity of the older material is present or the prolamellar body is absent. Like the *Euglena* proplastids, these develop normally into chloroplasts on light exposure. A comparison of plastid development in *Euglena* and in young and old leaves will be deferred until the protopigments have been discussed, but it is clear that in terms of structure the proplastid of *Euglena* and its structural development into a chloroplast on light exposure is far more similar to young etiolated bean leaves than to older etiolated material.

Earlier, the variation of plastid structure with phyllogeny was described (see Fig. 2) which indicated a trend from small plastids with a few unstacked thylakoids to larger plastids containing numerous stacked membrane systems. The proplastids of higher plants and *Euglena* resemble these more primitive plastids, or whole cells in the case of the prokaryotic blue–green algae or cyanobacteria, except that no chlorophyll pigments are present at this stage of development of the proplastid. We have already noted that many, perhaps most, photosynthetic cells contain constituitive chloroplasts, i.e., the chloroplasts are always present in the fully-developed form. It would seem that the arrested condition leading to proplastids and light-induced development came later as an adaptation to dark growth followed by light exposure as routinely occurs in germinating seeds which start underground and then emerge into the light. In *Euglena*, of course, we have an

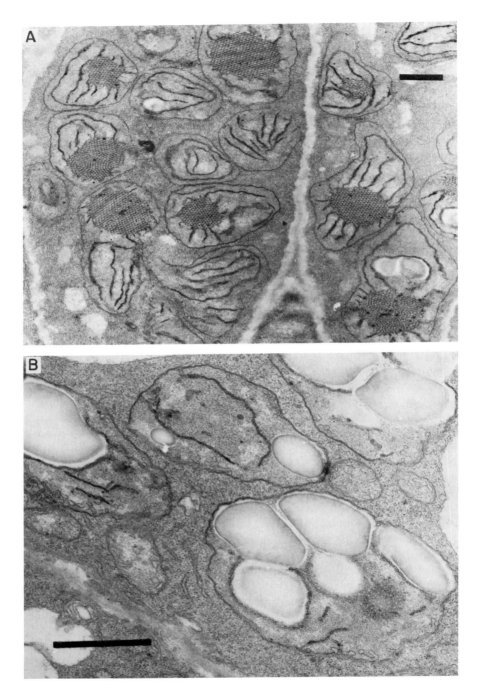

Fig. 5. A, Section through a cell from a primary leaf of a 7-day-old dark-grown bean seedling. The etioplasts contain fully developed prolamellar bodies and prothylakoids. B, Sections through proplastids from a primary leaf of a 3-day-old dark-grown bean seedling. Marker indicates 1 μm. From Klein and Schiff (1972).

organism that can exist as a phototroph in light and as an organotroph in darkness. In both cases it is advantageous to drop the excess baggage of a fully mature photosynthetic system during the dark phase. We know how easy it is through mutation to go from constituitive chloroplasts to a light-induced situation as in *Chlamydomonas reinhardi* Y_1. Pulling these threads together, ontogeny, to some extent may be recapitulating phyllogeny. The cells of cyanobacteria [which have been shown to contain rudimentary prolamellar-like bodies on occasion (Lang and Rae, 1967)] may resemble proplastids of higher forms. This is because when these higher forms are prevented from completing chloroplast development and must stop at a proplastid, they may fall back on an ancient program still present in their genes from the time when the mature plastid resembled the situation found in the cyanophytes (S. Klein *et al.,* 1972). This idea will be developed further when we come to consider the origin of the plastid.

4. Protopigments and Chlorophyll Formation

Older etiolated angiosperm leaves contain protochlorophyllide predominantly, the polar chlorophyll precursor in which the propionic acid side chain is unesterified. Since this pigment can be reduced on absorption of light to form chlorophyllide in the leaf followed by dark reactions that attach an ester group, this sequence has been viewed as the usual sequence of events (Fig. 6) (Boardman, 1966; Kirk and Tilney-Bassett, 1967; Rebeiz and Castelfranco, 1973). In extracted etioplast membranes, protochlorophyllide is thought to form a complex with an enzyme and NADPH in darkness; on illumination hydrogen would be transferred from the NADPH to the protochlorophyllide to form chlorophyllide (Griffiths, 1978). Perhaps this is the mechanism of phototransformation of the protochlorophyllide holochrome from beans and other plants, the holochrome being a protochlorophyllide–protein complex which is extracted from etiolated leaves with detergents; on illumination *in vitro* the protochlorophyllide in the holochrome is transformed to chlorophyllide (Schopfer and Siegelman, 1968). However, another possible route is one in which protochlorophyll (an esterfied form of protochlorophyllide) is photoconverted to an esterfied chlorophyll (Fig. 6) (Lancer *et al.,* 1976). Protochlorophyll and protochlorophyllide from higher plants cannot be distinguished spectroscopically (Fig. 7) since they have the same absorption spectrum and must be subjected to solvent partition or chromatography in order to detect the difference—the long chain fatty alcohol ester group making the protochlorophyll more nonpolar than protochlorophyllide which can form salts through its free carboxyl group.

Dark-grown *Euglena* cells contain both protochlorophyll and protochlorophyllide (Fig. 8) in about equal amounts and both pigments are transformable by light to the respective chlorophyll pigments (see Fig. 3 in C. Cohen and Schiff, 1976). Once again an investigation of the younger

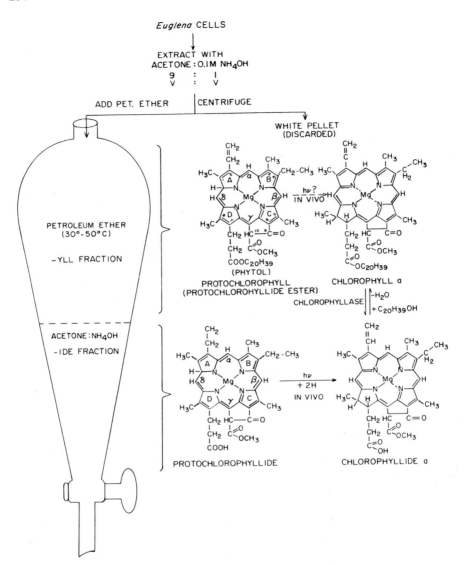

Fig. 6. Esterified chlorophyll and protochlorophyll are epiphasic on partition being more nonpolar than unesterified chlorophyllide and protochlorophyllide which can form salts at the free carboxyl group. The established path from work on older leaves of higher plants is shown as proceeding from protochlorophyllide through photoreduction to chlorophyllide followed by esterification forming chlorophyll a. However, work with *Euglena* and young leaves indicates a path from the esterified protochlorophyll followed by photoreduction to an esterified chlorophyll (see text for further detail).

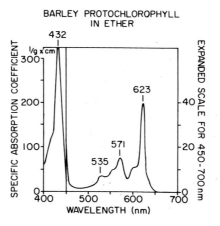

Fig. 7. Absorption spectrum of Barley protochlorophyllide in ether. After Boardman (1966). Protochlorophyll would show essentially the same absorption spectrum.

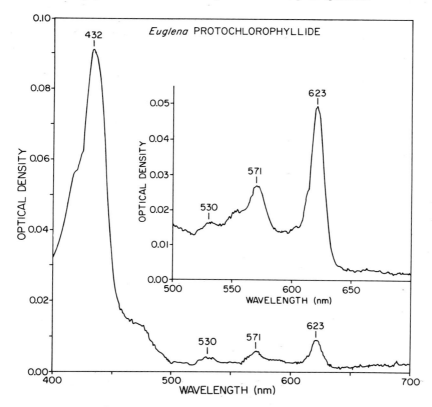

Fig. 8. Absorption spectrum of purified hypophasic pigment from dark-grown cells of *Euglena* which appears to be similar to or identical with protochlorophyllide of higher plants. From C. Cohen and Schiff (1976).

angiosperm leaves, in this case young bean leaves, shows that this situation is like *Euglena*, approximately equal amounts of esterified and unesterified protochlorophyll(ide) pigments are present which are transformable by light to their respective chlorophyll pigments (Lancer *et al.*, 1976). Studies of bean leaves undergoing development in the dark show that both protochlorophyll and protochlorophyllide are synthesized at early times; however, as development continues the synthesis of protochlorophyll stops and that of protochlorophyllide continues, leading to the predominance of protochlorophyllide in older leaves (Fig. 9).

The similarity between the young bean leaf and *Euglena* extends to the spectroscopic forms of the pigments found *in vivo* (Klein and Schiff, 1972; Kindman *et al.*, 1978). The etiolated older leaf material commonly employed shows protochlorophyll(ide) $[Pchl(ide)]_{650}$ predominantly (Fig. 10). Dark-grown *Euglena*, however, shows $Pchl(ide)_{635}$ to the exclusion of $Pchl(ide)_{650}$ (Fig. 11). Young beans, again, are like *Euglena*; $Pchl(ide)_{635}$ predominates although a small amount of $Pchl(ide)_{650}$ is also present. It should be noted that no correlation between protochlorophyll and protochlorophyllide on the one

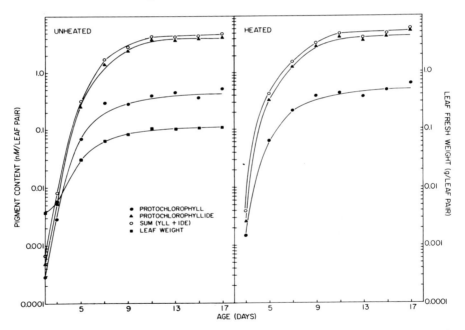

Fig. 9. Contents of protopigments during development of bean seedlings in the dark. The results are plotted semilogarithmically to allow the display of the wide range of concentrations on one graph. The two parts of the figure compare extraction in the cold of unheated leaves and leaves which had been steamed prior to extraction to inactivate chlorophyllase. From Lancer *et al.* (1976).

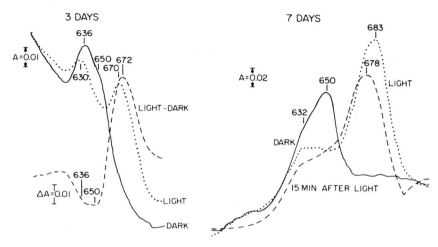

Fig. 10. Absorption spectra *in vivo* of 3-day-old and 7-day-old etiolated bean leaves showing predominance of Pchl(ide)$_{635}$ in young leaves and of Pchl(ide)$_{650}$ in older leaves. On light exposure of young leaves Pchl(ide)$_{635}$ is converted to Chl(ide)$_{670}$; the light minus dark difference spectrum shows a drop in absorption at 635–650 nm and a rise at 672 nm. On light exposure of older leaves, Pchl(ide)$_{650}$ is converted to Chl(ide)$_{683}$ which Shibata shifts to shorter wavelengths in darkness. Residual absorption in the 635–650 nm regions after light represents contributions of unconvertible Pchl(ide) plus Chl(ide). From Pardo and Schiff (1979).

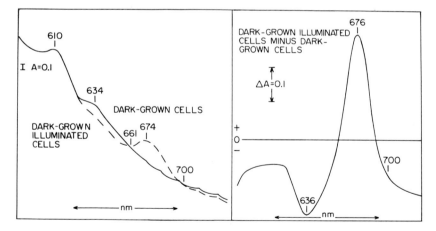

Fig. 11. Absorption spectra *in vivo* of dark-grown *Euglena* cells before and after illumination (red light, 125 W/m² for 2 min) and the corresponding difference spectrum. The cells were scanned 16 times in each case to produce the absorption spectra shown. Protochlorophyll(ide) absorbs at 634 nm; on illumination this is converted to chlorophyll(ide) at 674 nm. The residual absorption at 634 nm is due to unconvertible protochlorophyll(ide) plus some chlorophyll(ide) absorption. The light minus dark difference spectrum shows a drop at 636 nm and an increase at 676 nm. From Kindman *et al.* (1978).

hand and Pchl(ide)$_{650}$ and Pchl(ide)$_{635}$ on the other can be made. *Euglena* contains equal amounts of protochlorophyll and protochlorophyllide but has only Pchl(ide)$_{635}$; the older bean material has mainly Pchl(ide)$_{650}$ and contains mostly protochlorophyllide. The protochlorophyllide holochrome from beans absorbs at about 635–640 nm (Schopfer and Siegelman, 1968). It would seem that the explanation of the spectroscopic forms is to be sought in the way they are arranged in the thylakoid membranes, rather than in the chemical nature of the protopigments themselves. It is worth noting that systems having low amounts of Pchl(ide) have mainly Pchl(ide)$_{635}$ while those that have large amounts have mainly Pchl(ide)$_{650}$. Perhaps as Pchl(ide) is added to the membrane the pigment molecules are randomly distributed and lie far apart representing Pchl(ide)$_{635}$.. As the concentration of Pchl(ide) in the membrane increases, by chance more of them may lie close enough together to interact photochemically forming Pchl(ide)$_{650}$. If a mathematical model could be made perhaps the proportions of Pchl(ide)$_{635}$ and Pchl(ide)$_{650}$ would show such a relation to Pchl(ide) concentration.

The phototransformation of the various forms also differ (Kindman *et al.*, 1978). In the older leaves, Pchl(ide)$_{650}$ is phototransformed to Chl(ide)$_{685}$ followed by a Shibata shift in darkness to 675 nm. Other more transient shifts can also be distinguished spectroscopically (Bonner, 1969; Shibata, 1957; Gassman *et al.*, 1968; Bauer and Siegelman, 1972) often through low temperature stabilization. In younger bean leaves and *Euglena*, Pchl(ide)$_{635}$ is phototransformed to Chl(ide)$_{676}$. In *Euglena* (and possibly in the young bean leaf), there is then a small shift in darkness to Chl(ide)$_{673}$. Thus in all cases, the final position of the pigment is at about 673 nm, the region of absorption of the photosystem I antenna. Later P$_{700}$ and then 685 nm forms are formed in agreement with findings that during chloroplast development the formation of photosystem I preceeds that of photosystem II.

A summary of the principal protochlorophyll(ide) and chlorophyll(ide) forms *in vivo* is presented in Fig. 12. Pchl(ide) in ether absorbs at 623 nm (Fig. 7). Forms *in vivo* are found at 628 nm which are assumed to be unbound, perhaps dissolved in lipid. Pchl(ide)$_{635}$ formed in *Euglena* and young leaves and on aminolevulinic acid (ALA) feeding in older leaves was originally thought to be nontransformable, but more recent work has established that it is phototransformable *in vivo* (Boardman *et al.*, 1970; Sundquist, 1970; Kindman *et al.*, 1978). The nontransformable Pchl(ide)$_{635}$ formed from ALA feeding is converted to phototransformable Pchl(ide)$_{650}$ in darkness after a brief illumination. Pchl(ide)$_{650}$ can be converted to phototransformable Pchl(ide)$_{635}$ on freezing and thawing (Butler and Briggs, 1966). In general, single isolated pigment molecules absorb at shorter wavelengths when in solution (Shibata, 1957; Boardman *et al.*, 1970; Granick and Gassman, 1970; Kahn *et al.*, 1970). Association or binding to macromolecules or arrangement into crystals move the red absorption band to longer wavelengths.

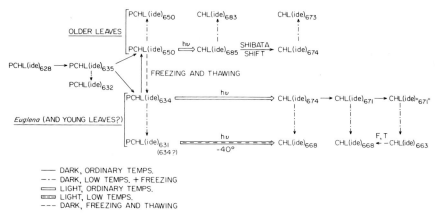

Fig. 12. Relationships among various spectroscopic forms of protochlorophyll(ide) and chlorophyll(ide) in various systems (see text for details). From Kindman *et al.* (1978).

If etiolated leaves or dark-grown *Euglena* cells are subjected to brief illumination to transform all of the protochlorophyll(ide), the protochlorophyll(ide) is regenerated in darkness to the same level it had preceding illumination. This can be done repeatedly (e.g., see Fig. 4 in Kindman *et al.,* 1978) indicating that protochlorophyll(ide) resynthesis is under tight control, perhaps through feedback inhibition. Since ALA is the first unique intermediate of the pathway of chlorophyll synthesis, it is reasonable to assume that the synthesis of ALA is the control point, although others may exist along the pathway. Control at the genome level through enzyme induction and turnover has been suggested for the ALA-synthesizing system of higher plants (Nadler and Granick, 1970).

There is one further ambiguity in the final steps of chlorophyll synthesis which is not yet resolved. Protochlorophyll (the esterified form of protochlorophyllide) and chlorophyll frequently contain long chain alcohols other than phytol such as geranylgeraniol (Rüdiger *et al.,* 1977; see C. Cohen and Schiff, 1976). Since geranylgeraniol is more unsaturated than phytol and since chlorophylls have been extracted from cells that contain various degrees of unsaturation in the side chain, one route of protochlorophyll and chlorophyll synthesis may be the esterification of protochlorophyllide or chlorophyllide with geranylgeraniol followed by successive reductions during chlorophyll synthesis, leaving only one double bond thereby forming phytol *in situ* (Rüdiger *et al.,* 1976, 1977; Schoch *et al.* 1977). Enzyme systems performing these reductions have been demonstrated in etioplast membrane preparations. Perhaps the one unsaturation is left in phytol because the tetrapyrrole ring sterically hinders the reductase from attacking the double bond closest to the ring.

Several properties of dark-grown *Euglena* and etiolated younger and older bean leaves are summarized in Table III. The young bean and *Euglena* are very similar and quite different from the older etiolated bean leaf. Since young bean leaves and *Euglena* carry out chloroplast development perfectly well when exposed to light, and since most seedlings under normal conditions emerge into the light rather rapidly, it would seem that the course of proplastid and chloroplast development in the young bean and *Euglena* represents the more normal mode of chloroplast development. The older bean leaf material shows characteristics that are more closely related to the pathology of prolonged etiolation. A plant that is growing in darkness for a long time is nearing the limits of its stored reserves and it is not surprising that evolutionary adaptation has brought about an increase in the size of the proplastid, the number and elaboration of membranes, and amount of protopigments; this leads to an etioplast that permits a more rapid development of photosynthetic function once the plant finally finds the light. Because the

TABLE III

Comparison of Properties of Dark-Grown *Euglena*, 2–3-Day-Old and 7–9-Day-Old Bean Leaves[a]

Property	Dark-grown *Euglena*	2–3 day etiolated bean	7–9 day etiolated bean
Structure	Proplastid	Proplastid	Etioplast
Size (μm)	~1–2	~2–3	~4
Prolamellar body	Small, noncrystalline	Absent, or small crystalline	Large, crystalline
Predominant absorption *in vivo*	Pchl(ide)$_{635}$	Pchl(ide)$_{635}$	Pchl(ide)$_{650}$
Total protopigment (pg/plastid)	~1×10^{-4}	~0.5×10^{-4}	~$7{-}10 \times 10^{-4}$
Protopigments present	Pchlide and Pchl	Pchlide and Pchl	Predominantly Pchlide
Ratio: moles -ide/moles -yll	3	1	6
First stable photoproduct *in vivo*	Chl(ide)$_{676}$ shifting to Chl(ide)$_{673}$	Chl(ide)$_{675}$	Chlide$_{685}$ Shibata shifting to 675 nm
Pigments produced directly on illumination	Chlide and Chl	Chlide and Chl	Chlide (predominantly)
Conversion Pchl(ide) to Chl(ide) (%)	10–50	40	80–90
Rate of Chlide esterification	Fast (no lag)	Fast (no lag)	Slow (lag)
Rate of protopigment regeneration	Fast (short lag)	Fast (no lag)	Slow (lag)

[a] From Lancer *et al.*, 1976.

older material is more easily obtained, most work has been done with this more specialized system of plastid development. It is time to turn our attention to the more normal course of events in the younger leaves and in microorganisms.

E. Evolution of Plastid Pigments in Response to Available Natural Light

Considering the wavelengths of light available at the surface of the earth (Fig. 13) (Collingbourne, 1966), it is paradoxical, at first sight, that chlorophyll a, the nearly universal photosensitizer of photosynthesis, has absorptions in the blue and the red regions of the spectrum, not in the middle of the spectrum where a considerable amount of energy is available. (In terms of numbers of available quanta, there is a constant plateau from about 550 to 800 nm with a steady decline from 550 to 400 nm.) One would think that life evolving under a yellow sun would evolve pigments to absorb light in the yellow–green region. Indeed, vision has evolved in this direction where rhodopsin and other retinal pigments absorb maximally in the middle of the spectrum with small shifts to accommodate various ecological situations (Lythgoe, 1972). In seeking reasons for the selection of chlorophyll we must

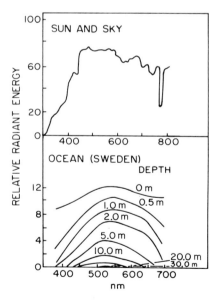

Fig. 13. Distribution of spectral energy from the sun at the earth's surface and at various depths in the ocean on relative scales. To compare upper and lower halves of figure, note 0 depth curve in lower half. After Collingbourne (1966); Levring (1966).

consider both the requirements of photosynthesis and the environmental pressures which selected the pigment systems found in contemporary organisms.

Another glance at Table II shows that all photosynthetic organisms contain plastid pigments other than chlorophyll a (or, in the case of the anoxygenic photosynthetic bacteria, bacteriochlorophyll a). These pigments have absorptions which fill the gaps in the middle of the spectrum left by chlorophyll a. Studies of the photochemical relationship of these pigments to each other and to the reactions of photosynthesis have shown that the light energy absorbed by these "accessory" pigments can be transferred efficiently to chlorophyll a (or bacteriochlorophyll a, where appropriate) (Butler, 1978; Blinks, 1964; Duysens, 1952; Myers and French, 1960; French, 1960). For this to occur several conditions must be satisfied.

1. The energy transfer must be from pigments absorbing at shorter wavelengths to those absorbing at longer wavelengths (the absorptions under consideration are those representing transitions from the ground state to the lowest excited singlet state). Since the energy in a photon or quantum of light is given by $E = hc/\lambda$ (where h is Planck's constant, c is the speed of light, and λ is the wavelength of light), it follows that if energy transfer is to occur it must do so in a manner that loses energy, since gaining energy from nowhere would violate the energy conservation laws. (This neglects small acquisitions of thermal energy which are small compared with photochemical quantities.) Thus in being transferred a quantum must stay the same or become smaller. If it becomes smaller, as it must do if the transfer process is less than 100% efficient, then the wavelength of the quantum must increase. Thus energy transfer is from shorter to longer wavelengths.

2. The two pigments involved must have energy states in common or, to put it another way, their absorption or fluorescence spectra must overlap.

3. The two pigments involved in energy transfer must be quite close to each other in space, usually less than 30–50 Å for efficient transfer to occur.

The thylakoid pigments satisfy these requirements and efficient energy transfer among these pigments has been measured in many systems (Butler, 1978; Blinks, 1964: Duysens, 1952). Several basic patterns exist depending upon the specific pigments present, but transfer is always from pigments whose lowest excited states result from absorptions at shorter wavelengths, to those whose transitions result from absorptions at longer wavelengths (Fig. 14 shows the basic schemes found in various groups). Although the series formulation shown in Fig. 14 is usual, studies based on the isolation of light harvesting pigment–protein complexes (Thornber *et al.*, 1979) from green and brown groups are beginning to provide evidence favoring parallel formulations. Thus carotenoids and chlorophyll b may transfer energy independently to chlorophyll a and fucoxanthin and chlorophyll c may transfer

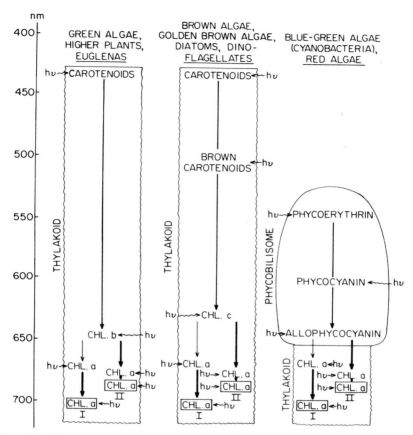

Fig. 14. Diagrammatic representation of energy transfer in representative groups of photosynthetic organisms. Any pigment can absorb light at its absorption maxima but energy transfer is always from the lowest excited singlet state of a pigment to a pigment whose lowest excited singlet state is at longer wavelengths; i.e., energy transfer is always from the top of the diagram downwards as indicated by the vertical arrows. The thickness of the arrows represents, roughly, the extent of energy transfer; the thicker the arrow the greater the energy transfer. The unboxed chlorophylls represent antenna pigments; the boxed chlorophylls represent the reaction centers of systems I and II which carry out the photodriven electron transport reactions with donors and receptors. All reactions are thought to occur in the thylakoids except for energy transfer among phycobilins in the red and blue–green algae which is thought to occur in phycobilisomes. Similar schemes involving bacteriochlorophylls can be written for energy transfer in the membranes of the photosynthetic bacteria. Although a series formulation is shown here, recent evidence indicates the possibility of parallel paths in green and yellow-brown groups (see text for details).

energy independently to chlorophyll a without transfer from carotenoids to chlorophylls b or c on the way to chlorophyll a (Prézlin and Alberte, 1978).

In chlorophyll-containing organisms, the ultimate recipient of the light energy is chlorophyll a serving as the reaction center of systems I and II of photosynthetic electron transport. In the photosynthetic bacteria, bacteriochlorophyll and bacteriopheophytin have been shown to be constituents of the reaction centers (Straley *et al.,* 1973; Arntzen and Briantis, 1975). The end result is that light energy absorbed in the region of the solar spectrum outside of the chlorophyll a absorptions themselves, is made available to the reaction centers and thus to the reactions of photosynthesis.

There are some clues as to how these accessory pigments have evolved. If we look at the way many of these groups of organisms are situated ecologically, for example, we find that they are sometimes stacked in a manner that takes advantage of energy transfer (Crisp, 1971). Some relationships among marine organisms which appear to result, at least in part, from light availability are shown in Fig. 15. Land plants and the green algae occupying the upper layers in ponds and ocean contain chlorophylls a and b besides carotenoids. The organisms living below them (e.g., the brown algae) have evolved special brown carotenoids which absorb blue–green light not absorbed by the green algae.

Photosynthetic bacteria can use far-red light not absorbed by other organisms through the bacteriochlorophyll long wavelength absorption. Blue–green and red algae are frequently the only photosynthetic organisms at very great depths. These organisms contain phycocyanin and phycoerythrin in phycobilisomes which transfer energy to chlorophyll a. These absorptions are in the green and orange regions of the spectrum not absorbed by the organisms above them. Absorption of light by other organisms is not the only factor which limits light at great depths to the blue–green region of the spectrum (as you look down into a body of water, the light reflected back through the water is blue–green). Water itself and yellow impurities in the water absorb red and blue light and the longer the column of water above the organism, the greener the light (Fig. 13) (Levring, 1966).

The red and blue–green algae are well adapted, through the phycoerythrin absorption, to make use of this residual green light. It should also be noted that these two groups have another dimension of adaptation. Cells of many blue–green algae and red algae can regulate the proportions of phycocyanin and phycoerythrin through a process called complementary chromatic adaptation (Bogorad, 1975; Haury and Bogorad, 1977; Fujita and Hattori, 1963; Vogelman and Scheibe, 1978; Björn, 1979; Ohki and Fujita, 1979). Organisms growing in light rich in green wavelengths form more phycoerythrin; those growing in light rich in longer wavelengths form more phycocyanin. Chromatic adaptation of this type seems to be restricted to organisms containing phycobilisomes; perhaps the phycobilisome evolved as a device to

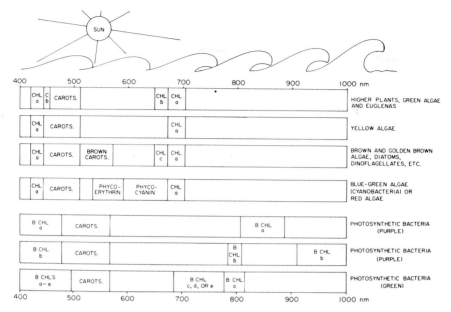

Fig. 15. Diagrammatic representation of absorption by photosynthetic pigments in various groups of organisms in relation to depth in the ocean. "Windows" at 510–600 nm and above 700 nm, where certain groups of organisms are transparent, allow light to penetrate to other organisms that have pigments in the right absorption range. Since water and substances in the water absorb light in the blue and red regions of the spectrum, green light predominates as depth increases. The photosynthetic bacteria are found at interfaces between aerobic and anaerobic environments such as mud and water where light (particularly far red light) can penetrate but an anaerobic environment containing reductants such as H_2S and H_2 can be maintained. chl a, b, c = chlorophyll a, b, c; Bchl a, b, c, d, e = bacteriochlorophyll a, b, c, d, e; carots = normal carotenoids (i.e., carotenes and xanthophylls); brown carots. = brown carotenoids (e.g., fucoxanthin).

facilitate modulation of the levels of phycobilin accessory pigments without changing the composition of the thylakoids and their reaction centers. Even organisms that cannot regulate the various forms of pigments they contain are able to control the total amount of pigments in response to varying light levels. Thus one finds greater amounts of photosynthetic pigments in algae (Halldal, 1970; Steeman-Nielsen, 1975; Ramus *et al.*, 1976) grown in dim light or in shaded leaves of higher plants (Bjorkman, 1973; Boardman, 1977). As the available quanta decrease, more antenna pigments are added to increase the probability of photon capture. Thus both evolutionary adaptation and short-term regulatory mechanisms ensure efficient capture of light and transfer to the reaction centers of photosynthesis.

It appears, then, that one way evolution has adapted organisms to the light available is to provide accessory pigments in response to available light which transfer their energy to the chlorophylls of the reaction centers. But what can we add about the evolution of chlorophyll itself?

We begin with a consideration of the biosynthetic pathway for heme and chlorophyll. (Fig. 16). (For chemical details of chlorophyll biosynthesis, consult Chapter 10 in Volume 8 of this series.) In animals (heme synthesis) and photosynthetic bacteria (bacteriochlorophyll synthesis) the pathway begins with an enzyme synthesizing aminolevulinic acid, ALA (Shemin and Russel, 1953), from succinyl-CoA and glycine (Shemin *et al.,* 1955; Kikuchi *et al.,* 1958; Gibson *et al.,* 1958). Although this system may participate in chlorophyll synthesis in cyanobacteria and algae (Klein and Senger, 1978; Harel, 1978), it has been extremely elusive in higher plants. Experiments in which labeled precursors are fed to algae and higher plants in which the utilization of ALA is blocked by administration of its analog, levulinic acid, the labeling of the accumulated ALA is consistent with compounds such as glutamate or alpha ketoglutarate being the precursors rather than succinyl-CoA and glycine (Jurgenson *et al.,* 1975; Beale, 1978; Harel, 1978); enzymatic investigations are in agreement (Kannangara and Gough (1978). In either case, condensation of two molecules of ALA yields the first pyrrole, porphobilinogen, and further condensation of four pyrroles yields the tetrapyrroles. Protoporphyrin IX is the branch point at which either iron can be inserted to yield heme and, eventually, when necessary, the open chain tetrapyrroles for phycocyanin and phycoerythrin formation by ring opening, or magnesium insertion is accomplished to begin the synthetic leg leading to chlorophyll. Granick first pointed out in his stimulating article (Granick, 1949) that the precursors to chlorophyll leading from magnesium protoporphyrin IX to chlorophyll showed both increasing shifts towards the red region of the spectrum and increasing strength of absorption (Fig. 17) of these bands. He further proposed that during evolution, each of these pigments might have served, transiently, as the sensitizer of photosynthesis. Thus an early photosynthetic pigment might have been Mg⁻protoporphyrin IX. When mutation brought about the formation of enzymes for the conversion of Mg protoporphyrin IX to Mg vinyl pheopophyrin a_5 or protochlorophyll(ide), this pigment was selected as the photosynthetic pigment and so on until the biosynthetic chain was lengthened to ultimately terminate in chlorophyll a. The longest wavelength of absorption of chlorophyll a is the farthest to the red of all of the precursors and this red band has the highest absorption coefficient of all of the precursors as well.

Now, if we combine all of these ideas, we can provide an explanation for the present pigment composition of photosynthetic organisms and its relation to available light. We must suppose that the evolution of chlorophyll for the reaction centers and the evolution of accessory pigments occurred concomitantly to adapt the organisms to the available light. If we adopt Granick's ideas, the long wavelength absorption of the early precursors to chlorophyll

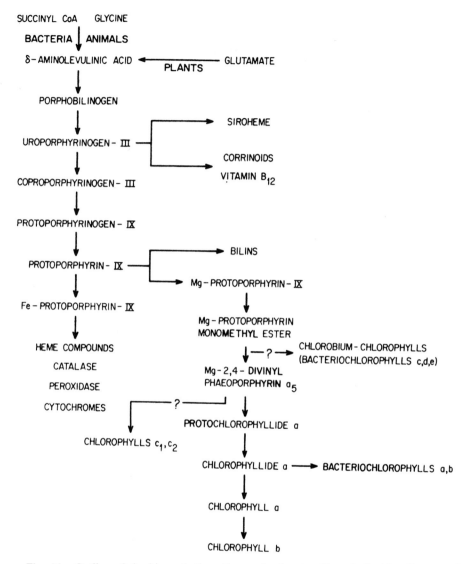

Fig. 16. Outline of the biosynthetic pathways leading to chlorophyll. After Ragan and Chapman (1978).

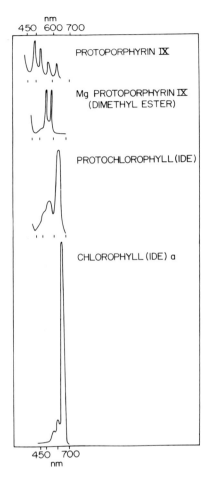

Fig. 17. Absorption spectra of intermediates in chlorophyll biosynthesis drawn roughly to scale for comparison of wavelength and relative absorbance. Insertion of Mg^{2+} into protoporphyrin IX increases long wavelength absorbance (Mg protoporphyrin IX). On closure of the isocyclic ring and other modifications the long wavelength peak moves into the red with an increase in absorbance [protochlorophyll(ide)]. On reduction of ring IV the peak moves further to the red with a further increase in absorbance [chlorophyll(ide) a]. After Granick (1948); Smith and Benitez (1955).

was more toward the yellow region of the spectrum where most light was available. As accessory pigments evolved, in response to each ecological situation (Fig. 15) the absorption of the primary sensitizer of photosynthesis had to move toward longer wavelengths because it would have to serve as the sink for energy transfer from the accessory pigments. We must suppose that the chlorophyll type of pigment was selected early in evolution for its chemical properties, i.e., for its ability to lose an electron to an acceptor when excited by light of the available energy. Thus chlorophyll (or its precursors) became the converter of photon energy to chemical energy and provided the interface between light reactions and electron transport. As accessory pigments filled in the middle of the spectrum in various organisms, the absorption of the chlorophyll pigments was steadily pushed to the red to serve both as the ultimate acceptor of photons from energy transfer by accessory pigments and as the primary step in photosynthetic electron transport. Accessory pigments and the blue peak of chlorophyll itself exist to capture the available photons; the red peak captures photons and also functions as the lowest energy sink for all of this energy and its absorption lies in the region of maximum available quanta from sunlight. We must suppose that evolution has rendered the structure of chlorophyll optimal for this function and for losing electrons to and gaining electrons from the primary electron transport carriers. A similar argument can be made for bacteriochlorophyll which has the added selective advantage of absorption at wavelengths so long that light is still available despite absorption by other organisms living above (Fig. 15). Photosynthetic bacteria are anaerobic when they are carrying out photosynthesis and are frequently found at the interface between anaerobic mud and water where both light and electron donors such as H_2S and H_2 are available. Since the photosynthetic bacteria lack system II of photosynthesis which needs a quantum with enough energy to split water, the lowest excited state of bacteriochlorophyll could be pushed by ecological selection to longer wavelengths where the energy per quantum is still enough to drive photosystem I, but may not be great enough for the splitting of water. The photoredox reactions of porphyrins in relation to the origins of photosynthesis have been discussed by Mauzerall (1978).

F. Organelle Transformations and Loss of Organelle Function

1. Organelle Transformations: Dedifferentiation, Chromoplasts, etc.

This section focuses on situations in which the form of the mature organelle is changed either through development into another type of organelle or through dedifferentiation to a precursor body. Mitochondria in facultatively anaerobic fungi such as yeast dedifferentiate to a promitochondrion when the cells are deprived of oxygen (Ephrussi and Slonimski, 1950; Slonimski, 1953; Watson *et al.*, 1971). Mature chloroplasts of certain algae (see preceding discussion) undergo dedifferentiation to proplastids on growth in darkness. This process has been carefully studied in *Euglena* (Ben-Shaul *et al.*, 1965). When green *Euglena* cells on a resting medium which prevents division through nutritional limitations are placed in darkness, the plastids remain in their fully mature form indefinitely. If the same cells are placed in a complete growth medium in darkness and allowed to divide, the plastid structure and pigment complement become progressively reduced until they are completely transformed into proplastids. This process resembles normal light-induced plastid development in reverse but at a considerably lower rate. During this process pDNA is conserved and when these cells are again returned to light either under dividing or nondividing conditions, the proplastids redifferentiate into chloroplasts. These results imply that the dedifferentiation of the plastid in *Euglena* is linked to processes of cell division and, perhaps, the reduction of the plastid requires plastid division in order to dilute the existing plastid structures and constituents among the daughter cells until the proplastid condition is reached and stabilized. This situation suits the peculiar ecological needs of *Euglena*. When growing in the light whether or not a utilizable carbon source is present, *Euglena* is always capable of making whatever use of photosynthesis the prevalent lighting conditions allow. In darkness, if no nutrients are available the cells do not divide and they retain their chloroplasts to make use of light as soon as it is available. When nutrients are available in darkness, the cells divide and rapidly reduce the excess baggage of chloroplasts to proplastids and live on the available nutrients as efficient organotrophs.

In higher plants the situation is quite different. The plastids are largely contained in a determinate structure, the leaf, which develops from a small protuberance on the stem apex, expands to the mature structure, and then undergoes senescence and death. Concomitantly plastids differentiate from small proplastids in the embryo and apex to the proplastid (or etioplasts) in darkness and to mature chloroplasts in light (Whatley, 1974; Klein and Schiff, 1972; Cran and Possingham, 1974). Although these maturing plastids may divide during leaf development, they never have an opportunity to return to the proplastid or reduced condition through cell division, since the leaf cells do not divide once they are mature and only face senescence and death. Instead, plastids are returned to the proplastid or reduced condition through the sexual cycle during the differentiation of egg cells and pollen. Once fertilization occurs and embryonic development forms new meristems, differentiation into a mature plastid can begin again. While all of the plastids in a microorganism such as *Euglena* are potentially immortal through division, only those plastids that are contained in sex cells are ultimately reduced and conserved through the entire life cycle of the higher plant. Of course, vegetative reproduction of various sorts can perpetuate plastids through division in some cases. [There is evidence in some systems such as the alga *Acetabularia* (Woodcock and Bogorad, 1970) that only a portion of the chloroplasts contain pDNA. Perhaps even in algae a "germ line" of plastids containing DNA and capable of division exists together with plastids formed from these but lacking DNA and serving only for photosynthetic function with no possibility of replication.]

The higher plants, however, have various forms of differentiated plastids which are not encountered in microorganisms. Among these are chromo-plasts, amyloplasts, and elaioplasts which accumulate carotenoids and other pigments, starch, and lipids, respectively (Kirk and Tilney-Bassett, 1967; Granick, 1960). Chromoplasts are structures which give color to flowers and fruits. Chromoplasts can be formed from proplastids, amyloplasts, or chloroplasts. Everyone is familiar with ripening in various fruits where the color changes from green to orange or yellow because chloroplasts are trans-formed into orange or yellow chromoplasts. The formation of chromoplasts from chloroplasts often entails extensive loss of chloroplast structure and the formation of new structures. Amyloplasts contain starch grains and serve as polysaccharide storage depots while elaioplasts accumulate oils, fats, or steroids. Although there is a fair amount of descriptive information, much more work is needed to understand the transformations among these plastids and how these transformations are controlled. All of the useful techniques of molecular biology which have been so effective in studying plastid develop-ment from the proplastid to the chloroplast await application to this problem.

2. Loss of Organelle Function

The loss of organelles can only be sustained in organisms that have evolved alternative means of survival which do not depend on the organelle's function. For example it is not possible to eliminate mitochondria or to reduce mitochondrial function in obligate aerobes such as *Euglena*. In facul-tatively anaerobic fungi such as yeast, however, where respiration can be re-placed by fermentation, it has been possible to obtain petit mutants (Ephrussi and Slonimski, 1950) which lack mitochondrial function, have reduced mitochondrial structure, and have lost mtDNA either partially or completely (Mahler *et al.*, 1975; Schatz, 1970; Nagley and Linnane, 1970; Slonimski and Lazowska, 1977). Similarly, it is only possible to study plastid reduction and loss in organisms which are not completely dependent on their chloroplasts for survival. Almost all higher plants, for example, are obligate phototrophs and are completely dependent on photosynthesis for survival. Since plastids can never be absent (they can be reduced to proplastids or etioplasts tran-siently in darkness only when nutrition is available from stored seed re-serves), essential functions other than those needed for photosynthesis have become lodged in the plastids. For example, starch synthesis (Kirk and Tilney-Bassett, 1967), sulfate reduction (Schmidt and Schwenn, 1971), and nitrite reduction (Ramirez *et al.*, 1966) are present in the plastids and could not be eliminated without killing the plant through nutritional deprivation. *Euglena,* however, is a facultative photo-organotroph. If light is not available the organism can use organic compounds as a source of reduced carbon. The plastid functions of *Euglena* can be greatly reduced yielding "bleached" strains which lack plastid pigments and pDNA (Edelman *et al.*, 1965). [It might be noted, as will be discussed subsequently, that the plastid and mitochondrial structures can never be completely eliminated because many of their functional components are coded in nuclear DNA (nDNA) and are synthesized outside the plastid.] Since *Euglena* can exist perfectly well without functional chloroplasts when given a reduced carbon source to replace photosynthesis, it may be inferred that this organism does not place essential cellular functions in its plastids other than those concerned directly with photosynthesis. Thus unlike the higher plants and green algae, storage polysaccharide synthesis occurs external to the plastid and sulfate reduction is in mitochondria and microbodies rather than in the plastids (Brunold and Schiff, 1976).

For this reason plastid loss is relatively easy to bring about and to study in *Euglena*. Among the "bleaching" agents known to cause plastid loss are ultraviolet light, nalidixic acid and other chemical agents (Ebringer, 1972), temperatures above 34° (Pringsheim and Pringsheim, 1952), and spontaneous mutation (Schiff *et al.*, 1971). Ultraviolet light causes a mass conversion of irradiated plated green cells to white colonies from which aplastidic strains can be cloned (such as W_3BUL) (Lyman *et al.*, 1961; Nicolas *et al.*, 1977). This ultraviolet effect has an action spectrum suggesting nucleoprotein (probably pDNA) as the uv absorber (Lyman *et al.*, 1961); in agreement, the effect of uv is fully photoreactivable through the action of a photoreactivating enzyme that opens pyrimidine dimers in the DNA (Schiff *et al.*, 1961; Diamond *et al.*, 1975). The level of this enzyme is raised by light and the effective wavelengths are the same as those for photoreactivation indicating, perhaps, that light acts as an inducing substrate for the photoreactivating enzyme. Mutants of *Euglena* (such as M_2^-BUL) are known which have normal chloroplasts when grown in the light. In darkness, however, the ability to form plastids is rapidly lost and white colonies result (Goldstein *et al.*, 1974). Unlike wild-type cells under the same conditions, the white cells cannot regain chloroplasts on light exposure. Although *Euglena* can grow at temperatures above 34°, the plastid is heat-sensitive and is rapidly lost above this temperature leading to wholesale production of bleached cells. Streptomycin, the first bleaching agent discovered (Provasoli *et al.*, 1948), produces permanent bleaching in *Euglena*, as does nalidixic acid (Lyman *et al.*, 1975) and various other chemical agents (Schiff *et al.*, 1971; Ebringer, 1972). Mutants, probably in pDNA, have been obtained which are resistant to the bleaching action of streptomycin (Diamond and Schiff, 1974); the plastids of these cells have 70 S ribosomes which fail to show the tight binding of streptomycin characteristic of wild-type cells (Schwartzbach and Schiff, 1974). This seems to be due to the alteration of several 70 S ribosomal proteins (Freyssinet, 1975).

In all cases the bleaching process leading to plastid loss requires division and results in the loss of pDNA from the bleached cells (Fig. 18). What appears to happen in the best-studied cases is that the bleaching agent brings about an arrest of the replication of the pDNA and as a result, the pDNA is

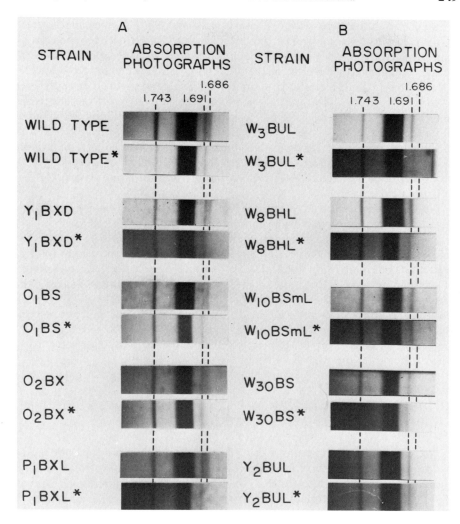

Fig. 18. The DNA's of *Euglena* mutants in cesium chloride gradients. Ultraviolet-absorption photographs of centrifuge cells loaded with 25–50 μg of DNA. Band of density 1.743 g/cm³ is a density standard. All strains in column A are capable of developing at least a partial chloroplast and possess both mitochondrial (1.691 g/cm³) and chloroplast (1.686 g/cm³) DNA. All strains in column B are incapable of even partial chloroplast development and contain only mtDNA (1.691 g/cm³). All strains, both in columns A and B, when subjected to bleaching doses of ultraviolet light (indicated by *) lack chloroplast DNA but retain mtDNA. The thick heavy band in all pictures represents nDNA. From Edelman *et al.* (1965).

diluted out as the cells divide (Goldstein *et al.,* 1974). Some destruction of pDNA may also accompany this process but its extent depends on the bleaching agent employed. The mechanism leading to the arrest of pDNA replication probably depends on the bleaching agent. Dimers are produced by uv in pDNA, which probably prevent replication, lead to errors in replication, or both. Streptomycin may prevent the synthesis of the enzymes necessary for pDNA replication on 70 S plastid ribosomes while nalidixic acid may act directly on the pDNA to prevent replication (Lyman *et al.,* 1975). Mutation, in the case of M_2^-BUL, appears to render the cell incapable of replicating its pDNA in darkness but not in light; perhaps different DNA polymerases or other enzymes associated with replication are active in mature plastids and in proplastids.

These bleached strains, particularly W_3BUL produced by uv irradiation, have served as important tools in determining where various proteins and nucleic acids are transcribed in the cell (Bovarnick *et al.,* 1974). Plastid DNA cannot be detected in this strain by ultracentrifugation on cesium chloride gradients (Edelman *et al.,* 1965) (Fig. 18), column separation of large amounts of cellular DNA (E. Stutz, personal communication), through sensitive hybridization experiments with cellular RNAs (Schwartzbach *et al.,* 1976; R. B. Hallick, personal communication) or by fluorescence microscopy of cells stained with a DNA-specific fluorochrome (Coleman, 1979). Since pDNA is undetectable in W_3, any protein or RNA in this strain must originate from a DNA other than the pDNA.

III. ORIGIN OF THE BIOCHEMICAL BUILDING BLOCKS FOR ORGANELLE DEVELOPMENT

There are several indications that the developing plastid does not require the products of photosynthesis. The optimal light intensity for plastid development is usually much lower than the optimum for photosynthesis (Stern *et al.,* 1964b); at the optimum for plastid development the rate of photosynthesis would be greatly reduced. Also, the developing chloroplast has a very low rate of photosynthesis at first when a great deal of synthesis of plastid constituents is occurring (Stern *et al.,* 1964a).

Euglena cells exposed to concentrations of DCMU which completely block photosynthetic carbon dioxide fixation by preventing the emergence of electrons from system II of photosynthetic electron transport still form normal amounts of chlorophyll, photosynthetic membranes, and plastids of normal size (see Fig. 2 in Schiff *et al.,* 1967). If these DCMU-treated cells are washed at the completion of plastid development, photosynthesis appears at once indicating that although the plastid has developed in the absence of photosynthesis, all of the machinery necessary for photosynthesis has formed.

In *Euglena* and other organisms, other organelles are found to be closely associated with the developing chloroplasts (see Fig. 8 in Schiff, 1973), such as mitochondria and microbodies. Although DCMU, an inhibitor of photosynthesis, does not prevent plastid development, inhibitors of mitochondrial electron transport and uncouplers of oxidative phosphorylation such as azide (Fong and Schiff, 1977) and dinitrophenol (Evans, 1971) are excellent inhibitors as is the removal of oxygen (Klein *et al.,* 1972).

This suggests that the energy, small molecules, and reducing power necessary for chloroplast development come from outside the developing plastid and are supplied by the rest of the cell. This seems to be fairly widespread. Many organisms such as members of the green algae and the cotyledons of gymnosperms are able to form chlorophyll and chloroplasts in darkness (Kirk and Tilney-Bassett, 1967); these organisms, too, must obtain the energy and other constituents for plastid development from processes other than photosynthesis, outside the developing plastid. This is also true of mutants of photosynthetic organisms blocked in some step in photosynthesis which make otherwise normal chloroplasts (Levine and Volkmann, 1961; Bishop, 1966).

Pyrenoids stand out as differentiated structures within many algal chloroplasts and are formed during plastid development (Griffiths, 1970) (Figs. 2 and 4). They frequently serve as centers about which the polysaccharide reserves of photosynthesis are organized and may be composed of enzymes participating in the dark reactions of photosynthetic carbon dioxide fixation. The pyrenoid of *Eremosphaera viridis* has been isolated and shown to contain ribulose bisphosphate carboxylase (Holdsworth, 1971); the same has been found for other organisms (Salisbury and Floyd, 1978; Kirby and Evans, 1978). When plastids are formed in *Euglena* in the presence of DCMU, pyrenoids fail to form (Schiff *et al.,* 1967). When the DCMU is washed out, pyrenoids rapidly appear without the need for protein synthesis and ribulose bisphosphate carboxylase activity increases (Schiff, 1973). It is possible that normal photosynthesis maintains an ionic environment inside the chloroplast which encourages a condensed form of ribulose bisphosphate carboxylase and, perhaps, other enzymes of CO_2 fixation in the form of a pyrenoid. When this ionic environment is altered through inhibition of photosynthesis and the consequent lack of ion pumping driven by photosynthetic electron transport, it may no longer favor the formation of condensed forms of the enzymes and no pyrenoids form. If the activities of the condensed and uncondensed forms are different this might serve to regulate photosynthesis. Another possible instance of the influence of ionic environment on photosynthetic structures is also found in *Euglena*. If light grown chlorophyll-containing cells are dark-adapted on an acidic medium under conditions of low aeration, the chlorophyll of the plastids is converted to pheophytin. Since Mg removal is known to be acid catalyzed, it is possible that under the conditions employed, hydrogen ions can no longer be pumped away from the chlorophyll leaving the pigments in a highly acidic environment that favors removal of Mg (Greenblatt and Schiff, 1959).

The source of constituents for mitochondrial development has not been worked out to my knowledge, but it would be unexpected if the energy, reducing power, and small molecules were not ultimately provided by fermentation through the glycolytic pathway (Perlman and Mahler, 1974).

If the rest of the cell provides the nutrition for organelle development, how is this organized and made available? We have some indications in the case of plastid development. In *Euglena,* for example, there is a light-induced increase in oxygen uptake associated with plastid development which may represent a light-induced respiration required for increased energy formation (Schiff, 1963). The formation of energy, small molecules, and reducing power begins with the light-induced breakdown of carbohydrate reserves. In *Euglena* paramylum is the reserve material which is broken down when the cells are illuminated to begin plastid development; this is localized outside the plastid since it is present in W_3BUL which lacks most plastid structure and pDNA (Schwartzbach *et al.,* 1975). Since paramylum breakdown can be induced in darkness by cycloheximide and levulinic acid, a model has been proposed based on negative control at the genome level with light and aminolevulinic acid acting to repress transcription (Schwartzbach *et al.,* 1975; Schiff, 1978). The reducing power formed during plastid development can also be detected through the reduction of the nitro group of chloramphenicol to an amino group (Vaisberg *et al.,* 1976). This requires electron transport from NADPH via NADPH-ferredoxin reductase; experiments with mutants and DCMU, however, show that the electrons do not come from photosynthesis, but rather from the central metabolism of the cell, probably from blue light-induced paramylum breakdown. Products of photosynthesis in higher plants appear to be altered in blue light (Voskresenskya, 1972) and in algae such as *Chlorella,* where carbon substrates can be fed, blue light brings about the mobilization of starch reserves in the manner described for *Euglena* (Kowallik, 1970).

IV. ORIGIN OF THE GENETIC INFORMATION FOR ORIGIN OF THE GENETIC INFORMATION FOR
ORGANELLE DEVELOPMENT

Eukaryotic cells containing mitochondria have two DNAs, one in the nucleus and one in the mitochondrion; plastid-containing cells have three, one each in the nucleus, mitochondrion, and plastid (Bücher *et al.*, 1976; Saccone and Kroon, 1976; Nigon and Heizmann, 1978; Schiff, 1973; Schatz, 1970) (see Fig. 18, for example). The nDNA represents some 95% of the cellular DNA with pDNA and mtDNA representing a few percent each. Nuclear DNA represents about 10^{12} daltons *in toto*. Plastid DNA is typically of the order of 10^8 daltons, only slightly smaller than a bacterial genome but mitochondrial DNAs are more variable. Many of these organelle DNAs have been shown to be circular, as in bacteria (Manning *et al.*, 1971; Schatz, 1970). The information, therefore, for plastid or mitochondrial formation could come from any of the three DNAs. There are, however, no well-documented cases confirming informational interractions between plastids and mitochondria. There is, however, a great deal of data that indicate that the chloroplast or mitochondrion relies both on its own DNA and the DNA of the nucleus for genetic information (Schiff, 1973, 1974; Schatz, 1970; Bücher *et al.*, 1976), a result consistent with the genetic experiments presented earlier.

It has already been noted that plastids and mitochondria resemble prokaryotic cells resident in a eukaryotic cell. Each of these cellular compartments has its own protein synthesizing system (Fig. 19). The nDNA codes for the (ribosomal) rRNAs of the 80 S-type ribosomes and the (transfer) tRNAs of the cytoplasm (Schwartzbach *et al.*, 1976). Another complete set of tRNAs and the rRNAs of the 70 S plastid ribosomes are coded in pDNA and these, as far as known, stay within the plastid (Schwartzbach *et al.*, 1976). Still another set may be coded mtDNA and remain within the mitochondrion (Buetow and Wood, 1978; Barnett *et al.*, 1978). To date, there does not seem to be any clear indication that RNAs or DNAs cross the boundaries of plastids and mitochondria.

The mutants lacking detectable pDNA in *Euglena* such as W$_3$BUL (Edelman *et al.*, 1965) have been extremely useful in deciding which proteins are coded in nDNA (Bovarnick *et al.*, 1974). Plastid proteins that are present in mutant W$_3$BUL cannot be coded in pDNA and are probably coded in nDNA. The synthesis of all of these enzymes is inhibited by cycloheximide, a specific inhibitor of translation on the cytoplasmic ribosomes of *Euglena*. Thus the messages for these proteins must be transcribed from nDNA and translated on cytoplasmic ribosomes. This is found to be the case in several other plastid systems. For example, it is possible to obtain mutants affecting chloroplast enzymes in *Chlamydomonas* which map to specific loci on the chromosomes of the nuclear genome (Levine and Goodenough, 1970; Hudock and Rosen, 1976; Adams *et al.*, 1976). There are several enzymes of yeast mitochondria which map to the nuclear genome (Schatz, 1970; Bücher *et al.*, 1976; Sherman *et al.*, 1968). In general, plastids and mitochondria have substantial cadres of proteins which are coded in nDNA and are translated (synthesized) on cytoplasmic 80 S-type ribosomes.

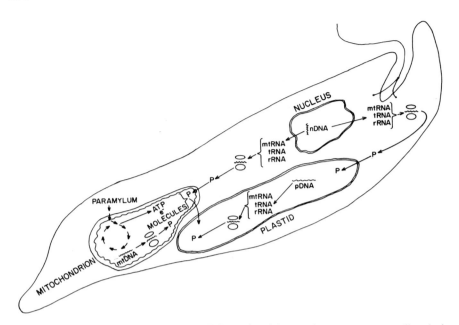

Fig. 19. Summary of nutritional and informational interactions among organelles during chloroplast development in *Euglena*. The DNAs of either the mitochondrion or the chloroplast code for that organelle's tRNAs and rRNAs and for certain proteins (P) which are translated on the ribosomes of that organelle and remain within the organelle. Nuclear DNA codes for rRNA and tRNAs of the cytoplasm and for proteins which are translated on cytoplasmic ribosomes. Some of these remain in the cytoplasm, others enter the developing organelles. Light-induced paramylum breakdown feeding mitochondrial respiration supplies small molecules, energy, and reducing power to the developing chloroplast. From Schiff (1978).

It has proven more difficult to find the organelle proteins which are coded in pDNA or mtDNA. Although certain proteins are thought to be coded in pDNA or mtDNA and synthesized on the 70 S-type organelle ribosomes, this is an area that requires more work. Interestingly enough, the most convincing evidence for organelle proteins of this type is obtained from studies of organelle enzymes which have subunits of nuclear-cytoplasmic origin *and* plastid or mitochondrial origin. For example, it seems clear that in most plastid-containing organisms, the small subunit of ribulose bisphosphate (RuBP) carboxylase, the first enzyme of carbon dioxide fixation in photosynthesis, is coded in nDNA (Kung, 1976; Chen *et al.*, 1976), translated on cytoplasmic ribosomes, and then is processed as it is transported into the plastid (Ellis and Barraclough, 1978; Highfield and Ellis, 1978; Dobberstein *et al.*, 1977; Schmidt and Chua, 1978). The large subunit of the RuBP carboxylase, on the other hand, is coded in pDNA (Kung, 1976; Chen *et al.*,

1976; Sagher *et al.*, 1976) and is translated on plastid ribosomes. A similar situation is found with cytochrome oxidase, a terminal oxidase of yeast mitochondria: here the enzyme consists of many different subunits, several of which are coded in nDNA and translated on cytoplasmic ribosomes and several that are coded in mtDNA and translated on mitochondrial ribosomes (see Bücher *et al.*, 1976).

An active field of study now is to prepare DNA from organelles and by fragmentation of this organelle genome DNA with specific endonucleases (restriction enzymes) to obtain segments that can by hybridized with (messenger) mRNA, rRNA, and tRNAs (see papers in Bücher *et al.*, 1976; Hallick *et al.*, 1978; Steinmetz *et al.*, 1978; Stutz, 1978, and other papers in Akoyunoglou and Argyroudi-Akoyunoglou, 1978). In this way it is becoming possible to map the organelle genomes to determine the relative placement of the DNA sites controlling the formation of various organelle RNAs and proteins.

Another current approach is to use isolated chloroplasts from higher plants (Ellis, 1978; Siddell and Ellis, 1975) and *Euglena* (Vasconcelos, 1976) or mitochondria from various sources (Buetow and Wood, 1978) to study protein synthesis. Radioactive amino acids are incorporated into plastid proteins at maximal rates in red light and probably represent protein synthesis on plastid polysomes already primed with messages which presumably originated in pDNA.

In addition to the soluble proteins of the plastid, there are the ribosomal and membrane proteins. In general, the cytoplasmic 80 S-type ribosomes have some 80 proteins while the plastid 70 S ribosomes are similar to their bacterial counterparts and have some 55 ribosomal proteins (Freyssinet and Schiff, 1974; deVries and van den Bogert, 1976; O'Brien *et al.*, 1976). At present it appears that some of the ribosomal proteins of the plastid may be coded in nDNA and others in pDNA (Freyssinet, 1977). Information on mitochondrial ribosomal proteins is also available (Buetow and Wood, 1978).

Using inhibitors of translation in various organisms, it has been inferred that some thylakoid membrane polypeptides are synthesized on cytoplasmic ribosomes and others on plastid ribosomes, including polypeptides which ordinarily carry pigments in the mature thylakoid membranes (Ohad, 1975; Bogorad and Weil, 1977). In *Euglena*, it has been possible to locate plastid thylakoid membranes from dark-grown cells and mutant W_3BUL in which plastid DNA is undetectable on sucrose gradients by using the sulfolipid of the thylakoids as a specific marker (the sulfolipid is synthesized by enzymes which are nuclear-coded) (Bingham and Schiff, 1976, 1979a,b). Surprisingly (see figures in references) there is an exact correspondence between the bands from W_3BUL and dark-grown wild-type cells indicating that all of the proplastid thylakoid polypeptides are coded in nDNA. During greening of the cells, most bands increase while others appear, including those which will ultimately carry the photosynthetic pigments. Thus far, no insoluble membrane polypeptides have been found which are coded in pDNA, although some of those which appear during greening may belong to this group. Cytochrome c-552, which is loosely associated with the *Euglena* thylakoid membrane and which serves in place of plastocyanin (as has been found for other algae) (Wood, 1976; Bengis and Nelson, 1977), may be coded in pDNA since it is undetectable in mutants lacking detectable pDNA such as W_3 and its synthesis is blocked by streptomycin indicating that it is synthesized on the 70 S plastid ribosomes (Freyssinet *et al.,* 1979). Some algae can synthesize both plastocyanin and the cytochrome c; with copper ions in the medium plastocyanin is made, in the absence of copper ions the cytochrome is made (Bohner and Böger, 1978).

There are indications that the formation of plastid thylakoid membranes is tightly regulated and coordinated. For example, if synthesis of carotenoids in *Euglena* (which, as in other organisms, protect the cells from photooxidation) is specifically blocked at the level of phytoene with the herbicide SAN 9789, the synthesis of chlorophyll and other membrane constituents is impeded and membrane synthesis stops (Vaisberg and Schiff, 1976). In higher plants such as beans, normal etioplast membranes and prolamellar bodies containing protochlorophyll(ide) are made in the absence of carotenoid synthesis when germination takes place in SAN 9789 (Pardo and Schiff, 1980; Frosch *et al.,* 1978). The case of *Euglena* is understandable, since there would be a strong evolutionary selection against making chlorophyll and membranes in the absence of carotenoids which would lead to cell death through photooxidation. Why this selection did not operate in the case of higher plants is puzzling.

The situation with respect to sources of energy, intermediates, and genetic information for organelle development (Schiff, 1978) is summarized in Fig. 19. The mitochondrion and the chloroplast each contain their own DNA which codes for the organelle tRNAs and rRNAs. It also codes for certain mRNAs which are translated on the 70 S-type organelle ribosomes to form proteins that remain in the respective organelles. The nDNA codes for the tRNAs and rRNAs of the cytoplasm with 80 S-type ribosomes as the functional translational units. Most of the proteins made on these ribosomes stay in the cell compartments outside of the mitochondrion and plastid but certain of them cross the organelle boundaries, usually with some processing enroute, to become proteins of the mitochondria and chloroplasts. The breakdown of the storage materials of the cell is under the control of light and the flow of these breakdown products through the metabolic pathways of the cell and the electron transport and oxidative phosphorylation routes of the mitochondria, provide the reducing power, energy, and small molecules for the formation of plastids (and, probably, the mitochondria).

V. COORDINATION OF ORGANELLE DEVELOPMENT THROUGH INDUCTION BY LIGHT AND OTHER SUBSTRATES

The picture we have, then, is of a plastid, or perhaps a mitochondrion, which must be supplied with reducing power, energy, small molecules, and proteins during substrate-induced development, oxygen in the case of the mitochondrion, light and other substrates in the case of the chloroplast. While light is the inducing substrate in *Euglena* and higher plants, the length of the lag phase in *Euglena* is governed by the previous nutritional history of the cells. Cells that are rich in paramylum by having been grown in media rich in carbon, have relatively short lags while cells grown in nitrogen-rich media and having low paramylum stores have very long lags (Freyssinet, 1976). In certain organisms such as *Chlorella prototosecoides* where chlorophyll and chloroplasts can form in darkness, the formation of the chloroplast is controlled in a similar manner by the carbon to nitrogen ratio (Hase, 1971). Utilizable reduced carbon substrates repress plastid development in *Euglena* (App and Jagendorf, 1963; Horrum and Schwartzbach, 1980).

From a considerable amount of information, chiefly from *Euglena*, a model has emerged for the sequence of functions concerned with plastid development. This model is summarized in Fig. 20. The lag period is thought to be the period during which the synthetic activities in the nonplastid compartments of the cell are mobilized to feed the developing proplastid with those materials that are provided from outside. This nonplastid contribution is under control of a blue light absorbing system in *Euglena* and in a mutant of *Scenedesmus* (Brinkmann and Senger, 1978), but can also be under control of a red-far red phytochrome system in other organisms such as higher plants (Virgin, 1972). When sufficient supplies have been imported into the developing plastid, the lag period ends and the plastid enters into a period of rapid development induced by a red–blue absorbing system which seems to be identical with the absorption of protochlorophyll(ide) (Schiff, 1978; Egan *et al.*, 1975). In *Euglena* it is possible to separate the early and late events of plastid development through preillumination. Normally, dark-grown cells exposed to light at zero time show a 12-h lag in chlorophyll synthesis. If 2 h of preilluminating light followed by a dark period is given previous to this zero time postillumination, the lag is eliminated (Holowinsky and Schiff, 1970). Thus given a brief preillumination, the cells can carry out in the ensuing dark period those processes which ordinarily take place in the lag period in continuous light. The action spectrum for preillumination shows a blue light type of action spectrum (see Fig. 11 in Schiff, 1974) while the action spectrum for postillumination shows a typical protochlorophyll(ide)

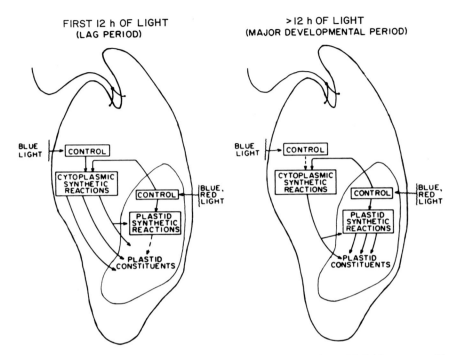

Fig. 20.   Temporal organization of plastid development in *Euglena*. Model for events of lag period and period of active development. The lag period is thought to be a period in which the proplastid receives energy, reducing power, small molecules, and proteins from the rest of the cell under control of the blue absorbing receptor system. When sufficient materials have been imported, synthesis within the plastid becomes predominant under control of the blue–red absorbing receptor system and the rapid period of plastid development ensues.

action spectrum. Similarly, transcription of cytoplasmic rRNAs from nDNA is induced by the blue light system while transcription of plastid rRNAs is controlled by the red–blue protochlorophyll(ide) system (D. Cohen and Schiff, 1976).

If one looks at the action spectrum for the formation of plastid proteins that are nuclear coded and cytoplasmically synthesized, one sees that a red and blue light photoreceptor like protochlorophyll(ide) is involved (Egan *et al.*, 1975). Since it is known that chloroplasts can develop normally given only red light, we must suppose that on red light induction of the protochlorophyll(ide) plastid system, a signal of some sort is sent to the nuclear genome which can replace the inductive effect of the blue light absorbing photoreceptor and turn on the nonchloroplast processes in the absence of blue light.

Remembering that light-induced breakdown of the storage carbohydrate in the cytoplasm of *Euglena* is necessary for chloroplast development, it is possible to form the scheme of Fig. 21 (Schiff, 1978). Transcription of the nuclear genome and paramylum breakdown are under control of the blue light photoreceptor system or the signal from the plastid. (Blue light induction of nonchloroplast events such as paramylum breakdown and transcription of cytoplasmic rRNAs from nDNA can be demonstrated in mutant W_3BUL lacking the plastid transcription system and protochlorophyll(ide) and, of course, in wild-type cells.) Formation of chlorophyll and control of plastid transcription are under control of the red–blue protochlorophyll(ide) photoreceptor. Induction of this system leads to a signal which operates to turn on nonchloroplast processes in the absence of blue light.

We have already discussed the nature of the protochlorophyll(ide) system and its involvement in chlorophyll synthesis. The blue light receptor system is less well defined and the responsible pigments are euphemistically called "cryptochromes." It encompasses those elusive pigment systems which are also involved in phototropism, phototaxis, spore germination, photorespira-

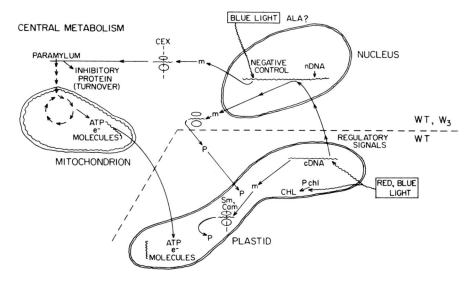

Fig. 21. Summary of photocontrol of plastid development in *Euglena*. Blue light is thought to induce paramylum breakdown and nuclear transcription, possibly through negative control. Blue and red light, through a protochlorophyll(ide)-type of photoreceptor, is thought to control plastid transcription and chlorophyll(ide) formation. Since those plastid enzymes which are nuclear-coded show protochlorophyll(ide)-type action spectra for induction, it is suggested that illumination of the plastid receptor results in a regulatory signal which passes to the nucleus to derepress the appropriate loci for transcription of these proteins in the absence of blue light. ALA, δ aminolevulinic acid; Cam, chloramphenicol; cDNA, plastid DNA; CEX, cycloheximide; Chl, chlorophyll; m, messenger RNAs; nDNA, nuclear DNA; P, proteins; Pchl, protochlorophyllide; Sm, streptomycin; W_3, mutant W_3 BUL (in which cDNA is undetectable); WT = wild-type cells (containing cDNA).

tion, etc. (Senger, 1980). There seem to be many blue light systems in plants which might be collected under the title "blue light syndrome" (Batra, 1971; Kowallik, 1970; Voskresenskya, 1972; Virgin, 1964; Carlile, 1970; Briggs, 1964; Clayton, 1964; Nultsch, 1970; Diehn, 1969). Blue light systems have been extracted from many organisms which show blue light-induced absorption changes in the appropriate redox environments (Poff and Butler, 1974; Schmidt and Butler, 1976; Zeldin and Schiff, 1975). These usually can be shown to consist of a flavoprotein interacting with a cytochrome of the b type. The flavoprotein is usually the photosensitizer bringing about redox changes in the cytochrome.

That this system is the underlying mechanism for blue light effects in cells is made uncertain by the fact that flavoproteins and even added flavins can sensitize photo-oxidations in extracts (Lewis *et al.*, 1961). Also, although the action spectra for the biological effects are frequently of the type expected for flavins, other action spectra are found which are more of a porphyrin type (Batra, 1971; Zeldin and Schiff, 1975). Perhaps either a flavoprotein or a porphyrin can act as the photosensitizer, or perhaps this is not the system acting *in vivo*. It is worth noting that both flavoprotein and porphyrin-type action spectra are found in closely related genera of bacteria where they control carotenoid biosynthesis (Batra, 1971). In any case, it seems wise to withhold judgment until it is possible to prove conclusively a connection between the systems *in vitro* and the processes they are thought to control *in vivo*. This might be achieved through comparison of more precise action spectra *in vitro* and *in vivo* and through mutants blocked in the photoreceptors (Schmidt and Lyman, 1974).

The phytochrome system is well studied in higher plants and certain filamentous green algae but the mechanism providing a connection of the pigment protein to the processes it is supposed to control has not yet been elucidated. It appears that the phytochrome system replaces the blue light system in many cases in multicellular plants, the cytoplasmic aspects of plastid development being one example (Virgin, 1972). Phytochrome-like biliprotein control pigments seem to have arisen very early in evolution in the blue–green algae or cyanobacteria (Björn, 1979); the control of phycocyanin and phycoerythrin ratios by a red–green system (Haury and Bogorad, 1977; Fujita and Hattori, 1963) and the control of morphogenesis in *Nostoc* by a red–green reversible system are examples (Lazaroff and Schiff, 1962). Perhaps they arose at the same time as the biliproteins whose formation they control in the blue–green algae. If so, the red–green system would be more primitive than the red-far red system which seems to have come later and has become pervasive in the multicellular plants.

While control at the level of metabolism by blue receptors, red-far red receptors, or protochlorophyll(ide) receptors is one possibility, we should not ignore another level of control. As we have noted previously, control of organelle development is an example of substrate control. Substrate control in bacteria involves repressors which bind to DNA and block the transcription of genes. For example, in *E. coli* grown on glucose the genes coding for the enzymes involved in lactose utilization are dormant because a repressor blocks their transcription. On presentation of lactose to the cells in place of glucose, the lactose binds the repressor of the lactose genes removing it from the DNA and allowing the transcription of the genes which code for the enzymes of lactose utilization. It is entirely possible that similar modes of control have evolved in organelle development. In the case of chloroplast development the inducing substrate is light. Perhaps pigmented repressors have evolved which recognize specific wavelengths of light and are removed from sites where they inhibit transcription by absorption of this light. One might further suppose that evolution has selected for pigmented repressors whose absorption spectra resemble the absorptions of the processes that they control. Thus plastid development is controlled by blue and red lights which are utilizable by photosynthesis. It would not be adaptive in *Euglena* and higher plants, for example, to have plastid development induced by green light since this light would not be effective for photosynthesis once the plastid was formed. Thus we have plastid development induced by blue and red light and the elusive photoreceptors involved may not be the readily measured flavin and protochlorophyll(ide) pigments but pigmented repressors present in exceedingly small amounts which control the cytoplasmic and chloroplast events through transcriptive control at the level of nuclear and chloroplast DNA. [The induction of carotenoid biosynthesis by light and oxygen which causes photodynamic damage protected against by carotenoids may be another example of this type of control (Batra, 1971).] In mitochondria we would suppose that substrate control of transcription is exerted by oxygen acting without light to bind a repressor. It would seem remarkable that having perfected a genomic mechanism of control of transcription which recognized inducing substrates of an organic or inorganic nature, that adaptation and selection did not extend this type of control to recognition of light as an inducing substrate.

VI. ORIGIN OF PRESENT INTERRELATIONSHIPS AMONG ORGANELLES DURING EVOLUTION

There seem to be two extremes of opinion about plastid origins. On one hand, it has been suggested that the plastid may have had an episomal origin (Raff and Mahler, 1972). Thus a portion of the main genome of the prokaryotic cell may have become detached at some point and may have established itself in the cytoplasm as the genetic unit for an independent organelle. Over the course of evolution these episomes would evolve into the organelles we see today. One argument against this is the fact that the base compositions of the organellar DNAs show no correlation with the nDNA of the cell in which they reside. On the other hand, it could be argued that a very nonrepresentative portion of the nDNA became detached to form the organelle, or that subsequent evolution has caused the organellular DNA to diverge from the nuclear base composition.

The other view begins with Schimper and Mereschkowsky who suggested that blue–green algae invaded eukaryotic cells and became established as endosymbionts (see Margulies, 1970). This is a very attractive hypothesis, since the organelles as we know them today are prokaryotic in nature, and it seems the simplest course to suppose that they started that way. We also know that blue–green algae form endosymbiotic relations with eukaryotic organisms (Tiffany, 1951), and there are the cases of *Cyanophora paradoxa* and *Glaucocystis nostichearum* (see Fogg *et al.*, 1973; and Trench, 1979) which are flagellates that harbor as endosymbionts organisms which resemble blue–green algae. Amoebae lacking mitochondria harbor endosymbiotic bacteria (Whatley, 1976). It would seem that it is a short way from symbiotic relationships of this sort to becoming cellular organelles.

What can be learned from the DNA base compositions of plastids and blue–green algae? When we first did our work in this area we had the hope that definitive relationships between the blue–green algae and plastids would emerge. We are sadder and wiser now. Both our own work (Edelman *et al.*, 1967) and that of Stanier *et al.*, (1971) have suggested that if a relationship exists it is quite complex (see Fig. 17 in Schiff, 1973).

The DNAs of the blue–green algae span such a wide range that it would be surprising if there were no overlap with the base composition of the plastids. Indeed, the base compositions of the *Nostocaceae* and Stanier's groups IIA and IIB of unicellular blue–greens do overlap those of the pDNAs from higher plants and green algae which are very close to each other as might be expected from their evolutionary affinities. Still, there are many blue–green algal groups that show no correspondence, and the pDNA of *Euglena* and the putative pDNA of *Ochromonas* are far outside the blue–green algal range. It seems likely that plastids in various groups might have their origins in different invaders. It certainly seems likely that the plastids of *Euglena* arose from a different invader than the plastids of the green algae and higher plants. Thus, the problem becomes, if we retain the invasion hypothesis, which prokaryotic groups gave rise to which plastids. We must be prepared for the possibility that the precursors of certain plastids might have not persisted to the present day and are lost in the fossil past.

Another dimension has opened with the discovery of a prokaryotic organism, *Prochloron* (Lewin, 1975, 1976, 1977; Lewin and Withers, 1975; Newcomb and Pugh, 1975; Thorne *et al.*, 1977; Whatley, 1977b; Perry *et al.*, 1978), having chlorophyll a and b but no phycobilins. Such organisms could have become the plastids of the green algae, and *Euglenas* through invasion and habituation.

Perhaps there was considerable evolutionary diversity among the primitive oxygenic prokaryotes with all of the photosynthetic pigments and structures familiar to us in the plastids of modern eukaryotic algal groups represented in one or another of the free-living prokaryotic organisms. These then invaded various eukaryotic cells to become established as chloroplasts. If this actually happened, then the diversity of plastid and pigment types we see today may have evolved in the free-living prokaryotes before they invaded and became established as plastids. We now have prokaryotes resembling the plastids of red algae (the blue–green algae) and the plastids of green algae, *Euglenas* and higher plants (*Prochloron*); it remains to be seen whether other free-living prokaryotes having the pigments and structures of the chloroplasts of other algal groups will turn up.

Despite the equivocal nature of the nucleic acid data, the endosymbiotic hypothesis is still attractive, given the large weight of evidence suggesting close similarities between the organelles and free-living prokaryotes (Table I). My own synthesis of existing ideas and information can be stated as follows.

We must suppose that during the anaerobic phase of the formation of life, organisms evolved which resembled the primitive anaerobic prokaryotes we know today. With the exhaustion of the original supply of reduced organic materials in the primordial soup, there was a selection pressure for photosynthesis which at first was anoxygenic and dependent on external reductants such as H_2S from the environment. With the exhaustion of these, organisms that could use water as a reductant were selected, and these first prokaryotic oxygen evolvers probably resembled primitive blue–green algae or prochlorophytes. With the appearance of free oxygen in the atmosphere as a result of oxygenic photosynthesis, aerobic bacteria could also evolve which carried out respiration. At this point some organisms existed which resembled primitive free-living plastids or mitochondria (blue–green algae and perhaps other prokaryotic oxygenic organisms, and bacteria, respectively).

At the same time, certain other cells were growing larger. Increased size and complexity implies that the genome was becoming larger to handle the structural and regulatory demands of such cells, and as the genome enlarged, mechanisms evolved to ensure that this genome was faithfully passed to daughter cells. As long as the primitive prokaryotes were small, a single circular DNA molecule was adequate as their genome, and this could be faithfully handed along through membrane attachment as in the modern bacteria (and, probably, blue–green algae). With an increase in the size of the genome and in regulatory complexity, chromosomes were evolved, and the necessary mitotic apparatus was formed to ensure regular transmission of the several chromosomes to the progeny. With it came the partitioning of this apparatus into a nuclear compartment surrounded by a membrane with extensions of endoplasmic reticulum to allow transmission of genetic information from the nucleus efficiently within the larger volume of this much enlarged cell. It seems obvious that most eukaryotic cells are much larger than prokaryotic ones and that they are, for the most part, obligate aerobes. There are of course facultatively anaerobic eukaryotes such as yeast, but we suppose that the ability to grow anaerobically in these organisms is a secondary adaptation, like the return of certain land organisms to water.

This enlarged eukaryotic cell (eukaryotic because it contains a nucleus partitioned from the rest of the cell by a membrane), would now be ready to obtain its respiratory and (in some cases) its photosynthetic machinery. The simplest means I suppose would be through phagocytosis of the already existing photosynthetic (blue–green algal) and respiratory (bacterial) prokaryotes. Most cells that engulf other cells have mechanisms to ensure that they will not be transformed into what they eat, i.e., they engulf the food organism into a vacuole and digestive enzymes are secreted into this vacuole from the engulfing cell to digest the food organism before the digested products are absorbed into the cell. It seems most important that the DNA of the ingested organism be destroyed to prevent the predator from being transformed into the prey. In the case of invasion by the putative organelles, we must suppose that the invaders had adapted to cope with the digestive apparatus of the engulfing organism, or that certain of the engulfing organisms became more tolerant of the prey. In any case, we suppose that once engulfed, the prokaryotes were not digested but continued to live within the food vacuoles. This would suggest that the outer membrane of contemporary organelles could be the membrane of the primitive food vacuole, while the inner membrane might be the cell membrane of the invading prokaryote. Having established themselves within the eukaryotic cell in this way, the prokaryotes could supply respiratory and photosynthetic energy to the cells and would divide along with them. One further series of adaptations is necessary, however, for a stable system to evolve. The invading prokaryotes have their own DNA and synthetic machinery and could reproduce faster than the host, thereby wrecking the system. The host must have some control over the reproduction, growth, and metabolism of the invader to ensure that the invader grows and divides along with the host in a coordinated manner and provides its contribution to the rest of the cell as needed. This seems to have been achieved by the invader ceding to the host certain synthetic prerogatives. We now know that organelles such as the chloroplast and mitochondrion require contributions of energy, small molecules, and even proteins from the rest of the cell in order to construct themselves. By regulating the flow of these molecules, the hose cell could regulate the rate of growth, division, and metabolism of the now endosymbiotic prokaryote, and thereby domesticate it, converting it into what we know now as a cellular organelle. In turn, the invader would gradually produce molecules that would regulate the production of necessary metabolites by the host leading to a highly efficient finely tuned relationship.

What might the original invaders have looked like in the case of the plastid? In work on the early development of bean proplastids (Klein and Schiff, 1972) and in an extensive study of proplastid and chloroplast development in *Euglena* (Klein *et al.*, 1972), we have noted that these proplastids resemble blue–green algae. On the basis of this evidence we suggested (Klein *et al.*, 1972) that if plastids arose by endosymbiotic invasion, the invading blue–green probably resembled a proplastid morphologically rather than a mature chloroplast. It was evident that the plastids of the red algae were very similar, and even the brown algal plastids have similar structures with somewhat more elaborate arrangements including thylakoid stacking. It is in the *Euglenophyta,* the yellow and green algae, and the line leading to the higher plants that the fully mature plastids have become much larger and more complex, with elaborate stacking of thylakoids as the rule. In these organisms the primitive condition is only evident when mutation (in nature or laboratory) has rendered the organism light-dependent for plastid development; the dark-grown organism then possesses a proplastid similar in morphology to the more primitive mature plastids. We suggest, then, that blue–green algae and perhaps other prokaryotes established themselves as endosymbionts which resembled the proplastids of today except that they had the full complement of photosynthetic machinery of plastids, or of the free living prokaryotes after invasion. We further suggest that continued evolution led to the types of mature plastids we see in the red and brown algae and ultimately, in the *Euglenas,* green algae, and higher plants. As already noted, Granick (1949) has proposed that the various precursors of chlorophyll in the present biosynthetic pathway at one time served as a less efficient photosynthetic pigment during the course of evolution; it is quite possible that the various structures between blue–green algae (or contemporary proplastids) and the mature green algal or higher plant plastid at one time served as the mature structure.

ACKNOWLEDGMENTS

The support of research grants from the National Institutes of Health (GM 14595) and the National Science Foundation (PCM 76-21486) are gratefully acknowledged.

REFERENCES

Adams, G. M. W., Vanwinkle-Swift, K. P., Gillham, N. W. and Boynton, J. E. (1976). *In* "Genetics of Algae" (R. Lewin, ed.), p. 69. Blackwell, Oxford.

Akoyunoglou, G., and Argyroudi-Akoyunoglou, J. H. (1978). "Chloroplast Development." Elsevier, Amsterdam.

App, A. A., and Jagendorf, A. T. (1963). *J. Protozool.* **10**, 340.

Arntzen, C., and Briantais J.-M. (1975). *In* "Bioenergetics of Photosynthesis" (Govindjee, ed.), p. 51. Academic Press, New York.

Barnett, W. E., Schwartzbach, S. D., and Hecker, L. I. (1978). *Prog. Nucleic Acid Res. Mol. Biol.* **21**, 143.

Batra, P. P. (1971). *In* "Photophysiology" (A. Giese, ed.), Vol. 1, p. 47. Academic Press, New York.

Bauer, S., and Siegelman, H. W. (1972). *FEBS Lett.* **20**, 352.

Beale, S. I. (1978). *Annu. Rev. Plant Physiol.* **29**, 95.

Bengis, C., and Nelson, N. (1977). *J. Biol. Chem.* **252**, 4564.

Ben-Shaul, Y., Epstein, H. T., and Schiff, J. A. (1965). *Can. J. Bot.* **43**, 129.

Bingham, S., and Schiff, J. A. (1976). *In* "Genetics and Biogenesis of Chloroplasts and Mitochondria" (T. Bücher, W. Neupert, W. Sebald, and S. Werner, eds.), p. 79. North-Holland Publ., Amsterdam.

Bingham, S., and Schiff, J. A. (1979a). *Biochem. Biophys. Acta* **547**, 512.

Bingham, S., and Schiff, J. A. (1979b). *Biochem. Biophys. Acta.* **547**, 531.

Bishop, N. I. (1966). *Annu. Rev. Plant Physiol.* **17**, 185.

Bjorkman, O. (1973). *In* "Photophysiology" (A. C. Giese, ed.), Vol. 8, p. 1. Academic Press, New York.

Björn, G. S. (1979). *Physiologia Plantarum* **46**, 281.

Blinks, L. R. (1964). *In* "Photophysiology" (A. C. Giese, ed.), Vol. 1, p. 199. Academic Press, New York.

Boardman, N. K. (1966). *In* "The Chlorophylls" (L. P. Vernon and G. R. Seely, eds.), p. 437. Academic Press, New York.

Boardman, N. K. (1977). *Ann. Rev. Plant Physiol.* **28**, 355.

Boardman, N. K., Anderson, J. M., Kahn, A., Thorne, S. W., and Treffry, T. E. (1970). *In* "Autonomy and Biogenesis of Mitochondria and Chloroplasts" (N. K. Boardman, A. W. Linnane, and R. M. Smillie, eds.), pp. 70–83. North-Holland Publ. Amsterdam.

Bogorad, L. (1950). *Bot. Gaz. (Chicago)* **111**, 221.

Bogorad, L. (1975). *Annu. Rev. Plant Physiol.* **26**, 369.

Bogorad, L. (1976). *In* "Chemistry and Biochemistry of Plant Pigments" (T. W. Goodwin, ed.), Vol. 1, p. 64. Academic Press, New York.

Bogorad, L., and Weil, J. H. (1977). "Nucleic Acids and Protein Synthesis in Plants." Plenum, New York.

Bohner, H., and Böger, P. (1978) *FEBS Lett.* **85**, 337.

Bonner, B. (1969). *Plant Physiol.* **44**, 739.

Bouck, B. (1965). *J. Cell Biol.* **26**, 523.

Bovarnick, J. G., Schiff, J. A., Freedman, Z. and Egan, J. M., Jr. (1974). *J. Gen. Microbiol.* **83**, 63.

Boynton, J. E., Gilham, N. W., Harris, E. H., Tingle, C. L., Van-Winkle-Swift, K., and Adams, G. M. W. (1976). *In* "Biogenesis of Chloroplasts and Mitochondria" (T. Bücher, W. Neupert, W. Seybald, and H. Werner, eds.), p. 313. North-Holland Publ., Amsterdam.

Briggs, W. R. (1964). *In* "Photophysiology" (A. Giese, ed.), Vol. 1, p. 223. Academic Press, New York.

Brinkmann, G., and Senger, H. (1978). *Plant Cell Physiol.* (in press).

Brunold, C., and Schiff, J. A. (1976). *Plant Physiol.* **57**, 430.

Bücher, T., Neupert, W., Sebald, W., and Werner, S. (1976). "Genetics and Biogenesis of Chloroplasts and Mitochondria." North-Holland Publ., Amsterdam.

Buetow, D. E., and Wood, W. M. (1978). *In* "Subcellular Biochemistry" (D. B. Roodyn, ed.), p. 1. Plenum, New York.

Butler, W. L. (1978). *Annu. Rev. Plant Physiol.* **29**, 345.

Butler, W. L., and Briggs, W. R. (1966). *Biochim. Biophys. Acta* **112**, 45.

Carlile, M. J. (1970). *In* "Photobiology of Microorganisms" (P. Halldal, ed.), p. 309. Wiley, New York.

Chen, K., Johal, S. and Wildman, S. G. (1976). *In* "Genetics and Biogenesis of Chloroplasts and Mitochondria" (T. Bücher, W. Neupert, W. Sebald, and S. Werner, eds.), p. 3. North-Holland Publ., Amsterdam.

Clayton, R. (1964). *In* "Photophysiology" (A. Giese, ed.), Vol. 2, p. 223. Academic Press, New York.

Cleland, R. E. (1962). *Adv. Genet.* **11**, 147.

Cohen-Bazire, G., Sistrom, W. R., and Stanier, R. Y. (1957). *J. Cell Comp. Physiol.* **49**, 25.

Cohen, C., and Schiff, J. A. (1976). *Photochem. Photobiol.* **24**, 555.

Cohen, D., and Schiff, J. A. (1976). *Arch. Biochem. Biophys.* **177**, 201.

Cohen, Y., Padan, E., and Shilo, M. (1975) *J. Bacteriol.* **123**, 855.

Coleman, A. W. (1979). *J. Cell Biol.* **82**, 299.

Collingbourne, R. H. (1966). *In* "Light as an Ecological Factor" (R. Bainbridge, G. C. Evans, and O. Rockham, eds.), p. 1. Blackwell, Oxford.

Cran, D. G., and Possingham, J. W. (1974). *Ann. Bot. (London)* **38**, 843.

Crisp, D. J., ed. (1971). *Eur. Marine Biol. Symp. 4th.*

Diamond, J., and Schiff, J. A. (1974). *Plant Sci. Lett.* **3**, 289.

Diamond, J., Schiff, J. A., and Kelner, A. (1975) *Arch. Biochem. Biophys.* **167**, 603.

Diehn, B. (1969). *Biochim. Biophys. Acta* **177**, 136.

Dobberstein, B., Blobel, G., Chua, N. (1977). *Proc. Natl. Acad. Sci. U.S.A.* **74**, 1082.

Dodge, J. D. (1975). *Phycologia* **14**, 253.

Dujon, B., and Slonimski, P. P. (1976). *In* "Biogenesis of Chloroplasts and Mitochondria" (T. Bücher, W. Neupert, W. Seybald, and H. Werner, eds.), p. 393. North-Holland Publ., Amsterdam.

Duysens, L. N. M. (1952). "Transfer of Excitation Energy in Photosynthesis," Thesis, Univ. of Utrecht, Utrecht, The Netherlands.

Ebringer, W. (1972). *J. Gen. Microbiol.* **71**, 35.

Edelman, M., Schiff, J. A., and Epstein, H. T. (1965). *J. Mol. Biol.* **11**, 769.

Edelman, M., Swinton, D., Schiff, J. A., Epstein, H. T., and Zeldin, B. (1967). *Bacteriol. Rev.* **31**, 315.

Edwards, M. K., Berns, D. S., Holt, S. C., and Ghiorse, W. C. (1968). *J. Phycol.* **4**, 283.

Egan, J. M., Jr., Dorsky, D., and Schiff, J. A. (1975). *Plant Physiol.* **56**, 318.

Ellis, R. J., and Barraclough, R. (1978). "Chloroplast Development" (G. Akoyunoglou *et al.*, eds.), p. 185. Elsevier, Amsterdam.

Ephrussi, B., and Slonimski, P. P. (1950). *C. R. Hebd. Seances Acad. Sci.* **230**, 685.

Erwin, J. A. (1968). *In* "The Biology of Euglena" (D. E. Buetow, ed.), Vol. II, p. 133. Academic Press, N.Y.

Evans, W. R. (1971). *J. Biol. Chem.* **246**, 6144.

Fogg, G. E., Stewart, W. D. P., Fay, P., and Walsby, A. E. (1973). "The Blue Green Algae." Academic Press, New York.

Fong, F., and Schiff, J. A. (1977). *Plant Physiol. Suppl.* **59**, 92.

French, C. S. (1960). The chlorophylls *in vivo* and *in vitro, In* "Encyclopedia of Plant Physiology" (A. Pirson, ed.), Vol. V/I, p. 252. Springer-Verlag, Berlin and New York.

Freyssinet, G. (1975). *Plant Sci. Lett.* **5**, 305.

Freyssinet, G. (1976). *Plant Physiol.* **47**, 824.

Freyssinet, G. (1977). *Physiol. Vegetale* **15**, 519.

Freyssinet, G., Harris, G., Nasatir, M., and Schiff, J. A. (1979). *Plant Physiol.* **63**, 908.

Freyssinet, G., and Schiff, J. A. (1974). *Plant Physiol.* **53**, 543.

Frosch, S., Bergfeld, R., and Mohr, H. (1978). *In* "Chloroplast Development" (G. Akoyunoglou and J. H. Argyroudi-Akoyunoglou, eds.), p. 781. Elsevier, Amsterdam.

Fujita, Y., and Hattori, A. (1963). *In* "Studies on Microalgal and Photosynthetic Bacteria" (Jpn. Soc. Plant Physiol., ed.), p. 431. Univ. of Tokyo Press, Tokyo.

Gannt, E., and Conti, S. F. (1966). *J. Cell Biol.* **29**, 423.

Gannt, E., Edwards, M. B., and Conti, S. F. (1968). *J. Phycol.* **4**, 65.

Gannt, F., Edwards, M. B., and Provasoli, L. (1971). *J. Cell Biol.* **48**, 280.

Gassman, M., Granick, S., and Mauzerall, D. (1968). *Biochem. Biophys. Res. Commun.* **32**, 295.

Gibbs, S. (1962). *J. Cell Biol.* **14**, 433.

Gibson, L. D. Laver, W. D., and Neuberger, A. (1958). *Biochem. J.* **61**, 618.

Glazer, A. N., Apell, G. S., Hixson, C. S., Bryant, D., Riwor, A., and Brown, D. M. (1976). *Proc. Natl. Acad. Sci. U.S.A.* **73**, 428.

Goldstein, N. H., Schwartzbach, S. D., and Schiff, J. A. (1974). *J. Protozool.* **21**, 443.

Goodenough, U. W. (1969). Ph.D. Dissertation, Harvard Univ., Cambridge, Massachusetts.

Grabowski, J., and Gannt, E. (1978). *Photochem. Photobiol.* **28**, 39 and 47.

Granick, S. (1948). *J. Biol. Chem.* **175**, 333.

Granick, S. (1949). *Harvey Lect. Ser.* **43**, 220.

Granick. S. (1960). *In* "The Cell" (J. Brachet and A. E. Mirsky, eds.), Vol. 2, p. 489. Academic Press, New York.

Granick, S. (1971). *In* "Methods in Enzymology: Vol. 23, Photosynthesis and Nitrogen Fixation, Part A" (A. San Pietro, ed.), p. 162.

Granick, S., and Gassman, M. (1970). *Plant Physiol.* **45**, 201.

Greenblatt, C. L., and Schiff, J. A. (1959). *J. Protozool.* **6**, 23.

Griffiths, D. J. (1970). *Bot. Rev.* **36**, 29.

Griffiths, W. T. (1978). *Biochem. J.* **174**, 681.

Gunsalus-Cadavid, N. F. (1974). *Sub-cell. Biochem.* **3**, 275.

Hagemann, R. (1976). *In* "Genetics and Biogenesis of Chloroplasts and Mitochondria" (T. Bücher, W. Neupert, W. Seybald, and H. Werner, eds.), p. 331. North-Holland Publ., Amsterdam.

Halldal, P. (1970). *In* "Photobiology of Microorganisms" (P. Halldal, ed.), p. 17. Wiley (Interscience), New York.

Hallick, R. B., Gray, P. W., Chelm, B. K., Rushlow, K. E., and Orozco, E. M., Jr. *In* "Chloroplast Development" (G. Akoyunoglou *et al.*, eds.), p. 619. Elsevier, Amsterdam.

Harel, E. (1978). *In* "Chloroplast Development" (G. Akoyunoglou *et al.*, eds.), p. 33. Elsevier, Amsterdam.

Hase, E. (1971). *In* "Autonomy and Biogenesis of Mitochondria and Chloroplasts" (N. K. Boardman, A. W. Linnane, and R. M. Smillie, eds.), pp. 434–446. North-Holland Publ., Amsterdam.

Haury, J., and Bogorad, L. (1977). *Plant Physiol.* **60**, 835.

Highfield, P. E., and Ellis, R. J. (1978). *Nature (London)* **271**. 420.

Hill, H. Z., Schiff, J. A., and Epstein, H. T. (1966). *Biophys. J.* **6**, 125.

Holdsworth, R. H. (1971). *J. Cell Biol.* **51**, 499.

Holowinsky, A. W., and Schiff, J. A. (1970). *Plant Physiol.* **45**, 339.

Horrum, M. A., and Schwartzbach, S. D. (1980). *Plant Physiol.* in press.

Hudock, G. A., and Rosen, H. (1976). *In* "Genetics of Algae" (R. Lewin, ed.). Blackwell, Oxford.

Jurgenson, J. E., Beale, S. I., Troxler, R. R., Bartolf, M. M., Fitzgerald, M. P. Ramus, J., and Schiff, J. A. (1975). *Biol. Bull.* **149**, 432.

Kahn, A. Boardman, N. K., and Thorne, S. W. (1970). *J. Mol. Biol.* **48**, 85.

Kannangara, C. G., and Gough, S. P. (1978). *Carlsberg Res. Commun.* **43**, 185.

Kessler, E. (1974). *In* "Algal Physiology and Biochemistry" (W. D. P. Stewart, ed.), p. 456. Univ. of California Press, Berkeley.

Kindman, L. A., Cohen, C. E., Zeldin, M. H., Ben Shaul, Y., and Schiff, J. A. (1978). *Photochem. Photobiol.* **27**, 787.

Kikuchi, G., Kumar, A., Talmage, P., and Shemin, D. (1958). *J. Biol. Chem.* **223**, 1214.

Kirby, N. W., and Evans, L. V. (1978). *Planta* **142**, 91.

Kirk, J. T. O., and Tilney-Bassett, R. A. E. (1967). "The Plastids." Freeman, San Francisco. (see also 2nd edition, 1978).

Klein, O., and Senger, H. (1978). *Plant Physiol.* **62**, 10.

Klein, S., and Schiff, J. A. (1972). *Plant Physiol.* **49**, 619.

Klein, S., Schiff, J. A., and Holowinsky, A. (1972). *Dev. Biol.* **28**, 253.

Kowallik, W. (1970). *In* "Photobiology of Microorganisms" (P. Halldal, ed.), p. 165.

King, S. D. (1976). *Science* **191**, 429.

Lancer, H. A., Cohen,, C. E., and Schiff, J. A. (1976). *Plant Physiol.* **57**, 369.

Lang, N. L., and Rae, P. M. M. (1967). *Protoplasma* **64**, 67.

Lazaroff, N., and Schiff, J. A. (1962). *Science* **137**, 603.

Ledbetter, M. C., and Porter, K. R. (1970). "Introduction to the Fine Structure of Plant Cells." Springer-Verlag, Berlin and New York.

Leedale, G. F. (1967). "Euglenoid Flagellates." Prentice-Hall, Englewood Cliffs, New Jersey.

Lefort, M. (1964). *C. R. Hebd. Seances Acad. Sci.* **258**, 4318.

Lefort-Tran, M. (1975). *In* "Les Cycles Cellulaires et Leur Blocage chez Plusieurs Protistes" pp. 297–308. Centre Nat. Rech. Sci., Paris.

Levine, R. P., and Goodenough, U. W. (1970). *Annu. Rev. Genet.* **4**, 397.

Levine, R. P., and Volkmann, D. (1961). *Biochem. Biophys. Res. Commun.* **6**, 264.

Levring, T. (1966). *In* "Light as an Ecological Factor" (R. Bainbridge, G. C. Evans, and O. Rackham, eds.), p. 305. Blackwell, Oxford.

Lewin, R. A. (1975). *Phycologia* **14**, 149.

Lewin, R. (1976). *Nature (London)*, **261**, 697.

Lewin, R. (1977). *Phycologia* **16**, 271.

Lewin, R. A. and Withers, N. W. (1975). *Nature (London)* **256**, 735.

Lewis, S. C., Schiff, J. A., and Epstein, H. T. (1961). *Biochem. Biophys. Res. Commun.* **5**, 221.

Linnane, A. W., and Crowfoot, P. D. (1975). *In* "Membrane Biogenesis Mitochondria, Chloroplasts, Bacteria" (A. Tzagaloff, ed.), p. 99. Plenum, New York.

Lucas, I. A. N. (1970). *J. Phycol.* **6**, 30.

Lütz, C. and Klein, S. (1979). *Planta,* in press.

Lyman, H., Epstein, H. T., and Schiff, J. A. (1961). *Biochim. Biophys. Acta* **50**, 301.

Lyman, H., Jupp, A., and Larrinua, I. (1975). *Plant Physiol.* **55**, 390.

Lythgoe, J. N. (1972). *In* "Handbook of Sensory Physiology: Vol. 7, Part I, Photochemistry of Vision" (H. J. Dartnall, ed.), p. 566. Springer-Verlag, Berlin and New York.

MacColl, R., and Berns, D. S. (1976). *Arch. Biochem. Biophys.* **177**, 265.

Mahler, H. R., Bastos, R. N., Flury, U. Lin, C. C., and Phan, S. H. (1975). *In* "Genetics and Biogenesis of Mitochondria and Chloroplasts" (C. W. Birky, Jr., P. S. Perlman, and T. J. Byers, eds.), p. 66. Ohio State Univ. Press, Columbus, Ohio.

Manning, J. E., Wolstenholme, D. R., Ryan, R. S., Hunter, J. A., and Richards, O. C. (1971). *Proc. Natl. Acad. Sci. U.S.A.* **68**, 1169.

Margulies, L. (1970). "Origin of Eucaryotic Cells." Yale Univ. Press, New Haven, Connecticut.

Mauzerall, D. (1978). *In* "Bioorganic Chemistry" (E. E. van Tamelen, ed.), Vol. 4, p. 303. Academic Press, New York.

Miller, K. (1978). *In* "Chloroplast Development" (G. Akoyunoglou *et al.*, eds.), p. 17. Elsevier, Amsterdam.

Myers, J., and French, C. W. (1960). *J. Gen. Physiol.* **43**, 723.

Nadler, K., and Granick, S. (1970). *Plant Physiol.* **46**, 240.

Nagley, P., and Linnane, A. W. (1970). *Biochem. Biophys. Res. Commun.* **39**, 989.

Newcomb, E. H., and Pugh, T. D. (1975). *Nature (London)* **253**, 533.

Nicolas, P., Innocent, J.-P., and Nigon, V. (1977). *Mol. Gen. Genet.* **155**, 123.

Nigon, V., and Heizmann, P. (1978). *Int. Rev. Cytol.* **53**, 211.

Nultsch, W. (1970). In "Photobiology of Microorganisms" (P. Halldal, ed.), p. 213. New York.

O'Brien, T. W., Matthews, D. E., and Denslow, N. D. (1976). *In* "Genetics and Biogenesis of Chloroplast and Mitochondria" (T. Bücher, W. Neupert, W. Sebald, and S. Werner, eds.), p. 741. North-Holland Publ., Amsterdam.

Ohad, I. (1975). *In* "Membrane Biogenesis Mitochondria, Chloroplasts, Bacteria" (A. Tzagaloff, ed.), p. 279. Plenum, New York.

Ohki, K., and Fujita, Y. (1979). *Plant and Cell Physiol.* **20**, 1341.

Padan, E. (1979). *Annu. Rev. Plant Physiol.* **30**, 27.

Pardo, A., and Schiff, J. A. (1980). *Can. Jour. Bot.*, **58**, 25.

Pearlman, P. S., and Mahler, H. R. (1974). *Arch. Biochem. Biophys.* **162**, 248.

Perry, G. J., Gillan, F. T., and Johns, R. B. (1978). *J. Phycol.* **14**, 369.

Poff, K. L., and W. L. Butler (1974). *Photochem. Photobiol.* **20**, 241.

Prézlin, B. B., and Alberte, R. S. (1978). *Proc. Natl. Acad. Sci. U.S.A.* **75**, 1801.

Pringsheim, F., and Pringsheim, O. (1952). *New Phytologist* **51**, 65.

Provasoli, L., Hutner, S. H., and Schatz, A. (1948). *Proc. Soc. Exp. Biol. Med.* **69**, 279.

Raff, R. A., and Mahler, H. R. (1972). *Science* **177**, 575.

Ragan, M. A., and Chapman, D. J. (1978). "A Biochemical Phylogeny of the Protists." Academic Press, New York.

Ramirez, J. M., Del Campo, F. F., Paneque, A., and Losada, M. (1966). *Biochim. Biophys. Acta* **118**, 58.

Ramus, J., Beale, S. I., and Mauzerall, D. (1976). *Marine Biol.* **37**, 231.

Rawson, J. R. Y. (1975). *Biochim. Biophys. Acta* **402**, 171.

Rebeiz, C., and Castelfranco, P. (1973). *Annu. Rev. Plant Physiol.* **24**, 129.

Rhoades, M. M. (1946). *Cold Spring Harbor Symp. Quant. Biol.* **11**, 202.

Rüdiger, W., Benz, J., Lempert, U., Schoch, S., and Steffens, D. (1976). *Z. Pflanzenphysiol.* **80**, 131.

Rüdiger, W., Hedden, P., Köst, H.-P., and Chapman, D. J. (1977). *Biochem. Biophys. Res. Commun.* **74**, 1268.

Saccone, C., and Kroon, A. M. eds. (1976). "The Genetic Function of Mitochondrial DNA." North-Holland Publ., Amsterdam.

Sager, R. (1958). *Brookhaven Symp. Biol.* **11**, 101.

Sager, R., and Palade, G. E. (1954). *Exp. Cell Res.* **7**, 584.

Sager, R., and Ramanis, Z. (1967). *Proc. Natl. Acad. Sci. U.S.A.* **58**, 931.

Sagher, D., Grosfeld, H., and Edelman, M. (1976). *Proc. Natl. Acad. Sci. U.S.A.* **73**, 722.

Salisbury, J. L., and Floyd, G. L. (1978). *J. Phycol.* **14**, 362.

Schatz, G. (1970). *In* "Membranes of Mitochondria and Chloroplasts" (E. Racker, ed.), p. 251. Van Nostrand-Reinhold, New York.

Schiff, J. A. (1963). *Carnegie Inst. Washington Yearb.* **62**, 375.

Schiff, J. A. (1973). *Adv. Morphogen.* **10**, 265.

Schiff, J. A. (1974). *Proc. Int. Congr. Photosynthesis 3rd*, p. 1691.

Schiff, J. A. (1978). *In* "Chloroplast Development" (G. Akoyunoglou *et al.*, eds.), p. 747. Elsevier, Amsterdam.

Schiff, J. A., and Hodson, R. C. (1970). *Ann. N.Y. Acad. Sci.* **175**, 555.

Schiff, J. A., Lyman, H., and Epstein, H. T. (1961). *Biochim. Biophys. Acta* **50**, 310.

Schiff, J. A., Zeldin, M. H., and Rubman, J. (1967). *Plant Physiol.* **42**, 1716.
Schiff, J. A., Lyman, H., and Russel, G. K. (1971). *In* "Methods in Enzymology: Vol. 23, Photosynthesis and Nitrogen Fixation, Part A" (A. San Pietro, ed.), p. 143. Academic Press, New York. (For addendum see Schiff, J. A., Lyman, H., and Russel, G. K. (1980). *In* "Methods in Enzymology: Vol. 69" (A. San Pietro, ed.). Academic Press, New York.
Schmidt, A., and Schwenn, J. D. (1971). *Proc. Int. Congr. Photosynthesis 2nd.*, p. 507.
Schmidt, G., and Chua, N. H. (1978). *Brookhaven Symp. Biol.*, in press.
Schmidt, G., and Lyman, H. (1974). *Proc. Int. Congr. Photosynthesis, 3rd*, p. 1755.
Schmidt, W., and Butler, W. L. (1976). *Photochem. Photobiol.* **24**, 71.
Schoch, S., Lempert, U., and Rudiger, W. (1977). *Z. Pflansenphysiol.* **83**, 427.
Schopfer, P., and Siegelman, H. W. (1968). *Plant Physiol.* **43**, 990.
Schwartzbach, S. D., and Schiff, J. A. (1974). *J. Bacteriol.* **120**, 334.
Schwartzbach, S. D., Schiff, J. A., and Goldstein, N. H. (1975). *Plant Physiol.* **56**, 313.
Schwartzbach, S., Hecker, L., and Barnett, W. (1976). *Proc. Natl. Acad. Sci. U.S.A.* **73**, 1984.
Senger, H. (1980). "The Blue Light Syndrome." Springer, Heidelberg, in press.
Senger, H., and Bishop, N. I. (1972). *Plant Cell Physiol.* **13**, 633.
Shemin, D., and Russel, C. S. (1953). *J. Am. Chem. Soc.* **75**, 4873.
Shemin, D., Russel, C. S., and Abramsky, T. (1955). *J. Biol. Chem.* **214**, 613.
Sherman, F., Stewart, J. W., Parker, J. H., Inhaber, E., Shipman, N. A., Putterman, G. J., Gardisky, R., and Margoliash, E. (1968). *J. Biol. Chem.* **243**, 5446.
Shibata, K. (1957). *J. Biochem. (Tokyo)* **44**, 147.
Siddell, S. G., and Ellis, R. J. (1975). *Biochem. J.* **146**, 675.
Slonimski, P. P. (1953). "La Formation des Enzymes Respiratoires Chez la Levure." Masson, Paris.
Slonimski, P. P., and Lazowska, J. (1977). *Mitochondria 1977, Proc. Colloq. Genet. Biogenesis Mitochondria*, p. 39.
Smith, J. H. C., and Benitez, A. (1955). *In* "Moderne Methoden der Pflanzenanalyse" (K. Paech and M. V. Tracey, eds.), Vol. 4, p. 142. Springer-Verlag, Berlin and New York.
Staehlin, L. A. (1976). *Brookhaven Symp. Biol.* **28**, 278.
Stanier, R. Y., Kunisawa, R., Mandel, M., and Cohen-Bazire, G. (1971). *Bacteriol. Rev.* **35**, 171.
Steeman-Nielsen, E. (1975). "Marine Photosynthesis," Oceanography Ser. 13. Elsevier, New York.
Steinmetz, A., Mubumbila, V., Keller, M., Burkard, G., and Weil, J. H. (1978). *In* "Chloroplast Development" (G. Akoyunoglou *et al.*, eds.) p. 573. Elsevier, Amsterdam.
Stern, A. I., Schiff, J. A., and Epstein, H. T. (1964a). *Plant Physiol.* **39**, 220.
Stern, A. I., Epstein, H. T., and Schiff, J. A. (1964b). *Plant Physiol.* **39**, 226.
Straley, S. C., Parson, W. W., Mauzerall, D. C., and Clayton, R. K. (1973). *Biochim. Biophys. Acta* **305**, 597.
Stutz, E. (1978). *In* "Chloroplast Development" (G. Akoyunoglou *et al.*, eds.), p. 609. Elsevier, Amsterdam.
Sundquist, C. (1970). *Physiol. Plant.* **23**, 412.
Surzycki, S. J., Goodenough, U. W., Levine, R. P., and Armstrong, J. J. (1970). *Symp. Soc. Exp. Biol.* **24**, 13.
Thornber, J. P., Markwell, J. P., and Reinman, S. (1979). *Photochem. Photobiol.* **29**, 1205.
Thorne, S. W., Newcomb, E. H., and Osmond, C. B. (1977). *Proc. Natl. Acad. Sci. U.S.A.* **74**, 575.
Tiffany, L. H. (1951). *In* "Manual of Phycology" (G. M. Smith, ed.), pp. 307–308. Ronald Press, New York.
Trench, R. K. (1979). *Annu. Rev. Plant Physiol.* **30**, 485.
Troxler, R. F. (1977). *In* "Chemistry and Physiology of Bile Pigments" (P. D. Berk and N. I. Berlin, eds.), p. 431. DHEW Publ. No. 77-100. Natl. Inst. Health, Washington D.C.

Troxler, R., and Bogorad, L. (1966). *Plant Physiol.* **41**, 491.

Vaisberg, A. J., and Schiff, J. A. (1976). *Plant Physiol.* **57**, 260.

Vaisberg, A. J., Schiff, J. A., Li, L., and Freedman, Z. (1976). *Plant Physiol.* **57**, 594.

Vary, M. J., Stewart, P. R., and Linnane, A. W. (1970). *Arch. Biochem. Biophys.* **141**, 430.

Vasconcelos, A. C. (1976). *Plant Physiol.* **58**, 719.

Virgin, H. (1964). *In* "Photophysiology" (A. Giese, ed.), Vol. 1, p. 273. Academic Press, New York.

Virgin, H. (1972). *In* "Phytochrome" (K. Mitrakos and W. Shropshire, Jr., eds.), p. 371. Academic Press, New York.

Vogelman, T. C., and Scheibe, J. (1978). *Planta* **143**, 233.

von Wettstein, D. (1958). *Brookhaven Symp. Biol.* **11**, 138.

Voskresenskya, N. (1972). *Annu. Rev. Plant Physiol.* **23**, 219.

de Vries, H., and van der Bogert, C. (1976). *In* "Genetics and Biogenesis of Chloroplasts and Mitochondria" (T. Bücher, W. Neupert, W. Sebald, and S. Werner, eds.), p. 271. North-Holland Publ., Amsterdam.

Waaland, J. R., Waaland, S. D., and Bates, G. (1974). *J. Phycol.* **10**, 193.

Watson, K., Haslam, J. M., Veitch, B., and Linnane, A. W. (1971). *In* "Autonomy and Biogenesis of Mitochondria and Chloroplasts" (N. K. Boardman, A. W. Linnane, and R. M. Smillie, eds.), p. 162. North-Holland Publ., Amsterdam.

Whatley, J. M. (1974). *New Phytol.* **73**, 1097.

Whatley, J. M. (1976). *New Phytol.* **76**, 111.

Whatley, J. M. (1977a). *New Phytol.* **78**, 407.

Whatley, J. M. (1977b). *New Phytol.* **79**, 309.

White, A. W. (1974). *J. Phycol.* **10**, 292.

Wildman, S. G., Hongladarom, T., and Honda, S. I. (1962). *Science* **138**, 434.

Wolk, C. P. (1973). *Bacteriol. Rev.* **37**, 32.

Wood, P. M. (1976). *FEBS Lett.* **65**, 111.

Woodcock, C. L. F., and Bogorad, L. (1970). *J. Cell Biol.* **44**, 361.

Zeldin, M., and Schiff, J. A. (1975). *Plant Physiol.* Suppl. **56**, 33.

Biochemistry of the Chloroplast

7

RICHARD G. JENSEN

The Biochemistry of Plants, Vol. 1
Copyright © 1980 by Academic Press, Inc.
All rights of reproduction in any form reserved.
ISBN 0-12-675401-2

I. INTRODUCTION

The unique and important feature of plants is their ability to grow using sunlight as the source of energy and CO_2 from the air as the carbon source with water and elements coming from the environment or soil. The process of photosynthesis with light capture, coupled to O_2 evolution and CO_2 fixation occurs in the chloroplast. This organelle operates in a semi-autonomous fashion with many of its metabolic processes apparently independent of direct cytoplasmic control. The chloroplast generates its own ATP and reducing power in the light, whereby CO_2 is assimilated. The carbon is either exported as triose phosphates to the cytoplasm or stored in the chloroplast as starch. In the dark, reducing power can be generated by a hexose monophosphate shunt in the chloroplast. Some amino acids but not all can be synthesized using amino nitrogen obtained by chloroplast nitrite reductase. Chloroplast mRNA is transcribed from chloroplast DNA for translation of chloroplast proteins. However, studies on this process indicate that only part of the chloroplast proteins are produced in the organelle while the rest are either partially or entirely coded in the nucleus and synthesized in the cytoplasm. This could explain why chloroplasts, after isolation and still capable of many vital functions such as photosynthesis or reduction of nitrate, have not been cultured as algae. In this chapter the emphasis will be on the biochemistry of the major metabolic functions unique to the chloroplast.

II. MORPHOLOGY OF HIGHER PLANT CHLOROPLASTS

The chloroplast as it appears in most published electron micrographs usually has a characteristic lens shape with a length of 4 to 10 μm (Fig. 1). This longitudinal profile is selected because it reveals best the stacking of the internal lamellar membranes. There are three major structural regions of the chloroplast: the double outer membrane or envelope; the mobile stroma containing the soluble enzymes for metabolism, protein synthesis, and starch storage; and the highly organized internal lamellar membranes containing chlorophyll and involved in the biophysical reactions of energy capture and conversion. The internal membranes are shaped like discs and are often stacked together like a pile of coins to form a granum. Each disc is vesiculated or saclike and is termed a thylakoid. If sectioned in a plane parallel to the thylakoid membrane both the chloroplast and the membranes appear disc-shaped. The outer envelope is a selectively permeable structure that regulates the movement of carbon intermediate products, reducing power and adenylates in and out of the chloroplast, while retaining starch for degradation at night. The stroma is mostly protein, consisting of about 50% of "fraction 1 protein" or ribulose-1,5-P_2 carboxylase/oxygenase.

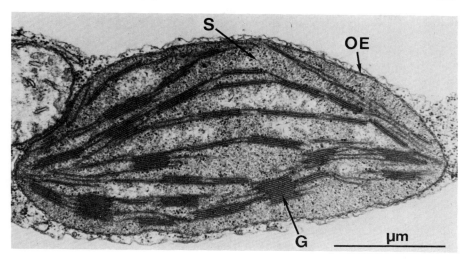

Fig. 1. Electron micrograph of a leaf mesophyll chloroplast of squash showing the double-layer outer envelope (OE), lamellar membranes stacked together to form grana (G), and the stroma space (S). The electron-dense dots in the stroma are probably caused by ribosomes. From W. M. Laetsch, unpublished.

The appearance of the chloroplast does differ between plants, e.g., between C_3 and C_4 plants. C_3 plants usually have similar appearing chloroplasts throughout the leaf, which is characteristic of the independence of photosynthesis or other metabolic functions between neighboring cells. C_4 plants have two prominent cell types, mesophyll and bundle sheath, whose chloroplasts may differ in appearance and function (Fig. 2). Carbon dioxide is fixed initially by phosphoenolpyruvate carboxylase in the mesophyll cell and rapidly transported as malate or aspartate to the neighboring bundle sheath cell for further utilization (see Section IV,C).

Chloroplasts from the leaf mesophyll cells of C_4 plants, such as corn or sugarcane, exhibit grana similar to other higher plant chloroplasts. In contrast, the chloroplast of the neighboring bundle sheath cells of these plants often appear without prominent grana. Most of the stored starch is also located in the bundle sheath chloroplast. However, if translocation of photosynthetic products is inhibited, large starch granules will appear in the mesophyll chloroplasts after the bundle sheath chloroplasts are first loaded. It was originally thought that the chloroplasts of C_4 bundle sheath cells were "agranal." Although earlier micrographs of C_4 bundle sheath cells showed primarily chloroplasts without grana, a closer evaluation of many C_4 plants showed that bundle sheath chloroplasts often do exhibit grana (Fig. 3). There appears to be no direct relationship between the presence or absence of grana in the chloroplast and C_4 photosynthesis. The degree of granal devel-

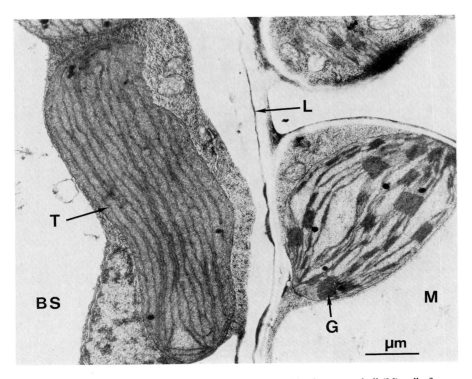

Fig. 2. Chloroplasts of a bundle sheath (BS) and neighboring mesophyll (M) cell of sugarcane, a C_4 plant. Note the unappressed agranal thylakoids (T) in the bundle sheath chloroplast and the grana (G) in the mesophyll chloroplast. The outer wall of the bundle sheath cell stops at an electron opaque layer (L). From Laetsch (1974). Reproduced with permission from the *Annual Review of Plant Physiology* **25**, 34; © 1974 by Annual Reviews Inc.

opment in bundle sheath chloroplasts may be related in some fashion with their function in synthesizing and storing starch.

III. FUNDAMENTAL ENERGY PROCESSES IN PHOTOSYNTHESIS

Photosynthesis as it operates in the chloroplast has two phases, the light reactions, which are directly dependent on light energy, and the dark reactions, which can occur without the direct influence of light. Research over the last 25 years has shown that the light reactions of photosynthesis are primarily responsible for converting light energy into chemical energy in the form of ATP and NADPH. These compounds in turn bring about the reduction of carbon dioxide to sugars and other products.

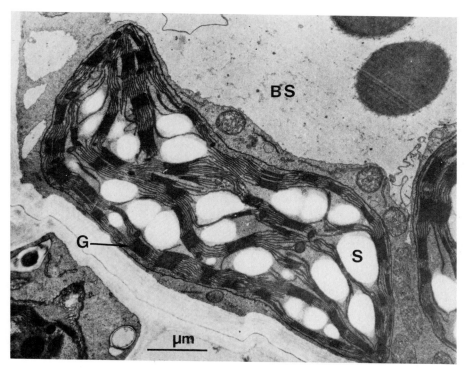

Fig. 3. Bundle sheath (BS) chloroplast of *Muhlenbergia racemosa,* a C_4 plant, showing prominent grana (G) and large amounts of starch (S). From W. M. Laetsch, unpublished.

Early observations indicated that the rate-limiting step in plant photosynthesis takes place in the dark. When photosynthetic organisms are subjected to intermittent illuminations with short flashes of light (milliseconds or less) followed by dark intervals of varying duration, evolution of O_2 after a single flash of 10^{-5} s was maximal if it was followed by a much longer dark period (greater than 0.06 s). The term "dark reactions" does not mean that they take place in the dark; in the living plants they function together with the light reactions in light. At night while the leaf respires many of the dark reactions of photosynthesis are inoperative. As explained later, the "dark reactions" communicate with the action of the light reactions not only by utilization of ATP and NADPH but by light-generated pH and Mg^{2+} gradients in the stroma in the presence of a reducing environment.

A. The Role of the Pigment Systems

The various photosynthetic pigments involved in light absorption from higher plants can be classified into two main groups: chlorophyll and carotenoids. The function of these pigments is to provide the plant with an efficient system of absorbing light throughout the visible spectrum. This energy is then transferred to reaction centers where it is utilized in a photochemical reaction. The bulk of the pigments are light-harvesting pigments involved in the process of light absorption and subsequent energy transfer.

There are two kinds of chlorophyll in higher plants, chlorophyll a and chlorophyll b. In algae and photosynthetic bacteria other chlorophyll pigments are also present. Chlorophyll a (Chl a) is the major pigment and is found in all photosynthetic organisms that evolve oxygen. In the plant, Chl a has various forms with different absorption maxima, due to unique environments, e.g., Chl 660, 670, 680, 685, 690, and 700–720 nm. The evidence for the existence of these various forms comes from derivative spectrophotometry, low-temperature absorption measurements, and the action spectra of various photochemical reactions. The short-wavelength Chl a forms are fluorescent and are predominantly present in photosystem II. The long-wavelength forms are weakly fluorescent and are mostly present in photosystem I.

Chlorophyll b (Chl b), also present in higher plants, has a major absorption maxima at 650 nm, with a minor component in some species at 640 nm. The major portion of Chl b is present in photosystem II.

Chlorophyll *in vivo* is noncovalently bound to protein in the thylakoid membrane. Upon treatment with organic solvents, the weak interaction between chlorophyll and the membrane components is eliminated and its absorption maxima shifts to a lower wavelength, depending on the solvent–chlorophyll interactions.

The carotenoids are the yellow and orange pigments found in most photosynthetic organisms. The two classes of carotenoids are (1) carotenes absorbing blue light of which β-carotene is the most common; and (2) carotenols or alcohols, commonly called xanthophylls. It is generally accepted that most of the carotenes are present in photosystem I while the xanthophylls are located in photosystem II. Both of these carotenoid pigments function by absorbing light mostly in the regions of the spectra not absorbed by chlorophyll and transfer the energy to Chl a. They also help protect chlorophyll from photo-oxidation.

Recently an orientation of the chlorophyll pools in the photochemical apparatus was proposed by Butler (1978) and co-workers using low temperature absorption and fluorescence measurements. There are three functional locations for Chl a: Chl a in photosystem I (Chl a_I), in photosystem II (Chl a_{II}), and a third known as the light-harvesting Chl complex (Chl LH) which can transfer energy to either of these systems. According to the model (Fig. 4), Chl a_I in photosystem I cannot transfer energy directly between neighboring photosystem particles. Transfer occurs by way of Chl a_{II} in photosystem

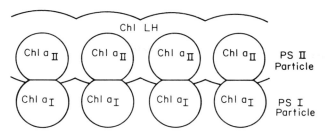

Fig. 4. Model of energy transfer occurring between photosystem units as mediated by Chl a. Adapted from Butler (1978).

II embedded in a large array of Chl LH. The Chl LH mediates the energy transfer. Earlier results indicated that energy transfer occurs from Chl a_{II} to Chl a_I in the absence of Chl LH indicating that photosystem I and II units are connected together directly.

An electron transport chain is connected by way of photochemical reaction centers to each pair of photosystems. As each photosystem is supplied by about 300 antenna Chl, each electron transport chain may pass a pair of electrons once every 15 ms in well-operating chloroplasts. Were a single Chl molecule to drive the reaction there would not be enough light quanta to suffice even if the molecule were exposed to bright sunlight. An average Chl molecule absorbs one quantum of light per 100 ms under bright sunlight, one per second under diffused daylight, and only one per 10 s on a cloudy day. An organized pool of Chl with several energy transfers occurring simultaneously is essential to match the rather low absorption rate of quanta per Chl to the higher rates of electron transport. A typical thylakoid disc from a mature spinach chloroplast contains at least 10^5 Chl molecules so that its membrane is covered by at least 200 electron transfer chains.

B. Flow of Electrons in Light

Light quanta absorbed by the chlorophyll and carotenoid pigments are funneled into a specific photochemical reaction center. The efficiency of this energy transfer is high implying that the probability for transfer of a quantum between two neighboring pigments is higher than the probability for any competing process such as fluorescence emission, formation of metastable states, wasteful photochemistry, and radiation-less deactivation. As these processes usually occur within nanoseconds, the transfer through the whole light-harvesting antenna pigment system to a reaction center must occur in a much shorter time. Rapid transfer of energy occurs via dipolar coupling between pigments which are tightly packed and communicate by resonance.

Because energy transfer is enhanced when the absorption spectra of neighboring pigments overlap, it is not unusual to note that the reaction centers have absorption maxima at longer wavelengths (lower energy).

The end result of photosynthetic electron flow in the chloroplast is the evolution of oxygen and the formation of ATP and NADPH necessary for the assimilation of CO_2. The currently accepted representation of photosynthetic electron transport is one of a cooperative interaction of two light reactions. This model originated with Hill and Bendall (1960). A representation of their hypothesis as it has evolved today is presented in Fig. 5. Their formulation was proposed primarily to account for three major experimental observations: (1) the decline in efficiency of photosynthesis at long wavelengths (greater than 685 nm) and the synergistic effect of shorter wavelengths on the photosynthetic action of far red illumination; (2) the presence in green tissues of two cytochromes, cytochrome f (Cyt f) and Cyt b_6, whose characteristic potentials differed about 400 mV as did their light-induced absorption changes; and (3) the stimulation of electron flow to $NADP^+$ when ATP formation occurred concurrently. According to the model (Fig. 5), photosystem II oxidizes water to free O_2 and reduces Q, while photosystem I reduces a low potential electron acceptor X and oxidizes P-700. Q may be

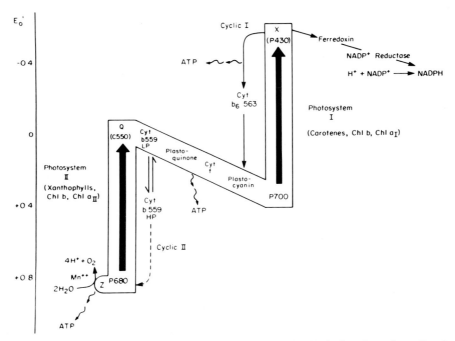

Fig. 5.   The Z-scheme for photosynthetic electron transport including sites of coupling for photophosphorylation. See text for details.

equivalent to a component producing an absorbance change at 550 nm, referred to as C-550. Similarly, X appears to be a pigment having an absorption change at 430 nm and is referred to as P-430. Oxidized P-700 is reduced by reduced Q via exergonic electron transport reactions that are coupled to the phosphorylation of ADP to ATP. The oxidation of water also provides protons and a membrane potential (Section III,F) to run a second phosphorylation of ADP. These two steps of ATP production occur during noncyclic electron flow and are called noncyclic photophosphorylation. The carriers catalyzing the electron flow reactions are Cyt b-559 (low potential), plastoquinone (PQ), Cyt f, and plastocyanin (PC), in that order. The site of phosphorylation is probably between PQ and Cyt f.

The low potential electron acceptor X for photosystem I can transfer energy to form NADPH via a ferredoxin reducing substance, ferredoxin (Fd) and the ferredoxin-NADP$^+$ reductase. Alternatively energy from the primary acceptor X can cycle back to Cyt f or PC by way of Cyt b_6. In this latter instance the electron transport traces a closed circuit utilizing only photosystem I. It is referred to as cyclic electron transport and the accompanying formation of ATP is designated cyclic photophosphorylation. The amount of cyclic photophosphorylation that occurs *in vivo* is still uncertain.

C. Photosystem II and the Evolution of Oxygen

Photosystem II of higher plants is associated with oxygen evolution to provide electrons for the subsequent reductive processes mediated by the electron transport chain and photosystem I. As with photosystem I, the pigment–enzyme complex of photosystem II is membrane bound and exhibits a high degree of structural and organizational integrity. According to various investigators, the trap Chl a of photosystem II exists in a reaction center complex with a primary electron donor Z and a primary acceptor Q. When the reaction center complex is in the proper redox state, the trapping of an exciton by the chlorophyll pool is funneled to the trap Chl a (P-680). This creates a photochemical product by a separation of charge between Z and Q.

$$Z\ P_{680}Q \xrightarrow{\quad\underset{Chl^*\quad Chl}{\overset{h\nu}{\frown}}\quad} Z\ P^*_{680}Q \qquad (1)$$

$$Z\ P^*_{680}Q \longrightarrow Z\ P^+_{680}Q^- \longrightarrow Z^+P_{680}Q^- \qquad (2)$$

Oxidized Z (Z^+) can then receive an electron by oxidation of water comprising a multistep procedure. The chemical nature of the primary Z complex is lacking at present. Tightly bound Mn^{2+} appears to be associated with Z, with Cl^- also shown to be essential.

Little is known about the biochemical mechanism of O_2 evolution. One of the functional problems is to understand how four photoreactions which correspond to the transfer of four electrons are able to cooperate to produce one O_2 molecule. Most of our present knowledge comes from kinetic studies. When dark-adapted chloroplasts were submitted to a series of short saturating flashes (10^{-5} s) the amount of O_2 evolved per flash oscillated with a periodicity of four. From these experiments Kok and Joliot concluded that the cooperation between four photoreactions occurs at the same photocenter, with each photocenter being independent of the other. In the electron donor side of photosystem II a multicomplex protein component containing Mn is proposed to successively collect and store four oxidizing equivalents which are used in a concerted oxidation of two water molecules to evolve one O_2. The precursor states for O_2 evolution are quite stable; under some conditions they can have a life time of several minutes. The O_2-yielding reactions and the reactions associated with trap recovery are fast relative to the rate-limiting steps of photosynthesis. Although the molecular entity of this photosystem II charge collector is unknown, some evidence suggests the participation of 4 Mn per charge collector. The role of Cl^- perhaps reflects the necessity of ion-pumping in O_2 evolution. Bicarbonate indirectly moderates the mechanism of O_2 evolution but is not the donor for O_2.

D. Photosystem I and the Reduction of $NADP^+$

Photosystem I has a characteristically longer absorption maxima than photosystem II. It is involved in moving electrons to reduce $NADP^+$ or provide for cyclic electron flow to give extra ATP without net electron transport. It has been hard to identify the primary electron acceptor for photosystem I. In 1971, a spectroscopic component, P-430, was discovered that exhibited properties necessary for the primary acceptor. Below 25° K an absorption component in the Soret region having the necessary kinetic characteristics was identified. It was associated with photosystem I, based on its enrichment upon isolation of photosystem I particles, and the quantum yield increased from one to two for both P-700 and P-430 when far-red light was substituted for red light. The chemical identity of this component has been speculated to possibly be bound Fd. At least one to two bound nonheme iron sulfur centers exist at the reducing end of photosystem I and a large amount of nonheme iron or Fd is bound to the thylakoid membrane. Studies with enriched photosystem I particles (prepared by Triton X-100 treatment) indicated a large pool of iron–sulfur protein of four to five times the amount of P-700, PC, or Cyt f.

The rest of the reducing side of photosystem I is one of the best understood segments of the photosynthetic electron transport pathway. There are two soluble proteins involved in the direct transfer of electrons from the reducing membrane-bound P-430 to $NADP^+$. The first is soluble Fd which has been shown to be photoreduced by isolated chloroplasts. Ferredoxin is a reddish brown protein having a potential of -430 mV. It contains 2 moles of nonheme iron and acid-labile sulfur per mole of protein and functions as a one electron carrier as shown by the ability of 1 mol of $NADP^+$ to oxidize 2 moles of reduced Fd. A second protein, ferredoxin-$NADP^+$ oxidoreductase is necessary for the collection and transfer of electrons, one at a time, from ferredoxin to the two election reduction of $NADP^+$. This enzyme contains one bound FAD per molecule; the nature of the binding of FAD to the protein is unknown. However, covalent binding has been excluded because FAD can be separated from the holoenzyme by reduction with $NADP^+$ in the presence of urea or in the presence of mercurials. The FAD prosthetic group must be hidden internally in the protein structure. Reduced X and/or reduced ferredoxin also appear to regulate the activities of some of the enxymes involved in carbon flow during photosynthesis (Section IV,B,1).

It is well established that P-700 is the primary donor of photosystem I. Its concentration in the chloroplast is about 1 per 400 Chl. Redox titrations have established it as a single electron carrier with a potential of 450 mV. Most likely P-700 itself is a chlorophyll localized in a special environment. Because its absorbance band is at a slightly longer wavelength than the bulk light-harvesting chlorophylls, most of the excitation energy captured by the bulk chlorophyll will be funneled to P-700. P-700 has been isolated in a Chl a–P-700 protein complex with 40 Chl a per P-700 and 90,000 MW.

E. Intermediates of Electron Transport

Identification of the primary electron acceptor of photosystem II (Q or C-550) has been complicated partially by the confusion concerning the roles of Cyt b-559 and P-680. It appears that Cyt b-559 can exist as two forms: a low potential form Cyt b-559, with a potential of about 80 mV interconvertible to a high potential form of 350 mV. The exact physiological role of these two forms is not clear but it seems likely that the low potential form interacts with the electron transport chain between the two photosystems via plastoquinone, while the high potential form is involved in a cyclic flow of electrons around photosystem II.

Plastoquinone (PQ) is the name given to a mixture of related electron transport quinone intermediates—the principle component is PQ A. The concentration of plastoquinone is much higher than that of the other electron transport intermediates. It is normally present in a concentration equal to

5–10% of the total chlorophyll or 25–50 molecules per photosystem I or photosystem II unit. The main pool of plastoquinone is located in the electron transport chain between the low potential Cyt b-559 and Cyt f.

Cytochrome f is the best known of the photosynthetic cytochromes. It can be released by gentle procedures from the photosynthetic membranes of several higher plants and algae. It has a characteristic absorption peak for the reduced form of 554 nm, with a redox potential of 365 mV. Although called Cyt f, it is actually a c-type cytochrome. Because of its association with the green part of the plant, f, for the latin *folium* (leaf), was used.

Plastocyanin (PC) is the electron donor to P-700. It is a copper protein which has a characteristic blue color in the oxidized form. Chloroplasts contain a Chl/plastocyanin ratio of about 300 with plastocyanin accounting for one-half the total copper in the chloroplast. The redox potential of spinach plastocyanin is 370 mV.

The second b-type cytochrome in photosynthetic tissue is Cyt b_6 or b-563. It has a potential of about zero volts and is auto-oxidizable. Although originally proposed to function in the main electron transport flow it has now been shown to mediate cyclic flow between X and the electron transport chain. Its interaction is probably with Cyt f or plastocyanin, or possibly even through the large plastoquinone pool.

Fig. 5 indicates the electron transport scheme in the characteristic Z shape, in order to represent the electrical potential of each step. Upon illumination a potential gradient is formed so that electrons can flow from Z to X by way of the indicated component. Physically these points may well lie close to one another across the thylakoid membrane in a characteristic fashion, and how this might occur is shown in Fig. 6. By use of antibody–antigen identification techniques to various components of the electron transport scheme, it has been determined which components are located on the outside and which are located on the inside of the thylakoid membrane. The ferredoxin and the flavoprotein are on the outside and are readily solubilized. The Z complex, where water is oxidized, appears to lie on the inside of the space. The donor side to photosystem I apparently is oriented toward the inside of the membrane, i.e., PC, Cyt f, and P-700. The acceptor side of photosystem II appears to lie on the outside of the thylakoid membrane (Trebst, 1974).

F. Photophosphorylation

Absorbed light energy is conserved in the formation of ATP. Chloroplasts in the light are capable of high rates of ATP formation. This process is very similar to the coupled conservation of energy during respiration with electron transport in mitochondria. In plants this light-activated process is called photophosphorylation.

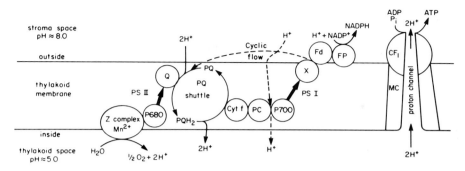

Fig. 6. Diagram of photosynthetic electron flow in the thylakoid membrane coupled to ATP formation. See text for details on intermediates. Modified from Trebst (1974) and Trebst and Avron (1977).

Both electron transport and photophosphorylation are said to be "coupled." No phosphorylation will occur unless electron transport is proceeding. Conversely there should be no electron transport unless ADP and Pi are present to permit simultaneous phosphorylation. Actually isolated chloroplasts always have a small amount of electron transport in the absence of added ADP or inorganic phosphate due to leaks in the system. The stimulation of this basal rate by added ADP and phosphate, expressed as a ratio, is often given the name "photosynthetic control" by analogy with the term "respiratory control" by ADP in mitochondrial oxidative phosphorylation. The degree of coupling varies depending on the conditions for measuring photophosphorylation and on the state of the isolated chloroplasts. When good chloroplasts, capable of high rates of light-dependent CO_2 fixation, are given a hypotonic shock at the last moment, photosynthetic controls of over four to six are observed.

As details of the electron transport chain became known, an important question was asked concerning how many of the electron transport steps conserved energy in ATP synthesis. The first attempts at determining the least number of sites were based on measurements of the $P/2e^-$ ratio—the number of ATP molecules formed per pair of electrons moving from water to the reducing side of photosystem I. This $P/2e^-$ ratio is used similarly with coupling of mitochondrial oxidative phosphorylation. The initial $P/2e^-$ ratios were observed at 1.0. Lower ratios are easily obtained with broken chloroplasts so that higher ratios are more significant. As preparation procedures improved and the use of the pH optimum 8.5 was discovered, ratios rose to 1.2 to 1.3. Using good chloroplasts after hypotonic shock, $P/2e^-$ ratios above 1.5 and approaching 2.0 have been obtained. It is now concluded that at least two phosphorylation sites must exist between water and photosystem I. Recent work using specific inhibitors or electron donors and acceptors also concur with this conclusion. The two phosphorylating sites in electron transport appear, first, between water and photosystem II and, second, between Q and Cvt f (Fig. 5).

Recent results by Krause and Heber (1976) have suggested that the coupling of phosphorylation to electron transport may not be tight. The photosynthetic apparatus can adjust itself to lower values, when lower stoichiometric amounts are needed for photoreduction. They suggest that the chloroplast *in vivo* operates with a flexible P/2e$^-$ ratio. The control of this flexible ratio may involve the concentration of ADP or the breakdown of the proton gradient across the thylakoid.

Photophosphorylation occurs in an intact vesicle similar to a bubble with an aqueous phase inside and out and whose membrane is relatively impermeable to protons or hydroxide ions. The electron transport intermediates are embedded anisotropically across the membrane. The hydrogen carriers are believed to be so oriented in the membrane that when an electron is passed the necessary proton to complement it comes from outside the vesicle. In turn, when giving up the electron it should be to an electron acceptor on the inside with a complimentary proton released on the inside. In this way, due to the geometry, electron transport through the chain is coupled obligatorily to vectorial proton translocation across the membrane. Specific ways in which H$^+$ translocation might occur in the chloroplast membrane are outlined in Fig. 6. The first internal protons are those released from water splitting and this constitutes site II for energy conservation. The second coupling site, site I, operates during the sequential reduction, with the reoxidation of plastoquinone (PQ). Because PQ is a pool of molecules, the shuttle may actually involve transfer between several or many molecules before complete oxidation and reduction across the pool has occurred.

Almost immediately when electron flow starts and protons move to the inside of the vesicle, a membrane potential will arise, positive on the inside forming what is known as an electrogenic "pump." In thylakoids this charge separation occurs during the primary photoact itself and is the source of the electric membrane potential. Once the membrane potential has been created by proton flux, other ions will diffuse passively across the thylakoid membranes, facilitated by ion carriers. In particular, Cl$^-$ ions are taken up with the protons and Mg^{2+} and K$^+$ leave in varying proportions depending on the medium in which the plastids are equilibrated. The counter ion flux, at least in part, collapses the membrane potential but leaves a hydrogen ion gradient undisturbed with the interior more acid than the exterior.

Thus, electron transport creates both an electrochemical gradient of protons and a membrane potential. These two parameters are additive, contributing to a protonmotive force differential across the membrane such that Δpmf = ΔpH + $\Delta\psi$, where pmf stands for protonmotive force, $\Delta\psi$ is the membrane potential, and ΔpH refers to the proton differential across the membrane. Present evidence suggests that a collapsing of both these components of the protonmotive force are the driving energy for ATP synthesis. This occurs by protons leaving the inner space of the thylakoid by the proton channel and passing by the ATP synthetase in such a manner that the energy of the H^+ gradient is conserved in ATP synthesis from ADP and inorganic phosphate.

The membrane-bound enzyme that is involved directly in photophosphorylation is called the chloroplast coupling factor one (CF_1). It sits on the surface of the thylakoid membrane and can be released by dilute EDTA. With an approximate 325,000 MW, it is composed of five different polypeptide subunits. It can move laterally in the membrane since antibodies make the knobs clump together. There appears to be one CF_1 per 500–850 Chl. The CF_1 acts as a proton translocator and, as well, as an ATP synthetase. There is no indication that CF_1 protein penetrates all the way through the membrane although the protons must move completely through. Apparently, there are highly hydrophobic proteins in the membrane which, as intrinsic components of the membrane, serve as a point of attachment of the CF_1 to facilitate ATP synthesis.

The magnitude of the pH gradient that illuminated thylakoids can reach is 3–3.5 pH units more acidic on the inside than on the outside. This pH gradient increases with increasing light intensity and electron flow rates and is decreased by uncouplers. ATP or ADP have been shown to decrease the rates of proton leakage and can enhance the pH gradient to a maximum value of 3.5 units. The measurements of the membrane potential are variable with most values reported between 5 and 50 mV. However, some estimates have suggested that the membrane potential can even approach 100 mV positive on the inside.

For many years when the maximum $P/2e^-$ ratios in intact chloroplasts were thought to be below 1.5, additional electron transfer steps for synthesis of extra ATP were considered. One of the most important has been cyclic phosphorylation supported by cyclic electron flow in photosystem I (Fig. 5). There is no compelling evidence to show that cyclic photophosphorylation plays a significant role in higher plants under natural aerobic conditions. Rather, in intact leaves or in chloroplasts, O_2 can readily react with an electron carrier beyond photosystem I. In an N_2 atmosphere intact leaves exhibit strong chloroplast shrinkage under far-red illumination (which excites preferentially photosystem I) indicative of photophosphorylation by cyclic electron transfer. As very low levels of O_2 (0.1%) reversed this effect, it seems that O_2 can easily drain electrons from the cyclic pathway and thereby inhibit cyclic photophosphorylation. Yet oxygen does support photophosphorylation in light using both photosystems. Apparently, electrons transferred through the two photosystems will either move to $NADP^+$ or, if this is reduced, can be transferred to molecular O_2 (Mehler reaction). In this reaction, H_2O_2 could leave the chloroplast by diffusion and be decomposed by catalase in the peroxisomes. Such a system would be self regulating with respect to photophosphorylation, if $NADP^+$ has a much greater affinity as the terminal electron acceptor than O_2. There are several indications that this is the case in intact plants. During induction of CO_2 fixation, ATP can become limiting whereupon NADPH accumulates and less $NADP^+$ is available for reduction. Electrons are then diverted to oxygen which results in additional photophosphorylation without $NADP^+$ reduction. This process is termed pseudocyclic electron transport as oxygen evolution and oxygen uptake balance each other and no net O_2 change is observed.

IV. CARBON METABOLISM IN THE CHLOROPLAST

The light produced intermediates, ATP and NADPH, are utilized in the chloroplast to fix CO_2 and reduce it to the level of carbohydrates. Part of the reduced carbon remains as starch in the chloroplast for utilization at night, while the rest is transported to the cytoplasm to form sucrose and organic and amino acids. Those products which are formed in the chloroplast and transported to the cytoplasm are mostly three-carbon compounds; 3-phosphoglycerate (glycerate-3-P), dihydroxyacetone phosphate, and glyceraldehyde-3-phosphate (glyceraldehyde-3-P). Glycolate and small amounts of pentose monophosphates can also leave the chloroplast. Depending on needs of the plant, 25–50% of the fixed carbon is stored as starch in the chloroplast.

A. Photosynthetic Carbon Reduction Pathway

The only pathway for net CO_2 fixation resulting in carbon incorporation into hexoses is the reductive photosynthetic carbon cycle. Even though C_4 plants initially fix CO_2 into oxaloacetate which is converted into malate and aspartate, these must be decarboxylated so that the CO_2 released can be refixed by way of the photosynthetic carbon cycle. Many of the reactions of the photosynthetic carbon cycle are similar to steps of the glycolytic pathway and the hexose monophosphate shunt and consist of three different phases of carbon metabolism (Fig. 7). The first phase is the production of ribulose-1,5-bisphosphate (ribulose-P_2) and its carboxylation, steps that are unique to the photosynthetic carbon cycle. The second is the reduction of glycerate-3-P to the level of an aldehyde, glyceraldehyde-3-P. The third phase involves the disproportionation of triose phosphates to produce pentose monophosphates, the precursors for ribulose-P_2, by way of tetrose, hexose, and heptose phosphates.

CO_2 is incorporated by carboxylation of ribulose-P_2, catalyzed by

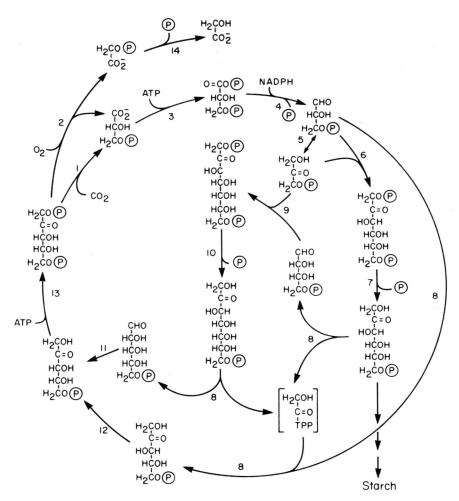

Fig. 7. Photosynthetic carbon reduction cycle or Calvin cycle. The enzymes involved are (1,2) ribulose-P_2 carboxylase/oxygenase; (3) glycerate-3-P kinase; (4) NADP-glyceraldehyde-3-P dehydrogenase; (5) triose phosphate isomerase; (6) aldolase; (7) fructose-P_2 1-phosphatase; (8) transketolase; (9) aldolase; (10) sedoheptulose-P_2 1-phosphatase; (11) ribose-5-P isomerase; (12) ribulose-5-P epimerase; (13) ribulose-5-P kinase; (14) P-glycolate phosphatase.

ribulose-1,5-P_2 carboxylase/oxygenase (ribulose-P_2 carboxylase). The initial products are two molecules of glycerate-3-P. The ribulose-P_2 carboxylase catalyzes a second important reaction, whereby molecular O_2 reacts with ribulose-P_2 to form P-glycolate and glycerate-3-P. P-glycolate phosphatase releases glycolate which diffuses from the chloroplast to be the substrate for part of photorespiration.

In the presence of ATP, glycerate-3-P is phosphorylated to 1,3-bisphosphoglycerate by action of glycerate-3-P kinase. Glycerate-1,3-P_2 is reduced by NADP-glyceraldehyde phosphate dehydrogenase and NADPH with release of inorganic phosphate to give glyceraldehyde-3-P. Glyceraldehyde-3-P quickly equilibrates with dihydroxyacetone phosphate by way of triose phosphate isomerase. These two triose phosphates are combined by the catalysis of aldolase to form fructose 1,6-bisphosphate (fructose-1,6-P_2). Then, fructose 6-phosphate (fructose-6-P) is formed by the action of fructose-P_2 phosphatase, a regulatory enzyme which controls the flow of carbon to subsequent pathways. The reverse step in glycolysis, by way of phosphofructokinase is also present in the chloroplast, but probably does not operate in the light. Fructose-6-P can be converted to glucose 6-phosphate (glucose-6-P), glucose 1-phosphate (glucose-1-P), and eventually starch. For continuation of the photosynthetic carbon cycle, trans-ketolase cleaves fructose-6-P into erythrose 4-phosphate and an enzyme-bound glycolaldehyde–thiamine pyrophosphate adduct. Erythrose-4-P and dihydroxyacetone phosphate combine to produce sedoheptulose-1, 7-bisphosphate which is split by sedoheptulose-P_2 phosphatase to form sedoheptulose-7-phosphate (sedoheptulose-7-P). In a second reaction catalyzed by transketolase, sedoheptulose-7-P is split to ribose-5-phosphate (ribose-5-P) and another bound glycolaldehyde–thiamine pyrophosphate. This activated aldehyde is transferred by transketolase to glyceraldehyde-3-P to produce xylulose-5-phosphate (xylulose-5-P). Ribose-5-P, by way of an isomerase, and xylulose-5-P, by way of an epimerase, are both converted to ribulose 5-phosphate (ribulose-5-P). Lastly, the cycle is completed by the formation of ribulose-P_2 from ATP and ribulose-5-P catalyzed by ribulose-5-P kinase.

The net result of the photosynthetic carbon cycle is to fix CO_2 to the oxidation level of a carbohydrate. This requires 3 ATP and 2 NADPH per CO_2 reduced. Thus, to produce one hexose from 6 CO_2, 18 ATP and 12 NADPH are required.

B. Regulation of CO_2 Fixation

Many factors are involved in regulating the rate of photosynthetic carbon assimilation. If CO_2 and ribulose-P_2 were present in saturating amounts, then the activity of ribulose-P_2 carboxylase would be limiting in the carboxylation phase. Indeed this appears partially true as the ribulose-P_2 carboxylase can be activated to varying degrees in the chloroplast, depending on light and CO_2 levels. A discussion on this and other properties of the ribulose-P_2 carboxylase will be given later. However, CO_2 is limiting for photosynthesis in the atmosphere and this condition can only be corrected by growing plants in enriched concentrations of CO_2. The usual atmospheric level of CO_2 is about 320 ppm, while 600–700 ppm CO_2 is necessary to saturate photosynthesis. The level of CO_2 at the site of carboxylation may be considerably less than 320 ppm, but equal or greater than the CO_2 compensation point (about 50 ppm with C_3 plants). This is due to limitations on CO_2 diffusion by the stomata and mesophyll cell resistances.

Levels of ribulose-P_2 can also be low and limiting for CO_2 fixation. This could occur under low light intensity or any condition that limits photophosphorylation and electron transport. Several of the photosynthetic carbon cycle enzymes needed to regenerate ribulose-P_2 are also regulated in the light and dark.

1. Enzyme Activity Regulated by Light

a. Glyceraldehyde 3-P Dehydrogenase. This was the first enzyme of the photosynthetic carbon cycle whose activity was shown to increase in the light. Its activation is caused by an increase in the NADPH-linked activity relative to its NADH-linked activity. The activation is apparently due to the dissociation of the enzyme by $NADP^+$ and NADPH from a high molecular weight oligomeric form to protomers. The enzyme from pea contains bound $NADP^+$ which locks it in the lower molecular weight form. The level of glycerate-1,3-P_2 and the amount of reduced pyridine nucleotides regulate the rate and the direction of carbon flow as catalyzed by glyceraldehyde-3-P dehydrogenase.

b. 3-Phosphoglycerate Kinase. This kinase can be controlled by the adenylate energy charge. In the light the energy charge in the chloroplast becomes greater than 0.9, which stimulates glycerate-3-P kinase in the direction of phosphorylation of glycerate-3-P. The enzyme does not appear to be stimulated by reduced ferredoxin.

c. Ribulose 5-P Kinase. The activity of this enzyme is increased in the light and reduced in the dark. The purified enzyme can be activated by dithiothreitol. It apparently is activated in the chloroplast either by reduced ferredoxin or by direct interaction with a component at the reduced end of photosystem I. Under conditions of full activity, its activity is considerably larger than the rate of the ribulose-P_2 carboxylase. If it were not regulated, then its high activity and the large negative free energy of the reaction would cause a large accumulation of ribulose-P_2. This is generally not the case. It has recently been shown that the ribulose-5-P kinase can also be influenced by the adenylate energy charge.

d. Fructose-1,6-P_2 1-phosphatase and Sedoheptulose-1,7-P_2 1-phosphatase. The reported activities of the fructose-P_2 phosphatase and the sedoheptulose-P_2 phosphatase are not much more than the rates of CO_2 fixation, hence the regulation of these enzymes are quite likely to influence the synthesis of ribulose-P_2. The fructose-P_2 phosphatase apparently exists as a dimer which is activated in the light in the presence of Mg^{2+} by reduced Fd, chloroplast thioredoxin, and ferredoxin–thioredoxin reductase. Dithiothreitol can also be used in place of reduced Fd. Upon dissociation at pH 8.5 the monomer loses the Mg^{2+} dependent activity and retains the Fd-dependent activity while acquiring a sedoheptulose-P_2 phosphatase activity dependent on reduced Fd, thioredoxin, and the reductase. The fructose-P_2 phosphatase responds sigmoidally with respect to fructose-P_2. These properties are consistent with the variable activity of this enzyme in the chloroplast in the light. Activity is dependent on pH and Mg^{2+}, with little or no activity in the dark.

e. Ribulose-1,5-P_2 Carboxylase/Oxygenase. Ribulose-P_2 carboxylase often comprises more than 50% of the protein in the chloroplast stroma. It is a protein of high molecular weight (560,000) existing as a aggregate of two types of subunits, large and small (L_8S_8). The larger of the two subunits has a molecular weight of 51,000 to 58,000 while the smaller subunit is 12,000 to 18,000 MW. The large subunit is catalytically active even in the absence of the small subunit. The function of the small subunit might well be regulatory, but how this is accomplished in the molecule remains to be determined. The structure of crystalline tobacco ribulose-P_2 carboxylase consists of a two-layered structure each having four large and four small spherical masses. The two layers are arranged about a fourfold axis with four twofold axes perpendicular to it. When viewed down the fourfold axis, the molecule is square.

The problem of how the ribulose-P_2 carboxylase operates and is regulated at air levels of CO_2 *in vivo* is now of considerable research interest. From earlier kinetic studies, it was apparent that the activity of the isolated enzyme, as exhibited by its apparent affinity for CO_2, was too low to account for the observed rates of photosynthetic CO_2 fixation. The $K_m(CO_2)$ for the purified ribulose-P_2 was considerably high, between 70 and 600 μM. In intact isolated spinach chloroplasts, the apparent $K_m(CO_2)$ for CO_2 fixation is of the order of 10–20 μM. Water, in equilibrium with 1 atm air with 0.03% CO_2, has 10 μM CO_2 at 25°C.

The $K_m(CO_2)$ of the ribulose-P_2 carboxylase assayed upon lysis of chloroplasts is 11–18 μM at pH 7.8, which is comparable to that for light-dependent CO_2 fixation by intact chloroplasts. At low CO_2 concentrations, the kinetics of the ribulose-P_2 carboxylase is not stable after release from the chloroplast. During assays up to 10 min a lower steady state rate is obtained after 3 min, which eventually displays a $K_m(CO_2)$ value of about 500 μM, comparable to that seen in most previous work with the purified carboxylase. Incubation of the enzyme in buffer alone or buffer plus ribulose-P_2 before adding CO_2 and

Mg^{2+} hastens the decline in activity. In addition it appears that the ribulose-P_2 carboxylase, while still in the intact chloroplast, is not fully activated and this degree of activation can be altered by incubating chloroplasts with various CO_2 concentrations.

As the rate of carboxylation of ribulose-P_2 in the plant defines the rate of gross photosynthesis, the rate of ribulose-P_2 oxygenation is the major determinant for glycolate production for photorespiration. O_2 has been shown to be a competitive inhibitor of carboxylation as CO_2 is of oxygenation. The relative rates of the two reactions are regulated by the concentration of O_2 and CO_2. Where both reactions have been measured under the same conditions, effector metabolites such as NADPH, gluconate-6-P and glycerate-3-P activate or inhibit both reactions to the same extent.

The time-dependent and order of addition-dependent kinetics of ribulose-P_2 carboxylase are a result of the activating effects of Mg^{2+} and CO_2 and the inactivating effects of ribulose-P_2 in addition to their roles as substrates. The initial activity of the enzyme responds to the concentration of CO_2 and Mg^{2+} during preincubation, indicating the reversible formation of an active enzyme-CO_2-Mg^{2+} complex. Kinetic analyses have indicated that the enzyme is activated by a slow, but reversible initial binding between enzyme and CO_2 followed by a rapid reaction with Mg^{2+}. The amount of activation is a function of both CO_2 and Mg^{2+} concentration and the final degree of activation at fixed CO_2 and Mg^{2+} is sharply pH dependent with a distinctly alkaline pK_a. The relationship between CO_2 and Mg^{2+} concentrations and pH during activation suggests that the protonated enzyme does not react with CO_2. When the enzyme is first incubated with ribulose-P_2 and the reaction is initiated with Mg^{2+} and CO_2, a marked lag is observed in the course of product formation. The process of activation and those reactions which compete with it are as follows:

Activating process

$$E_{(inactive)} + CO_2 \xrightleftharpoons{slow} E\text{-}CO_{2(inactive)}$$
$$E\text{-}CO_{2(inactive)} + Mg^{2+} \xrightleftharpoons{fast} E\text{-}CO_2\text{-}Mg^{2+}_{(active)}$$

Competing process

$$E + H^+ \rightleftarrows EH$$

Catalysis leading to carboxylation or oxygenation occurs as follows:

$E\text{-}CO_2\text{-}Mg^{2+} + \text{ribulose-}P_2 \rightleftarrows E\text{-}CO_2\text{-}Mg^{2+}\text{-ribulose-}P_2$
$E\text{-}CO_2\text{-}Mg^{2+}\text{-ribulose-}P_2 + CO_2 \rightarrow E\text{-}CO_2\text{-}Mg^{2+} + \text{two glycerate-3-P (carboxylation)}$
$E\text{-}CO_2\text{-}Mg^{2+}\text{-ribulose-}P_2 + O_2 \rightarrow E\text{-}CO_2\text{-}Mg^{2+} + \text{glycerate-3-P} + \text{P-glycolate (oxygenation)}$

Mg^{2+} and CO_2 are necessary for enzyme activation with the first CO_2 separate from the CO_2 involved in catalysis. When spinach chloroplasts are lysed in the presence of saturating amounts of ribulose-P_2, Mg^{2+}, and CO_2, the initial rate of CO_2 fixation can be about 200 μmoles/mg Chl·h. If upon lysis the extract is allowed to activate in the presence of Mg^{2+} and bicarbonate, then ribulose-P_2 added, the rate of fixation can be 300–400 μmoles/mg Chl·h. The amount of activation measured corresponds to the amount of the enzyme that exists as the E-CO_2 and E-CO_2-Mg^{2+} forms.

The activity of ribulose-P_2 carboxylase can be variable while the enzyme is in the chloroplast. Incubation of chloroplasts in high CO_2 levels activates the enzyme as does illumination. This can be reversed by removing the CO_2 and/or by dark. The light activation is inhibited by electron transport inhibitors, such as DCMU or uncouplers for photophosphorylation such as CCCP. The chloroplast carboxylase is inactivated in the dark under conditions where the E-CO_2 complex can dissociate in a CO_2-deficient media. Light activation and dark inactivation can be explained by light-dependent changes of Mg^{2+} and pH in the chloroplast.

The activation by lower pH values and the drop in Mg^{2+} concentration in the stroma are still too slow to fully explain the more rapid light to dark cessation of CO_2 fixation seen with intact chloroplasts. Most probably, the lower pH reached in the dark affects catalysis before it affects activation of the enzyme. At longer times, however, the activation of the enzyme does decrease in response to the conditions in the chloroplast in the dark. The activation of the enzyme in chloroplasts in the dark is usually about 20–50% of the fully activated ribulose-P_2 carboxylase.

2. Regulation of Photosynthesis by Ribulose-1,5-P_2 Levels

CO_2 fixation also depends on the operation of the photosynthetic carbon cycle to regenerate ribulose-P_2. Upon turning on the light with isolated chloroplasts, a brief initial lag in CO_2 fixation is noted. During this period the chloroplast ribulose-P_2 pool can increase to a peak followed by a steady decrease accompanying linear fixation of CO_2. When the lights are turned off, CO_2 fixation ceases and a small decrease in ribulose-P_2 occurs, yet significant ribulose-P_2 remains. The light-on peak in ribulose-P_2 levels suggests that more ribulose-P_2 is produced than is consumed and this condition holds until carboxylation of CO_2 equals the synthesis of ribulose-P_2.

The levels of ribulose-P_2 which sustain photosynthesis from intact spinach chloroplasts appears to be between 15–25 nmoles ribulose-P_2/mg Chl. This corresponds to a concentration in the chloroplast stroma (volume, 25 μl/mg Chl) of 0.62–1.0 mM ribulose-P_2. With a maximum of eight binding sites per ribulose-P_2 carboxylase, there are three to four mM binding sites for ribulose-P_2 present in the chloroplast. As the K_m (ribulose-P_2) is about 1 μM, essentially all of the chloroplast ribulose-P_2 would be bound to the ribulose-P_2 carboxylase. One might expect that catalysis by the ribulose-P_2 carboxylase would be limited by ribulose-P_2 if only part of the binding sites were occupied. Apparently not all the binding sites are available for ribulose-P_2

binding. When one out of four sites contains ribulose-P_2, then the enzyme in the chloroplast appears to be saturated with ribulose-P_2.

C. Carbon Metabolism in C_4 Photosynthesis

Those plants whose initial product of CO_2 fixation is glycerate-3-P are called C_3 plants. However, a second class of plants initially fix CO_2 into four-carbon organic acids, oxaloacetate, malate, and aspartate. These are called C_4 plants. C_4 photosynthesis is a cooperative process requiring operation of chloroplasts and cytoplasm of both the mesophyll cells and the bundle sheath cells. The compartmental localization of enzymes of this pathway and the movement of products between cells is well understood. Information is now appearing on the properties of the chloroplast to cytoplasm shuttle mechanisms in C_4 plants. The flow of carbon in C_4 plants will be discussed in this section, even though many of the events occur outside of the chloroplast.

Photosynthesis in C_4 plants has much practical value. It is these species which are capable of producing the highest yields of organic matter. They are better adapted to overcoming the negative effects of photorespiration and the low levels of atmospheric CO_2 and arid water conditions. The photosynthetic tissue is located in two distinct concentric layers. This Kranz-like (wreathlike) anatomical characteristic is discussed in Chapter 1.

In C_4 plants CO_2 is first assimilated in the cytoplasm of the outer mesophyll cells by carboxylation of phosphoenolpyruvate (P-pyruvate) to form oxaloacetate; the key enzyme involved is P-pyruvate carboxylase (Fig. 8). Oxaloacetate is reduced to malate in $NADP^+$-malic enzyme (NADP-ME) type species or converted to aspartate in P-pyruvate carboxykinase (PCK) type and NAD^+-malic enzyme (NAD-ME) type species. These three types are named after the different variations by which malate or aspartate are decarboxylated in the bundle sheath cell.

In the chloroplast of the bundle sheath cell of the NADP-ME type plants, malate is decarboxylated to CO_2, pyruvate, and NADPH. The CO_2 and NADPH are utilized by the reductive photosynthetic carbon cycle to form P-glycerate and to reduce it to triose phosphates. The pyruvate returns to the mesophyll cell. In contrast, in the PCK type of C_4 plants, aspartate in the cytoplasm of the bundle sheath is converted to oxaloacetate by transferring the amino group to glutamate. The oxaloacetate is decarboxylated in the chloroplast to P-pyruvate by P-pyruvate carboxykinase. The P-pyruvate may go to pyruvate, although this enzyme activity has not been established. Alanine is formed from pyruvate and returned to the mesophyll cell.

A third variation occurs with the NAD-ME type. In the bundle sheath mitochondria, aspartate is transaminated to oxaloacetate, then reduced to malate with NADH. The NAD^+-malic enzyme decarboxylates malate to

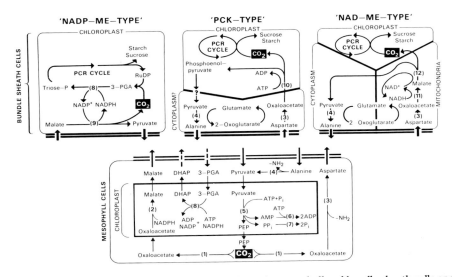

Fig. 8. Photosynthetic reactions of the C_4 pathway in mesophyll and bundle sheath cells and their intracellular location. Separate diagrams depict the different C_4 acid decarboxylation mechanisms for the NADP–ME type, PCK type, and the NAD–ME type species. The malate–pyruvate shuttle from the mesophyll cell operates with the NADP–ME type while the aspartate–alanine shuttle occurs with the other two types. The enzymes involved are (1) P-pyruvate carboxylase; (2) $NADP^+$-malate dehydrogenase; (3) aspartate aminotransferase; (4) alanine aminotransferase; (5) pyruvate–phosphate dikinase; (6) adenylate kinase; (7) pyrophosphatase; (8) glycerate-3-P kinase, $NADP^+$-glyceraldehyde-3-P dehydrogenase and triose phosphate isomerase; (9) $NADP^+$–malic enzyme; (10) P-pyruvate carboxykinase; (11) NAD^+–malate dehydrogenase; (12) NAD^+–malic enzyme. From Hatch (1978).

pyruvate and regenerates the NADH. In all three of these variations, the C_4 pathway has the effect of concentrating CO_2 from the mesophyll into the bundle sheath cells thereby increasing net CO_2 fixation during photosynthesis and reducing photorespiration.

The orientation of the C_4 pathway of carbon has been determined primarily from the location of the various enzyme activities in the two kinds of cells and their organelles. The complete photosynthetic carbon cycle appears to operate only in the bundle sheath chloroplast of all three types of C_4 plants. The ribulose-P_2 carboxylase is located only in the bundle sheath chloroplast. Of notable exception is the observation that P-glycerate kinase and $NADP^+$-glyceraldehyde-3-P dehydrogenase are about equally distributed between the mesophyll and bundle sheath chloroplasts. It has been suggested that this is necessary in agranal chloroplasts (NADP-ME type) to allow maximum photosynthetic CO_2 reduction to occur because these chloroplasts are deficient in photosystem II activity with reduced electron flow to form NADPH. Some of the P-glycerate formed in bundle sheath chloroplasts is thought to be transported to the mesophyll chloroplasts for reduction to dihydroxyacetone phosphàte. The decarboxylation of malate by the $NADP^+$-malic enzyme provides the rest of the NADPH needed for reduction of P-glycerate in bundle sheath chloroplasts. By this mechanism the responsibility for providing ATP and reducing power for CO_2 reduction is shared between the two cell types. To maintain the cycle nature of the C_4 pathway, a C_3 compound, pyruvate or alanine, must be returned to the mesophyll cell to regenerate P-pyruvate.

The reaction leading to P-pyruvate formation is catalyzed by pyruvate–phosphate dikinase, an enzyme common to all C_4 species but absent in C_3 species. It catalyzes the reaction of pyruvate with ATP and orthophosphate to give P-pyruvate plus AMP and pyrophosphate. High activities of adenylate kinase and pyrophosphatase are present to help photophosphorylation convert AMP and orthophosphate to ATP. For the NADP-ME type species, pyruvate is supplied directly to the pyruvate–phosphate dikinase from the bundle sheath cell. For the other two types the conversion of alanine to pyruvate via alanine aminotransferase is required.

The bundle sheath chloroplast is the preferential site for starch synthesis, although, after heavy deposition, starch will also appear in mesophyll chloroplasts. The enzymes involved in sucrose synthesis are either equally distributed or more prevalent in mesophyll cells. Sucrose for translocation is supplied by both cell types, although the bundle sheath cells may be the major donor.

Most of the carbon fixed by the photosynthetic carbon cycle in the bundle sheath chloroplast comes by way of the shuttling and decarboxylation of C_4 acids from the mesophyll cell. Only a minor amount of external CO_2 diffuses directly into the bundle sheath chloroplast. The high activity and affinity for CO_2 of the P-pyruvate carboxylase along with high amounts of carbonic anhydrase in the mesophyll cytoplasm form a strong sink for CO_2. The affinity for CO_2 of the PEP carboxylase is several times greater than that of the ribulose-P_2 carboxylase. The concentrating of CO_2 into the thick-walled bundle sheath cells causes increased CO_2 fixation, which competes with O_2 to reduce photorespiration. Any carbon possibly lost as CO_2 by photorespiration is refixed by the mesophyll P-pyruvate carboxylase. The concentrating mechanism increases the CO_2 in the bundle sheath cells to 0.6–2.0 mM bicarbonate, an amount that approaches saturation for ribulose-P_2 carboxylase.

V. ENERGETICS AND SHUTTLE MECHANISMS OF THE CHLOROPLAST

Intact chloroplasts are contained inside an envelope consisting of two bilayer membranes. Rupture of this outer envelope occurs by mechanical means, such as osmotic shock, and releases soluble constituents including enzymes and even the chlorophyll-containing thylakoids. When chloroplasts with intact outer envelopes are supplied with $NADP^+$, ADP, and phosphate and are illuminated, little oxygen evolution is observed in the absence of CO_2. These chloroplasts are unable to maintain light-driven phosphorylation of added ADP, nor are they capable of reduction of the exogenous $NADP^+$ or proton uptake from the medium as observed with naked thylakoid membrane systems. In the presence of substrates such as bicarbonate or P-glycerate, which need $NADP^+$ and ATP for reduction, illumination results in rapid oxygen evolution indicating the necessity for the functional integrity of the isolated organelles. The chloroplast outer envelope prevents free exchange of protons, adenylates, and pyridine nucleotides, but not of CO_2 and glycerate-3-P, between the reactive sites in the chloroplast interior and the suspension medium. The rapid reduction of CO_2 and glycerate-3-P inside intact chloroplasts is indicative of the rapid turnover of the internal adenine and pyridine nucleotide pools which are separated from the external media by the outer envelope.

A. Adenylates and Pyridine Nucleotides

The pool sizes of adenylates and pyridine nucleotides have been determined in chloroplasts either by aqueous isolation or by nonaqueous separation of them from the cell matrix. The *in vivo* concentration of adenylates (AMP + ADP + ATP) in spinach chloroplasts is between 1 and 3 mM. The levels of pyridine nucleotides, NAD(H), and NADP(H) are lower and usually between 0.5 and 1.5 mM. The light-driven increase in levels of NADPH and ATP are used to drive CO_2 assimilation, including the synthesis of ribulose-P_2 and the reduction of glycerate-3-P. Although light causes a large change in the level of NADPH, even in the dark about 5–25% of the endogenous NADP(H) is reduced. Upon illumination in the absence of electron acceptors such as glycerate-3-P, CO_2, or oxaloacetate, about 60%–90% of the $NADP^+$ becomes reduced. Importantly, a large part of the NADPH is bound, even in the dark, so that the changes in the redox state of the free NADP(H) occurring upon illumination is much larger than would be calculated on the basis of total NADP(H) content. Illumination of isolated chloroplasts also increases the ATP level. The rise in ATP is accompanied by a drop in ADP and AMP, indicating the presence of an active adenylate kinase. The total adenylate content remains constant.

The chloroplast envelope has a low permeability towards most ionic substances. A limited number of compounds do permeate by diffusion facilitated by specific "translocators" located in the inner membrane of the outer envelope. Although the chloroplast envelope does not permit transport of pyridine nucleotides, the reducing power can be shuttled out of the chloroplast. The permeability of the chloroplast envelope towards adenylates is small, limiting direct transport of adenylates to 2–4 μmoles of ADP phosphorylated by intact chloroplasts per milligram Chl · h. This compares to rates after rupture of the envelope of 60–120 μmoles ADP/mg Chl · h. This rate of transport with intact chloroplasts is insufficient to account for fast dark–light transit changes in ATP levels in the cytoplasm and suggests that cytoplasmic ATP produced during photophosphorylation leaves the chloroplast by mechanisms other than direct transfer of the adenylate moiety (see Section V,E).

B. pH, Magnesium, and Bicarbonate

Illumination causes the pH of the stroma to increase by one unit to about pH 8.0, while the pH in the intrathylakoid space drops by about two units. This alkalization of the chloroplast stroma in the light thermodynamically favors the ATP consuming reactions of the photosynthetic carbon reduction cycle. These pH changes are also accompanied by an increase of 1–3 mM Mg^{2+} in the stroma due to Mg^{2+} release from the thylakoids. The increase in pH of the stroma in the light causes changes in the levels of bicarbonate. CO$_2$ is rapidly transported across the outer envelope so that as pH changes occur inside the chloroplast there is also an accompanying change in the bicarbonate. Bicarbonate itself does not exhibit measurable rates of transport. Under conditions where the rates of CO$_2$ diffusion are not limiting, the CO$_2$ concentration on each side of the membrane is equal. This has been shown by experiments measuring the internal pH of the stroma using [^{14}C]dimethyloxazolidinedione, which distributes across the outer envelope depending on the internal pH, and comparing it to the distribution of ^{14}C-bicarbonate. As there appears to be ample carbonic anhydrase activity and rapid diffusion of CO$_2$, the pH-dependent increase of bicarbonate in the stroma probably does not alter the CO$_2$ concentration in the stroma compared to the cytoplasm. This is important as CO$_2$ rather than bicarbonate is both a substrate and effector for the ribulose-P$_2$ carboxylase.

C. The Phosphate Translocator

Of the various phosphate compounds involved in the photosynthetic carbon cycle, only glycerate-3-P, dihydroxyacetone phosphate, glyceraldehyde-3-P, and inorganic phosphate permeate the chloroplast through the inner membrane of the outer envelope at rates sufficient to support photosynthetic carbon flow. Measurements indicate these compounds are specifically translocated or exchanged across the inner membrane. This facilitated diffusion allows inorganic phosphate to enter while glycerate-3-P, dihydroxyacetone phosphate, or glyceraldehyde-3-P exit from the chloroplast (Fig. 9). In this manner the level of total phosphates in the

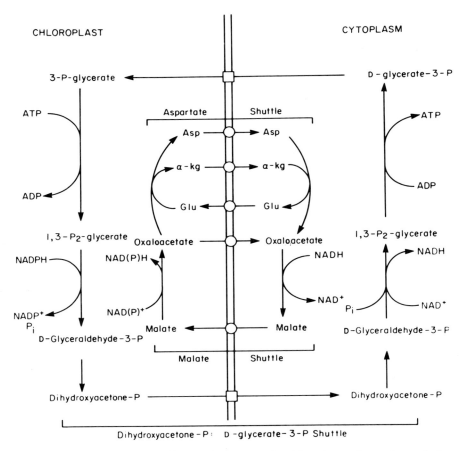

Fig. 9. Mechanism of metabolic transfer between the chloroplast and cytoplasm of triose phosphates and glycerate-3-P for indirect shuttle of ATP and NADH. Back transfer of NADH can occur by transfer of malate and oxaloacetate or malate, glutamate, α-ketoglutarate (αkg), and aspartate. Adapted from Heber (1975).

chloroplast remains relatively constant. Apparently, there is one specific carrier protein catalyzing this transport, called the "phosphate translocator." Other phosphates such as the pentose phosphates and erythrose-4-P are poorly translocated, while hexose phosphates are not moved.

D. Dicarboxylate Translocator and the Malate : Oxaloacetate Shuttle

Several dicarboxylic organic acids are also transported across the inner membrane of the chloroplast envelope by the "dicarboxylate translocator."

Because of limited binding sites, the shuttle mechanism shows substrate saturation with various dicarboxylates competing for transport into the stroma, e.g., the transport of malate is inhibited by fumarate, succinate, and aspartate. The dicarboxylate carrier facilitates a counter exchange of anions, but this counter exchange operates independent of transport by the phosphate translocator. Net transport in one direction is possible, but the rate is about an order of magnitude lower than the rate of counter transport.

The dicarboxylate translocator moves acidic amino acids (aspartate, glutamate) into the chloroplast. However, neutral amino acids are taken up by diffusion into the chloroplast stroma but not by the dicarboxylate translocator. The neutral amino acids moving most rapidly have a large hydrophobic moiety like phenylalanine or isoleucine, whereas the short amino acids like serine, alanine and, to some extent, glycine are taken up at considerably slower rates. There is no firm evidence for the existence of specific amino acid carriers in the spinach chloroplast envelope.

E. Dihydroxyacetone Phosphate: Glycerate-3-P Shuttle

During illumination, the cytoplasm uses ATP at rates severalfold higher than in the dark. The increased ATP/ADP ratio drives ATP consuming reactions of normal dark metabolism faster in the light. For example, the triose phosphates, glycerate-3-P and glycolate from the chloroplast are metabolized in the cytoplasm in energy requiring reactions. Sucrose is synthesized by consuming ATP in the cytoplasm transferred by triose phosphates produced in the chloroplast. The chloroplast does not form sucrose nor does the chloroplast envelope permit passage of sucrose. The major amount of the ATP for cytoplasmic sucrose synthesis comes from the chloroplast since mitochondrial respiratory activity is controlled and decreased by the rise in cytoplasmic ATP.

The required energy transfer from the chloroplast to cytoplasm exceeds by far the capacity for direct ATP transport. The most efficient indirect transport system for ATP appears to be the shuttle system involving the transport of triose phosphates and glycerate-3-P across the chloroplast envelope, mediated by the phosphate translocator (Fig. 9). In the light the two triose phosphates are exported from the chloroplast to form hexose phosphates which are then converted into sucrose. Part of the glyceraldehyde-3-P may be oxidized to glycerate-3-P with the formation of ATP and NADH as needed. Glycerate-3-P can return to the chloroplast to be reduced again. The result of this shuttle is the export of phosphorylation and reducing energy in a ratio of one ATP per NAPH per triose phosphate oxidized. In the dark the shuttle can operate in the opposite direction and serve for energy transport from the cytoplasm to the chloroplast. With isolated chloroplasts the transport rate has been measured at 40–50 μmoles/mg Chl · h for light-dependent phosphorylation of external ADP. The rate of energy transfer is controlled by mass action by the ATP/ADP and NADH/NAD$^+$ ratios in the cytoplasm.

A major obstacle for operation of the dihydroxyacetone phosphate : glycerate-3-P shuttle for energy export is the absolute stoichiometric coupling between phosphorylation and NAD$^+$ reduction (one ATP per NADH), producing potentially excess NADH in the cytoplasm. As the pyridine nucleotides are impermeable to the chloroplast envelope, an indirect transfer system to return reducing power to the chloroplast is required. The dicarboxylate translocator may well be involved in returning excess reducing power to the chloroplast. Chloroplasts are capable of taking up malate and oxidizing it back to oxaloacetate with the reducing power being used for support of the photosynthetic carbon cycle. This shuttle mechanism may be aided by the availability of a malate–aspartate system (see Fig. 9). Oxaloacetate is normally in low concentrations. In the chloroplast it can be transaminated to aspartate with the amino group coming from glutamic acid. The products are aspartate and α-ketoglutarate. These two compounds will react in the cytoplasm to produce glutamate and oxaloacetate which can be reduced again to malate for return to the chloroplast.

The chloroplast can also indirectly transfer NADPH using the dihydroxyacetone phosphate : glycerate-3-P shuttle. The cytosol of leaf cells contains a nonphosphorylating NADP-specific glyceraldehyde phosphate dehydrogenase which produces one NADPH per glyceraldehyde-3-P oxidized.

F. Transport of Glycolate

During photosynthesis in the presence of air levels of CO_2, glycolate is a major product of the chloroplast. It cannot be metabolized in the plastid and is excreted into the medium (see Chapter 9 on Microbodies). Most of the glycolate comes from P-glycolate, a product of the oxygenase reaction of the chloroplast. One of the products of peroxisomal metabolism is the formation of glycerate which is returned back to the chloroplast and phosphorylated to glycerate-3-P. Indeed, it may be that up to half of the carbon fixed during photosynthesis may flow to glycolate through the peroxisome to glycerate to be returned back to the photosynthetic carbon cycle. This system in the peroxisomes consumes energy but may be necessary to act as an "idle" so that when CO_2 is limiting or exhausted, the light reactions can still have glycerate-3-P as an electron sink, thereby reducing damage to the photosystems. If this view is correct that photorespiration is actually a "safety valve," attempts to abolish photorespiration by blocking of its pathway could well produce negative results.

VI. STORAGE OF ENERGY BY STARCH ACCUMULATION

The rate of leaf photosynthesis appears to respond proportionately to the rate at which the photosynthetic products are transported and utilized. CO_2 fixation during photosynthesis is often reduced when high accumulations of starch occur in the leaf chloroplast. This has led to the hypothesis of product inhibition on photosynthesis. If a cold night prevents the breakdown and translocation of starch, photosynthesis the next day is decreased. If the plants have a greater need for utilization or greater sink requirements, then little starch will accumulate during the day and the photosynthetic rates will be higher than those with less sink requirements. This relationship suggests that photosynthesis may be indirectly inhibited by accumulation of starch if translocation is limited. However, endogenous sucrose in leaves appears to have little adverse effect on the photosynthetic rate, most likely because sucrose is not formed nor located in the chloroplast where photosynthesis occurs. However, if leaves suspended in solution are fed sugars such as mannose or glucose, the starch content does increase in the light. This increase does not appear to be due to direct glucose incorporation into starch, but rather the results of sequestering cytoplasmic inorganic phosphate.

The drop in photosynthesis with high starch content in the chloroplast has been proposed to be due to the physical distortion of the chloroplast by the starch grains. A chloroplast, largely free of starch grains, is an extremely thin organelle in the living cell. The accumulation of starch between the thylakoids can distort the chloroplast so that it approaches the shape of a sphere. This could increase the effective path length of CO_2 diffusion or tend to bind Mg^{2+} and thus reduce the activity of the ribulose-P_2 carboxylase. The actual regulatory mechanisms by which chloroplast starch accumulation limits photosynthesis are still quite speculative and require more investigative research.

Starch biosynthesis in the chloroplast operates by the following reaction:

$$ATP + \alpha\text{-glucose-1-phosphate} \rightarrow ADP\text{-glucose} + \text{pyrophosphate}$$
$$ADP\text{-glucose} + \alpha1\text{-4 glucan} \rightarrow ADP + \alpha1\text{-4 glucosyl-glucan}$$

The first step catalyzed by ADP-glucose pyrophosphorylase primes the synthetic route by formation of ADP-glucose. Metabolic regulation of starch formation apparently occurs by control of this allosteric enzyme. The second step is catalyzed by an $\alpha1$-4 glucan (starch) synthetase and adds a glucosyl residue to the glucan primer. Usually, the enzyme is intimately associated with the starch granule.

Another enzyme, starch phosphorylase, is also capable of synthesizing starch:

$$\alpha\text{-glucose 1-phosphate} + \alpha1\text{-4 glucan} \rightleftarrows \alpha1\text{-4 glucosyl-glucan} + \text{phosphate}$$

Since both of the starch-synthesizing enzymes are located in the chloroplast, there is still some controversy as to which is of primary importance. The phosphorylase is most likely involved only in the degradation or breakdown of starch rather than in synthesis because of the high ratio of inorganic phosphate to α-glucose 1-phosphate in the chloroplast. With isolated chloroplasts, the levels of inorganic phosphate have been measured from a low value of around 4 mM to over 100 mM. That phosphate can be in high concentration in the chloroplast is borne out by the observation that during the development of etioplasts to chloroplasts, electron-dense inclusions containing phosphate and iron have been observed in the chloroplasts of tobacco cotyledons.

Most studies on the path of starch synthesis and breakdown have been deduced from experiments with purified chloroplast enzymes. They suggest that regulation resides with the ADP-glucose pyrophosphorylase. This enzyme is allosterically effected by intermediates of the photosynthetic carbon cycle, with positive activation by glycerate-3-P and inhibition by inorganic phosphate. Indeed with isolated chloroplasts, higher amounts of inorganic phosphate in the suspending media do reduce the amount of carbon going to starch. If glycerate-3-P is increased, then there is an increase in starch synthesis. Photosynthesis is also inhibited by high orthophosphate and reversed by glycerate-3-P and triose phosphates. The phosphate inhibition could be caused by a loss of intermediates of the carbon cycle including glycerate-3-P with depletion of the ribulose-P_2 pool. When triose phosphates are added, the phosphate-induced loss of sugar phosphates from the chloroplast is reversed, glycerate-3-P increases, and starch synthesis is enhanced.

The control of starch metabolism by levels of inorganic phosphate in the cytoplasm could well be operating in the leaf. Processes which increase inorganic phosphate in the cytoplasm, such as the hydrolysis of sucrose phosphate to sucrose, facilitate export of triose phosphates from the chloroplast in exchange for inorganic phosphate. If mannose is added to leaf discs of spinach beet, phosphate was sequestered in the cytoplasm as mannose 6-phosphate. Starch formation in these leaf discs was increased tenfold, yet the starch formed was not synthesized from the mannose carbon but from CO_2.

Starch stored in the chloroplast during the day is mobilized to soluble products and exported from the chloroplast at night. Isolated chloroplasts loaded with ^{14}C-starch in the light will remobilize this starch in the dark into glycerate-3-P and maltose as the major products. This mobilization is promoted by phosphate and inhibited by glycerate-3-P. Recent comparisons of enzyme activities of cytoplasm with chloroplast fractions of pea suggest that maltose comes from action of maltose phosphorylase rather than β-amylase during starch degradation. The results are consistent with a phosphorylytic mechanism of chloroplast starch breakdown.

VII. ASSIMILATION OF NITROGEN

A. Reduction of Nitrate to Ammonia

Nitrate reductase, which reduces nitrate to nitrite, is not located in the chloroplast, although some evidence does suggest that part may be associated with the outer membrane of the chloroplast envelope. It requires NADH which could be supplied during photosynthesis by the chloroplast and the dihydroxyacetone-P: glycerate-3-P shuttle.

Nitrite reduction in leaves is associated with the electron transport reactions of photosynthesis. It is inhibited by DCMU, is insensitive to uncouplers, and operates independent of CO_2 fixation. The physiological electron donor appears to be reduced ferredoxin. Nitrite reductase from spinach contains two Fe atoms per molecule, no flavin, and has a 63,000 MW. With isolated chloroplasts, nitrite can support noncyclic electron transport accompanied by the evolution of O_2 in the expected stoichiometric ratio of 1 mol of nitrate reduced per 1.5 mol O_2 evolved.

B. Assimilation of Ammonia and Biosynthesis of Amino Acids

Many of the reactions of amino acid biosynthesis are compartmentalized with some located in the chloroplast. This is essential as the photorespiratory pathway involved in CO_2 assimilation and metabolism also makes the amino acids, glycine and serine. Their involvement in this role is in addition to their utilization for protein synthesis. Likewise, part of the α-keto acids used for amino acid synthesis are separated from those used in mitochondrial respiration.

The major pathway by which ammonia is incorporated into amino acids begins with the enzyme, glutamine synthetase. This enzyme, located in the chloroplast stroma, utilizes glutamate and ATP to incorporate ammonia into glutamine.

$$\text{L-glutamate} + \text{ATP} + NH_3 \rightleftarrows \text{L-glutamine} + \text{ADP} + \text{phosphate}$$

Glutamine synthetase is markedly effected by Mg^{2+}, pH, and energy charge. Since all of these components change in the chloroplast stroma on transition from light to dark, these changes may be the mechanism by which the chloroplast glutamine synthetase is regulated. Thus, light favors increased glutamine synthetase activity as well as reduction of nitrate to ammonia, both being strongly light-activated in the leaf.

The production of glutamate from glutamine is catalyzed by glutamate synthetase, often referred to as GOGAT, in the stroma of chloroplasts. It catalyzes the reductive transfer of the amide–amino group of glutamine to α-ketoglutarate, producing two molecules of glutamate.

$$\text{L-glutamine} + \alpha\text{-ketoglutarate} + Fd_{red}(\text{NADPH}) \rightarrow 2 \text{ L-glutamate} + Fd_{ox} (\text{NADP}^+)$$

In the chloroplast this system uses reduced Fd rather than NADPH. The discovery of GOGAT in cooperation with ATP : glutamine synthetase provides the primary route for the net synthesis of glutamate from ammonia and α-ketoglutarate.

Glutamate dehydrogenase has only minor activity in the chloroplast. In addition glutamine synthetase has a considerably higher affinity for ammonia than glutamate dehydrogenase. Chloroplast glutamate dehydrogenase appears to be tightly bound to the thylakoid lamellae and is only released with difficulty. The activity of GOGAT in the chloroplast emphasizes that glutamine is the central organic nitrogen donor in the assimilation of nitrogen into amino acids and other organic compounds.

Biosynthesis of other amino acids does occur in the chloroplast. As most of the carbon fixed by photosynthesis exits from the chloroplast as triose phosphates and is metabolized outside to α-ketoglutarate, oxaloacetate, or pyruvate, these precursors must reenter the chloroplast for amino acid biosynthesis. Some of the enzymes involved have been identified in the chloroplast. In the aspartate family, aspartate kinase, diaminopimelate decarboxylase, and homoserine dehydrogenase are located in the chloroplast. The reactions of methionine biosynthesis also appear to be entirely localized in the chloroplast. In the branched chain amino acid pathway, acetolactate synthetase and threonine deaminase have been located in the chloroplast. With the aromatic amino acid pathway all of the enzymes required to convert chorismate to tryptophan are present in etioplasts. A range of aminotransferses are also located in the plastids, including two glutamate aminotransferases.

VIII. SULFATE REDUCTION

Sulfate is taken up through the envelope of intact chloroplast by way of the phosphate translocator. Cysteine is formed in intact chloroplasts with sulfate as the cource of sulfur, indicating that the complete sulfate-reducing cycle is localized inside the chloroplast. The rates obtained with intact chloroplasts are quite low, about 0.5 nmoles/mg Chl·h. With reconstituted chloroplasts, formed by adding the soluble extract from lysed chloroplasts to naked thylakoids, a rate of 12 nmoles cysteine formed per milligram Chl·h can be supported.

The intermediates of the reduction process have been identified as the activated sulfate, adenosine-5′-phosphosulfate (APS) and 3′-phosphoadenosine-5′-phosphosulfate (PAPS), membrane-bound form of sulfite $(X\text{-}S : SO_3H)$, as well as a membrane-bound thiolsulfide $(X\text{-}S : SH)$. A ferredoxin-dependent thiolsulfate reductase is known to catalyze the reduction of the thiol-bound sulfite to the thiol-bound sulfide, a six-electron requir-

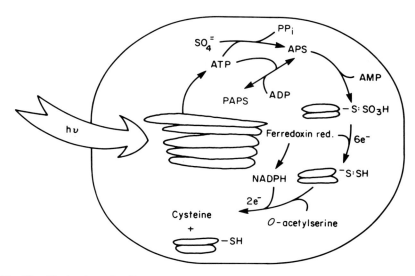

Fig. 10. Mechanism of sulfate reduction by intact chloroplasts and light. The involvement of PAPS (3′-phosphoadenosine-5′-phosphosulfate) is only hypothesized. Adapted from Schwenn and Trebst (1976).

ing process. Two more electrons are consumed in the formation of cysteine with the regeneration of the carrier X-SH. The electron transport system of photosynthesis provides this energy for sulfate reduction, which includes ATP, reduced ferredoxin, and NADPH (Fig. 10). The activated sulfur donor was assumed to be PAPS which is readily formed in chloroplasts. Recent evidence suggests that APS is actually the donor with PAPS only utilized after reconversion to APS. The most likely carbon precursor for chloroplast cysteine is O-acetylserine.

A current proposal suggests that the sulfur–sulfur bond of the bound X-S : SH is split by an intramolecular thiolysis with an adjacent sulfhydryl group:

$$X\genfrac{}{}{0pt}{}{S:SH}{SH} + O\text{-acetylserine} \quad X\genfrac{}{}{0pt}{}{S}{S} + \text{cysteine} + \text{acetate}$$

The NADPH would be required to regenerate the oxidized disulfide carrier "X" to the sulfhydryl to serve as the sulfo acceptor from APS.

IX. METABOLISM OF LIPIDS

A. Lipid Composition of the Chloroplast

Lipids are important constituents of chloroplast membranes and play a central role in their molecular organization and function. Some chloroplasts, as viewed by electron microscopy do contain some stored lipid as lipophilic granules; however, the bulk of the chloroplast lipids are as membranes. The main lipids in the chloroplast include three galactolipids: monogalactosyldiglyceride (MGDG), digalactosyldiglyceride (DGDG), and sulfoquinovosyldiglyceride (SL). These galactolipids are not uniformly distributed in chloroplast membranes. MGDG is concentrated in the internal thylakoid membrane, while DGDG is more prevalent in the chloroplast outer envelope (Table I). Phospholipids are also found in the chloroplast and include phosphatidylglycerol (PG), phosphatidylcholine (PC), and phosphatidylinositol (PI). PC is more concentrated in the chloroplast outer envelope. Phosphatidylethanolamine is not found within the chloroplast and is considered a negative marker for highly purified plant chloroplasts.

TABLE I

Lipid Composition of Higher Plant Chloroplast Membrane

Plant source	Total lipids (%)						References
	MGDG	DGDG	SL	PG	PC	PI	
Spinacia oleracea							
Chloroplast envelope	22	32	5.0	8.4	27	1.3	Douce *et al.*, 1973
Chloroplast lamellae	51	26	7.1	9.1	3.2	1.4	
Chloroplast envelope	8.4	29.3	5.5	13.2	27.5	—	Hashimoto and Murakami, 1975
Chloroplast lamellae	39.1	20.1	7.3	16.5	10.1	—	
Vicia faba							
Chloroplast envelope	29.1	32.4	—	8.9	29.6	—	Mackender and Leech, 1974
Chloroplast lamellae	65.4	26.2	—	5.5	2.8	—	
Beta vulgaris							
Whole chloroplasts	36	20	5.0	7.0	7.0	2	Nichols and James, 1968

	Total fatty acids (%)						
	16:0	16:1 (Δ^3 *trans*)	16:3	18:1	18:2	18:3	
Spinacia oleracea							
Whole chloroplasts							
MGDG	—	—	25	—	2	72	Allen *et al.*, 1966
DGDG	3	—	5	2	2	87	
PG	11	32	2	2	4	47	
SL	39	—	—	—	6	52	
PC	12	—	4	9	16	58	

The galactolipids and phospholipids from chloroplasts both have a high content of trienoic fatty acids (linolenic acid and hexadecatrienoic acid). The PG from chloroplasts is unique in that it contains a high amount of Δ^3-trans-hexadecenoic acid in the C_2-position of glycerol. The SL fraction is more saturated than the other chloroplast lipids and contains a high palmitic acid content.

B. Fatty Acid Biosynthesis

Noncyclic photophosphorylation provides ATP, NADPH, and O_2 as the essential components for fatty acid biosynthesis and desaturation. Chloroplasts can use [^{14}C]acetate as a substrate for biosynthesis of fatty acids in the light. The origin of acetate in leaf tissue is not understood although it might be generated outside of the chloroplast. Chloroplasts do contain an active acetyl-CoA synthetase, located as a stable enzyme in the stroma. The enzyme uses ATP and biotin to convert CO_2 and acetyl-CoA to malonyl-CoA. The CO_2 is first bound tightly to the thylakoid membrane in the light using ATP, apparently to a membrane-bound biotin carboxyl carrier protein. The binding is inhibited by avidin. Only CO_2, which is first bound to the thylakoid membrane, ends up in malonyl-CoA.

Fatty acids are synthesized by a chloroplast *de novo* fatty acid synthetase complex operating with acyl carrier protein (ACP) using acetyl-CoA and malonyl-CoA and the usual synthetase system of soluble enzymes [β-keto acyl ACP synthetase, β-keto acyl ACP reductase (requiring NADPH), 3-hydroxyacyl-ACP dehydratase, and enoyl-ACP reductase (requiring NADPH)]. Chloroplasts from developing spinach leaves have about 1.5 to 3 times as much protein per milligram chlorophyll as those from mature leaves. Incorporation of acetate into fatty acids by chloroplasts from developing leaves is also about three times higher than by those from mature leaves and the fatty acid synthetase on a protein basis is slightly larger.

Fatty acids are synthesized to the length of palmitoyl (16:0) ACP by the synthetase complex. They are elongated to stearyl (18:0) ACP by a system that requires palmitoyl ACP, malonyl ACP, and NADPH. Chloroplast membranes have a high composition of unsaturated fatty acids. Free oleic acid has been produced with chloroplast extracts using stearyl ACP in the presence of NADPH, ferredoxin, NADP-ferredoxin reductase, O_2, and the oleoyl ACP hydrolase, all present in the stroma. The principal dienoic acid in chloroplast membranes is linoleic acid (18:2). Its formation as well as the formation of α-linolenic acid is not well understood. For further information see Chapter 7, Vol. 4.

As chloroplasts are seldom engourged with fat droplets such as normally found in cotyledons of oil-containing seeds, the rate of lipid biosynthesis must be carefully regulated. Currently very little is known about the molecular control of lipid biosynthesis in the chloroplast, although the transfer of acetyl-CoA precursors into the chloroplast and the kinetic relationships of the enzymes would most likely be involved.

X. PROTEIN SYNTHESIS IN THE CHLOROPLAST

A fundamental feature of the organization of eukaryotic cells is that they contain organelles possessing genetic systems separate from the one located in the nucleus. In the chloroplast DNA, RNA, and ribosomes are present. The chloroplast genome has the potential capacity for encoding about 125 polypeptides, each of 50,000 MW. The chloroplast ribosomes represent up to 50% of the total ribosomal complement of the photosynthetic cell. With the discovery of such quantities of chloroplast DNA and ribosomes, the chloroplast was thought to be autonomous and perhaps capable of growing as isolated organelles in culture. With recent biochemical evidence, it is now quite apparent that such attempts would be ill founded. Although the chloroplast contains the four components necessary for biological autonomy—DNA, DNA polymerase, RNA polymerase, and a machinery for synthesis of proteins—these components neither encode nor synthesize all the chloroplast proteins. Many genes concerned with chloroplast structure and function are located in the nucleus and there is increasing evidence that the majority of chloroplast proteins are synthesized on cytoplasmic ribosomes.

Intact chloroplasts isolated from pea or spinach leaves can use light or added ATP to incorporate [^{35}S]methionine or $^{14}CO_2$ into protein. Analysis of these products by gel electrophoresis indicates at least 37 bands of radioactivity. There are two major bands containing 50% of the total label. One band is soluble and has been identified as the large subunit of ribulose-P_2 carboxylase. This subunit is released into the stroma in an aggregated form but does not form the native ribulose-P_2 carboxylase unless the small subunit is made available from the cytoplasm. The large subunit is made by unbound chloroplast ribosomes located in the chloroplast stroma. Further analysis by two-dimensional gel electrophoresis of the soluble fraction of proteins labeled in the stroma revealed up to 80 radioactive spots; these were, however, only faintly labeled, relative to the label in the large subunit. A comparison of this labeling pattern with that made from total stromal proteins as detected by staining indicated little or no correspondence between them. This suggests, at least, that the more abundant soluble proteins are probably synthesized outside the chloroplast. None of the minor soluble products of chloroplast protein synthesis have yet been identified.

The second major product of chloroplast protein synthesis is firmly attached to the thylakoid membranes. Antiserum prepared for the ATPase coupling factor precipitates most of the thylakoid-bound products of chloroplast protein synthesis. Evidence indicates that three of the five subunits of the coupling factor (CF_1) for the ATP synthetase complex are produced in isolated chloroplasts. The observation that the chloroplast contains both free and bound ribosomes suggests that a division of labor exists within the chloroplast. The free ribosomes seem to synthesize the large subunit of ribulose-P_2 carboxylase and other soluble polypeptides, while the bound ribosomes may synthesize components of the thylakoids, which are hydrophobic polypeptides.

The small subunit of ribulose-P_2 carboxylase is not labeled in isolated chloroplasts and has been detected as the product of protein synthesis by cytoplasmic polysomes. Studies indicate that it is synthesized as a precursor protein in the cytoplasm with an apparent molecular weight of 20,000, which is 6,000 greater than the molecular weights of the small subunit as separated from purified ribulose-P_2 carboxylase. The extra polypeptide piece may be involved in identification of the small subunit to the chloroplast envelope and transport across upon removal of the 6,000 MW polypeptide. Some evidence indicates that the small subunit must appear inside the chloroplast in order to initiate synthesis of the large subunit (see Fig. 11).

Although ribulose-P_2 carboxylase constitutes the greatest protein fraction in the stroma, many of the other chloroplast proteins may be totally or partially synthesized by cytoplasmic ribosomes. The transport of protein across the outer envelope must be on a rather large scale. Not only the small subunit of ribulose-P_2 carboxylase but the other proteins that are made on cytoplasmic ribosomes are probably crossing the chloroplast envelope at a rate of 8×10^4 molecules/plastid·h. The mechanism of this transport process

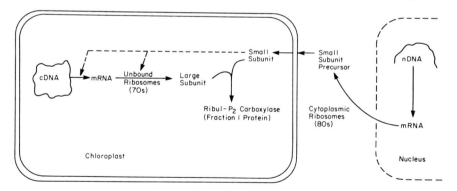

Fig. 11. Proposed model for the cooperative action of chloroplast and nuclear genomes in the synthesis of ribulose-P_2 carboxylase. cDNA and nDNA represent chloroplast and nuclear DNA, respectively. The dashed lines indicate sites of control at which the small subunit may regulate the synthesis of the large subunit. Adapted from Dobberstein *et al.,* (1977) and Ellis (1977).

is unknown, but it must be specific. It may be that proteins destined for the chloroplast all contain a terminal polypeptide piece which can recognize binding sites on the chloroplast outer envelope.

ACKNOWLEDGMENT

The author acknowledges The University of Arizona Experiment Station Paper No. 276.

REFERENCES

Allen, C. F., Hirayama, O., and Good, P. (1966). In "Biochemistry of Chloroplasts" (T. W. Goodwin, ed.), Vol 1, pp. 195–200. Academic Press, New York.

Amesz, J., and van Gorkom, H. J. (1978). Annu. Rev. Plant Physiol. 29, 47–66.

Avron, M. (1977). Annu. Rev. Biochem. 46, 143–155.

Barber, J., ed. (1976). "The Intact Chloroplast." Elsevier, Amsterdam.

Barber, J., ed. (1977). "Primary Processes of Photosynthesis." Elsevier, Amsterdam.

Bonner, J., and Varner, J. E., eds. (1976). "Plant Biochemistry," 3rd ed., Academic Press, New York.

Bottomley, W., Spencer, D., and Whitfeld, P. R. (1974). Arch. Biochem. Biophys. 164, 106–117.

Bucher, T. L., Neupert, W., Sebald, W., and Werner, S., eds. (1976). "Genetics and Biogenesis of Chloroplasts and Mitochondria." North-Holland Publ., Amsterdam.

Burris, R. H., and Black, C. C., eds. (1976). "CO_2 Metabolism and Plant Productivity." Univ. Park Press, Baltimore, Maryland.

Butler, W. L. (1978). Annu. Rev. Plant Physiol. 29, 345–378.

Cramer, W. A., and Whitmarsh, J. (1977). Annu. Rev. Plant Physiology 28, 133–172.

Dobberstein, B., Blobel, G., and Chua, N. (1977). Proc. Natl. Acad. Sci. U.S.A. 74, 1082–1084.

Douce, R., Holz, R. B., and Benson, A. A. (1973). J. Biol. Chem. 248, 7215–7222.

Ellis, R. J. (1977). Biochim Biophys. Acta 463, 185–215.

Gibbs, M., and Latzko, E., eds. (1979). "Encyclopedia of Plant Physiology [N.S.], Photosynthesis II: Photosynthetic Carbon Metabolism and Related Processes," Vol. 6. Springer-Verlag, Berlin and New York.

Govindjee. (1975). "Bioenergetics of Photosynthesis." Academic Press, New York.

Hashimoto, H., and Murakami, S. (1975). Plant Cell Physiol. (Tokyo) 16, 895–902.

Hatch, M. D. (1978). Curr. Topics Cell. Reg. 14, in press.

Hatch, M. D., and Osmond, C. B. (1976). In "Encyclopedia of Plant Physiology [N.S.], Transport in Plants III" (C. R. Stocking and U. Heber, eds.), Vol. 3, pp. 144–184. Springer-Verlag, Berlin and New York.

Heber, U. (1974). Annu. Rev. Plant Physiol. 25, 393–421.

Heber, U. (1975). Proc. Int. Congr. Photosynthesis, 3rd, 2, 1335–1347.

Heldt, H. W. (1976). In "Encyclopedia of Plant Physiology [N.S.], Transport in Plants III" (C. R. Stocking and U. Heber, eds.), Vol. 3, pp. 137–143. Springer-Verlag, Berlin and New York.

Heldt, H. W., Chon, C. J., Maronde, D., Herold, A., Stankovic, Z. S., Walker, D. A., Kraminer, A., Kirk, M. R., and Heber, U. (1977). Plant Physiol. 59, 1146–1155.

Hewitt, E. J. (1975). Annu. Rev. Plant Physiol. 26, 73–100.

Hill, R., and Bendall, F. (1960). Nature (London) 186, 136–137.

Jensen, R. G., and Bahr, J. T. (1977). Annu. Rev. Plant Physiol. 28, 379–400.

Kelly, G. J., Latzko, E.; and Gibbs, M. (1976). *Annu. Rev. Plant Physiol.* **27**, 181–205.

Kung, S. D. (1977). *Annu. Rev. Plant. Physiol.* **28**, 401–437.

Krause, G. H., and Heber, U. (1976). *In* "The Intact Chloroplast" (J. Barber, ed.), Vol. 1, pp. 171–214. Elsevier, Amsterdam.

Laetsch, W. M. (1974). *Annu. Rev. Plant Physiol.* **25**, 27–52.

McCarty, K. E. (1976). *In* "Encyclopedia of Plant Physiology [N.S.], Transport in Plants III" (C. R. Stocking and U. Heber, eds.), Vol. 3, pp. 347–416. Springer-Verlag, Berlin and New York.

Mackender, L. R. O., and Leech, R. M. (1974). *Plant Physiol* **53**, 496–502.

Mazliak, P. (1977). *In* "Lipids and Lipid Polymers in Higher Plants" (M. Tevini and H. K. Lichtenthaler, eds.), pp. 48–74. Springer-Verlag, Berlin and New York.

Miflin, B. J., and Lea, P. J. (1977). *Annu. Rev. Plant Physiol.* **28**, 299–329.

Nichols, B. W., and James, A. T. (1968). *In* "Progress in Phytochemistry" (L. Reinhold and Y. Liwschitz, eds.), Vol. 1, pp. 1–49. Wiley (Interscience), London.

Preiss, J., and Levi, C. (1979). *In* "Encyclopedia of Plant Physiology [N.S.], Photosynthesis II" (M. Gibbs and E. Latzko, eds.), Vol. 6. Springer-Verlag, Berlin and New York.

Schiff, J. A., and Hodson, R. C. (1973). *Annu. Rev. Plant Physiol.* **24**, 381–414.

Schwenn, J. D., and Trebst, A. (1976). *In* "The Intact Chloroplast" (J. Barber, ed.), Vol. 1, pp. 315–334. Elsevier, Amsterdam.

Siegelman, H. W., ed. (1978). "Photosynthetic Carbon Assimilation." Plenum, New York.

Stumpf, P. K. (1975). *In* "Recent Advances in the Chemistry and Biochemistry of Plant Lipids" (T. Galliard and E. I. Mercer, eds.), pp. 95–113. Academic Press, New York.

Thornber, J. P. (1975). *Annu. Rev. Plant Physiol.* **25**, 127–158.

Trebst, A. (1974). *Annu. Rev. Plant Physiol.* **25**, 423–458.

Trebst, A., and Avron, M., eds. (1977). "Encyclopedia of Plant Physiology [N.S.], Photosynthesis I," Vol. 5. Springer-Verlag, Berlin and New York.

Walker, D. A. (1976). *In* "Encyclopedia of Plant Physiology [N.S.], Transport in Plants III" (C. R. Stocking and U. Heber, eds.), Vol. 3, pp. 85–136. Springer-Verlag, Berlin and New York.

Zelitch, I. (1975). *Annu. Rev. Biochem.* **44**, 123–145.

Plant Mitochondria 8

J. B. HANSON

D. A. DAY

I. INTRODUCTION

The mitochondria of plants have the same functions as those of animals and fungi; they provide ATP as the principal energy source of the cell, and to varying degrees participate in intermediary metabolism. With these common requirements it is not surprising that eukaryote evolution has been remarkably conservative with respect to mitochondrial form and function. However, perhaps a billion years of divergent evolution separate the autotrophic eukaryotes from the heterotrophic, and it would be remarkable if some manifestation of this were not found in comparative studies.

Unfortunately, close comparison is not possible. Although there is now significant literature on plant mitochondria, much of it derives from physio-

The Biochemistry of Plants, Vol. 1

logical investigations. Seldom have plant mitochondria been used by biochemists as primary objects for investigation of oxidative phosphorylation, and rarely have they been directly compared with animal mitochondria. Experiments with plant mitochondria are most frequently devised and the results explained on the basis of information gained from studies of animal or fungal mitochondria. Enough work has been done in this vein, however, to make it clear that plant mitochondria differ only in detail, primarily in substrate utilization and the accessory functions unique to autotrophic metabolism.

The intent in this chapter is to provide a comprehensive overview of the occurrence, structure, and function of mitochondria in the plant cell. Volume 2 of this series provides additional detail on substrate oxidation, electron transport, and oxidative phosphorylation, plus participation in photorespiration, cyanide-resistant respiration, and tissue respiration. Volume 4 covers special aspects of membrane lipid metabolism, and Volume 6 the protein and nucleic acid synthesis of the organelle.

II. OCCURRENCE, FORM, AND DEVELOPMENT

Most higher plant cells contain several hundred simple mitochondria of spherical, ellipsoidal, or rodlike shape. Clowes and Juniper (1968) give an average number of 700 per cell with diameters of 0.5–1.0 μm and lengths of about 3 μm for the elongate forms. Numbers per cell can increase dramatically as cells grow (enlarge), but the number per cytoplasmic volume is more nearly constant (Juniper and Clowes, 1965). There are certain cells, especially in lower plants and motile spores, where one or a few large mitochondria are found; serial sectioning has revealed a reticulate mitochondrion in the alga *Chlorella fusci* (Atkinson *et al.*, 1974). Occasionally in higher plants there are reductions in number with increases in size and branching; there is a striking instance of this in the variegated leaves of *Ficus elastica* (Duckett and Toth, 1977). Exceptions aside, however, the number, size, and siting of plant mitochondria appears to be related to the cell's demand for ATP (Clowes and Juniper, 1968). Secretory cells of nectaries, phloem companion cells, and transfer cells are especially rich in mitochondria. In transfer cells mitochondria constitute up to 20% of the cytoplasm (Gunning and Steer, 1975). It is worth noting that these cells are believed to be heavily involved in transport processes.

Cinematography of leaf cells shows mitochondria to move freely in the streaming cytoplasm, appearing to divide and coalesce, and with transient adherence to the chloroplasts (Wildman *et al.*, 1962). Due to cytoplasmic streaming the mitochondria tend to be randomly distributed (Gunning and Steer, 1975). The streaming and transient contacts with other surfaces probably provides for rapid exchange of metabolites and obviates the need for the localized aggregations of mitochondria often found in cells with a more static cytoplasm.

In addition to greater numbers of mitochondria, growing cells have been frequently reported to show mitochondrial development. This development results in more cristae being formed (Lund *et al.,* 1958; Simon and Chapman, 1961; Clowes and Juniper, 1964; Chrispeels *et al.,* 1966) and an increase in substrate oxidation rates of isolated mitochondria (Lund *et al.,* 1958; Simon and Chapman, 1961; Van der Plas *et al.,* 1976). The balance observed between cell and mitochondrial growth may be achieved by mitochondrial products exerting influence over the expression of nuclear genes which code for mitochondrial proteins (Leenders *et al.,* 1974).

The correlation between cristae formation and respiratory activity is not obligatory and the synthesis of membranes as structural entities can proceed without the concomitant synthesis of respiratory chain enzymes. Öpik (1973) found that mitochondria in anaerobically germinating rice formed abundant cristae, but were deficient in cytochrome oxidase. During the cell cycle of *Chlorella,* the inner membrane area increases in parallel with cell growth, but succinic dehydrogenase and cytochrome oxidase are only synthesized during the last third of the cycle (Forde *et al.,* 1976). That is, mitochondrial growth involves insertion of spasmodically synthesized respiratory enzymes into preformed membranes. This imbalance between membrane and enzyme synthesis implies that the enzyme density of the inner membrane varies dramatically during cell growth. The inner and outer membrane grow in parallel, however, and the ratio of inner membrane to outer membrane area remains constant at 1.8 (Forde *et al.,* 1976).

Notwithstanding the lack of linkage between membrane and enzyme synthesis during cell growth, for mature tissues those cells with many mitochondria containing many cristae can be expected to have high respiration rates. The usual procedures for isolating plant mitochondria involve grinding plant material which can consist of several anatomically distinct tissues. The resulting yield of mitochondria is a bulked average. Lance and Bonner (1968), experimenting with mitochondria from five widely different types of material, found only modest variation in succinate oxidation rates (range of 750 – 1180 μl $O_2 \cdot h^{-1} \cdot$ mg N^{-1}). Tissue respiration rates, however, varied widely (40 to 4000 μl $O_2 \cdot h^{-1} \cdot$ g fr. wt^{-1}). This variability could be largely explained on the basis of tissue content of mitochondria.

Plant mitochondria persist throughout the life of the cell and appear to be the most resistant of the cytoplasmic structures to degradative changes during plant senescence (Butler and Simon, 1971).

III. STRUCTURE AND COMPOSITION

Figure 1 (from Öpik, 1974) is a diagram of a plant mitochondrion in trans-ection as revealed by thin sectioning and electron microscopy. Figure 2 gives electron micrographs of mitochondria typical of young and mature cells. The inner membrane of the plant mitochondrion encloses an aqueous matrix of solutes, soluble enzymes, and the mitochondrial genome. Invaginations of the membrane produce saclike cristae of variable shape and number, usually with a narrow neck; platelike, parallel cristae develop under anoxia (Moris-set, 1973). The whole is encased in the outer membrane, which *in vivo* is closely appressed to the inner, leaving a scarcely discernable intermembrane space (Öpik, 1974). Fixation and osmotic shrinkage of the matrix in sucrose solutions is largely responsible for much of the membrane separation and cristae dilation seen by electron microscopy, which clearly reveals the con-tinuity of the intermembrane and intercristal space but exaggerates the vol-ume of the space (Öpik, 1974).

The matrix sometimes appears to have an ultrastructure, but except for the demarkation of electron translucent areas containing DNA fibrils—the nucleoids (Gunning and Steer, 1975)—there is no clear resolution of its orga-nization. Inclusions of ribosomes and metal phosphate deposits are fre-quently seen (Figs. 1 and 2). The ribosomes appear smaller than those of the cytoplasm (Öpik, 1974; Gunning and Steer, 1975). This is inconsistent, how-ever, with the finding that the ribosomes of higher plant mitochondria have sedimentation coefficients of $78s$ (Leaver and Harmey, 1973; Pring, 1974).

Ribosomal RNA subunits are $24s$, $18.5s$, and $5s$; the $5s$ RNA is not found in animal mitochondria (Leaver and Harmey, 1976). Transfer RNA ($4s$) is also

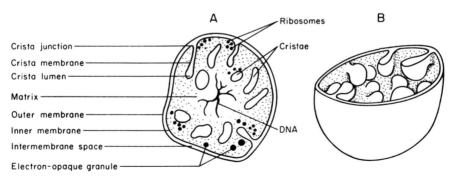

Fig. 1. Illustration of structure and organization of plant mitochondria. From Öpik (1974), "Mitochondria," *in* Dynamic Aspects of Plant Ultrastructure (A. W. Robards, ed.). Copyright © 1974, McGraw-Hill, New York.

found. Higher plant mitochondria DNA occurs as a large circular molecule of 30 μm length with a molecular weight of 70×10^6 (Kolodner and Tewari, 1972). Recently, several smaller classes of circular DNA have been reported for soybean mitochondria (Synenki *et al.*, 1978).

The solute content of the matrix is not accurately known. Grinding and washing plant tissues in media of high osmolarity (0.4–0.6 *M*) shrinks the matrix, concentrating the solutes and setting up steep gradients for outward diffusion. Respiration, which might serve to maintain the solute content, is suppressed by ice temperatures. Consequently, the solute content of isolated plant mitochondria is quite low and endogenous respiration is hard to detect. Bowman *et al.*, (1976) report washed mung bean mitochondria to contain 4 nmoles malate/mg protein with lesser amounts of pyruvate, citrate, α-ketoglutarate, aspartate, and glutamate. There were 3 nmoles NAD/mg protein and about 1 nmole of adenine nucleotides; the maximum reported value is 7 nmoles AdN/mg protein in corn mitochondria (Jung and Hanson, 1975). Phosphate content is high when isolations are made with phosphate buffer [115 nmoles/mg protein vs 45 with TES buffer (Day *et al.*, 1978)]. About 140 nmoles K^+/mg protein are retained when isolations are made in the absence of K^+, and magnesium content is in the range of 35–45 nmoles/ mg protein (Jung and Hanson, 1975). Sucrose inaccessible space (matrix volume) of corn mitochondria in 0.12 *M* sucrose is about 2 μl/mg protein (Kirk and Hanson, 1973).

A number of features in addition to the possession of cristae distinguish the inner membrane from the outer. Inner membranes stain more heavily (Gunning and Steer, 1975). With negative staining (Nadakavukara, 1964; Parsons et al., 1965) or acetate swelling (Zaar, 1974) they reveal the characteristic stalked particles on the matrix surface which are the sites of ATP synthesis.

Inner membranes are osmotic barriers, being relatively impermeable to most ions and hydrophilic organic solutes, but readily permeable to water. The matrix compartment bounded by the inner membrane thus becomes a nearly perfect osmometer in the range of -1 to -10 bars osmotic potential (Yoshida and Sato, 1968; Lorimer and Miller, 1969). Electron micrographs show that osmotic contraction produces a denser matrix and dilation of the cristae and intermembrane space (Baker *et al.*, 1968). However, not all mitochondria are equally responsive in this fashion. Malone *et al.* (1974) describe three types, with variable swelling–contraction responses, from etiolated corn shoots. Moreover, the osmotic properties of mitochondria may change with cell development (Malhotra and Spencer, 1970). Outer membranes, on the other hand, are freely permeable to most solutes up to 4,000 MW, but if intact will exclude cytochrome c (approximately 13,000

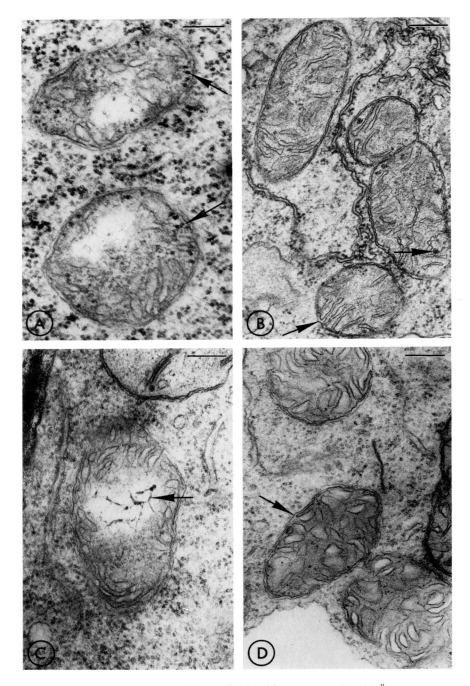

Fig. 2. Electron micrographs of plant mitochondria (courtesy of Helgi Öpik, University College of Swansea, Wales). (A) Immature mitochondria in tomato root tip meristem. Arrows

TABLE I

Lipid Composition of Cauliflower Mitochondria Membranes[a,b]

	Total PL (%)						Total FA (%)				
	PL	PC	PE	DPG	PG	PI	16:0	18:0	18:1	18:2	18:3
Inner membrane	32	41	37	14	3	5	10	1	7	13	69
Outer membrane	40	42	24	3	10	21	50	4	20	8	18
Microsomes	30	50	35	1	8	6	26	2	11	10	51

[a] PL, phospholipid; PC, phosphatidyl choline; PE, phosphatidyl ethanolamine; DPG, diphosphatidylglycerol; PG, phosphatidylglycerol; PI, phosphatidylinositol; FA, fatty acids.
[b] From Moreau et al., 1974.

MW), a fact which is used as a test of outer membrane integrity (Douce et al., 1972a; Day and Wiskich, 1974a). Outer membranes sometimes reveal a pitted surface (Parsons et al., 1965) and x-ray diffraction studies reveal an in-plane subunit structure which corresponds with the spacing of the pits (Mannella and Bonner, 1975b).

Inner membranes have a lower phospholipid/protein ratio (i.e., higher protein content) than outer membranes, have a much higher percentage of unsaturated fatty acids, and are enriched in diphosphatidyl glycerol (Mazliak, 1977; Table I). Diphosphatidyl glycerol (cardiolipin) serves as a marker for the inner membrane (Mazliak, 1977). In contrast, the outer membrane is relatively enriched in phosphatidyl glycerol and phosphatidyl inositol. Both membranes are rich in phosphatidyl choline and phosphatidyl ethanolamine, but the PC/PE ratio is lower than that of other plant organelles (Donaldson and Beevers, 1977). Galactolipids, characteristic of plastids, are missing (McCarty et al., 1973). Inner membranes are characterized by a high unsaturated fatty acid content, particularly linolenic (18:3) while outer membranes are rich in palmitic acid (16:0). The plasticity of the inner membrane relative to the outer is ascribed to the high content of unsaturated fatty acids (Moreau et al., 1974).

---

point to mitoribosomes. Translucent areas with DNA filaments constitute the nucleoids. (B) Mature mitochondria in the vascular bundle of a rice coleoptile tip with primarily tubular cristae. Lower arrow points to crista-inner membrane junction. Upper arrow designates a dense granule which is probably an insoluble metal phosphate deposit. (C) Mitochondrion in 3-day mung bean seedling leaf with coalesced DNA filaments in the central nucleoid (arrow). (D) Mature mitochondrion in 10-day mung bean seedling leaf. Note development of saccate cristae (arrow) extending through the mitochondria, and the loss of an obvious nucleoid zone (this is not due to sectioning plane). Scale = 0.2 μm.

Membrane lipids are believed to be of two classes: those of the fluid bimolecular leaflet, and those firmly bound to membrane proteins forming proteolipids (Mazliak, 1977). Differences in protein constituents may thus account for preferential association of different species of phospholipids with the membranes. The firmly bound enzymes of the inner membrane include the NADH dehydrogenases, succinate dehydrogenase, the electron transport chain, the coupling ATPase, and various transporting enzymes (discussed in Section V). Lectins are bound to castor bean mitochondria including the inner membrane (Bowles et al., 1976).

Outer membranes are characterized by a high proportion of phospholipid, appreciable amounts of sterols, and about 40 μg of galacturonic acid residues per milligram protein (Mannella and Bonner, 1975a). The breaking strength of the outer membrane is directly correlated with the uronide and divalent ion content. Polyacrylamide gel electrophoresis shows two major protein peaks corresponding to 30,000 and 50,000 MW (Mannella and Bonner 1975a). This study also showed that the proportion of unsaturated fatty acids in the outer membrane of mung bean mitochondria increased with lower growth temperatures. Fatty acid composition, and to some extent phospholipid composition, reflect environmental as well as genetic and developmental control (Lyons, 1973).

There appears to be no specific role for the outer membrane in the processes normally studied with isolated mitochondria. Mitochondria lacking outer membranes show normal respiratory control and ADP:O ratios (Wilson et al., 1973), but there is reason to doubt if these parameters are an adequate measure of "intactness" since respiration rates are much lower in vitro than in vivo (Bligny and Douce, 1976). It perhaps serves as a "screen" to prevent loss of intermembrane enzymes—or access of cytoplasmic enzymes—while allowing ready permeation of metabolites. To a limited degree it appears to have the restraining properties of a wall; Mannella and Bonner (1975a) report osmotic potential differences between mitochondria and medium of five to seven bars (depending on the uronic acid content of the membrane) before outer membrane rupture.

Plant mitochondria show ultrastructural states somewhat corresponding to the orthodox and condensed states of animal mitochondria both in situ (Thomson et al., 1972) and when isolated (Baker et al., 1968; Wilson et al., 1973). However, state 3–state 4 transformations in configuration are not found, and the transformation to the "orthodox" state upon addition of substrate can be ascribed to the uptake of solutes with osmotic swelling. Laties and Treffry (1969) found that the morphology of isolated potato mitochondria was highly conditioned by the suspending medium; inclusion of macromolecules of various kinds was essential to retention of the in vivo rodlike shape.

IV. ENZYMES AND ENZYMATIC ACTIVITIES

A. Occurrence

Those enzymes and enzyme systems that occur in plant mitochondria, and for which locations are known or indicated by investigations of fungal and animal mitochondria, are given in Table II. There is no assurance that the list is complete since there is incomplete evidence on mitochondrial participation in the intermediary metabolism of autotrophs. For example, phosphenol-pyruvate carboxykinase may be a constituent of some plant mitochondria (Benedict and Beevers, 1961; Dittrich *et al.*, 1973; Graesser and Wilson, 1977). Recently, Hunt and Fletcher (1977) established that pea leaf mitochondria contain isocitrate lyase, an enzyme previously known only for microbodies (glyoxysomes) of fatty seeds (Cooper and Beevers, 1969). Catalase is typically found in microbodies (Tolbert, 1971), and density gradient isolation of subcellular organelles indicates that mitochondrial catalase and peroxidase is a contaminant (Plesnicar *et al.*, 1976; Huang and Beevers, 1971; Tolbert, 1971). There are opinions that mitochondrial catalase may exist and have a functional role (Rich *et al.*, 1976).

Contaminants have been a problem in deciding just which enzymatic activities associated with isolated mitochondria truly represent mitochondrial enzymes. There are two sorts of contaminants: adhering soluble and solubilized enzymes, and inclusion of nonmitochondrial particles. Repeated washing by resuspension and recentrifuging can reduce the level of adhering soluble enzymes, and increasing the pH of isolation is sometimes helpful (Hanson *et al.*, 1965). For minimizing contaminating particulates, correct selection of centrifuge speeds is essential; often the speeds used to isolate mitochondria are too fast (Bonner, 1973).

There is also the hazard in the other direction of solubilizing mitochondrial enzymes during isolation. In pea epicotyls, for example, the soluble glutamate dehydrogenase appears to be derived from the mitochondria during isolation (Davies and Teixeira, 1975). This is not true for the glutamate dehydrogenase of pumpkin cotyledons, where there are distinct isoenzymes for the soluble and particulate phases (Chou and Splittstoesser, 1972). Isoenzymes for other mitochondrial enzymes have been described (Scandalios, 1974). A distinguishing criterion lies with cofactor requirements: the mitochondrial dehydrogenases are NAD^+, while the corresponding enzymes of cytoplasm and plastids tend to be $NADP^+$-linked (Bonner, 1973). Protein synthesis in plant mitochondria involves distinct tRNA and aminoacyl tRNA synthetases (Lea and Norris, 1977).

Continuous or discontinuous sucrose gradients have been used frequently to isolate or to purify isolated mitochondria (which band at about 1.18 g/cm^3).

TABLE II

Intramitochondrial Localization of Enzyme Systems

Location	Enzymes	References[a]
Outer membrane	NADH dehydrogenase	1, 2, 3
	Cytochrome b-555	1, 2
	Acid phosphatase	3
Intermembrane	Superoxide dismutase (Cu–Zn);	4
space	Malic enzyme[b]	5, 6
Inner membrane	Adenylate kinase (external)[c]	7
	NADH dehydrogenase (internal)	6, 8, 9
	NADH dehydrogenase (external)[c]	1, 2, 6, 9, 10
	NADPH dehydrogenase (external)[c]	11
	Succinate dehydrogenase	6, 8, 9
	Cytochromes b-556, b-560, b-565, c-551, c-550, a, and a_3	6, 8, 9
	Coupling ATPase	12, 13
	Alternate oxidase, cyanide insensitive	8, 9, 14
	Pyridine nucleotide transhydrogenase	10, 15, 16
	CDP-diglyceride synthetase	17
	Metabolite and ion transporters	18, 19
Matrix	NAD$^+$-linked dehydrogenases (malate, pyruvate, isocitrate, α-ketoglutarate)	8, 9, 15, 20
	Citric acid cycle enzymes	8, 9, 15, 20
	Malic enzyme, NAD$^+$-linked	15, 21
	Glutamate dehydrogenase	15, 22, 23, 24
	Amino transferase	22, 25
	Glycine decarboxylase (leaf)	26, 27, 28
	Isocitrate lyase (leaf)	29
	Phosphoenol pyruvate carboxykinase	30, 31, 32
	α-ketoglutarate-glyoxylate carboxyligase	33
	Glycolate-D lactate dehydrogenase (algae)	34
	Superoxide dismutase (Mn)	4
	DNA, RNA, and protein synthesis	35, 36

[a] References:
1. Moreau and Lance, 1972
2. Douce et al., 1973
3. Day and Wiskich, 1975
4. Arron et al., 1976
5. Palmer and Arron, 1976
6. Palmer, 1976
7. Arron et al., 1978
8. Ikuma, 1972
9. Bonner, 1973
10. Day and Wiskich, 1974a
11. Koeppe and Miller, 1972
12. Yoshida and Takeuchi, 1970
13. Takeuchi, 1975
14. Solomos, 1977
15. Davies, 1956
16. Wilson and Bonner, 1970
17. Douce et al., 1972b
18. Hanson and Koeppe, 1975
19. Wiskich, 1977
20. Beevers, 1961
21. Macrae and Moorhouse, 1970
22. Cooper and Beevers, 1969
23. Chou and Splittstoesser, 1972
24. Davies and Teixeira, 1975
25. Splittstoesser and Stewart, 1970
26. Tolbert, 1971
27. Bird et al., 1972
28. Woo and Osmond, 1976
29. Hunt and Fletcher, 1977

The disposition of enzyme activity along the gradient is of importance not only in determining which cytoplasmic organelle has a high concentration of enzyme, but also in allowing an estimation of whether a small amount of enzyme associated with the mitochondria represents a contaminant (e.g., the catalase problem above). Gradients have also been used to reduce bacterial contamination in studies of labeled precursor incorporation into protein and nucleic acids (Baxter and Hanson, 1968). Improved isolation techniques were instrumental in showing that β-oxidation of fatty acids in germinating seeds, initially thought to be mitochondrial, is in fact associated with microsomes and microbodies (Mazliak, 1973). However, Mazliak (1973) notes that subsequent to acylthioester formation, β-oxidation might occur in the mitochondria of carnitine-rich tissues. Mitochondria from avocado mesocarp (Panter and Mudd, 1973) and pea cotyledons (McNeil and Thomas, 1976) have been shown to oxidize palmitate and salts of other fatty acids. This oxidation is stimulated by addition of carnitine, and it is suggested (Thomas and McNeil, 1976) that carnitine acts by facilitating transport of fatty acids across the inner mitochondrial membrane, as it does in mammalian tissues.

B. Respiratory Substrates

It was established very early that plant mitochondria contain the complete complement of enzymes for the oxidation of pyruvate via the TCA (tricarboxylic acid) cycle (Beevers, 1961; Fig. 3), and pyruvate has been assumed to be the major substrate. Experimentally, pyruvate oxidation requires addition of a "sparker" TCA acid, and thiamin pyrophosphate is often found to be limiting. Recently, the pyruvate dehydrogenase complex has been shown to possess the component enzymes known for animal mitochondria; pyruvate dehydrogenase (decarboxylating), dehydrolipoate transacetylase, and dihydrolipoate dehydrogenase (Reid et al., 1977; Rubin and Randall, 1977). The complex is regulated by feedback inhibition from NADH and acetyl-CoA (Crompton and Laties, 1971; Rubin and Randall, 1977), and by phosphorylation by a specific kinase (Rubin and Randall, 1977).

In addition to pyruvate, all of the TCA cycle acids can be oxidized by intact mitochondria. However, the oxidation rates for the different substrate anions varies widely, especially with gradient purified mitochondria (Cooper and Beevers, 1969). Succinate is generally found to be the most rapidly oxidized, particularly in the presence of ATP which activates succinate dehydrogenase (Oestreicher et al., 1973). In intact mitochondria this is primar-

---

30. Benedict and Beevers, 1961
31. Dittrich et al., 1973
32. Graesser and Wilson, 1977
33. Davies and Kenworthy, 1970
34. Beezley et al., 1976

35. Leaver and Pope, 1977
36. Lea and Norris, 1977
   [b] postulated location (Ref. 6)
   [c] bound to external surface of inner membrane

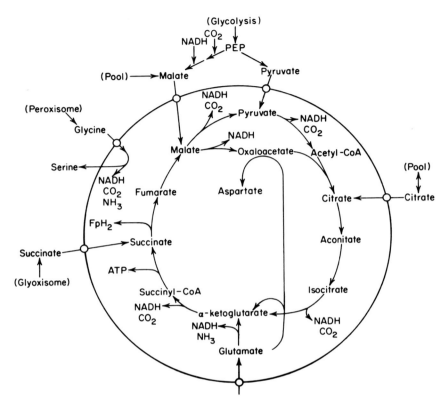

Fig. 3. Schematic diagram of the primary sources and pathways of respiratory carbon in plant mitochondria.

ily a one-step oxidation yielding malate (Bowman *et al.,* 1976), and malate oxidation is in turn regulated by turnover of NADH and oxaloacetate. Oxaloacetate removal can be achieved either by condensation with acetyl-CoA or transamination with glutamate. The disproportionately large levels of malate dehydrogenase found in plant mitochondria may reflect the adverse chemical equilibrium for oxaloacetate production ($K_{eq} = 10^{-5}$; Bowman and Ikuma, 1976). Oxaloacetate levels may regulate matrix NADH levels via malate dehydrogenase (Douce and Bonner, 1972; Palmer and Arron, 1976).

Added citrate and isocitrate are also oxidized by plant mitochondria, but much more slowly than other substrates, possibly due to limited transport rates (Bowman and Ikuma, 1976; Day and Wiskich, 1977b). Isocitrate dehydrogenase may be regulated by the $NAD^+/NADH$ ratio, but adenine nucleotides seem not to affect enzyme activity (Cox and Davies, 1967). Nonetheless, a role for adenine nucleotides in the regulation of NAD-linked substrate oxidation by some plant mitochondria has been suggested (Laties, 1973;

Sottibandhu and Palmer, 1976). ADP and AMP (but not ATP) stimulate NAD-linked substrate oxidation by potato and Jerusalem artichoke mitochondria, apparently by stimulating the respiratory linked NADH-dehydrogenase (Sottibandhu and Palmer, 1976). These effects are not seen with all plant mitochondria (Day and Hanson, 1977a), but different preparations contain different amounts of endogenous adenine nucleotide. The ADP effects observed with potato mitochondria (Laties, 1973) may account for the "conditioning" phenomena also seen with these (and other) mitochondria. Conditioning involves a gradual increase in state 3 (see Fig. 6) rates through successive state 3/state 4 cycles during oxidation of various substrates (Raison *et al.*, 1973).

It is rational that plant mitochondria should be adapted to oxidation of TCA cycle anions other than pyruvate. The glyoxylate cycle produces succinate, which can not be oxidized to malate except after transfer to the mitochondria (Cooper and Beevers, 1969). Plant cells frequently store large amounts of organic anions, notably malate and citrate. Malate pools become the depository for CO_2 fixed via PEP carboxylase or carboxykinase, especially in plants which are rapidly assimilating nitrate, absorbing excess K^+, or have C_4 or crassulacean acid metabolism (Osmond, 1976). Thus to varying degrees—dependent upon organic acid metabolism and compartmentation—plant mitochondria can be supplied with organic acid substrates in addition to pyruvate.

An essential adaptation here for complete TCA cycle operation is the Mn^{2+} or Mg^{2+} requiring NAD-malic enzyme of plant mitochondria which enables them to produce pyruvate directly from malate (Macrae and Moorhouse, 1970; Macrae, 1971a; Davies and Patil, 1975). In corn mitochondria, where rates of pyruvate transport are low, it appears that malate transport plus malic enzyme furnishes much of the pyruvate when respiration rates are high (Day and Hanson, 1977b). Although Wedding *et al.* (1976) found that pyruvate production via malic enzyme by sweet potato mitochondria required high malate concentrations ($K_m = 40$ mM), concentrations of 1–4 mM were sufficient for pyruvate formation in corn mitochondria (Day and Hanson, 1977b). A relatively low external pH (6.5–7.0) seems to favor intramitochondrial malic enzyme activity (Macrae, 1971b).

Another alternative to pyruvate as substrate is reduced pyridine nucleotide from the cytosol. Isolated plant mitochondria rapidly oxidize exogenous NADH (Palmer, 1976) and (in corn mitochondria) NADPH (Koeppe and Miller, 1972). An exception is found with mitochondria from fresh beet root, the dormant food-storing organ of the biennial plant. Only when beet slices have been aged (i.e., brought into active aerobic metabolism) by washing in aerated solutions do the mitochondria develop the capacity to oxidize NADH (Day *et al.*, 1976). Similarly, the low rate of NADH oxidation by mitochondria from fresh potato tissue is doubled by aging (Dizengremel and

Lance, 1976); oxidation of succinate or malate is not so affected. In general, however, mitochondria from actively respiring plant tissues rapidly oxidize external NADH.

NADH oxidation is stimulated by Ca^{2+} and other divalent cations (Hackett, 1961; Miller *et al.,* 1970; Coleman and Palmer, 1971). K^+ also stimulates NADH oxidation in the presence of Ca^{2+}, and Earnshaw (1975) attributes this to an increase in free internal Ca^{2+}. NADPH oxidation requires Ca^{2+} or phosphate, and is not mediated by transhydrogenation with NAD^+ (Koeppe and Miller, 1972). External NADH oxidation occurs at the outer surface of the inner membrane (Douce *et al.,* 1973), and interacts with the electron transport chain at the level of ubiquinone, by passing the first coupling site (Palmer, 1976). Oxidation is thus rotenone-insensitive but antimycin-sensitive. These are also characteristics of NADPH oxidation (Koeppe and Miller, 1972).

Oxidation of external NADH directly via the plant respiratory chain obviates the need for the complex anion shuttles known to operate in animal cells (Chappell, 1968; Klingenberg, 1970) for the reoxidation of glycolytic NADH. How this direct oxidation is regulated is not clear. Palmer (1976) suggests that Ca^{2+} levels or substrate acid oxidation may provide regulation; succinate and malate oxidation by isolated mitochondria inhibit exogenous NADH oxidation (Palmer, 1976; Day and Wiskich, 1977b).

NADH can also be oxidized at the outer membrane provided a suitable electron acceptor is present. Cytochrome c is such an acceptor, and activity is measured as NADH-cytochrome c reductase. Reduction involves a flavoprotein and cytochrome b-555 (Douce *et al.,* 1973). Potentially, intermembrane cytochrome c could act as a shuttle between the outer and inner membranes, transferring reducing equivalents from NADH to cytochrome oxidase, and Dizengremel (1977) suggests this may occur *in vivo.* He found that aging potato slices increased NADH oxidation by the outer membrane from 6 to 18% of the total.

In addition to the capacity for direct oxidation of cytoplasmic NADH, plant mitochondria also appear able to directly transfer reducing equivalents from the matrix to external NAD, probably via a transmembrane transhydrogenase (Day and Wiskich, 1974a,b, 1978). This transfer appears to be unidirectional and to be under the control of the extra mitochondrial NADH/NAD ratio.

Glutamate and aspartate are potential sources of respiratory carbon due to intramitochondrial glutamate dehydrogenase and amino transferase. However, as single substrates, glutamate is oxidized very poorly by intact mitochondria and aspartate not at all (Lance, 1974; Bowman *et al.,* 1976). Since the level of glutamate dehydrogenase and glutamate–oxaloacetate transaminase compare very favorably with TCA cycle enzymes, and since endogenous glutamate and asparate are present in millimolar concentrations in the matrix (Bowman *et al.*, 1976), there must be controls that limit the availability of dicarboxylate amino acids as substrates. Davies and Teixeira (1975) found ATP, ADP, and NADH to inhibit the deamination reaction of glutamate dehydrogenase. The enzyme has only a minor role in glutamate synthesis (Miflin and Lea, 1977a).

Glycine is a product of photosynthesis by way of photorespiration in C_3 plants (Tolbert, 1971). Intact leaf mitochondria possess an NAD^+-linked enzyme system which will oxidatively decarboxylate two molecules of glycine with the production of one molecule each of serine, CO_2, and NH_4^+ (Woo and Osmond, 1977). Oxidation of the NADH formed can be linked to oxaloacetate reduction (Woo and Osmond, 1976), or to ATP formation (Bird *et al.*, 1972; Douce *et al.*, 1977; Moore *et al.*, 1977). Thus mesophyll mitochondria of C_3 plants play a substantial role in photorespiration. The corollary of this is that glycine is a significant substrate for these mitochondria, and provides a pathway linking photorespiration with ATP production (Bird *et al.*, 1972; Moore *et al.*, 1977). It should be noted that studies of mung bean leaf mitochondria (Chapman and Graham, 1974a,b) indicate that the TCA cycle functions in the light at comparable rates to the dark, but with an increase in malate concentration of the tissue at the expense of phosphoenolpyruvate and aspartate. Light also increases the ATP : ADP and the NADH : NAD ratios.

C. Electron Transfer to Oxygen

With the exception of succinate dehydrogenase, the TCA cycle enzymes which initiate substrate oxidation produce NADH in the mitochondrial matrix. Due to a shift in fluorescence properties compared to extracted NADH, it is believed that the NADH becomes bound to the membrane (Bonner, 1973). Subsequent oxidation of the NADH and succinate by oxygen, the ultimate electron acceptor, is by means of a complex of oxidoreductases which are incorporated into the inner membrane. This complex constitutes the electron transport chain (Fig. 4), which despite intensive investigation remains poorly understood, especially with respect to its energy-conserving properties. The constituents of the complex in plant mitochondria are similar to those of animal mitochondria. However, there is clear evidence from studies of fluorescence and absorption spectra that the flavoproteins and cytochromes of plant mitochondria are distinct, albeit carrying out the same electron transport role (Ikuma, 1972; Bonner, 1973; Palmer, 1976). With minor exceptions the iron–sulfur proteins appear to be similar to those of animal mitochondria (Palmer, 1976; Rich *et al.*, 1977; Rich and Bonner, 1978).

The electron transport chain consists of three major subcomplexes in which electron transfer is linked to an electrogenic H^+ extrusion (Fig. 4). In

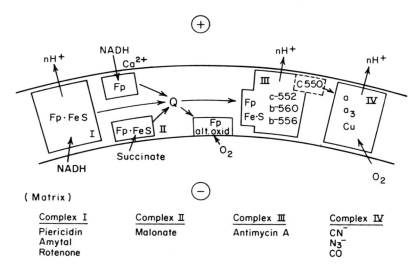

Fig. 4. Schematic diagram of the respiratory chain of plant mitochondria. Fp = flavoprotein; FeS = iron sulfur protein; Q = ubiquinone; b, c, and a are cytochromes.

addition, there are other complexes for reducing ubiquinone (coenzyme Q) at the expense of succinate and external NADH which are not linked to H^+ efflux. The diagram is based on a very broad extrapolation of what is known about complexes in animal mitochondria, with the addition of an externally located NADH-ubiquinone reductase to account for exogenous NADH oxidation. In animal mitochondria there are three inner membrane complexes which can be extracted, reinserted into membrane vesicles (liposomes), and demonstrated to carry out a redox reaction coupled to H^+ transport (Hinkle, 1976; Racker, 1976). These complexes are (I) NADH-ubiquinone reductase, (III) ubiquinone–cytochrome c reductase, and (IV) cytochrome c oxidase (Hatefi *et al.,* 1962). The succinate–ubiquinone reductase, complex II, does not produce H^+ transport.

Complexes I, II, and III contain specific flavoproteins plus iron–sulfur proteins. Complex III appears to be linked to ubiquinone (Q) through a flavoprotein (Storey, 1976). Complex III probably contains two b cytochromes (b-556 and b-560) and two c cytochromes (c-552, tightly bound, and c-550, dissociated at high ionic strength). A third b cytochrome (b-565) exists, but with debatable function (Ikuma, 1972; Bonner, 1973; Palmer, 1976). [The cytochromes are designated by their room termperature absorption maxima (Storey, 1976).] It is through studies of changes in the oxidized-reduced absorption spectra with inhibitors, uncouplers, and anoxia that participation in the electron transport chain is deduced. Complex IV has cytochromes a and a_3 plus copper.

Ubiquinone (UQ_{10} in mung bean mitochondria; Beyer *et al.*, 1968) serves as a lipid-soluble redox cofactor with a central position between the primary oxidations of NADH and succinate, and the reduction of complex III. Recent work with mammalian mitochondria (Saleno *et al.*, 1977; Yu *et al.*, 1977) has suggested that ubiquinone may be bound to protein(s) of the inner membrane, and may also be intimately involved in proton translocation by complex III (Mitchell, 1975; Rich and Moore, 1976).

In intact mitochondria respiring in the presence of ADP and phosphate, the passage of a pair of electrons through the chain from internal NADH to oxygen produces between two and three ATP. For succinate and external NADH oxidation between one and two ATP are formed. It has been customary to show electron transfer at sites corresponding to complexes I, III, and IV linked to ATP formation, at least schematically. However, the intensive investigations which have been made of ATP formation fail to disclose a direct linkage to ATP formation at these sites. The energy-conserving reaction detected is that illustrated in Fig. 4—electrogenic H^+ transport.

Mitchell (1966) proposed that H^+ efflux is the consequence of the electron transport chain in the region of these complexes being arranged in "loops" of alternating H-carriers and e^--carriers. The transfer from H to e^- occurs at the outer surface, releasing H^+, while the alternate transfer from e^- to H occurs at the inner surface, absorbing H^+. A two-electron redox reaction carried through one loop would thus transfer two H^+ from the matrix side to the outside. There is some evidence for "sidedness" in enzyme location. Cytochrome c (c-550) has an external location, and there are external and internal sites for NADH dehydrogenase (Douce *et al.*, 1973; Palmer, 1976). Succinate must enter the mitochondrion to be oxidized. Cytochrome oxidase appears to react with oxygen at the matrix surface. However, to date there is inadequate evidence for the presence of a loop in each complex, and the true mechanism of H^+ extrusion remains unknown. Furthermore, the stoichiometry may be as high as four H^+/complex (Lehninger *et al.*, 1977). Since the stoichiometry is not resolved, Fig. 4 indicates that electron transfer in each complex ejects nH^+ ($n = 2$ to 4, possibly not an integer) which are replaced from the matrix, leaving it negatively charged and alkaline with respect to the intermembrane space.

Inhibitors of each complex are shown in Fig. 4. In some plant tissues and plant mitochondria, blocking electron transfer in complexes III and IV, as with antimycin A or cyanide, does not completely inhibit respiration. This is the so-called cyanide-resistant respiration, or simply the alternative oxidase or alternative pathway. It is a highly variable component of mitochondrial respiration, depending on the plant species, organs, and stage of development (Solomos, 1977). The thermogenic spadices of Arum or skunk cabbage have mitochondria that are totally cyanide-resistant. Mitochondria from fresh potato tuber tissue are almost totally cyanide-sensitive; however, the aging of dormant storage tissue introduces a significant component of cyanide insensitivity (Ikuma, 1972; Solomos, 1977).

It now appears from kinetic analysis that the alternative pathway is operative *in vivo* with a K_m of 26–29 μM O_2 vs a K_m of 0.1 μM for cytochrome oxidase (Kano and Kageyama, 1977). It has been suggested that the alternate oxidase may have a functional role when mitochondria are participating heavily in intermediary metabolism (Ikuma, 1972; Palmer, 1976).

The point at which the alternative oxidase draws reducing equivalents from the electron transport chain is indicated to be ubiquinone (Rich and Moore, 1976; Storey, 1976). When skunk cabbage mitochondria are held anaerobic with CO to block the cytochrome oxidase, and pulses of oxygen are given to activate the alternative oxidase, ubiquinone and a distinct flavoprotein are oxidized (Storey, 1976). The flavoprotein has a midpoint redox potential 50 mV more negative than ubiquinone, which means that the alternative path will be favored only when the ubiquinone pool is highly reduced. Complex I is still operative with the alternative oxidase when NAD^+-linked stustrates are used, and ATP formation has been demonstrated (Solomos, 1977).

The oxidase of the alternative pathway is unknown. Previous suggestions on the participation of a b-cytochrome or autoxidizable flavoproteins have been discounted (Henry and Nyms, 1975; Solomos, 1977). Oxidation is strongly and specifically inhibited by substituted hydroxamic acids and certain other iron chelating agents (Ikuma, 1972; Bonner, 1973; Henry and Nyms, 1975; Solomos, 1977). The suggestion that nonheme iron proteins are implicated is viewed skeptically (Palmer, 1976).

D. Energy-Linked Processes

1. General Principles

The harnessing of respiratory energy to chemical and osmotic work will be treated here in terms of Mitchell's (1966) chemiosmotic hypothesis. Experimental observations with plant mitochondria can be explained most readily by this hypothesis, and this is the trend in the literature. However, as with the respiratory loops (above), it is not possible to make judgments on some of the postulated mechanisms.

It is undoubtedly true that operation of the electron transport chain drives proton efflux, but evidence has been meager for plants. Net proton efflux with corn mitochondria oxidizing NADH in a simple sucrose medium can be demonstrated, but only when salt (KCl) is added (Kirk and Hanson, 1973). Both H^+ efflux and respiration increase with increasing K^+ up to 20 mM.

Measurable net proton efflux requires a corresponding cation influx. Recently, protein ejection upon pulsing mung bean mitochondria with O_2 has been demonstrated (Moore, 1978) in experiments similar to those of Mitchell and Moyle (1967).

The inner membrane is considered to be impermeable to backflow of the extruded protons except in a controlled fashion through enzyme systems, or in association with lipid-soluble, proton-conducting uncouplers (Mitchell, 1966).

Proton extrusion by the respiratory chain is electrogenic and produces an electrochemical gradient of protons, or protonmotive force (Δp). Expressed in millivolts, $\Delta p = \Delta\psi - Z\Delta pH$, where $\Delta\psi$ is the electrical potential difference across the membrane, and $-Z\Delta pH$ is the chemical potential difference ($Z = 2.3\ RT/F$). Liver mitochondria respiring in state 4 (no ATP formation) have Δp values of 228 mV, with $\Delta\psi$ contributing about two-thirds depending on ion species in the medium (Nicholls, 1974). Plant mitochondria show Δp values in the range of 150–160 mV, with $\Delta\psi$ contributing 75–80 % (Moore *et al.*, 1978). Blocking the respiratory chain with antimycin A causes a sharp drop in Δp and a complete reversal of the pH gradient when succinate is the substrate, but not with malate. Stability with malate is attributed to electron transport through complex I and the alternate oxidase (Moore *et al.*, 1978).

Figure 5 illustrates a number of ways that Δp can be coupled to "chemiosmotic" work.

a. Backflow of protons through the coupling ATPase, forming ATP (see later). This is a reversible process, and the hydrolysis of ATP can drive H^+ efflux.

b. Influx or efflux of salts can be coupled to Δp through carrier enzymes, or "porters." For salt influx, $\Delta\psi$ drives an electrophoretic influx of the cation via a "uniport," while the chemical gradient (ΔpH) carries out a neutral exchange of OH^- for the anion via an "antiport" (an alternative would be a neutral cotransport of H^+ and anion via a "symport"). For salt efflux, the H^+ enters in exchange for the cation, and it is the anion which fluxes down the electrical gradient. The polarization of the process lies not with the porters, which can function in either direction, but with electrogenic H^+ efflux. According to this hypothesis the movement of an equivalent of salt is necessary to return an equivalent of extruded H^+. However, in the one report on the stoichiometry of K-salt influx and efflux, the K^+ transported per coupling site was about one, much lower than would be hypothesized (Kirk and Hanson, 1973).

There is an alternative mechanism which would return H^+ without net salt flux; cations could enter down the electrical gradient, followed by exit in neutral exchange for H^+ driven by the chemical gradient. In this case, cy-

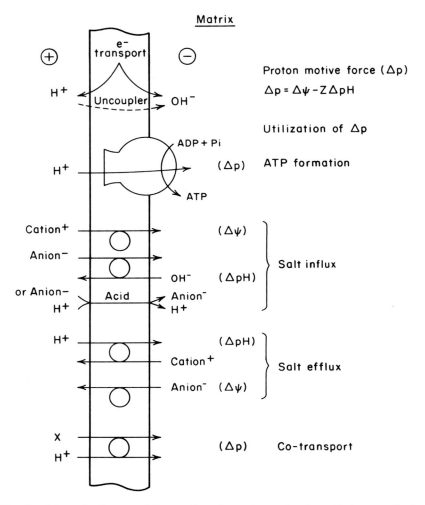

Fig. 5. Schematic diagram of the creation of a proton motive force (Δp) across the inner membrane during respiratory electron transfer, and the expenditure of Δp in chemiosmotic work.

cling of the cation would expend Δp. Obviously, cycling of anions would produce the same result. The evidence for respiring plant mitochondria is that steady-state salt loads are largely maintained by a balance of influx and efflux salt pumping (Hensley and Hanson, 1975).

c. Protons can be carried in by undissociated weak acids diffusing through lipid domains of the membrane. The pH gradient is discharged by the acid influx, and the cation uniport maintains electrical neutrality. Acetate salts are rapidly taken up by this mechanism; there is little discrimination as to cation (Wilson *et al.*, 1969).

d. Electrically neutral compounds (X in Fig. 5) could be cotransported with H^+ by the binding of both X and H^+ to the carrier. A similar mechanism could also be utilized to produce a partial neutralization of multivalent anions which are participating in an electrically neutral exchange (e.g., citrate^{3-} + H^+ ← carrier → HPO_4^{2-} or malate^{2-}). Note that this is not neutralization of the anion, but of the anion–carrier complex.

e. As with animal mitochondria, submitochondrial particles from plant mitochondria can carry out an energy-linked reversed electron flow from succinate to NAD^+, as well as an energy-linked transhydrogenase reaction, $NADH + NADP^+ \rightarrow NAD^+ + NADPH$ (Wilson and Bonner, 1970). In both cases the energy input can be derived from hydrolysis of ATP or by high energy intermediates derived from respiration. The high energy intermediate is now considered to be Δp (Racker, 1976). (No attempt is made to illustrate reversed electron flow in Fig. 5.)

f. The classical uncoupling agents, such as 2,4-dinitrophenol (DNP) are proposed to act as lipid-soluble proton carriers (Mitchell, 1966), allowing ready backflow of H^+ and collapsing Δp.

It is inherent in the chemiosmotic hypothesis that if there is no backflow of H^+, respiration will cease. That is, electron transport (and O_2 uptake) is controlled by Δp. If there are no proton "leaks," and if none of the above Δp-consuming systems is operative, the coupled electrogenic efflux of a trivial quantity of H^+ will bring Δp to a level where it comes to equilibrium with the redox energy of the electron transport chain complex. At this point it would be impossible for further electron transport to occur.

Experimentally, this does not happen. Plant mitochondria are notorious for carrying on an initial rapid rate of substrate oxidation in the absence of ADP. Bonner (1973) refers to this rate as a "substrate state," an additional respiratory steady state to those described by Chance and Williams (1956) for animal mitochondria (Fig. 6A). There are contentions that rapid substrate and state 4 rates are indicative of isolation procedures which damage the membranes or produce contaminants (Bonner, 1973). Damage and contamination are undoubtedly a factor, but it is noted that high substrate and state 4 rates still occur with carefully isolated and minimally contaminated preparations (Baker *et al.*, 1968; Ikuma, 1972; Palmer, 1976; Day and Hanson, 1978).

One factor in the high substrate and state 4 rates is the presence of phosphate (Fig. 6B). Addition of phosphate to mitochondria oxidizing exogenous NADH (which requires no energy expenditure in substrate transport) increases the respiration rate (Hanson *et al.*, 1972). Blocking further phosphate transport with mersalyl (see later) demonstrates that most of this "loose-coupled" respiration is due to the buildup of matrix phosphate. Addition of

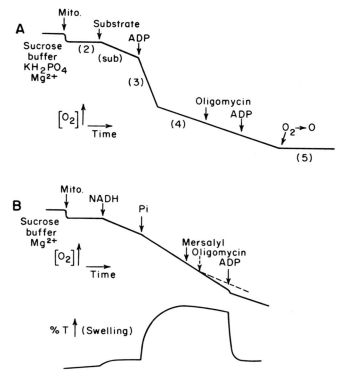

Fig. 6. Generalized tracings of oxygen concentration and light transmission obtained with plant mitochondria. (A) Illustration of respiratory states (Bonner, 1973). (B) Illustration of "loose coupling" introduced by matrix phosphate, and the loss of internal phosphate via ATP formation when phosphate transport is blocked with mersalyl (Hanson *et al.*, 1972).

oligomycin, which blocks the coupling ATPase (see later), reduces the respiration to about the "acceptorless" level. Withdrawal of the phosphate in ATP formation has the same effect. It is therefore argued that part of state 4 respiration is due to turnover of the coupling mechanism in the presence of high concentrations of internal phosphate. This is pronounced if arsenate is substituted (Bertagnolli and Hanson, 1973; Wickes and Wiskich, 1976). Recylcing of ions and salts plus energy expended in organic acid uptake must also be important. Simple electrophoretic backleakage of protons through the membrane cannot be excluded as an endogenous uncoupling factor, particularly if fatty acids are present, but the magnitude of leakage may be less than expected. The most likely avenue for H^+ leakage is the coupling ATPase (see below). Evidence for this is seen in the observations that oligomycin, which blocks the passage of H^+ through the ATPase, can increase the rate and extent of salt influx (Bertagnolli and Hanson, 1973). Blocking the leak increases the availability of Δp for ion transport.

2. ATP *Formation*

There is now much information on the coupling ATPase of animal and fungal mitochondria and the chloroplasts (Petersen, 1975; Racker, 1976) but scarcely any on plant mitochondria. One can reasonably assume, however, that the plant ATPase is basically the same, and examine the available data in this framework. For the purposes of this chapter it is again appropriate to use an illustrative diagram (Fig. 7).

The coupling ATPase forms the fifth enzyme complex of the inner membrane (Complex V). It consists of a hydrophilic knob (F_1) joined by a stalk piece (OSCP, or oligomycin sensitivity conferring factor), to a hydrophobic

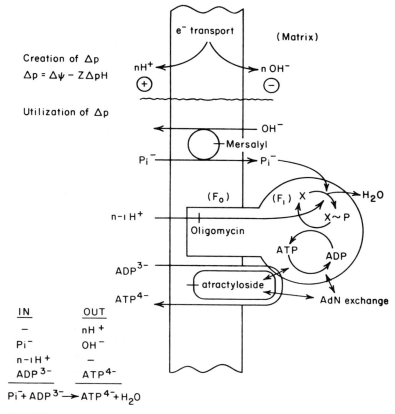

Fig. 7. Schematic diagram of ATP formation at the expense of Δp created at respiratory chain complexes (see Fig. 4).

base piece (F_0). F_1 preparations yield five polypeptide subunits, with a sixth polypeptide that functions as an inhibitor of ATP hydrolysis (Petersen, 1975). The stalk plus membrane sector yields another five polypeptide subunits. One of these, subunit 9, provides an oligomycin-inhibited avenue for proton flux in artificial phospholipid membranes (Criddle *et al.*, 1977). Isolated F_1, freed of the ATPase inhibitor, carries out an Mg-dependent, oligomycin-insensitive ATP hydrolysis, but only functions in ATP formation when bound to the membrane with the oligomycin-sensitive F_0. Insertion of isolated coupling ATPase into liposomes has shown that ATP hydrolysis drives a proton current as predicted by Mitchell, and addition of an electron transport complex provides an uncoupler-sensitive system for ATP formation at the expense of electron transport (Racker, 1976).

Figure 7 illustrates for plant mitochondria the probable pathway whereby the electrochemical potential associated with H^+ ejection is utilized in ATP formation (again the stoichiometry question is evaded, but in this case n is at least three). The diagram assumes a perfectly coupled flow of H^+ (no leaks) and emphasizes the connection between transport and phosphorylation. One proton is used in phosphate transport and the balance drive ATP formation. The need for more than one proton lies with the energy requirement for ATP formation and in observed stoichiometry (Nicholls, 1974, 1977; Lehninger *et al.*, 1977). The electrogenic exchange of ADP^{3-} for ATP^{4-} is driven by $\Delta\psi$ (Klingenberg, 1970), although more recent work suggests some compensatory H^+ efflux (Wulf *et al.*, 1978).

Phosphate is utilized only from the matrix, and inhibitors of phosphate transport, such as mersalyl, will block ATP synthesis. However, if mitochondria are preloaded with phosphate before addition of mersalyl, ATP synthesis to the level of accumulated phosphate can be demonstrated (Hanson *et al.*, 1972; see Fig. 6B). Thus ATP formation requires energy expenditure in phosphate influx, for which the Pi/OH antiporter must function.

The means by which phosphate enters into ATP formation are unknown for any mitochondrion. For plant mitochondria it has been suggested that the increased respiration rate of phosphate (or arsenate) loaded, mersalyl-blocked mitochondria ("loose coupled," Fig. 6B) is due to turnover of matrix phosphate with a labile intermediate ("$\sim$") which serves as a phosphoryl donor (Hanson *et al.*, 1972; Bertagnolli and Hanson, 1973). It was later recognized that "$\sim$" could be interpreted in terms of the chemiosmotic hypothesis as representing an activation of phosphorylation by Δp (Hanson and Koeppe, 1975). This view is adopted in Fig. 7, showing the formation of a phosphoryl donor, $X \sim P$. Recent work with animal mitochondria (Penefsky, 1977) discloses a distinct phosphate binding site in F_1 particles, and Racker (1976) notes that phosphorylated intermediates are not incompatible with the chemiosmotic hypothesis. However, such intermediates have not been identified.

An unexplained observation with plant mitochondria is that 2,4-dinitrophenol (the classical uncoupler) and fatty acids are competitive inhibitors of phosphate entry into ATP formation (Jackson *et al.*, 1962; Baddeley and Hanson, 1967). Recent investigations with animal mitochondria indicate an uncoupler-binding site associated with the coupling ATPase (Hanstein, 1976).

Figure 7 depicts bound ATP formed by F_1 being transferred to the adenine nucleotide translocase in exchange for an ADP from the medium. This transfer was postulated to explain the withdrawal of arsenate from mersalyl-blocked corn mitochondria upon addition of ADP (Bertagnolli and Hanson, 1973); if ADP–As were released to the aqueous matrix it should hydrolyze, and the arsenate would recycle without leaving the matrix. Active sites in F_1 are visualized to be in AdN exchange equilibrium with the matrix, which in turn can also exchange AdN with the translocase. Direct transfer between F_1 and the AdN translocase appears to have partial support from translocase studies with animal mitochondria (Vignais, 1976; Out *et al.*, 1976), although this is disputed (Klingenberg, 1977).

Whether directly coupled to F_1 or not, the exchange of ADP^{3-} for ATP^{4-} is largely electrogenic and constitutes a considerable share of the energy requirement for ATP synthesis (Klingenberg, 1970). Thus the electrical component of Δp is critical to oxidative phosphorylation. This would not be true for photophosphorylation where ADP/ATP exchange with the cytosol does not occur. Photophosphorylation can be driven by a pH gradient alone (Jagendorf and Uribe, 1966).

The coupling ATPase should be reversible, hydrolyzing ATP and producing Δp, which in turn can do osmotic work. This proves to be true in corn mitochondria where ATP hydrolysis will drive Ca^{2+} uptake (Hodges and Hanson, 1965; Elzam and Hodges, 1968) or KCl efflux (Stoner and Hanson, 1966) in an oligomycin-sensitive reaction. This could not be done with castor bean mitochondria (Yoshida, 1968). It was discovered that castor bean mitochondria possessed very little uncoupler-stimulated ATPase unless they were permitted to respire, in which case activity developed (Takeuchi *et al.*, 1969). The same proved true for sweet potato and cauliflower mitochondria (Carmelli and Biale, 1970; Jung and Hanson, 1973a, 1975). (The role of the uncoupler is to permit backflow of H^+, obviating the need for salt transport, etc.)

It was initially contended that respiration "primed" cauliflower mitochondria for ATP transport; a brief pulse of NADH oxidation in the presence of phosphate and Mg^{2+} activated subsequent dinitrophenol-stimulated ATPase, provided the ATP was added before the uncoupler (Jung and Hanson, 1973a, 1975). Sonication of cauliflower mitochondria, which inverts the membrane vesicles and exposes the ATPase directly to the ambient solution, or assay at high pH which increases membrane permeability, was adequate to induce high ATPase activity. However, an alternative interpretation is that respiration, alkaline treatment, and sonication may remove an inhibitor protein from the ATPase (Takeuchi, 1975). This view is supported by several studies showing trypsin digestion of submitochondrial particles to increase dramatically the ATPase activity (Takeuchi, 1975; Jung and Laties, 1976; Grubmeyer and Spencer, 1978). Thus in common with animal mitochondria (Van de Staat et al., 1973) it appears that in plant mitochondria a trypsin-sensitive ATPase inhibitor may be detached during respiration. This would explain the common observation that intact plant mitochondria without uncoupler-stimulated ATPase are quite effective in oxidative phosphorylation.

Disrupted monocotyledon mitochondria display oligomycin-resistant ATPase activity, while dicotyledonous mitochondria show oligomycin sensitivity (Jung and Hanson, 1973b; Sperk and Tuppy, 1977). Solubilized corn ATPase is cold stable and of low molecular weight (40,000–60,000) (Sperk and Tuppy, 1977) vs 380,000 for solubilized F_1 (Petersen, 1975). Grubmeyer and Spencer (1978) have shown that washed submitochondrial particles of corn have a normal oligomycin-sensitive ATPase. It is a solubilized fraction of the ATPase which produces the anomalous result. Solubilized ATPases from dicotyledon mitochondria resemble the F_1-ATPase of animal mitochondria (Yoshida and Takeuchi, 1970; Grubmeyer et al., 1977). Pea mitochondria ATPase is notable for the high degree of stimulation by NaCl and Ca_2Cl (Grubmeyer et al., 1977) and for the fact that chloride and bromide, as well as oxy-anions, give stimulation (C. Grubmeyer and M. Spencer, private communication).

3. Ion and Substrate Transport

a. General Aspects. Before examining the details of transport in plant mitochondria, it is worthwhile considering the general requirements for mitochondrial transport in vivo (Fig. 8). The guidelines for this come mainly from the extensive work done with mammalian mitochondria and now confirmed in many respects for plant mitochondria (Hanson and Koeppe, 1975; DeSantis et al., 1976; Wiskich, 1977). During oxidative phosphorylation there is influx of pyruvate, phosphate, ADP, and O_2, and efflux of CO_2, H_2O, and ATP (Fig. 8). Fluxes of the small, neutral molecules (O_2, CO_2, H_2O) are by diffusion through the inner membrane, and their transport is not rate limiting. However, the anions with their negative charge face accumulation against an electrical gradient, and this is bypassed by neutral exchange. The anion generated for exchange is OH^-, and the primary exchanges are P_i^-/OH^- and pyruvate/OH^-. Since pyruvate is rapidly oxidized, it is phosphate transport at the expense of the pH gradient which is fundamental to establishing and maintaining the anion content of the matrix and $\Delta\psi$ (Lehninger,

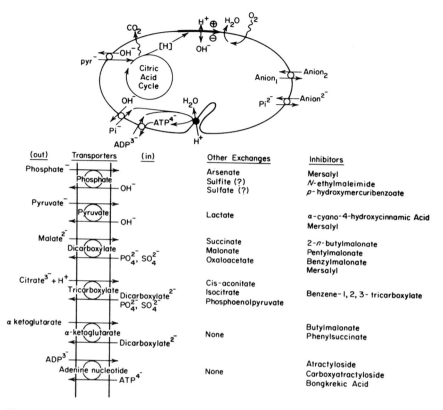

Fig. 8. Schematic diagram of the transport complex involved in oxidative phosphorylation and intermediary metabolism (above), with details on specific transporters described for plant mitochondria (below).

1974; Wiskich, 1977). Although other anions, such as acetate, can be transported *in vitro* it is phosphate that plays the physiological role.

The $ADP_{in}^{3-}/ATP_{out}^{4-}$ exchange is electrogenic and is driven by $\Delta\psi$, contributing to ATP formation and the high ATP:ADP ratio of the cytosol (Klingenberg, 1970; see preceding section). Hence, the cumulative effect of the neutral P_i^-/OH^- exchange may be viewed as maintaining an electrical potential favorable to $ADP_{in}^{3-}/ATP_{out}^{4-}$ exchange and ATP formation. However, the processes are closely integrated, and the electrogenic ADP/ATP exchange can be conversely considered as compensated by phosphate influx (McGiven *et al.*, 1971).

Substrates other than pyruvate (see Section I and Fig. 4) are supplied by exchange for phosphate or another anion. Figure 8 summarizes anion exchanges as they are known for plant mitochondria. The transporters are symmetrical in that they will exchange the same species of anion across the membrane (i.e., P_i/P_i, malate/malate, etc.) (DeSantis *et al.*, 1976).

b. Phosphate Transport. Perhaps the most extensively studied transport process in plant mitochondria is that of phosphate (and arsenate). In general, this transport conforms to that found in animal mitochondria, being readily inhibited by the hydrophilic sulfydryl reagents, mersalyl and *N*-ethylmaleimide. In respiring mitochondria the kinetics of potassium phosphate transport show a two-phase absorption curve, with the first phase half-saturated at about 0.25 m*M* phosphate (Hanson *et al.*, 1972). When respiration is inhibited or Δp collapsed with uncoupler, rapid passive efflux of phosphate occurs, and this is also inhibited by mersalyl (Hensley and Hanson, 1975; DeSantis *et al.*, 1975). Two simple but important points emerge from these studies: (1) phosphate transport is responsive to the phosphate gradient as well as the proton gradient; and (2) high matrix phosphate content (and hence other ion content) is maintained only so long as respiration maintains Δp.

Arsenate uncoupling involves cyclic arsenate transport (in by the phosphate transporter, out as ADP–As on the adenine nucleotide transporter), and mersalyl is as effective as oligomycin in blocking arsenate uncoupling (Bertagnolli and Hanson, 1973).

The exchange studies of De Santis *et al.* (1976) indicate that sulfate enters by the dicarboxylate transporter in exchange for phosphate or dicarboxylic acids. However, corn mitochondria oxidizing NADH in the absence of phosphate or dicarboxylate accumulate potassium sulfate by a process inhibited by mersalyl (Kimpel and Hanson, 1978).

Sulfite exchanges for phosphate with the same inhibitor sensitivity as phosphate/phosphate exchange (DeSantis *et al.*, 1976). However, DeSantis *et al.* (1976) suggest that SO_3/OH exchange occurs independently of the P_i transporter.

c. Substrate Transport. Pyruvate transport corresponds closely to that in animal mitochondria, with 2 μM α-cyano-4-hydroxycinnamic acid providing 50% inhibition (Day and Hanson, 1977b). About 10 nmoles of pyruvate per milligram protein can be accumulated if pyruvate oxidation is blocked; assuming a matrix volume of 1 μl/mg protein, this represents an internal concentration of approximately 10 m*M*. However, the rate of pyruvate transport (estimated at about 20 nmoles/min/mg protein) is too low to support vigorous respiration and under these conditions malate transport plus intramitochondrial malic enzyme furnishes a significant share of internal pyruvate to corn mitochondria (Day and Hanson, 1977b). Pyruvate does not enter into exchanges with phosphate or TCA cycle anions (DeSantis *et al.*, 1976), but lactate can use the pyruvate transporter and inhibits pyruvate accumulation (Day and Hanson, 1977b). Pyruvate stimulation of malate oxidation by corn mitochondria is dependent on the pyruvate transporter, and the apparent K_m for pyruvate transport is 0.53 mM (Day and Hanson, 1977b).

A number of swelling and respiration studies have shown malate and succinate transport to be dependent upon phosphate, and to be blocked by inhibitors of the phosphate and dicarboxylate transporter (Wiskich, 1977). About 20 nmoles malate/mg protein can be accumulated in respiring corn mitochondria, and malate is lost when the mitochondria are uncoupled (Day and Hanson, 1977a), accounting for the decline with time in the uncoupled oxidation rates. State 3 malate oxidation by corn mitochondria is dependent on malate transport via the dicarboxylate carrier, and the K_m for malate transport was estimated at 0.25 mM.

Exchange studies by DeSantis *et al.* (1976) with bean mitochondria have shown that oxaloacetate, malonate, and sulfate can participate in exchanges on the dicarboxylate transporter. Oxaloacetate is also transported on the α-ketoglutarate carrier, and the combined activity of the two transport systems may account for the observed rapid entry of oxaloacetate (Douce and Bonner, 1972; Day *et al.*, 1976). There is no evidence that fumarate transport occurs in plant mitochondria (Wiskich, 1977). The small amount of fumarate oxidation observed can be accounted for by diffusive entry; the slow rate of malate oxidation by completely uncoupled mitochondria appears to depend on diffusive entry (Day and Hanson, 1977a).

Oxidative and swelling studies show that citrate enters in exchange for a dicarboxylate anion, or phosphate, depending on the plant species (Phillips and Williams, 1973; Wiskich, 1974). Bean mitochondria show citrate exchange with cis-aconitate, malate, malonate, isomalate, and phosphoenolpyruvate, but not phosphate (DeSantis *et al.*, 1976). The exchange was inhibited by benzene-1,2,3-tricarboxylate but not by mersalyl or butylmalonate. A small amount of exchange was found with sulfate and α-ketoglutarate (10–17% vs 65–66% with malate), but this was scarcely inhibited by benzene-1,2,3-tricarboxylate. In corn mitochondria at pH 6.5, a combination of citrate plus sulfate or phosphate produces mersalyl-sensitive citrate uptake and oxidation (Kimpel and Hanson, 1977). Citrate transport in animal mitochondria is proton compensated, with citrate + H$^+$ exchanging for malate (Papa *et al.*, 1971; McGivan and Klingenberg, 1971). It is probable that proton compensation is required for plant mitochondria as well, and this may account for the phosphate- and sulfate-activated citrate transport at low pH. There is an α-ketoglutarate/dicarboxylate exchanger in bean mitochondria which differs from the dicarboxylate transporter in not utilizing phosphate and in being mersalyl-insensitive (DeSantis *et al.*, 1976).

Glutamate transport is perplexing. DeSantis *et al.* (1976) did not find glutamate to participate in any of the exchanges they studied. Yet it is clear that glutamate enters plant mitochondria since it is commonly used during malate oxidation to lower internal oxaloacetate by transamination. Swelling studies with glutamate in beetroot and cauliflower mitochondria show a requirement for phosphate plus dicarboxylate, suggesting transport on the tricarboxylate transporter (Day and Wiskich, 1977a). No reported attempts have been made to find the electrogenic glutamate–aspartate exchange transporter known for animal mitochondria (LaNoue and Tischler, 1976).

In summary, substrate transport is accomplished by a group of inner membrane anion exchanging porters or translocases which are presumably proteinaceous and able to discriminate between substrates. Net substrate influx requires a proton motive force. Substrate transport rates may at times limit respiration rates, but the site or nature of resistance to transport is unknown. Membrane lipids probably have a role here: low concentrations of detergents are reported to stimulate succinate oxidation with no impairment of respiratory control and ADP:O ratio (Mazliak and DeCotte, 1976).

Substrate transport is also involved when the mitochondria participate in intermediary metabolism. This subject is discussed in Section V.

d. Adenine Nucleotide Transport. Only in recent years has the adenine nucleotide translocator of plant mitochondria been studied, although the availability of the specific inhibitors atractyloside, carboxyatractyloside, and bongkrekic acid have made it the earliest and best studied translocator in animal mitochondria (Klingenberg, 1970; Vignais, 1976). Available evidence is that the translocator in plant mitochondria carries out a very rapid ADP/ATP exchange with low K_m values comparable to animal mitochondria (Janovitz *et al.*, 1976; Earnshaw, 1977). Exchange of internal ADP for external AMP is found, but is attributed to the action of adenylic kinase producing ADP for exchange (Earnshaw, 1977). Other than this, the AdN translocator is specific for ADP and ATP.

Atractyloside is a noncompetitive inhibitor of ATP binding and transport (Janovitz *et al.*, 1976) and thus an effective inhibitor of DNP-stimulated ATPase (Jung and Hanson, 1973a; Janovitz *et al.*, 1976). However, atractyloside is much less effective in inhibiting ADP binding and transport in plant mitochondria than in animal mitochondria (Jung and Hanson, 1973b; Passam and Coleman, 1975; Janovitz *et al.*, 1976; Vignais *et al.*, 1976; Earnshaw, 1977), with some variation between plant species. In *Vigna* mitochondria, 2.5 μM atractyloside produces 50% inhibition of state 3 respiration, and the K_m for ADP in state 3 is 25 μM (Silva Lima *et al.*, 1977).

ATP is also an inhibitor of ADP transport and the reversal of ATP inhibition by atractyloside leads Janovitz *et al.* (1976) to suggest that the transporter has an ATP-binding regulatory site. Earnshaw (1977) calculates that about 0.55 nmoles ADP/mg protein are bound to the translocator in corn mitochondria, a figure comparable to that for animal mitochondria.

Unlike atractyloside, carboxyatractyloside is an effective competitive inhibitor of ADP transport in plant mitochondria (Vignais et al., 1976; Abou-Khalil and Hanson, 1977). Bongkrekic acid is also effective in blocking transport (Passam and Coleman, 1975; Vignais et al., 1976). The reaction of ADP with the translocator causes a contraction of the membrane which is inhibited by atractyloside (Hanson et al., 1972; Earnshaw and Hughes, 1976). This is thought to represent a conformational change of the transporter (Stoner and Sirak, 1973; Scherer and Klingenberg, 1974; Earnshaw and Hughes, 1976).

Plant mitochondria have an energy-linked mechanism for net uptake of ADP which is insensitive to atractyloside and carboxyatractyloside but is sensitive to mersalyl (Jung and Hanson, 1975; Abou-Khalil and Hanson, 1977). There is a requirement for Mg^+, phosphate, and oligomycin (to prevent phosphorylation). The mechanism is not understood, but except for the lack of atractyloside sensitivity it resembles the net uptake of ADP accompanying Ca^{2+} phosphate accumulation in liver mitochondria (Carafoli et al., 1965).

e. Cation Transport. Net accumulation of anions must be charge compensated by uptake of cations (see Fig. 5), and in vivo these are primarily K^+ and Mg^{2+}. It is now widely accepted in animal mitochondria that cation uptake occurs via an electrophoretic "uniport" mechanism, while efflux is via a $cation^+/H^+$ antiporter (Brierley, 1976a,b). The same conclusion is drawn for plant mitochondria (Hanson and Koeppe, 1975).

For animal mitochondria it is questioned whether the uniport for electrophoretic entry of monovalent cations is a carrier protein, since there is little discrimination between cations during the rapid energized influx of acetate or phosphate (Brierley, 1976a). This is also true for plant mitochondria (Wilson et al., 1969). However, there is no critical basis for deciding between a simple electrophoretic cation penetration of the lipid domains and an interaction with a broad-spectrum cation carrier. The same applies to the electrophoretic efflux of anions during energy-linked salt extrusion (Fig. 5). The term uniport is used here without implication as to the nature of the transport site.

In studies of respiratory-linked potassium phosphate and acetate uptake by plant mitochondria, the uptake of K^+ via the uniport appears to be rate-limiting since addition of valinomycin or gramicidin (lipid soluble K^+ − mobilizing ionophores) greatly increases the rate and extent of salt uptake and osmotic swelling (Wilson et al., 1972; Hanson et al., 1972; Kirk and Hanson, 1973).

Unlike the cation uniport, the cation/H^+ antiport might be assumed to be a protein, but no specific inhibitor is known and no exchange carrier has been isolated. There is no evidence for plant mitochondria that Na^+ is strongly favored over K^+ as it is in animal mitochondria (Brierley, 1976b). The evidence for the existence of the antiport is largely based on the energized osmotic shrinkage of salt-loaded mitochondria, generally followed by absorbancy changes. The assumptions here are supported, however, by K^+ analyses (Kirk and Hanson, 1973) size changes by Coulter counter (Pomeroy, 1977) and ultrastructural changes (e.g., Pomeroy, 1977). Morphology, rather than size, may be a dominant factor in absorbancy changes (Pomeroy, 1977).

The two cation transport processes, outlined above, may be a means by which mitochondrial volume is controlled *in vivo* (Brierley, 1976b). In corn mitochondria during steady-state osmotic swelling in potassium phosphate, there is cyclic salt transport (Hensley and Hanson, 1975).

Valinomycin does not always react to increase the rate and extent of energy-linked salt influx. Corn mitochondria oxidizing NADH in 5 mM K phosphate respond to valinomycin with additional salt uptake and swelling; if 5 mM K_2SO_4 is substituted for K phosphate, the initial rate of swelling is much slower and there is a rapid shrinkage upon addition of valinomycin (Kimpel and Hanson, 1978). This sulfate response can be mimicked with phosphate if a limited amount of mersalyl is introduced to add a "resistance" to phosphate transport. It is concluded that the relative resistance of the anion$^-$/OH^- and K^+/H^+ antiports governs whether there is a net influx or efflux of salt (Kimpel and Hanson, 1978).

Divalent cation uptake (principally Ca^{2+}) is electrophoretic in mammalian mitochondria (Lehninger, 1974) probably by means of a lanthanum and ruthenium red sensitive Ca^{2+} carrier (Carafoli, 1976). Whether Ca^{2+} is taken up as the cation or as a phosphate complex is not resolved (Moyle and Mitchell, 1977a,b; cf., Reynafarje and Lehninger, 1977). For plant mitochondria there is no evidence for a Ca^{2+} uniport. Either there is no respiration-linked Ca^{2+} transport whatever, an exceptional result reported by Moore and Bonner (1977) for mung bean mitochondria, or Ca^{2+} uptake is linked to oligomycin-insensitive phosphate uptake (Hanson and Hodges, 1967; Chen and Lehninger, 1973; Wilson and Graesser, 1976). Sr^{2+} and Ba^{2+} are taken up as well as Ca^{2+} (Miller *et al.*, 1970; Wilson and Minton, 1974) and Mg^{2+} uptake occurs from high Mg^{2+} concentrations (Millard *et al.*, 1965). Accumulation of Sr^{2+} and Ca^{2+} in the presence of phosphate produces electron-dense precipitates in the matrix (Ramirez-Mitchell *et al.*, 1973; Peverly *et al.*, 1974). The Ca^{2+}/P_i accumulation ratio is 1.7 (Elzam and Hodges, 1968). Proton release accompanies calcium phosphate uptake with an H^+/Ca^{2+} ratio of 0.8, a result attributed to formation of calcium phosphate precipitates (Earnshaw *et al.*, 1973).

In the absence of added phosphate, there is an energized uptake or binding of about 100 nmoles Ca^{2+}/mg protein (Hanson and Miller, 1967). The amount of binding proves to depend on the level of endogenous phosphate (Earnshaw et al., 1973; Day et al., 1978). During Ca^{2+}-binding the endogenous phosphate moves from a readily-leached to a leaching resistant phase, with both the Ca^{2+} and phosphate rapidly released when respiration ceases (Earnshaw et al., 1973; Earnshaw and Hanson, 1973). Unlike the case with mammalian mitochondria, Ca^{2+} is not actively taken up with acetate (Truelove and Hanson, 1966; Day et al., 1978), and there is little or no evidence for high-affinity binding sites (Chen and Lehninger, 1973; Day et al., 1978). There is, however, H^+ release during Ca^{2+} binding with an H^+/Ca^{2+} ratio of 0.9 (Earnshaw et al., 1973; Day et al., 1978). It is believed that a Ca^{2+}-phosphate complex accounts for the Ca^{2+} binding and that the complex is the vehicle of Ca^{2+} transport during massive calcium phosphate uptake in the presence of phosphate (Wilson and Minton, 1974; Day et al., 1978). Calcium or strontium phosphate uptake is competitive with ATP formation, and is most active under state 4 conditions (Hanson and Miller, 1967; Johnson and Wilson, 1973).

V. CELLULAR AND PHYSIOLOGICAL INTERACTIONS

Most biochemical and biophysical studies of plant mitochondria have from necessity been made with isolated organelles, but there is a biological obligation to place the observed properties in the context of cellular function. In addition to the central role in energy metabolism, mitochondrial involvement in other areas of cellular metabolism, and in the physiological responses of the plant, are being studied. Mitochondrial participation in photorespiration has been discussed briefly. Further examples are provided in this section.

A. Intermediary Metabolism

1. C_4 Photosynthesis

The present concept of C_4 photosynthesis (Hatch and Osmond, 1976) is that the mesophyll cells produce malate and aspartate which are transferred to the bundle sheath cells. Here the C_4 acids are decarboxylated to produce CO_2 (which is reduced in photosynthesis) and pyruvate, which cycles back to the mesophyll where it serves as the C_3 skeleton for further photosynthetic C_4 production. In one group of C_4 plants, NAD-linked malic enzyme provides the necessary decarboxylation (e.g., Atriplex spongiosa). The bundle sheath mitochondria of these plants are especially adapted to carry out rapid C_4 decarboxylation, having high concentrations of NAD-linked malic enzyme and amino transferase. When optimal concentrations of phosphate and α-ketoglutarate are supplied, the rates of C_4 decarboxylation by these mitochondria far exceed the rates of respiration (Hatch and Kagawa, 1976). Current schemes (Hatch and Osmond, 1976) call for rapid entry of aspartate and α-ketoglutarate into the mitochondria, transamination, and reduction to produce malate, decarboxylation via malic enzyme, and exit of glutamate and pyruvate. Hence, bundle sheath mitochondria must possess transporters thus far unknown for other plant mitochondria (see previous section). In addition, there must be close regulation of pyruvate dehydrogenase and citrate synthetase.

2. Amino Acid Metabolism

Plant mitochondria are the presumed source of the α-ketoacids utilized in amino acid synthesis. Glutamate synthesis during N assimilation is now known to proceed through glutamine synthetase plus glutamate synthase, largely in the plastids (Miflin and Lea, 1977b). The assimilation of ammonia by this pathway requires a supply of α-ketoglutarate from the mitochondria.

3. Fat Synthesis

Many developing seeds synthesize and deposit triglycerides as food reserves for the embryo. It has been suggested (Nelson and Rinne, 1977) that the pathway from sugars to fat involves citrate synthesis by plant mitochondria, much as in animals. Pyruvate enters the mitochondrion and is oxidized to acetyl-CoA, which in turn condenses with oxaloacetate to form citrate. Citrate exits and is converted to oxaloacetate plus acetyl-CoA by cytoplasmic ATP-citrate lyase. Acetyl-CoA then enters into fatty acid synthesis while the oxaloacetate reenters the mitochondrion, possibly after reduction to malate, in exchange for citrate. No work has been reported on mitochondrial citrate transport in fat-synthesizing seeds.

B. Hormone Responses

As discussed early in this chapter, there is evidence for mitochondrial reproduction and development in dividing and growing cells, a process which is under hormonal control. Similarly, during seed germination there appear to be factors arising from the embryo which regulate the development of enzymatic activities in the cotyledons or endosperm, and in certain cases application of known hormones can substitute for the embryo (Mayer and Shain, 1974). There are many experiments showing the development of mitochondria subsequent to the hydration and metabolic activation of cotyledons or endosperm (Mayer and Shain, 1974). In the initial stages, this appears to be an assembly of normal mitochondria from particulates or pro-mitochondria preexisting in the dessicated cells (Sato and Asahi, 1975), with no known hormonal implication. There is no evidence for a direct effect of hormones (at physiological concentrations) on mitochondrial activities. Mitochondria from auxin-treated soybean seedlings, cleared of bacteria by gradient centrifugation, are slightly larger and incorporate labeled leucine into protein more rapidly than those from control tissue, but they are not more active in oxidative phosphorylation (Baxter and Hanson, 1968).

Phytochrome is a photoreversible chromoprotein that regulates a broad spectrum of red and far-red light responses in plants, possibly through membrane-mediated processes (Briggs and Rice, 1972; Marmé, 1977). Pulses of red light or continuous far-red light increase the fumarase, succinic dehydrogenase, and cytochrome oxidase activity of mustard cotyledon mitochondria, and the far-red exposure changes the inner membrane configuration from parallel *cristae* to the more common, random *sacculi* (Bajracharya *et al.,* 1976).

In addition to these developmental effects *in vivo,* red light *in vitro* is reported to activate $NADP^+$ reduction by gradient purified pea epicotyl mitochondria (Manabe and Furuya, 1974), and spectrophotometric determinations show red light to increase P_{fr} bound to mitochondria *in vivo* (Manabe and Furuya, 1975). Georgevich *et al.* (1977) have presented evidence on the binding of ^{125}I-labeled phytochrome to oat coleoptile mitochondria. Although Cedel and Roux (1977) could not confirm the phytochrome-activation of $NADP^+$ reduction reported by Manabe and Furuya, they did find 20% higher outer membrane NADH dehydrogenase with P_{fr} bound than with P_r bound.

Mitochondria isolated from etiolated *Avena* shoots which had been irradiated with far-red light were found to have altered permeability to certain organic acids (Hampp and Schmidt, 1977). Similar light treatment of the isolated organelles caused identical changes in permeability (Schmidt and Hampp, 1977), and the red light effects were fully reversible. It was concluded that phytochrome is bound to the mitochondria and causes reversible changes in the permeability of the inner membrane. Since similar effects were observed with etioplasts, it was postulated that phytochrome mediates changes in organelle permeability during chloroplast development, allowing coordinated exchange of metabolites between mitochondria and plastids.

In summary, it is probable that mitochondria participate in hormonal- and light-mediated shifts in metabolism, but there is little to indicate that they have more than an accessory role in the manifold physiological responses. In general, they appear to share in these responses, not initiate them, and the primary function of ATP production is remarkably stable throughout experimental treatments.

C. Environmental Stress

Due to the essentiality of mitochondria in supplying energy, there has been interest in the biochemical modifications of mitochondria associated with the tolerance of heat, cold, drought, etc.

Plants adapted to warm climates suffer chilling injury at temperatures below about 10°C, and mitochondria from species of these have been compared with chilling-resistant temperate species (Lyons, 1973). Mitochondria extracted from chilled tissues of sensitive species generally show injury, but it is difficult to determine if this is a direct effect of low temperature (Lyons, 1973). Mitochondria isolated from normal tissues of chilling-sensitive and chilling-resistant species, however, show an inherent difference in their response to a temperature range of about 2°–25°C (Lyons and Raison, 1970). In an Arrhenius plot of oxidation rates, chilling-resistant mitochondria have a constant activation energy, while those from chilling-sensitive plants show a discontinuity at 10°–12°, below which the activation energy is substantially increased. The discontinuity is attributed to a phase transition in membrane lipids from a flexible liquid-crystalline to a solid-gel structure, and it is reasonably correlated with a higher proportion of saturated fatty acids in chilling-sensitive membranes. However, this characteristic is not unique to the mitochondrial membranes (Lyons, 1973).

Mitochondria from wheat (a temperate species) grown at 24° show more distinct cristae and a denser matrix than those from plants grown at 2°, a temperature which hardens the plants against freezing injury and increases the proportion of unsaturated fatty acids (Pomeroy, 1977). Mitochondria from 24° plants show normal swelling in 0.2 M KCl, and contraction (due to operation of the efflux pump, Fig. 5) upon addition of succinate. Hardened mitochondria from 2° plants show slower swelling and no energy-linked contraction (Pomeroy, 1976, 1977). Yet such mitochondria have good substrate oxidation rates, respiratory control, and ADP:O ratios when tested at 24°C (Miller et al., 1974). Gel electrophoresis of mitochondrial proteins extracted from hardened wheat shows a greater diversity of proteins (Khoklova et al., 1975). Obviously there are shifts in metabolism under cold stress which change mitochondrial membranes in ways that cannot be determined by measuring state 3–state 4 respiration. In this respect, it should be noted that mitochondria from hardened plants are very resistant to damage even if the plant is severely damaged (Andrews and Pomeroy, 1977; Singh et al., 1977).

Adaptations to heat and drought stress are not as well studied. Mitochondria from beans, a heat resistant species, show less swelling in KCl at 35°–45° than do pea mitochondria, a cool season crop, and show better energized contraction (Andreeva, 1969). Mitochondria from heat-hardened corn seedlings also swell less and contract better at high temperatures than nonhardened mitochondria (Andreeva, 1969). Mitochondria from drought-stressed corn seedlings have lower substrate oxidation rates, more rapid swelling in KCl, and no capacity for energized contraction (Koeppe et al., 1973). Oddly enough, drought stress produced no marked damage in respiratory control and ADP : O ratios with succinate, malate plus pyruvate, or NADH as substrates.

Decreasing the osmotic potential of the assay medium lowers the state 3 respiration rate of plant mitochondria, especially with salt additions, but without altering ADP : O ratios (Flowers and Hanson, 1969; Campbell et al., 1976). Increasing the water potential by hydrostatic pressure will not reverse the inhibition, which is attributed to a salt effect on mitochondrial enzymes, not to water potential per se (Flowers and Hanson, 1969). High concentrations of sucrose may limit substrate uptake (Campbell et al., 1976). The water stress created in tissue by high concentrations of NaCl alters the configuration of the mitochondria in electron micrographs, decreasing the clarity of the cristae membranes but not reducing cytochrome oxidase activity (Nir et al., 1969, 1970). In developing bean leaves a salt-stress increases the number of mitochondria per cell as much as six-fold (Siew and Klein, 1968).

D. Pathological Stress

There is one clear case of a toxin produced by a pathogenic fungus affecting mitochondria. The toxin produced by Helminthosporium maydis, race T, specifically attacks the mitochondria of corn plants carrying a cytoplasmic gene for male sterility (Texas male sterile). Mitochondria from T-cytoplasm are extremely sensitive to the toxin compared to those from normal cytoplasm, showing loss of respiratory control and ATP formation, mitochondrial swelling, and stimulated ATPase activity (Miller and Koeppe, 1971; Gegenbach et al., 1973a,b; Bednarksi et al., 1977). Removal of the outer membrane from normal mitochondria makes them also susceptible to the toxin (Watrud et al., 1975), although caution is suggested in interpreting this finding as strictly an outer membrane phenomenon due to possible changes in the inner membrane during outer membrane removal (Bednarski et al., 1977). The toxin can be washed from the mitochondria, restoring activity (Bednarski et al., 1977). Although the action of the toxin is partially mimicked by uncouplers and monovalent cation ionophores, there are a number of dissimilarities, and the mode of action remains unknown.

In summary, plant mitochondria are part of the living cell and share in the developmental, adaptive, and genetic properties of the cell. However, there is little evidence that this sharing involves any changes in the fundamental parameters of ATP formation. Here the mitochondria prove quite conservative. Structural changes do accompany the dormancy or hardening of plant tissues, probably as part of the evolved mechanism for plant survival of adverse environments.

REFERENCES

Abou-Khalil, S., and Hanson, J. B. (1977). *Arch. Biochem. Biophys.* **183**, 581–587.
Andreeva, I. N. (1969). *Sov. Plant Physiol.* **16**, 182–187.
Andrews, C. J., and Pomeroy, M. K. (1977). *Plant Physiol.* **59**, 1174–1178.
Arron, G. P., Henry, L., Palmer, J. M., and Hall, D. D. (1976). *Biochem. Soc. Trans.* **4**, 618–620.
Arron, G. P., Day, D. A., and Laties, G. G. (1978). *Plant Physiol.* **61S**, 84.
Atkinson, A. W., Jr., John, P. C. L., and Gunning, B. E. S. (1974). *Protoplasma* **81**, 77–109.
Baddeley, M. S., and Hanson, J. B. (1967). *Plant Physiol.* **42**, 1702–1710.
Bajracharya, D., Falk, H., and Schopfer, P. (1976). *Planta* **131**, 253–261.
Baker, J. E., Elfvin, L. G., Biale, J. B., and Honda, S. I. (1968). *Plant Physiol.* **43**, 2001–2022.
Baxter, R., and Hanson, J. B. (1968). *Planta* **82**, 246–260.
Bednarski, M. A., Izawa, S., and Scheffer, R. P. (1977). *Plant Physiol.* **59**, 540–545.
Beevers, H. (1961). "Respiratory Metabolism in Plants." Row-Peterson, Evanston, Illinois.
Beezley, B. B., Graber, P. J., and Fredrick, S. E. (1976). *Plant Physiol.* **58**, 315–319.
Benedict, C. R., and Beevers, H. (1961). *Plant Physiol.* **36**, 540–544.
Bertagnolli, B. L., and Hanson, J. B. (1973). *Plant Physiol.* **52**, 431–437.
Beyer, R. E., Peters, G. A., and Ikuma, H. (1968). *Plant Physiol.* **43**, 1393–1400.
Bird, I. F., Cornelius, M. J., Keys, A. J., and Whittington, C. P. (1972). *Phytochemistry* **11**, 1587–1594.
Bligny, R., and Douce, R. (1976). *Physiol. Veg.* **14**, 499–515.
Bonner, W. D., Jr. (1973). *In* "Phytochemistry" (L. P. Miller, ed.), Vol. III, pp. 221–261. Van Nostrand-Reinhold, New York.
Bowles, D. J., Schnarrenberger, C., and Kauss, H. (1976). *Biochem. J.* **160**, 375–382.
Bowman, E. J., and Ikuma, H. (1976). *Plant Physiol.* **58**, 433–437.
Bowman, E. J., Ikuma, H., and Stein, H. J. (1976). *Plant Physiol.* **58**, 426–432.
Brierly, G. P. (1976a). *In* "Mitochondria: Bioenergetics, Biogenesis and Membrane Structure" (L. Packer and A. Gòmez-Puyou, eds.), pp. 3–21. Academic Press, New York.
Brierley, G. P. (1976b). *Mol. Cell. Biochem.* **10**, 41–62.
Briggs, W. R., and Rice, H. V. (1972). *Annu. Rev. Plant Physiol.* **23**, 293–334.
Butler, R. D., and Simon, E. W. (1971). *Adv. Gerontol. Res.* **3**, 73–129.
Campbell, L. C., Raison, J. K., and Brady, C. J. (1976). *Bioenergetics* **8**, 121–129.
Carafoli, E. (1976). *In* "Mitochondria: Bioenergetics, Biogenesis and Membrane Function" (L. Packer and A. Gómez-Puyou), pp. 47–60. Academic Press, New York.
Carafoli, E., Rossi, C. S., and Lehninger, A. L. (1965). *J. Biol. Chem.* **240**, 2254–2261.
Carmelli, C., and Biale, J. B. (1970). *Plant Cell Physiol.* **11**, 65–82.
Cedel, T. E., and Roux, S. J. (1977). *Plant Physiol.* **59**, S-101.
Chance, B., and Williams, G. R. (1956). *Adv. Enzymol.* **17**, 65–134.
Chapman, E. A., and Graham, D. (1974a). *Plant Physiol.* **53**, 879–885.
Chapman, E. A., and Graham, D. (1974b). *Plant Physiol.* **53**, 886–892.
Chen, C.-H., and Lehninger, A. L. (1973). *Arch. Biochem. Biophys.* **157**, 183–196.
Chou, K.-H., and Splittstoesser, W. E. (1972). *Plant Physiol.* **49**, 550–554.

Chrispeels, M. J., Vatter, A. E., and Hanson, J. B. (1966). *J. R. Microbiol. Soc.* **85**, 29–44.

Clowes, F. A. L., and Juniper, B. E. (1964). *J. Exp. Bot.* **15**, 622–630.

Clowes, F. A. L., and Juniper, B. E. (1968). "Plant Cells." Botanical Monographs, Vol. 8 (J. H. Burnett, ed). Blackwell, Oxford.

Coleman, J. O. D., and Palmer, J. M. (1971). *FEBS Lett.* **17**, 203–208.

Cooper, T. G., and Beevers, H. (1969). *J. Biol. Chem.* **244**, 3507–3513.

Cox, G. F., and Davies, D. D. (1967). *Biochem. J.* **105**, 729–734.

Criddle, R. S., Packer, L., and Shieh, P. (1977). *Proc. Natl. Acad. Sci. U.S.A.* **74**, 4306–4310.

Crompton, J., and Laties, G. G. (1971). *Arch Biochem. Biophys.* **143**, 143–150.

Davies, D. D. (1956). *J. Exp. Bot.* **7**, 203–218.

Davies, D. D., and Kenworthy, P. (1970). *J. Exp. Bot.* **21**, 247–257.

Davies, D. D., and Patil, K. D. (1975). *Planta* **126**, 197–211.

Davies, D. D., and Teixeira, A. N. (1975). *Phytochem.* **14**, 647–656.

Day, D. A., and Hanson, J. B. (1977a). *Plant Physiol.* **59**, 139–144.

Day, D. A., and Hanson, J. B. (1977b). *Plant Physiol.* **59**, 630–635.

Day, D. A., and Hanson, J. B. (1978). *Plant Sci. Lett.* **11**, 99–104.

Day, D. A., and Wiskich, J. T. (1974a). *Plant Physiol.* **53**, 104–109.

Day, D. A., and Wiskich, J. T. (1974b). *Plant Physiol.* **54**, 360–363.

Day, D. A., and Wiskich, J. T. (1975). *Arch. Biochem. Biophys.* **171**, 117–123.

Day, D. A., and Wiskich, J. T. (1977a). *Plant Sci. Lett.* **9**, 33–36.

Day, D. A., and Wiskich, J. T. (1977b). *Phytochem.* **16**, 1449–1451.

Day, D. A., and Wiskich, J. T. (1978). *Biochem. Biophys. Acta* **501**, 396–404.

Day, D. A., Rayner, J. R., and Wiskich, J. T. (1976). *Plant Physiol.* **58**, 38–42.

Day, D. A., Bertagnolli, B. L., and Hanson, J. B. (1978). *Biochim. Biophys. Acta* **502**, 289–297.

DeSantis, A., Borraccino, G., Arrigoni, O., and Palmieri, F. (1975). *Plant Cell Physiol.* **16**, 911–923.

DeSantis, A., Arrigoni, O., and Plamieri, F. (1976). *Plant Cell Physiol.* **17**, 1221–1233.

Dittrich, P., Campbell, W. H., and Black, C. C. (1973). *Plant Physiol.* **52**, 357–361.

Dizengremel, P. (1977). *Plant Sci. Lett.* **8**, 283–289.

Dizengremel, P., and Lance, C. (1976). *Plant Physiol.* **58**, 147–151.

Donaldson, R. L., and Beevers, H. (1977). *Plant Physiol.* **59**, 259–263.

Douce, R., and Bonner, W. D., Jr. (1972). *Biochem. Biophys. Res. Commun.* **47**, 619–624.

Douce, R., Christensen, E. L., and Bonner, W. D., Jr. (1972a). *Biochim. Biophys. Acta* **275**, 148–160.

Douce, R., Mannella, C. A., and Bonner, W. D., Jr. (1972b). *Biochem. Biophys. Res. Commun.* **49**, 1504–1509.

Douce, R., Mannella, C. A., and Bonner, W. D., Jr. (1973). *Biochim. Biophys. Acta* **292**, 105–116.

Douce, R., Moore, A. L., and Neuberger, M. (1977). *Plant Physiol.* **60**, 625–628.

Duckett, J. G., and Toth, R. (1977). *Ann. Bot.* **41**, 903–912.

Earnshaw, M. J. (1975). *FEBS Lett.* **59**, 109–112.

Earnshaw, M. J. (1977). *Phytochem.* **16**, 181–184.

Earnshaw, M. J., and Hanson, J. B. (1973). *Plant Physiol.* **52**, 403–406.

Earnshaw, M. J., and Hughes, E. A. (1976). *Plant Sci. Lett.* **6**, 343–348.

Earnshaw, M. J., Madden, D. M., and Hanson, J. B. (1973). *J. Exp. Bot.* **24**, 828–840.

Elzam, O. E., and Hodges, T. K. (1968). *Plant Physiol.* **43**, 1108–1114.

Flowers, T. J., and Hanson, J. B. (1969). *Plant Physiol.* **44**, 939–945.

Forde, B. G., Gunning, B. E. S., and John, P. C. L. (1976). *J. Cell Sci.* **21**, 329–340.

Gegenbach, B. G., Koeppe, D. E., and Miller, R. J. (1973a). *Physiol. Plant.* **29**, 103–107.

Gegenbach, B. G., Miller, R. J., Koeppe, D. E., and Arntzen, C. J. (1973b). *Can. J. Bot.* **51**, 2119–2125.

Georgevich, G., Cedel, T. E., and Roux, S. J. (1977). *Proc. Natl. Acad. Sci. U.S.A.* **74,** 4439–4443.

Graesser, R. J., and Wilson, R. H. (1977). *Plant Physiol.* **59,** 126–128.

Grubmeyer, C., and Spencer, M. (1978). *Plant Physiol.* **61,** 567–569.

Grubmeyer, C., Duncan, I., and Spencer, M. (1977). *Can. J. Biochem.* **55,** 812–818.

Gunning, B. E. S., and Steer, M. W. (1975). "Ultrastructure and the Biology of Plant Cells." Arnold, London.

Hackett, D. P. (1961). *Plant Physiol.* **36,** 445–452.

Hampp, R., and Schmidt, H. W. (1977). *Z. Pflanzenphysiol.* **82,** 68–77.

Hanson, J. B., and Hodges, T. K. (1967). *Curr. Top. Bioenerg.* **2,** 65–98.

Hanson, J. B., and Koeppe, D. E. (1975). *In* "Ion Transport in Plant Cells and Tissues" (D. A. Baker and J. L. Hall, eds.), pp. 79–99. Amer. Elsevier, New York.

Hanson, J. B., and Miller, R. J. (1967). *Proc. Natl. Acad. Sci. U.S.A.* **58,** 727–734.

Hanson, J. B., Wilson, C. M., Chrispeels, M. J., Kruger, W. A., and Swanson, H. A. (1965). *J. Exp. Bot.* **16,** 282–293.

Hanson, J. B., Bertagnolli, B. L., and Shepherd, W. D. (1972). *Plant Physiol.* **50,** 347–354.

Haustein, W. G. (1976). *Biochim. Biophys. Acta* **456,** 129–148.

Hatch, M. D., and Kagawa, T. (1976). *Arch. Biochem. Biophys.* **175,** 39–53.

Hatch, M. D., and Osmond, C. B. (1976). *In* "Encyclopedia of Plant Physiology, Transport in Plants III" (C. R. Stocking and U. Heber, eds.), Vol. 3, pp. 144–184. Springer-Verlag, Berlin and New York.

Hatefi, Y., Haavik, A. G., Fowler, L. R., and Griffiths, D. E. (1962). *J. Biol. Chem.* **237,** 2661–2669.

Henry, M. F., and Nyns, E. J. (1975). *Sub-Cell Biochem.* **4,** 1–65.

Hensley, J. R., and Hanson, J. B. (1975). *Plant Physiol.* **56,** 13–18.

Hinkle, P. (1976). *In* "Mitochondria; Bioenergetics, Biogenesis and Membrane Structure" (L. Packer and A. Gómez-Puyou, eds.), pp. 183–192. Academic Press, New York.

Hodges, T. K., and Hanson, J. B. (1965). *Plant Physiol.* **40,** 101–109.

Huang, A. H. C., and Beevers, H. (1971). *Plant Physiol.* **48,** 637–641.

Hunt, L., and Fletcher, J. (1977). *Plant Sci. Lett.* **10,** 243–247.

Ikuma, H. (1972). *Annu. Rev. Plant Physiol.* **23,** 419–436.

Jackson, P. C., Hendricks, S. B., and Vasta, B. M. (1962). *Plant Physiol.* **37,** 8–17.

Jagendorf, A., and Uribe, E. (1966). *Proc. Natl. Acad. Sci. U.S.A.* **55,** 170–177.

Janovitz, A., Chavez, E., and Clapp, M. (1976). *Arch. Biochem. Biophys.* **173,** 264–268.

Johnson, H. M., and Wilson, R. H. (1973). *Am. J. Bot.* **60,** 858–862.

Jung, D. W., and Hanson, J. B. (1973a). *Arch. Biochem. Biophys.* **158,** 139–148.

Jung, D. W., and Hanson, J. B. (1973b). *Biochim. Biophys. Acta* **325,** 189–192.

Jung, D. W., and Hanson, J. B. (1975). *Arch. Biochem. Biophys.* **168,** 358–368.

Jung, D. W., and Laties, G. G. (1976). *Plant Physiol.* **57,** 583–588.

Juniper, B. E., and Clowes, F. A. L. (1965). *Nature (London)* **208,** 864–865.

Kano, H., and Kageyama, M. (1977). *Plant Cell Physiol.* **18,** 1149–1153.

Khokhlova, L. P., Eliseeva, N. S., Stapishina, E. A., Bondar, I. G., and Suleimanova, I. G. (1975). *Sov. Plant Physiol.* **22,** 723–729.

Kimpel, J. A., and Hanson, J. B. (1977). *Plant Physiol.* **60,** 933–934.

Kimpel, J. A., and Hanson, J. B. (1978). *Plant Sci. Lett.* **11,** 329–335.

Kirk, B. I., and Hanson, J. B. (1973). *Plant Physiol.* **51,** 357–362.

Klingenberg, M. (1970). *In* "Essays in Biochemistry" (P. N. Campbell and F. Dickens, eds.), Vol. 6, pp. 119–159. Academic Press, New York.

Klingenberg, M (1977). *In* "Structure and Function of Energy Transducing Membranes" (K. van Dam and B. F. van Gelder, eds.), pp. 275–282, Elsevier, Amsterdam.

Koeppe, D. E., and Miller, R. J. (1972). *Plant Physiol.* **49,** 353–357.

Koeppe, D. E., Miller, R. J., and Bell, D. T. (1973). *Agron. J.* **65**, 566–569.

Kolodner, R., and Tewari, K. K. (1972). *Proc. Natl. Acad. Sci. U.S.A.* **69**, 1830–1834.

Lance, C. (1974). *Plant Sci. Lett.* **2**, 165–171.

Lance, C., and Bonner, W. D., Jr. (1968). *Plant Physiol.* **43**, 756–766.

LaNoue, K. F., and Tischler, M. E. (1976). *In* "Mitochondria: Bioenergetics, Biogenesis and Membrane Structure" (L. Packer and A. Gómez-Puyou, eds.), pp. 61–78. Academic Press, New York.

Laties, G. G. (1973). *Biochemistry* **12**, 3350–3355.

Laties, G. G., and Treffry, T. (1969). *Tissue Cell* **1**, 575–592.

Lea, P. J., and Norris, R. D. (1977). *Prog. Phytochem.* **4**, 121–167.

Leaver, C. J., and Harmey, M. A. (1973). *Biochem. Soc. Symp.* **38**, 175–193.

Leaver, C. J., and Harmey, M. A. (1976). *Biochem. J.* **157**, 275–277.

Leaver, C. J., and Pope, P. K. (1977). *In* "Nucleic Acids and Protein Synthesis in Plants" (L. Bogorad and J. H. Weil, eds.), pp. 213–237. Plenum, New York.

Leenders, H. J., Berendes, H. D., Helmsing, P. J., Derksen, J., and Koninkx, J. F. J. G. (1974). *Sub-Cell Biochem.* **3**, 119–147.

Lehninger, A. L. (1974). *Proc. Natl. Acad. Sci. U.S.A.* **71**, 1520–1524.

Lehninger, A. L., Reynafarje, B., and Alexandre, A. (1977). *In* "Structure and Function of Energy-Transducing Membranes" (K. van Dam and B. F. van Gelder, eds.), pp. 95–106. Elsevier, Amsterdam.

Lorimer, G. H., and Miller, R. J. (1969). *Plant Physiol.* **44**, 839–844.

Lund, H. A., Vatter, A. E., and Hanson, J. B. (1958). *J. Biochem. Biophys. Cytol.* **4**, 87–98.

Lyons, J. M. (1973). *Annu. Rev. Plant Physiol.* **24**, 445–466.

Lyons, J. M., and Raison, J. K. (1970). *Plant Physiol.* **45**, 386–389.

Macrae, A. R. (1971a). *Biochem. J.* **122**, 495–501.

Macrae, A. R. (1971b). *Phytochem.* **10**, 2343–2347.

Macrae, A. R., and Moorhouse, R. (1970). *Eur. J. Biochem.* **16**, 96–102.

Malhotra, S. S., and Spencer, M. (1970). *Plant Physiol.* **46**, 40–44.

Malone, C. P., Koeppe, D. E., and Miller, R. J. (1974). *Plant Physiol.* **53**, 918–927.

Manabe, K., and Furuya, M. (1974). *Plant Physiol.* **53**, 343–347.

Manabe, K., and Furuya, M. (1975). *Planta* **123**, 207–215.

Mannella, C. A., and Bonner, W. D., Jr. (1975a). *Biochim. Biophys. Acta* **413**, 213–225.

Mannella, C. A., and Bonner, W. D., Jr. (1975b). *Biochim. Biophys. Acta* **413**, 226–233.

Marmé, D. (1977). *Annu. Rev. Plant Physiol.* **28**, 173–198.

Mayer, A. M., and Shain, Y. (1974). *Annu. Rev. Plant Physiol.* **25**, 167–193.

Mazliak, P. (1973). *Annu. Rev. Plant Physiol.* **24**, 287–310.

Mazliak, P. (1977). *In* "Lipids and Lipid Polymers in Higher Plants" (M. Tevini and H. K. Lichtenthaler, eds.), pp. 48–74. Springer-Verlag, Berlin and New York.

Mazliak, P., and DeCotte, A. M. (1976). *J. Exp. Bot.* **27**, 769–777.

McCarty, R. E., Douce, R., and Benson, A. A. (1973). *Biochim. Biophys. Acta* **316**, 266–270.

McGivan, J. D., and Klingenberg, M. (1971). *Eur. J. Biochem.* **20**, 392–399.

McGivan, J. D., Grebe, K., and Klingenberg, M. (1971). *Biochem. Biophys. Res. Commun.* **45**, 1533–1541.

McNeil, P. H., and Thomas, D. R. (1976). *J. Exp. Bot.* **27**, 1163–1180.

Miflin, B. J., and Lea, P. J. (1977a). *Progr. Phytochem.* **4**, 1–26.

Miflin, B. J., and Lea, P. J. (1977b). *Annu. Rev. Plant Physiol.* **28**, 299–329.

Millard, D. L., Wiskich, J. T., and Robertson, R. N. (1965). *Plant Physiol.* **40**, 1129–1135.

Miller, R. J., and Koeppe, D. E. (1971). *Science* **173**, 67–69.

Miller, R. J., Dumford, S. W., Koeppe, D. E., and Hanson, J. B. (1970). *Plant Physiol.* **45**, 649–653.

Miller, R. W., de la Roche, I., and Pomeroy, M. K. (1974). *Plant Physiol.* **53**, 426–433.

Mitchell, P. (1966). "Chemiosmotic Coupling in Oxidative and Photosynthetic Phosphorylation." Glynn Research, Bodmin, Cornwall, England.

Mitchell, P. (1975). *FEBS Lett.* **56,** 1–6.

Mitchell, P., and Moyle, J. (1967). *Biochem. J.* **105,** 1147–1162.

Moore, A. L. (1978). *In* "Functions of Terminal Oxidases" (H. Degn, ed.), FEBS Colloquium B6, pp. 141–147. Pergamon, New York.

Moore, A. L., and Bonner, W. D., Jr. (1977). *Biochim. Biophys. Acta* **460,** 455–466.

Moore, A. L., Jackson, C., Halliwell, B., Dench, J. E., and Hall, D. O. (1977). *Biochem. Biophys. Res. Commun.* **78,** 483–491.

Moore, A. L., Bonner, W. D., Jr., and Rich, P. R. (1978). *Arch. Biochem. Biophys.* **186,** 298–306.

Moreau, F., and Lance, C. (1972). *Biochemie* **54,** 1227–1380.

Moreau, F., DuPont, J., and Lance, C. (1974). *Biochim. Biophys. Acta* **345,** 294–304.

Morisset, C. (1973). *C. R. Hebd. Seances Acad. Sci. Ser. D.* **276,** 311–314.

Moyle, J., and Mitchell, P. (1977a). *FEBS Lett.* **77,** 136–140.

Moyle, J., and Mitchell, P. (1977b). *FEBS Lett.* **84,** 135–140.

Nedakavukara, M. J. (1964). *J. Cell Biol.* **23,** 193–195.

Nelson, D. R., and Rinne, R. (1977). *Plant Cell Physiol.* **18,** 1021–1027.

Nicholls, D. G. (1974). *Eur. J. Biochem.* **50,** 305–315.

Nicholls, D. G. (1977). *Biochem. Soc. Trans.* **5,** 200–203.

Nir, I., Klein, S., and Poljakoff-Mayber, A. (1969). *Aust. J. Biol. Sci.* **22,** 17–33.

Nir, I., Poljakoff-Mayber, A., and Klein, S. (1970). *Plant Physiol.* **45,** 173–177.

Oestreicher, G., Hogue, P., and Singer, T. P. (1973). *Plant Physiol.* **52,** 622–626.

Öpik, H. (1973). *J. Cell Sci.* **12,** 725–739.

Öpik, H. (1974). *In* "Dynamic Aspects of Plant Ultrastructure" (A. W. Robards, ed.), pp. 52–83. McGraw-Hill, New York.

Osmond, C. B. (1976). *In* "Encyclopedia of Plant Physiology, Transport in Plants" (U. Lüttge and M. G. Pitman, eds.), Vol. IIA, pp. 347–372. Springer-Verlag, Berlin and New York.

Out, T. A., Valeton, E., and Kemp, A., Jr. (1976). *Biochim. Biophys. Acta* **440,** 697–710.

Palmer, J. M. (1976). *Annu. Rev. Plant Physiol.* **27,** 133–157.

Palmer, J. M., and Arron, G. P. (1976). *J. Exp. Bot.* **27,** 418–430.

Panter, R. A., and Mudd, J. B. (1973). *Biochem. J.* **134,** 655–658.

Papa, S., Lofrumento, N. E., Kanduc, D., Parradies, G., and Quagliariello, E. (1971). *Eur. J. Biochem.* **22,** 134–143.

Parsons, D. F., Bonner, W. D., Jr., and Verboon, J. G. (1965). *Can. J. Bot.* **43,** 647–655.

Passam, H. C., and Coleman, J. O. D. (1975). *J. Exp. Bot.* **26,** 536–543.

Penefsky, H. S. (1977). *J. Biol. Chem.* **252,** 2891–2899.

Petersen, P. L. (1975). *Bioenergetics* **6,** 243–275.

Peverly, J. B., Miller, R. J., Malone, C., and Koeppe, D. E. (1974). *Plant Physiol.* **54,** 408–411.

Phillips, M. L., and Williams, G. R. (1973). *Plant Physiol.* **51,** 667–670.

Plesnicar, M., Bonner, W. D., Jr., and Storey, B. T. (1967). *Plant Physiol.* **42,** 366–370.

Pomeroy, M. K. (1976). *Plant Physiol.* **57,** 469–473.

Pomeroy, M. K. (1977). *Plant Physiol.* **59,** 250–255.

Pring, D. R. (1974). *Plant Physiol.* **53,** 677–683.

Racker, E. (1976). "A New Look at Mechanisms in Bioenergetics." Academic Press, New York.

Raison, J. K., Lyons, J. M., and Campbell, L. C. (1973). *Bioenergetics* **4,** 397–408.

Ramirez-Mitchell, R., Johnson, H. M., and Wilson, R. H. (1973). *Exp. Cell Res.* **76,** 449–457.

Reid, E. D., Thompson, P., Lyttle, C. R., and Dennis, D. T. (1977). *Plant Physiol.* **59,** 842–848.

Reynafarje, B., and Lehninger, A. L. (1977). *Biochem. Biophys. Res. Commun.* **77,** 1273–1279.

Rich, P. R., and Bonner, W. D., Jr. (1978). *Biochim. Biophys. Acta* **501,** 381–395.

Rich, P. R., and Moore, A. L. (1976). *FEBS Lett.* **65**, 339–344.
Rich, P. R., Boveris, A., Bonner, W. D., Jr., and Moore, A. L. (1976). *Biochem. Biophys. Res. Commun.* **71**, 695–703.
Rich, P. R., Moore, A. L., Ingledew, W. J., and Bonner, W. D., Jr. (1977). *Biochim. Biophys. Acta* **462**, 501–514.
Rubin, P. M., and Randall, D. D. (1977). *Plant Physiol.* **60**, 34–39.
Saleno, J. C. Harmon, H. J., Blum, H., Leigh, J. S., and Ohnishi, T. (1977). *FEBS Lett.* **82**, 179–182.
Sato, S., and Asahi, T. (1975). *Plant Physiol.* **56**, 816–820.
Scandalios, J. C. (1974). *Annu. Rev. Plant Physiol.* **25**, 225–258.
Scherer, B., and Klingenberg, M. (1974). *Biochemistry* **13**, 161–170.
Schmidt, H. W., and Hampp, R. (1977). *Z. Pflanzenphysiol.* **82**, 428–434.
Siew, D., and Klein, S. (1968). *J. Cell Biol.* **37**, 590–596.
Silva Lima, M., Denslow, N. D., Fernandes de Melo, D. (1977). *Physiol. Plant.* **41**, 193–196.
Simon, E. W., and Chapman, J. A. (1961). *J. Exp. Bot.* **12**, 414–420.
Singh, J., de la Roche, A. I., and Simnovich, D. (1977). *Plant Physiol.* **60**, 713–715.
Solomos, T. (1977). *Annu. Rev. Plant Physiol.* **28**, 279–297.
Sottibandhu, R., and Palmer, J. M. (1976). *Biochem. J.* **152**, 637–645.
Sperk, G., and Tuppy, H. (1977). *Plant Physiol.* **59**, 155–157.
Splittstoesser, W. E., and Stewart, S. A. (1970). *Physiol. Plant.* **23**, 1119–1129.
Stoner, C. D., and Hanson, J. B. (1966). *Plant Physiol.* **41**, 255–266.
Stoner, C. D., and Sirak, H. D. (1973). *J. Cell Biol.* **56**, 51–64.
Storey, B. T. (1976). *Plant Physiol.* **58**, 521–526.
Synenki, R. M., Levings, C. S., III, and Shah, D. M. (1978). *Plant Physiol.* **61**, 460–464.
Takeuchi, Y. (1975). *Biochim. Biophys. Acta* **376**, 505–518.
Takeuchi, Y., Yoshida, K., and Sato, S. (1969). *Plant Cell Physiol.* **10**, 733–741.
Thomas, D. R., and McNeil, P. H. (1976). *Planta* **132**, 61–63.
Thomson, W. W., Raison, J. K., and Lyons, J. M. (1972). *Bioenergetics* **3**, 531–538.
Tolbert, N. E. (1971). *Annu. Rev. Plant Physiol.* **22**, 45–74.
Truelove, B., and Hanson, J. B. (1966). *Plant Physiol.* **41**, 1004–1013.
Van der Plas, L. H. W., Jobse, P. A., and Verlear, J. D. (1976). *Biochim. Biophys. Acta* **430**, 1–12.
Van de Staat, D., De Boer, B. L., and van Dam, K. (1973). *Biochim. Biophys. Acta* **292**, 338–344.
Vignais, P. V. (1976). *Biochim. Biophys. Acta* **456**, 1–38.
Vignais, P. V., Douce, R., Lauguin, G. J. M., and Vignais, P. M. (1976). *Biochim. Biophys. Acta* **440**, 688–696.
Walker, D. A., and Beevers, H. (1956). *Biochem. J.* **62**, 120–127.
Watrud, L. S., Baldwin, J. K., Miller, R. J., and Koeppe, D. E. (1975). *Plant Physiol.* **56**, 216–221.
Wedding, R. T., Black, M. K., and Papp, D. (1976). *Plant Physiol.* **58**, 740–743.
Wickes, W. A., and Wiskich, J. T. (1976). *Aust. J. Plant Physiol.* **3**, 153–162.
Wildman, S. G., Hongladarom, T., and Honda, S. I. (1962). *Science* **138**, 434–436.
Wilson, R. H., and Graesser, R. J. (1976). In "Encyclopedia of Plant Physiology, Transport in Plants" (C. R. Stocking and U. Heber, eds.), Vol. 3, pp. 377–397. Springer-Verlag, Berlin and New York.
Wilson, R. H., and Minton, G. A. (1974). *Biochim. Biophys. Acta* **333**, 22–27.
Wilson, R. H., Hanson, J. B., and Mollenhauer, H. H. (1969). *Biochemistry* **8**, 1203–1213.
Wilson, R. H., Dever, H., Harper, W., and Fry, R. (1972). *Plant Cell Physiol.* **13**, 1103–1111.
Wilson, R. H., Thurston, E. L., and Mitchell, R. (1973). *Plant Physiol.* **51**, 26–30.
Wilson, S. B., and Bonner, W. D., Jr. (1970). *Plant Physiol.* **46**, 31–35.

Wiskich, J. T. (1974). *Aust. J. Plant Physiol.* **1**, 177–181.
Wiskich, J. T. (1977). *Annu. Rev. Plant Physiol.* **28**, 45–69.
Woo, K. C., and Osmond, C. B. (1976). *Aust. J. Plant Physiol.* **3**, 771–785.
Woo, K. C., and Osmond, C. B. (1977). *Plant Cell Physiol.* **18**, 315–323.
Wulf, R., Kalstein, A., and Klingenberg, M. (1978). *Eur. J. Biochem.* **82**, 585–592.
Yoshida, K. (1968). *J. Fac. Sci. Univ. Tokyo* **10**, 63–82.
Yoshida, K., and Sato, S. (1968). *J. Fac. Sci. Univ. Tokyo* **10**, 49–62.
Yoshida, K., and Takeuchi, Y. (1970). *Plant Cell Physiol.* **11**, 403–409.
Yu, C. A., Yu, L., and King, T. E. (1977). *Biochem. Biophys. Res. Commun.* **78**, 259–265.
Zaar, K. (1974). *Bioenergetics* **6**, 57–68.

Microbodies—Peroxisomes and Glyoxysomes | 9

N. E. TOLBERT

I. INTRODUCTION

Microbodies are respiratory subcellular organelles found in all eukaryotic plant and animal tissue. In plants, microbodies range in size from 0.5 to 1.5 μm in diameter, have a single bounding tripartite membrane, and have a granular protein matrix without lamelular membranes, as shown in Figs. 1 and 2. They may contain amorphorus or crystalline inclusions. Biochemi-

The Biochemistry of Plants, Vol. 1

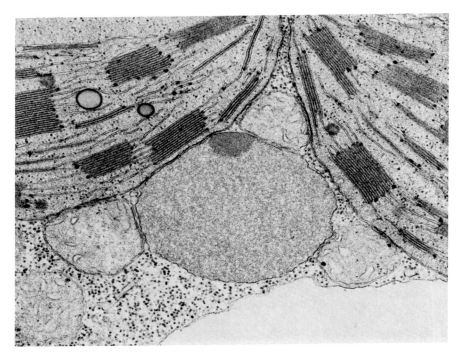

Fig. 1.   A leaf peroxisome among chloroplasts and mitochondria. This EM is from a tobacco leaf. Note the absence of microsomes in the peroxisomes, the single bounding peroxisomal membrane, and the granula peroxisomal matrix, × 40,000. From S. E. Frederick and E. Newcomb, University of Wisconsin.

cally microbodies are characterized by metabolic pathways associated with flavin-linked oxidases, which produce H_2O_2, and with catalase for removal of the H_2O_2. Cytologically they are detected by a stain for catalase. Microbodies from all biological tissues have many similarities and their general properties were summarized over a decade ago by de Duve and Baudhuin (1966), de Duve (1969), and Hruban and Rechcigl (1969). The metabolic pathways and enzymes in microbodies from leaves have been reviewed by Tolbert (1969, 1971a) and properties of microbodies from germinating fatty seeds were initially investigated by Beevers (1969, 1975). The morphology of plant microbodies was described by Newcomb's group (Frederick *et al.,* 1975). Other reviews on plant microbodies are to be found for algae (Tolbert, 1972), fungi (Maxwell *et al.,* 1977), protozoa (Müller, 1975), and in general (Vigil, 1973; Coleman, 1977; Tolbert, 1973, 1978; Gerhardt, 1978).

Metabolic pathways in microbodies are catabolic, yet the end products may be used in the cell for gluconeogenesis or other synthetic processes.

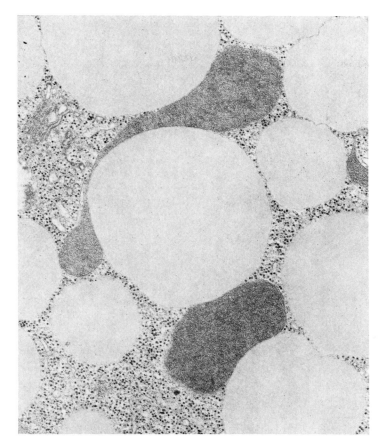

Fig. 2. Glyoxysomes associate with lipid bodies in a cotyledon cell of a tomato seedling. The appressed and distorted shape of the pliable glyoxysome is assumed to be the result of compression by the lipid bodies (Gruber *et al.*, 1970). × 29,000.

One of the reactions of the microbody, presumed to initiate, or control the rate of its metabolic pathway, is catalyzed by an irreversible flavin-containing oxidase coupled to oxygen uptake and H_2O_2 production. The H_2O_2 is destroyed by catalase with loss of energy by the system as heat. During mitochondrial oxidation, linked by electron transport to cytochrome c oxidase, part of the energy is biologically conserved in ATP synthesis. In contrast O_2 uptake in microbodies is due to the flavin oxidase located in its matrix, and there is no cytochrome-linked membrane-bound electron transport system for ATP synthesis. However, the energy of other oxidative steps in the microbody may be linked to NAD reduction and conserved by shuttles to other parts of the cell. In some cases only a portion of a pathway may be in the microbody and the rest is to be found elsewhere in the cell, i.e., in the cytoplasm or mitochondria. Although metabolic pathways associated with microbodies have been considered as degradative or catabolic, the products are part of pathways that may be essential for growth and development within the cell at other locations.

Nomenclature for Microbodies and Associated Respiration

Microbody is a name used for an organelle with the properties listed in Table I. Cytochemically a microbody is characterized by the DAB stain for catalase. On the basis of physiological and biochemical partial characterization of isolated microbodies from different tissues, they have been given other more specific names in addition, such as peroxisomes from leaves (Fig. 1) or glyoxysomes (Fig. 2) from germinating fatty seeds. The term, peroxisome, was proposed by de Duve (1966, 1969) for the particle that produced and consumed hydrogen peroxide. Their assay was based on the peroxidation by catalase of $HCOOH$ to CO_2 with added H_2O_2, which *in vivo* would be formed by a flavin oxidase. With this concept one substrate is oxidized by a flavin oxidase with O_2 to form H_2O_2 and a second substrate is oxidized with catalase by the H_2O_2. However, little *in vivo* evidence has been found for such a peroxidative type of metabolism in the peroxisomes of animals or plants. The term, leaf peroxisome, was adapted by Tolbert (1971a), because of morphological and enzymatic similarities of microbodies from leaves with de Duve's description of the organelle. Microbodies in germinating fatty seeds also met de Duve's characterization, but were called glyoxysomes by Breidenbach and Beevers (1967) and Beevers (1969), because one of their metabolic pathways was the glyoxylate cycle. Since then the term glyoxysome has been used only for microbodies with at least malate

TABLE I

General Characteristics of Plant Microbodies (Peroxisomes and Glyoxysomes)

Spheroides of about 0.5–1.5 μm diameter, but contorted and appressed among other particles
Single bounding membrane
Dense stroma or granular matrix which stains cytochemically for catalase
A core that may have a crystalloid structure.
Equlibrium sucrose density at 1.24–1.26 gm/cm^3
Apparently permeable to most small substrates
Contain catalase and different flavin oxidases that are part of specific metabolic pathways
About half as numerous as mitochondria and contain about 1–1.5% of the total leaf protein
Formed by budding from the smooth endoplasmic reticulum; contain no nucleic acids
Development and content may be regulated by substrate availability in some cases
In cotyledons glyoxysomes develop during germination and in leaves peroxisomes increase during greening

---

synthetase or isocitrate lyase, two unique enzymes of the glyoxylate cycle. Electron microscopists have observed microbodies from other plant tissues, such as roots, developing seeds, tubers, algae, and fungi, but generally enzymological characterization of these particles has not been extensive. As far as they have been examined, an α-hydroxy acid oxidase (or glycolate oxidase) and catalase are present, so they are often referred to either as microbodies or peroxisomes. Microperoxisomes of 0.1–0.5 μm diameter, as described in animal tissue (Novikoff and Novikoff, 1973), have not been reported in plant tissue. At present the terms in use are peroxisome or microbody for the organelle when it is known to contain glycolate oxidase and catalase; glyoxysome or microbody when it contains any part of the glyoxylate cycle, even though glycolate oxidase is also present; and microbody when referring to general properties of either peroxisomes or glyoxysomes or when the particle has not been biochemically characterized from the designated tissue.

There is no physiological nomenclature for microbody respiration. However, it must be significant and different from mitochondrial respiration or other O_2 uptake processes in the cell. The amount of total cellular respiration attributed to the microbodies has not been measured, but it should vary greatly in different tissues, at different stages of development, and at different periods of the day. The term, photorespiration, for glycolate biosynthesis during photosynthesis in the chloroplasts and its oxidation in the peroxisomes and mitochondria has been considered to be a manifestation of peroxisomal respiration (Tolbert, 1971a). However, since various parts of photorespiration or the glycolate pathway occur in the chloroplast, peroxisomes, and mitochondria, even this one respiratory process, which conceptually can be separated from other cellular processes, cannot be designated as simply peroxisomal respiration (see this series, Vol. 2, Chapter 12). The amount of oxygen uptake during seed germination that is attributed to glyoxysomal metabolism has not been quantitated, but since fatty acid degradation during germination occurs in the glyoxysomes, it too must be significant.

II. MORPHOLOGICAL DETECTION AND DESCRIPTION

After about 1954, with the introduction of osmium tetroxide as a fixative, the term microbody was used for a particle of unknown function bounded by a single membrane. Prior to that time the organelle went undetected because the microbody without internal membranes does not stand out in contrast in the cytoplasm, and the biochemist broke up the particle during homogenation of the tissue. Early cytological reports of them in plant tissue (i.e. Mollenhauer *et al.,* 1966) were not pursued until their function as glyoxysomes and peroxisomes was recognized. Detection of plant microbodies by electron microscopy (Frederick *et al.,* 1973, 1975; Gruber *et al.,* 1970, 1972; Vigil, 1973) is based on fixation in buffered 3% glutaraldehyde and post fixation in 2% OsO_4, which binds to unsaturated fatty acids in the lipid membranes and increases opacity. In the DAB cytological stain for catalase, the tissue is incubated after glutaraldehyde fixation with an alkaline buffer containing 3,3'-diaminobenzidine (DAB) and H_2O_2 (Novikoff and Goldfischer, 1969). Catalase inside the organelles is not inactivated by the glutaraldehyde and can peroxidatively oxidize the DAB to an insoluble compound, which is deposited in the microbody and results in a black or opaque organelle in the electron micrograph. There is also a cytological stain for α-hydroxyacid oxidase and for malate synthase (Burke and Trelease, 1975). Catalase and glycolate oxidase are distributed throughout the granular matrix of the microbody. Sometimes the core or crystalloid of the microbodies may be sliced by the plane of the field to reveal its crystalline structure, which has the DAB stain for catalase (Frederick and Newcomb, 1968). Thus, the core of plant microbodies is said to contain catalase, but what else may be present is not known. The core in microbodies from rats and mice contains urate oxidase, which is quite insoluble and apparently crystallizes out of solution. However, urate oxidase is not present in plant microbodies except in trace amounts. No explanation has been given for a core of catalase in plant microbodies. No physiological parameters have been established between the presence or absence of cores in leaf peroxisome. Fewer cores have been observed in glyoxysomes from germinating seeds than from leaf peroxisomes. The core may be a storage protein, yet there is no evidence for its accumulation or utilization as such.

It is not clear whether catalase is confined only to the microbodies of plant cells. The DAB stain for catalase occurs after mild glutaraldehyde fixation of the tissue in a manner that would inactivate cytoplasmic enzymes, but not catalase protected inside an organelle. Thus, it is known that catalase is not in chloroplasts and low levels of DAB peroxidation in the membranes of the cell, as in the mitochondria, could be explained by some peroxidative activity of the cytochrome system. Upon homogenization of plant tissue over half of the catalase is present in the soluble fraction, and some catalase activity is present in all subcellular fractions. This may be due to breakage of the fragile microbodies during homogenization and adherence of catalase to membranes, but it is also possible that there is a significant cytoplasmic pool of catalase. Similarly a large percentage of the activity of the other microbody enzymes cited in this chapter may be found at times in the soluble or cytosolic fraction after homogenation of plant tissue. Most investigators have assumed that the soluble activities were due to rupture of the particles, but this has not been proven.

III. ISOLATION PROCEDURES

Microbodies are not easily isolated because of their inherent fragility due to a single bounding membrane and the absence of inner membranes. At best only partial recovery of the particles after homogenization can be achieved. Destruction of the particle or partial loss of its matrix enzymes seems to occur during even the most gentle of grinding procedures. Chopping with razor blades or limited grinding with mortar and pestle or blender has been used for isolation from tissues that are easy to homogenize. Almost all of the work with leaves has been done with spinach plants, but for work on seeds, castor beans, watermelon, sunflower, flax, and wheat have been used. Microbody isolation from tough leaves, such as grasses and C_4 plants, has not been done extensively, and then only low yields were obtained (Tolbert et al., 1969). Even with spinach leaves, recovery of peroxisomes in the particulate fraction is a low 10–25%, so quantitation has been based on total activity of microbody marker enzymes in separate portions of tissue that have been completely homogenized. On the other hand, chloroplasts and mitochondria, isolated by simple differential centrifugation, are contaminated with peroxisomes, peroxisomal fragments, or peroxisomal enzymes adhering to their membranes. This fact is emphasized by the presence of catalase activity in nearly all particulate preparations, even though catalase has not been detected cytochemically in these particles.

Detailed procedures for isolating microbodies have been published (Leighton et al., 1968; Tolbert, 1971b, 1974; Beevers and Breidenbach, 1974). After limited, careful homogenization, the organelles in the homogenate or in a resuspension of all particles after an initial differential centrifugation are separated by equilibrium, or isopycnic, sucrose density gradient centrifugation. Microbodies band at a density between 1.24 and 1.26 g/cm³, whereas the other more lipid-rich particles with internal membranes, such as chloroplasts and mitochondria, band at lower sucrose densities in the upper part of the gradient (Table II). Nonlinear sucrose gradients are designed to provide maximum separation of the microbodies from broken chloroplasts and mitochondria by having minimal density changes per fraction over an extended volume of the gradient. Gradients of large volume are preferred by the use of 50-ml tubes or zonal rotors. Cytoplasmic proteins and enzymes lost from the broken microbodies remain at the top of the sucrose gradient.

Density gradient isolation of microbodies takes advantage of two properties of the microbodies. They are a protein-rich particle and they lose water to the concentrated sucrose gradient during centrifugation more rapidly than do the other organelles. As a result the microbodies move down in the gradient to a high sucrose density, but the isolated particles are dehydrated and have higher specific density than they would have *in vivo*. Prolonged

TABLE II

Density and Marker Enzymes for Subcellular Organelles as Separated on Sucrose Density Gradients

Organelle and marker enzymes	d_{10}° Density (g/cm^3)
Starch grains	Bottom of gradient
Protein bodies	$1.25 - 1.36$
Protein peak, protease activity	
Microbodies	$1.24 - 1.26$
Catalase, glycolate oxidase, NADH-hydroxypyruvate reductase, malate synthetase, isocitrate lyase	
Etioplasts and proplastids	$1.22 - 1.25$
Triose-P isomerase, dihydroxyphenylalanine oxidase, P-glycolate phosphatase	
Whole chloroplasts	$1.18 - 1.22$
Chlorophyll	
Mitochondria	$1.16 - 1.18$
Cytochrome c oxidase, glutamate dehydrogenase, succinate dehydrogenase	
Broken chloroplasts	$1.14 - 1.17$
Chlorophyll	
Microsomes	$1.12 - 1.14$
NADH : Cytochrome b_5 reductase	
Cytoplasm	At top of gradient

centrifugation for longer than 2–3 h or the use of too high centrifugational force result in microbody breakage, and the other particles will lose more of their bound water. These changes result in a broadening of the peaks for each organelle. Only sucrose gradients are used routinely to isolate microbodies because materials of higher molecular weight, such as Ficol, are not as effective for rapid extraction of the water from the microbodies, and as a result the microbodies do not equilibrate separately from the mitochondria and chloroplasts.

The immense bulk of chloroplasts poses special problems for isolation of leaf peroxisomes in that the large chloroplasts quickly form a dense, green, lipid band in the gradient during centrifugation, through which the smaller and slower moving peroxisomes must pass. To minimize this holdup the gradients cannot be overloaded and should be accelerated slowly to prevent early packing of the chloroplasts. Also the gradients must be broad through the area that retains the chloroplasts in order to keep them from forming too dense a band.

Electron microscopic examination of the isolated microbodies indicates a similar appearance to the particle *in situ,* except that the isolated particles are spherical, whereas in the cell the particle may be pushed into contorted shapes. The isolated microbody fraction is never 100% pure. Examination of the isolated microbody fraction by electron microscopy and assays by marker enzymes indicate that the main contaminants are protein bodies and lipid poor proplastids, as well as variable amounts of endoplasmic reticulum, perhaps attached or sticking to the microbodies (Schnarrenberger *et al.,* 1971). Consequently, reported specific activities for the enzymes in the so-called isolated microbody fractions are always considerably lower than their true values.

To locate the microbodies on the sucrose gradient and to quantitate the contaminating organelles, aliquots of each gradient fraction are analyzed for organelle-specific, marker enzymes. The markers most often used for the various organelles are cited in Table II. The peroxisomal fraction is delineated by its catalase activity and sometimes, in addition, glycolate oxidase or NADH:hydroxypyruvate reductase is measured for confirmation. The glyoxysomal marker enzymes are malate synthetase and/or isocitrate lyase along with catalase. Either glutamate dehydrogenase, cytochrome c oxidase, or succinate oxidase can be used as a marker enzyme for the mitochondria, and chlorophyll marks the chloroplasts. If the gradient is not overloaded at least two bands of chlorophyll are present due to broken (less dense) and whole (more dense) chloroplasts. Selection of the marker enzymes depends upon assay facilities and, of course, only organelle specific enzymes can be used. Thus characterization of a gradient to locate the peroxisome-enriched fraction requires measurements of sucrose density, the protein concentration for reporting specific activities, and assay profiles of the marker enzymes, in order to obtain microbody fractions with the highest enzymatic specific activity and lowest contamination.

After the microbodies have been partially isolated by a sucrose density gradient, they may be broken and recentrifuged by a second gradient into matrix, core, and membrane fractions. Methods of breakage have been combinations of osmotic shock, freeze-thaw, sonication, detergent, or dilution into salt solutions or into pyrophosphate buffer followed by standing overnight. In some cases particle ghosts are obtained which appear to have some of the protein of the matrix clumped or encapsulated inside the membranes. In addition the original contaminating ER remain with the microbody membrane fraction, so that studies of the composition of the microbody membrane are not as definitive as desired.

IV. ENZYMES IN MICROBODIES—CATALASE

Assays for enzymes so far reported in microbodies from plant tissue are in *Methods in Enzymology* (Tolbert, 1971b, 1974; Beevers and Breidenbach,

1974). Reported specific activities, where available, are low due to variable amounts of contaminating material in the microbody fraction from the sucrose gradients. Two groups of enzymes in plant microbodies represent the two metabolic pathways described in Figs. 3 and 4 and summarized in the following sections. Not all of these enzymes are present at one time in microbodies from the same tissue. Part or all of the enzymes of the glyoxylate cycle are in glyoxysomes from germinating seeds, algae, and fungi (yeast and molds). A portion of the enzymes for fatty acid beta oxidation are in the microbody, and some of the enzymes for purine and pyrimidine catabolism have been reported in microbodies. Generally the microbodies from a tissue with a specific function (e.g., leaves and photorespiration) will have the enzymes associated only with that metabolic pathway, and the other enzymes are either completely missing or repressed so that only traces of them are present. This changing composition of the microbody enzymes is discussed further in Section X.

A pH optimum of 8.5 ± 0.2 is a general characteristic of the flavin oxidases from microbodies. The pH optimum of catalase and some other microbody enzymes are broad ranging from below 6 to above 8. As a first approximation, the pH optima for cytoplasmic enzymes are near neutrality (7–7.5); enzymes in lysosome or vacuoles of plants have acidic pH optima around 5–5.5. In the chloroplast stroma the enzyme pH optima range from 7.8–8.3, and the highest pH optima of 8.5 and over are for the flavin oxidases in microbodies. Thus the different cellular compartments may function over a 10^3 range of H^+ concentration (pH 5.5–8.5), but proof for or a role for this change in regulating cellular metabolism is not known.

Catalase is present in microbodies in large amount relative to the activity of any other enzymes in the organelle or even in the cell. By the DAB cytological test catalase appears to be present only in the peroxisomes, where it is a soluble matrix component as well as the crystalloid core of plant microbodies. Of the many reviews on catalase, recent ones by Schonbaum and Chance (1976) refer to general properties, by Aebi (1974) to the assay, and by Sies (1974) to its function in microbodies. Isolated microbody fractions from leaves have a catalase specific activity of over 4000 μmoles/min/mg peroxisomal protein and catalase may represent 10–25% of the peroxisomal protein.

Catalase has four subunits and a molecular weight of 240,000 with heme as its prosthetic group. Catalase first forms a H_2O_2 complex, and then catalyzes a peroxidative reaction either with another molecule of H_2O_2 to form two H_2O and one half O_2, or a peroxidative reaction with an organic substance, such as the oxidation of formate to CO_2, or ethanol to acetaldehyde (Sies, 1974). Catalase activity is generally based on the rate of H_2O_2 use as measured by the loss of absorption at 240 nm. The specific activity of catalase is

10^3–10^4 times greater than other microbody enzymatic activities. This is due to the large amount of catalase in microbodies, and to its high rate constant of around 10^7 moles H_2O_2 per mole of hematin per second at pH 7 and 30°. Nonsaturating V_{max} with increasing H_2O_2 concentration and K_m (H_2O_2) values of over 1 M, make rather meaningless comparisons between catalase and other peroxisomal enzymes which generate only low levels of H_2O_2. The amount of catalase indicate that it is always present in great excess. Catalase is inhibited by cyanide, but as this is not too specific, aminotriazole inhibition of catalase is also sometimes used.

Wide variations in the amount of catalase in the cell have been reported. In unicellular algae, which do not have glycolate oxidase to form H_2O_2, the catalase content is less than 10% of that in plants (Frederick *et al.*, 1973) and the microbodies are not as numerous as in a leaf. In *Euglena* little catalase activity has been detected, and it has a few microbodies. On the other hand, some anaerobic photosynthetic bacteria with no known functional need for catalase, may contain as much as 25% of their protein as catalase (Clayton, 1960). The reason for an apparent great excess of catalase in most microbodies is unknown nor is it known whether catalase is confined only to the microbodies of plant cells. Although catalase is considered the universal marker enzyme for microbodies, the particles in some fungi yield negative DAB cytochemical tests, as if they did not contain catalase (Maxwell *et al.*, 1977). Microbodies that can be defined as glyoxysomes were isolated from Neurospora with no catalase (Theimer *et al.*, 1978), although another population of microbodies in the same cells does contain catalase. These results may reflect the fact that the glyoxylate cycle alone does not contain any flavin H_2O_2 generating reaction, and caution against the universal use of catalase to detect microbodies.

Peroxisomes from spinach leaves do not contain superoxide dismutase (SOD) (R. Gee and N. E. Tolbert, unpublished), which is rather ubiquitous enzyme in other parts of the cell for protection against O_2 toxicity. The product formed by SOD is H_2O_2, yet the catalase to remove the H_2O_2 is in the microbody. The reason for such apparent compartmentation is not understood.

V. METABOLIC PATHWAYS IN LEAF PEROXISOMES

As far as examined, there are two main carbon pathways in peroxisomes from all photosynthetic leaf tissue of C_3 plants and the bundle sheath cells of C_4 plants. One is the irreversible conversion of glycolate to glycine or the glycolate pathway, and the other is the reversible interconversion between glycerate and serine or the glycerate pathway. These interlocking reactions

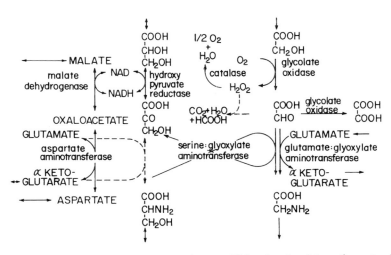

Fig. 3. Metabolic pathways in leaf peroxisomes. Right, the glycolate pathway to glycine; middle, the glycerate pathway between serine and glycerate; left, the malate shuttle. Only those reactions found in leaf peroxisomes are shown, and other reactions associated with photorespiration are in Vol. 2, Chapter 12.

and associated systems for transport in and out of the peroxisome are detailed in Fig. 3. They are a part of the overall process of photorespiration, which is further elaborated on in Vol. 2 of this series, Chapter 12.

A. Conversion of Glycolate to Glycine, Oxalate, or CO_2 and Formate

Peroxisomal glycolate oxidase in the leaf is the initiating flavin oxidase which forms H_2O_2 and which directs carbon flow to glycine during photorespiration. This oxidase, along with catalase, are the criteria of a microbody oxidase system. Glycolate oxidase has four equal subunits each with FMN as cofactor, a molecular weight of about 240,000, and a K_m for glycolate of 2×10^{-4} M. The active form is thought to have two subunits, since the tetramer as isolated has low activity. It catalyzes the oxidation of glycolate with O_2 to glyoxylate and H_2O_2 and can be assayed by the O_2 uptake. Glycolate oxidase has a high affinity for O_2 that is not saturated by aeration with 100% O_2. The reaction is often measured spectrophotometrically by coupling it with the reduction of 2,6-dichlorophenol indophenol under anaerobic conditions. In isolated peroxisomes from spinach or sunflower leaves its specific activity is about 1 μmole/min/mg of protein (Tolbert et al., 1969). The reaction with O_2 is irreversible with the loss of energy accounted for as H_2O_2 formation and its subsequent destruction by catalase. This enzyme is also called α-hydroxyacid oxidase, for it catalyzes the oxidation of L-lactate, but not D-lactate, at less than half the V_{max} for glycolate, as well as the oxidation of indole and phenylglycolate and lactate analogs. Longer chain α-hydroxyacids are not significantly oxidized except for α-hydroxyisocaproate which is oxidized as well as glycolate. The reason for this latter activity is unknown and the oxidation of L-lactate and the other substrates by leaf peroxisomes *in vivo* has not been documented.

Net oxygen uptake for the oxidation of one molecule of glycolate in the presence of catalase is one atom. In the absence of catalase, the H_2O_2 produced rapidly, nonenzymatically, and stoichiometrically further oxidizes glyoxylate to CO_2 from C_1 and formate from C_2. However, the excess of catalase is so great that only limited peroxidation of glyoxylate occurs in isolated, intact or broken, peroxisomes or in crude leaf preparations. Although de Duve's group originally used a peroxidase assay with $H^{14}COOH$ and H_2O_2 to detect peroxisomes, the extent of peroxidase activity *in vivo* by peroxisomal catalase is uncertain. Peroxisomes do not contain peroxidases, which are different heme proteins and located in the cell membrane. Since the flavin oxidases that generate H_2O_2 are in the microbodies with catalase, the source and regulation of H_2O_2 production for peroxidase located elsewhere in the cell is also uncertain. With isolated leaf peroxisomes oxidation of only a small part of the glycolate to CO_2 and formate is observed, and it has been proposed that perhaps as much as 10% of the CO_2 released during photorespiration may be generated by this mechanism (Halliwell and Butt, 1974). However, studies of the pool size and rate of formation of glycine and serine during photorespiration by Canvin's and Fock's groups (Mahon *et al.*, 1974; reviewed by Schnarrenberger and Fock, 1976), and the rate of conversion of [^{14}C]glycolate to glycine *in vivo* and by isolated peroxisomes indicate that most of the glycolate is converted to glycine (Tolbert, 1971a). In fact, the composition of the peroxisomal enzyme complex in the matrix seems to protect against peroxidation of glyoxylate. Excess oxidation of glyoxylate to CO_2 and formate would create a pool of formate which the leaf might be unable to handle, since there is only a low level of formate dehydrogenase in the mitochondria. Further, the limited extent that glycolate may be oxidized to glyoxylate and then on to CO_2 and formate, and the formate to CO_2, represents a complete loss of energy and carbon.

Glycolate oxidase also oxidizes glyoxylate to oxalate, since glyoxylate exists as the hydrated form, $CH(OH)_2$-$COOH$, in solution, and as such is an α-hydroxy acid analogue. In the absence of sufficient nitrogen for rapid conversion of glyoxylate to glycine, oxalate accumulates in spinach leaves, whereas heavy nitrogen fertilization, providing ample glutamate, results in a lowered production of oxalate.

Specific aminotransferase reactions between glyoxylate and an amine donor occur in the microbodies with the formation of glycine. In peroxisomes from leaves there are two different, active aminotransferases; one for glutamate : glyoxylate and the other for serine : glyoxylate. Activity of each is about 1–2 μmoles/min/mg of peroxisomal protein. Neither aminotransferase is absolutely substrate specific, but their much higher activity with the indicated substrates and location in the peroxisomes are strong evidence for functioning as shown in Fig. 3. In contrast to other aminotransferases, these two with glyoxylate are essentially unidirectional for glycine formation for reasons that are not clear. A coupling of glycine formation to serine conversion to hydroxpyruvate would move carbon through the glycolate and glycerate pathways of photorespiration (Tolbert, this series, Vol. 2, Chapter 12). However, two glycines are required for the formation of one serine, so that the second glycine must be generated from glyoxylate by the peroxisomal aminotransferase which utilized the cellular glutamate pool.

The oxidative conversion of glycolate to glycine in leaf peroxisomes is an active exothermic process and very unidirectional because both glycolate oxidase and the aminotransferases involved are physiologically irreversible reactions. Consequently in the plant during photosynthesis under normal conditions, the pool of glycolate is always small and the pool of glycine large. The O_2 and CO_2 concentrations cause large changes in these reservoirs by influencing both the rate of glycolate biosynthesis and glycolate oxidation.

A general hypothesis is that all reactions involving glyoxylate biosynthesis and metabolism are compartmentalized in microbodies in order to prevent undesired side reactions of glyoxylate. This is certainly a feature of the glycolate pathway in leaf peroxisomes and the glyoxylate cycle in seed glyoxysomes. Otherwise glyoxylate might participate in other aminotransferse reactions, it could be oxidized to CO_2 and formate, or two molecules of it could be dismutated by lactate dehydrogenase into oxalate and glycolate. Some lactate dehydrogenase is present in leaves (T. Betsche and B. Gerhardt, unpublished). Thus glyoxylate formation in the cytoplasm might result in these uncontrolled side reactions rather than its nearly complete conversion to either glycine in the leaf peroxisomes or malate in the glyoxysomes.

B. Interconversion between Glycerate and Serine

The peroxisomal part of the glycerate pathway (Fig. 3) consists of two reversible reactions, and as such can function for gluconeogenesis from serine or for serine synthesis via glycerate and P-glycerate derived from the photosynthetic carbon cycle or from glycolysis. The glycerate pathway in the peroxisomes seems essential for serine synthesis and would occur in both light and dark. Serine can be the precursor for essential glycine and C_1 units derived from the C_1-tetrahydrofolate complex. Synthesis of glycine and serine by the glycolate pathway only occurs during photorespiration. In C_4

plants or in algae with lower levels of photorespiration, serine and glycine formation from glycerate can be substantiatial even during photosynthesis.

In 1954 a glycerate dehydrogenase was isolated and characterized from leaves (Stafford et al., 1954), as well as a glyoxylate reductase (Zelitch, 1955). Similarities in the two activities, even then, suggested that they may be catalyzed by the same protein. It has a molecular weight of 240,000 and NAD as cofactor. Later it was recognized that this enzyme is in the peroxisomes (Tolbert et al., 1970). It catalyzes a NAD-linked glycerate dehydrogenase, a NADH: hydroxypyruvate reductase, and a NADH: glyoxylate reductase, but it does not catalyze the oxidation of glycolate. The K_m (hydroxypyruvate) is $2 \times 10^{-4} M$, but for glyoxylate the K_m is an unreasonable $2 \times 10^{-2} M$ and the pH optimum is 6. Consequently in Fig. 3 glyoxylate reductase activity is not shown and indeed none is predicted as there is no evidence for a glycolate–glyoxylate terminal oxidase system. Glycolate oxidation only occurs when it is produced during photosynthesis. Because peroxisomes contain much NADH: hydroxypyruvate reductase (≈ 1 μmol/min/mg protein for spinach leaf peroxisomes) and because the oxidation of NADH is a convenient and sensitive assay, this is an excellent marker enzyme for microbodies. It has been found in most plant microbodies. The peroxisomal pool of NAD/NADH must link this glycerate pathway to a malate shuttle and the rest of the cell as discussed in Section VIII.

The aminotransferase reaction between hydroxypyruvate and serine is catalyzed by at least two peroxisomal enzymes. The specific and irreversible serine: glyoxylate aminotransferase forms hydroxypyruvate and glycine, and is linked to glycolate metabolism and the flow of carbon from serine back toward hexose synthesis. However, this aminotransferase ought to function only during photorespiration in the direction of hydroxypyruvate synthesis. For serine formation from glycerate to occur at other times, additional aminotransferase activities in the leaf peroxisome are needed. Isolated leaf peroxisomes will catalyze a glutamate or alanine-linked transamination with hydroxypyruvate, but the specificity and certainty of this aminotransferase activity has not been clarified.

VI. METABOLIC PATHWAYS IN GLYOXYSOMES

During germination of fat-storing seeds there is a rapid conversion of the lipid reserve to sugar with about a 90% efficiency as measured by carbon retention (Beevers, 1969). After imbibition of water and initiation of germination, microbodies appear in the endosperm, and the enzymes of the glyoxylate cycle reach maximal activity at the time of rapid lipid degradation, radicle protrusion, and new shoot development. These glyoxysomes contain all the enzymes necessary for fatty acid beta oxidation to acetyl-CoA, as well as the glyoxylate cycle which condenses two acetyl-CoA molecules to form a C_4 acid, namely pools of succinate, malate, oxaloacetate and asparate (Beevers, 1969; Tolbert, 1971a). The C_4 products are used elsewhere in the cell of the cotyledon for synthesis of hexoses and sucrose for transport to the developing seedling. This gluconeogenesis is initiated by the synthesis of phosphoenol pyruvate from oxaloacetate (see Chapter 5). After the utilization of the lipid reserves, the glyoxysomes and the glyoxysomal enzymes disappear.

The complete process of fatty acid oxidation and the glyoxylate cycle, as shown in Fig. 4, has been studied most extensively in the glyoxysomes of the castor bean endosperm (Beevers, 1969), but it probably exists in its entirety in other fatty seeds. Some enzymes of the glyoxylate cycle have been reported in cotyledons from many fatty seeds during germination (extensive literature not cited). In tissue such as fungi, algae, and protozoan, the presence of part of the glyoxylate cycle has been established, but the fatty acid beta oxidation system in the organelle has not been extensively investigated. Graves and Becker (1974) reported the presence of crotonase, one of the

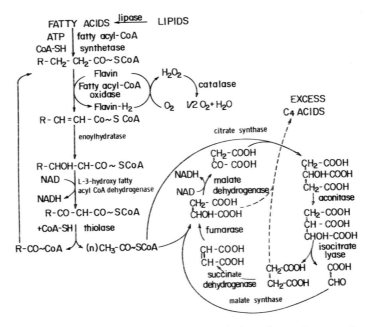

Fig. 4.   Metabolic pathways in glyoxysomes in cotyledon cell or endosperm of germinating seedlings. On the left is the fatty acid β-oxidation pathway to acetyl-CoA and on the right the glyoxylate cycle for converting two acetyl-CoA to one C_4 acid. The excess acids leave the glyoxysome to be utilized for gluconeogenesis in the cytoplasm.

enzymes of beta oxidation, in partially purified glyoxysomes from *Euglena*. It is logical for a biological system growing on C_2 or C_4 substrates to utilize the glyoxylate cycle for production of excess C_4 acids but not to require the fatty acid beta oxidation system. Glyoxysomes from *Tetrahymena* and yeast may contain only the two enzymes, isocitrate lyase and malate synthetase, which are not present in the citric acid cycle, so catalysis of the rest of the glyoxylate cycle must take place in the mitochondria (Hogg, 1969). In earlier research with some of these microbodies, the organelles were called peroxisomes, but for consistent nomenclature the organelle should now be called a glyoxysome.

A. Fatty Acid Beta Oxidation

Glyoxysomes from castor bean endosperm contain enzymes required for complete conversion of the fatty acids from lipids to acetyl-CoA (Cooper and Beevers, 1969). These glyoxysomes are found among the lipid bodies of a germinating seed at the time when the lipids are rapidly being utilized (Fig. 2) (Beevers, 1969; Frederick *et al.*, 1975). Likewise microbodies are abundant among the lipid bodies during germination of fungal spores (Maxwell *et al.*, 1977). A lipase in the glyoxysomal membrane and an ATP requiring fatty acyl-CoA synthetase have been reported with glyoxysomes from castor bean cotyledon (Cooper and Beevers, 1969). Fatty acyl-CoA oxidase is the unique glyoxysomal enzyme which fulfills the criteria of a microbody flavin oxidase linked to O_2 uptake and H_2O_2 production and which catalyzes an irreversible reaction with loss of energy to the system in order to direct carbon flow toward acetyl-CoA synthesis. The other enzymes of fatty acid beta oxidation in liver peroxisomes appear similar to their mitochondrial counterpart, although they have not yet been isolated and thoroughly characterized from plant glyoxysomes. Current research with enzymes for fatty acid beta oxidation from liver peroxisomes indicates that they are isoenzymic with those in the liver mitochondria (T. Osumi and T. Hashimoto, unpublished). In one turn of the beta oxidation cycle shown in Fig. 4, a fatty acyl-CoA is oxidized to one acetyl-CoA, and a fatty acyl-CoA two carbons shorter is generated. Repeated cycling is assumed to convert the fatty acid totally to acetyl-CoA. However, current research on peroxisomal beta oxidation in the liver suggests that the microbody beta oxidation from palmitoyl-CoA proceeds only about as far as octanoyl-CoA and four acetyl-CoA, and the shorter chain fatty acids are further oxidized in the mitochondria.

B. Glyoxylate Cycle

The glyoxylate cycle as shown in Fig. 4 is catalyzed by a complete set of enzymes in the glyoxysomes of the germinating seed by passing those in the mitochondria. Two of these, isocitrate lyase and malate synthetase, are unique to the glyoxylate cycle and have been considered to be always in glyoxysomes, even if the other steps occur in the mitochondria. However, exceptions even to this have been reported (see Section VII). The glyoxylate cycle and the enzymes, isocitrate lyase and malate synthetase, are also present in bacteria, which do not have defined organelles. Properties of both enzymes have been described after purification, as, for example, isocitrate lyase from flax seedlings (Kahn et al., 1977). Isocitrate lyase forms glyoxylate and succinate from isocitrate, and malate synthetase condenses glyoxylate with acetyl-CoA to form malate. The packaging of these two enzymes together in the glyoxysome is consistent with the concept mentioned in Section IV, that reactions involving glyoxylate are in a microbody compartment to prevent spurious side reactions of glyoxylate. The other enzymes of the glyoxylate cycle, if present in the glyoxysome, are presumed to be isoenzymic with those of the mitochondrial citric acid cycle. This is known for malate dehydrogenase (Curry and Ting, 1973), but not for the other enzymes.

The net reaction of one turn of the glyoxylate cycle in the glyoxysome is

$$2 \text{ acetyl-CoA} \rightarrow C_4 + 2 \text{ CoASH},$$

where as the net reaction of the citric acid cycle in the mitochondria is

$$\text{acetyl-CoA} \rightarrow 2 \text{ CO}_2 + \text{CoASH}.$$

The most likely C_4 acids to be excreted from the glyoxysome are succinate and malate, which are converted to oxaloacetate and then to phosphoenol pyruvate for glycolysis.

The other product of glyoxysomal metabolism is NADH, produced by two reactions: L-3-hydroxy fatty acyl-CoA dehydrogenase and malate dehydrogenase. The shuttle from the glyoxysomes or mechanism of reoxidizing NADH is not known. Because of the great efficiency in converting fats to sucrose during seed germination, the energy as NADH from the glyoxysomes must be conserved and transported to the cytoplasm. Only the loss of energy during fatty acyl-CoA oxidase activity need occur. Two ATP would be conserved had this step been catalyzed by the mitochondrial fatty acyl-CoA dehydrogenase. Since one palmitate, if totally oxidized to CO_2 in the mitochondria, forms a net of 129 ATP, the loss of only 2 ATP for each acetyl-CoA formed in the glyoxysomes or a total of 14 ATP per palmitate represents only a small (11%) loss of energy to be charged for accelerating fatty acid oxidation and rapid growth during germination.

VII. OTHER ENZYMES IN MICROBODIES

Besides the enzymes for the main metabolic pathways in leaf peroxisome and cotyledon glyoxysomes, as described in the previous sections, other groups of enzymes have been reported in microbodies. These other metabolic processes may predict what else is yet to be discovered about plant microbodies. Of these processes the most extensively investigated is the formation of microbodies containing methanol oxidase and catalase which develop in yeast grown on methanol (two of many references are Fukui *et al.*, 1975 and Veenhuis *et al.*, 1978). These yeast cells become filled with gigantic microbodies containing a flavin-linked methanol oxidase and catalase and as such fulfill de Duve's description of a peroxisome. In the use of these yeast as a food or protein supplement, the microbody protein is their main component. Growth of yeast on sugars severely represses microbodies and then mitochondria abound. Clearly the enzymatic composition and number of microbodies in yeast is very substrate-dependent.

When yeast are grown on long chain alkanes (as for degrading components of oil), they develop microbodies containing an oxidase to initiate the alkane oxidation (Osumi *et al.*, 1975), but this oxidase is not yet well characterized. This situation is akin to the development of microbodies for methanol utilization. In these cases, the flavin oxidase in the microbody with the catalase is the unique component, whose activity apparently initiates a catabolic pathway by the energy loss of the initial oxidation.

Some enzymes associated with amino acid oxidation are to be expected in microbodies. Low levels of D-amino acid and L-amino acid oxidase have been reported to be constituents of animal peroxisomes. These oxidases have not been extensively investigated in plant tissue. D-amino acid oxidase was detected in glyoxysomes from castor bean endosperm (Beevers and Breidenbach, 1974). Phenylalanine ammonia lyase has also been reported in plant microbodies (Ruis and Kindl, 1970), but because of its wide distribution in membrane fractions it is uncertain whether the observed activity was associated with the microbodies. Ruis and Kindl (1971) have also reported that plant microbodies metabolize aromatic amino acids and convert amino acids to α,β-unsaturated carboxylic acids.

Peroxisomes from liver and kidney in some cases have been reported to have one or more enzymes associated with purine and pyrimidine catabolism. Urate oxidase is abundant and is the core constituent in peroxisomes of those animals that degrade urate. Urate oxidase has not been studied in leaves, but it has been reported to be a minor component in glyoxysomes of germinating seeds (Theimer and Beevers, 1971). Xanthine dehydrogenase has not been found in plant microbodies, although it and allantoinase and allantoicase are located in hepatic peroxisomes from birds.

Other enzymes of liver peroxisomes, such as the carnitine acyl transferases and glycerol phosphate dehydrogenase have not been detected at all in plant microbodies (unpublished). Other enzymes have been reported in plant microbodies, but such claims have not been confirmed, and the enzyme may be in another compartment instead. Among these are nitrate reductase of the cytoplasm, polyphenol oxidase of the plastids or cytoplasm, phenylalanine ammonia lyase of the endoplasmic reticulum (ER) and formate dehydrogenase of the mitochondria.

VIII. MICROBODY MEMBRANE AND TRANSPORT

Many properties of plant microbodies are based on the fact that they have only a single outer membrane and no internal membrane structure, in contrast to the double outer membrane of chloroplasts and mitochondria. *In situ* microbodies appear contorted and appressed among the other organelles because of their apparently pliable nature (Fig. 2). In spite of this appearance there is no evidence of direct connections and transport between microbodies and other organelles, but rather all transport processes among the organelles probably pass through cytoplasmic pools.

Another consequence of the single membrane is that microbodies are fragile and hard to isolate. In fact preparations of leaf peroxisomes on a sucrose gradient are composed of few intact particles. Glyoxysomes from germinating cotyledons may be obtained in more intact condition. Because of particle rupture, the question of latency by the isolated fraction due to membrane transport systems is hard to assess. The microbodies are isolated in dense sucrose and must be assayed after dilution into a low density assay medium, which results in considerable particle disruption. Lower enzyme activity with isolated plant microbodies in the absence of detergent has been taken as criteria for latency and for restricted passage of substrates, especially NAD and NADH, into the organelle. Consequently, the enzyme assays for total activity are run in the presence of about 0.1% Triton X-100. Considerable activity in the absence of the detergent may be due in part to broken or cracked particles and cannot be quantitated to membrane transport.

A. Membrane Composition

The equilibrium sucrose density of the isolated microbody is about 1.25 g/cm^3, which is similar to that of a protein body, for the microbody has a low lipid to protein ratio. The lipids of the plant microbody bilayer membrane are similar to the ER membrane and are mainly phosphatidyl choline ($\approx 50\%$), phosphatidyl ethanolamine, and phosphatidyl inositol (Donaldson *et al.*, 1972; Donaldson and Beevers, 1977). The fatty acid composition of these phospholipids is similar to those found in other intracellular membranes. An antimycin a insensitive cytochrome b$_5$ reductase in the microbody fraction is presumed to be a part of the microbody membrane. However, it is nearly impossible by present methods to obtain microbodies free of attached or contaminating ER, so that both the lipid and cytochrome b$_5$ reductase composition must represent, in part, the occluded ER. From electron microscopic examination, this ER contamination is much more severe in leaf and liver peroxisomes, than in isolated seed glyoxysomes. Nevertheless, the membrane of the microbody seems similar in lipid composition and cytochrome b_5 reductase to the ER membrane. This is consistent with microbody formation by budding from the smooth ER (see Section VIII), which predicts that the membranes of the ER and microbody are similar.

Monoglyceride lipase activity has clearly been shown to be another enzymatic component of the glyoxysomal membrane (Muto and Beevers, 1974), but the enzymes for fatty acid beta oxidation are components of the matrix. It has been suggested, but not proven, that the matrix enzymes may be associated in complexes for rapid metabolism. Indeed a significant part of isocitrate lyase and malate synthetase remain with glyoxysomal ghosts after loss of catalase by osmotic shock upon dilution from 54 to 18% sucrose (Huang and Beevers, 1973; Köller and Kindl, 1977). These enzymes can be solubilized away from the membrane ghosts by treatment with 100 mM MgCl$_2$, as if they were in some larger internal complex closely associated with the membrane.

B. Malate Shuttle

Transport between mitochondria or chloroplast and the cytoplasm is controlled by specific shuttles and membrane-bound translocases. Since the microbodies have only a single membrane, translocase may not be as essential or even involved, but rather there may be only passive diffusion of the organic components of the shuttles. So far no translocases have been reported for microbodies. Since intact whole microbodies are difficult to obtain, definitive data on latency and transport between peroxisomes and other subcellular compartments are not available. Based on the presence of enzymes in the microbodies that are known to be shuttle components elsewhere in the cell, various microbody shuttles have been considered (Tolbert, 1973).

According to the metabolic pathways so far described in plant microbodies, a substantial transport system is required for NAD/NADH equivalents. Both the leaf peroxisomes and seed glyoxysomes contain a large amount of activity of an isoenzyme of NAD: malate dehydrogenase that is unique to microbodies (Yamazaki and Tolbert, 1969; Curry and Ting, 1973). Next to catalase the malate dehydrogenase activity of about 50 μmoles/min/mg protein is the most active enzyme in plant microbodies. A malate shuttle for leaf peroxisomes has been proposed by analogy with the mitochondrial malate shuttle and is included in Fig. 3. A similar malate shuttle for the glyoxysomes has not been detailed, but could be proposed. The oxidized component of the leaf peroxisomal malate shuttle is proposed to be aspartate rather than oxaloacetate, for three aspartate aminotransferases are present in the leaf peroxisomes (Rehfeld and Tolbert, 1972). This shuttle seems to be the only way to oxidize microbody NADH, since a microbody NADH oxidase has not been detected. The peroxisomal components should be linked through similar malate and aspartate pools in the cytoplasm, chloroplasts, and mitochondria. A malate shuttle out of the glyoxysome must transport substantial amounts of reducing power as continued operation of both the fatty acid oxidation and the glyoxylate cycle depend upon regeneration of NAD. The energy in the reduced NAD must be shuttled to the rest of the cell to account for the known efficient conversion of the lipids in the growth of the seedling. Likewise a very active malate shuttle for the leaf peroxisomes is required for serine reduction to glycerate during photorespiration. The maximum amount of reducing capacity to be shuttled into leaf peroxisomes by the malate shuttle during photorespiration may exceed the rate of mitochondrial respiration by several fold. The leaf peroxisomal malate shuttle must be connected either directly with the reducing capacity in the chloroplast (i.e., a chloroplast NADP : malate shuttle) or indirectly through the cytoplasmic pool of NADH generated by the chloroplast triose-P shuttle (see this series, Vol. 2, Chapter 12).

For transport of fatty acids in and out of liver mitochondria there are carnitine acyl transport systems, and liver peroxisomes contain carnitine acetyl and carnitine octanyl transferases but these have not been detected in plants (unpublished). The lipase in glyoxysomal membranes may facilitate fatty acid transport into the glyoxysome.

No transport systems for glycolate, glycerate, glycine, and serine in and out of leaf peroxisomes are known, but significant pools of each exist in the cell. The microbody is certainly a compartment for enzymes, but it is not yet known whether it is a substrate compartment controlled by membrane transport systems. A reason for proposing free movement of substrates and products through the microbody membrane is the fact that often only a part of an active metabolic pathway will be found in the microbody. The best examples are the presence of only part of the glyoxylate cycle in glyoxysomes from many tissues, and the fact that the glycolate pathway of photorespiration is only partly located in leaf peroxisomes. Large amounts of intermediates in these pathways have to move between the microbody and other compartments containing the enzymes for the rest of these metabolic pathways.

IX. DISTRIBUTION OF MICROBODIES IN PLANT TISSUES

Microbodies have been observed in angiosperms, gymnosperms, and bryophytes, and are probably universally present in plants, with the exception of algae. Leaf peroxisomes from pallisade and mesophyll cells of C_3

plants are similar in appearance and composition. Microbodies in bundle sheath cells and in mesophyll cells of C_4 plants and in CAM plants are also similar in appearance. Only Newcomb's group (Frederick *et al.*, 1975) has counted the number of microbodies per cell. They estimate the ratio of microbodies to mitochondria to chloroplasts in a cell of a C_3 plant to be about $1 : 2 : 3$. In the C_4 plant microbodies are mainly present in the bundle sheath cells and they are fewer in number and smaller in size in the mesophyll cells. Likewise there have been few surveys for the relative abundance of leaf peroxisomal enzymes as done by Tolbert's (1969, 1971a,b) group. It has only been estimated that microbodies in a C_3 plant contain about 1–2% of the total soluble protein. Based on total enzyme content in homogenates of C_4 plants obtained by differential grinding procedures, the peroxisomal enzymes in the bundle sheath cells may be similar to those in the C_3 plant. However, the function of enzymes in the few small microbodies of the mesophyll cells of C_4 plants is not known, since there should not be a glycolate pathway of photorespiration in these cells.

Although microbodies have been observed in many plant tissues, they have only been isolated and extensively investigated biochemically from spinach leaves and from the endosperm of germinating seeds of castor bean and some other fatty seeds such as sunflower and watermelon. Microbodies from other tissues have not been enzymatically characterized other than for the presence of catalase. Microbodies are abundant in the abcission zone of tobacco leaves. They are present in ripening fruit and in developing seeds. Preliminary reports indicate that most microbodies from other tissues contain NADH : hydroxypyruvate reductase and malate dehydrogenase in addition to catalase and a trace of glycolate oxidase (Huang and Beevers, 1971). Thus perhaps in nonphotosynthetic tissues, without glycolate production and photorespiration, the microbodies function in part for serine and glycine synthesis and metabolism via the glycerate pathway. Additional observations about microbodies in other plant tissues are cataloged under the following groups.

1. *Roots.* Some of the earliest cytological reports on the presence of microbodies in plants described their concentration in the meristem of root tips (Mollenhauer *et al.*, 1966). Only exploratory experiments on the isolation of microbodies from roots have been done because of the difficulty in homogenizing the tissue without also completely breaking up the microbodies. Microbodies from roots contain catalase and a little malate dehydrogenase and α-hydroxy acid oxidase, but otherwise their composition and function are unknown.

2. *Tubers.* There are several reports of microbodies in tubers which increase in number and enzyme activity during aging of sliced tissue (Tchang *et al.*, 1978). These microbodies have been found to contain catalase and a

low level of α-hydroxy acid oxidase, perhaps glycolate oxidase. Their function is unknown. The level of the α-hydroxy acid oxidase activity is far too low to be significant, as compared to the alternate respiration which develops in aging tuber slices.

3. *Algae.* Some algae, when grown heterotophically on acetate, malate, or other carbon sources, will develop abundant glyoxysomes in the same manner as yeast. When grown phototrophically, unicellular algae may contain only a few small microbodies, and the catalase content of the cell on a protein or chlorophyll basis is generally less than 10% of that found in a leaf from a C_3 plant with peroxisomes (Frederick *et al.,* 1973; Tolbert, 1972). The enzymatic composition and function of these algal microbodies are not well established, although it is conjectured that they are involved in glycolate and glycerate metabolism during photosynthesis as in higher plants. Hydroxypyruvate reductase or glycerate dehydrogenase is present; at least it is most often measured.

A major difference between plant and algal microbodies is the enzyme for oxidizing glycolate during photorespiration. Leaf peroxisome contain glycolate oxidase but in unicellular green algae, blue–green algae, and photosynthetic bacteria, glycolate is oxidized by a glycolate dehydrogenase, probably linked to a cytochrome c (Paul and Volcani, 1975). The activity of glycolate dehydrogenase in algae is very low and it may be confused with low levels of D-lactate dehydrogenase (Gruber *et al.,* 1974). This type of activity has been reported in mitochondria or chloroplasts from various algae, and in the case of *Euglena* and *Chlorella,* some of this dehydrogenase activity has been found on sucrose gradients at the density characteristic of peroxisomes. The excretion of glycolate by these algae during active photosynthesis has been taken as an indication of the production of more glycolate during photosynthesis than this limiting dehydrogenase can convert to glyoxylate and then to glycine.

There are as yet few definitive reports of microbodies in marine algae or plants, yet they too produce glycolate, presumably from the activity of ribulose-P_2 carboxylase/oxygenase. The symbiotic zooxanthaellae excrete some glycolate to their polyp. Most of these algae probably contain glycolate dehydrogenase (Tolbert, 1976), as do the fresh water algae, and would not be expected to have numerous peroxisomes.

4. *Fungi.* Microbodies in fungi are inducible organelles whose presence and enzymatic composition depend on the substrate and environment. When yeast are grown on methanol a few large microbodies will fill the whole cell and the isolated microbody contains a FMN-linked methanol oxidase and catalase (see Section VII). When yeast or neurospora are grown aerobically on C_2 or C_4 compounds they develop numerous glyoxysomes (Maxwell *et al.,* 1977; Thiemer *et al.,* 1978). When grown on *n*-alkanes yeast develop microbodies and catalase probably to initiate *n*-alkane oxidation (see Section

VII). However, when yeast are grown anaerobically or aerobically on glucose, microbodies are almost completely repressed, but many mitochondria are present.

Most fungi subsist on substrates, such as acetate and malate, from their host. Maxwell *et al.* (1977) have reviewed the importance of microbodies in plant pathogenic fungi, where they may be utilizing the glyoxylate cycle in the glyoxysomes for gluconeogenesis. Microbodies are found in zoospores, other spores, and hyphae, and apparently function for anaplerotic metabolism. In such diverse material as yeast, fungi, and algae, exceptions or perhaps just modifications to the normal distribution of microbody enzymes may be expected to occur. The presence of malate synthetase and isocitrate lyase in the mitochondria of a nematode may be such an exception (McKinley and Trelease, 1978).

5. Microbodies in other tissues. In order to understand the total concept of microbodies in biology, the plant scientists must also examine reviews about them from mammalian tissue, where they were first characterized. Peroxisomes from rat liver have been more thoroughly investigated than from any other source (de Duve and Baudhuin, 1966; de Duve, 1969). Microbodies are abundant in protozoa (Müller, 1975), trypanosomes, and trichomonads in which they have also been called glycerol phosphate oxidase bodies or hydrogensomes, because of their activity for glycerol phosphate metabolism and hydrogen production.

Metabolic pathways in plant microbodies were rapidly elucidated after their discovery because prior work on these metabolic pathways had already been done. The recent belated realization that liver peroxisomes (Lazarow and de Duve, 1976), also contain a fatty acid beta oxidation pathway similar to that in seed glyoxysomes, emphasizes a trend toward a type of subcellular organelle with universal properties, although all these metabolic pathways may not be expressed at one time in the particle from a given tissue. One major difference has been the absence of the glyoxylate cycle and glyoxysomes in mammals and in photosynthetic tissue of the higher plant, which form excess C_4 acids by carboxylation of P-enolpyruvate and which cannot utilize acetate gluconeogenically.

X. DEVELOPMENT AND BIOGENESIS OF PEROXISOMES AND GLYOXYSOMES

Development of glyoxysomes during seed germination and of peroxisomes during greening of the leaf are two striking phenomena that have been discussed in detail (Beevers, 1969; Tolbert, 1971a; Vigil, 1973; Gerhardt, 1978). After water imbibition, RNA, and ER proliferation occurs, and then on the second day of germination, microbody development begins and glyoxysomal enzymes appear. A rapid increase in the number of microbodies is accompanied by degradation of the lipid bodies. The ER attached to some of the microbodies has been observed in this period of development. There is *de novo* synthesis of microbody protein which can be inhibited by cycloheximide, an inhibitor of cytoplasmic protein synthesis. Subsequent decline in glyoxysomes occurs with the depletion of the stored lipids. In fatty cotyledons which will develop into cotyledonary leaves, there is initiated in the light an increase in leaf peroxisomal enzyme activities (i.e., glycolate oxidase) in the particle concurrent with a decrease in glyoxysomal malate synthetase and isocitrate lyase. Total catalase activity initially decreases along with glyoxysomal activity, but since catalase is also a constitutent of the leaf peroxisomes, the decrease in its activity is checked by the rise in its peroxisomal activity. These two biochemical classes of microbodies are morphologically similar and it has not been possible to separate the two microbody populations physically from these germinating tissues. Two working hypotheses for the period during greening and development of the cotyledon are that microbodies change in enzymatic composition or that there is a disappearance of glyoxysomes simultaneous with *de novo* formation of leaf peroxisomes. In liver there is a somewhat similar dilemma, where it has been impossible to isolate young or old, small or large peroxisomes, as marked by the incorporation of ^{14}C-labeled enzyme precursors. To explain this anomaly, de Duve (1973) proposed that liver peroxisomes are all interconnected through the ER channels, so that any alteration in their enzyme composition is freely mixed among all the peroxisomes.

During greening of etiolated leaves a large population of leaf peroxisomes develop. Prior to exposure to light, the etiolated leaf contains variable but always lower levels of peroxisomal activity. The development in the light seems to be controlled by phytochrome, in that it is reversibly stimulated by exposure to red and far-red light, at an intensity insufficient for photosynthesis or full chloroplast development (Feierabend, 1975). Leaf peroxisomal development thus seems to be another example of membrane development as influenced by photochrome. However, once a leaf has greened, there is as yet no evidence for a change in the composition or activity of the peroxisomal system, contrary to changes in microbody activity that occur in other systems. Although this problem needs to be further explored, the lack of change may be related to the close relationship between photosynthesis and photorespiration, which is associated with the peroxisomes.

There is little evidence that microbodies contain ribosomes or nucleic acids or can form proteins. Most investigators have reported that no ribosomes are seen in or bound to peroxisomes. Peroxisomes are closely associated with the ER and continuity between the smooth ER and the microbodies suggest that they are formed by budding from the ER. Rather than being autonomous, microbodies appear to be a part of the ER system. Molecular details of microbody development and turnover in plants have not been fully described, but analogies with peroxisomal biogenesis in the liver (de Duve, 1973) have established some working hypotheses. It is assumed that the protein for the microbody enzymes is formed by the rough ER and perhaps channeled down the inner cisternae to the developing microbodies. During early stages of development of glyoxysomes, the phospholipid components of the glyoxysomal membrane are made on the ER and the ER contains microbody enzymes (Beevers, 1975; Gonzalez and Beevers, 1976; Donaldson and Beevers, 1977), as if the enzymes were formed on the rough ER and transported to the developing microbody. It is not established whether cytoplasmic pools of these enzymes are artifacts of grinding procedures, or precursors of the microbodies, or representative of *in situ* degradation of microbodies. How specific enzymes are selected for packaging in the microbody is not understood. In liver the microbody and its catalase has a turnover time of about 1.5 days, which is rapid, relative to other organelles. Rapid biogenesis and then disappearance of glyoxysomes during seed germination suggests analogous properties. The turnover of glyoxysomal catalase has been estimated as being more rapid than the appearance and disappearance of the organelle during germination. The rate of turnover of the enzymes in leaf peroxisomes has not been reported.

Changes in number of microbodies per cell and changes in the enzyme content of the microbodies are influenced by several factors. That changes occur in microbody activity in the same tissue is contrasted to the predicted constancy of the mitochondria, and suggests that microbodies are involved in metabolic regulation. One phenomenon is the development of microbodies during growth such as glyoxysome formation during seed germination, leaf peroxisome formation during greening of an etiolated leaf, or postnatal development of liver peroxisomes. Another generality for regulating microbody activity is substrate induction of microbody formation. Microbodies are induced in yeast, other fungi, and algae, when grown on specific substrates, such as C_2 and C_4 compounds or methanol. Glyoxysomes develop and are active in seed germination only as long as there is lipid remaining for degradation. The formation of peroxisomal glycolate oxidase is associated with glycolate production by photosynthesis, although the development of glycolate oxidase can occur in far red light in the absence of chloroplast development. Thus for both cotyledon glyoxsomal and leaf peroxisomal development, direct substrate induction does not have to be the initiating factor. However, many ER metabolic reactions catalyzed by the P-450 system are substrate-induced, so it is possible that microbodies, as part of the ER system, may also be inducible and controlled in part by substrate availability.

The influence of hormones and growth regulators on plant microbody formation has hardly been explored. In animals, hypolipidemic agents, such as Clofibrate (p-chlorophenoxyisobutric acid), leads to the proliferation of liver peroxisomes. This compound is a mild plant antiauxin as well. In plants it also accelerates the development of glyoxysomes during seed germination, but it has no pronounced effect on the level of leaf peroxisomal activity (E. W. Smith and N. E. Tolbert, unpublished).

XI. FUNCTION

Several reasons for the distinct microbody subcellular compartment have been conjectured, but not proven. One biochemical reality is the need for a close association of flavin oxidases and catalase to quickly remove the H_2O_2. In metabolic reactions involving this sort of oxidation the chemical energy is lost as heat, but the flavin oxidase may serve for directing carbon flow that would otherwise not occur or not occur fast enough. Another concept is that two similar metabolic systems, one in the mitochondria with energy conservation by oxidative phosphorylation and one in the microbody with energy loss, may be used to balance growth. In this concept net growth may be regulated by the amount of metabolism that occurs in the microbody, while the mitochondrial system may be the more constant, as well as the indispensable system. Thus photorespiration, partially involving leaf peroxisomes, wastes a significant part of the newly acquired energy from photosynthesis (this series, Vol. 2, Chapter 12). As a better understanding of microbodies develops, the concept of a wasteful respiratory organelle is becoming less tenable. In the seed glyoxysomes the only energy loss during fatty acid conversion to C_4 acids (Fig. 3) may be that from the fatty acyl-CoA oxidase reaction, which is equivalent to only two ATP per turn of the cycle, if it had occurred in the mitochondria. In leaf peroxisomes there is only one oxidase step, glycolate oxidase, which represents only one of several steps of photorespiration. It seems likely that the packaging of the flavin oxidases with catalase in the microbody is essential for certain metabolic pathways.

De Duve (1969) thought that microbodies might represent a primitive respiratory organelle. However, they are present in aerobic eukaryotic cells, but absent in prokaryotes. Higher plants contain leaf peroxisomes with glycolate oxidase and catalase, whereas some algae do not have peroxisomes, but rather oxidize glycolate by a dehydrogenase not linked to O_2 and H_2O_2 production. Photorespiration and peroxisomes in the higher plant may have developed as a protective system against the photo-oxidative environment. Tolbert (1971a), Krause et al. (1978), and others have long proposed that for physiological stability of illuminated chloroplasts and cells, normal dissipation of excess photosynthetic reducing capacity by oxidative carbon metabolism is required. This protective process, in part in the peroxisomes, makes them essential.

REFERENCES

Aebi, H. (1974). *In* "Methods of Enzymatic Analysis" (H. U. Bergmeyer, ed.), Vol. 2, pp. 673–684. Academic Press, New York.

Beevers, H. (1969). *Ann. N.Y. Acad. Sci.* **168**, 313–324.

Beevers, H. (1975). *In* "Recent Advances in Chemistry and Biochemistry of Plant Lipids" (T. Galliard and E. I. Mercer, eds.), pp. 287–299. Academic Press, New York.

Beevers, H., and Breidenbach, R. W. (1974). *In* "Methods in Enzymology: Vol. 31, Biomembranes, Part A" (S. Fleischer, and L. Packer, eds.), pp. 565–571. Academic Press, New York.

Breidenbach, R. W., and Beevers, H. (1967). *Biochem. Biophys. Res. Commun.* **27**, 467–713.

Burke, J. J., and Trelease, R. N. (1975). *Plant Physiol.* **56**, 710–717.

Clayton, R. K. (1960). *Biochem. Biophys. Acta* **40**, 165–167.

Coleman, B. (1977). *In* "Cellular and Molecular Plant Physiology" (H. Smith, ed.), pp. 136–159. Blackwell, Oxford.

Cooper, T. G., and Beevers, H. (1969). *J. Biol. Chem.* **244**, 3507–3513; 3514–3520.

Curry, R. A., and Ting, I. P. (1973). *Arch. Biochem. Biophys.* **158**, 213–224.

Donaldson, R. P., and Beevers, H. (1977). *Plant Physiol.* **59**, 259–263.

Donaldson, R. P., Tolbert, N. E., and Schnarrenberger, C. (1972). *Arch. Biochem. Biophys.* **152**, 199–215.

De Duve, C. (1969). *Proc. R. Soc. London Ser. B.* **173**, 71–83.

De Duve, C. (1973). *J. Histochem. Cytochem.* **21**, 941–948.

De Duve, C., and Baudhuin, P. (1966). *Physiol. Rev.* **46**, 323–357.

Feierabend, J. (1975). *Planta* **123**, 63–77.

Frederick, S. E., and Newcomb, E. H. (1968). *Science* **163**, 1353–1355.

Frederick, S. E., Gruber, P. J., and Tolbert, N. E. (1973). *Plant Physiol.* **52**, 318–323.

Frederick, S. E., Gruber, P. J., and Newcomb, E. H. (1975). *Protoplasma* **84**, 1–29.

Fukui, S., Kawamoto, S., Yasuhara, S., and Tanaka, A. (1975). *Eur. J. Biochem.* **59**, 561–566.

Gerhardt, B. (1978). Microbodies/Peroxisomes Pflanzlicher Zellen. Springer-Verlag, Berlin and New York.

Gonzalez, E., and Beevers, H. (1976). *Plant Physiol.* **57**, 409–429.

Graves, L. B., and Becker, W. M. (1974). *J. Protozool.* **21**, 771–773.

Gruber, P. J., Trelease, R. N., Becker, W. M., and Newcomb, E. H. (1970). *Planta* **93**, 269–288.

Gruber, P. J., Becker, W. M., and Newcomb, E. H. (1972). *Planta* **105**, 114–138.

Gruber, P. J., Frederick, S. E., and Tolbert, N. E. (1974). *Plant Physiol.* **53**, 167–170.

Halliwell, B., and Butt, V. S. (1974). *Biochem J.* **138**, 217–224.

Hogg, J. F. (1969). *Ann. N.Y. Acad. Sci.* **168**, 209–381.

Huang, A. H. C., and Beevers, H. (1971). *Plant Physiol.* **48**, 637–641.

Huang, A. H. C., and Beevers, H. (1973). *J. Cell Biol.* **58**, 379–389.

Hruban, Z., and Rechcigl, Jr., M. (1969). "Microbodies and Related Particles: Morphology, Biochemistry and Physiology." Academic Press, New York.

Khan, F. R., Soleemuddin, S. M., and McFadden, B. A. (1977). *Arch. Biochem. Biophys.* **183**, 13–23.

Köller, W., and Kindl, H. (1977). *Arch. Biochem. Biophys.* **181**, 236–248.

Krause, G. N., Kirk, M., Heber, U., and Osmond, C. B. (1978). *Planta* **142**, 229–233.

Lazarow, P. B., and de Duve, C. (1976). *Proc. Natl. Acad. Sci. U.S.A.* **73**, 2043–2046.

Leighton, F., Poole, B., Beaufay, H., Baudhuin, P., Coffey, J. W., Fowler, S., and de Duve, C. (1968). *J. Cell Biol.* **37**, 482–513.

McKinley, M. P., and Trelease, R. N. (1978). *Protoplasma* **94**, 249–261.

Mahon, J. D., Fock, H., and Canvin, D. T. (1974). *Planta* **120**, 245–254.

Maxwell, D. P., Armentrout, V. N., and Graves, Jr., L. B. (1977). *Annu. Rev. Phytopathol.* **15**, 119–134.

Mollenhauer, H. H., Morre, D. J., and Kelley, A. G. (1966). *Protoplasma* **62**, 44–52.

Müller, M. (1975). *Annu. Rev. Microbiol.* **29**, 467–483.

Mutto, S., and Beevers, H. (1974). *Plant Physiol.* **54**, 23–28.

Novikoff, A. B., and Goldfischer, S. (1969). *J. Histochem. Cytochem.* **17**, 675–680.

Novikoff, A. B., and Novikoff, P. M. (1973). *J. Histochem. Cytochem.* **21**, 963–966.

Osumi, M., Fukuzumi, F., Teraniski, Y. Tanaka, A., and Fukui, S. (1975). *Arch. Microbiol.* **103**, 1–11.

Paul, J. S., and Volcani, B. E. (1975). *Plant Sci. Lett.* **5**, 281–285.

Rehfeld, D. W., and Tolbert, N. E. (1972). *J. Biol. Chem.* **247**, 4803–4811.

Ruis, H., and Kindl, H. (1970). *Hoppe Seyler's Z. Physiol. Chem.* **351**, 1425–1427.

Ruis, H., and Kindl, H. (1971). *Phytochem* **10**, 2627–2631; 2633–2636.

Schnarrenberger, C., and Fock, H. (1976). *In* "Encyclopedia of Plant Physiology [N.S.], Transport in Plants III" (C. R. Stocking and U. Heber, eds.), Vol. 3, pp. 185–234. Springer-Verlag, Berlin and New York.

Schnarrenberger, C., Oeser, A., and Tolbert, N. E. (1971). *Plant Physiol.* **48**, 566–574.

Schonbaum, G. R., and Chance, B. (1976). *In* "The Enzymes" (P. B. Boyer, ed.), Vol. 13: Oxidation-Reduction, Part C, pp. 363–408. Academic Press, New York.

Sies, H. (1974). *Angew. Chem. Int. ed. Engl.* **13**, 706–718.

Stafford, H. A., Magaldi, A., and Vennesland, B. (1954). *J. Biol. Chem.* **207**, 621–629.

Tchang, F., Mazliak, P., Catesson, A.-M., Kader, J.-C. (1978). *Biol. Cell.* **31**, 191–196.

Theimer, R. R., and Beevers, H. (1971). *Plant Physiol.* **47**, 246–251.

Theimer, R. R., Wanner, G., and Andig, G. (1978). *Cytobiologie* **18**, 132–144.

Tolbert, N. E. (1969). *Ann. N.Y. Acad. Sci.* **168**, 325–341.

Tolbert, N. E. (1971a). *Annu. Rev. Plant Physiol.* **22**, 45–74.

Tolbert, N. E. (1971b). *In* "Methods in Enzymology: Vol. 23, Photosynthesis, Part A"(A. San Pietro, ed.), pp. 665–682. Academic Press, New York.

Tolbert, N. E. (1972). *In* "Algal Physiology and Biochemistry" (W. D. P. Stewart, ed.), pp. 474–504. Blackwell, Oxford.

Tolbert, N. E. (1973). *Symp. Soc. Exp. Biol.* **27**, 215–239.

Tolbert, N. E. (1974). *In* "Methods in Enzymology: Vol. 31, Biomembranes, Part A" (S. Fleischer and L. Packer, eds.), pp. 734–746. Academic Press, New York.

Tolbert, N. E. (1976). *Aust. J. Plant Physiol.* **3**, 129–132.

Tolbert, N. E. (1978). *In* "Methods in Enzymology: Vol. 52, Biomembranes Part C" (S. Fleischer and L. Packer, eds.), pp. 493–505. Academic Press, New York.

Tolbert, N. E., Oeser, A., Kisaki, T., Hageman, R. H., and Yamazaki, R. K. (1969). *J. Biol. Chem.* **243**, 5179–5184.

Tolbert, N. E., Yamazaki, R. K., and Oeser, A. (1970). *J. Biol. Chem.* **245**, 5129–5136.

Yamazaki, R. K., and Tolbert, N. E. (1969). *Biochem. Biophys. Acta* **178**, 11–20.

Veenhuis, M., Van Dijken, J. B., Pilon, S. A., and Harder, W. (1978). *Arch. Microbiol.* **117**, 153–163.

Vigil, E. L. (1973). *Sub. Cell. Biochem.* **2**, 237–285.

Zelitch, I. (1955). *J. Biol. Chem.* **216**, 553–575.

The Endoplasmic Reticulum | *10*

MAARTEN J. CHRISPEELS

The Biochemistry of Plants, Vol. 1

I. STRUCTURE, ISOLATION, AND
   COMPOSITION OF THE ER

A. Structure

Around the turn of this century several light microscopists observed that secretory cells contain a lamellar body that stains intensely with basic dyes. They called this structure the *nebenkern* or ergastoplasm, because of its location next to the nucleus and its presumed role in secretion. A detailed analysis of its structure was impossible because of the low resolving power of the light microscope, and had to await the advent of the electron microscope. The first investigators of cellular ultrastructure noticed that many cells, and especially secretory cells, contain an extensive network of membranes that form interconnecting tubules and cisternae. This network was termed the endoplasmic reticulum (ER) and was first described in plant cells by Buvat and Carasso (1957).

In plant cells, as in animal cells, the membranes of the ER traverse the entire cytoplasm. The membranes delimit interconnecting channels that take the form of tubules or cisternae (flattened sacs). The membrane itself is somewhat thinner than the plasma membrane, measuring 5–6 nm in thickness, and has a "unit membrane" structure, i.e., two electron-dense layers separated by an electron translucent layer. The width of the tubules or cisternae can vary considerably depending on the cell type and its metabolic activity. The lumen (the space between the membranes) is usually electron translucent although electron-dense materials sometimes accumulate there.

The morphology of the ER and its abundance in the cell shows enormous variability depending on the cell type, its metabolic activity, or its stage of development. In cells that secrete proteins or sequester proteins in protein bodies or vacuoles, the reticulum consists of interconnected parallel cisternae studded with ribosomes attached to the cytoplasmic face of the membranes. This form of the ER is known as the rough endoplasmic reticulum (RER). The ribosomes are assembled into polysomes and the RER is a major site of protein synthesis. Cells that secrete lipophilic substances have an extensive network of tubules. The membranes that form these tubules do not bear ribosomes and this form of the ER is called the smooth ER (SER). Cells do not just contain RER cisternae and SER tubules, because many gradations exist between completely rough membranes with a high density of ribosomes and totally smooth membranes. Cisternae bearing ribosomes on one side but not on the other have been termed semirough. Examples of RER cisternae and SER tubules viewed by transmission electron microscopy in conventional thin sections and thick sections are shown in Figs. 1, 2, and 3.

The variety of form of the ER is matched by a multiplicity of functions. The extent to which the ER traverses the entire cell indicates that it may

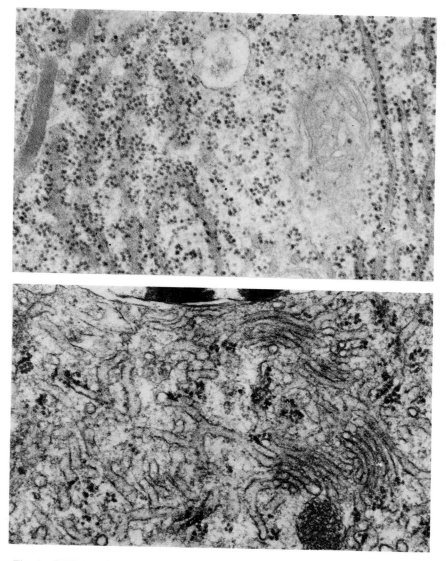

Fig. 1.   RER (top) in the tapetum of oat (*Avena sativa*) anthers. The tapetum, which lines the loculus of the anther, is a secretory tissue with a highly developed ER consisting of RER cisternae, smooth cisternae that ensheath the plastids and mitochondria, and smooth tubules with a wide lumen. Electron-dense material is found in the lumina of both the RER cisternae and the distended tubules. Magnification × 48,000.

Fig. 2.   SER (bottom) in a farina gland of a young petal of *Primula kewensis*. The labyrinths of tubules are thought to be associated with the synthesis of terpenoid substances. Magnification × 55,000. (Unpublished electron micrographs courtesy of B. E. S. Gunning and M. W. Steer.)

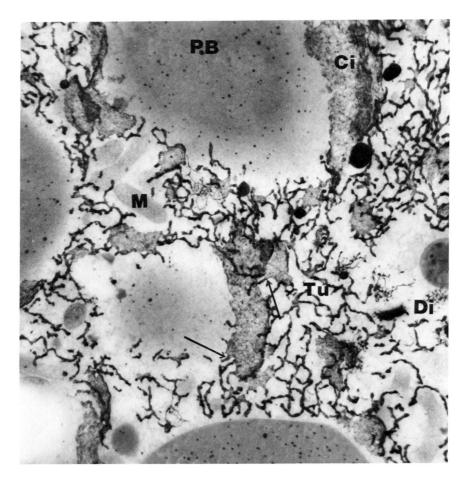

Fig. 3. Electron micrograph of a thick section (0.4 μm) of a cotyledon from a 4-day-old mung bean (*Vigna radiata*) seedling. The tissue was fixed/stained with zinc-iodide in osmium tetroxide, then dehydrated and embedded. This staining procedure makes the ER very conspicuous and tubules (Tu) and cisternae (Ci) are clearly visible as well as the connections between the two (arrows). Mitochondria (M), protein bodies (PB), and dictyosomes (Di) can be readily recognized. Magnification $\times$ 12,500. (Unpublished electron micrograph courtesy of Nick Harris.)

function as a communication or transport system throughout the cell, mediating the intracellular transport of small and large molecules. ER tubules often end at plasmodesmata indicating a possible role in intracellular communication. Yet the ER is more than a passive channel and contains a variety of enzymes which play important roles in cellular metabolism. It is the principal site of membrane synthesis and contributes to the formation of other subcellular organelles. It synthesizes proteins for export or for deposition in vacuoles or protein bodies. It may also play a regulatory role because it possesses binding sites for the hormone indoleacetic acid.

B. Isolation

The ER cannot be isolated in an intact form because it forms an interconnecting network of membranes throughout the cell. Homogenization of the tissue results in the vesiculation of the membranes, and pieces of the original membranes can then be recovered by differential centrifugation. Fragments of ER are a major component of the *microsomal fraction,* a collection of membranous vesicles which sediment when homogenates are centrifuged first at $15,000 \times g$ for 15 min (to remove larger organelles) and than at $100,000 \times g$ for 60 min. However, the microsomal fraction cannot be equated with the ER since the microsomal fraction also contains fragmented dictyosomes, tonoplasts, and plasma membranes. It was once thought that most of the ER could be recovered in the microsomal fraction. However, experiments by Larkins and Davies (1975) indicate that the RER sediments quite readily ($29,000 \times g$ for 10 min) in those isolation media formulated to give maximal yield of undegraded polysomes (high pH, high $MgCl_2$, and high KCl). As a result of such experiments differential centrifugation has been abandoned as the method of choice to obtain good preparations of ER.

A better way to isolate the ER from plant cells is by equilibrium density centrifugation on continuous sucrose gradients. The density of the ER depends on the homogenization medium and especially on the presence or absence of Mg^{2+} and EDTA. A low level of Mg^{2+} in the medium (2–3 mM) does not cause the organelles to aggregate and at the same time it does not allow the ribosomes to dissociate from the RER; EDTA (1 mM) will cause the ribosomes to dissociate making the ER less dense (Lord *et al.,* 1973; Ray *et al.,* 1976). When the tissue is homogenized in a medium containing 1 mM EDTA the ER bands at a density of $1.11–1.12$ g/cm^3. This identification is based on the location in the gradient of the ER-marker enzyme NADH-cytochrome c reductase. If the medium contains 2–3 mM $MgCl_2$, the ER forms a broad band with an average density of 1.17 g/cm^3. The presence or absence of ribosomes on the ER isolated by these two procedures has been confirmed by electron microscopy, but it is not known whether the difference in density can be solely attributed to the ribosomes.

C. Composition

The most detailed analysis and characterization of plant ER has been carried out by Philipp *et al.* (1976) who analyzed the constituents of microsomes isolated from onion root tips and onion stems. Purified rough micro-

somes from onion stems contain 16.4% RNA, 52.6% protein, 16.6% phospholipid, and 14.4% nonpolar lipids. When the ribosomes are removed (with
EDTA) protein accounts for 60–65% of the membrane. Membrane proteins
can be divided into peripheral and integral proteins. Peripheral proteins are
weakly bound to the membrane and can be dissociated by relatively mild
treatments which break ionic interactions. Integral proteins constitute the
bulk of the membrane proteins and can only be solubilized by detergents or
other denaturing agents. Analysis of the integral proteins of the ER of animal
cells by polyacrylamide gel electrophoresis indicates the presence of at least
30 different polypeptides. The ER of plant cells contains an equally diverse
population of proteins (Fig. 4). Some of the ER proteins have been studied in
detail and are known to be largely or exclusively associated with the ER.
These include components of an ER electron transport system (cytochrome
P-450, cytochrome b_5, NADH cytochrome b_5 reductase, and NADPH

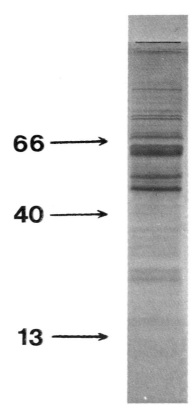

66 ⟶

40 ⟶

13 ⟶

Fig. 4.   SDS-polyacrylamide gel electrophoresis of the polypeptides present in purified ER
preparations obtained from mung bean (*Vignata radiata*) cotyledons. The numbers to the left
indicate molecular weights $\times\ 10^{-3}$.

cytochrome P-450 reductase) and enzymes involved in glycosyl transfer reactions, phospholipid biosynthesis, and hydroxylation or desaturation of organic compounds.

An analysis of the lipids present in the ER of castor bean endosperm showed that these membranes are rich in phosphatidylcholine and phosphatidylethanolamine and contain lesser amounts of phosphatidylinositol and phosphatidyl glycerol (Table I). The fatty acid composition of these phospholipids was palmitic acid (40%), stearic acid (8%), oleic acid (12%), linoleic acid (36%), and linolenic acid (4%). Free fatty acids, diacyl- and triacylglycerides, and small amounts of sterols were also found in the ER (Donaldson and Beevers, 1977).

Much less is known about the substances which are present in the lumina of the tubules and the cisternae, because some of these substances are lost from the ER vesicles during tissue homogenization. The lumina may contain ER-specific enzymes as well as various molecules synthesized or modified by ER-associated enzymes. For example, both animal and plant cells contain a hydroxylase which converts peptidyl–proline into peptidylhydroxyproline. This enzyme plays a role in the biosynthesis of collagen and extensin, extracellular glycoproteins that are secreted via the Golgi apparatus. The enzyme is readily solubilized during homogenization of the tissue, but immunocytochemical experiments show that it is present in the RER of chick embryo tendon cells (Olsen et al., 1973). Similarly, one of the forms of cellulase in indoleacetic acid-treated pea epicotyl tissue appears to be a "soluble" enzyme, yet immunocytochemical experiments show that it is located in the ER (Bal et al., 1976).

TABLE I

Lipid Composition of the ER Isolated from Castor Bean Endosperm[a]

Component	Quantity (nmoles/mg protein)
Phospholipids	1243
Phosphatidyl choline	502
Phosphatidyl ethanolamine	405
Phosphatidyl inositol	206
Phosphatidyl glycerol	33
Cardiolipin	4
Free fatty acids	213
Diacyl- and triacylglycerol	53
Stigmasterol	9
β-sitosterol	6

[a] Data from Donaldson and Beevers (1977).

II. BIOGENESIS AND DEGRADATION

A. The ER as a Dynamic System

The conformation and the amount of ER in a particular cell change with the cell's metabolic activity and stage of differentiation. Cells engaged in the synthesis of proteins which are transported—either to specialized protein storage vacuoles or to the outside of the cell—are almost invariably characterized by an extensive network of RER cisternae, the ergastoplasm of the light microscopists. The induction of enzyme secretion is usually accompanied by the proliferation of the RER. For example, feeding a carnivorous plant results in the proliferation of RER cisternae in the gland cells prior to the secretion of digestive enzymes by these cells (Schnepf, 1963). The tapetum that forms the inner lining of the anther loculus is another tissue with intense secretory activity. Its differentiation is accompanied by a rapid development of ER until the ER becomes the dominant component of the cytoplasm (Echlin and Godwin, 1968).

The development of the cotyledons of leguminous seeds is characterized first by the proliferation of RER cisternae and by the subsequent disappearance of these cisternae and the formation of tubules and vesicles of SER (Öpik, 1968). Morphometric measurements on the proliferation of RER cisternae in *Vicia faba* cotyledons indicate that the cytoplasmic volume occupied by the cisternae increases from 1% to 18% as the cotyledons increase in weight from 40 mg to 120 mg (Briarty, 1973). Selected electron micrographs showing the proliferation of ER do not constitute sufficient evidence that the amount of ER is increasing. Such information can only be obtained by morphometric analysis of numerous electron micrographs or by quantitative measurements of isolated ER. For example, germination of kidney beans and mung beans is accompanied by apparent increase in the RER of the cotyledons (Öpik, 1966; Chrispeels *et al.,* 1976); yet in mung beans, the total amount of ER in the cotyledons declines by more than 50% during the first 5 days of germination (Gilkes *et al.,* 1980). Do such changes involve the biogenesis of new ER and the breakdown of old ER, or is the ER of the cells transformed from one configuration to another?

There is considerable evidence that the ER is a dynamic system which is continuously synthesized and broken down. Incubation of tissues in the presence of radioactive precursors (e.g., choline, glycerol, phosphoric acid, amino acids) results in the rapid incorporation of these precursors into the ER macromolecules. The absence of an appreciable lag in the kinetics of incorporation indicates that the ER is capable of synthesizing its own macromolecules. Pulse-chase experiments show a decline in radioactivity associated with the ER indicating that newly synthesized molecules are degraded or transported to other organelles (Morré, 1970; Kagawa *et al.,* 1973; Moore,

1977). Different molecules turn over at different rates indicating that membrane turnover may be an expression of membrane differentiation.

Where and how in the cell is new ER made? The continuity between the ER and nuclear envelope, and the proximity of the RER to the nucleus (hence the old name *nebenkern*) led to the idea that the nuclear envelope is the site of ER biogenesis. This idea has been abandoned, however, and it is now generally accepted that in animal cells "new" ER arises from "old" ER, although we do not understand how this expansion of existing ER takes place. Since the membrane proteins can move laterally in the plane of the membrane, and lipids can be exchanged between membranes it seems unlikely that cells have old and new regions of ER. Rather, new molecules are probably inserted into existing ER, and each patch of membrane is a mosaic of old molecules and new molecules. The ER has the necessary enzymes to synthesize its own phospholipids (see below) and the RER at least has the capacity for protein synthesis and may synthesize its own proteins. The SER lacks ribosomes and is therefore not capable of protein synthesis. It probably arises from RER either by extension of existing membranes or after detachment of the ribosomes. Experiments with liver cells generally support the concept that there is a precursor–product relationship between RER and SER.

B. Biosynthesis of Specific Membrane Components

1. Fatty Acids

The synthesis of saturated fatty acids involves the action of at least three separate enzymes: fatty acid synthetase, which forms palmitic acid; palmitate elongase, which produces stearic acid; and another elongase which gives rise to the very long-chain fatty acids ($> C_{20}$). Saturated fatty acids can be further modified by hydroxylation and desaturation. (For a review see Harwood, 1975 and Vol. 4, Chapters 6 and 8 of this series.) In animal cells the synthesis of palmitic acid is carried out by large fatty acid synthetase complexes present in the cytosol, but in homogenates of plant tissues these enzymes are found in association with most subcellular organelles, including the microsomal fraction. In green leaves, the chloroplast is the major site of fatty acid synthesis. Careful investigations have shown that in nongreen tissues the plastids are the major or only site of fatty acid synthesis (Zilkey and Canvin, 1971; Vick and Beevers, 1978). The association of fatty acid synthesis with other subcellular fractions may be the result of the disruption of the plastids during homogenization and isolation.

Desaturation of common saturated fatty acids by microsomal fractions of animal cells is well documented, but in plants this reaction is carried out by a soluble enzyme using fatty acids attached to acyl carrier protein as a substrate. Further desaturation (from oleyl-CoA to linoleyl-CoA) is carried out by a microsomal mixed fraction oxidase (Vijay and Stumpf, 1966; Abdel-kader *et al.*, 1973). Other modifications of fatty acids such as the hydroxylation of oleyl-CoA to ricinoleyl-CoA by a mixed function oxygenase are also carried out by microsomal preparations (Galliard and Stumpf, 1966). Mixed function oxygenases which hydroxylate and desaturate a variety of organic compounds have been shown to be associated with the ER (see p. 409).

2. Triacylglycerides and Phospholipids

Phosphatidic acid (1,2-diacylglycerol *sn* phosphate) is a key intermediate in the synthesis of triacylglycerides and phospholipids. The acyl transferases which convert *sn* glycerol-3-P into phosphatidic acid occur where fatty acid synthetases are found: in the chloroplasts of green leaves (Cheniae, 1965; Marshall and Kates, 1974), in the oil droplets of fat storing seeds (Harwood and Stumpf, 1972), and in the ER of castor bean endosperm (Vick and Beevers, 1977). Phosphatidic acid phosphatase, the next enzyme in the biosynthesis of triacylglycerides and most phospholipids (see Fig. 5) is found in the microsomes of spinach leaves and the ER of castor bean endosperm.

The pathways for the biosynthesis of the major phospholipids starting from phosphatidic acid are shown in Fig. 5. The presence of the enzymes necessary for phospholipid synthesis has been demonstrated in several tissues. A careful analysis of the subcellular localization of these enzymes has been carried out by Beevers and collaborators. Using castor bean endosperm tissue they showed that these enzymes band at a density of 1.12 g/cm³

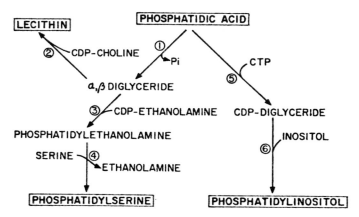

Fig. 5. Pathways of synthesis of three major phospholipids. Enzymes catalyzing reactions 1, 2, 3, 4, and 6 have been shown to occur in the membranes of the ER from castor bean endosperm. After Kennedy (1961); adapted by Beevers (1975).

in sucrose gradients and are associated with the ER (for a review see Beevers, 1975; also see Bowden and Lord, 1975).

3. Proteins

There is considerable evidence that the RER of plant cells, like that of animal cells, is a major site of protein synthesis especially in cells which synthesize proteins for secretion or sequestration within organelles (e.g., protein bodies). Whether the RER also synthesizes its own proteins has not been fully established for plant cells. Membrane proteins are generally more hydrophobic than soluble proteins and it is reasonable to assume that such proteins would be synthesized by membrane-bound rather than by free poly-somes. However, this remains to be demonstrated.

C. Degradation of the ER

At least two different mechanisms of ER degradation can be postulated. Membrane constituents (proteins and lipids) may continually dissociate from the membranes and be degraded in their dissociated state by normal intracel-lular degradative processes. Such a mechanism may account for the normal turnover of membrane constituents. Membranes of ER and of other organ-elles may also be internalized by autophagic vacuoles and digested by lysosomal enzymes. Ultrastructural studies indicate that the central vacuole plays a role in autophagy and autolysis of plant cells, and Matile (1975) has proposed that the vacuole is the major lytic compartment of plant cells. There is now good evidence that the central vacuole contains ribonuclease (Baumgartner and Matile, 1976; Butcher et al., 1977) and the presence of many other acid hydrolases in the vacuole has now been unequivocally demonstrated (Boller and Kende, 1979). There is as yet no evidence that vacuoles contain lipases. Rapid progress is now being made in the isolation of vacuoles from plant cells and this issue may be resolved in the near future.

Seedling growth of kidney beans and mung beans is accompanied by a rapid decline of the ER in the cotyledons. This may therefore be a useful system in which ER catabolism could be studied. These cells lack a central vacuole and the lytic compartment is made up of numerous protein bodies. Indeed, protein bodies contain not only storage proteins, but also a variety of acid hydrolases, especially ribonuclease, phosphatase, phosphodiesterase, protease, phospholipase D, and α-mannosidase (Van der Wilden et al., 1980). Thus they contain most of the enzymes necessary for protein and phos-pholipid breakdown. Ultrastructural evidence shows that protein bodies con-tain vesicles with cytoplasmic contents. It has been suggested that these are autophagic vesicles whose contents are being digested by protein body hy-drolases (Van der Wilden et al., 1980).

III. ROLE OF THE ER IN THE BIOGENESIS OF CYTOPLASMIC ORGANELLES

A. Mitochondria

Phosphatidylcholine and phosphatidylethanolamine are the most abundant membrane lipids in mitochondria as well as all other organelles (with the exception of the chloroplasts where galactolipids predominate). A survey of the biosynthetic properties of mitochondria reveals that they can synthesize saturated and monounsaturated fatty acids as well as acidic phospholipids such as phosphatidylinositol, phosphatidyl glycerol, and diphosphatidyl glycerol. However, they cannot synthesize polyunsaturated fatty acids or the neutral phospholipids phosphatidylcholine or phosphatidyl ethanolamine. To test the hypothesis that mitochondria may derive these components from the ER, Abdelkader and Mazliak (1968) and Kagawa *et al.* (1973) performed pulse-chase experiments with [^{14}C]choline on aged potato tuber slices and castor bean endosperm, respectively. They observed a decline in the radioactivity during the chase, and suggested that this was due to intermembrane lipid transfer. Subsequent experiments (Abdelkader and Mazliak, 1970) showed that such a transfer also occurred *in vitro,* and that nonradioactive mitochondria could accept phospholipids from radioactive liposomes. More recently a protein has been isolated from potato tuber cytoplasm which mediates this transfer of phospholipids (Abdelkader, 1973). These experiments indicate that the ER, as the major site of phosphatidylcholine and phosphatidylethanolamine synthesis in the cell, may supply these two phospholipids to other organelles via a lipid exchange process.

B. Golgi Apparatus

The Golgi apparatus in plant and animal cells is made up of dictyosomes each consisting of a stack of smooth membrane cisternae with associated secretory vesicles. Each dictyosome is a polarized structure with a forming face, in close proximity to the ER, and a maturing face characterized by the presence of mature secretory vesicles (see this volume, Chapter 12). It has been postulated that the dictyosome is a dynamic structure which is continuously formed at the expense of ER membranes by a process of membrane flow (for a review see Morré and Mollenhauer, 1974). The basis for this proposal comes from the following kinds of observations:

1. The forming face of the dictyosome is in close proximity to the ER and physical continuity is sometimes observed; "transition vesicles" are often present between the two.

2. There are many chemical similarities between ER membranes and dictyosome membranes and there is a gradual change in membrane properties from ER to Golgi to plasma membrane.

3. Proteins that are made by the RER of animal cells pass through the Golgi apparatus prior to secretion or sequestration.

4. When secretion is slowed down (by lowering the temperature), more physical connections between the ER and forming face of the dictyosome are seen (Mollenhauer et al., 1975).

While all these observations are consistent with the membrane flow hypothesis, it has been difficult to document experimentally that membrane flow from the ER to the dictyosome actually occurs.

C. Protein Bodies

Seeds contain reserve proteins which are stored in special organelles called protein bodies. These spherical organelles measure 2–10 μm in diameter and consist of a limiting membrane surrounding a protein matrix (usually amorphous, sometimes crystalline). In legume seeds, protein bodies are found in the cotyledons, while in cereals protein bodies are found in the tissues of the endosperm.

Cells of developing legume cotyledons contain numerous RER cisternae and these cisternae are the sites of reserve protein synthesis. Bailey et al., (1970) incubated slices of developing *Vicia faba* cotyledons with radioactive amino acids and demonstrated with autoradiography that the RER is a major site of reserve protein synthesis. They postulated that the membrane-bound polysomes were in the process of synthesizing reserve proteins. Recent experiments by Bollini and Chrispeels (1979) show that the RER is indeed the site of reserve protein synthesis in developing *Phaseolus vulgaris* cotyledons. Isolated free and membrane-bound polysomes synthesized different sets of polypeptides *in vitro,* and the polypeptides of the reserve protein vicilin were made only by the membrane-bound polysomes.

The reserve proteins of legumes are glycoproteins containing small amounts of mannose and N-acetylglucosamine. Recent experiments by Nagahashi and Beevers (1978) showed that the enzymes UDP-mannose mannosyl transferase and UDP-N-acetyl glucosamine N-acetylglucosaminyl transferase which glycosylate the reserve proteins are associated with the ER in pea cotyledons.

Several ultrastructural studies (Öpik, 1966; Harris and Boulter, 1976) on the deposition of reserve proteins in legume cotyledons show that these proteins accumulate in the central vacuole during the early stages of cotyledon development. The central vacuole fragments, giving rise to many irregularly shaped protein bodies. Later reserve proteins are deposited in protein bodies which arise *de novo*. There is circumstantial evidence that the transport of the reserve proteins from the RER to the protein bodies may be mediated by the Golgi apparatus. Electron-dense deposits, resembling in their staining properties the protein in the protein bodies, can often be seen in the secretory vesicles of the dictyosomes (Harris, 1979). This observation raises the possibility that new protein bodies originate from the Golgi-ER-lysosome (GERL) complex. Immunocytochemical experiments with ferritin-labeled antibodies against reserve proteins are needed to resolve the role of the Golgi apparatus in protein body formation.

Zein is the major protein present in the protein bodies of corn endosperm, and Larkins and Dalby (1975) demonstrated that the RER is the major site of zein synthesis in that tissue. They isolated free and membrane-bound polysomes from the endosperm of developing corn seeds and found that zein is made largely or perhaps exclusively by membrane-bound polysomes. Similar experiments were carried out by Burr and Burr (1976) who demonstrated that protein-body-associated polysomes synthesize zein. Zein synthesis probably occurs on both the RER and protein body membranes and the latter may simply be a differentiated form of the former. There is no evidence that the Golgi apparatus mediates the transport of reserve proteins in cereal endosperm and protein body membranes may arise directly from RER cisternae.

D. Glyoxysomes

Glyoxysomes are a special class of microbodies (see Chapter 9) present in the fat-storing tissues of young seedlings. Glyoxysomes have an amorphous protein matrix surrounded by a limiting membrane and contain a variety of enzymes, including those of the glyoxylate cycle (malate synthetase, isocitrate lyase, and citrate synthetase). The biogenesis of glyoxysomes has been investigated most extensively in castor bean endosperm. The endosperm cells of freshly imbibed castor beans contain numerous oil droplets and several large protein bodies, but few cytoplasmic organelles. Germination and seedling growth are accompanied by the biogenesis of mitochrondria, glyoxysomes, and ER. Experiments carried out in the laboratories of Beevers and Lord indicate that the ER plays a major role in the biogenesis of glyoxysomes. Gonzales and Beevers (1976) found that several enzymes of the glyoxylate cycle are associated with the ER before they appear in the glyoxysomes. They fractionated endosperm extracts on isopycnic sucrose gradients and found that malate synthetase—an enzyme normally found in glyoxysomes—banded with the ER as well as with the glyoxysomes. After 2 days of growth, when malate synthetase synthesis had just begun, half of the enzyme activity banded with the ER. This ratio shifted in favor of the glyoxysomes as growth proceeded and enzyme accumulated in the glyoxysomes (Fig. 6). Köller and Kindl (1978) demonstrated in a recent paper that the cosedimentation of malate synthetase with ER under the conditions used by Gonzales and Beevers (1976) may be fortuitous, and that ER and malate synthetase readily separate under other conditions of sedimentation. Further experimentation is needed to resolve this question.

Bowden and Lord (1977) prepared antisera against the glyoxysomal proteins and used them to study the synthesis of glyoxysomal proteins by the ER. Intact endosperm tissue of castor bean was incubated with [35S]methionine and the incorporation of radioactivity into soluble proteins of the ER and the glyoxysomes was followed. The radioactive proteins extracted from the ER and the glyoxysomes were precipitated with the anti-

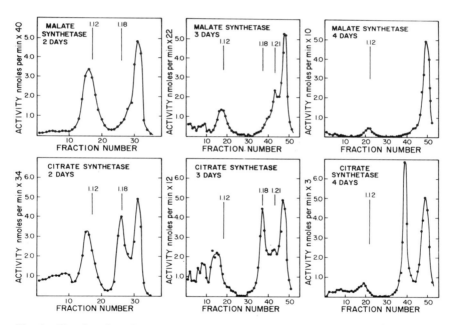

Fig. 6. Fractionation of organelles containing malate synthetase and citrate synthetase of castor bean endosperm. Castor beans were germinated for 2, 3, and 4 days and the endosperm homogenate layered on a 20–48% (wt/wt) sucrose gradient. The ER banded at 1.12 g/cm³ (marker enzyme NADPH-cytochrome c reductase), the mitochondria banded at 1.18 g/cm³ (marker enzymes, malate dehydrogenase, and citrate synthetase), and the glyoxysomes at 1.24 g/cm³ (marker enzyme, catalase). The glyoxysomal ghosts had a density of 1.21 g/cm³. At 2 days more than half of the malate synthetase was associated with the ER, but by 4 days most of it had "moved" to the glyoxysomes. From Gonzalez and Beevers, (1976), reprinted with permission.

sera against the glyoxysomal proteins. The results showed that ^{35}S-labeled antigens appeared without appreciable lag in the microsomal (ER) fraction, whereas a lag period preceded their appearance in the glyoxysomes. A chase with unlabeled methionine resulted in a rapid decrease in the ER associate ^{35}S-labeled antigens. These experiments are consistent with the interpretation that there is a precursor–product relationship between the ER and glyoxysomes.

A number of other experiments indicate that the glyoxysomal membrane is also derived from the ER, possibly as a result of membrane flow. The two types of membranes contain similar polypeptides and similar antigenic determinants, and the phospholipids of the glyoxysomal membrane are synthesized by ER associated enzymes (Kagawa *et al.*, 1973).

E. Vacuoles

One of the distinguishing features of most mature, living plant cells, is the presence of a large central vacuole. Vacuoles have a limiting membrane, called the tonoplast, and contain a variety of organic substances including proteins, tannins, water-soluble pigments, and organic acids. (For a more complete discussion see Chapter 16.) Vacuoles are thought to arise either directly or indirectly from ER cisternae. Numerous ultrastructural studies indicate that vacuoles may arise as local distensions of ER cisternae and suggest that the tonoplast is a differentiated membrane of the ER. Such an interpretation of vacuolar origin comes from studies showing physical continuity between ER cisternae and tonoplasts of small vacuoles (for a review, see Matile, 1975). Other investigators (Marty, 1973a,) suggest that the ER is only indirectly involved in vacuole formation. Marty examined thick sections of various tissues with the high voltage electron microscope and observed that vacuoles are formed when GERL-derived vesicles elongate into tubules and encircle an area of cytoplasm. These tubules then fuse to form an acid phosphatase rich prevacuole with a double membrane. He postulated that the acid hydrolases in the lumen digest the inner membrane and the enclosed cytoplasm, while the outer membrane becomes the tonoplast of the new vacuole. It is clear that further study is needed to show exactly how the ER contributes to vacuole formation.

That the ER and vacuole are ontogenetically related is also suggested by biochemical observations showing that the ER has enzymes which are involved in the biosynthesis of vacuolar substances. Recent evidence indicates that the cyanogenic glucoside, durrhin, present in sorghum seedlings is located in the vacuole (Saunders *et al.*, 1977a). Two mixed function oxidases involved in durrhin biosynthesis are associated with the ER (Saunders *et al.*, 1977b), indicating that durrhin needs to be transported from the ER to the vacuoles.

F. Oil Droplets

Oil droplets (also called oleosomes or spherosomes) consist of an amorphous mass of triacylglycerides surrounded by an osmiophilic coat. Yatsu and Jacks (1972) suggested that this osmiophilic coat is a half-unit membrane, but more recent analytical evidence indicates that the oleosomes do not contain enough phospholipid to be surrounded by a half-unit membrane (Kleinig *et al.*, 1978). Schwarzenbach (1971) suggested that oil droplets in castor bean endosperm originate from ER membranes by the gradual accumulation of triacyglyceridesides between the two monolayers of the unit membrane. Although it is now widely accepted that oleosomes are formed by the ER (for example, see Matile, 1975; Gunning and Steer, 1976), there is at present little evidence to substantiate that conclusion. Recent investigations into the origin of oleosomes in *Sinapis alba* cotyledons (Bergfield *et al.*, 1978) indicate that lipid droplets first appear in the cytoplasm near the surface of plastids (plastids are the site of fatty acid synthesis). The droplets become encircled by a cisterna of the endoplasmic reticulum and at the same time acquire an osmiophilic coat measuring 3 nm in thickness. Analytical evidence indicates that the coat is quite distinct from the ER and contains nine major polypeptides. These observations raise the possibility that the ER functions in the synthesis of these coat proteins and not in the synthesis of triacylglycerides as previously postulated.

IV. ROLE OF THE ER IN THE BIOSYNTHESIS AND SECRETION OF EXTRACELLULAR MOLECULES

A. Secretion of Proteins

The ER was first described in cells which secrete proteins, and it is in these cells that it takes on its most characteristic form i.e., stacks of long cisternae studded with ribosomes. Treatments that induce protein secretion often also induce the proliferation of the RER. However, biochemical evidence that the membrane-bound polysomes synthesize secretory proteins, and that the ER plays a role in the transport of secretory protein is still scant. The best researched case involves the gibberellic acid (GA_3)-mediated synthesis and secretion of α-amylase by the aleurone cells of cereal endosperm. When aleurone tissue is challenged with GA_3 it synthesizes and secretes a number of hydrolytic enzymes including α-amylase, ribonuclease, protease, β-glucanase, and xylanase. Synthesis and secretion start after an 8-h lag and are preceded and accompanied by marked ultrastructural changes: proliferation of RER cisternae, loss of protein body contents, and disintegration of the thick walls of the aleurone cells (Jones, 1969). Using autoradiography

Chen and Jones (1974) showed that the RER of GA_3-treated aleurone cells is a major site of amino acid incorporation and protein synthesis. Treatments that disrupt the normal configuration of the ER such as actinomycin D (Vigil and Ruddat, 1973) or water stress (Armstrong and Jones, 1973) also disrupt the synthesis and/or the secretion of the hydrolytic enzymes.

Is the RER involved in the synthesis and the transport of α-amylase and other hydrolases? Jones and Chen (1976) used an immunocytochemical approach to localize the enzyme within the cells. GA_3-treated aleurone tissue was challenged with fluorescent antibodies against α-amylase and most of the fluorescence was found in the perinuclear region, a region which is also rich in RER cisternae. Efforts to demonstrate that the α-amylase present in homogenates of aleurone tissue is associated with ER-derived vesicles have not been uniformly successful. Jones (1972) found that less than 10% of the total α-amylase in a tissue homogenate was particulate while Gibson and Paleg (1972, 1976) found that half the α-amylase was particulate. They characterized the "particles" and found them to be small vesicles (0.1–0.5 μM) with a density identical to the ER marker enzyme NADH-cytochrome c reductase. The α-amylase activity in the vesicles was latent and could be activated by treatment with Triton X-100. They called these vesicles lysosomes but did not rule out the possibility that they are secretory vesicles.

That these vesicles are indeed derived from the secretory system was recently shown by Locy and Kende (1978) who carried out pulse-chase experiments with aleurone layers of barley. Their experiments indicated that the radioactive α-amylase which was contained in the vesicles turned over rapidly (because the α-amylase was secreted) while the radioactivity which was in cytosolic α-amylase turned over much more slowly (probably because most of it was not secreted). Whether the α-amylase-containing vesicles are secretory vesicles (i.e. vesicles which carry α-amylase from the ER to the plasma membrane) or fragmented ER is still an open question. There is the possibility that the ER mediates secretion of enzymes because cisternae are continuous with the plasma membrane. Such a secretory mechanism may not involve secretory vesicles.

B. Secretion of Polysaccharides

Plant cell walls consist of cellulose microfibrils embedded in a matrix of pectin, hemicellulose, and glycoprotein (extensin). When tissues, which actively synthesize cell wall macromolecules are pulsed with [³H]glucose and examined by autoradiography, silver grains are found over the dictyosomes and the ER. These experiments indicated that both organelles may function in the synthesis or transport of cell wall polysaccharides. The role of the

Golgi apparatus in this process is well documented (see Chapter 12), but the role of the ER is not yet understood. Bowles and Northcote (1972, 1974) have proposed a direct role for the ER in hemicellulose biosynthesis. Their experiments with corn root tips support the view that as much as 90% of the hemicellulose is synthesized by the ER (or at least by a microsomal fraction). Ray and his collaborators, however, did not find any evidence for a direct role of the ER in hemicellulose biosynthesis. Their experiments with pea epicotyls support the view that the Golgi apparatus is the major site of hemicellulose synthesis and that the biosynthetic activities associated with the microsomes are due to the presence of secretory vesicles in that fraction (Ray *et al.,* 1969, 1976).

Cytochemical experiments generally do not support the interpretation that the ER plays a direct role in polysaccharide biosynthesis. Dictyosome cisternae and secretory vesicles stain positively for carbohydrate with either the silver-hexamine or the silver-proteinate stain. Cisternae of the ER, however, do not give a positive reaction, suggesting that they contain little polysaccharide material (for a review of these experiments, see Chrispeels, 1976). Yet, ultrastructural investigations of cell wall formation often show a characteristic distribution pattern of ER cisternae near the sites of deposition of cell wall macromolecules. Such observations are consistent with an indirect role of the ER in cell wall biogenesis. Such an indirect role might include the supply of precursors, primers, or enzyme complexes either to the dictyosomes for hemicellulose synthesis or to the plasma membrane for cellulose synthesis.

C. Secretion of Lipids

Plant cells synthesize and secrete a variety of specific lipids. For example, epidermal cells and cork cells, respectively, synthesize and secrete cutin and suberin into the cell wall. Oil glands synthesize and release terpenes and other volatile lipophilic substances. The cells of the tapetum in the anther secrete small lipid-rich globules (pro-orbicules) coated with sporopollenin, a polymerized form of carotenoid. Ultrastructural studies clearly show that cells which secrete lipids have an extensive ER. For example, the oil glands of *Arctium lappa* have an intricate network of SER tubules (Schnepf, 1969). This is also a property of the cells which secrete the lipophilic stigmatic fluid of *Petunia* (Konar and Linskens, 1966). In this respect these plant cells resemble animal cells active in steroid biosynthesis or drug detoxification. It is rarely possible, however, to detect the precursors of the lipophilic substances within the ER tubules. Dumas (1973) made a study of the stigmatic exudate of *Forsythia* and observed that substances which stained in the same way as the exudate were also present in the ER and the vacuole. However, the site of biosynthesis and the direction of transport could not be deduced from these experiments.

D. Secretion of Sugars

Involvement of the ER in sugar transport was first indicated by certain observations on mammalian liver cells in the process of mobilizing their glycogen reserves. Glycogen granules are normally surrounded by an extensive network of SER tubules which dilate and vesiculate when glycogen catabolism occurs. The observations are consistent with the hypothesis that the tubules collect the sugar and transport it throughout the cells. In plant cells a single ER cisterna is often found closely associated with a plastid. Evert *et al.* (1977) examined the ultrastructure of corn mesophyll and observed numerous cases in which the outer membrane of the chloroplast envelope appeared to be connected to a plasmodesma by means of a short piece of ER. Such observations indicate a role for the ER and the plasmodesmata in the intercellular transport of the products of photosynthesis. Other evidence that the ER may be involved in sugar transport in plant cells comes from ultrastructural observations on nectaries. Nectaries are small glands that secrete nectar, a sugar-rich liquid. The SER is a most conspicuous structure in the cells of nectaries and it normally proliferates at a time when nectar secretion begins. Unfortunately there is no cytochemical evidence that the ER tubules are involved in either the metabolism or the transport of the secreted sugars.

V. OTHER BIOSYNTHETIC AND REGULATORY FUNCTIONS OF THE ER

A. Binding of Auxin

Auxin (indoleacetic acid or other synthetic auxins) enhances the rate of cell elongation in many young plant tissues. While the mechanism of auxin action is not yet understood it is generally accepted that auxin interacts with a receptor, most probably a protein. A search for the intracellular location of this receptor in corn coleoptiles has recently led to the conclusion that the ER is the major site of auxin binding in that tissue. ER-derived vesicles bind α-naphthalene acetic acid with a K_D of $3.8 \times 10^{-7} M$ and the affinity of different auxins for this binding site is correlated with their relative auxin activity (Ray *et al.*, 1977a,b). This finding does not preclude the possibility that in other tissues auxin may also bind to receptors located on other membranes. However, the data do indicate that the ER may play an important regulatory function in the cell.

B. Geotropism

A regulatory role for the ER has also been suggested by investigators of geotropism in roots (for a review, see Juniper, 1976). When the young roots of certain plants are displaced from a vertical to a horizontal position a redistribution of cellular organelles occurs in the statocytes, the cells at the center of the root cap which respond to gravity. The amyloplasts fall to the "bottom" of the cells, while the ER changes its symmetrical distribution and accumulates at the "top." This rather rapid redistribution—it takes only 10 min—does not occur in all species, however, and may be a side effect of the gravitational stimulus rather than the cause of the redirection of growth.

C. Mono-Oxygenases or Mixed Function Oxygenases

Mono-oxygenases catalyze the insertion of one oxygen atom of O_2 into an organic substrate (hydroxylation) while the other oxygen atom is reduced to water. The enzymes require a second substrate to donate electrons to the oxygen atom which is reduced to water and this second substrate is ultimately NADH or NADPH. Different electron carriers are employed to transfer the electrons from NADPH or NADH to the oxygen atom. The ER of both plant and animal cells contains two electron carrier systems for the mono-oxygenases. One consists of the flavoprotein NADPH-cytochrome P-450 reductase and the microsomal cytochrome P-450, and the other one of the flavoprotein cytochrome b_5 reductase and cytochrome b_5. These electron transport components are integral membrane proteins which can only be solubilized when the membranes are treated with detergents (e.g. 1% deoxycholate). In liver cells the mono-oxygenases of the ER catalyze the hydroxylation of many different organic substrates, including steroids, fatty acids, certain amino acids, and a variety of drugs and carcinogenic hydrocarbons. In plant cells microsomal fractions catalyze the hydroxylation of kaurene, steroids, cinnamic acid, and fatty acids. Several recent studies indicate that these enzymes are associated with the ER and not with other membranous organelles present in the microsomal fraction. The enzymes which hydroxylate trans cinnamic acid to p-coumaric acid (Saunders et al., 1977b; Czichi and Kindl, 1977), methylate cycloartenol to form 24-methylene cycloartenol (Hartmann et al., 1977), demethylate N-methylaryl amines (Young and Beevers, 1976), and convert tyrosine into p-hydroxy-mandelonitrile (the precursor of dhurrin) (Saunders et al., 1977b) are all associated with the ER.

VI. CONCLUSION

The endoplasmic reticulum of plant cells displays a versatility of form matched by a diversity of function. Calculations show that 1 cm^3 of cytoplasm contains 1–10 m^2 of ER membrane (Gunning and Steer, 1976), and the ER thus provides the cell with an enormous surface area on which biochemical reactions can take place. The ER is the principal site of membrane synthesis in the cell and membrane components synthesized by the ER contribute to the formation of other membranous organelles. The ER participates in the biosynthesis of macromolecules which will be secreted or transported to vacuoles or protein bodies. The ER may play a role in the channeling of small molecules. It has binding sites for indoleacetic acid and may bind other regulatory molecules. In view of these many functions and structural forms the surface of the ER must be visualized as a mosaic of differentiated regions, a mosaic that is not static but that can be increased, decreased, or changed in response to external or internal stimuli. We have become accustomed to seeing it in electron micrographs as a static structure. We must discard that image and start viewing it as a dynamic organelle capable of performing many functions and able to take on many forms.

ACKNOWLEDGMENTS

I would like to thank Drs. Neil Gilkes, Russell Jones, and Harry Beevers for their critical review of the manuscript. Research in the author's laboratory has been consistently supported by the National Science Foundation.

REFERENCES

Abdelkader, A. B. (1973). *C. R. Hebd. Seances Acad. Sci.* **277**, 1455–1458.
Abdelkader, A. B. and Mazliak, P. (1968). *C. R. Hebd. Seances Acad. Sci.* **267**, 609–612.
Abdelkader, A. B., and Mazliak, P. (1970). *Eur. J. Biochem.* **15**, 250–262.
Abdelkader, A. B., Cherif, A., Demandre, C., and Mazliak, P. (1973). *Eur. J. Biochem.* **32**, 155–165.
Armstrong, J. E., and Jones, R. L. (1973). *J. Cell Biol.* **59**, 444–455.
Bailey, C. J., Cobb, A., and Boulter, D. (1970). *Planta* **95**, 103–118.
Bal, A. K., Verma, D. P. S., Byrne, H., and Maclachlan, G. A. (1976). *J. Cell Biol.* **69**, 97–105.
Baumgartner, B., and Matile, P. (1977). *Biochem. Physiol. Pflanz.* **170**, 279–285.
Beevers, H. (1975). *In* "Recent Advances in the Chemistry and Biochemistry of Plant Lipids" (T. Galliard and E. I. Mercer, eds.), pp. 287–299. Academic Press, New York.
Bergfeld, R., Hong, Y.-N., Kühnl, T., and Schopfer, P. (1978). *Planta* **143**, 297–307.
Boller, T., and Kende, H. (1979). *Plant Physiol.* **63**, 1123–1132.
Bollini, R., and Chrispeels, M. J. (1979). *Planta* **146**, 487–501.
Bowden, L., and Lord, J. M. (1975). *FEBS Lett.* **49**, 369–371.

Bowden, L., and Lord, J. M. (1977). *Planta* **134**, 267–277.
Bowles, D. J., and Northcote, D. H. (1972). *Biochem. J.* **130**, 1133–1145.
Bowles, D. J., and Northcote, D. H. (1974). *Biochem. J.* **142**, 139–144.
Briarty, L. G. (1973). *Caryologia* **25**, 289–301.
Burr, B., and Burr, F. A. (1976). *Proc. Natl. Acad. Sci. U.S.A.* **73**, 515–519.
Butcher, H. C., Wagner, G. J., and Siegelman, H. W. (1977). *Plant Physiol.* **59**, 1098–1103.
Buvat, R., and Carasso, N. (1957). *C. R. Hebd. Seances Acad. Sci.* **244**, 1532–1534.
Cheniae, G. M. (1965). *Plant Physiol.* **40**, 235–243.
Chen, R., and Jones, R. L. (1974). *Planta* **119**, 207–220.
Chrispeels, M. J. (1976). *Annu. Rev. Plant Physiol.* **27**, 19–38.
Chrispeels, M. J., Baumgartner, B., and Harris, N. (1976). *Proc. Natl. Acad. U.S.A.* **73**, 3168–3172.
Czichi, U., and Kindl, H. (1977). *Planta* **134**, 133–143.
Donaldson, R. P., and Beevers, H. (1977). *Plant Physiol.* **59**, 259–263.
Dumas, C. (1973). *Botaniste* **56**, 59–80.
Echlin, P., and Godwin, H. (1968). *J. Cell Sci.* **3**, 161–174.
Evert, R. F., Eschrich, W., and Heysen, W. (1977). *Planta* **136**, 77–89.
Galliard, T., and Stumpf, P. K. (1966). *J. Biol. Chem.* **241**, 5806–5812.
Gibson, R. A., and Paleg, L. G. (1972). *Biochem J.* **128**, 367–375.
Gibson, R. A., and Paleg, L. G. (1976). *J. Cell Sci.,* **22**, 413–425.
Gilkes, N. R., Herman, E. M., and Chrispeels, M. J. (1980). *Plant Physiol., in press.*
Gonzalez, E. and Beevers, H. (1976). *Plant Physiol.* **57**, 406–409.
Gunning, B. E. S., and Steer, M. W. (1976). "Ultrastructure and the Biology of Plant Cells." Arnold, London.
Harris, N. (1979). *Planta* **146**, 63–69.
Harris, N., and Boulter, D. (1976). *Ann. Bot. (London)* **40**, 739–744.
Hartmann, M. A., Fonteneau, P., and Benveniste, P. (1977). *Plant Sci. Lett.* **8**, 45–51.
Harwood, J. L. (1975). *In* "Recent Advances in the Chemistry and Biochemistry of Plant Lipids" (T. Galliard and E. I. Mercer, eds.), pp. 43–93. Academic Press, New York.
Harwood, J. L., and Stumpf, P. K. (1972). *Lipids* **7**, 8–19.
Jones, R. L. (1969). *Planta* **88**, 73–86.
Jones, R. L. (1972). *Planta* **103**, 95–109.
Jones, R. L., and Chen, R. (1976). *J. Cell Sci.* **20**, 183–198.
Juniper, B. E. (1976). *Annu. Rev. Plant Physiol.* **27**,
Kagawa, T., Lord, J. M., and Beevers, H. (1973). *Plant Physiol.* **51**, 61–65.
Kennedy, E. P. (1961). *Fed. Proc. Fed. Am. Soc. Exp. Biol.* **20**, 934–940.
Kleinig, H., Steinki, C., Kopp, C., and Zaar, K. (1978). *Planta* **140**, 233–237.
Köller, W., and Kindl, H. (1978). *FEBS Lett.* **88**, 83–86.
Konar, R. N., and Linskens, H. F. (1966). *Planta* **71**, 356–371.
Larkins, B. A., and Dalby, A. (1975). *Biochem. Biophys. Res. Commun.* **66**, 1048–1054.
Larkins, B. A., and Davies, E. (1975). *Plant Physiol.* **55**, 749–756.
Locy, R., and Kende, H. (1978). *Planta* **143**, 89–99.
Lord, J. M., Kagawa, T., Moore, T. S., and Beevers, H. (1973). *J. Cell Biol.* **57**, 659–667.
Marshall, M. O., and Kates, M. (1974). *Can. J. Biochem.* **52**, 469–482.
Marty, F. (1973a). *C. R. Hebd. Seances Sci.* **276**, 1549–1552.
Marty, F. (1973b). *C. R. Hebd. Seances Acad. Sci.* **277**, 1749–1752.
Marty, F. (1978). *Proc. Natl. Acad. Sci. U.S.A.* **75**, 852–856.
Matile, P. (1975). "The Lytic Compartment of Plant Cells." Springer-Verlag, Berlin and New York.
Mollenhauer, H. H., Morré, D. J., and Vanderwoude, W. J. (1975). *Mikroskopie* **31**, 257–272.
Moore, T. S. (1977). *Plant Physiol.* **60**, 754–758.

Morré, D. J. (1970). *Plant Physiol.* **45,** 791–799.

Morré, D. J., and Mollenhauer, H. H. (1974). *In* "Dynamic Aspects of Plant Ultrastructure" (A. W. Robards, ed.), pp. 84–137. McGraw-Hill, New York.

Nagahashi, J., and Beevers, L. (1978). *Plant Physiol.,* **61,** 451–459.

Olsen, B. R., Berg, R. A., Kishida, Y., and Prockop, D. J. (1973). *Science* **182,** 825–828.

Öpik, H. (1966). *J. Exp. Bot.* **17,** 427–439.

Öpik, H. (1968). *J. Exp. Bot.* **19,** 64–67.

Philipp, E., Franke, W. W., Keenan, T. W., Stadler, J., and Jarasch, E. (1976). *J. Cell Biol.* **68,** 11–29.

Ray, P. M., Shininger, T. L., and Ray, M. M. (1969). *Proc. Natl. Acad. Sci. U.S.A.* **64,** 605–612.

Ray, P. M., Eisinger, W. R., and Robinson, D. G. (1976). *Ber. Deutsch. Bot. Ges.* **89,** 121–146.

Ray, P. M., Dohrman, V., and Hertel, R. (1977a). *Plant Physiol.* **59,** 357–364.

Ray, P. M., Dohrman, V., and Hertel, R. (1977b). *Plant Physiol.* **60,** 585–591.

Saunders, J. A., Conn, E. E., Linn, C. H., and Stocking, C. R. (1977a). *Plant Physiol.* **59,** 647–652.

Saunders, J. A., Conn, E. E., Linn, C. H., and Shimada, M. (1977b). *Plant Physiol.* **60,** 629–634.

Schnepf, E. (1963). *Planta* **59,** 351–379.

Schnepf, E. (1969). *Protoplasma* **67,** 185–194.

Schwarzenbach, A. M. (1971). *Cytobiology* **4,** 145–147.

Van der Wilden, W., Herman E., and Chrispeels, M. J. (1980). *Proc. Natl. Acad. Sci. U.S.A.* (in press).

Vick, B., and Beevers, H. (1977). *Plant Physiol.* **59,** 459–463.

Vick, B., and Beevers, H. (1978). *Plant Physiol.* **62,** 173–178.

Vijay, I. K., and Stumpf, P. K. (1966). *J. Biol. Chem.* **241,** 5806–5812.

Vigil, E. L., and Ruddat, M. (1973). *Plant Physiol.* **51,** 549–558.

Yatsu, L. Y., and Jacks, T. J. (1972). *Plant Physiol.* **49,** 937–943.

Young, O., and Beevers, H. (1976). *Phytochemistry* **15,** 379–385.

Zilkey, B. F., and Canvin, D. T. (1971). *Can. J. Bot.* **53,** 323–326.

Ribosomes | 11

ERIC DAVIES
BRIAN A. LARKINS

I. INTRODUCTION

The term ribosome was first introduced about 20 years ago (Roberts, 1958) to describe a particle made up of approximately equal amounts of RNA and protein that was intimately involved in protein synthesis. At that time, more was known about the ribosomes from animals and plants than about those from bacteria, but that situation has changed considerably, as exemplified by the selection of articles in the most comprehensive review on the subject (Nomura *et al.*, 1974). At least part of the reason for the emergence of *E. coli* as a system for studies on ribosomes is its comparative simplicity. Bacteria (prokaryotes) have only one genome and produce only one type of ribosome

The Biochemistry of Plants, Vol. 1
Copyright © 1980 by Academic Press, Inc.
All rights of reproduction in any form reserved.
ISBN 0-12-675401-2

(70 S type), whereas animals and nonphotosynthetic plants have two genomes, with 70 S ribosomes in the mitochondria and 80 S ribosomes in the cytoplasm. Green plants have a third genome and an additional ribosome of the 70 S type in the chloroplast. There are obvious difficulties encountered when trying to isolate ribosomes when several different types are present in the same tissue, hence it is not altogether surprising that higher plant ribosomes are less well understood than bacterial or mammalian ribosomes. The problem is compounded in plants because of difficulties encountered in isolating ribosomes with intact RNA, and dissociating the ribosomes into subunits.

This chapter will deal primarily with the 80 S cytoplasmic ribosome from eukaryotes, although frequent references will be made to 70 S ribosomes from *E. coli*. Ribosomes from mitochondria and chloroplasts will be discussed by Edelman (this series, Vol. 6, Chapter 6). Some aspects of plant (and chloroplast) ribosomes have been reviewed earlier by Loening (1968a) and recently by Stutz (1976).

II. RIBOSOME STRUCTURE AND BIOGENESIS

Perhaps the major difference between eukaryotes and prokaryotes in regard to ribosome biogenesis is that the latter have no apparent subcellular compartmentalization of the sites of synthesis and assembly of their various ribosomal components. In contrast, strict compartmentalization exists in eukaryotes. The RNA component of ribosomes is synthesized directly from a DNA template in the fibrillar region of the nucleolus, whereas the ribosomal proteins are made on cytoplasmic polyribosomes and must be transported into the nucleolus for assembly (Warner *et al.*, 1973). The subcellular compartmentalization of various facets of ribosome biogenesis is represented diagrammatically in Fig. 1. However, eukaryotes are similar to prokaryotes in that ribosomal proteins aggregate with ribosomal RNA before the rRNA has been completely processed and perhaps before it has been completely transcribed (Hadjiolov, 1977; Krakow and Kumar, 1977). In the normal situation (i.e., in the absence of inhibitory experimental conditions) the synthesis of ribosomal RNA and of ribosomal proteins seems to be closely coordinated, even though synthesis occurs in separate regions of the cell (Perry, 1973).

A. Ribosomal RNA

In all eukaryotes studied so far, ribosomal RNA is synthesized as a large precursor molecule by a specific enzyme, RNA polymerase I. This precursor (35 S in yeast, 45 S in HeLa cells) is cleaved by highly specific RNases

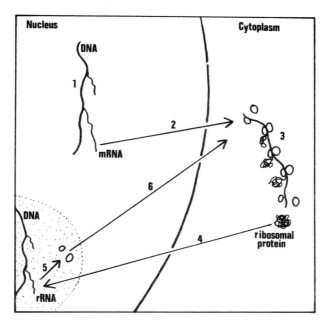

Fig. 1. Diagrammatic representation of ribosome biogenesis in eukaryotes. (1) Transcription of mRNAs for ribosomal proteins in the nucleus; (2) transport of these mRNAs into the cytoplasm; (3) translation on cytoplasmic ribosomes; (4) transport of ribosomal proteins into the nucleolus; (5) transcription and processing of ribosomal RNA with concomitant addition and modification of ribosomal proteins in the nucleolus; (6) maturation of ribosomal precursor into ribosomal subunits and their transportation into cytoplasm.

(Perry, 1976; Hadjiolov, 1977; Krakow and Kumar, 1977) into the 18 S RNA of the small subunit and the 25 S–28 S RNA plus the 5.5–5.8 S RNA of the large subunit (Hadjiolov, 1977). The RNA of the small subunit appears to have been strongly conserved during evolution (Hagenbüchle *et al.*, 1978), whereas the 25–28 S RNA of the large subunit increases from about 1.3×10^6 MW in plants (including algae and fungi) to 1.4, 1.5, 1.6, and 1.7×10^6 in sea urchins and insects, amphibians, birds, and mammals, respectively (Loening, 1973; Wool and Stöffler, 1974). The molecular weights of ribosomal RNA from the large and small subunits from a variety of organisms are given in Table I (from Loening, 1968b). The spacer regions of the precursor molecule, which are not found in the mature RNA, are thought to be degraded rapidly, although recent evidence from *E. coli* suggests that the spacer region contains cistrons for at least three different transfer RNAs. An additional component of the large ribosomal subunit, 5 S RNA, is synthesized by a different enzyme, RNA polymerase III, from an entirely different genetic locus apparently without cleavage from a precursor molecule

TABLE I

Molecular Weights of Ribosomal RNA[a]

Species	Molecular weight ($\times 10^{-6}$)	
	Large subunit	Small subunit
Animals		
HeLa	1.75	0.70
Rat (liver)	1.75	0.70
Mouse (liver)	1.71	0.70
Rabbit (reticulocytes)	1.72	0.70
Chick (liver)	1.58	0.70
Xenopus (tadpole)	1.51	0.70
(liver, ovary)	1.54	0.69
Drosophila	1.40	0.73
Arbacia	1.40	0.68
Plants and protozoa		
Amoeba	1.53 ⎫ Unstable	0.89
Euglena	(1.3?) ⎭	0.85
Tetrahymena	1.30	0.69
Paramecium	1.31	0.69
Higher plants		
Pea, bean, radish, corn	1.27–1.31	0.70–0.71
Algae		
Chorella	1.28	0.69
Chlamydomonas	1.30	0.69
Fern		
Dryopteris	1.34	0.72
Fungi		
Aspergillus	1.30	0.73
Botrytis	1.30	0.68
Chaetomium	1.30	0.71
Rhyzopus	1.28	0.72
Saccharomyces	1.30	0.72
Prokaryotic cells		
Bacteria		
E. coli	1.07	0.56
Rhodopseudomonas	1.08	0.59
Actinomycetes		
Streptomyces	1.11	0.56
Other species	1.13	0.56
Blue–green algae		
Anabaena	1.07	0.55
Nostoc	1.07	0.56
Oscillatoria	1.07	0.56
Higher plant chloroplasts	1.07–1.11	0.56

[a] From Loening (1968b).

(Erdmann, 1976). It appears as though 5 S RNA from *E. coli* is functionally related to eukaryotic 5.8 S RNA and not to eukaryotic 5 S RNA. Further aspects of ribosomal RNA synthesis and processing in plants will be discussed more fully in Vol. 6, Chapter 3 of this series and have been reviewed recently by Dalgarno and Shine (1977).

B. Ribosomal Proteins

The last 5 years have seen a vast increase in the detailed knowledge concerning ribosomal proteins, especially those of *E. coli* and, to a lesser extent, those of yeast and mammals. This results primarily from the development and refinement of a number of techniques (Brimacombe *et al.*, 1976; Kurland, 1977a,b; Stöffler and Wittmann, 1977). The first, upon which most of the others are ultimately dependent, is two-dimensional electrophoresis, especially with the use of SDS in the second dimension. This technique has led to the separation, purification, and identification of the numerous individual proteins. The second, reconstruction of ribosomes from rRNA and ribosomal proteins, has led (primarily in prokaryotes) to a deeper understanding of the function of many of the individual proteins. The third, immunochemical reaction to antibodies raised to specific ribosomal proteins, has led to an appreciation of the similarities and dissimilarities of ribosomes and ribosomal proteins from different sources and of the ribosomal subunits from the same source. Cross-linking of (presumed) adjacent ribosomal proteins with bifunctional reagents has led to ambiguities, because the proteins are not globular and are dispersed about the ribosomal RNA (Kurland, 1977a).

The 80 S ribosomes of eukaryotes are both larger and more complex than the 70 S ribosomes of bacteria. Not only are the RNAs longer, but the proteins are both more numerous and (on the average) larger. This increased complexity of the 80 S ribosome is presumably related to the greater reliance on translational control in eukaryotes, although there is yet no conclusive evidence that this exists (Wool and Stöffler, 1974). Eukaryotic ribosomes contain at least 30 proteins in the small (40 S) subunit and 40 in the large (60 S) subunit (Hadjiolov, 1977) whereas *E. coli* ribosomes contain 21 proteins in the small (30 S) subunit and 30 in the large (50 S) subunit (Kaltschmidt and Wittmann, 1970). They are denoted S_1 to S_{30} and L_1 to L_{40} depending on whether they arise from the small or large subunit and on their mobility in gels. The estimated number of ribosomal proteins in a wide variety of organisms is listed in Table II.

A major difficulty is encountered in defining exactly what a ribosomal protein is. There are proteins involved in ribosome biogenesis in the nucleolus which are associated with precursor ribosomes in that organelle, but do not accompany the mature particle into the cytoplasm (Martini *et al.*,

TABLE II

Number of Proteins in Ribosomes from Various Origins[a,b]

| | Number of proteins | | | | | | |
| | | | | Acidic proteins | | Basic proteins | |
Ribosome origin	Small subunit	Large subunit	Mono-some	Small subunit	Large subunit	Small subunit	Large subunit
Rat liver	30	40	73	2	4	28	36
Rabbit (reticulocyte)	30	46		1	4	29	42
Nicotiana (cytoplasm)			70–80				
Chlamydomonas reinhardtii (cytoplasm)	26	39					
	31	44		4	6	27	38
Euglena gracilis (cytoplasm)	27–32	39–46	66–78				
			77–85				
	20–21	20–21					
	33–36	37–43		0	3	34	39
Escherichia coli	21	34	54[b]	4	7	17	27
Nicotiana (chloroplast)	20–24	34–38					
C. reinhardtii (chloroplast)	22	26					
	25	34		8	5	17	29
	23	21		9		14	
Euglena gracilis (chloroplast)			35–42				
	14	16					
	22–24	30–34	56–60	7	4	15	27

[a] From Freyssinet (1977).
[b] S20 and L26 are identical.

1973). There are other proteins involved in protein synthesis which are found associated with the particle in the cytoplasm but are not present in the nucleolus. Thus one operational definition of permanent ribosomal components would include only those associated with the particle in both compartments (Martini *et al.*, 1973). Another operational definition would include those proteins that remain tenaciously bound to the ribosome even after extensive washing. However, it is not known exactly what conditions should be used to discriminate between tenaciously and transiently-bound proteins. Repeated washing in solutions of high KCl is the most frequently employed method of obtaining "genuine" ribosomal proteins. Since this technique often yields equimolar amounts of some of the proteins, it is assumed to be satisfactory, although one would only expect such yields if all the ribosomes were identical, and this may not always be the case. However, ribosomal proteins can be classified on a stoichiometric basis into at least three classes; "unit" proteins which occur in 0.8–1.2 copies per ribosome; "marginal" proteins which occur in 0.5–0.8 copies per ribosome; and "fractional" proteins which occur in less than 0.5 copies per ribosome. Although these groupings are convenient, they are also arbitrary. The unit protein may represent core proteins and the fractional or marginal proteins may represent functional proteins involved in different stages of mRNA translation (Cox and Godwin, 1975).

C. Ribosome Assembly

The reconstruction of ribosomes *in vitro* has met with much greater success with prokaryotes than eukaryotes (Hadjiolov, 1977). Perhaps successful assembly of proteins can only occur on precursor RNA and different proteins can only assemble during specific stages of the maturation process (Wool and Stöffler, 1974). Conversely, it seems as though the assembly of ribosomes is a prerequisite for the correct processing of the RNA (Perry, 1973). There seems, therefore, to be a high degree of cooperativity between the RNA and protein components. Similarly, cooperativity occurs between different proteins in the assembly (and perhaps the function) of *E. coli* ribosomes (Kurland, 1977a). With *E. coli*, it has been found that at least 18 of the 21 proteins in the small (30 S) subunit may be linked to RNA and that the order in which these proteins are added is crucial (Kurland, 1977a). A recent observation that pinpoints the essential role of RNA/protein and protein/protein interactions in the ribosome assembly process is the ability of 16 S rRNA to exist in at least two conformations which can be changed by proteins S_4, S_7, S_8, and S_{15}. The presence or absence of these proteins partially determines which other proteins can associate with the rRNA (Hochkeppel and Craven, 1977). It is now thought that the rRNA and the maturing ribosomal particles with their current complement of proteins can all undergo conformational transitions which, in turn, can permit or prevent the assembly of subsequent proteins (Hochkeppel and Craven, 1977). These "assembly clusters" may be topographically located in a functional domain (Kurland, 1977a).

D. Ribosome Structure

Information concerning ribosome structure has come primarily from studies employing X-ray diffraction, neutron scattering, electron microscopy, cross-linking with bifunctional reagents, RNA/protein interacting species remaining after limited hydrolysis, visualization of specific antibodies on ribosomes by electron microscopy, and affinity labeling (Kurland, 1977a,b; Pellegrini and Cantor, 1977; Stöffler and Wittmann, 1977). Kurland (1977a,b) seems less willing than others to accept that the vast amount of information obtained by these techniques has markedly advanced our understanding of ribosome structure and function. The apparent dimensions of ribosomal subunits and intact particles vary using different techniques. For instance, X-ray scattering yields dimensions of 220 Å × 220 Å × 55 Å for the *E. coli* 30 S subunit (Van Holde and Hill, 1974), whereas recent electron micrographs yield dimensions of 190 Å × 100 Å × 80 Å (Lake *et al.*, 1974). It would appear that information might only be valid for those instances where different techniques yield similar results. However, since the same techniques (e.g., electron microscopy of immunochemically reacted ribosomes) can yield similar results and yet be interpreted entirely differently by different groups (e.g., Lake, 1976; *cf.* Stöffler and Wittmann, 1977), a real understanding of ribosome structure lies in the future (Kurland, 1977a). Earlier experiments in which ribosomes were reconstructed from component parts, and the deficiency of a particular protein was linked to a deficiency in a particular function, may not be so straightforward to interpret as was thought. This is especially true in light of the recent finding that a number of different proteins can cause changes in rRNA conformation (Hochkeppel and Craven, 1977). It is further heightened by the realization that many ribosomal proteins are not isolated globules on the surface of the ribosome, but may exist in a linear conformation and stretch over much of the ribosomal subunit's surface. Thus one protein could cause changes in the conformation of other proteins, of the rRNA, or of the entire particle so that the modification of a specific protein may not be directly involved in the modification of a particular function. The acceptance of this kind of cooperativity and interaction has led to the concept of "assembly clusters" or "functional clusters" of ribosomal proteins interacting with rRNA to give topographical and functional domains (Kurland, 1977a). Except for some diagrammatic representations of *E. coli* ribosomes shown in Fig. 2 (from Lake, 1976) and EM pictures of crystalline sheets of membrane-bound polyribosomes from lizard ooctyes in Fig. 3 (Unwin, 1977) and the observation that ribosomes seem to be far from the "roughly spherical particles" as they are routinely described; we will offer no more on the presumed details of ribosome structure.

E. Control of Ribosome Content in Eukaryotes

Since one copy of the 18 S, 25–28 S, and 5.8 S RNAs is present in the precursor RNA, this ensures an equimolar production of each. It is not known how the synthesis of 5 S RNA is coordinated. Attainment of a balanced amount of all the rRNAs and ribosomal proteins can be accomplished through control of the rate of synthesis, processing, or degradation of the individual macromolecules. The synthesis of the RNA and protein components seems to be coordinated. When the synthesis of ribosomal proteins is

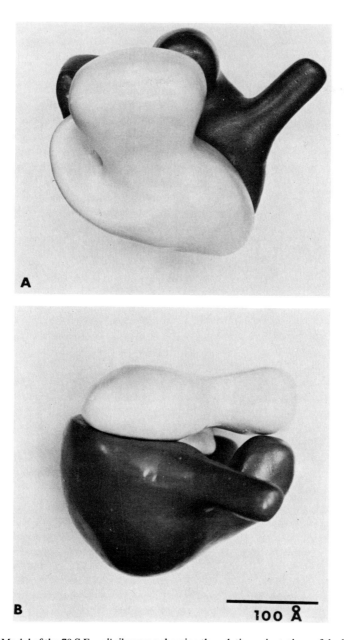

Fig. 2. Model of the 70 S *E. coli* ribosome showing the relative orientations of the large (dark) and small (light) subunits. (A) View of the model corresponding to the mirror image (in order to illustrate the position of the small subunit) of the overlap orientation of the ribosome. (B) View of the model corresponding to the nonoverlap orientation of the ribosome. From Lake (1976).

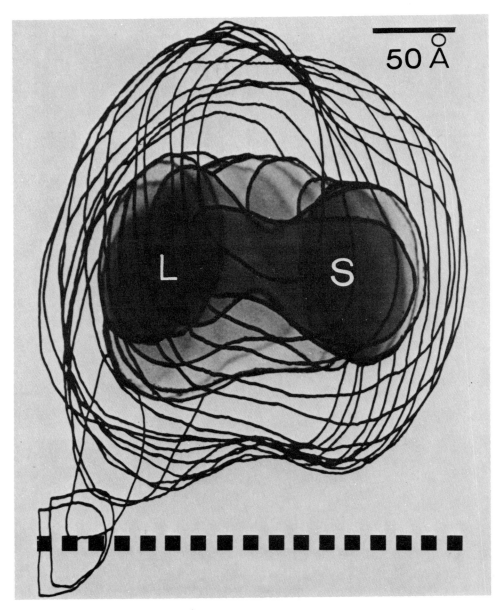

Fig. 3. View of a single ribosome emphasizing the dominant features of the three-dimensional map. L refers to the large subunit, S to the small subunit. The feature protruding from the large subunit may anchor the ribosome to the membrane (dotted line). From Unwin (1977).

inhibited, processing (and further synthesis) of precursor RNA halts (Perry, 1973), but it is uncertain whether it is the presence of the ribosomal proteins themselves or other nucleolar proteins involved in RNA processing that is essential for the continued maturation of ribosomal RNA.

In contrast, the synthesis of ribosomal proteins may continue in the absence of rRNA synthesis, but these proteins are degraded and do not accumulate (Warner, 1977). Their instability may be a result of their flexible, partly helical conformation that makes them especially sensitive to proteases unless they are fixed in ribosomes (Martini et al., 1973). The general observation that ribosomal proteins do not accumulate in the absence of ribosome biogenesis, and the parallel observation that mature ribosomes with permanently-associated proteins are transported from the nucleus to the cytoplasm have met with only occasional contradiction. Dice and Schimke (1972) reported that as much as 70% of the total ribosomal proteins could be found free in the cytoplasm and that these proteins could exchange with those in ribosomes. In a specific attempt to confirm or deny this finding, Wool and Stöffler (1974) were unable to detect any ribosomal proteins free in the cytoplasm even though they used the highly sensitive and specific technique of immunoprecipitation with antibodies raised to individual ribosomal proteins and to whole ribosomes. However, Berger (1977) reported the accumulation (by exchange) of at least three large subunit proteins onto preexisting ribosomes in certain mutants of *Drosophila*. In a rather detailed study, Subramanian and Van Duin (1977) showed that in *E. coli* most proteins do not exchange either *in vivo* or *in vitro,* some exchange equally *in vivo* and *in vitro* (during extraction), whereas L_9 and perhaps L_{33} and S_{21} exchange much more *in vivo*. Whether ribosomes can exist in different functional states cannot be inferred from these results.

Finally, ribosome degradation seems to occur when the rate of protein synthesis declines and monosomes accumulate (Perry, 1973) and in at least some instances, the 60 S subunit turns over more rapidly than the 40 S subunit.

F. Interaction of Ribosomes with Other Subcellular Components

Ribosomal RNAs, ribosomal proteins, or both, must necessarily interact with other macromolecules during protein synthesis. Such macromolecules include tRNA (Letham and Wettenhall, 1977), mRNA (Wettenhall and Clark-Walker, 1977), and proteins involved in initiation, translocation, and termination during protein synthesis (Kurland, 1977a). Some of these aspects will be discussed more fully by Weeks in Vol. 6, Chapter 10 of this series. Interactions also occur between ribosomes and membranes and these will be discussed in Section II,C,2 of this chapter.

III. POLYRIBOSOMES

The sole known function of ribosomes is to provide an environment conducive to accurate translation of the information contained in the base sequence of a mRNA into the amino acid sequence of a polypeptide. Such functional units, with more than one ribosome traversing the same mRNA, are known as polyribosomes or polysomes. Since protein biosynthesis in plants will be discussed in depth in Vol. 6, Chapter 12 of this series, and since a major review of the molecular aspects of protein biosynthesis has appeared recently (Weissbach and Pestka, 1977), this chapter will contain only a brief outline of ribosome function. We will concentrate on other aspects of polyribosome metabolism.

A. Polyribosome Function

The three major steps in protein biosynthesis are initiation, translocation, and termination. The first involves the formation of an initiation complex composed of the initiation codon (base triplet) on mRNA specific for the initiating transfer RNA and the small (40 S) ribosome subunit. After attachment of the 60 S subunit, the ribosome moves relative to the messenger, a second codon is exposed (within the groove between the subunits) and the appropriate tRNA bearing its amino acid is brought into position. With the aid of numerous factors, including many of the ribosomal proteins, various soluble factors, and an energy source, the initiating amino acid is linked (by peptide bond) to the second amino acid. This process, translocation, is repeated with the sequential addition of amino acids to make an increasingly longer nascent polypeptide chain. Termination occurs when the terminating codon is reached and the completed polypeptide is released from the polysome. The ribosome is released (possibly as subunits) and may then reinitiate on the same or a different messenger RNA.

Normally, the longer the messenger, the more ribosomes will be traversing it, each making identical copies of the same polypeptide. Various modifications of the protein (e.g., glycosylation, methylation, cleavage of some amino acids) may take place in a variety of subcellular locations before or after it has been released from the polysome.

B. Polyribosome Isolation

Frequently a major aim in isolating polysomes from a tissue is to provide information concerning the total amount, the subcellular location, and the distribution of various size-classes (e.g., dimers, hexamers, decamers, bearing two, six, or ten ribosomes, respectively) of the polysome population *in vivo*. Briefly, this entails extracting the polysomes, separating them into different size-classes by gradient centrifugation, and monitoring the absorbance. A diagrammatic representation of such a polysome profile is depicted in Fig. 4. The information gained by these techniques can, in turn, yield insights into the metabolic state of the tissue from which the polysomes were derived. However, for such information (and hence for the insights) to be valid, the isolated polysomes must reflect the situation *in vivo* and artifacts must be prevented. This is not easy. The most common artifacts encountered are changes in apparent state of aggregation (i.e. number of ribosomes per polysome); inadequate extraction, causing the total yield to be underestimated; and redistribution between subcellular compartments.

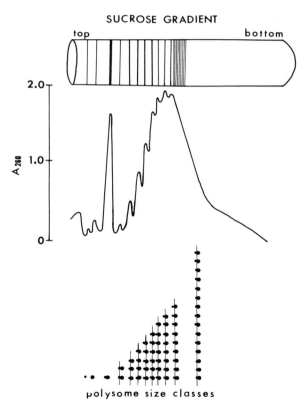

Fig. 4. Diagrammatic representation of a polyribosome profile. A mixture of polyribosomes was layered on a sucrose gradient and centrifuged. Top, the hypothetical banding achieved by the different ribosome size-classes. Center, the actual profile of their uv absorbance. Bottom, each size-class; from the left, the 40 S subunit, the 60 S subunit, monosome, dimer, trimer, etc.

1. Changes in Polysome Aggregation during Extraction

The average number of ribosomes per polysome can increase (aggregation) or decrease (disaggregation). Disaggregation is more prevalent and can arise from hydrolysis of inter-ribosomal bonds of mRNA by ribonuclease; from mechanical shearing; or from release of ribosomes after chain termination, i.e., ribosome "runoff" (Noll, 1969). These processes can be distinguished by computer analysis (Vassart *et al.,* 1970), or visually, from displays of ribosome distribution. Ribonuclease action, which is by far the most severe problem normally encountered, is characterized by an initial shift from larger to smaller polysomes followed much later by an accumulation of monosomes (Davies and Larkins, 1974). Mechanical shearing results in preferential cleavage of large polysomes, with little or no effect on small polysomes or monosomes, whereas ribosome runoff causes rapid accumulation of monosomes with a gradual shift from larger to smaller polysomes (Noll, 1969; Vassart *et al.,* 1970).

Mechanical shearing is rarely considered to be a major difficulty, and ribosome runoff can usually be prevented by rapid cooling of the tissue (and perhaps pulverizing it) in liquid nitrogen or by inclusion in the extraction medium of inhibitors of ribosome translocation or termination (e.g., cycloheximide). In contrast, degradation of polysomes through hydrolysis of interribosomal bonds of mRNA by endogenous or exogenous RNase is not so easily circumvented.

Ribonucleases are ubiquitous in plant tissues and, even though they may be compartmentalized *in vivo,* the physical forces needed to break the cell wall will also disrupt organelles that may contain RNase. A wide array of additives has been incorporated into extraction buffers to minimize RNase action and these include, but are by no means limited to, bentonite, exogenous RNA, heparin, diethylpyrocarbonate (DEP), ethylene glycol bis (2-aminoethyl ether)-tetraacetic acid (EGTA), KCl, $MgCl_2$, and buffers of high ionic strength and pH. Since these agents, while possibly protecting against ribonuclease action, may also have deleterious side effects, we will attempt to assess their relative usefulness.

Bentonite is a clay mineral which is thought to adsorb RNase and was used earlier in ribosome research; however, even when prepared to exacting directions it can cause ribosome precipitation (Loening, 1968a). Excess RNA provides an alternative substrate for the enzyme, thus rendering protection to polysomal mRNA, but care must be taken to ensure that is contains no RNases. Heparin, a sulphated glycoprotein, is a naturally occurring inhibitor from liver which is effective against plant ribonucleases (Akalehiywot *et al.,* 1977). However, in some tissues where it liberates DNA from nuclei and makes the extract unmanageably viscous it can not be used for initial homogenization. DEP is a highly effective inhibitor of RNases and has been used frequently for polysome isolation from plants (Weeks and Marcus, 1969). However, it inactivates a wide variety of enzymes and its main usefulness may be in treating glassware and solutions for work with isolated RNA. EGTA is a rather specific chelator of calcium and other divalant metal ions. It has been shown to be valuable for polysome isolation because it inhibits Ca^{2+}-activated RNases and prevents polysome precipitation by heavy metal ions (Larkins and Davies, 1973; Jackson and Larkins, 1976). KCl seems to be effective by removing ribosome-bound riboncleases (Breen *et al.*, 1971), but the concentration frequently employed (0.4 *M*) is similar to the concentration (0.5 *M*) used to dissociate monosomes, to remove ribosomes from membranes, and to wash ribosomes of nonribosomal proteins. It can, therefore, create artifacts in polysome aggregation, subcellular distribution, and function. Magnesium also inhibits RNase (Larkins and Tsai, 1977) but, since it can also cause aggregation and precipitation of ribosomes (Akalehiywot *et al.*, 1977), artifacts can arise. Buffers of high pH and ionic strength seem highly effective at inhibiting RNase action in a wide variety of tissues (Davies *et al.*, 1972), but such buffers can cause loss of ribosome-associated factors needed for protein synthesis (Ramagopal and Hsaio, 1973). Despite this, these buffers are generally suitable, although in certain cases they may need supplementing with heparin, RNA, or EGTA. The complete homogenizing medium would also include Mg^{2+} at 10–30 m*M*, K^+ at at least twice that concentration, an osmoticum such as sucrose, perhaps a sulfhydryl reducing agent, and a detergent, if the total polysomes were to be isolated in one fraction. It is advisable to use RNase-free sucrose at all stages.

Artificial aggregation of ribosomes can occur through interactions between ribosomes, between nascent proteins, or perhaps between both. Polysomes isolated from membranes may also tend to aggregate because of incomplete removal of membrane proteins. Ribosome–ribosome interactions are enhanced by divalent cations such as Mg^{2+}, and ribosomes pelleted by Mg^{2+} should be resuspended in buffers low in Mg^{2+} to reverse this aggregation. Nascent protein interactions are rarely reported, but they do occur with zein-synthesizing polysomes and can be prevented by mild protease K treatment (Larkins and Tsai, 1977). Margulies and Michaels (1975) found that it was necessary to treat polysomes from *Chlamydomonas* membranes with protease before they were susceptible to RNase.

2. Subcellular Distribution and Incomplete Extraction of Polysomes

Polyribosomes of the cytoplasmic (80 S) type can exist free in the cytoplasm or attached to membranes. In some earlier studies, this distinction was not recognized and incomplete extraction resulted. Frequently, homogenized tissue was centrifuged at low speed to remove debris and at higher speeds to pellet the mitochondria, and both of these pellets were normally discarded. The polysomes in the supernatant were purified by pelleting through concentrated sucrose; occasionally membrane-bound polysomes were collected from the interface of the sucrose pad. More recently it has been shown (Larkins and Davies, 1975) that the majority of membrane-bound polysomes cosediment with the mitochondria and can be released from the membranes by treatment with detergent. It has also been shown (Leaver and Dyer, 1974; Larkins and Davies, 1975) that sedimenting polysomes through a sucrose pad discriminates against the smaller polysomes, monosomes, and subunits, unless inordinately long periods of centrifugation are employed. Therefore, in most earlier studies the content of membrane-bound polysomes was grossly underestimated and a distorted distribution of free polysomes was presented. Occasionally, however, underestimation of the membrane-bound polysome content was avoided by use of detergents in the original homogenization, but this method prevented isolation of free and membrane-bound polysomes as discrete populations.

C. Free and Membrane-Bound Polysomes

The original suggestion (Siekevitz and Palade, 1960), that polysomes free in the cytoplasm synthesize proteins for internal use by the cell and polysomes attached to the endoplasmic reticulum synthesize proteins for export, has needed modification. First, even though the endoplasmic reticulum is the main site for ribosome attachment, polysomes do bind to membranes of other organelles; second, the eventual destination of the proteins is more varied than was thought; and third, the existence of discrete populations of polysomes bound to different membranes implies a more specific polysome–membrane interaction than was originally envisaged. Even though the latter two aspects have not been investigated in as much detail with plant systems as with animals, we will discuss them here since their implications for intracellular and extracellular compartmentalization are so important. Major reviews on membrane-bound polysomes in animal tissues include those of McIntosh and O'Toole (1976), and Shore and Tata (1977).

1. Organelles with Associated Ribosomes

Ribosomes of the 70 S type found in mitochondria and chloroplasts have been reviewed recently by Stewart (1977) and Whitfield (1977), respectively, and will be discussed by Edelman in Vol. 6, Chapter 6 of this series.

Ribosomes of the 80 S type have been found associated with Golgi, and on the outer membrane (or on endoplasmic reticulum closely appressed to the outer membrane) of mitochondria and chloroplasts. The great majority are, however found associated with the endoplasmic reticulum and many of these ER-bound polysomes are involved in the synthesis of proteins which are channeled through the smooth endoplasmic reticulum/Golgi system for processing (e.g., glycosylation) prior to secretion (Chrispeels, this volume, Chapter 10.

2. Interactions between Polysomes and Membranes

The components of the polysome likely to be involved in attachment to membranes are limited to the ribosome (ribosomal protein and rRNA), the messenger RNA, or the nascent polypeptide; components of the membrane likely to bind ribosomes are the protein and lipid. Evidence exists to support a role for each of them (Shore and Tata, 1977).

Interaction between the ribosome and the membrane is through the 60 S subunit, which binds to stripped rough endoplasmic reticulum much more readily than does the 40 S subunit. The binding is maintained in the presence of divalent cations such as Mg^{2+} and can be partially reversed by EDTA or high concentrations of KCl (McIntosh and O'Toole, 1976). Although some evidence exists to implicate ribosomal RNA, binding appears to involve proteins, either in the ribosome, in the membrane, or both. It is not known whether ribosomes from free polysomes contain an extra protein(s) which inhibits binding or lack protein(s) required for binding, although it has been shown that ribosomal-binding proteins do exist in the membrane and they can migrate within its plane (Ojakian et al., 1977). However, it is entirely conceivable that a membrane protein involved in ribosome attachment (or conversely, a ribosomal protein involved in membrane attachment) could be isolated attached to the membrane, attached to the ribosome, or partitioned between both, depending on the conditions used to strip the ribosomes from the membranes. Even though ribosome–membrane interactions undoubtedly occur, it is unlikely that they alone could provide sufficient specificity to determine whether a particular protein would be made on free or on bound polysomes.

A role for nascent proteins in the binding of polysomes to membranes is well established. Puromycin, which causes premature termination of nascent polypeptides, causes release of ribosomes from membranes and concomitant vectorial discharge of nascent proteins into the lumen of the endoplasmic reticulum (or into the interior of microsomal vesicles in vitro). It has been shown recently that a number of different proteins made on membrane-bound polysomes have a hydrophobic sequence of up to 30 amino acids at the N-terminal end (Blobel and Dobberstein, 1975). It is thought that this hydrophobic sequence can act as a "signal" which binds to, or passes through, the membrane, thereby anchoring the nascent protein to, or pulling the protein through, the membrane. The sequence might be cleaved before or after translation of the protein is completed (Blobel and Dobberstein, 1975). Such a system could obviously confer the specificity necessary to determine which proteins will be synthesized on membrane-bound polysomes.

Even after extensive removal of polysomes from membranes by puromycin and KCl treatment, some RNA remains associated with the membranes.
This RNA, which can be removed by mild RNase treatment, is enriched in
poly(A) content compared with the original polysome fraction and is presumably enriched in mRNA. The attachment of this mRNA to the membranes could be mediated through its poly(A) tract or through a neighboring
oligonucleotide region. The average chain lengths of poly(A) regions from
free and membrane-bound polysomes are usually similar, although one report suggests that, in peas, the mean length of the poly(A) tract from bound
polysomes is about 30% shorter than that from free polysomes (Verma and
Maclachlan, 1976). Unless there are discrete differences between poly(A)
regions of free and bound polysomes, it is difficult to envisage how they
could confer specificity for the site of translation of the message. This lack of
specificity is further supported by the finding that the poly(A)-binding proteins from free and bound polysomes are similar (Janssen *et al.,* 1976). It is
possible, however, that oligonucleotide sequences close to the poly(A) region could confer specificity in binding. This would imply that the 3' end of
the mRNA confers specificity directly through its nucleotide sequence,
whereas the 5' end confers specificity through its translation product (signal).

It is highly probable that in different tissues any one or any combination of
these components (60 S subunits, nascent protein, mRNA) are involved in
binding polysomes to membranes. Polysomes attached through all three
components are likely to be the most tightly bound and exhibit the greatest
specificity in their binding. In contrast, those polysomes bound solely by
mRNA (dangling polysomes) are likely to be most loosely bound, whereas
those bound through the 60 S subunit alone will be the least specific in their
attachment. The type of binding will also influence the site of initiation of
protein synthesis. Binding achieved through the 60 S subunit might require
that initiation commences when a (correct) 40 S subunit–mRNA complex
combines with a 60 S subunit already attached to the membrane, whereas
polysomes bound solely through nascent chains might form the initiation
complex free in the cytoplasm and only become bound when their hydrophobic signal becomes sufficiently long to interact with the membrane.

One of the few studies on polysome–membrane interactions in plants
(Dobberstein *et al.,* 1974) has shown that binding occurs through the 60 S
subunit, the nascent protein and the mRNA and the relative contribution of
each binding mechanism is more similar to that occurring in liver cells than
to animal cells in culture.

3. Types of Protein Synthesized on Membrane-Bound Polysomes

The two major classes of protein synthesized on bound polysomes are the
intrinsic proteins of the membranes and the proteins found within the cister-

nae of organelles (Shore and Tata, 1977). Both classes are intimately involved in organelle biogenesis.

a. **Intrinsic Membrane Proteins.** Membrane proteins exist on the cisternal (noncytoplasmic) face of the membrane (ectoproteins), traversing the membrane (transmembrane proteins), or on the cytoplasmic face of the membrane (endoproteins). Synthesis of ectoproteins and transmembrane proteins is thought to take place on bound polysomes (so that the hydrophobic signal on the nascent protein can be inserted into the membrane during synthesis), whereas endoproteins could be synthesized on free polysomes and then inserted into the cytoplasmic face of the membrane after release of the completed protein (Shore and Tata, 1977). Since the intrinsic membrane proteins of the various endomembranous systems are not identical, some subcellular partitioning must occur either in their site of synthesis, in their redistribution after synthesis, or in their selective loss from different membranes. Partitioning of the sites of synthesis is supported by the recent finding (Elder and Morré, 1976) that polysomes attached to Golgi vesicles synthesize some of the membrane proteins specific to that organelle. On the other hand, redistribution after synthesis is supported by evidence that newly synthesized intrinsic membrane proteins can migrate within the plane of the endoplasmic reticulum to the Golgi membranes (Autuori et al., 1975). We are not aware of any evidence to support a selective loss of specific proteins from different membranes, although such a mechanism could explain the relative paucity of ribosomal binding proteins on organelles other than the endoplasmic reticulum.

b. **Cisternal Proteins.** These proteins may eventually be secreted (presumably via the endoplasmic reticulum/Golgi system) or they may remain within the cell inside organelles such as lysosomes, microbodies, and protein bodies. There are a number of recent examples from plant systems, where it has been shown that cisternal proteins are synthesized primarily on membrane-bound polysomes and where specific mRNAs are located predominantly in the membrane fraction. These examples include: buffer-soluble and buffer-insoluble cellulase (Verma et al., 1975), both of which must be secreted through the plasma membrane in order to reach their substrate in the cell wall; zein, the major storage protein produced by developing corn kernels, which is stored in protein bodies within the endoplasmic reticulum (Larkins et al., 1976); the glyoxysomal isozyme of malate dehydrogenase (Walk and Hock, 1978); the small subunit of the photosynthetic enzyme, ribulose biphosphate carboxylase (Cashmore et al., 1978; Highfield and Ellis, 1978), which is made on 80 S-type ribosomes and must be secreted into the chloroplast; and perhaps, α-amylase and other hydrolytic enzymes

(Ho and Varner, 1974), which are synthesized in aleurone cells of barley and other grasses and secreted into the endosperm. In at least some of these cases, there is evidence that the product formed *in vitro* is longer than that formed *in vivo*. This suggests that the normal cleavage (possibly of the hydrophobic signal) does not take place *in vitro*, in the absence of endoplasmic reticulum membranes, as has also been found with a number of animal systems (Shore and Tata, 1977).

D. Changes in Polysome Aggregation

Changes in aggregation, especially the conversion of monosomes and small polysomes into large polysomes, have been reported for a wide variety of tissues exposed to various treatments or undergoing different stages of development. Such treatments include hormones, e.g., auxins (Davies and Larkins, 1973), gibberellins (Evins and Varner, 1972), and cytokinins (Fosket *et al.*, 1977), as well as light (Smith, 1976) and hydration (Bewley, 1973); and the developmental stages include germination (Marcus *et al.*, 1966); dormancy breaking (Leaver and Key, 1967) and different periods in the cell cycle (Baumgartel and Howell, 1977).

Results of such experiments are commonly reported as the ratio of polysomes (or larger polysomes) to total ribosomes. A frequent assumption in such instances is that the formation of large polysomes is indicative of an increased rate of protein synthesis *in vivo*. This assumption may not be valid. The number of ribosomes per polysome is a function of both transcription, i.e., the length of the translated portion of the message and of translation, i.e., the spacing of ribosomes on the message. Spacing is governed by the relative rates of initiation, translocation, and termination. Using a computer simulation study, Vassart *et al.* (1971) showed that with a constant population of mRNAs, polysome aggregation could result from either an increased rate of initiation or a decreased rate of termination. The former would, indeed, lead to an increased rate of protein synthesis, but the latter would lead to a lower rate. Aggregation could also arise from an increase in the number or length of the complement of mRNAs, in the absence of any change in initiation or termination rates. Consequently, the expression of changes in polysome aggregation as ratios or percentages alone provides no information on how aggregation occurred, nor on the rate of protein synthesis *in vivo*. Additional information is required to answer these questions.

One of the more complete attempts to provide such information using a plant system is the recent report (Baumgartel and Howell, 1977) on changes in protein synthesis and polyribosome aggregation during different stages of the cell cycle of light/dark synchronized *Chlamydomonas reinhardi*. These workers were unable to make any definite statements on the rate of protein synthesis *in vivo* solely on the basis of amino acid uptake and incorporation.

Using mainly pulse-labeling of polysomes *in vivo,* they were, however, able to demonstrate (1) that the relative elongation rate (ribosome translocation) per cell varied two-fold during the cell cycle with a maximum occurring at the light/dark transition period; (2) that the proportion of ribosomes as polysomes was consistently high (about 70%) during the light period, but declined steadily to about 15% during the dark; and (3) that the relative initiation rate (ribosome recruitment) varied at least 25-fold with a steady increase during the light period and a decline during the dark. They were able to conclude that the rate of protein synthesis (and polysome formation) *in vivo* in *C. reinhardi* is governed primarily by the rate of initiation. Similar effects of light at the posttranscriptional level have been reported for higher plants (Smith, 1976).

REFERENCES

Akalehiywot, T., Gedamu, L., and Bewley, J. D. (1977). *Can. J. Biochem.* **55,** 901–904.

Autuori, F., Svensson, H., and Dallner, G. (1975). *J. Cell Biol.* **67,** 700–714.

Baumgartel, D. M., and Howell, S. H. (1977). *Biochemistry* **16,** 3182–3189.

Berger, E. (1977). *Mol. Gen. Genet.* **155,** 35–40.

Bewley, J. D. (1973). *Plant Physiol.* **51,** 285–288.

Blobel, G., and Dobberstein, B. (1975). *J. Cell Biol.* **67,** 852–862.

Breen, M. D., Whitehead, E. I., and Kenefick, D. G. (1971). *Plant Physiol.* **49,** 733–739.

Brimacombe, R., Nierhaus, K. H., Garrett, R. A., and Wittmann, H.-G. (1976). *Prog. Nucleic Acid Res. Mol. Biol.* **18,** 1–44.

Cashmore, A. R., Broadhunt, M. K., and Gray, R. E. (1978). *Proc. Natl. Acad. Sci. U.S.A.* **75,** 655–659.

Cox, R. A., and Godwin, E. (1975). *MTP Int. Rev. Sci. Biochem. Ser One* **7,** 179–253.

Dalgarno, L., and Shine, J. (1977). *In* "The Ribonucleic Acids" (P. R. Stewart and D. S. Letham, eds.), pp. 195–232. Springer-Verlag, Berlin and New York.

Davies, E., and Larkins, B. A. (1973). *Plant Physiol.* **52,** 339–345.

Davies, E., and Larkins, B. A. (1974). *Anal. Biochem.* **61,** 155–164.

Davies, E., Larkins, B. A., and Knight, R. H. (1972). *Plant Physiol.* **50,** 581–584.

Dice, J. F., and Schimke, R. T. (1972). *J. Biol. Chem.* **247,** 98–112.

Dobberstein, B., Volkmann, O., and Klämbt, D. (1974). *Biochim. Biophys. Acta* **374,** 187–196.

Elder, J. H., and Morré, D. J. (1976). *J. Biol. Chem.* **251,** 5054–5068.

Erdmann, V. A. (1976). *Prog. Nucleic Acid Res. Mol. Biol.* **18,** 45–90.

Evins, W. H., and Varner, J. E. (1972). *Plant Physiol.* **49,** 348–352.

Fosket, D. E., Volk, M. J., and Goldsmith, M. R. (1977). *Plant Physiol.* **60,** 554–562.

Freyssinet, G. (1977). *Biochimie* **59,** 597–610.

Hadjiolov, A. A. (1977). *Trends Biochem. Sci.* **2,** 84–86.

Hagenbüchle, V., Santer, M., Steitz, J. A., and Mans, R. J. (1978). *Cell* **13,** 551–563.

Highfield, P. E., and Ellis, R. J. (1978). *Nature (London)* **271,** 420–424.

Ho, D. T.-H., and Varner, J. E. (1974). *Proc. Natl. Acad. Sci. U.S.A.* **71,** 4783–4786.

Hochkeppel, H.-K., and Craven, G. R. (1977). *J. Mol. Biol.* **113,** 623–634.

Jackson, A. O., and Larkins, B. A. (1976). *Plant Physiol.* **57,** 5–10.

Janssen, D. B., Counotte-Potman, A. D., and Van Venrooij, W. J. (1976). *Mol. Biol. Rep.* **3,** 87–95.

Kaltschmidt, E., and Wittmann, H.-G. (1970). *Anal. Biochem.* **36**, 401–412.

Krakow, J. S., and Kumar, S. A. (1977). *In* "Comprehensive Biochemistry" (M. Florkin, A. Neuburger, and L. L. M. Van Deenen, eds.), Vol. 24, pp. 105–184. Elsevier, Amsterdam.

Kurland, C. G. (1977a). *Annu. Rev. Biochem.* **46**, 173–200.

Kurland, C. G. (1977b). *In* "Molecular Mechanisms of Protein Biosynthesis" (H. Weissbach and S. Pestka, eds.), pp. 81–116, Academic Press, New York.

Lake, J. A. (1976). *J. Mol. Biol.* **105**, 131–159.

Lake, J. A., Sabatini, D. D., and Nonomura, Y. (1974). *In* "Ribosomes" (M. Nomura, A. Tissières, and P. Lengyel, eds.), pp. 543–557. Cold Spring Harbor Lab., Cold Spring Harbor, New York.

Larkins, B. A., and Davies, E. (1973). *Plant Physiol.* **52**, 655–659.

Larkins, B. A., and Davies, E. (1975). *Plant Physiol.* **55**, 749–756.

Larkins, B. A., and Tsai, C. Y. (1977). *Plant Physiol.* **60**, 482–485.

Larkins, B. A., Jones, R. A., and Tsai, C. Y. (1976). *Biochemistry* **15**, 5506–5511.

Leaver, C. J., and Dyer, J. A. (1974). *Biochem. J.* **144**, 165–167.

Leaver, C. J., and Key, J. L. (1967). *Proc. Natl. Acad. Sci. U.S.A.* **57**, 1338–1344.

Letham, D. S., and Wettenhall, R. E. H. (1977). *In* "The Ribonucleic Acids" (P. R. Stewart and D. S. Letham, eds.), pp. 129–193. Springer-Verlag, Berlin and New York.

Loening, U. E. (1968a). *In* "Plant Cell Organelles" (J. B. Pridham, ed.), pp. 216–227, Academic Press, New York.

Loening, U. (1968b). *J. Mol. Biol.* **38**, 355–365.

Loening, U. E. (1973). *Biochem. Soc. Symp.* **37**, 95–104.

Marcus, A., Feeley, J., and Volcani, T. (1966). *Plant Physiol.* **41**, 1167–1172.

Margulies, M. M., and Michaels, A. (1975). *Biochim. Biophys. Acta* **402**, 297–308.

Martini, O. H. W., Gould, H. J., and King, U. W. S. (1973). *Biochem. Soc. Symp.* **37**, 51–68.

McIntosh, P. R., and O'Toole, K. (1976). *Biochim. Biophys. Acta* **457**, 171–212.

Noll, H. (1969). *In* "Techniques in Protein Biosynthesis" (P. N. Campbell and J. R. Sargent, eds.), Vol. 2, pp. 101–179, Academic Press, New York.

Nomura, M., Tissières, A., and Lengyel, P. (1974). "Ribosomes." Cold Spring Harbor Lab., Cold Spring Harbor, New York.

Ojakian, G. K., Kriebich, G., and Sabatini, D. D. (1977). *J. Cell Biol.* **72**, 530–551.

Pellegrini, M., and Cantor, C. R. (1977). *In* "Molecular Mechanisms of Protein Biosynthesis" (H. Weissbach and S. Pestka, eds.), pp. 203–244. Academic Press, New York.

Perry, R. P. (1973). *Biochem. Soc. Symp.* **37**, 105–116.

Perry, R. P. (1976). *Annu. Rev. Biochem.* **45**, 605–629.

Ramagopal, S., and Hsiao, T. C. (1973). *Biochim. Biophys. Acta* **299**, 460–467.

Roberts, R. B. (1958). *In* "Microsomal Particles and Protein Synthesis" (R. B. Roberts, ed.), p. viii. Pergamon, Oxford.

Shore, G. C., and Tata, J. R. (1977). *Biochim. Biophys. Acta* **472**, 197–236.

Siekevitz, P., and Palade, G. E. (1960). *J. Biophys. Biochem. Cytol.* **7**, 619–630.

Smith, H. (1976). *Eur. J. Biochem.* **65**, 161–170.

Stewart, P. R. (1977). *In* "The Ribonucleic Acids" (P. R. Stewart and D. S. Letham, eds.), pp. 271–295. Spring-Verlag, Berlin and New York.

Stöffler, G., and Wittmann, H.-G. (1977). *In* "Molecular Mechanisms of Protein Biosynthesis" (H. Weissbach and S. Pestka, eds.), pp. 117–202. Academic Press, New York.

Stutz, E. (1976). *In* "Plant Biochemistry" (J. Bonner and J. E. Varner, eds.), pp. 15–35. Academic Press, New York.

Subramanian, A.-R., and Van Duin, J. (1977). *Mol. Gen. Genet.* **158**, 1–9.

Unwin, P. N. T. (1977). *Nature (London)* **269**, 118–122.

Van Holde, K. E., and Hill, W. E. (1974). *In* "Ribosomes" (M. Nomura, A. Tissières, and P. Lengyel, eds.), pp. 53–91, Cold Spring Harbor Lab., Cold Spring Harbor, New York.

Vassart, G. M., Dumont, J. E., and Cantraine, F. R. L. (1970). *Biochim. Biophys. Acta* **224,** 155–164.

Vassart, G., Dumont, J. E., and Cantraine, F. R. L. (1971). *Biochim. Biophys. Acta* **247,** 471–485.

Verma, D. P. S., and Maclachlan, G. A. (1976). *Plant Physiol.* **58,** 405–410.

Verma, D. P. S., Maclachlan, G. A., Byrne, H., and Ewings, D. (1975). *J. Biol. Chem.* **250,** 1019–1026.

Walk, R. A., and Hock, B. (1978). *Biochem. Biophys. Res. Commun.* **81,** 636–643.

Warner, J. R. (1977). *J. Mol. Biol.* **115,** 315–333.

Warner, J. R., Kumar, A., Udem, S. A., and Wu, R. S. (1973). *Biochem. Soc. Symp.* **37,** 3–22.

Weeks, D. P., and Marcus, A. (1969). *Plant Physiol.* **44,** 1291–1294.

Weissbach, H., and Pestka, S. (1977). "Molecular Mechanisms of Protein Biosynthesis." Academic Press, New York.

Wettenhall, R. E. H., and Clark-Walker, G. D. (1977). *In* "The Ribonucleic Acids" (P. R. Stewart and D. S. Letham, eds.), pp. 233–264. Springer-Verlag, Berlin and New York.

Whitfield, P. R. (1977). *In* "The Ribonucleic Acids" (P. R. Stewart and D. S. Letham, eds.), pp. 247–332. Springer-Verlag, Berlin and New York.

Wool, I. G., and Stöffler, G. (1974). *In* "Ribosomes" (M. Nomura, A. Tissières, and P. Lengyel, eds.), pp. 417–460. Cold Spring Harbor Lab., Cold Spring Harbor, New York.

The Golgi Apparatus | *12*

HILTON H. MOLLENHAUER
D. JAMES MORRÉ

The Biochemistry of Plants, Vol. 1

I. INTRODUCTION

A Golgi apparatus for plant cells was reported from light microscope preparations by Bowen in 1928, but the interpretations of the light microscope image were to be the subject of considerable controversy. Golgi apparatus can be seen in electron micrographs published by Hodge *et al.* in 1956. In the following 2 years, Golgi apparatus were noted in a variety of plant cells by Porter, Buvat, Heitz, Perner, Setterfield and Bayley, Sitte, Dalton and Felix, Rouiller and Faure-Fremiet, Chardard and Rouiller, and Sager and Palade (see Whaley 1975, for details). However, it was not until the following decade that ultrastructural details of the plant Golgi apparatus became generally recognized.

II. DEFINITION

The recognition of a common architecture among Golgi apparatus is one of the important generalizations resulting from electron microscopy (Mollenhauer and Morré, 1966a; Morré et al., 1971a). It is morphology that serves as the basis for the definition of Golgi apparatus. A biochemical definition, although desirable, is not possible because biochemical markers unique to the plant Golgi apparatus are unknown.

The Golgi apparatus is a component of the endomembrane system of the cell and appears to serve as an intermediate between endoplasmic reticulum and plasma membrane (see Section III). The architecture of the Golgi apparatus will be considered at three levels of organization, i.e., cisternae, dictyosomes, and Golgi apparatus. Additionally, coated and/or smooth-surfaced vesicles of various types may be associated with Golgi apparatus. The structural features characteristic of each level of organization are seen consistently in electron micrographs even though they may vary somewhat with cell type and the metabolic state of the cell.

A. Cisterna

A cisterna is a sac or cavity within a cell or organism. The cisternae of the Golgi apparatus are bounded by smooth-surfaced membranes (lacking ribosomes), are flattened, and consist of a central platelike region continuous with a peripheral system of fenestrae, tubules, and vesicles (Mollenhauer and Morré, 1966a; Cunningham *et al.,* 1966; Fig. 1). In some cisternae, the platelike regions predominate; others consist mainly of fenestrae or tubular elements. Both types of cisternae may exist within a single dictyosome. The platelike regions of the cisternae are typically 0.5–1.0 μm in diameter. The tubules are 300–500 Å in diameter and may extend for several microns from the dictyosome. Both smooth and coated vesicles may be attached to the tubules or to the fenestrated peripheries of the cisternae. The term saccule is synonomous with cisterna.

B. Dictyosome

When cisternae are organized into stacks, the stacks are called dictyosomes (Mollenhauer and Morré, 1966a). There are usually five to eight cisternae per dictyosome (Fig. 2), but 20 or more are not unusual for dictyosomes of some lower organisms.

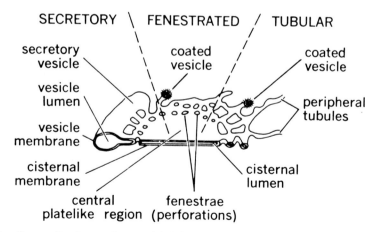

Fig. 1. Composite of several types of Golgi apparatus cisternae. Coated vesicles are distinct from secretory vesicles and appear to be a consistent feature of all Golgi apparatus cisternae. From Mollenhauer and Morré (1971).

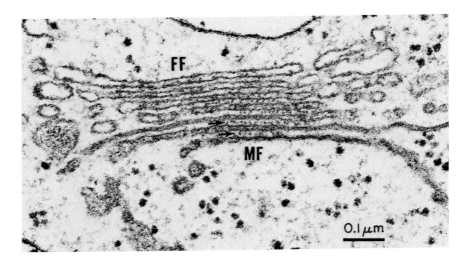

Fig. 2. Most dictyosomes of higher plants consist of a stack of 5-8 cisternae separated from one another by about 90–150 Å. The cisternae appear to develop sequentially across the stack from the forming face (FF) to the maturing face (MF) of the dictyosome. Intercisternal elements (arrows) are present between some of the cisternae and generally increase in number toward the mature pole of the dictyosome. The appearances of the cisternal membranes as well as biochemical analyses of isolated dictyosomes indicate that the membranes of the forming cisternae are like endoplasmic reticulum while those of the mature cisternae are like plasma membrane. The dictyosome of this illustration is from the maize root tip. Glutaraldehyde-osmium tetroxide fixation.

The term dictyosome was used originally by Perroncito (see Whaley, 1975) to designate a component of the Golgi apparatus that was visible following cell division and that had a definite pattern of distribution in the daughter cells. The term was also used (see Bowen, 1928, 1929; Whaley, 1975) to indicate a form of Golgi apparatus characteristically found in invertebrates, which was more lamellar than the reticular apparatus described by Golgi. When the Golgi apparatus was finally demonstrated in plant cells, it was of the dictyosome form. It is interesting to note also that the term dictyosome, as derived from the Greek, means "net body," a definition that closely fits the modern concept of dictyosome structure.

Dictyosomes are polarized structures in that cisternae at one pole or face of the cisternal stack differ from those at the opposite pole or face. In many animal cells, algae, and fungi, the proximal pole, or forming face, of each dictyosome is associated with the nuclear envelope or endoplasmic reticulum in a characteristic manner. In most higher plants, however, such an endoplasmic reticulum—Golgi apparatus association is not as apparent (see Figs. 3, 7, 11, 14, 19). Yet, the membranes of the forming cisternae of all Golgi

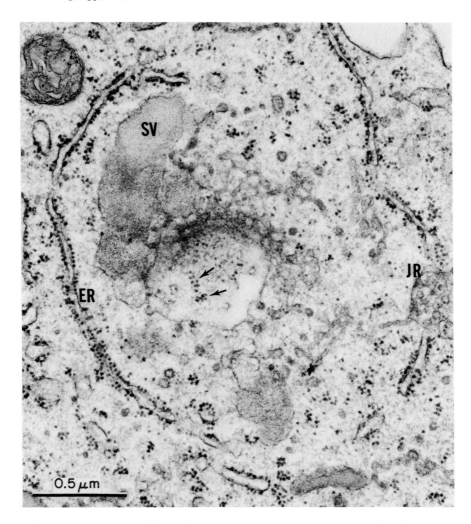

Fig. 3. Dictyosomes of higher plant cells are not closely associated with endoplasmic reticulum (ER) as is common in many animal cells (e.g., Brunner's gland; see Friend, 1965). Yet, associations between ER and plant cell dictyosomes clearly exist either as a loosely-associated amplexis of ER as in this figure (see also Figs. 6 A–F of Morré *et al.,* 1971a) or as a junctional region (JR) near the edges of the dictyosome cisternae (see also Mollenhauer and Morré, 1976a). The dictyosome illustrated here is from an outer root cap cell of maize. Note the polysomes (arrows) on the forming face of the dictyosome. The forming cisterna can be identified easily because it is swollen by OsO_4 fixation and its luminal contents are light. Secretory vesicle (SV). Glutaraldehyde-osmium tetroxide fixation.

apparatus are morphologically (Section VII,A) and cytochemically similar to the membranes of the endoplasmic reticulum. Toward the opposite pole (distal pole or maturing face) the morphology and staining characteristics of the cisternae become progressively more like plasma membrane.

The cisternae within a dictyosome are separated from one another by a minimal space of about 100–150 Å. In most plant cells, layers of parallel fibers, called intercisternal elements (Mollenhauer, 1965a; Turner and Whaley, 1965), are present within the intercisternal regions midway between the surfaces of adjacent cisternae (Figs. 2, 7, 18). The function and/or composition of the intercisternal elements is not known although a role in shaping secretory vesicles has been indicated (Mollenhauer and Morré, 1975).

The peripheral tubules of the cisternae apparently serve as a means of interconnecting adjacent dictyosomes into Golgi apparatus (see below). Additionally, the tubules may help segregate activities of the cisternal lumina from those of the forming vesicles.

C. Golgi Apparatus

The dictyosomes of most cells are interassociated so that they function synchronously. Interassociated dictyosomes form a Golgi apparatus. The number of dictyosomes within a Golgi apparatus may range from none [as in some fungal cells where single cisternae or tubules function as Golgi apparatus equivalents (Franke et al., 1971a; Bracker, 1974)] to more than 25,000 in rapidly growing hyphal tips or pollen tubes (Rosen, 1968; Grove and Bracker, 1970). When cells contain a single dictyosome, then that dictyosome is the Golgi apparatus. In most instances, rudimentary Golgi apparatus can be differentiated from the other membranous components of the cell by the zones of exclusion (see Section III,E) and the kinds of vesicles associated with them. It should be noted that the word apparatus may denote either singular or plural.

III. ASSOCIATION WITH OTHER CELL COMPONENTS: THE ENDOMEMBRANE CONCEPT

The concept of an endomembrane system was proposed to explain the functional continuum that exists between the membranous compartments of the eukaryotic cell (Morré and Mollenhauer, 1974). Included within the endomembrane system are the nuclear envelope, rough and smooth endoplasmic reticulum, Golgi apparatus, and various cytoplasmic vesicles. Plasma membranes, vacuolar membranes, and lysosomes are considered as end products of the system. Organelles such as mitochondria and chloroplasts are not usually included as part of the endomembrane system even though their outer membranes may be closely associated with, or directly connected to, the endoplasmic reticulum (Franke, 1971; Bracker *et al.*, 1971).

A functional continuum among the components of the endomembrane system has been established by documentation of direct membrane continuity like that between nuclear envelope and endoplasmic reticulum, by tubular transfer elements like those between endoplasmic reticulum and Golgi apparatus, or by vesicular transfer elements like those between Golgi apparatus and plasma membrane. The relationship of the Golgi apparatus to the endomembrane system is discussed in more detail in the sections that follow.

A. Endoplasmic Reticulum

Functional continuity between endoplasmic reticulum and Golgi apparatus was first deduced from observations of the morphological relationships between the two structures. These relationships are generally expressed as an alignment of endoplasmic reticulum cisternae near the forming face of a dictyosome (see Figs. 6 A–F of Morré *et al.*, 1971a; Figs. 3, 11). Presumably, continuity between the two structures is provided by small smooth-surfaced vesicles that bleb from the endoplasmic reticulum, move to the forming face of the dictyosome, and fuse to form new cisternae. Both membrane and vesicle content are thus transferred from the endoplasmic reticulum to the Golgi apparatus. However, as pointed out in the previous section, the endoplasmic reticulum is not in strict juxtaposition to the forming faces of the dictyosomes in most higher plants. Therefore, other criteria such as membrane appearance and size (see Section VII,A), biochemical analyses (see Section VII,C), and cytochemical analyses (see Section VII,B) must be used to deduce the relationship. Other observations, including those with higher plants, suggest that the endoplasmic reticulum is at least intermittently continuous with the peripheral tubules of the dictyosomes (Benbadis and Deysson, 1975; Mollenhauer *et al.*, 1975; Mollenhauer and Morré, 1976a) and with the GERL complex at the mature face (Novikoff *et al.*, 1971; Whaley, 1975; Marty, 1978). This relationship between Golgi apparatus and endoplasmic reticulum is most easily visualized in thick sections of tissues impregnated with heavy metals (Fig. 4).

Functional continuity between endoplasmic reticulum and Golgi apparatus has been demonstrated as well by autoradiographical studies showing sequential incorporation and transfer of product through the cell (Jamieson and Palade, 1967a,b; Flickinger, 1974a,b; Chang *et al.*, 1977). Additionally, cytochemical and biochemical analyses confirm the similarity between endoplasmic reticulum and parts of the Golgi apparatus (Holtzman and Dominitz, 1968; Morré, 1975).

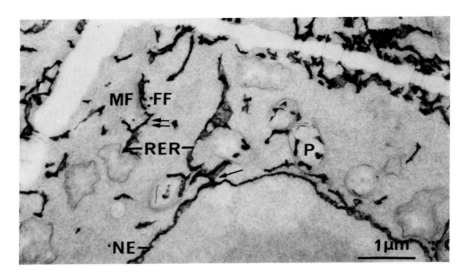

Fig. 4. The endoplasmic reticulum and Golgi apparatus are easily visualized following prolonged exposure to aqueous solutions of osmium tetroxide. Deposits of reduced osmium fill the lumina of the rough endoplasmic reticulum (RER), the nuclear envelope (NE), and the forming cisternae (FF) of the dictyosomes. The endoplasmic reticulum is most closely associated with the peripheral edges of the dictyosomes in the manner illustrated at the double arrows. Cortical cells of *Ricinus* (castor bean). Maturing face of dictyosome (MF), plastid (P). From Morré and Ovtracht (1977).

B. Plasma Membrane

Continuity between Golgi apparatus and plasma membrane is provided by secretory vesicles that move from the dictyosomes to the cell surface (Mollenhauer and Morré, 1966a; Whaley, 1975). Here, the membranes of the secretory vesicles fuse with the plasma membrane and the contents of the secretory vesicles are discharged from the cell (Fig. 5). This secretory route accounts for the direct transfer of plasma membrane, or plasma membrane constituents, from the Golgi apparatus to the cell surface. In fast growing cells such as pollen tubes and hyphal tips, almost all of the plasma membrane appears to be derived from the Golgi apparatus via this route (Morré and VanDerWoude, 1974; Grove *et al.,* 1970).

C. Vacuoles

An origin of the tonoplast from vesicles of the Golgi apparatus has been suggested in numerous reports. Most authors, however, suggest the endoplasmic reticulum as the primary source of the vacuolar membranes (see

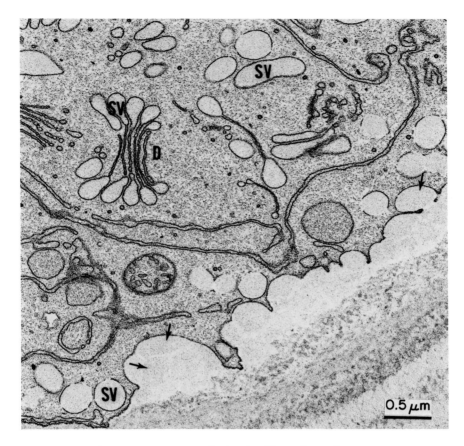

Fig. 5. Part of an outer root cap cell of maize illustrating a dictyosome (D), secretory vesicles (SV), and the contents of secretory vesicles (arrows) that have accumulated between the plasma membrane and cell wall. Potassium permanganate fixation.

Morré, 1975; Morré and Mollenhauer, 1976 for literature). Nonetheless, transfer of Golgi apparatus-derived product into the vacuoles of some cells seems likely. By following an electron-dense substance induced in maize roots by 2,4,6-trichlorophenoxyacetic acid, Mollenhauer and Hanson (1976) showed that secretory product from the Golgi apparatus was incorporated into the vacuolar system by an autophagic mechanism. Here, the membranes of the secretory vesicles did not fuse with the tonoplast (Fig. 6). However, much additional work needs to be done to establish the exact relationship between Golgi apparatus and plant vacuoles.

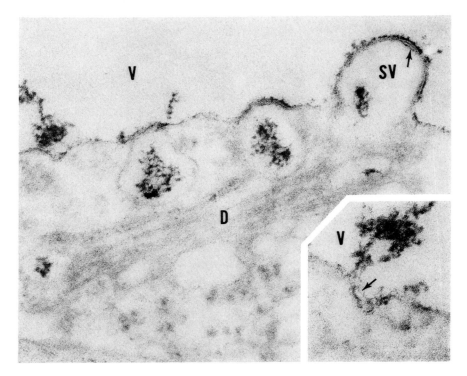

Fig. 6. Vesicles from a maize root dictyosome (D) are secreted into the central vacuole (V) by pinocytosis. The membranes (arrows) of the secretory vesicles (SV) do not fuse with the tonoplast membrane (insert). Glutaraldehyde-osmium tetroxide fixation. From Mollenhauer and Hanson (1976).

D. Lysosomes

In animal cells, the Golgi apparatus appears to play a major role in the formation of primary lysosomes and in the packaging of hydrolytic enzymes contained therein (deDuve and Wattiaux, 1966; Cook, 1973). The primary lysosomes are formed by a system of transition elements that includes smooth endoplasmic reticulum and Golgi apparatus (Essner and Novikoff, 1962). An equivalent to the acid phosphatase-positive GERL complex of animal cells at the distal faces of Golgi apparatus may also be present in plant cells (Novikoff *et al.*, 1962; Marty, 1978).

Lysosomes of the type found in animal cells are not a general feature of plant cells. In plants, acid hydrolases may be present in Golgi apparatus, vacuoles, and storage bodies (Poux, 1963, 1970; Dauwalder *et al.*, 1969; Jelsema *et al.*, 1977). Vacuoles and protein bodies are superficially related to one another in that the protein bodies of some seeds are derived by compartmentalization of the central vacuole during the latter stages of seed maturation and revert to a central vacuole during seed germination (Poux, 1963; Öpik, 1968). However, evidence is lacking to show that Golgi apparatus secrete acid hydrolases into the vacuole (see Section III,C) or play a role in the formation of lipid or protein storage bodies which may contain acid hydrolases.

E. Zones of Exclusion and Microfilaments

A zone of exclusion is a differentiated region of cytoplasm in which ribosomes, glycogen, and organelles such as mitochondria and plastids are scarce or absent (Morré et al., 1971a; Mollenhauer and Morré, 1978). Zones of exclusion are particularly conspicuous around dictyosomes (Fig. 7), mi-

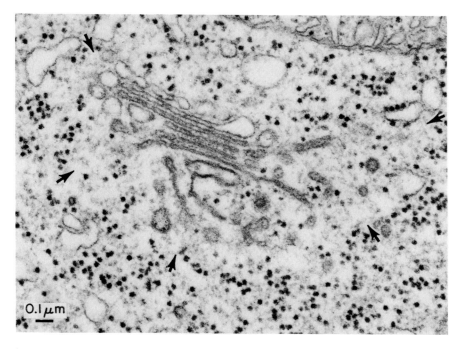

0.1 μm

Fig. 7. The Golgi apparatus zone of exclusion illustrated here (outlined by arrows) is from a maize root cortical cell and is typical of other plant cells as well. Occasionally, however, a zone of exclusion may become much more sharply differentiated from the rest of the cytoplasm (see Mollenhauer and Morré, 1978). From Mollenhauer et al. (1973). Glutaraldehyde-osmium tetroxide fixation.

crotubules, centrioles, flagellar bases, and the cell surface. In electron micrographs, zones of exclusion appear fibrillar or granular with a density similar to that of the cytoplasmic ground substance. The endoplasmic reticulum within a zone of exclusion is usually smooth surfaced (lacking attached ribosomes). Golgi apparatus zones of exclusion were recognized in 1954 by Sjöstrand and Hanzon who referred to them as Golgi ground substance. However, there is little evidence regarding the composition or function of the Golgi apparatus zone of exclusion.

Since the zone of exclusion surrounds the Golgi apparatus, it undoubtedly functions, at least in part, in the selective transfer of product into and out of the Golgi apparatus. Other postulates for its function suggest a role in dictyogenesis, as a center from which dictyosomes may be regenerated during seed germination, and as a pool of intracellular aggregates for the formation and transformation of dictyosome cisternae (see Morré *et al.*, 1971a).

Zones of exclusion, similar to those of the Golgi apparatus, are present at the cell surface of most, if not all, eukaryotic cells (Bluemink, 1971). These cell surface zones of exclusion contain actin-rich proteins that are capable of binding heavy meromysin and thus may be expected to contain special forms or arrangements of microfilamentous components (Wessells *et al.*, 1971; Pollard and Weihing, 1974; Franke *et al.*, 1976a; Tilney, 1976a,b). The presence of microfilamentous-like structures imparts to this zone of exclusion a potential role in cell contractility and perhaps the lateral movement of membranous constituents (Nicolson, 1976). Because of certain structural similarities between the zones of exclusion of the cell surface and of Golgi apparatus, it might be expected that filamentous material containing actin is also a part of the Golgi apparatus zone of exclusion. This, however, remains to be demonstrated.

A potential function for the filamentous component of the Golgi apparatus zone of exclusion was indicated in maize root cap and epidermal cells exposed to cytochalasin B (Mollenhauer and Morré, 1976b). In this study, cytochalasin B prevented the transfer of secretory vesicles from the Golgi apparatus to the cell surface (Fig. 8). Cytochalasin B disrupts the filamentous components of the cell, perhaps by acting on the membrane at the filament attachment site (Bluemink, 1971; Miranda *et al.*, 1974). Thus, inhibition of secretory vesicle transport in the maize root cap and epidermal cells could be interpreted as a disruption of a filamentous component associated with the migration of secretory vesicles.

The relationship, if any, between the intercisternal elements and the filamentous material of the zone of exclusion is unknown. However, the location of the intercisternal elements and their sometimes globular appearance (Kristen, 1978) suggests that the two filamentous components are not equivalent.

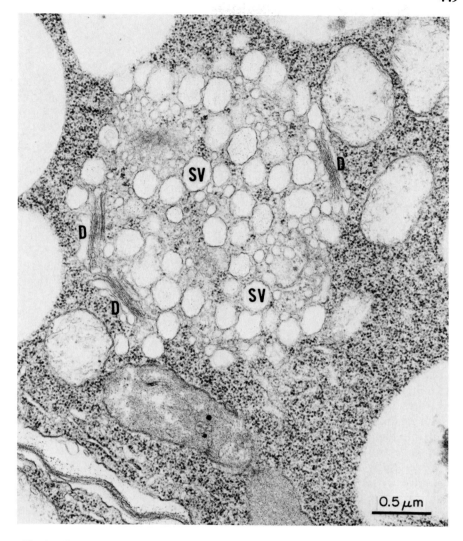

Fig. 8. Treatment of maize root tips with cytochalasin B prevents the transfer of secretory vesicles (SV) from the Golgi apparatus to the cell surface. When this occurs, the secretory vesicles accumulate in large masses around parts of the Golgi apparatus as illustrated here. These results imply that a microfilament system is involved in the movement of the Golgi apparatus vesicles to the cell surface. This illustration is from a maize root epidermal cell. Dictyosome (D). Glutaraldehyde-osmium tetroxide fixation.

F. Microtubules

Microtubules within the vicinity of a dictyosome have been reported in many tissues including liver and pancreas (Moskalewski *et al.*, 1975; Thyberg et al , 1976; Morré, 1977a), *Acanthamoeba* (Bowers and Korn, 1968), and *Euglena* (Mollenhauer, 1974). Additionally, a close association between dictyosomes and the microtubules of centrioles, flagellar bases, and rhizoplasts or between dictyosomes and microtubule bundles have been reported (Heath and Greenwood, 1971; Bouck and Brown, 1973).

In animal tissues, microtubule inhibitory drugs such as colchicine, colcemid, vinblastine, vincristine, griseofulvin, and agent R 17934 have been shown to cause structural aberrations of Golgi apparatus cisternae (Moskalewski *et al.*, 1976; Thyberg *et al.*, 1976) or to inhibit Golgi apparatus-mediated secretion (e.g., Redman *et al.*, 1975). However, an inhibition of Golgi apparatus-mediated secretion was not observed in the root cap cells of maize (Mollenhauer and Morré, 1976b) or in pollen tubes (Franke *et al.*, 1972) following treatment with microtubule inhibiting drugs. Thus, an unequivocal structural or functional relationship between microtubules and Golgi apparatus has yet to be demonstrated for plants.

IV. BIOCHEMISTRY OF PLANT GOLGI APPARATUS

The biochemistry of plant Golgi apparatus lags far behind that of animal Golgi apparatus. Primarily, this reflects the difficulty of isolating plant Golgi apparatus free of contaminating membranes and organelles (see Section IV,A). At the present time, a Golgi apparatus fraction of about 50% purity is attainable from some plant tissues. In many instances, glutaraldehyde fixation has been utilized to achieve higher fraction purity (Morré *et al.*, 1965). However, such chemically altered Golgi apparatus fractions are of only limited value for biochemical investigations because glutaraldehyde inactivates many enzymes. In spite of these obstacles, however, some progress has been made toward biochemically characterizing the plant Golgi apparatus. These preliminary studies show similarities as well as marked differences between plant and animal Golgi apparatus and point out the need for further study.

A. Isolation Methods

The Golgi apparatus presents some formidable technological difficulties to the usual isolation procedures. Dictyosomes unstack easily during homogenization, apparently by dissolution of the bonding constituents between the cisternae. If the dispersed cisternae fragment, the smooth membranous vesicles so derived are difficult to distinguish from plasma membrane or smooth endoplasmic reticulum. The polarized nature of the dictyosome causes most markers to be asymmetrically distributed through the dictyosome. If the dictyosome remains intact during isolation, then marker asymmetry is not a problem. However, if the dictyosome becomes unstacked, then only the dictyosome part carrying the marker can be assayed. This problem offers a compelling reason for keeping dictyosomes intact during isolation. Dictyosomes that remain intact sediment with plasma membrane, tonoplast vesicles, mitochondria, and plastids. Nonetheless, dictyosomes can be isolated from selected animal tissues in purities exceeding 90%. One factor that seems particularly important to Golgi apparatus isolation is the use of low sheer homogenization (Morré and Mollenhauer, 1964; Morré, 1971; Morré *et al.*, 1974c). Procedures for isolating plant dictyosomes have been described for onion stem (Morré and Mollenhauer, 1964; Cunningham *et al.*, 1966; Powell and Brew, 1974), pea epicotyl (Ray *et al.*, 1969; Shore and MacLachlan, 1975; Shore *et al.*, 1975), soybean hypocotyl (Hardin *et al.*, 1972), carrot root (Gardiner and Chrispeels, 1975), and maize root (Bowles and Northcote, 1972, 1974; Paull and Jones, 1976). An additional problem for those isolating plant Golgi apparatus is that well-documented and specific marker enzymes for plant Golgi apparatus are not yet available. Therefore, structural analyses by electron microscopy are indispensible for the qualitative and quantitative assay of isolated Golgi apparatus fractions.

Contamination of plant dictyosome fractions by plasma membrane can be estimated through use of a cytochemical stain [i.e., phosphotungstic acid acidified with chromic aid (PACP), see Section VII,B] specific for plasma membrane (Roland *et al.*, 1972). The PACP-positive component of the membrane is soluble in ethanol and in some of the low viscosity embedding resins. Therefore, the best differentiation between membranes is obtained when the tissues are dehydrated in acetone and embedded in Epon or Epon-Araldite resins. Contamination of plant dictyosome fractions by endoplasmic reticulum may be estimated by NADPH-cytochrome c reductase. The classic ER marker of mammalian cells, glucose-6-phosphatase, is largely a soluble enzymatic activity in plants.

B. Enzyme Composition and Biochemical Markers

Enzymatic activities common to plant dictyosomes include those that catalyze the hydrolysis of nucleoside diphosphates at neutral pH, including inosine diphosphatase (IDPase) and thiamine pyrophosphatase (TPPase). The catalysis of IDPase is especially pronounced in plant dictyosomes and IDPase has been used as a marker enzyme for plant Golgi apparatus (see Morré *et al.*, 1977). However, substantial quantities of IDPase may be present also in vacuoles and endoplasmic reticulum (Poux, 1967; Dauwalder *et al.*, 1969) and, therefore, the specificity of IDPase as a general marker for plant Golgi apparatus remains to be established. The IDPase of plant Golgi apparatus is characterized by an unusual form of latency (Ray et al., 1969; Powell and Brew, 1974; Morré et al., 1977) in that it shows a dramatic increase in specific activity and a change in pH optima from six to seven when the tissue homogenate is cold-stored for 2–3 days. According to Ray et al. (1969), the increased IDPase activity parallels a decline in glucan synthetase activity suggesting that the IDPase of Golgi apparatus represents an inactivated polysaccharide synthetase. A similar IDPase latency is not known to be characteristic of animal Golgi apparatus.

Glycosyl transferases as a general class are present in both plant and animal Golgi apparatus but specific enzymes may vary markedly with respect to donor and acceptor specificities (Schachter, 1974). Perhaps the best characterized is the galactosyl transferase of rodent liver (Morré et al., 1969; Fleischer and Fleischer, 1970; Schachter et al., 1970). Galactosyl transferase is localized exclusively within the Golgi apparatus and, when measured by the transfer of galactose from UDP-galactose to N-acetylglucosamine with the formation of N-acetylactosamine, is a useful marker enzyme. A galactyosyl transferase with the same donor and acceptor specificities has been reported from onion stems (Powell and Brew, 1974). Lercher and Wojciechowski (1976) showed that Golgi apparatus-rich fractions from onion stem and *Calendula officinalis* seedlings contained a UDP–glucose : sterol–glucosyl transferase activity.

β-Glucan synthetase activity has been identified in membrane fractions from *Acetobacter zylinum* (Glaser, 1958; Cooper and Manley, 1975a,b), *Acannthamoeba* (Potter and Weisman, 1971), and higher plants (Ray et al., 1969; VanDerWoude et al., 1974; Helsper et al., 1977). The β-glucan synthetase activity from the *Petunia* pollen tube differed from that in other plants in its inability to incorporate glucose from GDP–glucose (Helspar et al., 1977). This characteristic was also noted in membranes from *Lilium longiflorum* pollen tubes (Southwood and Dickinson, 1975) and may be a common feature of haploid cells or pollen tubes (Helspar et al., 1977). The β-glucan synthetase activity of dictyosomes indicates a capacity for the synthesis of cellulose; yet, the higher plant Golgi apparatus does not appear to be a primary site of cellulose synthesis. In these cells, the β-glucan synthetases may function in the synthesis of the pectic and hemicellulosic polysaccharides (VanDerWoude et al., 1974).

C. Lipid and Protein Composition

Golgi apparatus membranes consist of lipid and protein in about equal proportions with a relatively high sugar content (Franke et al., 1976b; Morré and Ovtracht, 1977). Densitometer scans of proteins of stripped Golgi apparatus- and plasma membrane-rich fractions from onion (*Allium cepa*) stem, soybean (*Glycine max*) hypocotyl, and rat liver are shown in Fig. 9 (Morré, 1977b). There are at least five major bands common to both Golgi apparatus and plasma membrane fractions for both onion and soybean but only two major bands that appear common to two species. There seem to be even fewer similarities when plant and animal Golgi apparatus fractions are compared. Thus, the composition of Golgi apparatus may be characterized by marked differences in proteins among different species.

Plant and animal Golgi apparatus share four classes of phospholipids: phosphatidylcholine, phosphatidylserine, phosphatidylethanolamine, and phosphatidylinositol (Fig. 10) (Keenan and Morré, 1970; Morré and Ovtracht, 1977). Yet, plant Golgi apparatus (and other plant membranes) lack sphingomyelin, one of the major phospholipids of mammalian Golgi apparatus and plasma membranes. The phospholipid compositions of the Golgi apparatus, like the protein compositions, are expected to be intermediate between endoplasmic reticulum and plasma membrane within a species but may differ markedly among species.

The sugars of plant membranes consist mainly of hexoses, whereas animal membranes are built around sialic acid and hexosamines in addition to

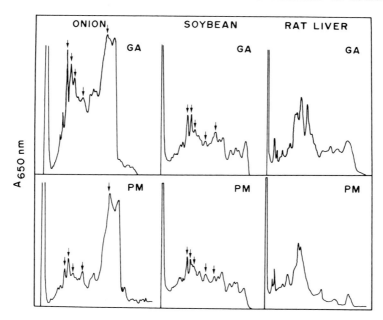

Fig. 9. Densitometer tracings of sodium dodecyl sulfate (SDS) polyacrylamide gel electrophoretograms of purified and stripped membranes comparing Golgi apparatus (GA) and plasma membrane (PM) fractions of onion (*Allium cepa*) stem, soybean (*Glycine max*) hypocotyl, and rat liver. Arrows indicate bands common to both onion and soybean fractions. Gels were stained with Coomassie blue and scanned at 650 nm. From Morré (1977b).

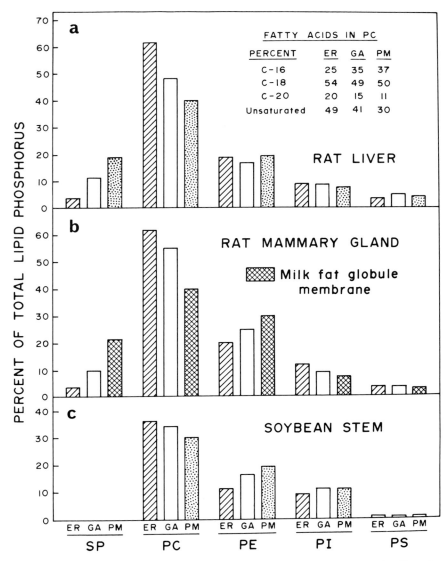

Fig. 10. Phospholipid composition of animal and plant endomembranes. (a) Rat liver. Inset gives characteristics of the fatty acid composition of phosphatidylcholine (PC); Keenan and Morré (1970). (b) Rat mammary gland. Golgi apparatus (GA) and endoplasmic reticulum (ER) as compared with membranes of milk fat globules. The milk fat globule membrane is a plasma membrane derivative that is probably similar to plasma membrane (PM); Keenan *et al.* (1974). (c) Elongating hypocotyls of etiolated soybeans (*Glycine max*). Data for endoplasmic reticulum (ER) based on total rough microsomes. SP = sphingomyelin; PC = phosphatidylcholine; PE = phosphatidylethanolamine; PI = phosphatidylinositol; PS = phosphatidylserine. From Morré and Ovtracht (1977).

hexoses (Morré, 1977b). Plant membranes, including Golgi apparatus, lack sialic acid, whereas this sugar accounts for about 20% of the total membrane carbohydrate of the Golgi apparatus of rat liver and plasma membrane (Franke and Kartenbeck, 1976). Pentoses (e.g., xylose and arabinose) are also found in plant Golgi apparatus but are absent from Golgi apparatus of rat liver (Morré, 1977b). Though both plant and animal Golgi apparatus contain glucosamine, galactose, glucose, mannose, and fucose, it is expected that the linkages will be different (Morré, 1977b).

V. CYTOCHEMISTRY OF PLANT GOLGI APPARATUS

The discovery of Golgi apparatus has its origins in a histological method known as the black reaction (Inferrera and Carrozza, 1975; Whaley, 1975). The technique included hardening of tissues with a mixture of potassium dichromate and ammonia followed by prolonged reaction in a solution of 0.5% or 1.0% silver nitrate. Two reticular apparatus were made visible. The inner eventually became known as the apparatus of Golgi or Golgi apparatus. It was soon discovered that results were improved if solutions of osmium tetroxide replaced the silver nitrate. A form of the black reaction called postosmication is still used in electron microscopy to highlight parts of the endoplasmic reticulum and the Golgi apparatus (see Fig. 4).

Investigators found that reticular apparatus were present in all higher cells in striking and confusing patterns. Many controversies developed regarding the meaning and even the reality of the apparatus, and it was many years before a reasonably common understanding was established. This period of the Golgi controversy is of considerable interest because it graphically illustrates the merits as well as the pitfalls of the histochemical–cytochemical approach. Literature about the histochemical period of Golgi apparatus discovery and controversy can be found in the papers by Inferrera and Carrozza (1975), Whaley (1975), and Beams and Kessel (1968).

The cytochemical methods available today to the electron microscopist are more varied and some are capable of visualizing direct molecular interactions. However, most cytochemical studies of plants have been directed toward illucidating the general pattern of distribution of a particular enzyme activity or reactive species such that few sophisticated questions have been asked. The discussion that follows is limited to electron microscope cytochemistry and no attempt is made to further evaluate evidence from light microscope preparations.

A. Enzyme Markers

Most of the plant cytochemical literature is directed to determining the presence and distribution of nucleoside diphosphatases, particularly inosine diphosphatase (IDPase) and thiamine pyrophosphatase (TPPase) (see Dauwalder *et al.,* 1969; Goff, 1973; Goff and Klohs, 1974; Carasso *et al.,* 1971). IDPase and TPPase are present in most plant Golgi apparatus and both have been used as marker enzymes for identification of isolated dictyosomes. However, both enzymes may be found in the endoplasmic reticulum. IDPase is most often associated with the maturing cisternae of the dictyosomes, the peripheral tubules of the cisternae, and the secretory vesicles (Dauwalder *et al.,* 1969). In maize root tip cells, the IDPase activity is most evident in secretory Golgi apparatus and may be absent from Golgi apparatus not producing conspicuous secretory vesicles (Dauwalder *et al.,* 1969). As pointed out by Dauwalder *et al.* (1969), the IDPase activity appears to correlate with the differentiation of the apparatus for polysaccharide synthesis. TPPase is also associated with the maturing cisternae but is usually not as obvious a constituent of the secretory vesicles as is IDPase (Dauwalder *et al.,* 1969).

Acid phosphatase is sometimes distributed through the entire dictyosome in plants as well as parts of the endoplasmic reticulum. However, it may also be localized near the mature pole of the dictyosome in the fashion described by Novikoff and co-workers as the GERL complex (see also Marty, 1978).

The accuracy of cytochemical procedures depends in large measure on the conditions of tissue preparation, particularly on the extent of enzyme inactivation by the fixative and the degree of enzyme inhibition by the lead salts added in the incubation media (Lake and Ellis, 1976; Dauwalder *et al.,* 1969). These factors limit cytochemical analyses almost entirely to qualitative observations.

BED oxidase and BAXD oxidase have been localized in the nuclear envelope and Golgi apparatus, respectively, of some root tip cells (Nir and Seligman, 1971). Peroxidase is present in all of the endomembrane components of cucumber root tip cells (Poux, 1969).

B. Impregnation Methods

Osmium tetroxide may be selectively reduced in the endomembrane system when tissues are incubated in aqueous solutions of osmium tetroxide for periods of 1–2 days at temperatures of 35–40°C (usually called postosmication; see Friend and Murray, 1965). The osmium deposits are primarily associated with the forming poles of the Golgi apparatus and, in some instances, endoplasmic reticulum. The deposits usually fill the lumina of the cisternae and are visible in both light and electron microscropy. The selective deposition indicates fundamental differences between the components of the endomembrane system but the chemical basis of the reaction is unknown.

Osmium impregnation has been of particular benefit to the high voltage electron microscopists who use sections 1–5 μm thick. The heavy osmium deposits are readily visible in the thick sections even under low magnification and have allowed an evaluation of gross Golgi apparatus architecture (see Carasso *et al.*, 1971; Marty, 1973a). The results clearly demonstrate that the Golgi apparatus is a very elaborate structure with extensive tubular networks connecting adjacent dictyosomes. In some instances, endoplasmic reticulum–tubule–cisternae associations are indicated (Fig. 4).

A mixture of osmium tetroxide and zinc iodide (OZI) has also been used to stain Golgi apparatus (Elias *et al.*, 1972; Marty, 1973b; Marty and Buvat, 1973; Dauwalder and Whaley, 1973). The results are similar to those obtained by postosmication.

C. Localization of Carbohydrates

A relatively precise localization of carbohydrates is possible using cytochemical methods. A variety of techniques are available for its visualization including selective stains (ruthenium red, colloidal metals, and alcian blue for acidic groups), reactions of periodic acid followed by a complexing reagent combined with a heavy metal, and lectins coupled with ferritin or other markers specific for certain sugar linkages (see Rambourg, 1969, 1971; Roland, 1973 for literature). Cytochemical methods for detecting carbohydrate components show what may be interpreted as progressive elaboration of polysaccharides across the stacked cisternae of the Golgi apparatus. These results (an increasing gradient of reaction product from the forming face to the maturing face of the dictyosome and from immature vesicles to mature vesicles) suggest stepwise assembly within the Golgi apparatus. At the plasma membrane, carbohydrate materials are concentrated on the outer or external membrane leaflet. In endomembrane components such as Golgi apparatus, the sugar residues are oriented toward the luminal surfaces. Upon fusion with the plasma membrane, the inner surfaces of the cisternae or vesicles become equivalent to the external surfaces of the plasma membrane. In plant cells, the staining procedure involving phosphotungstic acid at low pH has been useful to show that secretory vesicles of the Golgi apparatus progressively acquire cytochemical characteristics of the plasma membrane (Roland, 1969; Vian and Roland, 1972).

VI. ROLE IN SECRETION

A role for the Golgi apparatus in cellular secretion was suggested as early as 1923 by Nassonov (see Whaley, 1975). It was noted that primary secretory granules and small mucus drops became visible in the reticular regions of the Golgi apparatus, increased in size while still associated with the Golgi apparatus, and then migrated to the surface of the cell. The accumulated products from the Golgi apparatus were subsequently secreted from the cell. Bowen (1929) further illucidated the role of the Golgi apparatus in the secretory process and emphasized that the formation of secretory product by the Golgi apparatus was an intracellular process distinct from the transfer of accumulated products out of the cell. An important contribution of Bowen was the recognition that the Golgi apparatus was only an intermediary component in a much larger system of cellular components. An insight into the early development of the hypotheses on Golgi apparatus secretion are well presented in the reviews by Bowen (1929) and Whaley (1975).

Mammalian cells are known to produce and export a wide variety of macromolecular products through the endoplasmic reticulum–Golgi apparatus–secretory vesicle pathway including hormones, digestive enzymes, mucins, glycoproteins, surface coats, connective elements, and lipoproteins (Mollenhauer and Morré, 1966a; Whaley, 1975; Beams and Kessel, 1968; Favard, 1969; and references cited therein). Plant Golgi apparatus are not known to secrete large quantities of proteins in the manner of the animal tissues. Yet they might be expected to participate to some extent in the secretion of cell wall enzymes and in the secretion of digestive enzymes in carnivorous plants. Secretion of proteins via Golgi apparatus-derived vesicles has been suggested for ribonuclease (Jones and Price, 1970), proteases (Schwab et al., 1969), and certain phosphatases (see Dauwalder et al., 1969, 1972 for reviews). Some extracellular enzymes, however, appear to be derived directly from endoplasmic reticulum. These include α-amylase (Vigil and Ruddat, 1973), glucanase (Cortat et al., 1972), wall-degrading enzymes (Bal and Payne, 1972), and other extracellular materials (see Franke et al., 1972). Secretory products such as nectar and stigmatic exudates are probably mediated through the endoplasmic reticulum and do not follow the Golgi apparatus pathway.

The secretory activities of plant cells are well documented only for slimes and mucilages and some wall constituents. Slimes and mucilages are polysaccharides or polysaccharide–protein complexes of high viscosity which occur throughout the plant kingdom. Mucilages closely resemble the hemicelluloses and pectic substances of the cell wall. Ordinarily, only individual cells or groups of cells within a plant produce slimes and mucilages (Schnepf, 1969). These polysaccharides are often acidic and have a wide range of chemical compositions.

The above secretory activities refer primarily to the internal constituents of the secretory vesicles, i.e., secretory activity is usually equated to the products within the hypertrophied Golgi apparatus cisternae or the enlarged Golgi apparatus-derived vesicles. Only recently has it been recognized that membranes derived from Golgi apparatus vesicles or cisternae may constitute a significant part of the secretory activity and that the elaboration of membranes may well be the universal function of the Golgi apparatus (Whaley *et al.,* 1972; Morré, 1975, 1977a; Whaley, 1975; Morré and Ovtracht, 1977). Membrane formation may occur even when hypertrophied cisternae or secretory vesicles are inconspicuous.

A. Structural Pattern of Vesicle Formation and Secretion

The pattern of Golgi apparatus secretion in plants has been most extensively studied in the outer cells of the maize root cap. The secretory vesicles of these cells were conspicuous and easily recognized through their entire development. In addition, the quantity of product produced was so massive that simple electron microscopical observations were sufficient to convince early investigators that the contents of the secretory vesicles were the precursors of a slime droplet that forms on the root tip. Subsequently, the pattern of secretion in the maize root cap was found to be a useful model applicable to other plant cells.

In the outer cells of the maize root cap, the secretory vesicles develop progressively from one face of the dictyosome to the other (Fig. 11). These changes are evidenced in the physical form of the secretory vesicles as well as in the density and fibrillar appearance of the vesicle contents (Fig. 11). In these cells, there is usually only one, and at the most two, secretory vesicles per cisterna (Fig. 11) and these vesicles are elongated and curved to correspond to the peripheral edge of the cisterna. The secretory vesicles are connected to the cisternae by tubules (Mollenhauer and Morré, 1966b) (Fig. 12). When the secretory vesicles are mature, they, as well as the cisternae, are sloughed from the dictyosome (Figs. 11, 13, and 19). The secretory vesicles separate from the cisternae and the sloughed cisternae lose their identity (Mollenhauer, 1971). The separated secretory vesicles assume spherical shapes and, usually, show an increase in the density of the matrix substances. The membranes of the secretory vesicles eventually fuse with the plasma membrane, thus moving the contained secretory product to the cell exterior (Fig. 5).

The secretory vesicles of the maize epidermal cells, pollen tubes, and most other cell types are more spherical in form (Fig. 14). There are usually several to many secretory vesicles per cisterna, yet the vesicles are attached to the cisternae by tubules (Mollenhauer and Morré, 1966b). The secretory vesicles move to the cell surface and their contents are secreted out of the cell in the same manner as for the outer root cap cells described above.

Scale secretion in *Pleurochrysis scherffelii* as described by Brown and co-workers (Brown *et al.,* 1973) is also sequential from the forming to the maturing face of the dictyosomes. However, the secretory pattern differs from that of the maize root cap cells in that the entire cisterna becomes the secretory

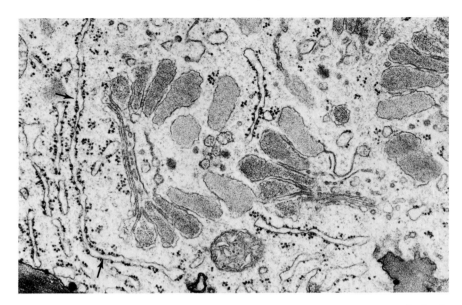

Fig. 11. A transverse section through two dictyosomes from an outer root cap cell of maize. These cells are highly secretory and produce a polysaccharide slime that covers the root tip. The slime is synthesized, at least in part, within the Golgi apparatus where it accumulates in large elongated vesicles around the periphery of the cisternae. There are only one or two sets of secretory vesicles per dictyosome and these vesicles are attached to the cisternae by tubules (see Mollenhauer and Morré, 1966b; Fig. 12). Ribosomes are associated with the forming cisternae. Amplexes of endoplasmic reticulum (arrows) are often present around parts of the dictyosome. Glutaraldehyde-osmium tetroxide fixation.

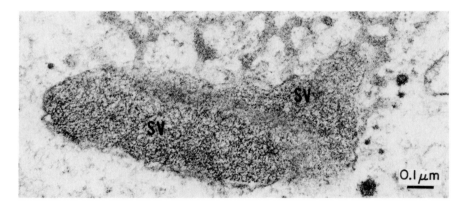

Fig. 12. The secretory vesicles (SV) of plant cell Golgi apparatus are attached to the cisternae by tubules as illustrated here. The tubules act, perhaps, to segregate functional activities and as transitional elements between the cisternae and the secretory vesicles. This illustration is from an outer root cap cell of maize. Glutaraldehyde-osmium tetroxide fixation.

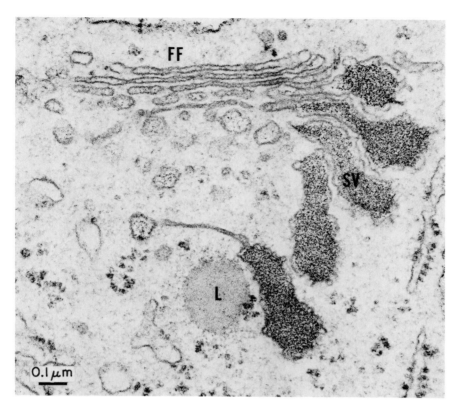

Fig. 13. Micrograph of a dictyosome from an outer cell of the maize root cap illustrating the developmental pattern of the secretory vesicles. The secretory vesicles (SV) begin to form along the periphery of the cisterna on the forming face (FF) of the dictyosome. The secretory vesicles enlarge and maturate sequentially across the stack and are ultimately sloughed (along with the attached cisternae) from the dictyosome. The secretory vesicles then separate from the cisternae and assume a spherical form. The sloughed cisternae break down and become unidentifiable. Lipid vesicle (L).

vesicle. When the scale is mature, the membrane of the cisterna fuses with the plasma membrane and the scale is released to its final position on the cell surface (Fig. 15).

B. Autoradiographic Pattern of Product Migration

Peterson and Leblond (1964a,b) and Neutra and Leblond (1966a,b) first showed that tritiated labeled sugars were incorporated into Golgi apparatus of goblet cells of the rat intestine within 5 min after intraperetoneal injection. The sugars appeared in the forming cisternae of the Golgi apparatus and then

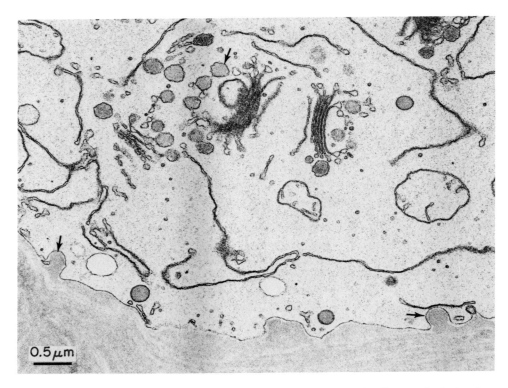

Fig. 14. The secretory vesicles (arrows) of maize root epidermal cells are spherical and about 0.1 μm in diameter. They are attached to the cisternae by tubules in much the same way as that described for the outer cells of the root cap (see Fig. 12) and follow the same pattern of secretion as that seen in the outer cap cells (see Fig. 5). From Mollenhauer and Mollenhauer (1978). Potassium permanganate fixation.

sequentially in maturing cisternae and secretion vesicles (mucigen granules). The entire sequence through the Golgi apparatus required less than 40 min. It was assumed that proteins were synthesized on the rough endoplasmic reticulum, transferred to the Golgi apparatus, and there glycosylated and sulfated to form mucin. Studies with a variety of plant and animal cells confirm the participation of Golgi apparatus in the synthesis and secretion of polysaccharides and mucopolysaccharides (see Whaley, 1975; Morré, 1977a,b for references).

Polysaccharide secretion in plant cells follows the pattern established for mucopolysaccharides of animal cells. For example, radioautography of wheat root cap cells exposed to tritiated sugars is interpreted to show that

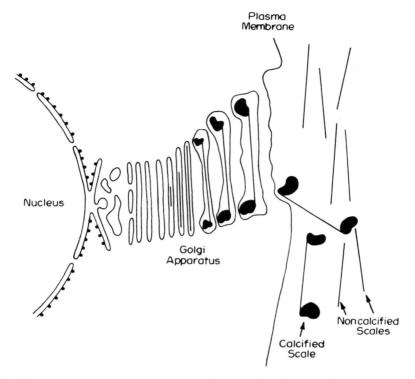

Fig. 15. In chrysophycean algae, scales are formed within cisternae of the Golgi apparatus. Calcification occurs at the scale margin (solid black projections) prior to discharge of the scale to the cell surface. From Morré and Mollenhauer (1976).

labeled polysaccharides are formed in the Golgi apparatus, passed into the secretory vesicles of the Golgi apparatus, and then moved through the plasma membrane into the extracellular environment (Northcote and Pickett-Heaps, 1966; Pickett-Heaps, 1967a,b). The secreted polysaccharides impregnate the cell wall and/or accumulate as a slime droplet covering the root tip. The same secretory mechanism occurs in the outer cells of the maize root cap (Rougier, 1976). Analyses show that the secreted polysaccharides from the maize root cap cells contain glucose, galactose, fucose, and galacturonic acid residues with smaller quantities of mannose, arabinose, xylose, and rhamnose (Jones and Morré, 1973; Harris and Northcote, 1970). This general pattern of polysaccharide synthesis and secretion has been confirmed by numerous structural, cytochemical, radioautographic, and organelle isolation studies in other species as well.

C. Secretion of Cell Wall Constituents

Except in a few instances, the cells of plants are surrounded by biphasic walls consisting of a microfibrillar phase assembled predominantly from polysaccharides (β1-4 glucan, β1-3 glucan, β1-3 xylan, β1-4 mannan, or chitin) and a matrix phase of pectins and hemicelluloses derived predominantly from mixed polymers of uronic acids, pentoses, and hexoses (Preston, 1974). The fibrillar phase is probably assembed at the cell surface by a process involving plasma membrane-bound enzymes and membrane-based assembly and orientation mechanisms. The matrix phase may be secreted by components of the endomembrane system and is probably distinct from the extraneous wall components of similar composition such as slimes and mucilages. A fibrillar appearance does not necessarily indicate the presence of cellulosic fibrils.

1. Cell Plate Formation

During anaphase, or perhaps earlier, small membrane-bounded vesicles move toward the equitorial regions of the cell where clusters of them aggregate and then fuse to initiate cell plate formation (Whaley and Mollenhauer, 1963; Roberts and Northcote, 1970; O'Brien, 1972; Whaley, 1975 for references). Since cell plate formation begins in the midregion of the cell and extends outward from this point, formation of new plasma membrane or new matrix substance by extension of an existing wall does not occur (see Whaley, 1975). Thus, the cell plate is assembled almost entirely from components supplied by the endomembrane system. At least some of these components appear to come from the Golgi apparatus. A role for coated vesicles in this process has also been indicated (Franke and Herth, 1974).

In most cells, the small size of the cell plate vesicles and their lack of definitive staining limits the amount of information that can be obtained from ultrastructural studies. Even radioautography does not always give conclusive information about the source of the cell plate vesicles due to the limits of resolution of autoradiographic grains. Conclusive evidence for the involvement of Golgi apparatus in cell plate formation has come primarily from studies of specialized cells such as those of the maize root epidermis (Mollenhauer and Mollenhauer, 1978), the root tips of *Phalaris canariensis* (Frey-Wyssling *et al.,* 1964), and the cross walls in desmids (Drawert and Mix, 1962), where Golgi apparatus sequester substances into vesicles that can be characterized by size and density and differentiated from other types of vesicles. In maize epidermis, the distinct Golgi apparatus vesicles are incorporated into the cell plate (Figs. 16A,B) where they contribute at least 12% of the new plasma membrane and perhaps as much as 50% of the wall matrix substance (Mollenhauer and Mollenhauer, 1978). Unfortunately, these data place only a lower limit on the extent of Golgi apparatus participa-

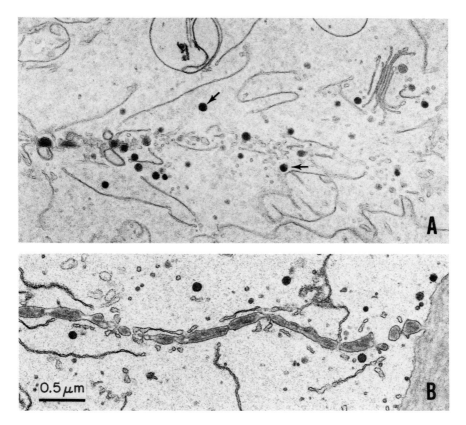

Fig. 16. The secretory vesicles of the maize epidermal cells are unique and can be identified even when separated from the Golgi apparatus. Analyses show (see Mollenhauer and Mollenhauer, 1978) that the secretory vesicles are preferentially incorporated into growing walls and almost none enter walls that are fully formed. In the cell plate, the secretory vesicles account for at least 12% of the new plasma membrane and perhaps as much as 50% of the wall matrix substances. (A) shows an accumulation of secretory vesicles near the forming plate (arrows). After the secretory vesicles are incorporated into the forming wall, their contents become evenly dispersed through the wall (B). From Mollenhauer and Mollenhauer (1978). Potassium permanganate fixation.

tion since only those vesicles with distinctive form or density are recorded. In these and other cells, many small Golgi apparatus vesicles lacking distinctive contents may also be present and may contribute to the cell wall.

2. Tip Growth

Rapid tip growth in cells such as pollen tubes, rhizoids, fungal hyphae, plant hairs, and cotton fibers is correlated with primary cell wall formation and Golgi apparatus secretion (see Sievers, 1963; Rosen, *et al.,* 1964;

VanDerWoude *et al.*, 1971; Ramsey and Berlin, 1976; Westafer and Brown, 1976; Ryser, 1977 for examples and literature). Dashek and Rosen (1965) site cytochemical evidence that both Golgi apparatus-derived secretory vesicles and the tip wall of pollen tubes contain pectic substances. Morré and VanDerWoude (1974) estimate that in pollen tubes of Easter lily, the Golgi apparatus must produce and export more than 1000 secretory vesicles per minute to generate the 300 μm^2 of new plasma membrane and corresponding volume of matrix substances necessary for each minute of steady-state growth.

3. Secondary Wall Formation

There is some evidence, mostly structural, suggesting a role for the Golgi apparatus in secondary wall formation. For example, in *Hibiscus esculentis* (okra) pods, the presence of Golgi apparatus secretory vesicles closely parallels wall thickening, i.e., secretory vesicles are first apparent at the initiation of wall thickening and disappear after the secondary wall is complete (Mollenhauer, 1967b; Figs. 17A,B). Moreover, profiles of secretory vesicles are visible along the cell surface suggesting a transfer of product into the wall. The products secreted into the wall then undergo further changes as indicated by a gradual increase in electron density and fibrillar appearance (Fig. 17B).

4. Scale Formation

Cell wall components consisting of discrete scales were initially observed in the haptophycean alga *Chrysochromulina* by Manton and co-workers (Parke *et al.*, 1955). Scales were subsequently found in other haptophycean species, in Chrysophyceae, Prasinophyceae, Chlorophyceae, and other protists (see Brown and Romanovicz, 1976 for review). The scales are synthesized within the Golgi apparatus and then secreted out of the cell. The scales can be seen by electron microscopy and their passage through the Golgi apparatus has been described in great detail. In addition, cell-free fractions of scales have been obtained for chemical and physical analyses (Green and Jennings, 1967; Herth *et al.*, 1972).

Scale formation has been detailed by Brown and co-workers (Brown, 1969; Brown and Romanovicz, 1976; Brown *et al.*, 1973) for *Pleurochrysis scherffelii*. Two kinds of scales are produced. In the Pleurochrysis phase of the growth cycle, the scales consist of (1) a radial system of noncellulosic microfilaments upon which a cellulosic (alkali-insoluble, β1-4 glucan) microfilament system is spirally arranged and covalently linked to protein (Herth *et al.*, 1972); and (2) an amorphous matrix deposited upon and within the filamentous network. In the Criscophaera phase of the growth cycle, the scales are formed with a peripheral network of calcium carbonate deposited on the rims. The calcified scales are called coccoliths. Assembly of both forms of scales occurs within the Golgi apparatus cisternae in a stepwise

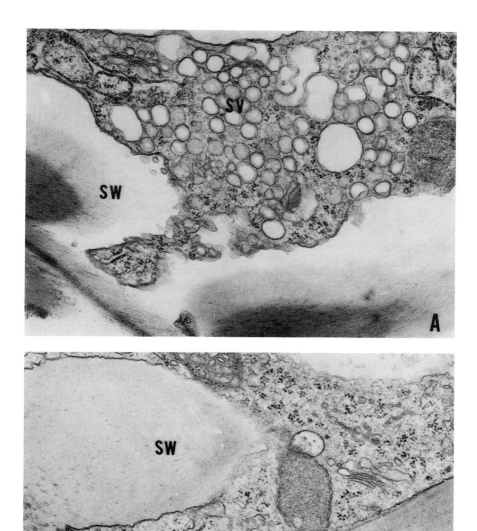

Fig. 17. Golgi apparatus are probably involved in the synthesis and secretion of secondary wall substances. This is illustrated here in cells of the okra pod associated with an immature (A) and mature (B) secondary wall (SW). Golgi apparatus secretory vesicles (SV) are abundant during wall formation (A) but are almost absent after the wall has been completed (B). Glutaraldehyde-osmium tetroxide fixation.

fashion (Fig. 15). The radial microfilaments of the scales are assembled first in a folded configuration which next unfolds. Spiral bands of cellulosic microfilaments are then deposited onto the distal surface of the radial microfilaments. Finally, the network of radial and concentric microfilaments is covered with an amorphous material. To form coccoliths, calcium is deposited during the latter stages of scale formation. Completed scales are secreted in the manner described in Section VI,A except that the whole cisterna acts as a secretory vesicle by fusing with the plasma membrane.

VII. ROLE IN MEMBRANE DIFFERENTIATION

Most Golgi apparatus exhibit a strong polarization from one face of each dictyosome to the other (Mollenhauer and Morré, 1966a; Whaley, 1975). This polarity is expressed in both physical and biochemical parameters and presumably represents a transformation of secretory product and/or membrane across the dictyosome. Almost all recent data indicate that the Golgi apparatus is a component of the endomembrane system with properties intermediate between endoplasmic reticulum and plasma membrane (Mollenhauer and Morré, 1966a; Whaley et al., 1972; Morré and Ovtracht, 1977). It is presumed that the parts of the Golgi apparatus in proximity to the endoplasmic reticulum (i.e., the forming faces) are like endoplasmic reticulum, whereas the parts of the Golgi apparatus associated with the secretory vesicles (i.e., the maturing faces) are like plasma membrane. However, even nonsecretory Golgi apparatus show membrane transformations across the dictyosomes. Thus, membrane transformations are, perhaps, a universal feature of the plant Golgi apparatus (Whaley et al., 1972; Whaley, 1975; Morré and Mollenhauer, 1976; Morré, 1977a; Morré and Ovtracht, 1977). Membrane transformations also occur where Golgi apparatus and endoplasmic reticulum are sometimes structurally continuous as might occur at the peripheral edges of the cisternae (Mollenhauer and Morré, 1976a). The transformation of one type of membrane to another, or the structural and chemical modification of existing membranes is *membrane differentiation* (Morré et al., 1971b; Morré and Mollenhauer, 1976).

A. Morphological Evidence

Plant dictyosomes characteristically show linearly progressive changes in membrane thickness (Table I) and staining intensity from one face to the other (Fig. 18). The membranes of the forming cisternae are structurally similar to those of endoplasmic reticulum whereas the membranes of the mature cisternae and secretory vesicles are thicker and show the dark–

TABLE I

Membrane Differentiation in Golgi Apparatus of Animals and Plants[a]

	Membrane thickness (nm)			
Membrane type	Rat liver	Rat mammary gland	Onion stem	Soybean hypocotyl
Nuclear envelope	65 ⎱	60	56	56
Endoplasmic reticulum	65 ⎰		53	56
Golgi apparatus				
Cisterna 1	65 ⎫		53	56
Cisterna 2	68 ⎪	70	60	58
Cisterna 3	72 ⎬		65	61
Cisterna 4/5	80 ⎭		75	69
Secretory vesicle	83	85	88	78
Plasma membrane	85	97	93	88

[a] Determined from measurements of photographically enlarged electron micrographs of glutaraldehyde-osmium tetroxide fixed materials. From Morré (1977b).

light–dark pattern characteristic of plasma membrane. These differences are accentuated in tissues that have been block-stained with uranyl acetate.

Membrane changes were particularly well demonstrated by Grove *et al.* (1968) in the dictyosomes of the fungus *Pythium ultimum*. The membrane differences were enhanced by using a $Ba(MnO_4)_2$ poststain on the thin section. Changes in the number and distribution of intramembranous particles was also used by Vian (1974) as an indicator of membrane transformation. Using the freeze–fracture technique, Vian showed progressive increases in numbers of particles from endoplasmic reticulum, Golgi apparatus, secretory vesicles, and plasma membranes of root tip cells of pea (*Pisum sativum*). Particle densities characteristic of the plasma membrane were already evident in the secretory vesicles of the Golgi apparatus. Similar results were obtained by Staehelin and Kiermayer (1970) for dictyosomes of the algae *Microsterias denticulata*.

B. Cytochemical Evidence

Cytochemistry permits the visualization of gradients within the compartments of the cell and has provided important contributions to the concept of membrane differentiation. For example, in animal tissues, carbohydrate complexing reagents show an increase in reactivity across the Golgi apparatus from the forming to the maturing faces of the dictyosomes (Ram-

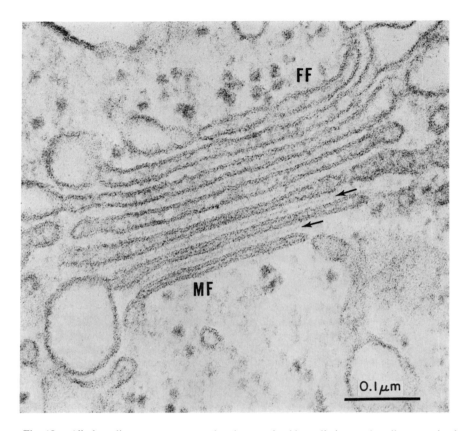

Fig. 18. All plant dictyosomes seem to be characterized by a distinct and easily recognized polarity. Moreover, the changes from the forming (FF) to the maturing (MF) faces of the dictyosome are linearly progressive across the stack. These changes include a decrease in the width of the cisternal lumina from the forming to the maturing faces, a change in the appearance of the membranes, and the gradual appearance of intercisternal elements (arrows) toward the maturing pole of the dictyosome. Note polyribosomes near forming face. Glutaraldehyde-osmium tetroxide fixation.

bourg, 1971). Transformations within the Golgi apparatus are shown also by the acquisition of acidic groups only on the last one to three of the most mature cisternae (Berlin, 1968; Stockem, 1969). In plants, Roland and co-workers (e.g., Roland, 1969, 1973; Roland and Sandoz, 1969; Roland and Vian, 1971) used an acidified phosphotungstic acid stain (PACP) to show that a unique plasma membrane constituent is probably first acquired in the mature cisternae and secretory vesicles of the Golgi apparatus and then transferred to the plasma membrane. Using the same cytochemical method, Frantz (1973) demonstrated that in isolated soybean hypocotyl dictyosomes, the PACP-positive component is acquired over a period of time with the staining reaction spreading from the mature toward the forming faces of the dictyosomes.

Additional evidence for membrane differentiation has come from enzyme cytochemistry. Nucleoside diphosphatase, acid phosphatase, inosine diphosphatase, and thiamine pyrophosphatase are the enzymes most often studied in plant cells; all are usually present within the Golgi apparatus and endoplasmic reticulum and occasionally present in or on plasma membrane (see Section V).

C. Biochemical Evidence

The biochemical basis for membrane differentiation has been sought through studies that compare endoplasmic reticulum, Golgi apparatus, and plasma membrane fractions isolated from rat liver (see Morré *et al.*, 1974a,b for references and summary of methodology). If Golgi apparatus function in the conversion of endoplasmic reticulum membranes to plasma membranes, the composition of Golgi apparatus membranes should reflect this transformation (Keenan and Morré, 1970). Similarly, a comparison of endoplasmic reticulum membranes and plasma membranes will indicate the biochemical changes required to effect the transformation. The bulk of the information available showing biochemical transformations within the Golgi apparatus has been derived from animal tissues. Nonetheless, indications are that both plant and animal Golgi apparatus have biochemical characteristics intermediate between endoplasmic reticulum and plasma membrane. This transitional nature of Golgi apparatus membranes is reflected in the lipid and protein compositions of the membranes for liver (Morré, 1975, 1977a; Morré and Ovtract, 1977) and mammary gland (Keenan *et al.*, 1972). Phospholipids and fatty acids of the major lipid classes of Golgi apparatus are intermediate between those of the endoplasmic reticulum (or nuclear envelope) and plasma membrane (Keenan and Morré, 1970; Fig. 10). At present levels of resolution, all endomembrane fractions (rough endoplasmic reticulum, smooth endoplasmic reticulum, Golgi apparatus, and plasma membrane) from a single tissue have at least some major protein bands in common, based on analyses by polyacrylamide disc gel electrophoresis comparing apparent molecular weights. Enzymatic activities characteristic of plasma membrane, i.e., plasma membrane marker enzymes, appear to be acquired at the Golgi apparatus, whereas enzyme activities characteristic of endoplasmic reticulum membranes appear to be lost (Morré and Mollenhauer, 1974; Morré and Ovtracht, 1977).

VIII. ROLE IN MEMBRANE FLOW

Membrane flow is the physical transfer of membrane from one compartment of the endomembrane system to another (Franke *et al.*, 1971b). The concept was originally applied to endocytosis by Bennett (1956) but applies equally well to all of the transfer processes associated with membrane biogenesis and differentiation (Morré and Mollenhauer, 1974). Membrane flow is not necessarily a random process (Morré *et al.*, 1971b; Franke and Kartenbeck, 1976), but appears to be highly selective for specific membrane components or constituents.

Physical transfer of membrane from the Golgi apparatus to plasma membrane via secretory vesicles (see Sections VI,A, VI,C, and VIII,A) is perhaps the clearest example of a membrane flow mechanism. Membrane flow from endoplasmic reticulum to Golgi apparatus is more difficult to document although it is strongly implied especially where secretion involves the gradual maturation of membrane from endoplasmic reticulumlike to plasma membranelike across a dictyosome with the periodic loss of entire cisternae at the maturing face (e.g., maize root cap cells and scale-forming algae).

A. Kinetic Estimates of Flow Rates in Plant Cells

Perhaps the most accurate determination of flow rate through a Golgi apparatus was determined by real time visual observation of the alga *Pleurochrysis scherffelii* (Brown, 1969). The Golgi apparatus of these cells is a single dictyosome of about 30 cisternae which is visible by light microscopy. Synthesis of each scale is sequential across the dictyosome beginning at the forming face. Secretion of each scale is accompanied by the loss of an entire cisterna. Replacement cisternae must form at the same rate as scale secretion to account for the continued functioning of the dictyosome. Therefore, at a secretory rate of about one scale per minute (Brown, 1969), the turnover rate for the entire dictyosome is approximately 30 min.

Mastigonemes are also useful markers for determining the time scale of secretion. Mastigonemes are the microtubulelike hairs attached to the flagella of certain algae (Bouck, 1969). Mastigonemes are synthesized in the endomembrane system of the cell and synthesis can be induced by deflagellating the algae by mechanical or osmotic shock. In *Ochromonas minute* (Hill and Outka, 1974), mastigonemes appear within the nuclear envelope and endoplasmic reticulum within 10 min of deflagellation. They are then transferred to the Golgi apparatus where they reach a maximum concentration about 30 min after deflagellation. The concentration of mastigonemes then drops as the mastigonemes are secreted to the forming flagellar surface. However, some synthesis of mastigonemes continues until the flagella are complete. The synthesis and passage of the initial mastigonemes through the endoplasmic reticulum and Golgi apparatus takes about 30 min with about 20 min required for the passage through the Golgi apparatus.

Radioautographic analyses of the secretory products in the root cap cells of wheat and maize (Northcote and Pickett-Heaps, 1966; Rougier, 1976) show that tritiated sugar is incorporated almost immediately into elements of the Golgi apparatus. The label is subsequently transferred to the secretory vesicles and into the wall in a period of about 30 min. Similarly, Robinson *et al.* (1976) using tritiated glucose and cell fractionation techniques showed that incorporation of glucose into the Golgi apparatus required about 20 min to reach a steady-state value and also that a chase period of about 20 min was required to remove the glucose label from the dictyosomal and microsomal fractions. Other estimates of product and membrane turnover time in both plant and animal Golgi apparatus generally fall within the range of 10–40 min. However, Bowles and Northcote (1974) calculate markedly shorter times for turnover based on the labeling of polysaccharide materials of maize root tip cells. These relatively short turnover times may reflect a synthesis of polysaccharides within free secretory vesicles as well as within vesicles still attached to the mature faces of the Golgi apparatus.

The above estimates of turnover times were obtained predominantly from secretory cells and little comparable information is available for nonsecretory cells. However, one might suspect that nonsecretory Golgi apparatus would turn over very slowly, perhaps at a rate equal to that for the natural decay of membranes generally. An indication of this trend was obtained in a study of the maize root tip epidermal cells (Mollenhauer and Mollenhauer, 1978). Though the Golgi apparatus of these cells are secretory and supply both plasma membrane and wall matrix material, their secretory pattern is highly selective. Secretory product is incorporated almost exclusively into walls undergoing expansion and almost no secretory product enters fully developed walls. In these cells, the differential in the numbers of secretory vesicles entering growing and nongrowing walls may vary by several orders of magnitude. Thus, the rate of secretion and dictyosome turnover appear to be variable parameters closely coupled to the rates of synthesis required for wall formation or modification.

B. Membrane Recycling

The plasma membranes of nongrowing mucus- or slime-secreting cells do not increase in surface area or thickness as a result of the influx of secretory vesicle membranes. Thus, compensatory mechanisms must exist within these cells to remove plasma membrane at a rate that equals the influx of new plasma membrane from the secretory vesicles. During periods of maximum secretion, turnover of plasma membrane is high. The incorporation of secretory vesicles into plasma membrane of slime-secreting cells of *Mimulus tilingii* has been estimated at over 500 vesicles per minute per cell with a net increase in surface area of 3% per minute (Schnepf and Busch, 1976). Yet, there is no morphological evidence that the flow of membrane from Golgi apparatus to plasma membrane is compensated for by backflow in the form of vesicles or other organized membrane structures (Morré and Mollenhauer, 1974; Schnepf and Busch, 1976).

C. Multiple Pathways of Membrane Flow

The physical and biochemical properties of the Golgi apparatus are intermediate between endoplasmic reticulum and plasma membrane and a major function of the Golgi apparatus is the transformation of membranes from endoplasmic reticulumlike to plasma membranelike (Whaley *et al.,* 1972; Whaley, 1975; Morré and Mollenhauer, 1974). Membranous constituents, either in molecular configuration or in preformed packets, must be incorporated into the Golgi apparatus at some stage in its pattern of functioning. Similarly, membranous constituents must move from the Golgi apparatus into other sites of incorporation.

The secretory vesicle provides the clearest example of bulk flow of membrane from the Golgi apparatus to the plasma membrane. Vesicular transfer of membrane to the vacuole of the plant cell may occur also (see Section III,C) but, here, there is little evidence that the membranes of the secretory vesicles fuse with those of the tonopast (Fig. 6). The transfer of product from the Golgi apparatus to a lysosomal system in plants like that in animal cells has not yet been documented. Neither has the transfer of preformed membranous constituents from the endoplasmic reticulum into the Golgi apparatus been well documented. Presumably, vesicles from endoplasmic reticulum juxtaposed to the forming pole of the dictyosome transfer membrane to the Golgi apparatus (see Beams and Kessel, 1968; Mollenhauer and Morré, 1966a; Favard, 1969; Whaley, 1975 for references). Vesicles from the endoplasmic reticulum accumulate approximately 100 Å from the forming pole of the dictyosome where they appear to fuse to form new cisternae.

Tubular projections from the endoplasmic reticulum are also found at the periphery of the dictyosome cisternae (Fig. 2A of Morré and Ovtracht, 1977) and are at least intermittently continuous with the cisternal tubules (Mollenhauer *et al.,* 1975). However, no unequivocal evidence is available to show product or membrane transfer between endoplasmic reticulum and Golgi apparatus in plants.

IX. PHYSIOLOGY OF THE PLANT GOLGI APPARATUS

The Golgi apparatus emerges as one of the more versatile cellular components with a remarkably diverse range of functions. In its capacity as a membrane transformation device, it may be required to provide bulk quantities of plasma membrane in rapidly expanding cells, provide a secretory vesicle membrane that will fuse with the plasma membrane and allow the transfer of secretory product out of the cell, or provide for the insertion of specific informational molecules into the plasma membrane. In its capacity as a secretory organelle, it may be required to transfer enzymes into the wall matrix or to provide various extracellular carbohydrate constituents of the wall matrix or specific kinds (e.g., scales, mastigonemes) of wall constituents. However, the details of how Golgi apparatus are programmed for a particular function remain to be determined.

The Golgi apparatus of both plant and animal cells appear to have some polyribosomes associated with their forming poles (Mollenhauer and Morré, 1974; Fig. 19), thus allowing for the synthesis of a limited number of proteins (Elder and Morré, 1976). The unusual position of the polyribosomes suggests that their function is the synthesis of specific Golgi apparatus enzyme complexes, perhaps enzymes such as the glycosyl transferases. These polyribosomes may be important in determining, or regulating, Golgi apparatus and cell function.

Golgi apparatus can change functional states in times that are relatively short. For example, an osmotic shock to the maize root cap will temporarily

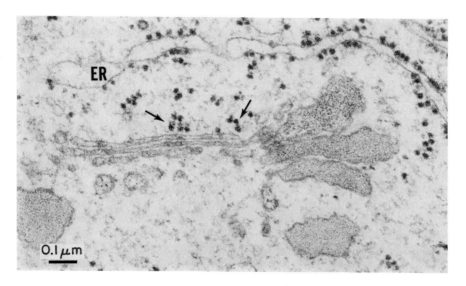

Fig. 19. Polyribosomes (arrows) are associated with the forming faces of both plant and animal dictyosomes. The polyribosomes are not intimately associated with the cisternal membrane though they remain attached even during dictyosome isolation. An "amplexis-like" segment of endoplasmic reticulum (ER) is often present around part of the dictyosome. Glutaraldehyde-osmium tetroxide fixation.

block Golgi apparatus secretion, i.e., secretory vesicles disappear from the dictyosomes within seconds and then begin to reappear in about 30 min after the cells have presumably become acclimated to the new osmolarity (H. H. Mollenhauer, unreported results). Exogenous sugars have also been shown to induce changes in the pattern and amount of Golgi apparatus secretion in the outer rootcap cells of maize (Jones and Morré, 1973).

A. Hormone Receptors

Binding of auxin hormones has been ascribed to various plant fractions including endoplasmic reticulum (Ray, 1977), plasma membranes (Batt and Venis, 1976; Kasamo and Yamaki, 1976; Williamson et al., 1977), and Golgi apparatus (Batt and Venis, 1976) or possibly even tonoplast (see also Williamson et al., 1977). Insulin binding activity has been reported for Golgi apparatus of rat liver (Bergeron et al., 1973). Changes in the Golgi apparatus secretory pattern was noted in maize root cap cells exposed to 2,4,6-trichlorophenoxyacetic acid (Mollenhauer and Hanson, 1976). In these cells, the Golgi apparatus vesicles became dense by electron microscopy and more easily visible.

B. Geotropism

The physical translocation of a cell component has long been considered as the mechanism for gravity perception (Schröter et al., 1973; Shen-Miller and Hinchman, 1974; Sievers and Volkman, 1977). The prime candidate for this role has been the amyloplast or "starch grain" so easily visible by light microscopy. Classically, the statoliths (i.e., the amylopasts) have been considered as behaving like a group of ball bearings rolling through the viscus medium of the cytoplasm. Displacement of the statoliths was thought to cause hormonal imbalance which in turn induced the observed changes in the rate of cell wall elongation.

More recent studies indicate that a simple translocation of amyloplasts is not sufficient in itself to perceive gravitational changes and trigger an appropriate response. Changes in gravitational accelerations of only 10^{-3}–10^{-4} g and angular deviations of only 5° are perceived by the gravity-sensing elements of the cells in time spans of only fractions of a second (Sievers and Volkmann, 1977; Volkmann and Sievers, 1979). Thus, current concepts favor a more integrated perceptual mechanism involving the interaction of several organelles and including endoplasmic reticulum and Golgi apparatus (Schröter et al., 1973; Shen-Miller and Hinchman, 1974; Volkmann, 1974; Sievers and Volkmann, 1977).

Golgi apparatus may function in geotropism through changes in the secretory pattern of cell wall constituents. For example, Sievers (1967) has shown that Golgi apparatus secretory vesicles are preferentially incorporated into the upper wall of *Chara* roots as compared to the lower wall when the roots are changed from a vertical to a horizontal position. Similarly, Shen-Miller and co-workers (Shen-Miller, 1971; Shen-Miller and Miller, 1972; Shen-Miller and Hinchman, 1974) have shown that the dictyosomes of oat coleoptiles move toward the bottom of the cell when the coleoptiles are tilted to a horizontal position. In addition, the number of secretory vesicles associated with the dictyosomes (a presumptive measure of secretory activity) increases as the dictyosomes fall toward the bottom of the cell. These changes may reach significant levels after only 6 min of gravity stimulation.

C. Abscission

Leaf abscission usually results from dissolution of cementing substances within the cell wall coupled with internal shear forces generated by differential growth and hydrostatic pressure within the cell (Morré, 1968). The process can be initiated by several conditions including senescence, injury, disease, environmental change, or chemicals such as ethylene. Initiation of the abscission process results in the synthesis of wall-degrading enzymes in the cells adjacent to the separation layer (Morré, 1968; Sexton *et al.,* 1977). In most instances, abscission involves only the middle lamella of the cell wall, although more extensive wall lysis may occur in some instances. The enzymes that degrade the wall constituents are moved into the wall by a secretory process (Sexton and Hall, 1974; Sexton *et al.,* 1977). The plasma membranes of the cells bordering the separation layer remain intact and do not lose their osmotic properties (Morré. 1968; Sexton *et al.,* 1977). Cells of the abscission zone are characterized by increased amounts of rough endoplasmic reticulum, increased numbers of dictyosomes, and greater numbers of dictyosome vesicles (Sexton and Hall, 1974; Sexton *et al.,* 1977) during the periods when wall-hydrolyzing enzyme secretion is occurring. In addition, there appears to be a significant increase in the number of vesicle profiles continuous with plasma membrane (Sexton and Hall, 1974; Sexton *et al.,* 1977). Thus, the observations suggest a role for Golgi apparatus in the transport and secretion of wall-dissolving enzymes.

D. Cell Wall Constituents

In the outer cells of the maize root cap, the Golgi apparatus secrete large quantities of polysaccharides into the cell wall (Mollenhauer and Whaley, 1963). Under fully hydrated conditions, the secreted polysaccharides pass into and through the cell wall and cover the root tip with slime. When not fully hydrated, however, the secreted polysaccharides accumulate between the cell wall and the plasma membrane (Fig. 5). A similar situation occurs with the slime-secreting cells of the okra pod (Mollenhauer, 1967b), the aerial roots of maize, and the aerial roots of orchids (Mollenhauer, 1967a). If water is added to the aerial roots of maize or orchid, then the secreted polysaccharides quickly hydrate and pass through the cell wall to the root exterior where they hold large quantities of water to form a slime droplet. It is of interest that root tip cells of water hyacinth, which are always under water, show no Golgi apparatus secretory activity of the type described above (Mollenhauer, 1967a). Secretory products from Golgi apparatus are also present in epidermal and phloem cells of maize roots (Mollenhauer and Mollenhauer, 1978). These products also penetrate the cell wall but are morphologically distinct from those of the maize root cap. In some cells, Golgi apparatus first produce epidermallike secretory products and then change to produce root cap type secretory products (Mollenhauer, 1965b). During the transformation, the Golgi apparatus secretory vesicles appear as epidermal-root cap hybrids (Fig. 20).

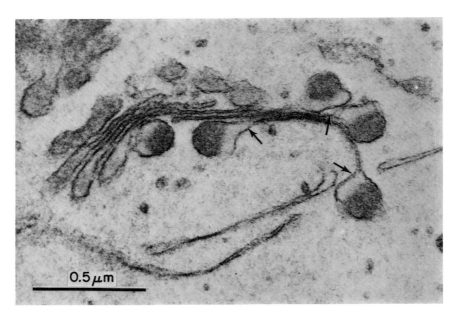

Fig. 20. Golgi apparatus of inner cells of the maize root cap form secretory vesicles that are structurally identical to those of the epidermal cells (see Fig. 14). The product of these vesicles is electron dense. As the cells mature, the secretory pattern changes abruptly to that characteristic of the outer cells of the root cap (see Figs. 5, 11, and 13). That is, the secretory vesicles become elongate and the secretory product is electron lucent. The newly synthesized secretory product is associated with the inner (cisternal) edge of the secretory vesicles (arrows) indicating, perhaps, that the secretory product is synthesized within the central parts of the cisternae and then transferred through the connecting tubules (see Fig. 12) to the forming vesicles. Potassium permanganate fixation.

Epidermal secretions are incorporated predominantly into the outer walls of the cells; however, some of the secreted product also impregnates the inner and side walls of the epidermal cells. The inner epidermal secretions penetrate only the epidermal walls and do not pass into the underlying cortical walls. The implication is that the middle lamella of these cells acts as a barrier to the secreted product. The same situation occurs in the walls of the maize phloem cells (see plate 13 of Leech *et al.,* 1963). The above examples indicate that a major function of the Golgi apparatus is to produce wall-impregnating agents that may impart special characteristics to the wall rather than to produce the microfibrillar phase of the wall.

E. Response to Injury

The study of the reaction of plant Golgi apparatus to physiological stress has been limited to morphological manifestations. Even the animal literature offers little or no insight into the biochemical changes accompanying stress.

There are three structural responses that characterize stressed Golgi apparatus: (1) change in secretory pattern; (2) change in number of cisternae per dictyosome; and (3) change in cisternal architecture. In addition, where extreme physical stress occurs (i.e., cell disruption), the Golgi apparatus may undergo rapid breakdown and disappear. In secretory cells, the most immediate response to adverse stress is the cessation of secretory activity.

An increase in number of cisternae has been noted in root meristem cells of *Allium cepa* following 6-azauracil treatment (Hall and Witkus, 1964), in root tip cells of *Allium* following barbital treatment (Benbadis and Deysson, 1975), and in maize root tip cells exposed to cold temperatures (Mollenhauer *et al.,* 1975), potassium cyanide, or an atmosphere of nitrogen (H. H. Mollenhauer, unreported results). Curling of cisternae has been observed in maize root tip cells during isolation (Morré and Mollenhauer, 1964) and following exposure to potassium cyanide or an atmosphere of carbon dioxide.

When cells are disrupted by cutting or homogenization, the Golgi apparatus tends to degenerate. The rate of breakdown may be very rapid (5–10 s) where suitable protective media are not provided. The breakdown pattern appears to be by a transformation of cisternae into tubules and then a breakdown of tubules into vesicles (Whaley and Mollenhauer, 1963; Fig. 21).

An inference of Golgi apparatus response to physiological shock is implied by changes in plasma membrane, lysosomes, and other Golgi apparatus-derived constituents. For example, an excess of calcium ion causes visible changes in the plasma membranes of wheat root tip cells and in transformed animal cells. There may even be direct changes in the width/length ratios of the dictyosomes (McCarthy *et al.,* 1974). However, a true "Golgi apparatus

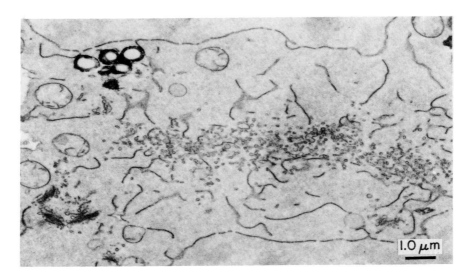

Fig. 21. The dictyosomes of mechanically injured cells break down rapidly (often within 5–20 s) into small vesicles. In dividing cells, similar appearing vesicles accumulate around the forming plate. The implication is that vesicles from the Golgi apparatus are transferred rapidly and selectively to this site of vesicle utilization. From Whaley and Mollenhauer (1963). Potassium permanganate fixation.

disorder'' caused wholly by Golgi apparatus disfunction, has not yet been recognized either in plants or animals.

X. GOLGI APPARATUS MULTIPLICATION

The literature is replete with suggestions as to how dictyosome replication might occur (see reviews by Whaley, 1966, 1975; Morré *et al.,* 1971a). Possibilities include formation by flow and reorientation of smooth membranes (Novikoff *et al.,* 1962; Morré *et al.,* 1971a), *de novo* synthesis from lipid spherulites (Mercer, 1962), perinuclear clear bodies (Werz, 1964), division by constriction (Buvat, 1958, 1963; Grassé, 1957; Grassé and Carasso, 1957), and replication by fragmentation and resynthesis of new cisternae (Dougherty, 1964).

Concern about Golgi apparatus replication dates back to at least the early 1900s when Perroncito applied the term ''dictyokinesis'' to what he assumed to be a process of division (see Whaley, 1966). With light microscropy, however, the dictyosomes were barely perceptible and the details of the replicative process could not be adequately observed. With the advent of electron microscropy, dictyosome structure became visible but the dynamics of the replicative process were lost. Thus, it was still not possible to define the replicative process. The problem remains today; we still do not know exactly how a dictyosome replicates.

Several general observations may be useful in reconsidering the problem of dictyosome replication. (1) Replication must be a subtle process as evidenced by the lack of definitive information about it. Replication does not necessarily require cleavage and splitting of an existing dictyosome into daughter structures. Such a process would be easy to recognize even in the static pictures obtained by electron microscropy. (2) Dictyosomes probably cannot replicate without endoplasmic reticulum to provide at least part of the membrane proteins of the new cisternae. In this sense, the endoplasmic reticulum and nuclear envelope would have considerable control over the replicative process. In dormant seeds, dictyosome regeneration does not occur to a recognizable extent until the endoplasmic reticulum has been at least partly reestablished. Similarly, dictyosome multiplication in the central root cap cells of maize clearly parallels endoplasmic reticulum extension (Whaley *et al.*, 1964). (3) Secretory dictyosomes must be in a continuous state of turnover and thus require the periodical replication of new cisternae to replace those lost by secretion (Morré, 1977a; Morré and Ovtracht, 1977). Cisternae may be formed and lost in intervals of less than 30 min (see Section VIII,B). Yet, the process is almost completely unobservable. Moreover, the number of cisternae per dictyosome may increase or decrease rapidly depending on the rate of secretion.

The above observations perhaps indicate that the fundamental unit for dictyosome replication is the cisterna or some replicative center associated with the cisterna. The latter idea is based upon the observation that new cisternae are formed upon a "replicative surface" associated with the forming pole of the dictyosome. That is, the new cisterna forms on a surface about 100 Å removed from the first cisterna of the stack. New cisternae may form without demonstrable direct membrane continuity with existing cisternae. This "replicating surface" is perhaps similar to the bonding constituents present between adjacent cisternae (Mollenhauer *et al.*, 1973). Thus, rapid dictyosome formation could occur by the tubular extensions of "replicating surfaces" from the peripheral edges of the cisternae, from individual cisternae, or from specialized regions of endoplasmic reticulum.

XI. CONCLUDING COMMENTS

The homology in form and function of plant and animal Golgi apparatus was one among many of the important concepts to arise from the applications of electron microscopy to plant cell biology. Both plant and animal Golgi apparatus are composed of dictyosome subunits and are surrounded in the cytoplasm by a special zone. Plant dictyosomes are usually more dispersed and often appear as discrete organelles while animal dictyosomes are closely spaced into structures more easily recognized as aggregate arrays. Both show a polarity across the stacked cisternae from endoplasmic reticulumlike at the forming (immature, proximal, or cis) face to plasma membranelike at the opposite (mature, distal, trans, or exit) face. Plant and animal Golgi apparatus differ structurally in the thickness of cisternae and in the intercisternal region. Plant dictyosomes are characterized by small fibers called intercisternal elements within the intercisternal region. Also, plant dictyosomes are easily disassembled into component cisternae (unstacked). Additionally, plant and animal Golgi apparatus differ markedly in their protein and phospholipid composition. This is all the more remarkable since their morphologies are very similar.

A major function of plant and animal Golgi apparatus is in the packaging of materials for export to the cell's exterior. In plants, polysaccharide slimes and matrix substances dominate as secretory products. A definitive role in formation of primary walls of rapidly elongating plant cells, however, has been established only for rhizoids, pollen tubes, fungal hyphae, and certain other tip-growing cells. Evidence is lacking or incomplete for other forms of wall growth. A second major function may be in the formation of new plasma membrane to support growth or to replace that lost to turnover. The latter function would explain why Golgi apparatus, even in nonsecretory cells, show evidence of membrane differentiation and exhibit plasma membrane characteristics.

An important characteristic of Golgi apparatus is their transitional nature. Golgi apparatus appear to depend on endoplasmic reticulum (or nuclear envelope) for their formation. Golgi apparatus contribute to the formation of plasma membrane. A contribution of Golgi apparatus to vacuole membranes has also been suggested, but definitive evidence is lacking. The transition nature of the Golgi apparatus is reflected in the overall composition of its membranes and in the functional and structural polarity gradient observed across the stacked cisternae. While certain partial reactions of membrane flow and differentiation, i.e., glycosylations, are carried out readily by isolated apparatus, it may be necessary to maintain structured relationships to endoplasmic reticulum or nuclear envelope and with the Golgi apparatus zone of exclusion in order for more complex manifestations of Golgi apparatus function to proceed. In this regard, Golgi apparatus should not be regarded as autonomous, or even semiautonomous, organelles but rather as components within a functional endomembrane system that includes endoplasmic reticulum, nuclear envelope, and various other transitional membranes and vesicles.

REFERENCES

Bal, A. K., and Payne, J. F. (1972). *Z. Pflanzenphysiol.* **66,** 265–272.
Batt, S., and Venis, M. A. (1976). *Planta* **30,** 15–21.
Beams, H. W., and Kessel, R. G. (1968). *Int. Rev. Cytol.* **23,** 209–276.
Benbadis, M.-C., and Deysson, G. (1975). *Planta* **123,** 283–290.
Bennett, H. S. (1956). *J. Biophys. Biochem. Cytol.* **2** (Suppl.), 99–103.
Bergeron, J. J. M., Evans, W. H., and Geschwind, I. I. (1973). *J. Cell Biol.* **59,** 771–776.
Berlin, J. D. (1968). *Radiat. Res.* **34,** 347–356.
Bluemink, J. G. (1971). *Z. Zellforsch. Mikrosk. Anat.* **121,** 102–126.
Bouck, G. B. (1969). *J. Cell Biol.* **40,** 446–460.
Bouck, G. B., and Brown, D. L. (1973). *J. Cell Biol.* **56,** 340–359.
Bowen, R. H. (1928). *Z. Zellforsch. Mikroskop. Anat.* **6,** 689–725.
Bowen, R. H. (1929). *Quart. Rev. Biol.* **4,** 299–324; 484–519.
Bowers, B., and Korn, E. (1968). *J. Cell Biol.* **39,** 95–111.
Bowles, D. J., and Northcote, D. H. (1972). *Biochem. J.* **130,** 1133–1145.
Bowles, D. J., and Northcote, D. H. (1974). *Biochem. J.* **142,** 139–144.
Bracker, C. E. (1974). *Electron Microsc. Proc. Int. Congr. 8th* **2,** 558–559.
Bracker, C. E., Grove, S. N., Heintz, C. E., and Morré, D. J. (1971). *Cytobiologie* **4,** 1–8.
Brown, R. M., Jr. (1969) *J. Cell Biol.* **41,** 109–123.
Brown, R. M., Jr., and Romanovicz, D. K. (1976). *Appl. Polym. Symp.* **28,** 537–585.
Brown, R. M., Jr., Herth, W., Franke, W. W., and Romanovicz, D. (1973). *In* "Biogenesis of Plant Cell Wall Polysaccharides" (F. Loewus, ed.), pp. 207–257. Academic Press, New York.
Buvat, R. (1958). *Ann. Sci. Nat. Bot. Biol. Veg.* **19,** 121–161.
Buvat, R. (1963). *Int. Rev. Cytol.* **14,** 41–155.
Carasso, N., Ovtracht, L., and Favard, M. P. (1971). *C. R. Hebd. Seances Acad. Sci.* **273,** 876–879.
Chang, P. L., Riordan, J. R., Moscarello, M. A., and Sturgess, J. M. (1977). *Can. J. Biochem.* **55,** 876–885.
Cook, G. M. W. (1973). *In* "Lysosomes in Biology and Pathology" (J. T. Dingle, ed.) Vol. 3, pp. 236–277. Elsevier North-Holland, Amsterdam.
Cooper, D., and Manley, R. St. J. (1975a). *Biochem. Biophys. Acta* **381,** 97–108.
Cooper, D., and Manley, R. St. J. (1975b). *Biochem. Biophys. Acta* **381,** 109–119.
Cortat, M., Matile, P., and Wiemken, A. (1972). *Arch. Mikrobiol.* **82,** 189–205.
Cunningham, W. P., Morré, D. J., and Mollenhauer, H. H. (1966). *J. Cell Biol.* **28,** 169–179.
Dashek, W. V., and Rosen, W. G. (1965). *Plant Physiol.* **40,** Suppl. XXIX.
Dauwalder, M., and Whaley, W. G. (1973). *J. Ultrastruct. Res.* **45,** 279–296.
Dauwalder, M., Whaley, W. G., and Kephart, J. E. (1969). *J. Cell Sci.* **4,** 455–497.
Dauwalder, M., Whaley, W. G., and Kephart, J. E. (1972). *Sub-Cell. Biochem.* **1,** 225–276.
DeDuve, C., and Wattiaux, R. (1966). *Annu. Rev. Physiol.* **28,** 435–492.
Dougherty, W. J. (1964). *J. Cell Biol.* **23,** 25a.
Drawert, H., and Mix, M. (1962). *Planta* **58,** 448–452.
Elder, J. H., and Morré, D. J. (1976). *J. Biol. Chem.* **251,** 5054–5068.
Elias, P. M., Park, H. D., Patterson, A. E., Lutzner, M. A., and Wetzel, B. K. (1972). *J. Ultrastruct. Res.* **40,** 87–102.
Essner, E., and Novikoff, A. (1962). *J. Cell Biol.* **15,** 289–312.
Favard, P. (1969). *In* "Handbook of Molecular Cytology" (A. Lima-de-Faria, ed.), pp. 1130–1155. North-Holland Publ., Amsterdam.
Fleischer, B., and Fleischer, S. (1970). *Biochem. Biophys. Acta* **219,** 301–319.
Flickinger, C. J. (1974a). *Anat. Rec.* **180,** 407–426.

Flickinger, C. J. (1974b). *Anat. Rec.* **180,** 427–448.

Franke, W. W. (1971). *Exp. Cell Res.* **66,** 486–489.

Franke, W. W., and Herth, W. (1974). *Exp. Cell Res.* **89,** 447–451.

Franke, W. W., and Kartenbeck, J. (1976). *In* "Progress in Differentiation Research" (N. Müller-Bérat, ed.), pp. 213–243. Amer. Elsevier, New York.

Franke, W. W., Eckert, W. A., and Krien, S. (1971a). *Z. Zellforsch. Mikrosk. Anat.* **119,** 577–604.

Franke, W. W., Morré, D. J., Deumling, B., Cheetham, R. D., Kartenbeck, J., Jarasch, E. D., and Zentgraf, H. W. (1971b). *Z. Naturforsch.* **26b,** 1031–1039.

Franke, W. W., Herth, W., VanDerWoude, W. J., and Morré, D. J. (1972). *Planta* **105,** 317–341.

Franke, W. W., Luder, M. R., Kartenbeck, J., Zerban, H., and Keenan, T. W. (1976a). *J. Cell Biol.* **69,** 173–195.

Franke, W. W., Morré, D. J., Herth, W., and Zerban, H. (1976b). *In* "Progress in Botany" (H. Ellenberg, K. Esser, H. Merxmüller, E. Schnepf, and H. Ziegler, eds.), Vol. 38, pp. 1–12. Springer-Verlag, Berlin and New York.

Frantz, C. E. (1973). M. S. Thesis, Purdue University, Lafayette, Indiana.

Frey-Wyssling, A., López-Sáez, J. F., and Mühlethaler, K. (1964). *J. Ultrastruct. Res.* **10,** 422–432.

Friend, D. S. (1965). *J. Cell Biol.* **25,** 563–576.

Friend, D. S., and Murray, M. J. (1965). *Am. J. Anat.* **117,** 135–149.

Gardiner, M., and Chrispeels, M. J. (1975). *Plant Physiol.* **55,** 536–541.

Glaser, L. (1958). *J. Biol. Chem.* **232,** 627–636.

Goff, C. W. (1973). *Protoplasma* **78,** 397–416.

Goff, C. W., and Klohs, W. D. (1974). *J. Histochem. Cytochem.* **22,** 945–951.

Grassé, P. -P. (1957). *C. R. Hebd. Seances Acad. Sci.* **245,** 1278–1281.

Grassé, P. -P., and Carasso, N. (1957). *Nature (London)* **179,** 31–33.

Green, J. C., and Jennings, D. H. (1967). *J. Exp. Bot.* **18,** 359–370.

Grove, S. N., and Bracker, C. E. (1970). *J. Bacteriol.* **104,** 989–1009.

Grove, S. N., Bracker, C. E., and Morré, D. J. (1968). *Science* **161,** 171–173.

Grove, S. N., Bracker, C. E., and Morré, D. J. (1970). *Am. J. Bot.* **57,** 245–266.

Hall, W. T., and Witkus, E. R. (1964). *Exp. Cell Res.* **36,** 494–501.

Hardin, J. W., Cherry, J. H., Morré, D. J., and Lembi, C. A. (1972). *Proc. Natl. Acad. Sci. U.S.A.* **69,** 3146–3150.

Harris, P. J., and Northcote, D. H. (1970). *Biochem. J.* **120,** 479–491.

Heath, J. B., and Greenwood, A. D. (1971). *Z. Zellforsch. Mikrosk. Anat.* **112,** 371–389.

Helsper, J. P. F. G., Veerkamp, J. H., and Sassen, M. M. A. (1977). *Planta* **133,** 303–308.

Herth, W., Franke, W. W., Stadler, J., Bittiger, H., Keilich, G., and Brown, R. M., Jr. (1972). *Planta* **105,** 79–92.

Hill, F. G., and Outka, D. E. (1974). *J. Protozool.* **21,** 299–312.

Hodge, A. J., McLean, J. D., and Mercer, F. V. (1956). *J. Biophys. Biochem. Cytol.* **2,** 597–607.

Holtzman, E., and Dominitz, R. (1968). *J. Histochem. Cytochem.* **16,** 320–336.

Inferrera, C., and Carrozza, G. (1975). *Golgi Centennial Symp.* pp. 13–38.

Jamieson, J. D., and Palade, G. E. (1967a). *J. Cell Biol.* **34,** 577–596.

Jamieson, J. D., and Palade, G. E. (1967b). *J. Cell Biol.* **34,** 597–615.

Jelsema, C. L., Morré, D. J., and Ruddat, M. (1977). *Bot. Gaz. (Chicago)* **138,** 138–149.

Jones, D. D., and Morré, D. J. (1973). *Physiol. Plant.* **29,** 68–75.

Jones, R. L., and Price, J. M. (1970). *Planta* **94,** 191–202.

Kasamo, K., and Yamaki, T. (1976). *Plant Cell Physiol.* **17,** 149–164.

Keenan, T. W., and Morré, D. J. (1970). *Biochemistry* **9,** 19–25.

Keenan, T. W., Morré, D. J., and Huang, C. M. (1974). *In* "Lactation: A Comprehensive Treatise" (B. L. Larson and V. R. Smith, eds.), Vol. 2, p. 191. Academic Press, New York.

Keenan, T. W., Huang, C. M., and Morré, D. J. (1972). *Biochem. Biophys. Res. Commun.* **47**, 1277–1283.

Kristen, U. (1978). *Planta* **138**, 29–33.

Lake, B. D., and Ellis, R. B. (1976). *Histochem. Journal* **8**, 357–366.

Leech, J. H., Mollenhauer, H. H., and Whaley, W. G. (1963). *Symp. Soc. Exp. Biol.* **17**, 74–84.

Lercher, M., and Wojciechowski, Z. A. (1976). *Plant Sci. Lett.* **7**, 337–344.

McCarthy, P., Richardson, C. L., Merritt, W. D., Morré, D. J., and Mollenhauer, H. H. (1974). *Proc. Indiana Acad. Sci.* **84**, 160–165.

Marty, M. F. (1973a). *C. R. Hebd. Seances Acad. Sci.* **277**, 2681–2684.

Marty, M. F. (1973b). *C. R. Hebd. Seances Acad. Sci.* **277**, 1317–1320.

Marty, M. F. (1978). *Proc. Nat. Acad. Sci. U.S.A.* **75**, 852–856.

Marty, M. F., and Buvat, M. R. (1973). *C. R. Hebd. Seances Acad. Sci.* **277**, 2681–2684.

Mercer, E. H. (1962). *In* "The Interpretation of Ultrastructure" (R. J. C. Harris, ed.), pp. 369–384. Academic Press, New York.

Miranda, A. F., Godman, G. C., Deitsch, A. D., and Tanenbaum, S. W. (1974). *J. Cell Biol.* **61**, 481–500.

Mollenhauer, H. H. (1965a). *J. Cell Biol.* **24**, 504–511.

Mollenhauer, H. H. (1965b). *J. Ultrastruct. Res.* **12**, 439–446.

Mollenhauer, H. H. (1967a). *Am. J. Bot.* **54**, 1249–1259.

Mollenhauer, H. H. (1967b). *Protoplasma* **63**, 353–362.

Mollenhauer, H. H. (1971). *J. Cell Biol.* **49**, 212–214.

Mollenhauer, H. H. (1974). *J. Cell Sci.* **15**, 89–97.

Mollenhauer, H. H., and Hanson, J. B. (1976). *Proc. Electron Microsc. Soc. Am.* **34**, 44–45.

Mollenhauer, H. H., and Mollenhauer, B. A. (1978). *Planta* **138**, 113–118.

Mollenhauer, H. H., and Morré, D. J. (1966a). *Annu. Rev. Plant Physiol.* **17**, 27–46.

Mollenhauer, H. H., and Morré, D. J. (1966b). *J. Cell Biol.* **29**, 373–376.

Mollenhauer, H. H., and Morré, D. J. (1971). *In* "Encyclopedia of Science and Technology" (D. N. Lapedes, ed.), pp. 262–264. McGraw-Hill, New York.

Mollenhauer, H. H., and Morré, D. J. (1974). *Protoplasma* **79**, 333–336.

Mollenhauer, H. H., and Morré, D. J. (1975). *J. Cell Sci.* **19**, 231–237.

Mollenhauer, H. H., and Morré, D. J. (1976a). *Cytobiologie* **13**, 297–306.

Mollenhauer, H. H., and Morré, D. J. (1976b). *Protoplasma* **87**, 39–48.

Mollenhauer, H. H., and Morré, D. J. (1978). *In* "Subcellular Biochemistry" (D. S. Roodyn, ed.), Vol. 5, pp. 327–359. Plenum, New York.

Mollenhauer, H. H., and Whaley, W. G. (1963). *J. Cell Biol.* **17**, 222–225.

Mollenhauer, H. H., Morré, D. J., and Totten, C. (1973). *Protoplasma* **78**, 443–459.

Mollenhauer, H. H., Morré, D. J., and VanDerWoude, W. J. (1975). *Mikroskopie* **31**, 257–272.

Morré, D. J. (1968). *Plant Physiol.* **43**, 1545–1559.

Morré, D. J. (1971). *In* "Methods in Enzymology: Vol. 22, Enzyme Purification and Related Techniques" (W. B. Jakoby, ed.), pp. 130–148. Academic Press, New York.

Morré, D. J. (1975). *Annu. Rev. Plant Physiol.* **26**, 441–481.

Morré, D. J. (1977a). *In* "The Synthesis, Assembly and Turnover of Cell Surface Components" (G. Poste and G. L. Nicolson, eds.), pp. 1–83. Elsevier North-Holland Biomed. Press, Amsterdam.

Morré, D. J. (1977b). *In* "International Cell Biology" (B. R. Brinkley and K. R. Porter, eds.), pp. 293–303. Rockefeller Univ. Press, New York.

Morré, D. J., and Mollenhauer, H. H. (1964). *J. Cell Biol.* **23**, 295–305.

Morré, D. J., and Mollenhauer, H. H. (1974). *In* "Dynamic Aspects of Plant Ultrastructure" (A. W. Robards, ed.), pp. 84–137. McGraw-Hill, New York.

Morré, D. J., and Mollenhauer, H. H. (1976). *In* "Encyclopedia of Plant Physiology [N.S.], Transport in Plants, III" (C. R. Stocking and U. Heber, eds.), Vol. 3, pp. 288–344. Springer-Verlag, Berlin and New York.

Morré, D. J., and Ovtracht, L. (1977). *Int. Rev. Cytol., Suppl.* **5**, 61–188.

Morré, D. J., and VanDerWoude, W. J. (1974). *In* "Macromolecules Regulating Growth and Development" (E. D. Hay, T. J. King, and J. Papconstantinou, eds.), pp. 81–111. Academic Press, New York.

Morré, D. J., Mollenhauer, H. H., and Chambers, J. E. (1965). *Exp. Cell Res.* **38**, 672–675.

Morré, D. J., Merlin, L. M., and Keenan, T. W. (1969). *Biochem. Biophys. Res. Commun.* **37**, 813–819.

Morré, D. J., Mollenhauer, H. H., and Bracker, C. E. (1971a). *In* "Results and Problems in Cell Differentiation. II. Origin and Continuity of Cell Organelles" (J. Reinert and H. Ursprung, eds.), pp. 82–126. Springer-Verlag, Berlin and New York.

Morré, D. J., Franke, W. W., Deumling, B., Nyquist, S. E., and Ovtracht, L. (1971b). *Biomembranes* **2**, 95–104.

Morré, D. J., Keenan, T. W., and Huang, C. M. (1974a). *Adv. Cytopharmacol.* **2**, 107–125.

Morré, D. J., Yunghans, W. N., Vigil, E. L., and Keenan, T. W. (1974b). *In* "Methodological Developments in Biochemistry" (E. Reid, ed.), Vol. 4, pp. 195–236. Longmans, London.

Morré, D. J., Lembi, C. A., and VanDerWoude, W. J. (1974c). *In* "Biochemische Cytologie der Pflanzenzelle" (G. Jacobi, ed.), pp. 147–172. Georg Thieme, Stuttgart.

Morré, D. J., Lembi, C. A., and VanDerWoude, W. J. (1977). *Cytobiologie* **16**, 72–81.

Moskalewski, S., Thyberg, J., Lohmander, S., and Friberg, K. (1975). *Exp. Cell Res.* **95**, 440–454.

Moskalewski, S., Thyberg, J., and Friberg, U. (1976). *J. Ultrastruct. Res.* **54**, 304–317.

Neutra, M., and Leblond, C. P. (1966a). *J. Cell Biol.* **30**, 119–136.

Neutra, M., and Leblond, C. P. (1966b). *J. Cell Biol.* **30**, 137–150.

Nicolson, G. L. (1976). *Biochem. Biophys. Acta* **457**, 57–108.

Nir, I., and Seligman, A. M. (1971). *J. Histochem. Cytochem.* **19**, 611–620.

Northcote, D. H., and Pickett-Heaps, J. D. (1966). *Biochem. J.* **98**, 159–167.

Novikoff, A. B., Essner, E., Goldfischer, S., and Heus, M. (1962). *In* "The Interpretation of Ulstrastructure" (R. J. C. Harris, ed.), pp. 149–192. Academic Press, New York.

Novikoff, P. M., Novikoff, A. B., Quintana, N., and Hauw, J. -J. (1971). *J. Cell Biol.* **50**, 859–886.

O'Brien, T. P. (1972). *Bot. Rev.* **38**, 87–118.

Öpik, H. (1968). *J. Exp. Bot.* **19**, 64–76.

Parke, M., Manton, I., and Clarke, B. (1955). *J. Marine Biol. Assoc. U.K.* **34**, 579–609.

Paull, R. E., and Jones, R. L. (1976). *Plant Physiol.* **57**, 249–256.

Peterson, M., and Leblond, C. P. (1964a). *J. Cell Biol.* **21**, 143–148.

Peterson, M., and Leblond, C. P. (1964b). *Exp. Cell Res.* **34**, 420–423.

Pickett-Heaps, J. D. (1967a). *J. Histochem. Cytochem.* **15**, 442–455.

Pickett-Heaps, J. D. (1967b). *J. Ultrastruct. Res.* **18**, 287–303.

Pollard, T. D., and Weihing, R. R. (1974). *CRC Crit. Rev. Biochem.* **2**, 1–65.

Potter, J. L., and Weisman, R. A. (1971). *Biochem. Biophys. Acta* **237**, 65–74.

Poux, N. (1963). *J. Microsc.* **2**, 557–568.

Poux, N. (1967). *J. Microsc.* **6**, 1043–1050.

Poux, N. (1969). *J. Microsc.* **8**, 855–866.

Poux, N. (1970). *J. Microsc.* **9**, 407–434.

Powell, T., and Brew, K. (1974). *Biochem. J.* **142**, 203–209.

Preston, R. D. (1974). "The Physical Biology of Plant Cell Walls." Chapman and Hall, London.

Rambourg, A. (1969). *C. R. Hebd. Seances Acad. Sci., Ser. D.* **269**, 2125–2127.

Rambourg, A. (1971). *Int. Rev. Cytol.* **31**, 57–114.

Ramsey, J. C., and Berlin, J. D. (1976). *Am. J. Bot.* **63**, 868–876.

Ray, P. M. (1977). *Plant Physiol.* **59**, 544–547.

Ray, P. M., Shininger, T. L., and Ray, M. M. (1969). *Proc. Natl. Acad. Sci. U.S.A.* **64**, 605–612.

Redman, C. M., Banerjee, D., Howell, K., and Palade, G. E. (1975). *J. Cell Biol.* **66**, 42–59.
Roberts, K., and Northcote, D. H. (1970). *J. Cell Sci.* **6**, 299–321.
Robinson, D. G., Eisinger, W. R., and Ray, P. M. (1976). *Ber. Deutsch. Bot. Ges.* **89**, 147–162.
Roland, J.-C. (1969). *Compt. Rend. Hebd. Seances Acad. Sci.* **269**, 939–942.
Roland, J.-C. (1973). *Int. Rev. Cytol.* **36**, 45–92.
Roland, J.-C., and Sandoz, D. (1969). *J. Microsc.* **8**, 263–268.
Roland, J.-C., and Vian, B. (1971). *Protoplasma* **73**, 121–137.
Roland, J.-C., Lembi, C. A., and Morré, D. J. (1972). *Stain Technol.* **47**, 195–200.
Rosen, W. G. (1968). *Annu. Rev. Plant Physiol.* **19**, 435–462.
Rosen, W. G., Gawlik, S. R., Dashek, W. V., and Siegesmund, K. A. (1964). *Am. J. Bot.* **51**, 61–71.
Rougier, M. (1976). *J. Microsc. Biol. Cell* **26**, 161–166.
Ryser, U. (1977). *Cytobiologie* **15**, 78–84.
Schachter, H. (1974). *Adv. Cytopharmacol.* **2**, 207–218.
Schachter, H., Jabbal, I., Hudgin, R. L., Pinteric, L., McGuire, E. J., and Roseman, S. (1970). *J. Biol. Chem.* **245**, 1090–1100.
Schnepf, E. (1969). Sekretion und Exkretion bei Pflanzen. *Protoplasmatologia,* **8**, 1–181.
Schnepf, E., and Busch, J. (1976). *Z. Pflanzenphysiol.* **69**, 62–71.
Schröter, K., Rodrigues-Garcià, M. I., and Sievers, A. (1973). *Protoplasma* **76**, 435–442.
Schwab, D. W., Simmons, E., and Scala, J. (1969). *Am. J. Bot.* **56**, 88–100.
Sexton, R., and Hall, J. L. (1974). *Ann. Bot. (London)* **38**, 849–854.
Sexton, R., Jamieson, G. G. C., and Allan, H. I. L. (1977). *Protoplasma* **91**, 369–387.
Shen-Miller, J. (1971). *Horm. Regul. Plant Growth Dev., Proc. Adv. Study Inst. 1971,* pp. 365–376.
Shen-Miller, J., and Hinchman, R. R. (1974). *BioScience* **24**, 643–651.
Shen-Miller, J., and Miller, C. (1972). *Plant Physiol.* **49**, 634–639.
Shore, G., and MacLachlan, G. A. (1975). *J. Cell Biol.* **64**, 557–571.
Shore, G., Raymond, Y., and MacLachlan, G. A. (1975). *Plant Physiol.* **56**, 34–38.
Sievers, A. (1963). *Protoplasma* **56**, 188–192.
Sievers, A. (1967). *Protoplasma* **64**, 225–253.
Sievers, A., and Volkmann, D. (1977). *In* "Plant Growth Regulation" (P. E. Pilet, ed.), pp. 208–217. Springer-Verlag, Berlin and New York.
Sjöstrand, F. S., and Hanzon, V. (1954). *Exp. Cell Res.* **7**, 415–429.
Southwood, D., and Dickinson, D. B. (1975). *Plant Physiol.* **56**, 83–87.
Staehelin, L. A., and Kiermayer, O. (1970). *J. Cell Sci.* **7**, 787–792.
Stockem, W. (1969). *Histochemie* **18**, 217.
Thyberg, J., Moskalewski, S., and Nilson, S. (1976). *J. Ultrastruct. Res.* **54**, 490.
Tilney, L. G. (1976a). *J. Cell Biol.* **69**, 51–72.
Tilney, L. G. (1976b). *J. Cell Biol.* **69**, 73–89.
Turner, F. R., and Whaley, W. G. (1965). *Science* **147**, 1303–1304.
VanDerWoude, W. J., Morré, D. J., and Bracker, C. E. (1971). *J. Cell Sci.* **8**, 331–351.
VanDerWoude, W. J., Lembi, C. A., Morré, D. J., Kindinger, J. I., and Ordin, L. (1974). *Plant Physiol.* **54**, 333–340.
Vian, B. (1974). *Compt. Rend. Hebd. Seances Acad. Sci.* **278**, 1482–1486.
Vian, B., and Roland, J.-C. (1972). *J. Microsc.* **13**, 119–136.
Vigil, E. L., and Ruddat, M. (1973). *Plant Physiol.* **51**, 549–558.
Volkmann, D. (1974). *Protoplasma* **79**, 159–183.
Volkmann, D., Sievers, A. (1979). *In* "Physiology of Movements" (W. Haupt and M. E. Feinleib, eds.), pp. 573–600. Springer-Verlag, Berlin and New York.
Werz, G. (1964). *Planta* **63**, 366–381.
Wessels, N. K., Spooner, B. S., Ash, J. F., Bradley, M. O., Luduena, M. A., Taylor, E. L., Wrenn, J. T., and Yamada, K. M. (1971). *Science* **171**, 135–143.

Westafer, J. M., and Brown, R. M. Jr. (1976). *Cytobios* **15**, 111–138.

Whaley, W. G. (1966). *In* "Probleme der Biologischen Reduplication" (P. Sitte, ed.), pp. 340–371. Springer-Verlag, Berlin and New York.

Whaley, W. G. (1975). "The Golgi Apparatus." Cell Biol. Monographs: Vol. 2. Springer-Verlag, Berlin and New York.

Whaley, W. G., and Mollenhauer, H. H. (1963). *J. Cell Biol.* **17**, 216–221.

Whaley, W. G., Kephart, J. E., and Mollenhauer, H. H. (1964). *In* "Cellular Membranes in Development" (M. Locke, ed.), pp. 135–173. Academic Press, New York.

Whaley, W. G., Dauwalder, M., and Kephart, J. E. (1972). *Science* **175**, 596–599.

Williamson, F. A., Morré, D. J., and Hess, K. (1977). *Cytobiologie* **16**, 63–71.

The Plant Nucleus | *13*

E. G. JORDAN
J. N. TIMMIS
A. J. TREWAVAS

The Biochemistry of Plants, Vol. 1

I. INTRODUCTION

In prokaryotes the intracellular space is essentially a single compartment and is relatively homogeneous in terms of its soluble molecules. In contrast the eukaryotic cell contains numerous compartments consisting of organelles bounded by membranes. Of these the most characteristic is the nucleus, an essential feature of all animal and plant cells. The presence of a nuclear envelope results in the establishment of a discrete compartment, the nucleoplasm, from which cytoplasmic organelles are excluded.

The nucleus is composed mainly of chromatin defined as the nuclear material taking basic stains. The acidic components of the nucleus, the nucleic acids, are responsible for this staining characteristic. Furthermore, local variations in the density of the nucleic acids in the chromatin are mirrored by equivalent variations in the density of staining. On this basis chromatin can be divided roughly into two categories—diffuse chromatin which is weakly staining, and dense chromatin which is heavily staining. These two forms of chromatin may perform different functions in the nucleus, with active RNA transcription associated with diffuse chromatin rather than the dense chromatin. Examination of chromatin at high magnification under the electron microscope reveals a mass of fibrils giving rise to the picture of a nucleus as an enclosed mass of tangled fibrils variously aggregated or dispersed.

The nucleus as an organelle is unique in that it undergoes a set of striking but reversible structural changes during mitosis. The interphase chromatin, much of which is diffuse, condenses on the approach of mitosis into a set of densely staining, well-defined chromosomes. This process of cell division enabled Weisman in the late nineteenth century to formulate the first chromosomal theory of inheritance, thereby ascribing to the chromosome the importance that is now known to be justified. The work of Avery (DNA is the genetic material) and Watson/Crick (on the structure of DNA) when they were published confirmed amply what had been surmised for many years concerning the chemical nature of genes. The accurate replication and transmission of genetic information in mitosis and reproduction was mirrored by a molecular precision in the DNA molecule based on hydrogen bonding between the two chains.

The presence of the genetic material in the nucleus is the major reason for the interest in this organelle. About 95% of the cell DNA is in the nucleus and this accounts for 10–40% of its total dry mass. Plant development and the generation of form are the result of a sequential expression of information from the genes. If it is understood how cells control the flow of information from DNA to messenger RNA to protein, a major step in appreciating the bases of these phenomena will have been made. There is, however, a two-way traffic of material across the nuclear membrane. Signals from the cytoplasm enter the nucleus where they may initiate the transcription of new genes or modify those already in activity. The idea that the nuclear membrane exerts a controlling influence on this inward and outward movement of molecules is an attractive but still highly speculative possibility.

The presence of a nuclear envelope is the major distinguishing feature of the eukaryotic cell. Its function must therefore be related to some fundamental feature of eukaryotic organization. This function may be the separation of transcription from translation. In the prokaryote, transcription and translation take place concomitantly. The eukaryotes have sequences of "control" DNA adjacent to the genes. This introduces much greater flexibility into the way in which the genetic apparatus can be regulated in eukaryotes. However, these controlling sequences are also transcribed as part of the initial gene product which may be 10–20 times the size of the final messenger RNA. By the use of selective nucleases and ligases these extra pieces are removed during a stage of processing in the nucleus. Only then is the messenger RNA moved to the cytoplasm for translation. Concomitant transcription and translation in eukaryotes can not occur, therefore. The presence of the nuclear envelope emphasizes this separation of transcription from translation.

II. THE NUCLEUS AT INTERPHASE

A. Types of Interphase Nuclei

Figure 1 is an electron micrograph of a typical plant nucleus (from a young leaf cell of *Zea mays*) which illustrates the main anatomical features. The chromatin is enclosed by two membranes; the outer is studded with ribosomes and sometimes can be seen to be continuous with the rough endoplasmic reticulum. The gap between the membranes is termed the perinuclear space. The occasional dark staining areas that cross the two membranes are nuclear pores in transverse section. Internally the large spherical body is the nucleolus, the cellular site of ribosome synthesis. This is attached by the lighter staining nucleolus-organizing region to dense chromatin. The remainder of the nucleus is filled with numerous dense chromatin strands interspersed with regions of finely dispersed chromatin fibrils and granules. Since the dense chromatin strands often appear to form a reticulum, this type of nuclear structure is termed reticulate and can be seen in many plants.

In contrast to this, nuclei from other plants have very little internal structure except for a nucleolus and occasional peripheral dense chromatin

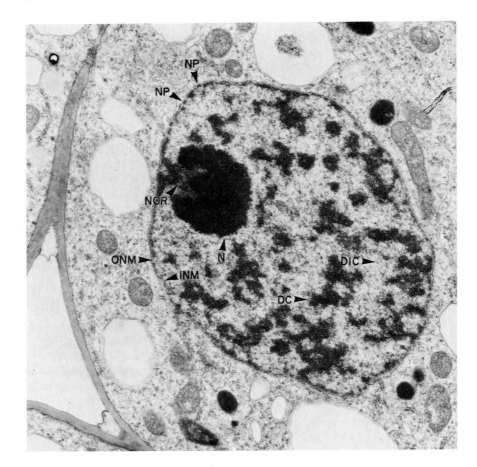

Fig. 1. Electron micrograph of a reticulate nucleus from a young leaf cell of *Zea mays*. ONM, outer nuclear membrane; INM, inner nuclear membrane; NP, nuclear pores; N, nucleolus; NOR, nucleolus-organizing region; DC, dense chromatin; DiC, diffuse chromatin. Magnification × 24,000. (Photograph courtesy of J. Pacy.)

patches called chromocenters. A typical chromocentric nucleus from *Spirodela* is shown in Fig. 2 and comparison of Fig. 1 and 2 immediately highlights the differences. Lacking the dense chromatin reticulum such nuclei have only occasional regions of dense chromatin. In this particular example the nucleolus contains a vacuole possibly indicating a high rate of ribosome synthesis.

The reasons for the gross difference in nuclear structure between reticulate and chromocentric plants have been discussed at some length by Lafontaine (1974b). Plants containing reticulate nuclei such as *Pisum, Maize,* or *Allium* in

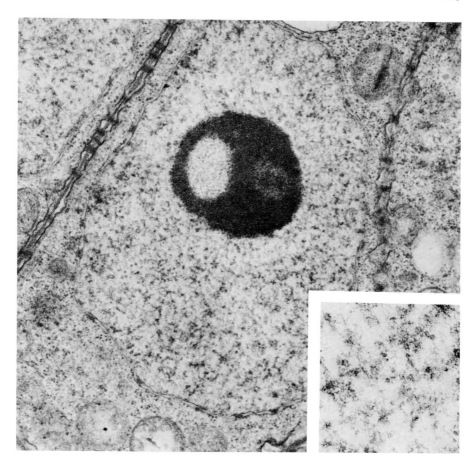

Fig. 2. Electron micrograph of a chromocentric nucleus from *Spirodela polyorhiza*. Magnification × 48,000. Inset shows chromatin magnified 120,000 × showing granule and fibril composition.

general have much longer chromosomes as well as more DNA/chromosome and more DNA/cell. The difference in structure cannot result from simply packaging larger amounts of DNA into the nucleus, however, because there is a linear relationship between DNA content and nuclear volume. An alternative possibility is suggested by the observation that higher DNA contents/cell are often associated with higher levels of intermediate repetitive DNA which may be packaged or condensed in a different way (Chooi, 1971) (see Section III,A,). Any attempt to understand the difference between reticulate and chromocentric nuclei is complicated by the observation that in differing metabolic states a nucleus may acquire a different internal structure. An example of this can be seen in Fig. 14 where the dense chromatin of a reticulate nucleus almost totally disappears at times of high RNA synthesis.

B. RNP Particles, Fibrils, and Puffs in the Nucleoplasm

The nucleoplasm contains a number of structures of which the majority can be convincingly seen in reticulate nuclei but are much less obvious in chromocentric nuclei. Most of these structures are likely to be concerned with the major function of the nucleus, the synthesis and processing of RNA.

Autoradiographic work has shown that the highest rates of RNA synthesis in the nucleus are found in the diffuse chromatin (Bouteille *et al.*, 1974). Very short term pulse-chase experiments have located the highest RNA specific activities as being in the diffuse chromatin areas surrounding the dense chromatin, i.e., in the so-called perichromatin region. Only later does the labeled RNA move into the main body of diffuse chromatin. Both fibrils (perichromatin fibrils) (Bouteille *et al.*, 1974) and granules (perichromatin granules) (Jordan and Chapman, 1971; Chaly and Setterfield, 1975) have been observed in this region.

On a biochemical basis the form of RNA with the highest specific activities is the so-called heterogenous nuclear RNA, (hnRNA). This RNA is of high molecular weight and is usually 10–20 times the size of messenger RNA of which it is a precursor. Normally it is processed extremely quickly by appropriate nucleases in the nucleus. The perichromatin fibrils which range from 30–200 Å in diameter have cytochemical characteristics which suggest that they may be composed of RNA. They have been tentatively identified as hnRNA (Monneron and Bernhard, 1969).

The perichromatin granules (shown in Fig. 3) are individual particles 350–400 Å in diameter which may be composed of densely packed fibrils. Some work has indicated that they might contain RNA and also that they might migrate to the cytoplasm (Bernhard, 1969). Their function is unknown although it has again been suggested that the RNA might be messenger RNA (Bouteille *et al.*, 1974).

Other granules have been observed in the main area of diffuse chromatin and an example is shown in Fig. 4. These interchromatin granules (Bouteille *et al.*, 1974) are usually found in clumps and may form part of a network which is composed of coiled filaments stretching from the nucleolus to the nuclear membrane. The granules are generally of the order of 200–250 Å in diameter although they can reach 400 Å in some plants (Heywood, 1976). It seems likely that they contain both RNA and protein. The morphology is distinctly different from ribosomes and it has been suggested that they could be informosomes (Spirin, 1969) which are about the same size as interchromatin granules. Informosomes contain both RNA and protein and may

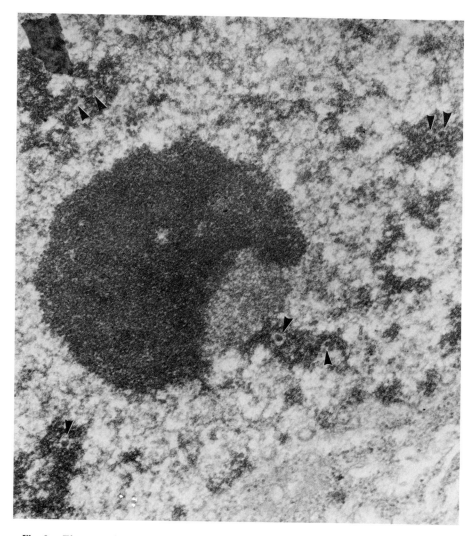

Fig. 3.   Electron micrograph of artichoke nucleus showing perichromatin granules (arrows). Magnification × 70,000.

be concerned with the processing and transport of mRNA to the nuclear envelope.

Structures again characteristically seen in reticulate nuclei are called micropuffs (Lafontaine 1965; Lafontaine and Lord, 1969) (Fig. 4). These are spherical and consist of a loose meshwork of fine twisted fibers which are unmistakably joined to the chromatin reticulum. Their staining density is

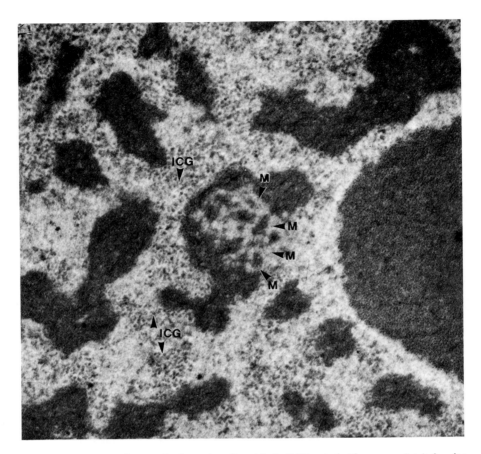

Fig. 4.   Electron micrograph of a nucleus from bluebell (*Hyacinthoides non-scripta*) showing a micropuff (M) and interchromatin granules (ICG). Magnification × 19,000. (Photograph courtesy of J. Pacy.)

heterogeneous with both diffuse and dense regions. They seem to be more apparent at times of high RNA synthesis but disappear as this declines (Chaly and Setterfield, 1975; Nagl, 1977). The more recent evidence of Lafontaine *et al.* (1979) shows that they contain DNA and indicates that they are regions of centromeric chromatin but centromeres do not always seem to be characterized by heterogeneity (Church and Moens, 1976).

   Other large spherical bodies in the nucleus have been observed and have recently been described in detail by Jordan (1976, and references cited therein). According to the internal structure these may be referred to as loose nuclear bodies, sometimes called micronucleoli or karyosomes, and dense nuclear bodies. Examples of both are shown in Figs. 5a and b. These

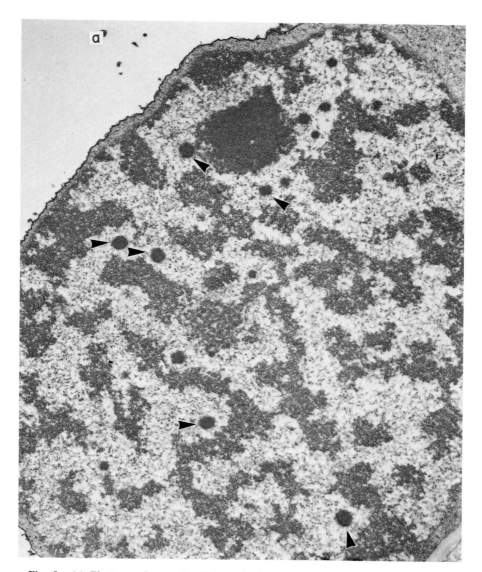

Fig. 5. (a) Electron micrograph of dense bodies (arrows) in nucleus of *Hyacinthoides non-scripta*. Magnification × 30,000. (b) Electron micrograph of karyosome in carrot nucleus. Magnification × 70,000.

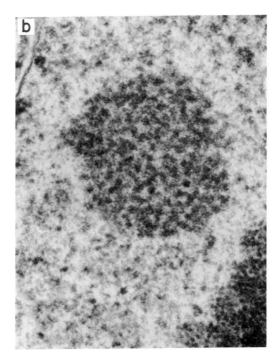

Fig. 5. (*Continued*)

structures are 0.5–1 μm across and are often found adjacent to nucleoli. Their function is unknown although again there appears to be a relationship with nuclear activity (Chaly and Setterfield, 1975).

Figure 6 summarizes the most distinctive nuclear features of reticulate nuclei in the form of a diagram.

C. The Interphase Nucleus and Its Defined Internal Architecture

Because the interphase nucleus has few anatomical landmarks (usually only a nucleolus) it is often considered that the chromosomes are not oriented in any way. In contrast to this, a view, common in the early part of this century (see references listed in Fussell, 1975) was that chromosomes (chromatin) were nonrandomly arranged in the interphase nucleus, which therefore had a defined internal order. An article by Comings (1968) and another by Vogel and Schroeder (1974) revived this point of view and discussed the evidence. A growing body of evidence, much of it derived from work in plants, now supports this idea and will be discussed.

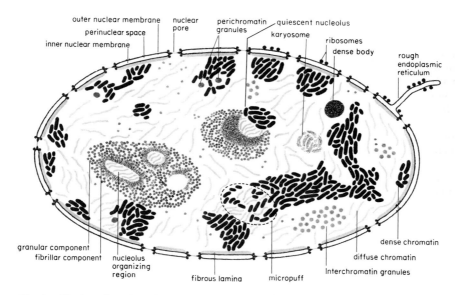

outer nuclear membrane   nuclear   perichromatin   quiescent nucleolus
perinuclear space   pore   granules   karyosome
inner nuclear membrane   ribosomes
dense body   rough
endoplasmic
reticulum

granular component   dense chromatin
fibrillar component   nucleolus   diffuse chromatin
organizing
region   Interchromatin granules
fibrous lamina   micropuff

Fig. 6. Ideal section of a reticulate nucleus showing all the main components. The nucleus is surrounded by an outer and inner nuclear membrane that encloses the perinuclear space which is one with the rough endoplasmic reticulum and has ribosomes attached to it. Between the chromatin and the inner membrane lies the fibrous lamina which is joined to the annuli of the nuclear pores. The chromatin is found as condensed chromatin and diffuse chromatin. The nucleolus shows fibrillar and granular components and a nucleolus-organizing region. The two nucleoli show the arrangement of components characteristic of low (smaller) and higher (larger) transcriptional activity. Around the borderline of the dense chromatin are shown perichromatin granules while the diffuse chromatin contains karyosomes, dense bodies, and interchromatin granules. A centromere or micropuff consisting both of diffuse and dense chromatin is also shown. All structures are to scale but the whole nucleus is drawn artificially small.

1. Attachment of Chromosomes to the Nuclear Membrane

That the nucleus is not simply a bag of chromosomes is shown by the fact that they appear to be attached to the nuclear membrane at a number of different sites. Early cytological reports observing this in plants are summarized in Van Der Lyn (1948) and a more dramatic centrifugal demonstration of membrane attachment in meiotic cells was made by Pusa in 1963.

Electron microscope studies have shown that during mitosis the nuclear envelope often does not break down completely but that large sections of it remain in intimate contact with the chromosomes throughout division. The onion photographs published by Porter and Machado (1960) underscore this point, and evidence in animal systems, which includes excellent pictures, is found in Franke and Scheer (1974a).

More recent work has shown that isolated nuclear membranes have attached DNA. Spread preparations of membrane-attached DNA have been photographed by Franke and Scheer (1974a) and leave little doubt on this point. In cotton seedling nuclei, Clay *et al.* (1975) showed that this DNA was newly synthesized as it is in some animal nuclei (Comings 1968; Berezne and Coffey, 1974a) inviting comparisons with the bacterial replicon. This, however, is a controversial point and at the present time it can only be concluded that the replication kinetics of membrane-bound DNA are different from that which is unbound (Vogel and Schroeder, 1974). The DNA membrane attachment is strong and resistant to high salt concentrations, urea, mild detergents, and even sonication. Its exact chemical nature is unknown.

Chromosomes are attached to the nuclear membrane at a number of sites that are probably specific in their location (Comings and Okada, 1970a,b). The most common are the centromeric and telomeric regions of the chromosome composed of densely staining heterochromatin (see Section V,A for this terminology). The adherence of these regions to the membrane accounts for the peripheral dense chromatin frequently observed in nuclear sections (Franke and Scheer, 1974a). The clearest case of obligatory telomeric attachment to a membrane in plants is that observed by Pusa (1963) in meiotic prophase. Fussell (1975) has recently shown in *Allium* that the centromeric regions of interphase chromosomes are clustered on one side of the nucleus while the telomeric regions are scattered in an arc on the other side.

2. Evidence for a Nuclear Skeleton

Although it used to be considered that the two nuclear membranes acted like a bag to retain the interphase chromosomes, if these membranes are removed the nucleus still retains its spherical shape (Riley *et al.*, 1975). If these nuclei are then further treated to remove the chromatin, a spherical ghost or matrix remains (Riley *et al.*, 1975; Berezney and Coffey 1974b). This nuclear ghost is composed of three discrete polypeptides—small amounts of RNA, phospholipid, and DNA which may be newly synthesized. A structure called the fibrous lamina (Fig. 7) has been observed a number of times in electron microscope pictures of animal nuclei. This structure lies just under the inner nuclear membrane and is about 150–180 Å thick (Bouteille *et al.*, 1974). It was initially thought to be identical with the nuclear ghost. Aaronson and Blobel (1975) have recently reported a preparation of nuclear pores attached to this lamina material.

Microscopic evidence by Comings and Okada (1976) and Busch and co-workers (Narayan *et al.*, 1967; Steele and Busch, 1966; Smetana *et al.*, 1963), however, has convincingly shown that the nuclear ghost is composed not only of a nuclear pore/fibrous lamina complex, but also of fibrillar nucleoli and an intranuclear matrix. The photograph of a nuclear ghost in Fig. 7a clearly shows the densely staining matrix material surrounded by a nuclear pore lamina complex. The intranuclear matrix appears to radiate outwards through the nucleoplasm to the fibrous lamina. Higher magnification of the ghost structure (Fig. 7b) shows the matrix to be composed of a network of protein fibers 20–30 Å thick and these may associate in places to form larger 100–300 Å thick fibers. It is possible that cross sections of the strands of this network give rise to the clusters of interchromatin granules (Section II,B). The function of the matrix is unknown but Comings and Okada (1976) have published evidence indicating that it may serve as sites of attachment for the DNA. Matrix-attached DNA exists in the form of a rosette (Fig. 7).

3. Nonrandom Arrangement of Homologous Chromosomes in the Nucleus

It is of course well known that homologous chromosomes pair up during meiosis. Such a phenomenon suggests either the presence of an accurate recognition system or alternatively, and more simply, that homologous chromosomes may remain attached throughout their somatic existence. Evidence for the latter possibility has been obtained several times. Somatic chromosomes are normally observed during metaphase and usually by the disruptive and randomizing method of squashing. Despite this a number of observations indicate that homologous chromosomes at metaphase lie much closer together than would be expected by chance. Results observing this in seven different plants were summarized by Kitani (1963); an excellent recent report is that by Werry *et al.* (1977) in *Happlopappus* who used direct observations of undisturbed metaphase plates. They reported that homologues always lay adjacent to each other. Convincing microscopic evidence for actual interchromosomal attachments between homologous chromosomes of *Ornithogalum vireus* has been published by Godin and Stack (1976) (Fig. 8A). The connections usually involve the dense telomeric chromatin (Ashley and Wagenaar, 1974) and could represent fragments of nuclear membrane. Maguire (1967) reported in maize that there were tendencies for homologous sections of dense chromatin to lie close to each other in the germ cells providing evidence for the possible pairing of homologues in the interphase nucleus.

4. Nonrandom Arrangement of Other Chromosomes in the Nucleus

Treatment of interphase nuclei with radiation or certain chemicals can cause chromosome breakage and rejoining of the new ends to different chromosomes. If interphase chromatin were randomly arranged it would be expected that there would be a random distribution of reunion. That this is not the case has been shown in *Tradescantia, Vicia,* and *Happlopappus* (Sax, 1940; Evans, 1961; Werry *et al.,* 1977; Wolff, 1959; Rao and Natarajan, 1967). There is often a marked preference for exchange between homologues and usually between the same regions of dense chromatin.

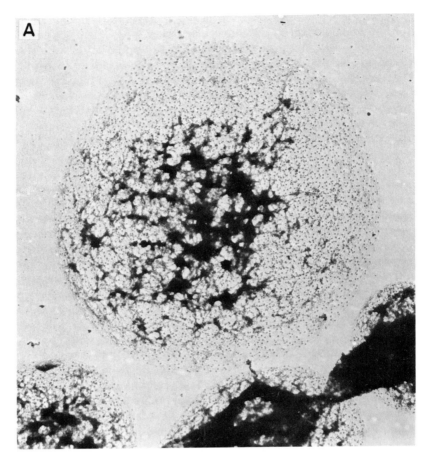

Fig. 7. Nuclear ghost structure and matrix-attached DNA. (A) Mouse liver nucleus washed with 2 *M* NaCl, triton X-100, treated with DNase and RNase and then water spread. The dense areas represent a fibrillar intranuclear matrix and spread around this is the nuclear pore-lamina complex. (B) A high magnification of a platinum shadowed nuclear pore-lamina complex showing the fibrous strands between the nuclear pores. (C). Rosettes of DNA-nuclear matrix complex from nuclei disrupted in dilute tris, solubilized in 2 *M* NaCl and centrifuged at 120,000 *g* for 12 h. The DNA pellet was spread, picked up on grids, and shadowed. There are multiple sites of DNA supercoiling indicated by arrows. Photographs courtesy of D. Comings. From Comings and Okada (1976).

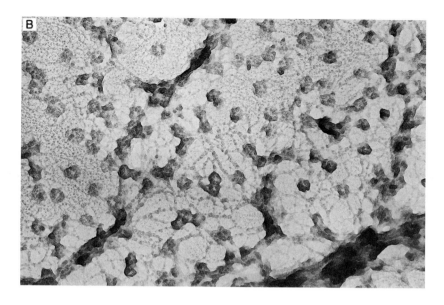

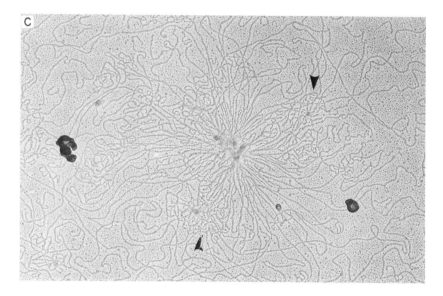

Fig. 7. (*Continued*)

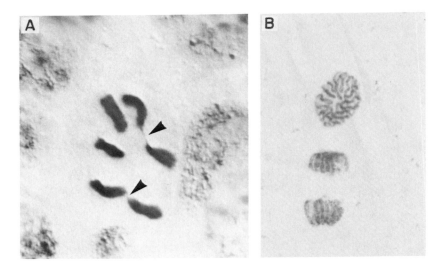

Fig. 8. A. Photograph of chromosomes of *Ornithogalum virens* showing telomeric attachments (arrows) between different chromosomes. Photographs were the gift of Stephen Stack. From Godin and Stack (1976). B. Mitosis in *Allium* root tips showing the polarized arrangement of the chromosomes within the nuclear envelope, × 600. At telophase (side view, lower cell) the centromeres aggregate in the polar regions of the two daughter nuclei. At prophase (polar view, upper cell) the same arrangement can be seen with both chromosome arms of each chromosome extending away from the almost circular group of centromeres. Apparently the polarized chromosome arrangement is maintained throughout interphase and either the orientation of the last division of the lower cell was perpendicular to the adjacent telophase or the lower nucleus has rotated through 90° during interphase. Photograph the gift of A. Dyer.

Interchange of chromatin between different chromosomes may occur spontaneously during interphase in somatic cells. Sax (1940) showed that over 80% of the exchanges which occurred in *Tradescantia* involved exchanges at loci which correspond to each other in respect of their centromeres. Similar results were reported by Evans (1961) in *Vicia* and imply that the chromosomes maintain a fixed orientation to each other in the interphase nucleus.

The telophase orientation of chromosomes has been reported on a number of occasions to be maintained through to the next prophase. Fussell (1975), who lists some of the early observations, clearly showed this in *Allium* by a study of heterochromatin distribution. Such results again imply that the position of the chromosomes is fixed in the interphase nucleus. Ghosh and Roy (1977) studied Giemsa C band orientation in *Allium* and clearly confirmed this point. An example of polarized arrangements of chromosomes within the nuclear envelope is shown in Fig. 8B.

5. Somatic Reduction in Plants.

This phenomenon has been described on a number of occasions (Storey, 1968a,b). In this process the normal diploid number of chromosomes may be reduced to the haploid number or less in the absence of a genuine meiotic process. The most informative example comes from Storey (1968b) who studied the spider plant (*Chlorophytum*). The somatic chromosome numbers of all members of this group have diploid numbers which are multiples of seven ranging from 14 to 56. This suggests that the genus arose as an autotetraploid in which the basic number was originally seven. Some mitotic reductions have been observed which separate seven chromosomes to one daughter cell and 21 to another. This ability to separate out a single genomic set is unusual and suggests a high degree of order among the chromosomes in an interphase nucleus.

6. Summary

That the interphase nucleus has a degree of internal order is now beginning to become apparent. Chromosomes are attached to the nuclear membrane at their centromeric and telomeric regions. Homologous chromosomes lie close to each other and may even have interchromosomal connections. Other chromosomes are held in position in the interphase nucleus and this position may be specified at each interphase period. Du Praw (1970) has even suggested that a haploid set of chromosomes may form a single giant genome; all the chromosomes may be connected by interchromosomal fibrils to form a large circle of DNA folded in places into the appropriate chromosomes. The evidence both for and against this notion has been discussed by Nagl (1976b). The suggestion is an intriguing one but it will require more substantiation before it is generally accepted.

III. STRUCTURE AND FUNCTIONAL ASPECTS OF INTERPHASE CHROMATIN

A. Chemical Constituents of Chromatin

1. DNA

Prokaryotes possess a genome which is compatible in complexity with the number of polypeptides they produce. Eukaryotic nuclei contain a much larger amount of DNA, an amount which is in vast excess of even the most liberal estimate of their requirements for making proteins. In addition, eukaryotes show a wide range of genome size which is essentially unrelated to phenotypic complexity. This phenomenon is most obvious in the higher plants, whose nuclei may show several hundredfold variation in DNA content. *Arabidopsis thaliana* has a haploid DNA value of 0.2 pg whereas *Fritillaria davisii* contains 89.5 pg. Even within the same family this variation may be observed; the Ranunculacea shows an eighty-fold variation. In the genus *Lathyrus,* where the chromosomes are similar in both number and morphology, the genomes show several hundred percent variation.

DNA has been analyzed by a variety of methods which determine the base composition and complexity, or size, of sequences comprising the genome. The size of the genome may be determined from the rate at which denatured DNA strands reassociate to form a duplex molecule. The simpler the sequence, the faster the renaturation reaction.

When a sample of DNA is rendered single stranded and incubated in solution under suitable conditions, the molecules collide with each other and will reform hydrogen bonds between complementary, or near complementary strands. The rate at which the reaction occurs is dependent upon the DNA concentration, the molarity of sodium ions in solution, and the base composition of the DNA sample, as well as on the kinetic complexity of the sequences renaturing. A useful method of presenting renaturation data is therefore to plot the percentage of DNA remaining single stranded against the product of the initial concentration of DNA (C_0, in moles of nucleotides per liter) and the time (t, in seconds) during which the reassociation reaction has proceeded. This is the C_0t curve of Britten and Kohne (1968) (Fig. 9).

Homogeneous genomes, such as those of bacteria or viruses show the expected second order kinetics, their renaturation spreading over two orders of C_0t value. Genomes such as T_4 or *E. coli* may therefore be described by their characteristic $C_0t_{1/2}$ value (the C_0t at which 50% of the DNA is renatured) (Curves a and b in Fig. 9). The genomes of higher organisms may not follow second order kinetics (curve c) indeed, if they did the times and concentrations of DNA required to achieve reassociation would be enormous in species with very large DNA contents. Instead eukaryote DNAs produce C_0t curves which indicate that the genomes are composed of a heterogeneous mixture of sequences varying from very simple and often highly repetitious, to much more complex tracts which are represented only once or a few times in the haploid nucleus. Some eukaryotes contain fractions that do not follow second order kinetics but renature so fast that a small proportion of the genome is thought to be composed of contiguous palindromic nucleotide sequences (Smith and Flavell, 1975). While the purpose of much of the repeated DNA is unclear its presence is characteristic of eukaryotes, most particularly of plants (Table I). There would seem to be a general if rather tenuous correlation between nuclear DNA content and the content of repetitive DNA. This correlation is very strong in diploid members of the genus *Lathyrus* where the increase in repetitive DNA accounts in large measure for the observed variation in the nuclear DNA amount (Narayan and Rees,

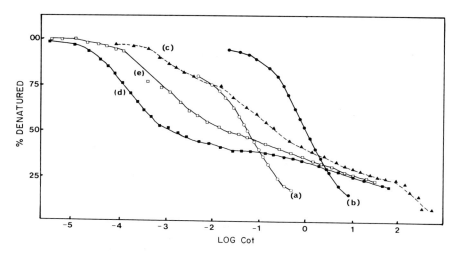

Fig. 9. Renaturation kinetics of cucumber DNAs. Denatured DNA samples, sheared to approximately 0.8×10^6 daltons single-stranded molecular weight, were renatured at appropriate concentrations and in suitable salt molarities. Renaturation of cucumber satellite I ($\square$), satellite II ($\blacksquare$), bacteriophage T_4 ($\bigcirc$), *E. coli* ($\bullet$), and rye total ($\blacktriangle$) DNA was followed optically at 270 nm corrected to $2 \times$ SSC and 35% formamide. C_0t value represents the product of the concentration of DNA (moles of nucleotides per liter) and the length of time in seconds during which the renaturation reaction has taken place. No correction was made for G + C content. Modified from Timmis and Ingle (1977). Data for rye previously unpublished.

1976), and is also closely linked with the amount of DNA in heterochromatin. The genus *Anemone* however, does not appear to behave similarly (Cullis and Schweizer, 1974).

Certain genes, such as those which code for ribosomal RNA, tRNA, and histone proteins, are known to be present in multiple copies in tandem. In plants ribosomal RNA genes may be present in very large numbers; up to about 32,000 copies are present in a telophase nucleus of some hyacinth varieties (Ingle *et al.*, 1975). Other repeated sequences are interspersed with unique sequences. These sequences are about 1,000 nucleotides long in wheat (Flavell and Smith, 1976) and are of about the size expected to code for proteins. Based on similar findings in animal systems it has been proposed that the genes of eukaryotes consist of the main, unique coding sequence with attached repeated regions, the latter having a role in the regulation of gene transcription (Britten and Davidson, 1969). In wheat there is also a small unique fraction of the genome which is much more complex and not interspersed with reiterated sequences (Flavell and Smith, 1976). It is notable, however, that even the unique portion of plant genomes thus far described is in excess in information content of the cell's probable requirements for proteins.

TABLE I

The Proportion of Repeated Nucleotide Sequences in Nuclear DNA of Some Plants

		Haploid DNA content (daltons)	Proportion of repetitive DNA (%)
Fungi	Torulopsis candida	1.16×10^{10}	5.3^a
	Saccharomyces exiguus	1.13×10^{10}	10.8^a
	Neurospora crassa	$2.20 \times 10^{10\ b}$	12.3^c
	Coprinus lagopus	$2.50 \times 10^{10\ b}$	15.0^c
	Torulopsis holmii	1.36×10^{10}	16.2^a
	Phycomyces blakesleeanus	1.90×10^{10}	35.0^b

		2C DNA content (pg)	
In a range of higher plants[d]	Lamium purpureum	2.7	60
	Daucus carota	2.1	62
	Beta vulgaris	2.7	63
	Helianthus annuus	10.7	69
	Pisum sativum	9.9	75
	Zea mays	11.0	78
	Vicia faba	29.3	85
	Secale cereale	18.9	92
In *Lathyrus* spp.[e]	L. articulatus	12.5	56
	L. cicera	14.2	58
	L. nissolia	13.2	59
	L. ochrus	14.0	60
	L. tingitanus	17.9	60
	L. clymenum	13.8	62
	L. aphaca	14.0	63
	L. sativus	17.2	66
	L. hirsutus	20.3	70
In *Anemone* spp.[f]	A. coronaria	19.9	53
	A. blanda	32.0	57
	A. cylindrica	21.9	65
	A. riparia	21.0	67

[a] Christiansen *et al.* (1971).
[b] Dusenberg (1975).
[c] Dutta (1974).
[d] Flavell *et al.* (1974).
[e] Narayan and Rees (1976).
[f] Cullis and Schweizer (1974).

It may be fortuitous that the highly reiterated DNA is sometimes sufficiently different in composition, and therefore in buoyant density, to allow
separation and subsequent purification by equilibrium centrifugation in
heavy metal salt gradients. The minor components resolved by these methods have been called satellite DNAs and they occur in a wide evolutionary
range of organisms from fairly primitive to the most advanced. The best
studied satellite DNA is that from the mouse where it comprises 10% of the
genome and differs in buoyant density from the bulk of the DNA by 0.010 g
cm^{-3}. In guinea pig the satellite DNAs are composed of very simple sequences with a unit of only six nucleotide pairs repeated millions of times
(Southern, 1970).

In many organisms renaturation is often inaccurate, indicating that the
satellite sequences are not highly conserved, but subject to considerable
evolutionary divergence. Hybridization of radioactive satellite DNA or *in
vitro* labeled complementary RNA copies of satellite DNA to preparations of
chromosomes denatured *in situ,* indicate that the sequences are present near
the centromeres of all or most of the chromosomes of the mouse complement
(Jones, 1970; Pardue and Gall, 1970).

In plants satellite DNAs may comprise up to 40% of the genome, as in
cucumber (Ingle *et al.,* 1973), but they have been analyzed in detail in relatively few species (Timmis and Ingle, 1977). They all contain very simple,
highly repetitious nucleotide sequences, comparable with those of animals,
which are not sufficiently complex to code for polypeptides. The plant satellite DNAs differ from those of animals in containing more complex portions
which are potentially able to code for proteins.

Curves d and e of Fig. 9 show C_0t plots for the two satellite DNAs from
cucumber. Clearly a major fraction of each is composed of a sequence much
simpler than T4 DNA, having a complexity of about 0.2×10^6 daltons
(Timmis and Ingle, 1977). A smaller heterogeneous portion of each satellite
DNA renatures at much greater C_0t values approaching the complexity of *E.
coli* DNA. Whether any RNA products are transcribed remains an open
question.

The simple sequences from *Scilla sibirica* have been located on the
chromosomes by *in situ* hybridization (Timmis *et al.,* 1975). The sequence is
present on all the chromosomes of the complement and is localized in regions
of cold sensitive heterochromatin at both interphase and metaphase (Fig.
10). The type of heterochromatin in *Scilla* could have a very different function from the centromeric form of mouse nuclei and yet it is also associated
with repetitious simple sequences (Jones, 1970). One suggestion for a function for satellite DNA is that it confers a mechanical advantage on the
chromosomes, an idea that is based on the centromeric hybridization observed in mouse. The very different patterns found in *Scilla,* where the
heterochromatin is not centromeric, do not support this suggestion. The

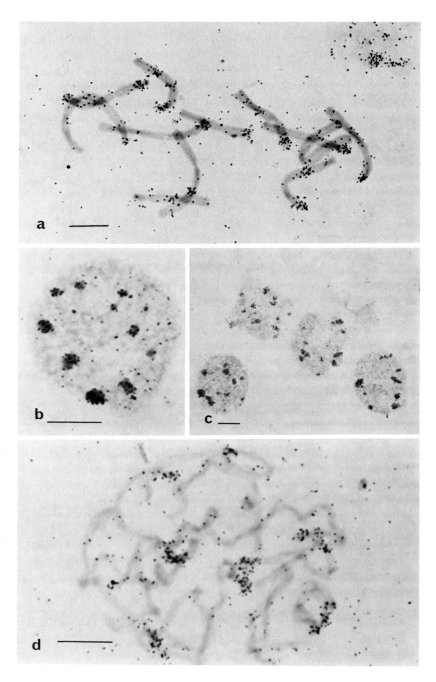

Fig. 10. *In situ* hybridization of *Scilla sibirica* satellite DNA. RNA complementary to pure satellite DNA was transcribed *in vitro* and hybridized to denatured squash preparations from

fairly constant association between heterochromatin and reiterated DNA does imply a role for satellite DNA in chromosome coiling and supercoiling. This possibility has been apparently undermined by the extreme localization of fast renaturing sequences on chromosomes.

2. Histones

DNA in eukaryotic cells is generally associated with large amounts of protein. The types of protein are divided operationally into acid-soluble proteins or histones and the acid-insoluble proteins or nuclear acidic proteins.

The histones can be classified into five groups on the basis of their lysine and arginine content; the very lysine-rich histones, H1; the slightly lysine-rich histones, H2A; the moderately lysine-rich histones, H2B; and the arginine-rich histones H3 and H4. All five groups have been detected in plant chromatin. Some of the information on plant histones has been summarized in Tables II and III.

Histones are not detectable in bacterial organisms and the simplest plants in which they have been clearly demonstrated are *Chlorella* and *Euglena,* both eukaryotic green algae. It is often considered that histones evolved at the same time as other eukaryote characteristics but this conclusion is probably premature. Although histones have not been detected in the chromatin of a number of blue–green algae a recent report indicates that there may be three histone-like basic proteins in the chromatin of *Anacystis*. These proteins are not ribosomal, but their cellular content is extremely low and at the levels detected they are unlikely to play a significant structural role in *Anacystis* chromatin. However, it is possible that they represent the primitive ancestors of present-day histones. Histones are also absent in the dinoflagellates; chromatin from these organisms contains only one basic protein in very small amounts. In agreement with this, dinoflagellates also lack the 100 Å and 250 Å eukaryotic chromosome fibrils, typical of DNA complexed with histone. Instead, like blue–green algae, they contain the smaller 20–25 Å fibril typical of bacterial nucleoids and representing naked DNA. However, the dinoflagellates are eukaryotes in that they possess a nuclear envelope. Such results suggest that the origin of histones and other eukaryote characteristics are separable evolutionary events.

---

colchicine treated root tips. The slides were coated with Ilford K2 emulsion and exposed for 5 weeks at 4°C. After developing, the nuclear material was stained with Geimsa. a, Mitotic prophase showing about half the number of groups of silver grains as in metaphase chromosomes. b,c, Interphase nuclei showing groups of silver grains over the more dark staining, heterochromatic chromocenters. d, Mitotic metaphase cell ($2n = 12$), showing mainly distal localization of the hybridized complementary RNA. The bars represent 10 μm. Adapted in part from Timmis *et al.* (1975).

The presence of histones in another group of primitive organisms, the fungi, is still very controversial. Early reports (e.g., Leighton *et al.,* 1971) indicated that chromatin from *Neurospora, Phycomyces,* and *Microsporus* had an extremely low acid-soluble protein content and no detectable histones. Later reports (Table II) failed to confirm this but instead demonstrated a nearly normal complement of histones in these and other fungi with only occasional omissions. Possible reasons for this conflict of views have been

TABLE II

Distribution of Histone Species throughout the Plant Kingdom

Organism	Histone					Reference
	H1	H2A	H2B	H3	H4	
Blue–green algae 4 species	None detected but three basic proteins detected in Anacystis which may have histone similarities					Gofschtein *et al.* (1975); Makino and Tsuzuki (1971)
Algae						
Dinoflagellates (two species)	None detected					Rizzo and Nooden (1974)
Euglena	1	1	1	1	1	Netrawali (1970)
Chlorella	1	1	1	1	1	Kanazawa and Kanazawa (1968)
Fungi						
Neurospora	1	1	1	1	1	Goff (1976)
Phycomyces	1	1	1	0	1	Cohen and Stein (1975)
Aspergillus	1	1	1	1	1	Felden *et al.* (1976)
Achlya	1	1	1	1	0	Horgen *et al.* (1973)
Yeast	0	1	1	1	1	Thomas and Furber (1976)
Blastocladiella	None detected					Horgen *et al.* (1973)
Microsporum	None detected					Leighton *et al.* (1971).
Moss						
Polytrichum	3	1	0	1	1	Spiker (1975)
Pteridophytes						
Psilotum	2	1	0	1	1	Spiker (1975)
Equisetum	0	1	1	1	1	Spiker (1975)
Polypodium	2	1	1	1	1	Spiker (1975)
Gymnosperms						
Four species	2–4	1	1	1	1	Berkofsky and Roy (1976) Spiker (1975)
Dicotyledonous Angiosperms 14 species	3–4	1–2	1	1	1	Nadeau *et al.* (1974) Spiker (1975)
Monocotyledonous Angiosperms Seven species	1–5	1–2	1	1	1	Nadeau *et al.* (1974) Spiker (1975)

TABLE III

A Comparison of Some Properties of Plant and Animal Histones[a]

Organism	Molecular wts.					Lysine/arginine ratio				
	H1	H2A	H2B	H3	H4	H1	H2A	H2B	H3	H4
Calf thymus	21,500 ⎱ 20,500 ⎰	12,500	13,700	14,000	11,000	17	1.34	2.61	0.74	0.71
Fungi										
Neurospora[b]						4.9	1.3	3.6	0.9	0.8
Phycomyces	19,000[c]	12,500	13,700	0	11,000	5.3	—	—	—	—
Aspergillus	17,500[c]	13,000	13,500	14,000	11,000	—	1.32	2.96	—	0.76
Pea	25,000 ⎱ 24,000 ⎰	16,500 ⎱ 15,000 ⎰	16,500	14,000	11,000	9.1	2.2	4.5	0.66	0.71
Wheat	25,500 25,200 24,700 24,000 22,000	16,200 15,200	16,800	14,000	11,000	6.8	—	—	0.69	0.69

[a] Data taken from Cohen and Stein (1975); Fambrough et al. (1968); Felden et al. (1976); Goff (1976); Nadeau et al. (1974); Panyim and Chalkley (1971); Sommer and Chalkley (1974); Spiker and Isenberg (1977); Spiker (1975).

[b] Majority have same molecular weight as mouse histones.

[c] Value estimated from publication.

discussed at some length by Goff (1976). These may hinge around the difficulties of purifying the rather small nuclei from these organisms and the rather active proteases often found associated with fungal chromatin. Leighton et al. (1976), in defense of their original work, have shown that fungal chromatin may be grossly contaminated with ribosomal proteins, some of which are deceptively histone-like and that this may explain the reported detection of histones in some fungi. Chromatin with low RNA/DNA ratios may have little or no acid-soluble protein. However, the analytical data provided by Goff (1976) for the five histone fractions in Neurospora seem very convincing and the detection of nucleosomes in some fungi, particularly Aspergillus and Neurospora, must place the balance of opinion in favor of those who believe fungal chromatin contains the normal complement of histones (Kornberg, 1977).

In the higher plants the situation is much clearer. Apart from occasional omissions in the mosses and pteridophytes, other plants possess representatives of the five histone groups. The amino acid sequences of pea histones H3 and H4 have been determined and they differ by only four and two residues, respectively, from their calf counterparts indicating exceptional evolu-

tionary conservation (Elgin and Weintraub, 1975). The remaining plant histones H1, H2A, and H2B do show marked dissimilarities to the appropriate
calf histones and their generally greater size indicates that at the best there
can only be partial conservation of sequence. These modifications in histone H2A and H2B are unlikely to impair their ability to form nucleosomes
(Spiker and Isenberg, 1977). The number of histone H1 components in higher
plants can be extremely variable but the majority of plants examined do
seem to have one major and two minor components. Fambrough *et al.* (1968)
have shown that in peas both the number and the percentage of histone as
histone H1 varies with the tissue examined. A putative tissue-specific function is therefore possible for plant histone H1 as it is in animals. However,
the total histone/DNA ratio of chromatin also varies with the tissue source in
peas and an inverse correlation between this ratio and *"in vitro"* RNA synthesis by the chromatin has been noted (Bonner, 1976).

The polypeptide structure of histones can be modified in a number of
ways. In plant systems, methylation, acetylation, and phosphorylation have
all been detected. The modifying enzymes may be attached to the chromatin
and S-adenosyl methionine, acetyl-CoA, and ATP are the appropriate donors
of the modifying group (Berkofsky and Roy, 1976; Patterson and Davies,
1969; Chapman *et al.*, 1975). Phosphorylation of histones H1 and H2B have
been detected in artichoke (Trewavas and Stratton, 1976) and phosphorylation of H1 is carried out by a specific histone kinase (Lin and Key, 1976) on
both serine and threonine residues. Acetylation and methylation takes place
on lysine residues (Nadler, 1976; Patterson and Davies, 1969).

3. Nuclear Acidic Proteins

The function of the histones is to initiate the condensation and packaging
of the DNA into the interphase nucleus and to nonspecifically repress transcription. Other nuclear properties including enzymatic activities and specific
genetic restriction are the responsibility of the non-histone proteins.

The abundance of individual non-histone proteins is certainly variable.
Non-histone nuclear proteins from ungerminated barley embryos have been
separated by two-dimensional gels (A. J. Trewavas, unpublished). The major
protein species have approximate molecular weights of 70,000, 68,000,
57,000, 50,000, 42,000, 30,000, and 20,000. The major species in rat liver, as
determined by Douvas and Bonner (1977) and Comings and Okada (1976),
have molecular weights of 68,000, 67,000, 65,000, 50,000, 45,000, 34,000,
32,000, and 28,000, which are not dissimilar. The proteins from 50,000 to
28,000 have been identified in liver as tubulin, actin, tropomyosin, and
myosin. The three remaining proteins, 65,000 to 68,000 MW, have been identified as components of the structural matrix (Comings and Okada, 1976).

Nuclear non-histone proteins share certain physical characteristics with
cytoplasmic proteins. They are weakly acidic (isoelectric points between pH
5 and 7) but in general have higher molecular weight. When nuclear non-his-

tone proteins from artichoke are labeled during a 3-h incubation and then separated two-dimensionally and autoradiographed, nearly all the proteins have over a 40,000 MW and a substantial number have over 100,000 MW. Cytoplasmic proteins in the artichoke are between 20,000 and 60,000 MW (D. Melanson and A. J. Trewavas, unpublished). Several hundred proteins can be detected on the original autoradiographs of artichoke and similar numbers have been seen in nuclei from barley embryo and pea root (A. J. Trewavas, unpublished). Longer labeling periods result in the appearance of more protein in the artichoke nucleus suggesting a heterogeneity of turnover rates. About 500–1000 non-histone proteins have been observed in mammalian nuclei (Peterson and McConkey, 1976). Most plant nuclear non-histone proteins can be solubilized in 8 M urea implying some sort of hydrophobic attachment in the nucleus (Trewavas, 1976b).

As far as is known nuclear proteins are synthesized in the cytoplasm and are then moved to the nucleus. Although isolated nuclei have frequently been reported to incorporate amino acids into protein (Kuehl, 1974), a recent analysis using two-dimensional gel electrophoresis indicate that isolated barley nuclei fail to make discrete protein products (A. J. Trewavas and C. J. Leaver, unpublished). If nuclei are incubated with an *in vitro* wheat germ system a number of recognizably cytoplasmic proteins are synthesized. The protein synthetic capacity of the nucleus is therefore probably limited to the ribosomes attached to the nuclear membranes.

A substantial number of the nuclear acidic proteins in barley, artichoke, and pea can be modified by phosphorylation. The numbers of proteins range from 50 to 80 and they are generally the most abundant proteins in the nucleus. In all three plants the predominant phosphorylated proteins are 40,000 to 50,000 and 90,000 to 110,000 MW. These proteins are phosphorylated by nuclear-based protein kinases using ATP and the group modified is usually serine (Van Loon *et al.,* 1975).

B. Chromatin Composed of Fibrils and Fibrils Composed of Strings of Nucleosomes

As indicated by the inset in Fig. 2, nucleoplasmic chromatin is composed of fibrils. Since these fibrils can be shown both enzymatically and by specific staining to contain both DNA and protein (mainly histone), much interest has centered on their actual molecular structure. Although early work involved the use of the electron microscope it has been generally considered that the results were too subject to artifact to be of general use. Instead an alternative technique developed by Gall (1963) has been used. Chromatin was spread on an air–water interface, negatively stained or air dried, and then examined microscopically.

The initial work (reviewed by Solari, 1974) indicated that the chromatin and chromosome fibrils were about 200–300 Å in diameter averaging about 250 Å. Occasional areas up to 500 Å could also be observed giving the fibers a distinctly bumpy appearance. Subsequent work extended this and showed that variation in fiber width could be anywhere from 25–500 Å but with a mean value still about 250 Å (Wolff, 1959). Tissue variations could also be detected. For example, barley root chromatin had an average fibril diameter of 170 Å while barley endosperm was 244 Å (Wolff, 1968). In chromatin actively synthesizing RNA, fibrils 35–50 Å at their thinnest could be seen. Metaphase chromosomes had fibrils of about 300 Å diameter compared to interphase fibrils averaging 250 Å (Dupraw, 1970).

The picture that emerged from this work was of a fairly basic fibril structure of 250 Å in thickness which could become thicker (when chromosomes were being condensed in mitosis), or thinner when they were being actively transcribed. Treatment with chelating agents such as EDTA (a technique that results in chromatin swelling) caused thinning of many 250 Å fibers to 100 Å (Solari, 1974). Examination of EDTA-treated chromatin by negative staining revealed some areas of fibrils with diameters of 25–30 Å which is very close to the DNA double helix of 20 Å. Since the effect of EDTA is to cause unfolding it was considered that the 100 Å and 250 Å fibril were produced by folding of the DNA duplex.

Measurement of the dry mass of the 250 Å fibril and its DNA content showed that DNA represented some 40% of the dry mass. Each micron of fibril contained some 60 μm of DNA (packing ratio 60). This packing ratio was increased to 150:1 during mitosis when the metaphase fibrils had a thicker diameter and greater mass/unit length. These measurements have very important implications for the packing of DNA into the fibrils. This can only be achieved in the 100-Å fibril by supercoiling the DNA coil and in the 250-Å fibril by supercoiling the supercoil.

Recent studies have greatly clarified the internal molecular architecture of the chromatin fibrils. These studies have shown that fibrils are composed of a repeating structure unit called a nucleosome and in turn each nucleosome is composed of a globular histone core around which the DNA is wrapped (Kornberg and Thomas, 1974). For chromatin that is not undergoing transcription or replication, Kornberg (1977) proposed that each nucleosome comprises a short length of DNA containing about 200 base pairs associated with two molecules each of histones H2A, H2B, H3, and H4 (histone octamer) and one molecule of H1. The experimental basis for the nucleosome concept began with studies by Hewish and Burgoyne (1973) on the digestion of animal nuclei with a micrococcal endonuclease which yielded 200 base pair DNA fragments and higher multiples thereof. Precisely similar results have been shown for plants (McGhee and Engel, 1975). At the same time electron microscope studies of stretched chromatin fibrils showed the presence of beadlike particles on threads (Olins and Olins, 1974; Woodcock *et al.*, 1976).

Brief micrococcal nuclease treatment of nuclei or chromatin releases nucleosomes containing approximately 200 base pairs of DNA. Extended digestion reaches a kinetic pause at about 140 base pairs where the product contains all the histones listed above except H1 (Kornberg, 1977). The limit digest particle is referred to as the core particle. The nucleosome core length is 140 base pairs in Aspergillus, Yeast, Neurospora, HeLa cells, and chicken (Morris, 1976; Compton *et al.*, 1976; Noll, 1976; Lohr *et al.*, 1977). Core particles are joined in intact chromatin by a piece of DNA continuous with that in the particle and comprising the remainder of the 200 base pairs (i.e., about 60) called the linker DNA. Histone H1 is primarily attached to the linker DNA and its presence stabilizes the packing of core particles into a continuous thread. This is shown diagrammatically in Fig. 11. In intact chromatin acidic protein may be associated with the linker DNA.

Under suitable conditions chromatin can be disaggregated into DNA and histone and then reaggregated to a structure in which the nucleosomes seem identical to those in the original chromatin (Oudet *et al.*, 1975). The reconstruction of nucleosomes in this case is a self-assembly process and requires no molecules which are not present in the final structure. The histones do not seem to recognize specific DNA sequences and the nucleosomes are randomly distributed along DNA.

Nucleosome cores have recently been crystallized and studied by X ray diffraction and electron microscopy (Finch *et al.*, 1977). The core is roughly disc-shaped with a diameter of 100 Å and a thickness of 57 Å. It is divided into two symmetrically-arranged halves along the short axis and it is believed that each half contains one molecule each of histones H2A, H2B, H3, and H4. The path of the DNA cannot be precisely located but the number of turns of DNA around the core can be estimated. Figure 12 shows nucleosome cores and the approximate diameter of DNA if it were wrapped once or twice around the outside. There must obviously be almost two complete turns of DNA around each core. Each turn probably involves 80 DNA base pairs with a pitch of 28 Å (Finch *et al.*, 1977). The dimensions of the core are summarized in Fig. 11.

DNA strand separation, an essential prerequisite of both transcription and replication, is impossible without some degree of structural disassembly of the nucleosome. Since transcriptionally active genes are in the form of nucleosomes (Section III,C,3), this suggests that such a process is induced by the polymerase as it reads the base sequence. Weintraub and Groudine, 1976) have described a model in which the two halves of the nucleosome separate during transcription, rather like a fully open oyster.

The exact arrangement of the histones in the core is not known. Nucleo-

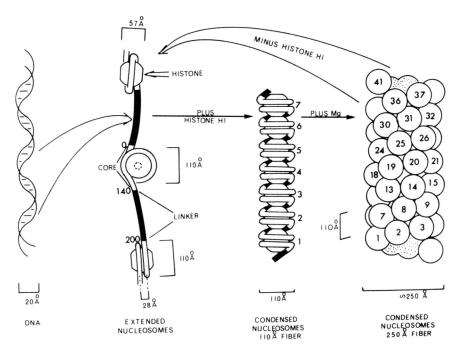

Fig. 11. A diagrammatic scheme of various levels of chromatin structure. The diagram shows the essential features of the extended nucleosome structure, includes the various measurements of the nucleosome core (Finch *et al.*, 1977), and indicates the position of the core DNA (140 base pairs) and the linker DNA (base pair number 140 to 200). Addition of histone H1 results in folding of the linker and abjunction of the cores of successive nucleosomes to form the 110-Å fibril. As drawn the linker DNA is folded behind the fibril. Further addition of low concentrations of Mg^{2+} ion (1 mM) causes condensation of the nucleosomes into a solenoidal structure having about six nucleosomes per turn and equivalent to a 250-Å fibril. For clarity the DNA has been omitted from the drawing of the 250-Å fibril and only the position of the nucleosome cores is shown. Removal of histone H1 from the 250 Å fibril results in formation of extended nucleosomes. Diagram based on Finch and Klug (1976); Finch *et al.* (1977); Thoma and Koller (1977).

somes can be prepared using just histones H3 and H4 but not with just histones H2A and H2B (Kornberg, 1977). Thus histones H3 and H4 may form a basic skeletal structure whilst the other two act as fillers agreeing with the strict evolutionary conservation of the former but not the latter (Weintraub and Groudine, 1976) (Section III,C,2). Attempts using chemical cross-linking agents to clarify the histone arrangement in the core have not led to any definite proposals (Kornberg, 1977) but two interesting facts have emerged. Extensive cross-linking of chromatin leads initially to a cross-

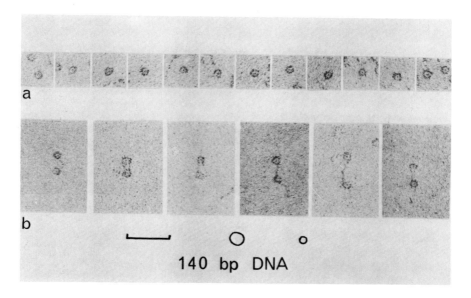

Fig. 12. Electron micrographs of (a) single and (b) binucleosomes. The scale underneath shows the length of 140 base pairs of DNA drawn to form either one circle or two. The size of the nucleosome core is clearly closer to two circles of DNA around the histone core. The photographs show the presence of a central hole in the nucleosome. Photograph from C. Woodcock.

linked octamer and then later to multiples of this octamer. At the same time histone H1 is converted to a homopolymer. Obviously the histones composing the core octamer are in contact with each other and other nucleosome cores in chromatin. Histone H1 in the linker region is not in direct contact with the core histones but is in contact with histone Hl molecules in other linker regions.

The length of the linker region, in contrast to the core, can vary by up to 30 base pairs in different chromatins (Lohr *et al.*, 1977). The linker region is somehow folded by its association with the histone H1 so that higher orders of chromatin structure can occur. This can be seen convincingly in Fig. 13 which are electron micrographs of chromatin in the presence and absence of histone H1 (Thoma and Koller 1977). A diagrammatic scheme of the various stages of chromatin structure has been drawn in Fig. 11. Association of histone H1 with extended nucleosomes at very low salt concentrations leads to the formation of the classical 100 Å fibril (Brasch, 1976). Addition of low concentrations of Mg^{2+} ion (1 mM) to the 100 Å fibril leads to the formation of a solenoidal structure with six to seven nucleosomes per turn and an

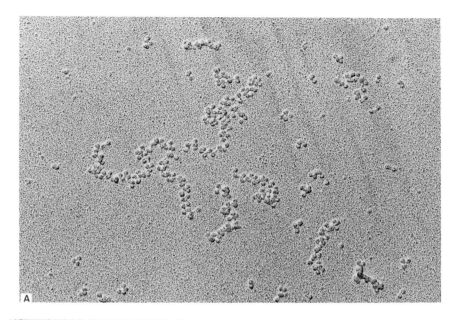

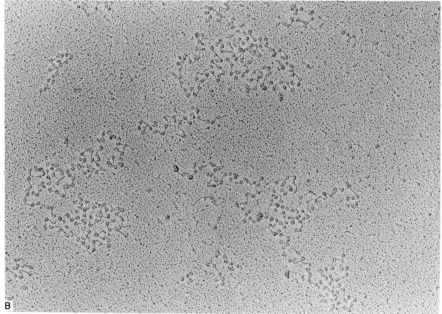

Fig. 13. Electron micrographs of chromatin and chromatin depleted of histone H1. A, chromatin fixed in 10 mM sodium acetate showing individual nucleosomes packed into a filament about 250 Å thick; B, chromatin depleted of histone H1 and fixed in 10 mM sodium chloride showing extended nucleosome subunits joined by linker DNA. Photographs from F. Thoma. From Thoma and Koller (1977), with permission of MIT Press.

approximate width of 250 Å (Finch and Klug, 1976; Thoma and Koller, 1977), representing the classical chromatin fibril diameter. Removal of histone H1 from this structure results in a return to extended nucleosomes.

C. The Structure of Transcribable Chromatin and the Control of Genetic Restriction

1. The Structure of Transcriptionally Active Chromatin

In the normal interphase nucleus variable degrees of chromatin condensation can be observed. The diffuse lightly staining areas of chromatin are associated with high rates of RNA synthesis (Frenster, 1969) and contain ribonucleoprotein particles. The dense chromatin areas are inactive in RNA synthesis. In nuclei that are totally inactive in RNA synthesis the majority of chromatin is extremely dense as may be seen in nuclei of maturing sieve elements (Burr and Evert, 1973). Most nuclei retain a balance between the two forms with perhaps a conversion of dense chromatin to diffuse chromatin at times of high RNA synthesis and a return when nuclear activity subsides (Frenster, 1974). An example is shown in Fig. 14 of nuclei from the dividing, elongating, and mature zones of the maize root. The highest rates of RNA synthesis are found in cells in the elongating zone and nuclei in these regions show higher amounts of diffuse chromatin. Comparable results have been described for the pea root (Chaly and Setterfield, 1975).

The fibril diameter of diffuse and dense chromatin is different. Dense or inactive chromatin which represents 90% of the DNA has a fibril diameter of about 250 Å which in mitotic chromosomes is compacted to about 300 Å (DuPraw, 1970). Diffuse chromatin, on the other hand, has a fibril diameter of 100 Å or less (Frenster, 1969) and is also strongly depleted in histone H1 (Gottesfeld et al., 1975). The transition may therefore be of the form described by Finch and Klug (1976), Oudet et al. (1975), and Thoma and Koller (1977) in which a tightly coiled 250 Å chromatin fibril is unwound to extended nucleosomes by removal of histone H1. This change is diagrammed in Fig. 11 and discussed in Section III,B. This unwinding phenomenon is confirmed by physical measurements. Optical techniques show the DNA of diffuse chromatin to be in a more unraveled state (Gottesfeld et al., 1975) and this is supported by the much lower melting temperature of diffuse chromatin (Frenster, 1969). Along with this change in fibril diameter the formation of diffuse chromatin is accompanied by an unraveling of the whole chromatin structure. This can be clearly visualized by examining the pictures of polytene chromosomes shown in Fig. 15. In the inactive state the chromosome is condensed and banded. As the RNA synthetic capacity of the chromosome increases the chromatin becomes diffuse and eventually an unraveling and a looping out of portions of the chromosome occurs.

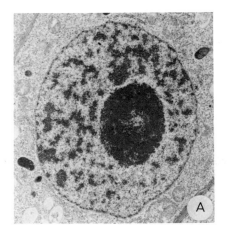
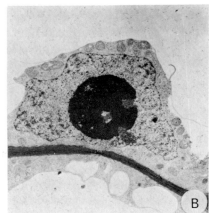
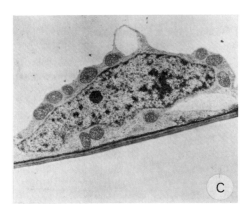

Fig. 14. Nuclei from three developmental stages in the maize root. A, from cell in the dividing zone (0.5 mm from tip); B, from cell in the elongating zone (4 mm from tip); C, from cell in the mature zone (10 mm from tip). Note the disappearance of much of the dense chromatin from stage A as cells go to stage B and with some return in stage C (A. J. Trewavas, unpublished).

2. Transcriptionally Active Chromatin versus Inactive Chromatin

An important consequence of the unraveling of dense chromatin to diffuse chromatin is that the latter now becomes accessible to many molecules. This has been shown in a number of different ways. By using cell fusion techniques, nuclei of cells which are inactive in RNA synthesis (e.g., mature chick erythrocyte) can be reactivated (Harris, 1970). As these nuclei recommence RNA synthesis they swell dramatically increasing some 20- to 30-fold in volume. There is a direct linear relationship between nuclear volume and

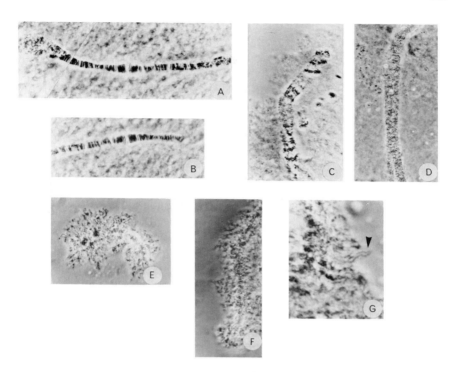

Fig. 15. Photographs of polytene chromosomes of Phaseolus suspensor in various stages of activity. A and B, chromosomes inactivated by chilling. Note the well-defined band structure. C, polytene chromosome in partly active state showing some heterochromatic and euchromatic bands; D, granular appearance of active polytene chromosome; E and F, hyperactive polytene chromosomes showing lamp brush appearance; G, detail of loop of lampbrush chromosome. Photographs from W. Nagl (Nagl, 1970, 1974, 1977), with permission of the publisher.

resultant RNA synthesis indicating a conversion of dense to diffuse chromatin. This dispersion of the chromatin is accompanied by a marked increase in the affinity of the chromatin for DNA-intercalating dyes such as acridine orange and for DNA-binding antibiotics such as actinomycin D. At the same time the nuclear protein content is increased 3- to 4-fold. Nuclear enlargement therefore loosens the chromatin and renders it accessible not only to proteins but to small molecules such as acridine orange.

The accessibility of diffuse chromatin to macromolecules has been strikingly demonstrated by the use of nucleases. Very mild treatment of chromatin with such enzymes leads to the selective release of diffuse chromatin (Marushige and Bonner, 1971; Gottesfeld et al., 1975; Tata and Baker, 1978). Clearly the DNA of diffuse chromatin is directly accessible to enzymatic hydrolysis while that of dense chromatin is not. This differential nuclease sensitivity is also demonstrated by individual genes in the process of transcription. The globin gene can be transcribed in the immature chick erythrocyte nucleus but not in the fibroblast nucleus. Weintraub and Groudine (1976) using nucleases showed that the globin gene in erythrocyte chromatin was very susceptible to digestion whilst that in fibroblast chromatin was very resistant.

Further indications of accessibility differences have followed from the use of specific antibodies to DNA (Frenster, 1969). These are readily able to bind to the DNA in diffuse chromatin but cannot bind to that of dense chromatin.

Finally, the accessibility differences also include RNA polymerase. Considerable *in vitro* transcription work has been conducted using RNA polymerase from *E. coli*. Despite the fact that the enzyme is wholly alien to eukaryote chromatin it very faithfully mimics the normal restriction pattern shown in gene expression. For example, it will transcribe the ovalbumin gene in chromatin from estrogen-treated oviduct tissue but not in the chromatin from untreated tissue; likewise the histone and globin genes in the appropriate systems (Stein *et al.*, 1975; Axel *et al.*, 1973). Since it is unlikely that *E. coli* RNA polymerase recognizes specific promoter regions on the eukaryotic DNA, its ability to transcribe some genes and not others must be the result of simple structural, i.e., accessibility differences.

In summary, the conversion of dense to diffuse chromatin is the result of simple structural alterations. The DNA of dense chromatin which is normally tightly packed in the 250-Å fibril (Fig. 11) is relatively inaccessible to a number of small and large molecules including most importantly RNA polymerase. The conversion to diffuse chromatin takes place by an unwinding or unraveling of the 250-Å fibril to the 100-Å fibril or extended nucleosomes possibly by loss of histone H1. As a consequence the DNA becomes freely accessible to RNA polymerase and can be transcribed.

3. Nuclear Non-Histone Proteins as Controlling Elements

Although very early work suggested that the histones played a specific role in genetic restriction this is no longer thought to be the case. Besides being highly conserved molecules in evolution they are very limited in complexity and apart from histone H1 are identical throughout the various tissues of an organism. Differential gene transcription that accompanies tissue and organ formation could not therefore be explained by changes in histones. Furthermore, the DNA/histone ratios of diffuse and dense chromatin differ only slightly (Frenster, 1969; Gorovsky and Woodward, 1967) although there is a very marked depletion of histone H1 in diffuse chromatin (Gottesfeld *et al.*, 1975).

In contrast the nuclear acidic proteins are qualitatively complex and variable between different tissues and organisms. Furthermore the non-histone protein to DNA ratio is 2- to 4-fold higher in diffuse than dense chromatin (Frenster, 1969). Unequivocal evidence to support a specific role for this group of proteins in genetic restriction has come from the chromatin reannealing experiments. They are best understood by reference to an example. The globin gene can be transcribed in reticulocyte chromatin but not in liver chromatin. These two types of chromatin were purified and each was divided into its three-component fractions—DNA, histone, and non-histone protein. These were then mixed together in various combinations and reannealed to form heterologous chromatins, e.g., DNA from liver, histone from reticulocyte, and non-histone protein from liver. After addition of *E. coli* RNA polymerase the capability of such heterologous chromatins to synthesize globin mRNA could be tested. The results of these experiments were unequivocal. Specific genetic restriction of the globin gene was a property of the tissue source of the non-histone protein and not the DNA or histone (Gilmour and Paul, 1975). Similar conclusions concerning the control of the histone genes (Stein *et al.*, 1975) and ovalbumin genes (Tsai *et al.*, 1976) have also been drawn; the specific controlling elements are in the nonhistone proteins.

It should not be understood from this that histones perform no function in genetic restriction. Early results (Huang and Bonner, 1962) showed that histones strongly inhibit the template activity of DNA. This inhibition is nonspecific. It is the function of the acidic proteins to specifically release genes from their inactive state induced by histones.

4. How Do Acidic Proteins Modify the Structure of Chromatin to Make It Available for Transcription?

The answer to this question is still very incomplete. A number of ideas have dominated the research and three of these will be briefly considered.

In the first theory it has been supposed that histones coat the DNA and sterically interfere with the movement of the RNA polymerase along the DNA. It was thought that acidic proteins promoted transcription by preventing the binding of histones to DNA (Stein *et al.*, 1974). To support this it was shown in some pea and other tissues that there was variation in the DNA/histone ratios. It was highest in the tissues most active in RNA synthesis (Bonner, 1976) implying some loss of histones upon transcription. The development of the nucleosome structure for chromatin has led to a modification of these views since it is clear that the majority of histones do not bind with the DNA in the way originally visualized. Recent work has shown unequivocally that transcriptionally active chromatin is structurally in the form of nucleosomes (Tata and Baker, 1978). Furthermore a minimum of six nucleosomes in length is essential for the RNA polymerase to engage. The demonstration has also been extended to globin, ovalbumin, and ribosomal RNA genes. While they are being transcribed they are in the form of nucleosomes (Axel, 1976; Garel *et al.,* 1977; Reeve, 1977; McKnight and Miller, 1976; Grainger and Ogle, 1978). (See Section VI,D,3 for further discussion on ribosomal RNA genes.) The length of the linker region may be increased during transcription (Lohr *et al.,* 1977; Grellet *et al.,* 1977), but this may simply reflect modification or loss of histone H1. The possibility that nuclear proteins specifically interfere with the binding of histone H1 molecules to DNA is still an open question.

In an adaptation of the above idea it has been proposed that phosphorylation or some other post-translational change in the non-histone protein may initiate its gene-controlling properties. It was thought phosphorylation would increase the ability of such weakly acidic proteins to bind to the basic positively charged histones (Stein, *et al.,* 1974; Kleinsmith, 1975) removing them from DNA. This view developed from the demonstration that many non-histone proteins are phosphorylated and that there are much higher levels of phosphorylated non-histone proteins in diffuse than in dense chromatin (Frenster, 1969). Although much evidence has been adduced for the proposal of phosphoproteins as specific gene regulators, on critical examination it stands up rather badly (Trewavas 1976a). Furthermore since the development of diffuse chromatin results in increased accessibility to enzymes this is very likely to include protein kinases which would then phosphorylate available proteins. The function of non-histone protein phosphorylation at the present time remains unknown although two of the phosphorylated proteins have been identified as enzymes (Trewavas, 1976a).

More enlightening information has come from work on the action of steroid hormones. The outline of their mode of action in controlling transcription is now clear. In the target tissue the steroid combines with a specific receptor protein in the cytoplasm. After a conformational change this complex moves to the nucleus. Here the steroid receptor binds *not* to DNA but to *other* acidic proteins in the nucleus (Spelsberg *et al.,* 1975). This nuclear receptor site is initially in part of the chromatin which is inaccessible to nucleases as is the ovalbumin gene itself (Buller *et al.,* 1975; Weintraub and Groudine, 1976). After binding of the steroid/receptor protein complex both the nuclear binding site and the ovalbumin gene become accessible to nuclease digestion (Garel *et al.,* 1977; Senior and Frankel, 1978) indicating a conformational change in the chromatin. As a consequence the ovalbumin gene becomes available to RNA polymerase and the number of chromatin binding sites for this enzyme are greatly increased. Transcription of the ovalbumin gene then commences (O'Malley and Means, 1976).

There are approximately 5000 binding sites for the steroid/protein complex in the receptive nucleus (O'Malley and Means, 1976). The simple bacterial model of one gene/one regulatory protein clearly does not fit this situation since there are only one to two ovalbumin genes/chick genome (O'Malley and Means, 1976). This figure of 5000 has important implications particularly if it is considered, as discussed earlier, that the change from inactive to active chromatin involves a change from a 250-Å fibril to extended nucleosomes. Since the 5000 nuclear receptor sites are composed of non-histone proteins these must already be attached to the 250-Å fibril possibly at specific but repetitive DNA regions. Upon interaction with the steroid/receptor complex the interactions holding the 250-Å fibril together (possibly histone Hl-histone Hl interactions) (Thoma and Koller, 1977) are broken and the structure unwinds permitting transcription. Perhaps phosphorylation of some of these non-histone proteins may then occur in part fixing the chromatin in its open transcribable structure.

It has been shown by restriction enzyme analysis that repetitive DNA regions in DNA can occur of the order of one per nucleosome (Musich *et al.,* 1977). This suggests that perhaps the 5000 repetitive nuclear binding sites for the steroid/receptor complexes could occur with the same frequency perhaps adjacent to each of 5000 histone H1 molecules. If this is so it would occupy a considerable portion of DNA; approximately 10^6 base pairs. Since the chick genome size is of the order of 2×10^9 base pairs the binding sites in total would occupy a region about 1/2000th of the genome. The ovalbumin gene itself is unlikely to be much bigger than 3000–6000 base pairs (Mandel *et al.,* 1978). How can this discrepancy be reconciled?

A possible answer can be deduced from studies on Dipteran polytene chromosomes. These are banded structures in which the bands can be seen under the microscope (Fig. 15) (D'Amato, 1977). There is very good evidence to associate each band with only one gene and significantly there are approximately 2000 bands. When a band becomes active in transcription it becomes diffuse (forms puffs) or may even loop out (Fig. 15). The important point is that every band contains very much more DNA than could possibly be accounted for by a single gene and yet the whole band must be activated and unwound during transcription.

This suggests then that the nuclear binding sites for the steroid receptor complex may represent not only the ovalbumin gene but very considerable portions of the DNA surrounding the ovalbumin gene. Furthermore this DNA should have some form of repeat sequence to which acidic proteins can bind to form the nuclear binding site. These repetitious sequences may be adjacent to or on the linker region where histone H1 binds. If this repeat sequence occurs once every nucleosome then the total length of DNA involved in binding of the steroid receptor complex is about 1/2000th of the genome which could credibly correspond with one whole chromosome band.

IV. DYNAMIC ASPECTS OF CHROMOSOMES DURING DIVISION AND MEIOSIS

A. Mitotic Division

The majority of nuclei in mature plant somatic tissues are found in highly differentiated cells. In these cells the function of the nucleus is primarily to direct metabolism; they are the end products of development and no longer divide. Other populations of nuclei, which in mature plants are found in a minority of cells, are responsible for cell proliferation. These cells include the root and shoot apical meristems and the cambial tissues. Here nuclear division occurs by mitosis.

The long fibers associated with interphase chromatin are clearly incompatible with exact division of DNA into identical daughter nuclei because entanglement during separation would be inevitable. Preparation for separation therefore involves shortening by coiling and super-coiling of the DNA of the chromosomes. Under the light microscope these highly coiled structures are individually visible during prophase, metaphase, and anaphase but then begin to disappear and are diffuse at interphase (Fig. 16). The chromosomes of dividing cells go through alternate periods of DNA synthesis (S) and division (M) separated by "gaps," G1 and G2. A diploid nucleus therefore has a 2C DNA value at the onset of synthesis and a 4C value on completion. At anaphase the nucleus gives rise to two identical nuclei each with a 2C DNA value. The C-value is therefore the amount of DNA in an unreplicated haploid nucleus.

The nuclear membrane is broken down during prophase and the chromosomes, which consist of two chromatids, are attached to the spindle by a constriction called the centromere. They collect in a plane at the center of the cell at metaphase. The centromeres are attracted to opposite poles of the cell at anaphase. Much work on chromosome complements utilizes the drug colchicine which destroys the spindle and causes metaphase chromosomes to spread throughout the cell (Fig. 16a,c).

Eukaryote species have characteristic and remarkably constant chromosome complements, with related species often showing similarities in number and morphology. This general rule is broken by a large number of species of plants which may contain a variable number of additional chromsomes (Jones, 1971). These chromosomes are often smaller (Fig. 16a), replicate their DNA late in the S phase, and may sometimes remain as condensed "chromocentres" during interphase (Fig. 16b). Because of their distinctive appearance and behavior these are called B-chromosomes to distinguish them from the normal A-chromosome complement.

Although the B-chromosomes may account for a very large proportion of the genome (up to 80% in *Zea mays* is reported), they are dispensable to the

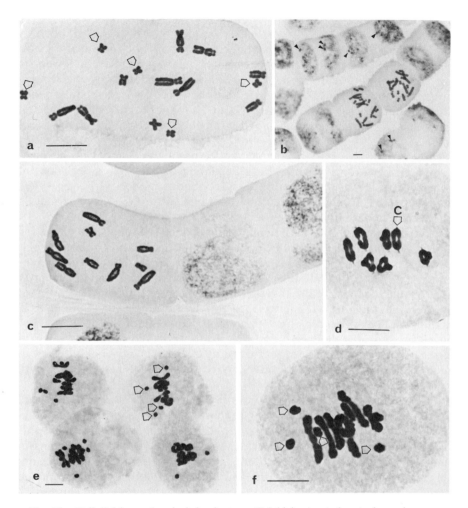

Fig. 16. Cell division and meiosis in plants. a, Colchicine treated metaphase chromosomes from a root tip cell of *Puschkinia libanotica* (2*n* = 10) with five additional B chromosomes. b, Low power photomicrograph of *P. libanotica* cells with six B chromosomes which are visible in interphase nuclei as chromocenters. c, Colchicine treated metaphase chromosomes of *P. libanotica* with no B chromosomes and consequently no chromocenters at interphase. d, Meiosis in *Secale ceriale* (2*n* = 14) showing the paired homologous chromosomes as they appear at first metaphase. In this cell there are seven ring bivalents some with two and some with three chiasmata holding the homologues together. The centromeres (c) are clearly visible. e,f, First meiotic metaphase from *S. cereale* with four B chromosomes which may pair with each other, but not with any of the A chromosome complement. In these cells the B chromosomes appear mainly as univalents. The group of cells in (e) illustrates the synchrony with which meiosis occurs in a population of pollen mother cells within an anther. B chromosomes are indicated by open arrows (△), B-chromosome associated chromocenters by closed arrows (▲) (J. N. Timmis, unpublished). The bars represent 10 μm.

plant life cycle and their phenotypic effects are not obvious. No major genes have been located on B-chromosomes but increasing numbers may reduce vigor and fertility. Yet B-chromosomes are widespread in many species and there are strong grounds for believing that they are of adaptive significance under certain stress conditions (Rees and Hutchinson, 1973). The genetic consequences of B-chromosomes will be further discussed in the next section.

The biochemical events accompanying some of the changes in the plant nucleus during the mitotic cycle are beginning to emerge. A fairly recent summary of some of them can be found in Yeoman and Aitchison (1975), Peaud-Lenoel (1977), and the book by Rost and Gifford (1977). This section includes additional information.

The contraction of chromatin which accompanies chromosome formation in mitosis seems to be initiated by a massive phosphorylation of histone H1. This was convincingly demonstrated in *Physarum* by Bradbury and co-workers (1973, 1974) and some confirmation has been obtained in partly synchronous artichoke cells (Trewavas and Stratton, 1976). The level of phosphorylated histone H1 increases 3- to 6-fold during middle and late G2 and then declines dramatically at M and the beginning of the next G1. Histone H1 is phosphorylated by a specific histone H1 kinase which itself increases in activity some 20- to 40-fold during late G2. The time of mitosis can actually be manipulated in *Physarum* by treating plasmodia externally with purified histone H1 kinase.

As the chromatin is condensed in prophase, RNA synthesis ceases and cannot be detected again until early telophase. This has been shown cytologically in both *Vicia* and *Pisum* (Das *et al.*, 1965; Davidson, 1964; Van't Hof, 1963). The observations are good evidence that highly condensed chromatin cannot be transcribed. However, both the nuclear membrane and the nucleolus also break down at this time and certainly the disintegration of the latter would have a marked effect on ribosomal RNA synthesis. The breakdown of the nuclear envelope appears to be dependent upon continued transcription in early prophase (Gimenez-Martin *et al.*, 1977). If inhibitors such as ethidium bromide are added at early prophase the chromosomes continue their cycle up to an advanced stage of condensation, beyond normal metaphase, but inside an intact nuclear envelope.

Along with the doubling of DNA which characterises the S phase the nuclear content of histones also doubles. Woodward *et al.* (1961) have carefully characterized this in *Vicia*. Whether this is the only phase in the cell cycle in which histones are synthesized is very much an open question. Turnover of histones can be detected in both G1 and G2 in artichoke and onion (Trewavas and Stratton, 1976) and in other cell cycle phases of *Vicia* cells synchronized by amino uracil. Histone synthesis also continues when

DNA synthesis is inhibited by FUDR. Thus the coupling between these two processes is obviously weak. In Hela cells transcription of the histone genes only occurs during S phase and not in other phases of the cell cycle (Stein *et al.*, 1975).

The acquisition of new acidic proteins in the nucleus during the cell cycle has received extensive study in animal systems. Summaries of the information can be found in Kleinsmith (1975), Jeter and Cameron (1974), Platz *et al.* (1975), and Stein *et al.*, (1974). The nuclear acidic proteins vary qualitatively throughout the phases of the cell cycle as does their phosphorylation. Among these proteins are some which regulate the activity of specific genes such as the histone genes (Stein *et al.*, 1975) and enzymes catalyzing relevant stages of the cycle such as DNA polymerase. The synthesis of new acidic proteins during S is not strongly coupled to DNA synthesis as has been shown using inhibitors. The little work available in plants on this topic is in basic agreement with that outlined above. In barley embryo cells approaching mitosis, about six phosphorylated proteins show a transient appearance in the nucleus leaving after cell division (Trewavas, 1978). In dormant artichoke tuber cells induced to divide by auxin interesting results have been obtained by analyzing the nuclear proteins using two-dimensional gel electrophoresis (D. Melanson and A. J. Trewavas, unpublished). The changes in acidic proteins were found to be sequential and progressive in the Gl and S phases of the cycle. Four novel acidic proteins appear 3 h after auxin treatment in early Gl, this is increased to 8, and then 20 by late Gl (12–15 h after auxin treatment) until at S there are about 40 novel proteins. Phosphorylated nuclear proteins only showed qualitative differences at the commencement of DNA synthesis. Such results are strongly suggestive of a progressive change during Gl and S with the protein pattern in one stage modifying subsequent stages.

B. Meiotic Division

The chromosomes of diploid organisms are present in two similar, homologous sets. One homologous set originates in the male parent and the other in the female. At meiosis, the cell division which gives rise to gametes or to an alternate generation, a single round of DNA replication, and chromosome duplication, is followed by two nuclear divisions. Very soon after DNA replication comes the onset of meiotic prophase, the gradual process of chromosome shortening and thickening in preparation for separation. Various stages of prophase may be recognized as the morphology and behavior of the chromosomes change. Although DNA has been replicated the first visible structures which appear at *leptotene* appear as single threads under the light microscope. At *zygotene* the homologous chromosomes begin to pair and this pairing is completed by *pachytene*. During pachytene the chromosome strands still appear single and it is at this stage that chiasmata are formed. Formation of a chiasma involves the breakage and cross-wise rejoining of homologous chromatids and represents the visible site of genetic recombination. The coincidence of chiasmata and crossing over has been established either by using a morphologically aberrant chromosome as one homologue or by autoradiographic techniques (Jones, 1971). At *diplotene* the paired chromosomes, or bivalents, are held together by chiasmata and each individual chromosome is clearly seen to be composed of two chromatids. The nuclear membrane and the nucleoli disappear, the spindle is formed, and the much coiled and contracted bivalents are orientated at the equator of the cell with the centromeres attracted to opposite poles (Fig. 16). At first anaphase the homologous chromosomes are separated to opposite poles. At the second anaphase the two chromatids of each chromosome are separated, giving rise to four haploid nuclei. Specifically, each daughter receives a haploid set of chromosomes, consisting of one member of each homologous pair. In heterozygous individuals these four products all differ from each other because of random orientation of the chromosomes during separation.

The genetic variation between the four haploid nuclei is increased by structural exchange of segments of homologous chromatids at chiasmata, occurring before the chromosomes separate. Fusion of the haploid male and female gametes restores the diploid complement in the zygote. The segregation and recombination of genes which result from the meiotic process and fertilization are a major source of genetic diversity in diploid organisms.

The meiotic chromosome is organized to allow recognition and pairing of homologues, the formation of a synaptonemal complex, and the exchange of genetic material at chiasmata. The light microscope shows that homologous chromosomes are unpaired at leptotene, begin to pair at zygotene, and are tightly paired at pachytene. A tripartite structure, the synaptonemal complex, is seen in electron micrographs between the paired pachytene chromosomes (La Cour and Wells, 1977; Moses, 1956). It is made up of two electron-dense lateral elements associated with the axes of the paired chromosomes and separated by a transparent region containing a central element of medium electron-density. The presence of the synaptonemal complex is universal in eukaryotes and shows considerable uniformity of structure (Westergaard and von Wettstein, 1972). The synaptonemal complex is complete during zygotene and pachytene (Fig. 17) and disintegrates thereafter. It is apparently more stable at the chiasmata suggesting that the synaptonemal complex is directly involved in chiasma formation and in the mechanism of chromosome recognition and pairing. Enzyme studies (Comings and Okada, 1970a,b; Westergaard and von Wettstein, 1972) show that the complex is unaffected by DNAse but is digested by RNAase and trypsin, indicat-

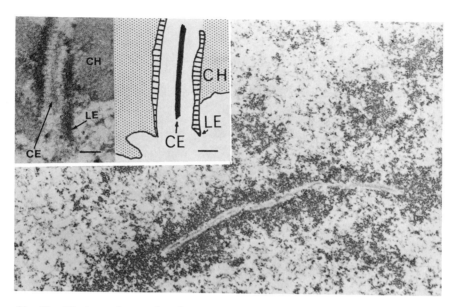

Fig. 17. Electron micrographs of synaptonemal complexes. Synaptonemal complex from *Tradescantia virginia*. The central element and chromosome fibers are clearly visible. × 33,000 (G. Venot and E. G. Jordan, unpublished). Inset: Electron micrograph of a small section of the synaptonemal complex at pachytene in a pollen mother cell of *Fritillaria lanceolata*. Section stained by the preferential RNA-staining method of Bernhard, showing high contrast in the lateral elements (LE). The fine filament in the central space is the central element (CE) and the stippled area represents the chromosome fibers (CH). The bar represents 100 nm. Photograph from L. La Cour. From La Cour and Wells (1977).

ing that it is composed of ribonucleo-proteins. The presence of a synaptonemal complex is obligatory for chiasma formation but it may also exist between nonhomologous chromosomes in haploid plants where no chromosome exchange occurs.

The B-chromosomes mentioned earlier have profound effects upon meiosis (Fig. 16) and perhaps here lies their significance. They do not pair with themselves, but affect the frequency and distribution of chiasmata in the A-chromosomes (Ayonadu and Rees, 1968; Cameron and Rees, 1967; Vosa and Barlow, 1972), and they restrict pairing in diploid and tetraploid hybrids to true homologous chromosomes. The latter phenomenon has been demonstrated in crosses between *Lolium perenne* and *Lolium temulentum* (Evans and Macefield, 1972) and may be considered a far-reaching property of B-chromosomes in view of the high incidence of polyploidy in wild and agricultural plants.

C. Replication of DNA

The variation in nuclear DNA content noted in Section III,A,1 is correlated with the duration of the mitotic cell cycle (Van't Hof and Sparrow, 1963; Bennett, 1972). The greater the DNA content the longer the cycle. There are equally well-defined correlations between the nuclear DNA content and the duration of meiotic cell cycle (Bennett, 1977) and again a major reason for these correlations appears to be the length of time spent synthesizing DNA. One difference between meiosis and mitosis is that premeiotic S phase is substantially longer than S for mitosis. In wheat, meiotic S phase is 8 h compared with the normal mitotic S phase of 3.5 h (Bennett *et al.*, 1971) and in *Lilium* the periods are 192 and 24 h, respectively (Bennett, 1971).

A specific ordered pattern of DNA replication for individual chromosomes and for different chromosome segments throughout the S period in mitosis has been repeatedly shown in plants (Taylor, 1958; Wimber, 1961; Evans, 1964; Evans and Rees, 1966; Barlow and Vosa, 1969); however, the pattern is species-specific. In *Crepis* (Taylor, 1958) replication begins at the chromosome ends and proceeds to centromeres. In *Tradescantia* (Wimber, 1961) this situation is reversed. A general conclusion that can be drawn, however, is that the DNA in heterochromatin replicates later than that in euchromatin irrespective of the position of the heterochromatin along the chromosome.

Replication of DNA takes place by a semi-conservative mechanism in which the double-stranded DNA of each chromosome directs formation of two new double-stranded DNA molecules which compose the daughter chromosomes. Replication does not take place continuously but is achieved by the simultaneous action of many replicating forks (Huberman and Riggs, 1968).

The units of replicating DNA in pea vary between 20 and 140 μm in length, averaging just over 40 μm (Van't Hof, 1975, 1976a,b). They consist of two replicating forks moving bidirectionally from the origin at the same time in opposite directions. Each fork, therefore, moves 10–70 μm from origin to end. Approximately 40,000 replicon units function during the S phase in this tissue. The rate of fiber growth varies throughout the S period being 5 μm/h/fork in early S and 20–30 μm/h/fork in late S. These values are considerably lower than those of prokaryotes which may be up to 900 μm/h.

Because the DNA polymerase only replicates in the 5′ to 3′ direction replication is discontinuous (Bryant, 1976) and intermediates of various sizes have been detected. In *Vicia* the earliest intermediates that can be detected are of the order of 10 S and are double stranded (Sakamaki *et al.*, 1976). Later intermediate sizes are 19–22 S and then 28 S and presumably these represent the joining together of the smaller pieces; 28 S represents a DNA length of about 6 μm. Considerable nicking must occur in the DNA of the replicon units to enable isolation of intermediates to be carried out. The joining together of fragments implies the presence of ligating enzymes.

Replication may well be initiated on the nuclear membrane. In *Antirrhinum* the rate of DNA synthesis is porportional to the surface area of the nucleus (Alfert and Das, 1969) and nuclear membrane preparations of cotton contain newly synthesized DNA (Clay *et al.*, 1976). Replication is carried out by a complex of enzymes of which the following have been detected in plant systems; an unwinding protein (Stern and Hotta, 1978); an endonuclease to nick the original strands (Stern and Hotta, 1978); DNA polymerases of which two sorts have been purified (McLennan and Keir, 1975; Tarrýo-Litwak, *et al.*, 1975; Gardener and Kado, 1976); a reannealing or ligating enzyme and a single-stranded DNA-dependant ATPase (Stern and Hotta, 1978) whose function is not clear at the present time but may be part of a ligase complex.

Initiation of S phase is almost certainly accomplished by proteins initially present in the cytoplasm. Jakob (1972) found that treatment of *Vicia* meristem cells in Gl with actinomycin D did not affect subsequent replication provided the drug was given within 3 h of S. A similar treatment 5 h before S markedly inhibited replication. Rather more direct evidence has been obtained in transplantation experiments in animals in which resting nuclei injected into the cytoplasm of a cell synthesizing DNA initiate DNA synthesis (Gurdon, 1974). The initiation of S phase in artichoke cells is accompanied by the acquisition of four new phosphorylated proteins in the nucleus (D. Melanson and A. J. Trewavas, unpublished). One of these becomes the most abundant protein in the nucleus at that time. Phosphorylated proteins regulating DNA replication in animals including the very abundant unwinding protein have also been reported (Sinha and Snustad, 1971; Tegtmeyer *et al.*, 1977).

Certain events are specific to the replication of DNA in meiosis. In *Lilium* microsporocytes (Stern and Hotta, 1978) about 0.3–0.4% of the chromosomal DNA remains unreplicated at the end of premeiotic S. This DNA is called Z-DNA because it is replicated at zygotene, when chromosome pairing begins. If the cells are induced to fully replicate their DNA during premeiotic S, mitosis rather than meiosis ensues. At zygotene, chromosome pairing is accompanied by three virtually meiosis-specific events (Stern and Hotta, 1978): (a) the appearance of a lipoprotein complex; (b) the appearance of a DNA-binding protein; and (c) the replication of the remaining 0.3–0.4% Z-DNA. The DNA-binding protein catalyzes single-stranded DNA renaturation and may therefore constitute the required agent for matching and alignment of homologous chromosomes. The significance of the Z-DNA is uncertain; it consists almost entirely of unique sequences of about 10^4 base pairs and is probably not localized but scattered along the chromosomes. The Z-DNA segments have different susceptibilities to nuclease, are relatively high in GC content and, after replication, remain unligated to the body of the chromosomal DNA until chromosome disjunction.

At pachytene, specific nucleases produce nicks in the DNA at loci which contain intermediately repeated sequences. These nicks are subsequently repaired. The DNAs synthesized during zygotene and pachytene are therefore distinct—unique sequences in the former and repetitive in the latter.

V. DYNAMIC ASPECTS OF CHROMOSOMES IN DIFFERENTIATION

A. Endopolyploidy and the Polytene Chromosome

The cells of the meristematic tissue pass through a number of mitotic cycles. When a cell starts to differentiate it usually leaves the cycle and it may become arrested in G1 or G2 (Evans and Van't Hof, 1974). In most angiosperms, however, differentiating cells that have left the mitotic cycle continue to synthesize DNA and to double their chromosome number in endomitotic and endoreduplication cycles passing from G1 to S and back again to G1. These cycles result in what are termed endopolyploid nuclei. Endopolyploidy is the rule rather than the exception for plant cells. Butterfass (1966) has estimated that up to 80% of the somatic cells of a plant may be polyploid.

The degree of endopolyploidy can be extensive. The highest levels have been reported in the ovular tissue of *Arum maculatum* [24,576 C, Erbrich, (1965)] the suspensor of *Phaseolus* [8192 C, Brady (1973)], and in gall tissue induced in *Poa* [4096 C, Hesse (1969)]. Most polyploid somatic plant cells, e.g., in root, will range from 4 to 64 C.

Because of the prevalence of endopolyploidy it has been presumed to play an important function in the life of the plant. Since increases in nuclear volume generally result in larger cells, endopolyploidy is a way of generating increased cell (and organ) volume without having to go through the metabolically wasteful events in mitosis. Since growth in plants is generally by cell expansion, endopolyploidy becomes an important feature of this process (Capesius and Stohr, 1974). Alternative functions for endopolyploidy have been suggested and these may be found in Nagl (1976a,b) and D'Amato (1977). The significance of endopolyploidy has recently been questioned by Evans and Van't Hof (1975). They were able to demonstrate that polyploidy was nonexistent in the petals or leaves of pea and the roots of Triticum but present in other tissues of these plants. Polyploidy could not be detected in any tissue of *Helianthus*. Certainly in the latter case endopolyploidy is not essential for the life cycle of the plant and its role in development then can only be a matter for speculation.

Often the chromatin of endopolyploid nuclei appears to be in the interphase condition but in certain cells the chromosomes are organized as polytene chromosomes which may have undergone up to 11 rounds of DNA replication without separation of the centromeres. Polytene chromosomes have been observed in a variety of cells in several plant species, but most notably in the suspensor cells of *Phaseolus* (Nagl, 1976b). Figure 15 (see p. 523) contains photographs of polytene chromosomes from *Phaseolus* in various states of activity. In the inactive state (Fig. 15a and b) the chromosome is highly condensed and banded. In a partly active state (Fig. 15c) some of the bands disappear and are said to "puff" by analogy with a similar phenomenon in salivary gland chromosomes in insects. Puffing in salivary gland chromosomes is a reflection of the unraveling of the chromatin fibers which makes them available for transcription. The puffs are active in RNA synthesis and specific chromosome regions are associated with specific developmental periods. They are the visible sites of transcription at specific loci (Ashburner, 1969).

A fully active polytene chromosome has a highly granulated appearance (Fig. 15d). The granules may represent ribonucleoprotein particles. Figure 15e and f shows hyperactive or lampbrush type polytene chromosomes from *Phaseolus*. In these, the whole chromosome structure is diffuse and expanded and portions are looped out (Fig. 15g). Lampbrush chromosomes were originally described in amphibian meiotic cells. These chromosomes though not endopolyploid, are of a far greater length than usual for the organism, and show tightly coiled DNA in chromomeres connected by very thin threads rather like beads on a string. Some of the chromosomes extrude characteristic symmetrical loops of DNA which are transcriptionally active and surrounded by a matrix of ribonucleoprotein. The loops are far too long to encode the information for single polypeptides and it was from observations of lampbrush chromosomes that the first ideas about gene multiplicity were formulated (Callon and Lloyd, 1960).

The study of polytene and lampbrush chromosomes indicates that chromatin must be diffuse rather than condensed for transcription to occur. Other terminology has used euchromatin for diffuse and heterochromatin for dense and this has led to confusion as pointed out by Nagl (1977). Euchromatin is the form of chromatin which undergoes the normal cycle of condensation–decondensation during the mitotic cycle. Euchromatin is normally active in RNA synthesis and contains both single genes and unique as well as intermediate repetitive DNA sequences. Heterochromatin, on the other hand, remains condensed throughout interphase except for a short period in prophase (so-called dispersion or Z phase). Normally this heterochromatin, which may be better termed constitutive heterochromatin, contains highly repetitive DNA sequences (Jones, 1971; Timmis *et al.*, 1975) and is often associated with centromeric and telomeric regions of chromosomes (see Section III,A,1). It is inactive in RNA synthesis and replicates later than euchromatin. In a number of monocotyledonous species, constitutive heterochromatin can be detected by growing the plants at low temperature producing the so-called cold-sensitive heterochromatin (Vosa, 1973). Otherwise heterochromatin may be detected by Giemsa or fluorescence banding techniques. The exception to the general rule is the nucleolus-associated heterochromatin. This is not genetically inert but contains the ribosomal genes and it can form a pufflike structure, the intranucleolar chromatin, with high rates of RNA synthesis.

B. Variation in Specific Portions of the Genome

For the maintenance of genetic information and its transfer from one generation to the next, it is essential that all the DNA is perfectly replicated and conserved in at least the germ line cells. In other somatic cells this strict conservation may not be necessary and there are a number of examples known of differential replication of DNA. Synthesis of DNA usually takes several hours, and chromosome segments may replicate at different times during this period. Heterochromatin characteristically replicates late in synthesis compared with euchromatin and during endopolyploid DNA replication this difference may lead to over or underreplication of the heterochromatin. The euchromatin may have undergone a greater or lesser number of rounds of replication than the heterochromatin. This phenomenon appears to be fairly widespread in animals and has been observed cytologically (Fox, 1971) and biochemically by analysis of satellite DNA content (Endow and Gall, 1975). Fruit tissues of melon and cucumber (Pearson *et al.*, 1974) contain 15% satellite DNA, only about half as much as meristematic tissues such as root tips. The main band and satellite DNAs in these different tissues also show differences in buoyant density perhaps indicating more widespread changes in composition of many genome fractions. The amount of heterochromatin found in the majority of fruit nuclei was also reduced compared with meristematic cells, but in a minority (5%) the proportion of heterochromatin was increased. In melon and cucumber it therefore appears that euchromatin and heterochromatin DNA replication are independent with an overall reduction in heterochromatin in mature fruit nuclei. Strikingly similar results have been obtained in cultured *Cymbidium* protocorms (Nagl, 1972; Schweizer and Nagl, 1976). Other numerous examples of specific variation in DNA or genome regions can be found in Nagl (1977) and D'Amato (1977). A discussion of ribosomal DNA variation can be found in Section VI,D.

Chromosomes themselves can also show considerable variation. In a hybrid of two *Nicotiana* species (Gerstel and Burns, 1966), giant chromosomes up to 30 times the normal size may be produced during a few somatic divisions. The B chromosomes of certain species may be entirely lost from dividing cells. Those of *Crepis capillaris* regularly undergo nondisjunction in shoot meristems so that their frequency is increased in the germ line, and in *Aegilops speltoides* and *Haplopappus gracilis,* B chromosomes are found only in the shoot system.

The DNA of the nucleus, particularly in plants, does not comprise all the cellular DNA as plastids and mitochondria also possess functional genomes. Estimates of the contribution of chloroplast DNA to total DNA are equivocal and widely variable. Siegel (1974) suggests that leaf tissue contains about five-fold more copies of chloroplast DNA than root tissue and this in itself represents an example of differential replication of a specific sequence of cellular DNA. Evidence is also present (Siegel, 1974) which surprisingly indicates that the majority of the chloroplast DNA of green tissues is located in the nucleus, not in chloroplasts. The contamination of nuclear with organelle DNA is a major problem in deciding whether chloroplast DNA sequences do in fact reside in the nucleus. Chloroplast DNA itself, on the other hand, may be prepared with a high degree of purity, and in Fig. 18 are autoradiographs of isolated nuclei and chloroplasts after *in situ* hybridization with complementary RNA to pure chloroplast DNA. The preparations show hybridization to chloroplasts, nuclei, and chromosomes and support the argument that chloroplast DNA is indeed present in the nucleus.

It is clear from the above examples that genetic constancy, while a characteristic of the replication and division mechanisms of the eukaryotic cell, is by no means their invariable consequence.

C. Transcription of Chromomeres: The Synthesis and Processing of Messenger RNA

Nearly all the chromosomal DNA is located in beadlike structures called chromomeres. These are most clearly seen in amphibian lampbrush chromosomes but it is assumed that this structure is representative of chromosome structure in general. Fusion of chromomeres in adjacent daughter chromosomes of a polytene chromosome gives rise to the typical band structure seen in Fig. 15.

It is thought that the chromomere may be produced by extensive folding and clustering of the 250-Å chromatin fibril. One possible interpretation of this structure is in Fig. 19. The initial stage is a clustering of the fibril into a set of loops held together by a fragment of membrane or matrix. This forms a rosette type of structure clearly seen in photographs of membrane-bound and matrix-bound DNA by Comings and Okada (1976) and Franke and Scheer (1974a). Clustering of the rosettes is then assumed to form the beadlike chromomere and fusion of adjacent chromomeres produces the polytene band.

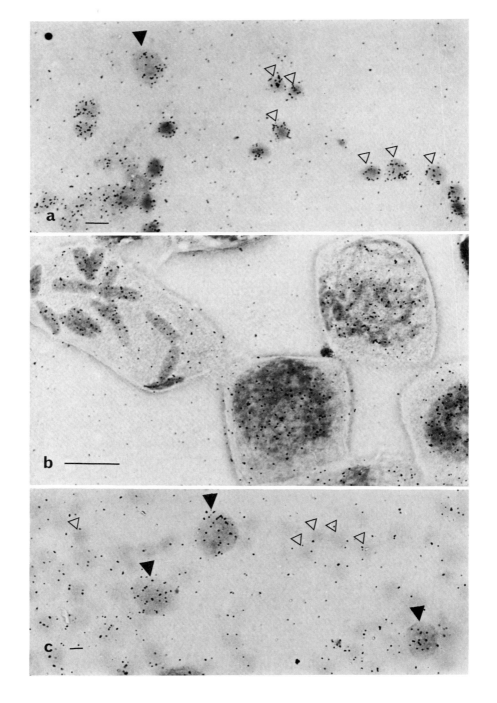

The chromomere is assumed to be the functional subunit of the chromosome for transcription and possibly DNA replication. Thus each replicon could be represented by a loop of the rosette structure and the replication fork would move in both directions from the membrane or matrix attachment of the various loops of the rosette simultaneously.

Chromomeres functionally represent single genes in certain organisms. This conclusion follows from extensive studies on polytene chromosomes. These have shown that deletion of a single band is phenotypically equivalent to a single point mutation (Beerman, 1972); that the numbers of complementation groups on polytene chromosomes are equivalent to the number of bands (Judd and Young, 1973); that puffing (i.e., transcription) of a single band results in the appearance of a single discrete gene product (Grossbach, 1973); and that rearrangement breakpoint studies show no evidence of more than one structural gene being present in one band (Lefevre, 1973). The amount of DNA in individual chromomeres varies from 7000 to 150,000 base pairs, averaging 25,000. An average structural gene could be expected to be represented by about 2000 base pairs. Clearly there is an excess of DNA in each chromomere and this seems likely to be functional in two different categories.

Recent observations (e.g. Mandel *et al.*, 1978), and other references herein) have shown that single genes are not represented by a contiguous nucleotide sequence on the DNA but by a set of fragments. The ovalbumin gene, for example, has seven coding fragments that are spaced along a piece of DNA 6000 base pairs in length. This is three times bigger than ovalbumin mRNA. Whether the interspersed nonstructural fragments perform a function is unknown but it seems likely that they are transcribed.

A second function for the remaining extra DNA seems likely to be transcriptional regulation (Georgiev, 1969). DNA hybridization studies have shown that unique sequences (structural genes) may be spaced by short repetitive nucleotide sequences. Bonner and Wu (1973) have pointed out that if all the copies of a single repeated sequence are physically linked together

---

Fig. 18. *In situ* hybridization of *Beta vulgaris* (swiss chard) chloroplast DNA. RNA complementary to DNA from pure chloroplasts was transcribed *in vitro* and hybridized to colchicine treated root tip squashes, or to isolated nuclei and chloroplasts. The slides were coated with Ilford K2 emulsion and exposed at 4°C for 4–8 weeks. After developing, the material was stained in Geimsa. Inset: a, hybridization to nuclei (▲) and chloroplasts (△) prepared from whole green leaf tissue of *B. vulgaris;* b, hybridization to root tip cells of *Vicia faba,* indicating dispersed chloroplast sequences in interphase nuclei and metaphase chromosomes; c, hybridization to nuclei (▲) and chloroplasts (△) isolated from whole green leaf tissue of *V. faba.* In all cases hybridization was carried out in the presence of a large excess of nonradioactive cytoplasmic rRNA to prevent any cross-hybridization of chloroplast rRNA sequences to the nuclear rRNA genes. The nuclei are distinguished from the chloroplasts by being larger, containing visible nucleoli, and by being slightly more densely stained in the original preparation. (J. N. Timmis, M. Pascoe, and J. Ingle, unpublished.) The bars represent 10 μm.

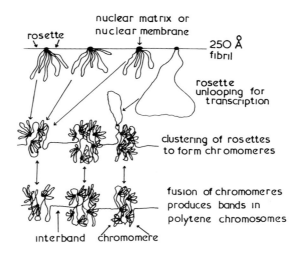

Fig. 19. Diagrammatic illustration of possible structure of chromomeres and polytene chromosome bands. The 250-Å chromatin fibril is clustered into rosettes by attachment to the nuclear matrix or nuclear membrane. This rosette is unlooped when transcription commences. Rosettes are then further clustered to give the chromomere. Alignment and fusion of chromomeres in polytene chromosomes produces the chromosome bands.

with an interspersed unique sequence, the DNA length generated is in close agreement with the average size of the chromomere. Repeated base sequences surrounding the structural gene were proposed in Section III,C and would act as a repetitive binding site for nuclear proteins forming the nuclear binding site for regulatory proteins.

Chromomeres are probably transcribed in total (Judd and Young, 1973). The initial RNA product, heterogenous nuclear RNA (hnRNA), has been shown to contain mRNA by *in vitro* translation and kinetic and hybridization studies (Perry *et al.,* 1976). The hnRNA is extremely variable in size and can contain repetitive sequences as well as unique ones (Holmes and Bonner, 1974). It seems likely that each hnRNA molecule contains only one or a few mRNA sequences. In some cases the precursors may be considerably larger (10- to 20-fold) than mRNA, in others the pre-mRNA is similar in size to mature mRNA. In the former case a stage of processing involving both ligases and nucleases is undergone to synthesize the mRNA. In the latter case the transcriptional units may be a similar size to the final mRNA or, alternatively, processing might occur so quickly that it cannot be detected. The surplus RNA fragments are further degraded in the nucleus. Probably the best model for the processing stage is that described for ribosomal RNA (Section VI,D).

Nuclear processing of mRNA precursors is completed by addition to the 5′ terminus of the cap sequence m⁷Gppp (Busch, 1976) which seems likely to be essential for mRNA translation. For most mRNAs, methylation of internal adenylate residues also occurs (S-adenosyl methionine is the donor and the reaction is catalyzed by a protein methylase) followed by addition of a polyadenylate sequence 100–300 bases long on the 3'OH terminus (Harris and Dure, 1976).

Transcription of the chromomere is catalyzed by RNA polymerase II. This molecule is large, reaching molecular weights of 5 to 6 $\times$ 10^5 in a number of plants (Duda, 1976). Some plants appear to have more than one polymerase in this category differing in template preferences and sensitivity to inhibition.

Although majority opinion favors the location of the structural gene as being in the polytene band, Crick (1971) suggested instead that the structural gene may be located in the interband region while the chromomere itself contained sequences concerned with the regulation of transcription. Recent studies by Jamrich et al. (1977) have shown that RNA polymerase is mainly located in the interband regions and puffs. Autoradiographic analysis has shown that low rates of RNA synthesis also occur in these regions (Zhimulev and Belyaeva, 1975). Interband DNA may also be euchromatic and early replicating (Comings, 1974). Transcription of the interband DNA obviously does then occur. However, there is insufficient DNA in this region (2–5% of the total) to account for the 20–25% which is known to be actually transcribed (Nagl, 1976b). As an alternative Paul (1972) has suggested that the interband region may function as a site of initiation for the unraveling of the adjacent chromomere which occurs when transcription commences.

VI. THE NUCLEOLUS

A. Role of the Nucleolus in the Life of the Cell

The nucleolus, discovered by light microscopists in the 18th century, was recognized as having a special relation to a chromosomal region in 1931 and shown to be involved in RNA metabolism by 1940. The related chromosomal region was found to contain many genes for ribosomal RNA with the use of nucleic acid hybridization techniques in Drosophila by Ritossa and Speigelman in 1965. This observation was confirmed by Wallace and Birnstiel in 1966 using the "Oxford mutant" of Xenopus laevis, discovered by Elsdale et al. in 1958. The spreading technique of Miller and Beatty in 1969, permitted the visualization of the transcription of these genes and their arrangement. Biochemical studies by Perry (1966) elucidated the essentials of the processing steps while the application of the methods of conventional electron microscopy (Bernhard, 1968) and EM autoradiography has provided a detailed structural model of nucleolar function.

Many lines of research which have been extensively reviewed (Vincent and Miller, 1966; Birnstiel, 1967; Lafontaine, 1965; Perry, 1969; Miller, 1966; Busch and Smetana, 1970; Smetana and Busch, 1974) indicate that the nucleolus is the cellular site of ribosome manufacture. This is an over-simplification because ribosomes are made in the cytoplasm from two parts, which are preformed in the nucleus at the nucleolus. The assembly of ribosomal subunits in the nucleolus involves the addition of proteins to the ribosomal RNA molecules during the course of their synthesis and later modification. The primary function of the nucleolus may then be said to be the transcription of ribosomal RNA genes. The transcripts however, undergo a complex series of modifications which may be termed "processing." At the same time they become complexed with many protein molecules. These proteins being the result of cytoplasmic synthesis and subsequent transloca-tion to the nucleus.

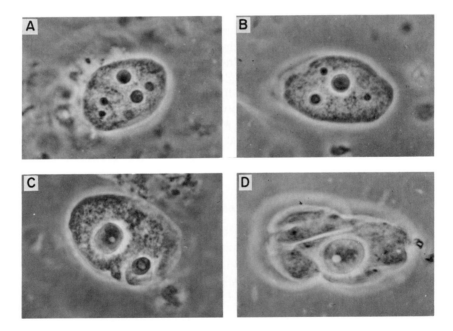

Fig. 20.   Isolated nuclei from *Helianthus tuberosus* tuber tissue viewed by phase contrast microscopy. Explants of dormant tissue were shaken in distilled water for 0 h, A; 3 h, B; 10 h, C; and 24 h, D. The stored cells at 0°C have up to six typically inactive nucleoli, A. With activation caused by removal of explant and elevation of temperature to 25°C the nuclei and nucleoli enlarge, B–D. With increasing activity the nucleoli fuse together and the nuclei become lobed and irregular, D (A and D, Severs and Jordan, 1975b; B and C courtesy of N. J. Severs). (× 1,800).

The nucleolus, therefore, although occupying a central position in the scheme for ribosome manufacture, represents only a part of the process. The proteins for the ribosomal subunits are synthesized in the cytoplasm, the RNA molecules are synthesized and modified in the nucleolus, while the ribosomes themselves are not complete until they are assembled in the cytoplasm. Cytoplasmic ribosome assembly results in polysomes through the addition of small and then large subunits to messenger RNA.

The nucleolus is a feature of eukaryotic cells and it may be asked how much the nuclear envelope contributes, by the separation of transcription and translation into separate compartments, to the morphological expression of a nucleolus. Prokaryotes may be said to have a "cryptic nucleolus" in the sense that they can be shown to synthesize ribosomes and show similar transcriptional features to eukaryotes in spread preparations. The presence of the nucleolus is due to the accumulation of the molecules, which are being fashioned into ribosomes, into a large conglomerate mass which can be visualized. It is quite possible to conceive of nucleolar functions being performed in a dispersed fashion, as in the case of prokaryotes. The question remains whether the noncryptic or conglomerate nature of the nucleolus of eukaryotes confers some special functional capacity on the overall process or whether it is only a consequence of the increased production characteristic of the eukaryotic nucleus. Without knowing the explanation, we may say that the nucleolus essentially results from the dense packing of the various molecules involved in the assembly of ribosomal subunits in the eukaryotic nucleus, into a clearly definable structure—the nucleolus.

B. Gross Structure of the Nucleolus

1. Size and Number of Nucleoli per Nucleus

The nucleolus is easily demonstrated by light microscopy using conventional stains for protein or RNA and is readily visualized by phase contrast microscopy as a dark refractile body especially in isolated nuclei (Fig. 20).

It is common for diploid organisms to have either one large or two smaller nucleoli. Nucleoli frequently fuse but the maximum number of nucleoli encountered per nucleus is an indication of the number of organizers present. There seems to be a regulation of nucleolar size which is reflected in the total surface area rather than volume, so that the surface area of a single nucleolus is the same as that of a pair (Barr and Hildegard, 1965; Gimenez-Martin *et al.*, 1977). Nucleoli tend to fuse with increasing time in the cell cycle or with increased activity of the cell (Fig. 20) (Jordan and Chapman, 1971; Anastassova-Kristeva and Nocoloff, 1978). The total volume of nucleolar material is higher in cells which are at the early stages of differentiation like young root hair cells (Rothwell, 1964; Wimalaratna, 1976) and drops to a minimum in cells which have finished their differentiation or are simply dormant or quiescent (Hyde, 1966).

2. Nucleolus Organizers

From a study of 33 species of *Vicia,* Heitz (1931) concluded that nucleoli arose at specific chromosomal sites, the secondary constrictions, also called the SAT regions of satellited chromosomes. The organizers are easily identified when they occur as secondary constrictions but sometimes part or all of the organizer may be as fully condensed as the rest of the chromosome. In *Phaseolus* the secondary constriction is flanked on both sides in this way by nonconstricted organizer as shown by *in situ* hybridization techniques (Avanzi *et al.,* 1972). In *Zea mays* it occurs on one side only of the constriction.

These chromosome regions are essential for the formation of a new nucleolus or the "organization" of a new nucleolus following its dispersal at nuclear division. For this reason they are called organizers, nucleolus organizers, or nucleolar-organizing regions. It does not seem to be necessary to discriminate between these different ways of referring to the same structure though attempts have been made to distinguish the external organizer or nucleolar organizer from the chromosomal site of organization, the nucleolar organizing region. We shall use any of these terms to mean no more than nucleolus organizer as defined by the committee on nucleolar nomenclature (Nucleolus Nomenclature Committee, 1966), the specific chromosomal site at which the nucleolus arises, but with the knowledge that the ribosomal RNA genes have now been located there.

The maximum number of nucleoli per nucleus found in any organism, sometimes called the primary number, is a reflection of the number of organizers (Fig. 20). This may be affected by aneuploidy or polyploidy (De Mol, 1927; Darvey and Driscoll, 1977). Some reports show that there is an increased total nucleolar volume (Phillips *et al.,* 1971) with consequent increase in cellular RNA and dry weight when extra organizers are introduced (Longwell and Svihla, 1960).

Organizers may vary in their capacity to form nucleoli, some being stronger. This gives rise to the concept of competition between organizers, especially noted in hybrids (Nawaschin, 1934). Latent or silent organizers may also occur, which in the absence of those normally present prove capable of nearly 100% compensation (Longwell and Svihla, 1960).

3. The Nucleolar Organizer Track

The arrangement of the organizer in the fully organized nucleolus of interphase is not discernable in all organisms. In certain algae, however, with large nucleoli, a well-defined structure is apparent (Van Wisselingh, 1898). Godward (1950) showed that this structure was formed from the extended nucleolar organizers which were covered in a layer of material clearly distinguishable from the rest of the nucleolus (Fig. 21). By following the early stages of mitosis she showed that the satellited chromosomes derived their organizer regions from the central filament within this material by a process of shortening and thickening (Fig. 22). She further showed that an organizer which was only 2 μm long at metaphase had condensed from one which was 25 μm long in the interphase nucleolus. The thick coating of material which surrounded the organizer has been called the organizer track. This structure has been shown clearly in higher plants by Lafontaine and Lord (1973), Lafontaine (1974a), and La Cour and Wells (1967), who call it a nucleolonema and nucleolar loop, respectively. Nucleolonema is used generally to describe any filamentous or reticular structure which can be distinguished from an amorphous background material in nucleoli, but which is not necessarily related to the organizer. The nucleolar material may display ordered structure in thick threads or sheets, even in the absence of the organizers

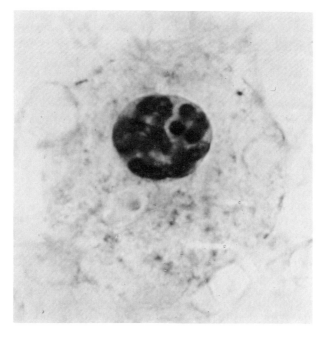

Fig. 21. Nucleolus showing the coiled and folded structure of the organizer track. *Spirogyra crassa*. From Godward (1950).

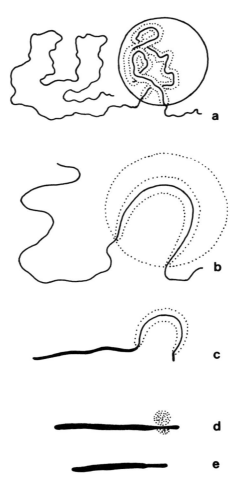

Fig. 22. Diagrammatic representation of the changes occurring in the nucleolus-organizing regions of nucleolar chromosome during the time of chromosome condensation in mitosis in *Spirogyra*. The nucleolus disperses in two stages, first the cortex, and second the material of the organizer track. The nucleolus organizer changes from its extenuated configuration in the interphase nucleolus to appear as a constricted region in the condensed chromosome. From Godward (1950).

(Jordan and Godward, 1969) and the term nucleolonema can be used in a noncommital way for these (Nucleolus Nomenclature Committee, 1966).

There is a greater clarity of nucleolar organizer tracks in plant nucleoli which is not thought to represent any fundamentally different organization, but only a difference related to the size and number of organizers. If a nucleolus results from the fusion of ten small nucleoli each having only a short organizer, as in man, then a long, sinuous, organizer track would not be expected in the interphase nucleolus.

C. Fine Structure of the Nucleolus

1. Fibrillar and Granular Components

The use of the electron microscope has revealed that the nucleolus is a mass of 150-Å granules which surround irregularly shaped regions lacking these granules. This nongranular part is composed of packed fibrils about 50 Å in diameter. These two parts are called the granular and fibrillar components and can usually be identified in nucleoli from all organisms (Bernhard, 1966; Lafontaine, 1958; Lafontaine and Chouinard, 1963). There is a matrix of amorphous material to both these components besides other fibrils and granules which may be distinguished from those providing the chief characteristics of the two components (Marinozzi, 1963).

Autoradiographic evidence shows that the fibrillar component is the first to receive newly synthesized RNA which later passes to the granular component (La Cour and Crawley, 1965). New nucleoli in dormant, quiescent, or inactive cells are almost entirely composed of the fibrillar component, (Chouinard, 1975; Jordan and Chapman, 1971).

Work with inhibitors on animal cells provides evidence that the proportion of fibrillar to granular component is a reflection of the rate of nucleolar function as expressed in the speed of processing of the large ribosomal precursor RNA (Smetana et al., 1966; Smetana and Busch, 1974). RNP particles of various sizes have been isolated from nucleoli and it is possible that a thorough and more objective analysis of particle size in electron micrographs may reveal their nucleolar location. However, the major component, the mass of 150-Å granules, would appear to be nearly mature large ribosomal subunits awaiting transport to the cytoplasm (Liau and Perry, 1969).

2. Nucleolar Vacuoles

Not infrequently nucleoli have more or less spherical inclusions of low density, looking like nucleoplasmic inclusions. These are the nucleolar vacuoles. They have been studied carefully by some workers who report that they fill slowly, discharge rapidly, and characterize active nucleoli (Johnson, 1967, 1969; Erdelska, 1973; Barlow, 1970). They have been assigned the role of RNA transport (Rose et al., 1972) and though they seem to contain very few particles, it has been calculated that the number of particles is sufficient to permit such an explanation. The absence of nucleolar vacuoles from normal active cells in many circumstances and their variable size and frequency in otherwise similar tissues (Chaly and Setterfield, 1975) do not seem to support any hypothesis assigning them an essential role.

3. *Dense Chromatin Associations*

The nucleolus in animal cells is very frequently surrounded by or associated with large quantities of dense chromatin. This has been called the nucleolus-associated chromatin (Bernhard, 1966). It may penetrate the main body of the nucleolus even forming a complex meshwork within it. The relationship of this chromatin to the organizer is not clear though a substantial part of it may be centromeric heterochromatin containing satellite DNA (Jones, 1970). Plant nucleoli are not ensheathed in dense chromatin in this way. However, it is very common for the nucleolus to show a clear junction with one or two blocks of dense chromatin (Fig. 23) (Jordan and Chapman, 1971, 1973). The dense chromatin may form a kind of pedicel from the nuclear envelope terminating at the nucleolus (see Fig. 1 and Franke and Scheer, 1974a). The common association of chromatin with the nuclear envelope might be expected to produce this arrangement without implying any special structural or functional significance. In *Zea mays,* it has been possible to identify a nucleolus-associated heterochromatic region as part of the nucleolar organizer (Gillies, 1973). In animal cells some non-nucleolar heterochromatin is associated with nucleoli and it may prove that some of the dense masses of chromatin juxtaposed to nucleoli in plants also are not part of the organizer.

4. *The Nucleolus-Organizing Region*

In favorable sections of *Spirogyra* the interpretation of the fine structure of the nucleolus is easily made. The organizer takes a meandering course ensheathed by the fibrillar component. The granular component accounts for the rest of the nucleolus (Jordan and Godward, 1969). However this simple picture which can easily be related to the earlier light microscopic work becomes complicated by the various different states in which the organizer chromatin may occur.

The nucleolus organizer can exist in at least three different configurations or states of condensation, and this fact has led to some confusion in the interpretation of conventional thin section electron micrographs.

a. Transcriptionally Active Fully Dispersed Organizer. The location of the organizer when it is in a fully dispersed condition may be inferred from two approaches. The rRNA genes in transcription have been clearly shown in spread preparations by the Miller and Beatty (1969) technique and since this is a spread out fibrillar component it demonstrates that it contains the dispersed organizer (see Section VI,D,3,c). Transcription has also been shown to occur in the fibrillar component by autoradiographic work. But the view

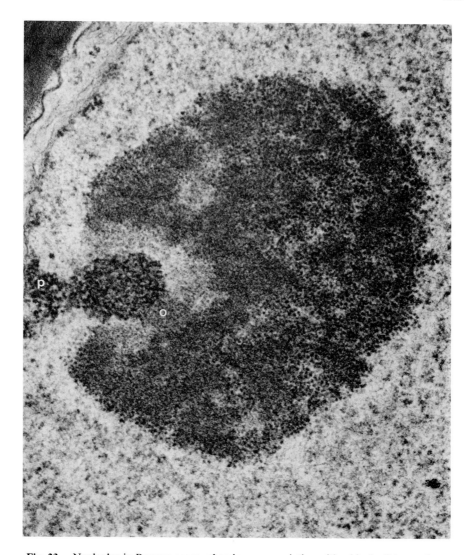

Fig. 23. Nucleolus in *Daucus carota* showing an association with a block of dense chromatin which is also in contact with the nuclear envelope in the form of a pedicel (p). The smaller fibrils and lower staining intensity characteristic of organizer chromatin (o) can be seen between the condensed chromatin and the fibrillar component of the nucleolus. From Jordan and Chapman (1973). × 80,000.

obtained of the fibrillar component in conventionally stained thin sections does not permit the visualization of the dispersed chromatin hidden in the dense mass of RNP resulting from transcription and addition of protein (Fig. 24). In a nucleolus with all its ribosomal genes in full operation, therefore, no other manifestation of the organizer, at least as a discrete condensed chromatin region, would be expected. Recent work, employing new staining techniques specific for DNA at the EM level has demonstrated DNA fibrils in this zone in thin sections, confirming the interpretation (Moyne *et al.*, 1975; Mirre and Stahl, 1976).

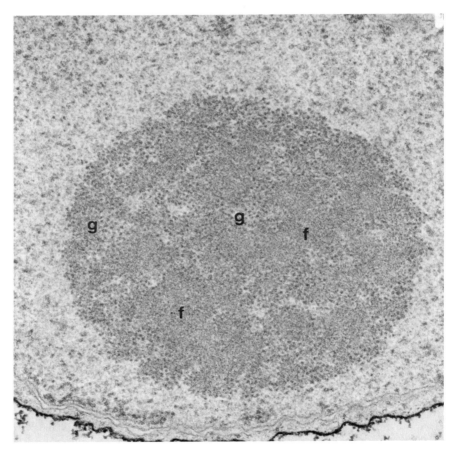

Fig. 24. Nucleolus from *Daucus carota*. Only two components are discernable: the granular (g) and fibrillar (f). This is the characteristic appearance of a fully active nucleolus where no manifestation of the organizer occurs except as the massed fibrils of the fibrillar component where the rRNA genes of the organizer are obscured by the mass of fibrils resulting from RNA transcription and processing. × 75,000.

b. Inactive Fully Condensed Organizer. The number of genes coding for ribosomal RNA in plants is high in comparison with most animals, especially in some plants where there may be as many as 20,000 per nucleus (Ingle and Sinclair, 1972). All these genes might not be expected to be working to full capacity in all cells so that some structure corresponding to condensed or at least partially condensed organizer might be expected. In maize there are a large number of rRNA genes which appear to be surplus to the normal requirements (Phillips *et al.,* 1974), at least for some cells, and these occur in the form of heterochromatin (Gillies, 1973; Phillips, 1976; McClintock, 1934). This represents the opposite extreme for the configuration of the genes from that found in the organizer in transcription within the fibrillar component. The condition of the organizer containing these apparently extra genes is indistinguishable from other condensed chromatin regions showing densely stained 200-Å fibrils. This is also the view obtained of a fully condensed organizer in mitotic chromosomes when no secondary constriction is formed.

c. Nucleolus Organizer with an Intermediate State of Condensation. The above situations, though representing both the active and inactive states of the organizers, are not the complete picture because the nucleolus organizer, especially in plants, can often be seen to be in an intermediate state of condensation. Such a condition can be seen in both interphase nucleoli and in chromosomes at nuclear division. The difference of this state of the organizer is not only one of staining intensity, but the constituent fibrils measure about half the diameter (100 Å) of those found in dense chromatin (200 Å) (Figs. 1, 3, and 23).

5. Micronucleoli and Cytoplasmic Nucleoloids

Besides the nucleoli that are situated on the organizers some cells have additional "free" nucleoli. These are usually smaller and hence they can be called micronucleoli they are also called secondary or accessory nucleoli. In animals that produce extra copies of the ribosomal genes in the process of gene amplification, many small nucleoli result each with their own ribosomal genes. Structures identified as micronucleoli have been seen in many plants, but the evidence for gene amplification or the presence of DNA in the micronucleoli is sparse at present (Avanzi *et al.,* 1972).

Micronucleoli in animal cells have been shown to have both fibrillar and granular components (Miller, 1966). They should be distinguished from karyosomes or nuclear bodies for these do not display nucleolar fine structure (Jordan, 1976) or any relationship with the nucleolus organizer. Some micronucleoli arise in meiotic prophase (Dickenson and Heslop-Harrison, 1970; Williams *et al.,* 1973; Jordan and Luck, 1976; Luck and Jordan, 1977). They are composed solely of fibrillar component and such structures are often seen enclosed within or adjacent to the disorganizing nucleolus (Gillies and Hyde, 1973). The explanation for the lack of a granular component in these meiotic micronucleoli may be that they result from synthesis of RNA in the absence of concurrent processing, a situation apparently occurring at this time.

Nucleolus-like bodies have been found in the cytoplasm at meiosis and these have been designated nucleoloids. They are thought to arise by a reconstitution of the material deriving from the dispersal of micronucleoli at the end of prophase (Williams *et al.*, 1973).

D. Chemical and Structural Aspects of Ribosomal RNA Synthesis

1. The Mechanism of Ribosomal RNA Synthesis

The genes that code for ribosomal RNA (rRNA) are associated with nucleoli, where they may be observed undergoing transcription (Trendelenburg *et al.*, 1974). These genes in eukaryotes are known to be in multiple copies with plants showing particularly high redundancies, from 1250 in a telophase nucleus of orange to 31,900 copies in a tetraploid hyacinth (Ingle *et al.*, 1975). They are arranged in tandem, clustered together at one or a few chromosomal secondary constrictions or nucleolus-organizing regions.

The rRNA cistrons consist of a transcription unit and a nontranscribed spacer region. The transcription unit encodes not only the large and small ribosomal subunit RNAs but other DNA regions which are adjacent to and contiguous with the ribosomal RNA genes. These other regions are transcribed but are split from the ribosomal RNA components and degraded by specific nucleases during a stage of processing.

A generalized outline of the production of mature rRNA is in Fig. 25. The first rRNA gene product, detectable by radioactive labeling and polyacrylamide gel electrophoresis, is a large, short-lived precursor molecule of 2.2–2.6×10^6 daltons. This molecule has been shown to contain the sequences for both the nominal 1.3×10^6 and 0.7×10^6 dalton rRNA species from mature ribosomes. The processing of this precursor rRNA to mature stable rRNA is well understood (Grierson, 1977) and occurs by methylation, cleavage, and loss of specific sequences from the molecule. This process may occur in two stages yielding further precursor rRNAs, 1.4×10^6 and 1.0×10^6 daltons in molecular weight, before processing is completed. It is stressed that various plants show variations from this generalized scheme, some showing multiple precursors and many showing minor deviations from the nominal molecular weights stated. Furthermore, the sizes of the predominant class of transcription units is different for different organisms and this is considerably influenced by the amount of nonconserved RNA transcribed. It is minimal in *Acetabularia* where almost the total length of the precursor is required to account for the two ribosomal RNA molecules leaving very little

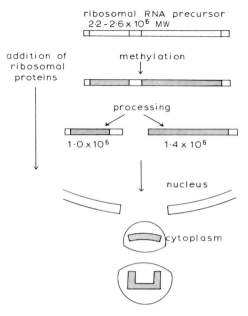

Fig. 25. The synthesis and processing of ribosomal RNA in plants. Modified from Grierson (1977).

for excision in the nucleolar maturation process (Spring *et al.*, 1974, 1976). However, in some animals the situation is very different where up to half the RNA transcribed is removed in the processing to leave only about half the primary transcript as rRNA, the rest being nonconserved regions.

Although the variation in length of the ribosomal precursor genes is greatly influenced by the proportion coding for the nonconserved part of the precursor, that coding for the conserved part is also different in different organisms and species, as evidenced by the different molecular weights of rRNA.

2. Utilization and Variability of Ribosomal RNA Gene Number

The ribosomal RNA genes are of particular interest because it is possible to accurately determine their number in a particular genome by hybridization of radioactive rRNA to denatured total DNA. It is also possible to study the amount of product of this specific gene as rRNA (Timmis and Ingle, 1975b). In a number of animal systems the rRNA gene redundancy per cell may be varied according to the demand for protein synthetic machinery (Birnsteil *et al.*, 1971), but this phenomenon of amplification has never been satisfactorily demonstrated in plants. It may be that maintaining high redundancies of rRNA genes in all plant tissues is an alternative method of supplying even the greatest demand.

It is clear that many plants are utilizing their rRNA genes with less than the maximum possible efficiency. Ribosomal RNA accumulation remains unaffected in aneuploid hyacinth varieties with only 16,000 instead of the usual 24,000 gene copies (Timmis and Ingle, 1975b). Furthermore, calculation of the rate of rRNA synthesis in pea roots suggests that each cell, which contains 8000 copies of the gene, is synthesizing 30,000 molecules of rRNA at any one time (Timmis *et al.*, 1972). Making the assumption that, as in *Acetabularia,* up to 100 molecules of precursor rRNA may be in the process of transcription from a single gene (Trendelenburg *et al.*, 1974), then the pea needs to use only about 4% of its available genes.

The fact remains, however, that many plants show intervariety and even interplant variation in rDNA (Flavell, 1975; Timmis *et al.*, 1972). Flavell (1975) demonstrated at least a twofold variation in the percentage of the wheat genome complementary to rRNA. The variety "Cheyenne" has 0.111% hybridization to rRNA compared with only 0.055% in "Holdfast." In rye the variety "Petkus" shows 0.174% hybridization compared with 0.710% in "King II." The rRNA gene redundancy in wheat is clearly under genetic control (Mohan and Flavell, 1974). When the nucleolus organizing chromosomes of hexaploid wheat are removed from or added to the genome the aneuploid forms do not always behave additively in terms of their rRNA gene dosage. The redundancy at the NORs appears to vary depending on the chromosomal environment. Addition or removal of the 1B NOR increases or decreases the rRNA gene redundancy as would be expected on a dosage basis (Table IV). In contrast, removal of the NOR on chromosome 1A certainly decreases the redundancy but its addition has practically no effect.

TABLE IV

Ribosomal RNA Genes in *Triticum aestivum* (var. Chinese Spring)[a]

Nucleolus organizers (1B, 1A, 5D)			rRNA genes (% of control)
2	2	2	100
4	2	2	128
0	2	2	84
2	4	2	101
2	0	2	58
2	2	4	76
2	2	0	105

[a] Data taken from Mohan and Flavell (1974).

Addition of the 5D chromosome actually reduces the overall redundancy whereas its removal causes little change. Somewhat similar results were observed in hyacinth aneuploid varieties (Timmis *et al.*, 1972), where the aneuploid forms show a reduced number of rRNA genes at each NOR compared with the euploids.

One of the most interesting examples of rDNA variation is found in flax (Timmis and Ingle, 1973, 1975a; Cullis, 1976), where changes may be induced by different nutritional environments. The same environments also induce large and small plants (Durrant, 1974) with the large plants having 16% more DNA per nucleus than the small (Evans, 1968). After the initial differences have been induced, they may either be maintained or lost in subsequent generations, according to the conditions of growth. This maintenance and reversion has given rise to a series of plants bearing various combinations of induced characters. In all the types studied thus far there is a correlation between plant size and rRNA gene redundancy with a phenotypically large plant having 60–70% more genes than a small plant. The results suggest that these flax types, unlike most plants (Timmis and Ingle, 1975b) may be phenotypically susceptible to rRNA gene changes. It has been proposed that the low inherent rDNA amount in flax compared with other plants (Ingle *et al.*, 1975) may allow this susceptibility and be responsible for the unusual environmental influence on inheritance. The 16% difference in total DNA (Evans, 1968) is much larger than may be accounted for by rDNA variation ($<1\%$) and the flax genome must be able to change quantitatively at numerous chromosomal sites.

In none of the examples of rDNA variation in plants can it be concluded that the variation is unique to rDNA. The same mechanisms which regulate rDNA could also control sequences elsewhere in the genome as they do in *Xenopus* (Amaldi *et al.*, 1973). Unless such changes involve relatively large proportions of the genome or involve sequences which have easily purified RNA products, it will be difficult to detect what could be a very widespread phenomenon.

3. Microscopic Observations of Ribosomal RNA Synthesis and the Ribosomal RNA Genes

a. Structure of Spread Nucleoli. The technique devised by Miller and Beatty (1969b) for displaying genetic transcription has provided elegant confirmation of the understanding of the nucleolus built up from other approaches and opened the way for a detailed analysis of the ribosomal RNA genes. Miller and Beatty (for procedures see Miller and Bakken, 1972) isolated nucleoli and then allowed them to disperse for a short time in a low ionic environment and then centrifuged the expanded transcriptional complexes through a stabilizing solution of fixative onto a carbon film supported on an

EM grid. After dipping in a solution containing a wetting agent to minimize the rearrangement of the material and assist in the spreading, the grids were dried.

The technique was first used on nucleoli from animal oocytes but has now also been applied to plants (Trendelenburg *et al.*, 1974; Woodcock *et al.*, 1975; Grainger and Olgle, 1978). It was found that the granular region of the nucleolus migrated away in the spreading solution to leave the loosened fibrillar component which proved to be the region containing the transcription complexes. The degree of spreading is variable and together with adjustments in time and conditions it is possible to find granular components (Fig. 26) or aggregated groups of transcription complexes from the fibrillar component (Fig. 27A) or well-separated transcriptional complexes from the fibrillar component (Fig. 27B). The granular component yields a network of fibrils with intermingled granules (Fig. 26). The significance of the fibrils is not clear, whether they correspond to some nucleolar counterpart of the nuclear matrix (Section II,C,2), or whether they correspond to inactive chromatin cannot yet be decided. It would not be expected that there is a significant amount of chromatin in the granular region, but there is a report of a feulgen reaction in this component in *Acetabularia* (Spring *et al.*, 1974).

The spread fibrillar component is more easily interpreted. Here the ribosomal precursor RNA genes can be identified by the mass of RNA transcripts splayed out to either side of the chromatin axis. The high multiplicity of ribosomal genes is very apparent. The individual genes seem to be involved in near maximal transcriptional activity for the polymerases and the growing RNA transcripts are very closely packed [110 per gene of 1.8 μm in *Acetabularia* (Spring *et al.*, 1974)].

Close packing seems to be a general characteristic of ribosomal transcripts (Puvion-Dutilleul *et al.*, 1977) though the transcripts on the silk fibroin gene show a similar close packing in a messenger RNA gene (McKnight *et al.*, 1976).

The transcripts show increasing length with distance along the gene making it possible to tentatively identify the initiation and termination sites and so measure the length of the precursor gene and the apparent spacer segments between adjacent transcriptional complexes.

The RNA transcripts appear to be shorter than the coding sequence and it must be concluded that they are being folded up through the formation of double stranded regions or perhaps by being complexed with protein, a suggestion supported by the presence of a terminal knob on the longer transcripts (Fig. 28).

The lengths of the spread genes correspond fairly well with what is expected for the coding sequences for the ribosomal precursor RNA genes (Scheer *et al.*, 1973, Spring *et al.*, 1976). Furthermore Franke *et al.* (1976a) have argued that such spread out DNA could not be coiled up into the

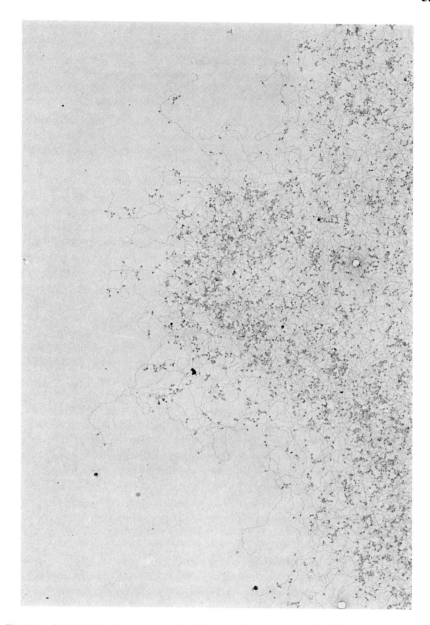

Fig. 26. Spread granular component of *Xenopus laevis* oocyte. Besides scattered granules there is a filamentous background material. (Courtesy of O. L. Miller and B. R. Beatty). × 20,000.

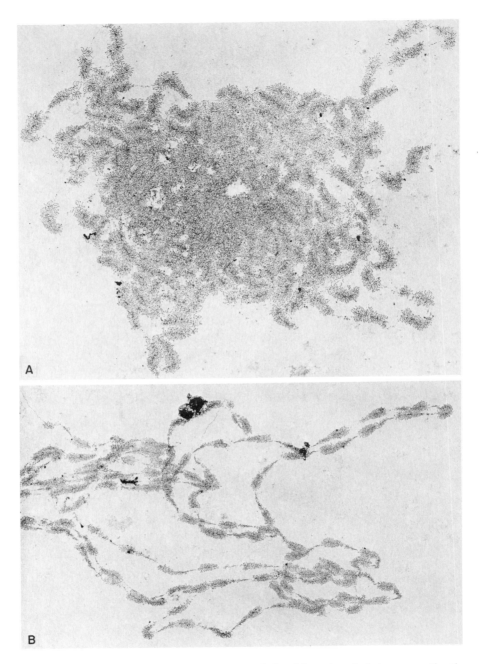

Fig. 27. A, Group of transcriptional complexes isolated from *Acetabularia* representing the dispersed fibrillar component. B, well spread transcriptional complexes from *Acetabularia*. From Trendelenburg *et al.* (1974).

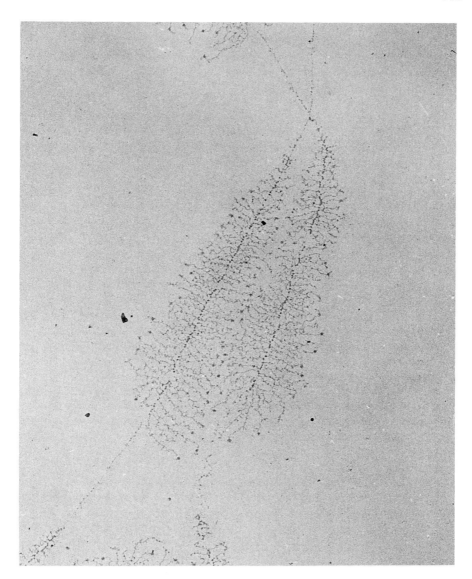

Fig. 28.   Well spread matrix units from amphiban oocyte nucleolus. The axis with granules is the DNA and transcribing polymerases. The lateral fibrils are folded nascent chains of precursor ribosomal RNA. The RNA molecules show a terminal granule in the longer transcripts. From Miller and Hamkalo (1972). × 63,000.

nucleosomal configuration and this may be consistent with the apparent absence of nucleosomes from spread transcribing ribosomal cistrons. The granules at the junction of the RNA transcripts and the gene axis have been identified as polymerases, and after treatment with the detergent sarkosyl, which removes the RNA, the genes can be seen marked out only by the closely packed polymerases (Fig. 29) (Scheer *et al.*, 1977).

From slightly spread nucleolar subunits in *Acetabularia* it has been possible to count the ribosomal genes and compute the number per nucleus as 17,000 in *Acetabularia mediterranea* and 21,000 in *Acetabularia major* (Spring *et al.*, 1974).

b. Possible Length Heterogeneity of Ribosomal RNA Genes. The spreading technique has permitted an assessment of length heterogeneity in a polulation of repeated genes (Spring *et al.*, 1976). Differences are greatest in the length of the spacer and this has been confirmed in *Xenopus* by techniques employing nucleases which have shown that the spacers contain certain short highly repetitious AT-rich sequences, the number of which is variable (Botchan *et al.*, 1977). The transcribed regions are not so variable but there are reports of different classes having different mean lengths (Spring *et al.*, 1976). The arrangement of the cistrons along the strand is variable though the uniform "head to tail" pattern is most abundant.

An interesting situation exists in *Physarum* where the nonchromosomal genes exist in palindromic sequences (Vogt and Braun, 1976) and the spreading technique reveals that the spacers come between the "facing" initiation points of the paired genes (Fig. 30) (Grainger and Ogle, 1978).

There are reports of transcription from the spacers either near the initiation point (prelude pieces) or in the central region of the spacer (spacer transcripts). The various arrangements are illustrated in Fig. 31 and the sizes of matrix units in a range of organisms is shown in Table V.

c. Interpretation of the "Christmas Tree" Pattern. The spread transcription complexes present a satisfyingly complete picture of the transcriptional process but there are some underlying assumptions in this interpretation. Firstly, it is assumed that the start of the gene is the tip of the Christmas tree whose branches are the lateral fibrils, but the prelude transcripts, spacer transcripts, and the occasional extra long lateral fibril indicate that initiation may occur at an earlier site (Franke *et al.*, 1976b). It is not impossible that polymerases join the axis earlier and proceed to the apparent initiation point making transcripts that are unstable or that are rapidly digested. The lack of polymerase molecules in most spacer regions makes this rather unlikely, but we need more information about such preliminary events.

A further complication to the interpretation would arise if processing or trimming of the RNA transcripts occurred while they were still growing and

Fig. 29.   Ribosomal precursor RNA genes from *Xenopus* marked out by the polymerase molecules after the RNA transcripts have been released by treatment with sarkosyl. From Scheer *et al.* (1977). × 18,000.

Fig. 30. Spread rDNA of *Physarum* showing the long spacers between the two genes of a palindromic sequence. From Grainger and Ogle (1978). × 12,000.

TABLE V

Comparison of Matrix Unit and Spacer Lengths in a Range of Organisms

Matrix unit (μm)	Spacer (μm)	Organism	References[a]
Plants			
1.3		*E. coli*	1
2.0	0.5	*Acetabularia mediterranea*	2,3
1.85	0.8	*Acetabularia mediterranea*	2,3
3.7		*Acetabularia mediterranea*	2,3
8.5		*Acetabularia mediterranea*	2,3
1.84	1.03	*Acetabularia major*	2,3
4.0		*Acetabularia major*	2,3
8.7		*Acetabularia major*	2,3
1.9	1.0	*Chlamydomonas reinhardi*	4
4.2	3.0	*Physarum polycephalum*	5
Animals			
2.6	2.1	*Xenopus leavis*	6
3.1	1.9	*Xenopus mulleri*	6
2.3	0.75	*Triturus viridescens*	7
2.89	2.2	*Triturus alpestris*	8
3.9		*Triturus alpestris*	8
	7.3	*Triturus alpestris*	8
	8.3	*Triturus alpestris*	8
	9.6	*Triturus alpestris*	8
2.47	2.22	*Triturus cristatus*	8
2.63	2.18	*Triturus helviticus*	8
5.58	5.49	*Acheta domesticus*	6
2.2	0.45	*Chironomus thummi*	6
2.5	0.8	*Drosophila melanogaster*	9
	1.4	*Drosophila melanogaster*	9
5.2	0.5	Rat liver	6
3.8	6.4	Chinese hamster ovary	10
3.5	3.5	HeLa cells	8

[a] Key to references: 1. Miller *et al.* (1970), 2. Spring *et al.* (1974), 3. Spring *et al.* (1976), 4. Woodcock *et al.* (1975), 5. Grainger and Ogle (1978), 6. Trendelenburg *et al.* (1973), 7. Hamkalo and Miller (1973), 8. Scheer *et al.* (1973), 9. Laird and Chooi (1976), 10. Puvion-Dutilleul *et al.* (1977).

joined to the polymerase. This appears to happen in the procaryotic ribosomal genes and could account for some of the apparent shortening of transcript length; any differences between precursor RNA molecular weight and gene length could be accounted for in this way.

The apparently extra large transcription units which have been reported (Spring *et al.*, 1976) may contain more than one gene and be indicative of a read-through and failure of termination at the first gene. However, it is not impossible that some nonribosomal RNA genes are intercalated in the or-

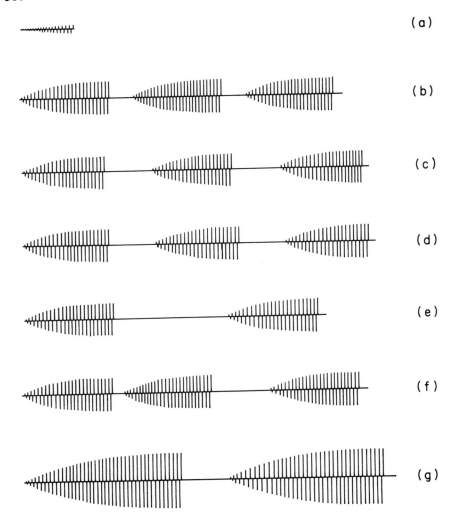

Fig. 31. Diagram showing the different arrangement of rDNA transcription units seen in plants. Length and spacing of transcripts, i.e., RNA side branches approximate and diagrammatic. The regular arrangement of matrix and spacers is most frequent but in *Acetabularia* various arrangements of spacer region have been found, consistent with the less stable highly repetitious sequences now known to be in these regions. The shortest spacers, b, are found in *Acetabularia mediterranea. Acetabularia major*, c, and *Chlamydomonas reinhardi*, d, have very similar matrix unit and spacer sizes. Large and variable spacers are indicated by e and f, respectively. Large and giant matrix units are shown at g and h. Prelude and spacer transcripts are shown in i. Facing initiation points are seen in some *Acetabularia* spreads, j, and in the *Physarum* palindromes, l. Termination to termination arrangements have been reported in *Acetabularia*, k. (Data from Miller *et al.*, 1970; Spring *et al.*, 1974, 1976; Woodcock *et al.* 1975; Grainger and Ogle, 1978). × 20,000.

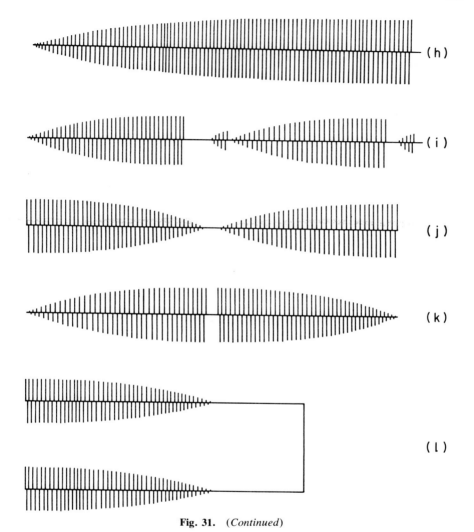

Fig. 31. *(Continued)*

ganizer, although it would be surprising to find such close packing of transcripts and polymerases on nonribosomal RNA genes.

All measurements from spread preparations are subject to slight variations according to the degree of spreading but in well-spread examples the DNA corresponds with what would be expected for the B conformation (Scheer *et al.*, 1973). The size of the polymerases is around 130 Å and the center to center spacing in *Acetabularia* (Spring *et al.*, 1974; Franke *et al.*, 1976a) is 150–250 Å. The width of the axis is between 50 and 150 Å which corresponds to fully dispersed DNA. If histones are present on the axis of the transcripts they must be spaced out, perhaps as half nucleosomes. However, it has been calculated in *Physarum* that the shortening of the axis of the spacer region from that of the B form of DNA is 1.8-fold, while there is very little, if any, in the transcribed regions (Grainger and Ogle, 1978).

VII. THE NUCLEAR ENVELOPE AND NUCLEAR PORE COMPLEXES

A. Introduction

The nuclear envelope and especially its pore complex may hold the key to our understanding of the cell control systems essential to the eukaryotic level of organization. The possession of a barrier between the cellular compartments of transcription and translation has permitted the development of a whole series of regulatory processes concerned with gene expression. The demonstration by Miller of growing polysomes on nascent message in prokaryotes emphasizes the different eukaryotic organization (Miller *et al.,* 1970).

The inner and outer membranes of the nuclear envelope are perforated by the pores (Fig. 32). Though the membranes themselves can be understood mainly as a barrier with certain synthetic and skeletal functions, the nuclear pores and their associated nonmembranous components have a highly organized architecture. Their location and structure is suggestive of a critical role in the life of a eukaryotic cell. The most interesting parts of the pore complex are tantalizingly near the limits of resolution obtainable in electron microscopy of biological specimens and the detail of their ultrastructure is still uncertain (Figs. 33 and 34). Biochemical appreciation of the pore complex is in its infancy but with methods now becoming available for the purification of the proteins of the nuclear pore complexes we can look forward to a better understanding in the near future.

The nuclear envelope and nuclear pore complexes and related annulate lamellae have been the subject of extensive reviews. All show the need for more refined biochemical and microscopic methods to improve the description of the molecules of the pore complex (Kay and Johnston, 1973; Kessel, 1973; Franke, 1974; Franke and Scheer, 1974a,b; Wunderlich *et al.,* 1976; Maul, 1977).

B. The Structure of the Nuclear Pore Complex

1. The Pore Orifice

A good deal of agreement has been reached on the main features of the nuclear pore complexes though differences exist in the interpretation of the

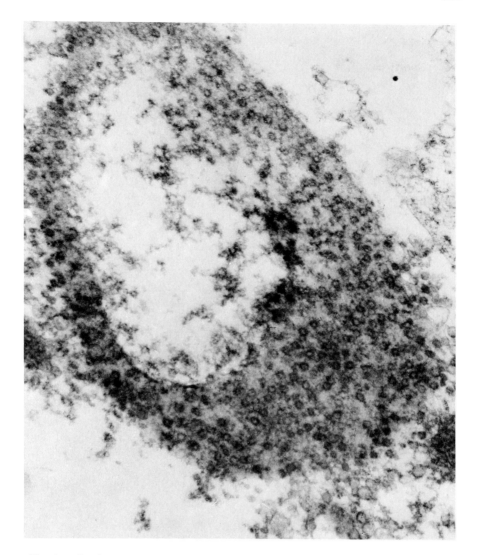

Fig. 32.   Section tangential to the nuclear envelope showing the dark stained annuli of the nuclear pores in *Spirogyra*. × 43,000.

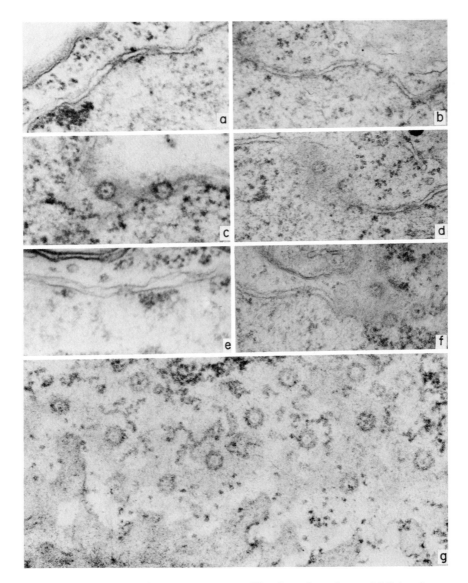

Fig. 33. Nuclear pores from *Daucus carota*. The three tiers of material lining the pore lumen are clear in a and b. Annulus subunits can be seen in side view in a, b, and f and in surface view in c, d, and g. Central granules can be identified in g and features representing central granules in side view are suggested in b, e, and f. The central features in b and f can be understood better as fibrils. From Jordan and Chapman (1973). × 60,000.

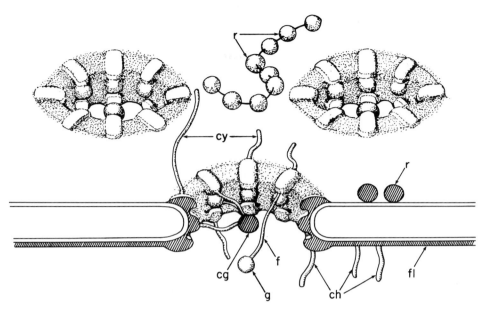

Fig. 34. Diagram of a nuclear pore complex. The discontinuities in the double membraned nuclear envelope show a lumen lined with three rings of material each showing eight smaller subunits. The inner and outer rings on the rims are called annuli. The one at the equator is described as composed of peripheral granules or projecting tips. Granules, g, of different sizes can be seen above and below the central plane often connected to fibrils, f, but these are not envisaged as having a fixed location. In the center of the pore a central granule, cg, can often be seen. Fibrils have been reported in various positions within the pores, projecting into the cytoplasm from the annulus granules, cy; on the nucleoplasmic annulus granules projecting into the nucleus, n; between one pore complex and another pore-connecting fibrils (not shown); and between pore complexes and chromatin (not shown).

The nuclear envelope shows a fibrous lamina, fl, on the nucleoplasmic side (especially in animal cells), ribosomes, r, on the cytoplasmic or outer membrane, and connections with chromatin, ch, on the inner or nucleoplasmic membrane.

The three rings, i.e., the inner and outer annuli and the central ring of material at the equator are all part of a continuous matrix which is more easily lost in preparative techniques than the smaller subunit components.

finer details. The nuclear pore complexes are simple membrane-lined perfo-rations through the nuclear envelope. The membranous walls of the pore which are formed by fusion of the inner and outer nuclear membranes en-close a circularly-shaped lumen about 700 Å in diameter. It has occasionally been reported that the size of the perinuclear space (the cisterna between the two membranes) is increased around the pore (Roberts and Northcote, 1970). The shape of the lumen has sometimes been described as octagonal (Gall, 1967; Speth and Wunderlich, 1970; Maul, 1971) but the argument that octagonal outlines are artifacts has not been adequately refuted and the common appearance of circular outline in unfixed unglycerinated frozen cells and the low frequency of noncircular or polygonal outlines favors this description (Franke and Scheer, 1974a).

2. The Annulus Structures

The nonmembranous parts of the pore complex consist of material which accentuates the pore perimeter and appears as rings lying on top of the nucleoplasmic and cytoplasmic pore margins. Each of these two ring structures which have been termed "annuli" are composed of eight subunits embedded in a matrix which may or may not be preserved according to the procedures used for specimen preparation (Franke and Scheer, 1970) (Fig. 33). In whole mounts of nuclei, viewed by high resolution scanning microscopy, the fully rounded doughnut-shaped profile of the annulus is very evident showing the subunits as slight distensions (Kirschner *et al.,* 1977). The annulus structure is shown diagramatically in Fig. 34.

The shape of the subunits of the annulus is most often described as spherical but in plants there is evidence that they may be flattened ovoids (Roberts and Northcote, 1970). Their ultrastructure has sometimes been described as fibrillar (La Cour and Wells, 1972). Fibrils clearly enter the annuli on both the cytoplasmic and nucleoplasmic sides (Fig. 35).

The narrowest part of the lumen, midway between the two annuli, is lined with material forming another ring at the center of the pore. This also has an eightfold substructure and the subunits have been described variously as projecting tips or peripheral granules (Franke and Scheer, 1970; Roberts and Northcote, 1970). Some reports describe this central ring at the pore equator as being the most prominent part of the pore-complex substructure (Wunderlich, *et al.,* 1976) but usually it is seen as a less pronounced feature.

The pore can then be envisaged as composed of a lumen, lined by three bands or rings. The three rings of material are not separate structures and may at times all be hidden in a darkly staining, thick lining throughout the pore lumen. This whole structure may be described as the annulus though this term is more usually restricted to the rings at the two ends of the lumen.

Some workers envisage the subunits of all three rings as connected into tubules running between the nucleus and cytoplasm equidistantly spaced and lying against the membrane bounding the pore lumen (Vivier, 1967; Abelson and Smith, 1970). The evidence for this view is that the subunits appear to have less electron density in their center in some preparations. It is possible, however, to understand this image as having been derived from collapsed fibrils.

3. Nuclear Pore Granules

Another characteristic of nuclear pore complexes is the frequent occurrence of central granules (Fig. 33). These do not seem to be obviously related to the functional capacity of pores (Jordan and Chapman, 1973). The appearance of the central granules is variable; they may be dense and small or large and diffuse or apparently formed of several smaller fibrils (Fig. 33). This central part of the pore has been unequivocally demonstrated to be the channel for transport of nuclear granules of various types—Balbiani ring granules (Stevens and Swift, 1966); injected polyvinyl pyrrolidone coated gold particles (Feldherr, 1965), or viral nucleic acid (Summers, 1969, 1971) and this may explain the variation in appearance.

Ribosomes on the nuclear envelope are often seen close to nuclear pores and the similarity of polysomes and their juxtaposition to the annulus granules has given rise to the suggestion that nuclear pore complexes are somehow related to the formation of polysomes (Swift, 1958; Mepham and Lane, 1969) perhaps even polysomes in formation. However, the stainability of ribosomes is always greater than annulus granules and size differences are also apparent (Franke and Scheer, 1974a). However, it is known that all ribosome subunits must cross the nuclear envelope and since they travel as separate subunits, only becoming ribosomes in the cytoplasm, the nuclear pore occupies a site which could easily be involved in their final assembly and possibly also with the supply or even selection of the mRNA. The identification of a nuclear envelope-associated poly A polymerase is interesting in this connection.

4. Nuclear Pore Fibrils

A further feature of the pore complexes is the presence of various fibrils which run between the central granules and between the annuli or peripheral granules of the equatorial region of the pore. These have been considered as structural components of the pore complex but they may be only transiently associated fibrils. Sections of the pore cut parallel to the nuclear envelope frequently show a cartwheel type of structure but this does not necessarily indicate that the "spokes" of this cartwheel are all in the same plane. Sections in the vertical plane show how fibrils in various positions could give this effect. In freeze–fracture studies granules appearing in the lumen may be these same fibrils giving the appearance seen for proteins of the intramembranous particles in cleaved membranes.

The pore complexes also have fibrils which radiate outwards from the annuli in the plane of the envelope and these have been described as pore-connecting fibrils within the membrane (Scheer et al., 1976) or skeletal support fibrils of the fibrous lamina (Aaronson and Blobel, 1975), the submembrane nuclear layer seen clearly in some animal cells.

Connections between the pores and chromatin have also been reported (Du Praw, 1965; Engelhardt and Pusa, 1972). Chromatin fibrils are thicker than the other types (Fig. 35). However, the nuclear pores are usually situated in chromatin-free areas of the nuclear envelope (Fig. 35), yet close

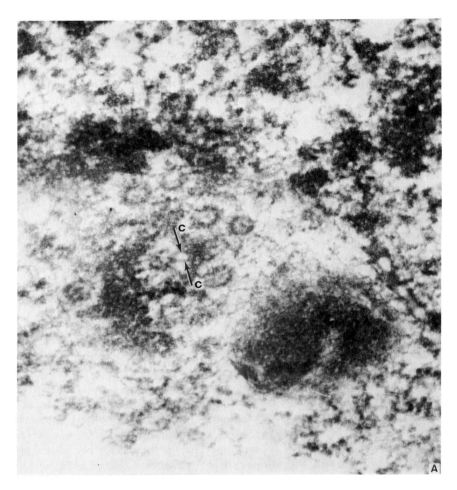

Fig. 35. A, tangential section through nuclear pores showing a chromatin free channel around the pore and fibrillar connections between chromatin and annulus granules, c. The differences in the central granules between pores are also apparent and some pores show evidence of a polygonal outline which may be a shrinkage artefact. *Helianthus tuberosus,* × 180,000. B, cross section through a nuclear pore showing the channel in the chromatin on the nucleus side and fibrils extending from the pore complex into both the cytoplasm and the nucleus. *Helianthus tuberosus,* × 100,000.

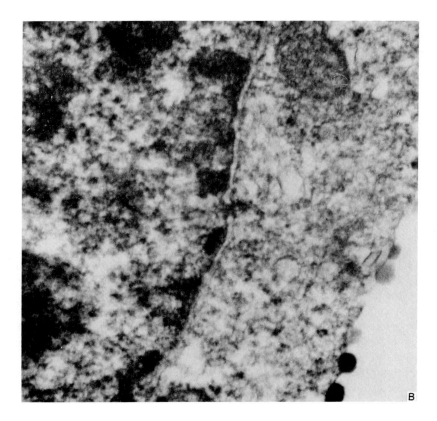

Fig. 35. (*Continued*)

association of chromosomes and the nuclear envelope is known to occur in mitosis and especially in meiosis (Hecht and Stern, 1969; Moens, 1969) and there are reports of chromosome-associated pore complexes even where no membrane component accompanies them (Maul, 1977).

C. Nuclear Pore Formation and Distribution

The process of pore formation remains a mystery. Apart from reports of minor deformations in the nuclear membranes almost nothing is known. From this it may be concluded that any intermediate stages in pore formation are rapid and unstable.

The size of the pore lumen seems to be constant within a given nuclear type although some reports of variably-sized pores have been published (Willison and Rajaraman, 1977). The variation in size reported in freeze–

fracture studies has been interpreted as resulting from different fracture planes (Severs and Jordan, 1975a) or as the variation encountered in a population of pores showing a normal distribution of diameters (Severs and Jordan, 1978). The idea that pores may open and close or change in size cannot be dismissed while the possibility also exists for very short-lived transient states not figuring significantly in normal preparations.

The positioning of pores in the nuclear envelope is frequently ordered or nonrandom indicating controlling elements in their distribution (Maul *et al.*, 1971). It is very common to find pores in short or long rows or clustered (Northcote and Lewis, 1968; Cole and Wynne, 1973; Scott *et al.*, 1971; Fabri and Bonzi, 1975) and sometimes in very regular octagonal or spiral distributions (Thair and Wardrop, 1971; Roberts and Northcote, 1971). Although the idea that pores can move or flow about on the nuclear envelope is suggested by their apparent absence in regions of organelle apposition (La Cour and Wells, 1974; Severs *et al.*, 1976), their links with the nuclear matrix and fibrous lamina and their nonrandom distribution would make other explanations for their absence from certain areas of the nuclear envelope more likely. For example, their formation may be impossible where one membrane of the nuclear envelope is occluded or adpressed to another structure.

The number of nuclear pores per nucleus or per unit area might be thought to give some indication of the processes with which they are involved. Much work has been concerned with this and it appears that a general correlation may be found between the amount of RNA exported from the nucleus and the total number of nuclear pores (Franke and Scheer, 1974b). Nuclear pores are formed rapidly in the new postmitotic nucleus and slower rates of formation then continue through the rest of the cell cycle (Scott *et al.*, 1971; Maul *et al.*, 1972; Jordan *et al.*, 1976, 1977). In animal cells a burst of pore synthesis was found in some cells to be related to the time of DNA synthesis (Maul *et al.*, 1972) but this does not occur in the yeast S-phase (Jordan *et al.*, 1976, 1977), making a chromosome anchorage role for nuclear pores less likely. However, the presence of DNA tightly bound to the nuclear envelope (Franke *et al.*, 1973; Zentgraf *et al.*, 1975) is strong evidence that the chromosomes are structurally integrated into the nuclear envelope somewhere, even if not at the pores.

D. The Nuclear Envelope

1. Structural Aspects of the Envelope

The nuclear envelope consists of two membranes 50 to 70 Å wide enclosing the perinuclear space or cisterna which is about 200 Å wide. The spacing between the membranes is very regular except in the following situations: (1) where the envelope is disintegrating (La Cour and Wells, 1972); (2) the envelope is involved in blebbing (Kessel, 1973); (3) the synthesis and accumulation of material in the perinuclear space (Bouck, 1971; Severs, 1976; Krutrachue and Evert, 1978); (4) some other developmental situation, e.g., spermiogenesis (Bell, 1978).

The outline of the nucleus becomes very irregular in some cells, especially those which are becoming increasingly active (Jordan and Chapman, 1973; Severs and Jordan, 1975b). This indicates an increase in surface area and may also be accompanied by an increase in nuclear pores per unit area making a considerable increase in nuclear pore number per nucleus (Jordan and Chapman, 1973).

The nuclear envelope possesses ribosomes on its outer membrane and frequently shows indications of being involved with synthetic activity like that of the endoplasmic reticulum. Biochemical analysis supports this analogy (Franke and Scheer, 1974b).

The nuclear envelope can often be found forming either invaginations or evaginations (Camefort, 1968; Bell, 1972, 1975; Dickenson and Bell, 1972; Jordan, 1974). These involve both the inner and outer membranes unlike the more usual connections with endoplasmic reticulum. These have been closely studied in the developing eggs of some Pteridophytes where the impression has been gained that they detach and give rise to cytoplasmic organelles (Bell, 1972, 1975).

Fingerlike nuclear projections have been seen in both male and female pronuclei in *Spirogyra* zygotes (Jordan, 1974). These show a band running under the inner nuclear membrane (Fig. 36) but whether this is part of the fibrous lamina or chromatin is unknown. Such projections and evaginations seem to occur at generation change-over points in the plant life cycle and could therefore be involved in the control of differentiation. The occurrence of similar fingerlike projections on the side of the nucleus towards the pole of the future rhizoid in *Fucus* zygotes is confirmation for this view (Quatrano, 1972).

2. Possible Biochemical Functions of the Envelope

The suggestion that the envelope may discriminate between the various diffusible cytoplasmic molecules only permitting entry of some of them follows from two basic observations. Loewenstein (1964) demonstrated the presence of a sizeable electrical potential difference between the inside and outside of the nucleus. This indicates a probable selectivity in ion uptake and shows that the nuclear pores do not permit free transit of small molecules. Secondly, the complement of nuclear proteins is markedly different to those in the cytoplasm. This has been shown using two-dimensional gel analysis for several plants (Trewavas, 1978) and for HeLa cells (Peterson and McConkey, 1976). Furthermore, oocytes microinjected with labeled nuclear or cytoplasmic proteins selectively accumulate the nuclear proteins (Bonner,

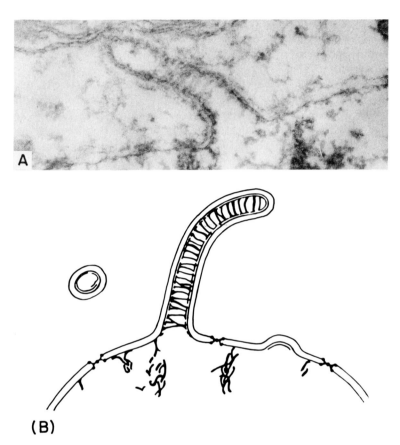

Fig. 36.   A, an evagination involving both membranes of the nuclear envelope. Irregular nuclear envelopes with evaginations have been seen to characterize critical phases of the life cycles in a number of plants. These evaginations with regular width and material associated in parallel lines with the inner nuclear envelope are found on both male and female pronuclei in the zygotes of *Spirogyra* before nuclear fusion. From Jordan (1974). × 60,000. B, diagram from serial sections showing the regularity of the width of the evagination and the presence of the fibrillar material just beneath the inner nuclear membrane. No nuclear pores are found on these evaginations.

1975). That this may represent an active accumulation is illustrated by the histones; the oocyte nucleus can accumulate up to 700-fold the amount of histone normally complexed with DNA. Perhaps more intriguing is the evidence for a pool of nuclear proteins resident in the cytoplasm. On receipt of the appropriate signals some of these may be moved to the nucleus where they initiate the expression of new genes or replication of DNA (Fansler and

Loeb, 1972; Johnson *et al.*, 1974; Merriam, 1969; Okita and Zardi, 1974; Seale and Aronson, 1974).

The nuclear envelope then can be envisaged not only as the regulator of nuclear-cytoplasmic exchange but also as a site of synthesis, similar to the endoplasmic reticulum, as a place for chromosome anchorage (especially during synapsis in meiosis) and as a possible site for membrane synthesis and rearrangements in developing cells especially at significant stages in the life cycle.

The biochemical analysis of the nuclear envelope has now reached the point where some of the molecules of the nuclear pore are being identified. These are large proteins of 150,000 and 73,000 MW (Krohne *et al.*, 1978).

Reviews of the biochemistry of the nuclear envelope (Franke, 1974; Franke and Scheer, 1974a; Wunderlich *et al.*, 1976) list many molecules of the nuclear envelope and discuss their possible distribution between membrane, nuclear pore, fibrous lamina, and nuclear matrix.

E. The Fibrous Lamina

In studies on animal nuclei a layer of finely fibrous material has been identified closely applied to the inside of the nuclear envelope. This has been called the fibrous lamina or zonula nucleum limitans (Patrizi and Poger, 1967). In detergent extracted preparations of nuclear envelope this fibrous lamina can be seen as a network of fibrils between the pores (Aaronson and Blobel, 1975). In plant cells there have not been clear demonstrations of such a structure though fibrous accumulations have been seen below the envelope in *Bryopsis* (Burr and West, 1971) and in *Spirogyra* it is possible to see a granule free zone of nucleoplasm in this region (E. G. Jordan, unpublished observations). The presence of darkly staining fibrils beneath the membrane after Bernhard's regressive staining technique indicate a nonchromatin layer in this position in plants.

ACKNOWLEDGMENTS

Unpublished work by A. Trewavas was carried out during receipt of a grant from the Science Research Council.

REFERENCES

Aaronson, R. P., and Blobel, G. (1975). *Proc. Natl. Acad. Sci. U.S.A.* **72**, 1002.
Abelson, H. T., and Smith, G. H. (1970). *J. Ultrastruct. Res.* **30**, 558.
Alfert, M., and Das, N. K. (1969). *Proc. Natl. Acad. Sci. U.S.A.* **63**, 123.
Amaldi, F., Lava-Sanchez, P. A., and Buongiorno-Nardelli, M. (1973). *Nature (London)* **242**, 615.

Anastassova-Kirsteva, M., and Nocoloff, H. (1978). *Biol. Zentralbl.* **97,** 83.

Ashburner, M. (1969). *Chromosomes Today* **2,** 99.

Ashley, T., and Wagenaar, E. B. (1974). *Can. J. Genet. Cytol.* **16,** 61.

Avanzi, S., Durante, M., Cionini, P. G., and D'Amato, F. (1972). *Chromosoma* **39,** 191.

Axel, R. (1976). *Prog. Nucleic Acid Res. Mol. Biol.* **19,** 355.

Axel, R., Cedar, H., and Felsenfeld, G. (1973). *Proc. Natl. Acad. Sci. U.S.A.* **70,** 3440.

Ayonadu, U., and Rees, H. (1968). *Genetika* **39,** 75.

Barlow, P. (1970). *Caryologia* **23,** 61.

Barlow, P. W., and Vosa, C. G. (1969). *Chromosoma* **28,** 457.

Barr, H. J., and Hildegard, E. (1965). *Exp. Cell Res.* **31,** 211.

Beermann, W. (1972). *In* ''Results and Problems in Cell Differentiation'' (W. Beermann, ed.),
    Vol. 4, p. 1. Springer-Verlag, Berlin and New York.

Bell, P. R. (1972). *J. Cell Sci.* **11,** 739.

Bell, P. R. (1975). *Endeavour* **34,** 19.

Bell, P. R. (1978). *J. Cell Sci.* **29,** 189.

Bennett, M. D. (1971). *Proc. R. Soc. London Ser. B.* **178,** 277.

Bennett, M. D. (1972). *Proc. R. London Ser. B.* **181,** 109.

Bennett, M. D. (1977). *Phil. Trans. R. Soc. London Ser. B.* **277,** 201.

Bennett, M. D., Chapman, V., and Riley, R. (1971). *Proc. R. Soc. London Ser. B.* **178,** 259.

Berezney, R., and Coffey, D. S. (1974a). *Biochem. Biophys. Res. Commun,* **60,** 1410.

Berezney, R., and Coffey, D. S. (1974b). *Proc. Natl. Acad. Sci. U.S.A.* **60,** 1410.

Berkofsky, J., and Roy, R. M. (1976). *Can. J. Bot.* **54,** 663.

Bernhard, W. (1966). *Natl. Cancer Inst. Monogr.* **23,** 13.

Bernhard, W. (1969). *J. Ultrastruct. Res.* **27,** 250.

Bernhard, W., Frayssinet, C., Lafarge, C., and Le Breton, E. (1965). *C. R. Hebd. Seances
    Acad. Sci.* **261,** 1785.

Birnstiel, M. (1967). *Annu. Rev. Plant Phys.* **18,** 25.

Birnsteil, M. L., Chipchase, M., and Spiers, J. (1971). *Progr. Nucleic Acid Res. Mol. Biol.* **11,**
    351.

Bonner, J. (1976). *In* ''Plant Biochemistry'' (J. Bonner and J. E. Varner, ed.), p. 38. Academic
    Press, New York.

Bonner, J., and Wu, J. R. (1973). *Proc. Natl. Acad. Sci. U.S.A.* **70,** 535.

Bonner, W. M. (1975). *J. Cell Biol.* **64,** 421.

Botchan, P., Reeder, R. H., and Dawid, I. B. (1977). *Cell* **11,** 599.

Bouck, G. B. (1971). *J. Cell Biol.* **50,** 362.

Bouteille, M., Laval, M., and Dupuy-Ioin, A. M. (1974). *In* ''The Cell Nucleus'' (H. Busch, ed.)
    Vol. 1, p. 3. Academic Press, New York.

Bradbury, E. M., Inglis, R. J., Matthews, H. R.,, and Sarner, N. (1973). *Eur. J. Biochem.* **33,**
    131.

Bradbury, E. M., Inglis, R. J., and Matthews, H. R. (1974). *Nature (London)* **247,** 257.

Brady, T. (1973). *Cell Dif.* **2,** 65.

Brasch, K. (1976). *Exp. Cell Res.* **101,** 396.

Britten, R. J., and Davidson, E. H. (1969). *Science* **165,** 349.

Britten, R. J., and Kohne, D. E. (1968). *Science,* **161,** 529.

Bryant, J. (1976). *In* ''Molecular Aspects of Gene Expression in Plants'' (J. Bryant, ed) p. 1.
    Academic Press, New York.

Buller, B. E., Toft, D. O., Schrader, W. T., and O'Malley, B. W. (1975). *J. Biol. Chem.* **250,** 801.

Burr, F. A., and Evert, R. F. (1973). *Protoplasma* **78,** 71.

Burr, F. A., and West, J. A. (1971). *J. Phycol.* **7,** 108.

Busch, H. (1976). *Perspect. Biol. Med.* **19,** 549.

Busch, H., and Smetana, K. (1970). "The Nucleolus." Academic Press, New York.

Butterfass, T. (1966). *Mitt. Max-Planck Inst. Ges.* **1**, 47.

Callan, H. G., and Lloyd, L. (1960). *Phil. Trans. R. Soc. London Ser. B.* **243**, 135.

Camefort, H. (1968). *C. R. Hebd. Seances Acad. Sci., Ser. D.* **261**, 4479.

Cameron, F. M., and Rees, H. (1967). *Heredity* **22**, 446.

Capesius, I., and Stohr, M. (1974). *Protoplasma* **82**, 147.

Chaly, N. M., and Setterfield, G. (1975). *Can. J. Bot.* **53**, 200.

Chapman, K., Trewavas, A., and Van Loon, K. (1975). *Plant Physiol.* **55**, 293.

Chooi, W. (1971). *Genetics* **68**, 213.

Chouinard, L. A. (1975). *J. Cell Sci.* **19**, 85.

Christiansen, C., Leth Bak, A., Standerup, A., and Christiansen, G. (1971). *Nature (London) New Biol.* **231**, 176.

Church, K., and Moens, P. B. (1976). *Chromosoma* **56**, 249.

Clay, W. F., Catterman, F. H. R., and Bartels, P. G. (1975). *Proc. Natl. Acad. Sci. U.S.A.* **72**, 3134.

Clay, W. F., Bartels, P. G., and Katterman, F. H. (1976). *Proc. Natl. Acad. Sci. U.S.A.* **73**, 3220.

Cohen, R. J., and Stein, G. S. (1975). *Exp. Cell Res.* **96**, 247.

Cole, G. T., and Wynne, M. J. (1973). *Cytobios.* **8**, 161.

Comings, D. E. (1968). *Am. J. Human Genet.* **20**, 440.

Comings, D. E. (1974). *In* "The Nucleus." (H. Busch, ed.), Vol. 1, p. 537. Academic Press, New York.

Comings, D. E. and Okada, T. A. (1970a). *Chromosoma* **30**, 269.

Comings, D. E. and Okada, T. A. (1970b). *Exp. Cell Res.* **63**, 62.

Comings, D. E. and Okada, T. A. (1976). *Exp. Cell Res.* **103**, 341.

Compton, J. L., Bellard, M., and Chambon, P. (1976). *Proc. Natl. Acad. Sci. U.S.A.* **73**, 4382.

Crick, F. (1971). *Nature (London)* **234**, 25.

Cullis, C. A. (1976). *Heredity* **36**, 73.

Cullis, C. A., and Schweizer, D. (1974). *Chromosoma* **44**, 23.

D'Amato, F. (1977). "Nuclear Cytology in Relation to Development." Dev. Cell Biol. Ser. 6. Cambridge Univ. Press, London and New York.

Darvey, N. L., and Driscoll, C. J. (1977). *Chromosoma* **36**, 131.

Das, N. K., Siegel, E. P., and Alfert, M. (1965). *Exp. Cell Res.* **40**, 178.

Davidson, D. (1964). *Exp. Cell Res.* **35**, 317.

De Mol, W. E. (1927). *Cellule* **36**, 1.

Dickenson, H. G., and Bell, P. R. (1972). *Dev. Biol.* **27**, 425.

Dickenson, H. G., and Heslop-Harrison, J. (1970). *Protoplasma* **69**, 187.

Douvas, A. S., and Bonner, J. (1977). *In* "Mechanisms and Control of Cell Division" (T. L. Rost and E. M. Gifford, eds.), p. 44. Dowden, Philadelphia, Pennsylvania.

Duda, C. T. (1976). *Annu. Rev. Plant Physiol.* **27**, 119.

Du Praw, E. J. (1965). *Proc. Natl. Acad. Sci. U.S.A.* **53**, 161.

Du Praw, E. J. (1970). "DNA and Chromosomes." Holt, New York.

Durrant, A. (1974). *Heredity,* **32**, 133.

Dusenberg, R. L. (1975). *Biochim. Biophys. Acta* **378**, 363.

Dutta, S. K. (1974). *Nucleic Acids Res.* **1**, 1411.

Elgin, S., and Weintraub, H. (1975). *Annu. Rev. Biochem.* **44**, 725.

Elsdale, T. R., Fischberg, M., and Smith, S. (1958). *Exp. Cell Res.* **14**, 642.

Endow, S. A., and Gall, J. G. (1975). *Chromosoma* **50**, 175.

Engelhardt, P., and Pusa, K. (1972). *Nature (London)* **240**, 163.

Erbrich, P. (1965). *Oesterr. Bot. Wochenbl.* **112**, 197.

Erdelska, O. (1973). *Protoplasma* **76,** 123.

Evans, H. J. (1961). *Genetics* **46,** 257.

Evans, H. J. (1964). *Exp. Cell Res.* **35,** 327.

Evans, G. M. (1968). *Heredity* **23,** 25.

Evans, G. M., and Macefield, A. J. (1972). *Nature (London) New Biol.* **236,** 110.

Evans, H. J., and Rees, H. (1966). *Exp. Cell Res.* **44,** 150.

Evans, L. S., and Van't Hof, J. (1974). *Exp. Cell Res.* **87,** 259.

Evans, L. C., and Van't Hof, J. (1975). *Am. J. Bot.* **62,** 1060.

Fabri, F., and Bonzi, L. M. (1975). *Caryologia* **28,** 548.

Fambrough, D., Fujimura, F., and Bonner, J. (1968). *Biochemistry* **7,** 575.

Fansler, B., and Loeb, L. A. (1972). *Exp. Cell Res.* **75,** 433.

Felden, R. A., Sanders, M. M., and Morris, N. R. (1976). *J. Cell Biol.* **68,** 43.

Feldherr, C. M. (1965). *J. Cell Biol.* **25,** 43.

Finch, J. T., and Klug, A. (1976). *Proc. Natl. Acad. Sci. U.S.A.* **73,** 1897.

Finch, J. T., Lutter, L. C., Rhodes, D., Brown, R. S., Rushton, B. R., Levitt, M., and Klug, A. (1977). *Nature (London)* **269,** 29.

Flavell, R. B. (1975). *In* ''Modification of the Information Content of Plant Cells'' (R. Markham, D. R. Davies, D. A. Hopwood, and R. W. Horne, eds.), p. 53. North-Holland/Am. Elsevier, Amsterdam.

Flavell, R. B., and Smith, D. B. (1976). *Heredity* **37,** 231.

Flavell, R. B., Bennett, M. D., Smith, J. B., and Smith, D. B. (1974). *Biochem. Genet.* **12,** 257.

Fox, D. P. (1971). *Chromosoma* **33,** 183.

Franke, W. W. (1974). *Int. Rev. Cytol. Suppl.* **4,** 71.

Franke, W. W., and Scheer, U. (1970). *J. Ultrastruct. Res.* **30,** 288.

Franke, W. W., and Scheer, U. (1974a). *In* ''The Cell Nucleus'' (H. Busch, ed.), Vol. 1, p. 219. Academic Press, New York.

Franke, W. W., and Scheer, U. (1974b). *Symp. Soc. Exp. Biol.* **28,** 249.

Franke, W. W., Trendelenburg, M. F., and Scheer, U. (1973). *Planta* **110,** 159.

Franke, W. W., Scheer, U., Trendelenburg, M. F., and Krohne, G. (1976a). *Exp. Cell Res.* **100,** 233.

Franke, W. W., Scheer, U., Trendelenburg, M. F., Spring, H., and Zentgraf, H. (1976b). *Cytobiologie* **13,** 401.

Frenster, J. H. (1969). *In* ''Handbook of Molecular Cytology'' (A. Lima-de-Faria ed.). p. 251–276. North-Holland Publ., Amsterdam.

Frenster, J. H. (1974). *In* ''The Nucleus'' (H. Busch ed.), Vol. 1, pp. 565–581. Academic Press, New York.

Fussell, C. P. (1975). *Chromosoma* **50,** 201.

Gall, J. G. (1967). *J. Cell Biol.* **32,** 391.

Gall, S. G. (1963). *Science* **139,** 320.

Gardener, J. M., and Kado, C. I. (1976). *Biochemistry* **15,** 688.

Garel, A., Zolan, M., and Acel, R. (1977). *Proc. Natl. Acad. Sci. U.S.A.* **74,** 4867.

Georgiev, G. P. (1969). *J. Theor. Biol.* **25,** 47.

Gerstel, D. U., and Burns, J. A. (1966). *Chromsomes Today.* **1,** 41.

Ghosh, S., and Roy, S. C. (1977). *Chromosoma* **61,** 49.

Gillies, C. B. (1973). *Chromosoma* **43,** 145.

Gillies, C. B., and Hyde, B. B. (1973). *Hereditas* **74,** 137.

Gilmour, S., and Paul, J. (1975). *In* ''Chromosomal Proteins and Their Role in the Regulation of Gene Expression'' (G. Stein and L. Kleinsmith, eds.), p. 19, Academic Press, New York.

Gimenez-Martin, G., Delatorre, C., Lopez-Saez, J. F., and Esponda, P. (1977). *Cytobiologie* **14,** 421.

Godin, D. E., and Stack, S. M. (1976). *Chromosoma* **57,** 309.

Godward, M. B. E. (1950). *Ann. Bot.* (*London*) **14,** 39.

Goff, C. C. (1976). *J. Biol. Chem.* **251,** 4131.

Gofshtein, L. U., Yurina, N. P., Romashkin, V. I., and Oparin, A. I. (1975). *Biokhimiya* **40,** 1104.

Gorovsky, M. A., and Woodward, J. (1976). *J. Cell Biol.* **33,** 723.

Gottesfeld, J., Kent, D., Ross, M., and Bonner, J. (1975). *In* "Chromosomal Proteins and Their Role in the Regulation of Gene Expression" (G. Stein and L. Kleinsmith, eds.), p. 227. Academic Press, New York.

Grainger, P. M., and Ogle, R. C. (1978). *Chromosoma* **65,** 115.

Grellet, F., Elseny, M. D., and Guitton, Y. (1977). *Nature* (*London*) **267,** 724.

Grierson, D. (1977). *In* "The Molecular Biology of Plant Cells" (H. Smith, ed.) Bot. Monog. 14, p. 213. Blackwell, Oxford.

Grossbach, U. (1973). *Cold Spring Harbor Symp. Quant. Biol.* **38,** 619.

Gurdon, J. B. (1974). "The Control of Gene Expression in Animal Development. Oxford Univ. Press, London and New York.

Hamkalo, B. A., and Miller, O. L. (1973). *Ann. Rev. Biochem.* **42,** 379.

Harris, B., and Dure, L. (1976). *Prog. Nucleic Acid Res. Mol. Biol.* **19,** 113.

Harris, H. (1970). "Nucleus and Cytoplasm," 2nd ed. Oxford Univ. Press (Clarendon), London and New York.

Hecht, N. B., and Stern, H. (1969). *J. Cell Biol.* **43,** 51a.

Heitz, E. (1931). *Planta* **15,** 495.

Hesse, M. (1969). *Cesterr. Bot.* **2, 115,** 34.

Hewish, D. R., and Burgoyne, L. A. (1973). *Biochem. Biophys. Res. Commun.* **52,** 504.

Heywood, P. (1976). *Cytobios* **66,** 79.

Holmes, D. S., and Bonner, J. (1974). *Proc. Natl. Acad. Sci. U.S.A.* **71,** 1108.

Horgen, P. A., Nagao, R. T., Chia, L. S. Y., and Key, J. L. (1973). *Arch. Mikrobiol.* **94,** 249.

Huang, R. G., and Bonner, J. (1962). *Proc. Natl. Acad. Sci. U.S.A.* **48,** 1216.

Huberman, J. A., and Riggs, A. D. (1968). *J. Mol. Biol.* **32,** 327.

Hyde, B. B. (1966). *Natl. Cancer Inst. Monogr.* **23,** 13.

Ingle, J., and Sinclair, J. (1972). *Nature* (*London*) **235,** 30.

Ingle, J., Pearson, G. G., and Sinclair, J. (1973). *Nature* (*London*) **242,** 193.

Ingle, J., Timmis, J. N., and Sinclair, J. (1975). *Plant Physiol.* **55,** 496.

Jamrich, M. J., Greenleaf, A. L., and Bautz, E. K. F. (1977). *Proc. Natl. Acad. Sci. U.S.A.* **74,** 2079.

Jakob, K. M. (1972). *Exp. Cell Res.* **72,** 370.

Jeter, J. R., and Cameron, I. L. (1974). *In* "Acidic Proteins of the Nucleus" (I. L. Camerson and J. R. Jeter, eds.), p. 213. Academic Press, New York.

Johnson, E. M., Karn, J., and Allfrey, V. G. (1974). *J. Biol. Chem.* **249,** 4990.

Johnson, M. J. (1967). *Am. J. Bot.* **54,** 189.

Johnson, M. J. (1969). *J. Cell Biol.* **43,** 197.

Jones, K. W. (1970). *Nature* (*London*) **225,** 912.

Jones, G. H. (1971). *Chromosoma* **34,** 367.

Jordan, E. G. (1974). *Protoplasma* **79,** 31.

Jordan, E. G. (1976). *Cytobiologie* **14,** 171.

Jordan, E. G., and Chapman, J. M. (1971). *J. Exp. Bot.* **22,** 627.

Jordan, E. G., and Chapman, J. M. (1973). *J. Exp. Bot.* **24,** 197.

Jordan, E. G., and Godward, M. B. E. (1969). *J. Cell. Sci.* **4,** 15.

Jordan, E. G., and Luck, B. T. (1976). *J. Cell Sci.* **22,** 75.

Jordan, E. G., Severs, N. J., and Williamson, D. H. (1976). *In* "Progress in Differentiation Research" (N. Müller-Berat, ed.), p. 77. North-Holland Publ., Amsterdam.

Jordan, E. G., Severs, N. J., and Williamson, D. H. (1977). *Exp. Cell Res.* **104,** 446.

Judd, B. H., and Young, M. W. (1973). *Cold Spring Harbor Symp. Quant. Biol.* **38,** 573.

Kanazawa, T., and Kanazawa, K. (1968). *Plant Cell Physiol.* **9,** 701.

Kay, R. R., and Johnston, I. R. (1973). *Sub-Cell Biochem.* **2,** 127.

Kessel, R. G. (1973). *Prog. Surf. Membr. Sci.* **6,** 243.

Kirschner, R. H., Rusli, M., and Martin, T. E. (1977). *J. Cell Biol.* **72,** 118.

Kitani, Y. (1963). *Jpn. J. Gen.* **38,** 244.

Kleinsmith, L. (1975). *In* "Chromosomal Proteins and Their Role in the Regulation of Gene Expression" (G. Stein and L. Kleinsmith, eds.), p. 45. Academic Press, New York.

Kornberg, R. (1977). *Annu. Rev. Biochem.* **46,** 931.

Kornberg, R. D., and Thomas, J. O. (1974). *Science* **184,** 865.

Krohne, G., Franke, W. W., and Scheer, U. (1978). *Exp. Cell Res.* **116,** 85.

Krutrachue, M., and Evert, R. F. (1978). *Ann. Bot. (London)* **42,** 15.

Kuehl, L. (1974). *In* "The Nucleus" (H. Busch, ed.), Vol. 3, p. 345. Academic Press, New York.

La Cour, L. F., and Crawley, J. W. C. (1965). *Chromosoma* **16,** 124.

La Cour, L. F., and Wells, B. (1967). *Z. Zellforsch. Mikrosk. Anat.* **82,** 25.

La Cour, L. F., and Wells, B. (1972). *Z. Zellforsch. Mikrosk. Anat.* **123,** 178.

La Cour, L. F., and Wells, B. (1974). *Phil. Trans. R. Soc. London Ser. B.* **268,** 95.

La Cour, L. F., and Wells, B. (1977). *Phil. Trans. R. Soc. London Ser. B.* **277,** 259.

Lafontaine, J. G. (1958). *J. Biophys. Biochem. Cytol.* **4,** 777.

Lafontaine, J. G. (1965). *J. Cell Biol.* **26,** 1.

Lafontaine, J. G. (1974a). *In* "Dynamic Aspects of Plant Ultrastructure" (A. W. Robards, ed.), p. 1. McGraw-Hill, New York.

Lafontaine, J. G. (1974b). *In* "The Cell Nucleus" (H. Busch, ed.), Vol. 1, p. 149. Academic Press, New York.

Lafontaine, J. G., and Chouinard, L. A. (1963). *J. Cell Biol.* **17,** 167.

Lafontaine, J. G., and Lord, A. (1969). *In* "Handbook of Molecular Cytology" (A. Lima-de-Faria, ed.), p. 381. North-Holland Publ., Amsterdam.

Lafontaine, J. G., and Lord, A. (1973). *J. Cell Sci.* **12,** 369.

Lafontaine, J. G., Luck, B. T., and Dontigny, D. (1979). *J. Cell Science* **39,** 13.

Laird, C. D., and Chooi, W. Y. (1976). *Chromosoma* **58,** 193.

Lefevre, G. (1973). *Cold Spring Harbor Symp. Quant. Biol.* **38,** 591.

Leighton, T. J., Dill, B. C., Stock, J. J., and Philips, C. (1971). *Proc. Natl. Acad. Sci. U.S.A.* **68,** 677.

Leighton, T. J., Leighton, F., Dill, B., and Stock, J. (1976). *Biochim. Biophys. Acta* **432,** 381.

Liau, M. C., and Perry, R. P. (1969). *J. Cell Biol.* **42,** 272.

Lin, P. P. C., and Key, J. L. (1976). *Biochem. Biophys. Res. Commun.* **73,** 396.

Loewenstein, W. (1964). *Protoplasmatologia* **5,** 26.

Lohr, D., Corden, J., Tatchell, K., Kovacic, R. T., and Van Holde, K. E. (1977). *Proc. Natl. Acad. Sci. U.S.A.* **74,** 79.

Longwell, A. C., and Svihla, G. (1960). *Exp. Cell Res.* **20,** 294.

Luck, B. T., and Jordan, E. G. (1977). *J. Cell Sci.* **25,** 111.

McClintock, B. (1934). *Z. Zellforsch. Mikrosk. Anat.* **21,** 294.

McGhee, J. D., and Engel, J. D. (1975). *Nature (London)* **254,** 449.

McKnight, G. L., and Miller, D. L. (1976). *Cell* **8,** 305.

McKnight, S. L., Sullivan, N. L., and Miller, O. L. (1976). *Prog. Nucleic Acid Res. Mol. Biol.* **19,** 313.

McLennan, A. G., and Keir, H. M. (1975). *Biochem. J.* **151,** 227.

Maguire, M. (1967). *Chromosoma* **21,** 221.

Makino, F., and Tsuzuki, J. (1971). *Nature (London)* **231,** 446.

Mandel, J. L., Breathnach, R., Gerlinger, P., Lemeur, M., Gannon, F. and Chambon, P. (1978). *Cell* **14,** 641.

Marinozzi, V. (1963). *J. Prog. Microsc. Soc.* **81**, 141.

Marushige, K., and Bonner, J. (1971). *Proc. Natl. Acad. Sci. U.S.A.* **68**, 2941.

Maul, G. G. (1971). *J. Cell Biol.* **51**, 558.

Maul, G. G. (1977). *Int. Rev. Cytol. Suppl.* **5**, 73.

Maul, G. G., Price, J. W., and Lieberman, M. W. (1971). *J. Cell Biol.* **51**, 405.

Maul, G. G., Maul, H. M., Scogna, J. E., Lieberman, M. W., Stein, G. S., Hsu, B. Y., and Borun, T. W. (1972). *J. Cell Biol.* **55**, 433.

Mepham, R. H., and Lane, G. R. (1969). *Nature (London)* **221**, 288.

Merriam, R. W. (1969). *J. Cell Sci.* **5**, 333.

Miller, O. L. (1966). *Natl. Cancer Inst. Monogr.* **23**, 53.

Miller, O. L., and Bakken, A. H. (1972). *Acta. Endocrinol. Suppl.* **168**, 155.

Miller, O. L., and Beatty, B. R. (1969). *In* "Handbook of Molecular Cytology" (A. Lima-de-Faria, ed.), p. 605. North-Holland Publ., Amsterdam.

Miller, O. L., and Hamkalo, B. A. (1972). *Int. Rev. Cytol.* **33**, 1.

Miller, O. L., Beatty, B. R., Hamkalo, B. A., and Thomas, C. A. (1970). *Cold Spring Harbor Symp. Quant. Biol.* **35**, 505.

Mirre, C., and Stahl, A. (1976). *J. Ultrastruct. Res.* **56**, 186.

Moens, P. B. (1969). *Chromosoma* **28**, 1.

Mohan, J., and Flavell, R. B. (1974). *Genetics* **76**, 33–44.

Monneron, A., and Bernhard, W. (1969). *J. Ultrastruct. Res.* **27**, 266.

Morris, N. R. (1976). *Cell* **8**, 357.

Moses, M. J. (1956). *J. Biophys. Biochem. Cytol.* **2**, 215.

Moyne, G., Bertaux, O., and Puvion, E. (1975). *J. Ultrastruct. Res.* **52**, 362.

Musich, P. R., Maio, J. J., and Brown, F. L. (1977). *J. Mol. Biol.* **117**, 657.

Nadeau, P., Pallotta, D. and Lafontaine, J. G. (1974). *Arch. Biochem. Biophys.* **161**, 171.

Nadler, K. D. (1976). *Exp. Cell Res.* **101**, 283.

Nagl, W. (1970). *J. Cell Sci.* **6**, 87.

Nagl, W. (1972). *Cytobios* **5**, 145.

Nagl, W. (1974). *Z. Pflanzenphysiol.* **73**, 1.

Nagl, W. (1976a). *Nature (London)* **261**, 614.

Nagl, W. (1976b). *Annu. Rev. Plant Physiol.* **27**, 39.

Nagl, W. (1977). *In* "Mechanisms and Control of Cell Division" (T. L. Rost and E. M. Gifford, eds.), p. 166. Dowden, Philadelphia, Pennsylvania.

Narayan, K. S., Steele, W. J., Smetana, K., and Busch, H. (1967). *Exp. Cell Res.* **46**, 65.

Narayan, R. K. J., and Rees, H. (1976). *Chromosoma* **54**, 141.

Nawaschin, M. (1934). *Cytologia* **5**, 169.

Netrawali, M. S. (1970). *Exp. Cell Res.* **63**, 422.

Noll, M. (1976). *Cell* **8**, 349.

Northcote, D. H., and Lewis, D. R. (1968). *J. Cell Sci.* **3**, 199.

Nucleolus Nomenclature Committee (1966). *Natl. Cancer Inst. Monogr.* **23**, 573.

Okita, K., and Zardi, L. (1974). *Exp. Cell Res.* **86**, 59.

Olins, A. L., and Olins, D. E. (1974). *Science* **183**, 330.

O'Malley, B. W., and Means, A. R. (1976). *Prog. Nucleic Acid Res. Mol. Biol.* **19**, 403.

Oudet, D., Gross-Bellard, M., and Chambon, P. (1975). *Cell* **4**, 281.

Panyim, S., and Chalkley, R. (1971). *J. Biol. Chem.* **246**, 755.

Pardue, M. L., and Gall, J. G. (1970). *Science* **168**, 1356.

Patterson, B. D., and Davies, D. D. (1969). *Biochem. Biophys. Res. Commun.* **34**, 791.

Patrizi, G., and Poger, M. (1967). *J. Ultrastruct. Res.* **17**, 127.

Paul, J. (1972). *Nature (London)* **238**, 444.

Peaud-Lenoel, C. (1977). *In* "Plant Growth Regulation" (P. E. Pilet, ed.), p. 240. Springer-Verlag, Berlin and New York.

Pearson, G. G., Timmis, J. N., and Ingle, J. (1974). *Chromosoma* **45**, 281.

Perry, R. P. (1969). *In* "Handbook of Molecular Cytology" (A. Lima-de-Faria, ed.), p. 620. North-Holland Publ., Amsterdam.

Perry, R. P., Bard, E., Hames, B. D., Kelley, D. E., and Schibler, U. (1976). *Prog. Nucleic Acid Res. Mol. Biol.* **19,** 275.

Peterson, J. L., and McConkey, E. H. (1976). *J. Biol. Chem.* **251,** 548.

Phillips, R. L. (1976). *In* "Maize Breeding and Genetics" (D. B. Walden, ed.), p. 221. Wiley, New York.

Phillips, R. L., Kleeze, R. A., and Wang, S. S. (1971). *Chromosoma* **36,** 79.

Phillips, R. L., Weber, D. F., Kleeze, R. A., and Wang, S. S. (1974). *Genetics* **77,** 285.

Platz, R. D., Grimes, S. R., Hord, G., Meistrich, M. L., and Anilica, L. S. (1975). *In* "Chromosomal Proteins and Their Role in the Regulation of Gene Expression" (G. S. Stein and L. Kleinsmith, eds.), p. 67, Academic Press, New York.

Porter, K., and Macado, R. D. (1960). *J. Biophys. Biochem. Cytol.* **7,** 167.

Pusa, K. (1963). *Proc. Int. Cong. Genet. 11th,* p. 1.

Puvion- Dutilleul, F., Bachellerie, J. P. Zalta, J. P., and Bernhard, W. (1977). *Biol. Cell.* **30,** 183.

Quatrano, R. (1972). *Exp. Cell Res.* **70,** 1.

Rao, R. N., and Nataratan, A. T. (1967). *Genetics* **57,** 821.

Rees, H., and Hutchinson, J. (1973). *Cold Spring Harbor Symp. Quant. Biol.* **38,** 175.

Reeve, R. (1977). *Eur. J. Biochem.* **75,** 545.

Riley, D. E., Keller, J. M., and Byers, B. (1975). *Biochemistry* **14,** 3005.

Rittossa, F. M., and Spiegelman, S. (1965). *Proc. Natl. Acad. Sci. U.S.A.* **53,** 737.

Rizzo, P. J., and Nooden, L. D. (1974). *Biochim. Biophys. Acta* **249,** 415.

Roberts, K., and Northcote, D. H. (1970). *Nature (London)* **228,** 385.

Roberts, K., and Northcote, D. H. (1971). *Microsc. Acta* **71,** 102.

Rose, R. J., Setterfield, G., and Fowke, L. C. (1972). *Exp. Cell Res.* **71,** 1.

Rost, T. L., and Gifford, E. M., eds. (1977). "Mechanisms and Control of Cell Division." Dowden, Philadelphia, Pennsylvania.

Rothwell, N. V. (1964). *Am. J. Bot.* **51,** 172.

Sakamaki, T., Takahashi, N., Takaiwa, F., and Tanifuji, S. (1976). *Biochim. Biophys. Acta* **447,** 76.

Sax, K. (1940). *Genetics* **25,** 41.

Scheer, U., Trendelenburg, M. F., and Franke, W. W. (1973). *Exp. Cell Res.* **80,** 175.

Scheer, U., Kartenbeck, J., Trendelenburg, M. F., Stadler, B. J., and Franke, W. W. (1976). *J. Cell Biol.* **69,** 1.

Scheer, U., Trendelenburg, M. F., Krohne, G., and Franke, W. (1977). *Chromosoma* **60,** 147.

Schweizer, D., and Nagl, W. (1976). *Exp. Cell Res.* **98,** 411.

Scott, R. E., Carter, R. L., and Kidwell, W. R. (1971). *Nature (London) New Biol.* **233,** 219.

Seale, R. L., and Aaronson, A. I. (1974). *J. Mol. Biol.* **75,** 633.

Senior, M. B., and Frankel, F. R. (1978). *Cell* **14,** 857.

Severs, N. J. (1976). *Cytobios* **16,** 125.

Severs, N. J., and Jordan, E. G. (1975a). *J. Ultrastruct. Res.* **52,** 85.

Severs, N. J., and Jordan, E. G. (1975b). *Experientia* **31,** 1276.

Severs, N. J., and Jordan, E. G. (1978). *Experientia* **34,** 1007.

Severs, N. J., Jordan, E. G., and Williamson, D. H. (1976). *U. Ultrastruct. Res.* **54,** 374.

Siegel, A. (1974). *In* "Modification of the Information Content of Plant Cells" (R. Markham, D. R. Davies, D. A. Hopwood, and R. W. Horne, eds.), p. 15. North-Holland/Am. Elsevier, Amsterdam.

Sinha, N. K., and Snustad, D. P. (1971). *J. Mol. Biol.* **62,** 267.

Smetana, K., and Busch, H. (1974). *In* "The Cell Nucleus" (H. Busch, ed.), Vol. 1, p. 75. Academic Press, New York.

Smetana, K., Steele, W. J., and Busch, H. (1963). *Exp. Cell Res.* **31,** 198.

Smetana, K., Shankar Napayan, K., and Busch, H. (1966). *Cancer Res.* **26,** 286.

Smith, D. B., and Flavell, R. B. (1975). *Chromosoma* **50**, 223.

Solari, A. J. (1974). *In* "The Nucleus" (H. Busch, ed.), Vol. 1, p. 493. Academic Press, New York.

Sommer, K. R., and Chalkley, R. (1974). *Biochemistry* **13**, 1022.

Southern, E. M. (1970). *Nature (London)* **227**, 794.

Spelsberg, T. C., Webster, R., and Pikler, F. M. (1975). *In* "Chromosomal Proteins and Their Role in the Regulation of Gene Expression" (G. Stein and L. Kleinsmith, eds.), p. 153. Academic Press, New York.

Speth, V., and Wunderlich, F. (1970). *J. Cell Biol.* **47**, 772.

Spiker, S. (1975). *Biochim. Biophys. Acta* **400**, 461.

Spiker, S., and Isenberg, I. (1977). *Biochemistry* **16**, 1819.

Spirin, A. G. (1969). *Eur. J. Biochem.* **10**, 20.

Spring, H., Trendelenburg, M. F., Scheer, U., Franke, W. W., and Herth, W. (1974). *Cytobiologie* **10**, 1.

Spring, H., Krohne, G., Franke, W. W., Scheer, U., and Trendelenburg, F. (1976). *J. Microsc. Biol. Cell.* **25**, 107.

Steele, W. J., and Busch, H. (1966). *Biochim. Biophys. Acta* **129**, 54.

Stein, G. S., Spelsberg, T. C., and Kleinsmith, L. J. (1974). *Science* **183**, 817.

Stein, G.. Stein, J., Thrall, C., and Park, W. (1975). *In* "Chromosomal Proteins and Their Role in the Regulation of Gene Expression" (G. Stein and L. Kleinsmith, eds.), p. 1. Academic Press, New York.

Stern, H., and Hotta, Y. (1978). *Phil. Trans. R. Soc. London Ser. B.* **277**, 277.

Stevens, B. J., and Swift, H. (1966). *J. Cell Biol.* **31**, 55.

Storey, W. B. (1968a). *Science* **159**, 648.

Storey, W. B. (1968b). *J. Heredity* **29**, 51.

Summers, M. D. (1969). *J. Virol.* **4**, 188.

Summers, M. D. (1971). *J. Ultrastruct. Res.* **35**, 606.

Swift, H. (1958). *In* "A Symposium of the Chemical Basis of Development" (W. D. McElroy and B. Glass, eds.), p. 174. John Hopkins Press, Baltimore, Maryland.

Tarragó-Litvak, L., Castroviejo, M., and Litvak, S. (1975) *FEBS Lett.* **59**, 125.

Tata, J. R., and Baker, B. (1978). *J. Mol. Biol.* **118**, 249.

Taylor, J. H. (1958). *Exp. Cell Res.* **15**, 350.

Tegtmeyer, P., Rundell, K., and Collins, J. (1977). *J. Virol.* **21**, 647.

Thair, B. W., and Wardrop, A. B. (1971). *Planta* **100**, 1.

Thoma, F., and Koller, T. H. (1977). *Cell* **12**, 101.

Thomas, J. O., and Furber, V. (1976). *FEBS Lett.* **66**, 274.

Timmis, J. N., and Ingle, J. (1973). *Nature (London) New Biol.* **244**, 235.

Timmis, J. N., and Ingle, J. (1975a). *Biochem. Genet,* **13**, 629.

Timmis, J. N., and Ingle, J. (1975b). *Plant Physiol.* **56**, 255.

Timmis, J. N., and Ingle, J. (1977). *Biochem. Genet.* **15**, 1159.

Timmis, J. N., Sinclair, J., and Ingle, J. (1972). *Cell Dif.* **1**, 335.

Timmis, J. N., Deumling, B., and Ingle, J. (1975). *Nature (London)* **257**, 306.

Trendelenburg, M. F., Scheer, U., and Franke, W. W. (1973). *Nature (London) New Biol.* **245**, 167.

Trendelenburg, M. F., Spring, H., Scheer, U., and Franke, W. W. (1974). *Proc. Natl. Acad. Sci. U.S.A.* **71**, 3626.

Trewavas, A. J. (1976a). *Annu. Rev. Plant Physiol.* **27**, 349.

Trewavas, A. J. (1976b). *Phytochem.* **15**, 363.

Trewavas, A. J. (1979). *Symp. Phytochem. Soc.* **16**, 175.

Trewavas, A., and Stratton, B. R. (1976). *In* "Nucleic Acids and Protein Synthesis in Plants" (L. Bogorad and J. Weil, eds.), p. 309. Plenum, New York.

Tsai, G. Y., Harris, G. E., Tsai, M. J., and O'Malley, B. W. (1976). *J. Biol. Chem.* **251**, 4713.

Van Der Lyn, L. (1948). *Bot. Rev.* **14**, 270.
Van Loon, K., Trewavas, A., and Chapman, K. (1975). *Plant Physiol.* **55**, 288.
Van't Hof, J. (1963). *Cytologia* **28**, 30.
Van't Hof, J. (1975). *Exp. Cell Res.* **93**, 95.
Van't Hof, J. (1976a). *Exp. Cell Res.* **99**, 47.
Van't Hof, J. (1976b). *Exp. Cell Res.* **103**, 395.
Van't Hof, J., and Sparrow, A. H. (1963). *Proc. Natl. Acad. Sci. U.S.A.* **49**, 897.
Van Wisselingh, C. (1898). *Bot. Z.* **56**, 197.
Vincent, W. S., and Miller, O. L. (1966). *Natl. Cancer Inst. Monogr.* **23**.
Vivier, E. (1967). *J. Microsc. (Paris)* **6**, 371.
Vogel, F., and Schroeder, T. M. (1974). *Human genetik* **25**, 267.
Vogt, and Braun (1976). *J. Mol. Biol.* **106**, 567.
Vosa, C. G. (1973). *In* "Chromosome Identification" (T. Casperson and L. Zech, eds.), Nobel
     Symp. 23, p. 152. Academic Press, New York.
Vosa, C. G., and Barlow, P. W. (1972). *Caryologia* **25**, 1–8.
Wallace, H., and Birnsteil, M. L. (1966). *Biochem. Biophys. Acta.* **114**, 296.
Weintraub, H., and Groudine, M. (1976). *Science* **193**, 848.
Werry, P. A. Th. J., Stoffelsen, K., Engles, F. M., Van Der Lan, F., and Spanjers, A. W. (1977).
     Chromosoma **62**, 93.
Westergaard, M., and Von Wettstein, D. (1972). *Annu. Rev. Genet.* **6**, 71.
Williams, E., Heslop-Harrison, J., and Dickenson, H. G. (1973). *Protoplasma* **77**, 79.
Willison, J. H. M., and Rajaraman, R. (1977). *J. Microsc.* **109**, 183.
Wimalaratna, S. D. (1976). *Cytobiologie* **12**, 189.
Wimber, D. E. (1961). *Exp. Cell Res.* **23**, 402.
Wolff, S. (1959). *Radiat. Res. Suppl.* **1**, 453.
Wolff, S. (1968). *J. Cell Biol.* **37**, 610.
Woodcock, C. L. F., Stanchfield, J. E., and Gould, R. R. (1975). *Plant Sci. Lett.* **4**, 17.
Woodcock, C. L. F., Safer, J. P., and Stanchfield, J. E. (1976). *Exp. Cell Res.* **97**, 101.
Woodward, J., Rasche, E., and Swift, H. (1961). *J. Biophys. Biochem. Cytol.* **9**, 445.
Wunderlich, F., Berezney, R., and Kleinig. H. (1976). *In* "Biological Membranes" (D. Chap-
     man and D. F. H. Wallach, eds.), p. 241. Academic Press, New York.
Yeoman, M. M., and Aitchison, P. A. (1975). *In* "Cell Division in Higher Plants" (M. M.
     Yeoman, ed.), p. 111. Academic Press, London.
Zentgraf, H., Falk, H., and Franke, W. W. (1975). *Cytobiologie* **11**, 10.
Zhimulev, I. F., and Belyaeva, E. S. (1975). *Chromosoma* **49**, 219.

Protein Bodies | *14*

JOHN N.A. LOTT

I. INTRODUCTION

Protein deposits, either free in the cytoplasm or enclosed by a bounding membrane, have been reported in a variety of plant tissues. Protein storage deposits in seed tissues have been of greatest interest to scientists and thus the bulk of scientific literature on plant protein deposits relates to seed protein bodies. The main reason behind the considerable interest in seed protein bodies is the fact that protein bodies are of great importance as a source of food for mankind and many other animals. Protein bodies of seeds are also of great importance to most angiosperms and gymnosperms because protein bodies provide nitrogenous compounds and minerals necessary for establishment of seedling plants. While this chapter will devote itself to a consideration of seed protein bodies, readers should remember that proteinaceous

The Biochemistry of Plants, Vol. 1

deposits do occur in nonseed tissues, including roots (Rothwell, 1966; R. L. Peterson, personal communication) leaves (Shumway *et al.*, 1970), stems (Kidwai and Robards, 1969; Shumway *et al.*, 1972) and flowers (Shumway *et al.*, 1972).

Seed protein bodies are single membrane-bound subcellular structures that are rich in protein and are often approximately spherical in shape. Protein bodies have been reported to vary in size from 0.1 μm to 22 μm in diameter (Ashton, 1976) and are known to occur in a variety of seed tissues irrespective of ploidy level. In conifer seeds the haploid megagametophyte tissue may contain protein bodies. The diploid perisperm tissue of some angiosperm seeds may contain protein bodies as do many embryo tissues. In most dicotyledonous plants, the bulk of the seed's protein storage is contained in the cotyledons. Polyploid endosperm tissues, which are especially common in monocotyledonous plants, contain protein bodies. In cereals the aleurone cells of the endosperm contain prominent protein bodies. Since protein bodies may be considered a specialization of the plant cell's vacuole system readers interested in protein bodies may also wish to read Chapter 15 in this volume, Matile and Wiemken (1976), and portions of Matile (1975).

II. PROTEIN BODY STRUCTURE

This chapter employs the terminology outlined in Ashton (1976), Lott *et al.* (1971), and Lott and Buttrose (1978b) and uses the term ''protein body'' rather than ''aleurone grain.'' While some workers (Altschul *et al.*, 1966) prefer the term aleurone grain, that term is increasingly being abandoned. The more general term protein body seems preferable to the term aleurone grain, especially when one is dealing with protein-rich particles located in seed tissues other than the aleurone layer of monocot seeds. Within a protein body, proteinaceous material that is structurally amorphous will be termed proteinaceous matrix. A protein body region that consists of protein in a definite crystalline lattice arrangement is termed a protein crystalloid. Inclusions, which are often approximately spherical in shape, are called globoids. In some cases globoid regions may be further subdivided into regions that are very electron-dense and regions that are electron-transparent. The electron-dense portion of a globoid is called a globoid crystal while the electron-transparent portion is called a soft globoid. Both globoids and protein crystalloids are surrounded by proteinaceous matrix material.

Protein bodies in seeds appear to have a limiting membrane. However, this membrane is sometimes difficult to observe clearly in electron micrographs of mature seed tissue. Evidence from developing seeds and germinating seeds supports the concept that each protein body is separated from other cell constituents by a membrane. Protein crystalloid regions do not appear to be surrounded by a membrane. While most workers have not observed the presence of a membrane around globoids there are some reports to that effect. Where chemical fixation is involved it may be difficult to determine if a stain deposit occurs because of the presence of a membrane or whether it occurs at the boundary of two chemically different regions. Freeze–fracture studies of barley (Buttrose, 1971) and wheat (Swift and Buttrose, 1972) support the presence of a membrane around the globoid in these species. Particle arrays, indicative of a globoid envelope, were found in wheat.

Protein bodies differ in structural complexity. Some differences in structure may be found within one cell or within different regions of a seed. Considerable differences in protein body structure may occur between species in different families (see Table I). Classification of protein bodies into different structural types provides a useful framework for considering the protein body system. Such a classification, as begun by Rost (1972) and extended by Lott and Buttrose (1978b), will be used in this chapter.

The simplest structural arrangement possible for seed protein bodies consists of proteinaceous matrix material surrounded by a limiting membrane (Fig. 1a). The proteinaceous matrix is structurally homogenous both in freeze–fracture preparations and in sections of well fixed tissue. Protein bodies of legumes are often thought to have this simple structural arrangement. New studies plus a review of the literature (see Lott and Buttrose, 1978a) indicate that this generality with regard to legume protein bodies is untrue.

A second structural arrangement possible in protein bodies consists of one to many globoids surrounded by proteinaceous matrix material. The globoids, which often are only electron-dense globoid crystal material, vary both in number and size (Figs. 1b–d, 2–7). At present there is no explanation as to why some species have a few large globoid crystals in their protein bodies (e.g., *Cassia* in Figs. 6 and 7) while the protein bodies of other species contain numerous smaller globoid crystals (e.g., soybean and sunflower in Figs. 2–5). The fraction of the total protein body volume occupied by globoid crystals clearly varies from species to species. In addition to situations where the globoids are entirely globoid crystal material there are situations where electron-transparent regions are evident. Whether these electron-transparent regions are artifacts is not certain. Further discussion of methodological problems is given in Section VI.

A third structural type of seed protein body contains one or more globoids, one or more protein crystalloids, and proteinaceous matrix (Figs. 1e–g, 8–11). In section the edges of the protein crystalloids may be angular (1e),

TABLE I

Electron Microscopic Studies of Seed Protein Bodies[a]

Family	Plant	Method	Protein crystalloid	Globoid crystal	Source of micrographs
A. Class Coniferopsida					
Pinaceae	*Pinus sylvestris*	T	++, M	++, M to L	Simola, 1974
	Pinus banksiana	T	++, L	++, M to L	Durzan et al., 1971
B. Class Angiospermopsida, Subclass Dicotyledonae					
Anacardiaceae	*Anacardium occidentale*	F,T,X	–	++, S to M	Buttrose and Lott, 1978a
	Pistacia vera	F,T,X	–	++, S	Buttrose and Lott, 1978a
Betulaceae	*Corylus avellana*	F,T,X	+, S to M	++, S to L	Lott and Buttrose, 1978c
Buxaceae[b]	*Simmondsia chinensis* =(*S. californica*)	F,T,X	–	++, S to L	Buttrose and Lott 1978b; Muller et al. 1975
Cannabinaceae	*Cannabis sativa*	T	++, L	++, S?	St. Angelo et al., 1968
Chenopodiaceae	*Kochia childsii*	T	–	++, S to L	Marin, 1975
Compositae	*Bidens cernua*	T	–	++?, S to M	Simola, 1973
	Bidens radiata	T	–	++?, S to M	Simola, 1969
	Helianthus annuus	F,T,X	–	++, S to M	Gruber et al., 1970; Buttrose and Lott, 1978a
	Helichrysum bracteatum	T,X	–	++, S to L	Buttrose and Lott, 1978a
	Lactuca sativa	T	–	++, S to M	Paulson and Srivastava, 1968
Cruciferae	*Capsella bursa-pastoris*	T	–	++, S to M	Dieckert and Dieckert, 1976a
	Crambe abyssinica	T,X	–	++, S to L	Hofsten, 1973; Smith, 1974
	Sinapsis alba =(*Brassica hirta*)	T	–	++, M to L	Kirk and Pyliotis, 1976; Rest and Vaughan, 1972; Werker and Vaughan, 1974
	Brassica campestris	T	–	++, S to L	Hofsten 1970, 1974
	Brassica napus	T	–	++, S to L	Hofsten 1970, 1974
Cucurbitaceae	*Apodanthera undulata*	T	++, L	+?	Hensarling et al., 1974
	Cucumis sativus	T	++, L	++, L	Poux, 1965
	Cucurbita maxima	T,F,X	++, L	++, L	Lott, 1975; Lott and Vollmer, 1973a,b; Lott et al., 1971, 1978b

Family	Species				Reference
	Cucurbita foetidissima	T	++, L	+?	Hensarling et al., 1974
	Cucurbita pepo	T	++, L	++, L	Hensarling et al., 1974
	Cucurbita palmata	T	+?	+?	Hensarling et al., 1974
	Cucurbita digitata	T	+?	+?	Hensarling et al., 1974
Euphorbiaceae	*Euphorbia helioscopia*	T	−	−	Gori, 1976
	Manihot esculenta	T	++?, L?	++?, S to M?	Nartey et al., 1973, 1974
	Ricinus communis	T	++, L	++, L	Sobolev, 1966; Sobolev et al., 1968; Suvorov et al., 1971; Tully and Beevers, 1976; Youle and Huang, 1976
Juglandaceae	*Juglans regia*	F,T,X	++, S to M	++, M to L	Lott and Buttrose, 1978c
Lecythidaceae	*Bertholletia excelsa*	F,T,X	++, L	++, S to M	Lott and Buttrose, 1978b
Leguminosae	*Acacia conferta*	F,T,X	−	++, S	For legume information consult data plus literature review found in Lott and Buttrose, 1978a
	Arachis hypogaea	I,T	−	++, S	
	Cassia artemisioides	F,T,X	−	++, M to L	
	Clianthus formosus	F,T,X	−	++, S	
	Glycine max	F,T,X	−?	+, S to M	
	Lablab purpureus	T	−	−?	
	Medicago sativa	T	−	++, S to M	
	Phaseolus aureus =(*Vigna radiata*)	T	−	+, S	
	Phaseolus lunatus	T	−	+, S	
	Phaseolus vulgaris	I,T	−	−?	Singh, 1977
	Pisum sativum	F,I,T	−(+? in radicle)	+, S	
	Vicia faba	F,T,X	−	+, S	
	Vigna unguiculata	T	−	+, S	

TABLE I (*Continued*)

Family	Plant	Method	Protein crystalloid	Globoid crystal	Source of micrographs
Linaceae	*Linum usitatissimum*	T	++, L	++, L	Poux, 1965
Malvaceae	*Gossypium hirsutum*	I,T	—	++, M	Engleman, 1966; Hensarling et al., 1970; Lui and Altschul, 1967; Yatsu, 1965; Yatsu and Jacks, 1968
Myrtaceae	*Eucalyptus erythrocorys*	F,T,X	—	++, S to L	Buttrose and Lott, 1978b
Oleaceae	*Fraxinus excelsior*	T	—	++, ?	Villiers, 1971
Onagraceae	*Clarkia rubicunda*	T	—	++, L	Dengler, 1967
Proteaceae	*Macadamia integrifolia*	T,F,X	—	+, M to L	Lott and Buttrose, 1978c
	Protea compacta	T,X	+, L	++, L	Van Staden et al., 1975a,b; Van Staden and Comins, 1976
Rosaceae	*Prunus dulcis*	T,F,X	+, S	++, S to L	Lott and Buttrose, 1978b
Santalaceae	*Santalum acuminatum*	T,F,X	++, M to L	++, S to M	Lott and Buttrose, 1978b
Scrophulariaceae	*Paulownia tomentosa*	T	—	++, M	Rickson, 1968
Solanaceae	*Lycopersicon esculentum*	T	—?	+, ?	Eggers and Geisman, 1976
Tropaeolaceae	*Tropaeolum majus*	T	—	++, M to L	Nougarède, 1963
C. Class Angiospermopsida, Subclass Monocotyledonae					
Gramineae	*Bromus inermis*[c]	T	—	+?, M to L?	MacLeod et al., 1964
(Poaceae)					
	Hordeum vulgare[c]	F,T	++, S	++, M to L	Buttrose, 1971; Jacobsen et al., 1971
	Oryza sativa[a]	F,I,T,X	—	++, S to L	Bechtel and Pomeranz, 1977; Buttrose and Soeffky, 1973; Ogawa et al., 1977; Tanaka et al., 1973, 1977

594

Species / Family	Tissue			References
Setaria lutescens[e]	T,I	—	—	Rost, 1971, 1972
Sorghum bicolor	F,T	—	+?, S?	Adams and Novellie, 1975
Triticum aestivum[d]		—	++, S to L	Poux, 1963; Morrison *et al.*, 1975; Swift and Buttrose, 1972; Swift and O'Brien, 1972
Zea mays[f]	T	-?	-(+?,S? in scutellum)	Burr and Burr, 1976; Khoo and Wolf, 1970; Kyle and Styles, 1977; Zamski, 1973
Liliaceae *Yucca brevifolia* *Yucca glauca* *Yucca schidigera* *Yucca whipplei*	T	Core-type ++, L Meshwork-type —	Core-type ++, L to S Meshwork-type ++, L to S -?	Horner and Arnott, 1965, 1966
Palmae *Cocos nucifera*	T	++, L	-?	Dieckert and Dieckert, 1976a

[a] This table, modified from Buttrose and Lott (1978a), contains the author's interpretation of the published electron micrographs available in the references quoted. The protein body type listed in this table is the one commonly found in storage tissues of the species or in the only tissue studied. In some cases, for example cereal grains, there are considerable tissue-to-tissue differences that cannot be covered adequately in a table of this type. Key to symbols: F = freeze-fractured; I = isolated particles; T = thin-sectioned; X = x-ray analyzed; ++ = present and common; + = present but infrequent; – = not present; L,M,S = large, medium, small-sized in relation to protein body size; ? = observation uncertain.

[b] Takhtajan (1969) places *Simmondsia* in the monotypic family Simmondisiaceae.

[c] Aleurone tissue.

[d] Aleurone, coleoptile, and scutellum tissues.

[e] Embryo and endosperm tissue.

[f] Aleurone, starchy endosperm, and scutellum epithelium.

595

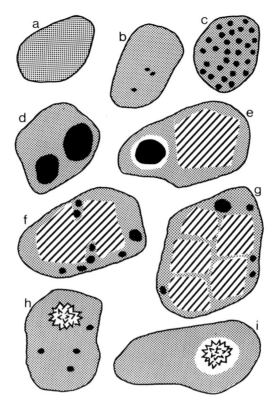

Fig. 1. Models of dry seed protein bodies illustrating some of the structural arrangements that are possible in different species. (a), Section of a protein body that contains only protein-aceous matrix (stippled) material inside the bounding membrane; (b)–(d), protein bodies in which there are globoid crystals (solid black) as well as proteinaceous matrix (stippled). The globoid crystals vary considerably in size and number. Complex protein bodies, as illustrated in (e), (f), and (g), contain protein crystalloids (black and white stripes), proteinaceous matrix (stippled) and globoid crystals (black). In (e) the globoid crystal is surrounded by a soft globoid (white) region. Protein bodies may contain druse crystals thought to be calcium oxalate. In (h) both the druse crystal and some globoid crystals (black) are surrounded by proteinaceous matrix (stippled). In (i) a region of unknown composition (white) separates the druse crystal from the proteinaceous matrix (stippled). The thick line around each protein body represents a membrane.

more irregular (1f) or even rounded. Protein bodies may contain one (1e, 1f), to many (1g) protein crystalloids. The number of protein crystalloids per protein body may vary within one cell or between cells of an organ. Freeze–fracture studies of protein bodies from some species suggest that a region which probably would appear as one protein crystalloid in fixed and sectioned material, may actually be composed of several constituent crystals

(Lott and Buttrose, 1978b). Such a situation occurs in Brazil nut protein bodies (Fig. 11). Protein crystalloids are composed of subunits arranged in a distinctive lattice. While the subunits and lattice structure are often not clearly visible in fixed and sectioned protein bodies, the freeze–fracture method may reveal them clearly (Fig. 11). In most cases that have been examined the presence of globoid crystals does not appear to have had any influence on protein crystalloid shape. In Brazil nut protein bodies (Fig. 9) however, the protein crystalloid is irregular in shape and usually contains globoid crystals in pockets (Lott and Buttrose, 1978b). While this situation has not as yet been studied developmentally a tentative hypothesis is that the presence of globoid crystals alters the way in which the protein crystalloid forms. Protein bodies containing protein crystalloids may contain various sizes and numbers of globoids. These globoids may consist of electron-dense globoid crystal material exclusively (Figs. 1f, 1g) or there may be both electron-transparent regions and electron-dense regions (Fig. 1e). In *Cucurbita* (Lott *et al.,* 1971; Lott and Vollmer, 1973a) and in *Ricinus* (Tully and Beevers, 1976) there is evidence for an electron-transparent region surrounding the globoid crystal (Fig. 8). This electron-transparent region was termed the soft globoid by Lott *et al.* (1971). The composition of the soft globoid is uncertain, but may be hydration dependent.

Protein bodies in a few species may contain groups of crystals in an arrangement commonly called druse or rosette crystals (Figs. 1h, 1i, 12). The druse crystal is generally surrounded by proteinaceous matrix material but may be surrounded by a region that is structurally distinct from the proteinaceous matrix (Buttrose and Lott, 1978b). Globoid crystals may or may not be evident in protein bodies that do contain druse crystals. The number of protein bodies in a seed containing druse crystals varies. In hazelnut cotyledon protein bodies druse crystals are very rare (Lott and Buttrose, 1978c) whereas in *Eucalyptus* cotyledon protein bodies the druse crystals are more common (Buttrose and Lott, 1978b). However, even in *Eucalyptus* there are more protein bodies containing no druse crystals than there are protein bodies with druse crystals. Crystals are also commonly found in protein bodies of umbelliferous seeds (Cutter, 1978).

Protein bodies containing concentric rings of stain deposit have been reported from cereal endosperm tissue or protein body isolates of cereal grains (Adams and Novellie, 1975; Adams *et al.,* 1976; Mitsuda *et al.,* 1969; Rost, 1971, 1972). Whether this is a real phenomenon or an artifact is an unresolved question since the alternating light–dark rings have generally been reported when protein bodies have been isolated first and/or in cases where fixation is marginal. In the author's lab it has been observed that excess poststain deposits may adhere to protein bodies, especially in certain species (E. Spitzer, personal communication). Perhaps drying down of water adhering to the protein body regions could result in the approximately spherical stain

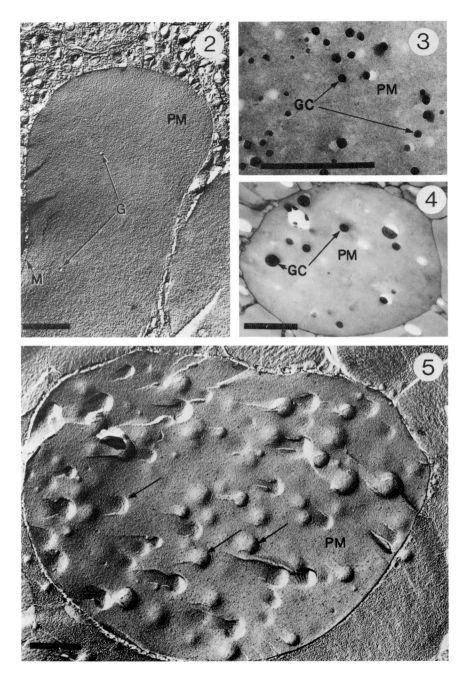

deposits. Sorghum protein bodies illustrated by Adams and Novellie (1975) contain dense deposits that are random in Fig. 1 and approximately concentric in Fig. 2. In Adams *et al.* (1976) the concentric deposits in Fig. 4 are not centered over each protein body. In Rost (1971) not only is there considerable variation in the density of the deposit on the protein bodies in Fig. 3 but the dense deposit is not centered over the protein bodies in every case.

The major storage tissue of a seed is likely to have protein bodies of the same structural complexity. When compared on this basis the protein bodies from different species within a family show a great deal of similarity (see Table I and Lott and Buttrose, 1978a). Table I, which shows tabulated results from most available electron microscopic studies of seed protein bodies, illustrates well just how few species have been studied carefully. Since most of the seeds studied with the electron microscope are from plants of current or potential economic importance it is likely that more structural diversity will be discovered as protein bodies in seeds of a wider range of species are investigated.

Classification of protein bodies into various levels of structural complexity, such has been done here, is useful. However, there is danger that such classifications will lead to an underestimation of the diversity of protein body structure and chemical composition that occurs within a seed. While it is probable that the majority of protein bodies of a given cell type in an organ will be structurally similar, there is considerable evidence showing that variation does occur within one cell, between cells in a tissue, or between different tissues (Graham and Gunning, 1970; Lott *et al.*, 1971; Lui and Altschul, 1967; Rost, 1972).

Fig. 2.   Freeze–fracture replica of dry soybean (*Glycine max*) cotyledon tissue demonstrating the presence of small globoids (G) in the proteinaceous matrix (PM) of a protein body. Position of the membrane that surrounds the protein body is indicated (M). Bar = 1 μm. Reproduced by permission of CSIRO from Lott and Buttrose (1978a).

Fig. 3.   Thin section of a protein body region from glutaraldehyde-OsO₄ fixed soybean cotyledon tissue. Numerous small electron-dense globoid crystals (GC) can be seen surrounded by the proteinaceous matrix (PM). Protein bodies in soybean cotyledon tissue rarely contain this many globoid crystals. Bar = 1 μm. Reproduced by permission of CSIRO from Lott and Buttrose (1978a).

Fig. 4.   Section of a protein body from glutaraldehyde-OsO₄ fixed sunflower (*Helianthus annuus*) cotyledon tissue. The globoid crystals (GC) are surrounded by proteinaceous matrix (PM). Bar = 1 μm. Micrograph courtesy of M. S. Buttrose, reproduced with permission of the National Research Council of Canada from Buttrose and Lott (1978a).

Fig. 5.   Freeze–fracture replica of a protein body from ungerminated sunflower cotyledon tissue. Depressions or elevated bumps (arrows) resulted from the fracture plane passing over or under globoid inclusions. Proteinaceous matrix (PM). Bar = 1 μm. Micrograph courtesy of M. S. Buttrose, reproduced with permission of the National Research Council of Canada from Buttrose and Lott (1978a).

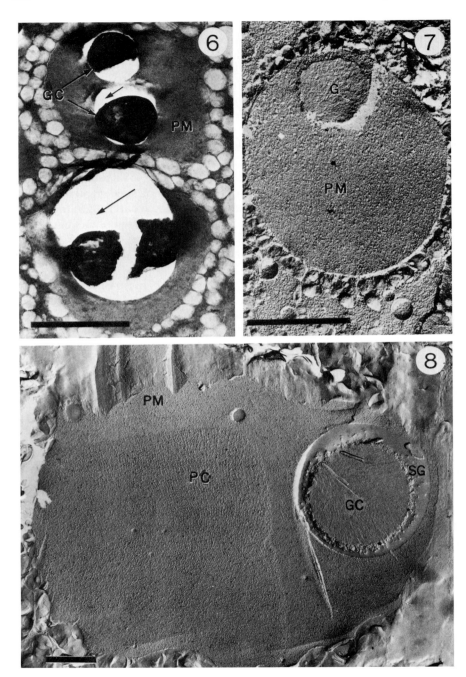

III. PROTEIN BODY COMPOSITION

A. Proteins

In this chapter no attempt will be made to review the field of seed protein chemistry, synthesis, and degradation. Readers are referred to reviews by Altschul *et al.*, 1966; Ashton, 1976; Millerd, 1975; and to Chapter 11 in Vol. 6 of this series. For information on protein quality in nutrition and genetic improvement of seed proteins, readers are referred to Inglett (1972), Sylvester-Bradley and Folkes (1976), and the National Academy of Sciences (Washington, D.C.) book *Genetic Improvement of Seed Proteins* (1974).

A given species normally has one or a few major storage proteins located in protein bodies. These proteins occur largely in proteinaceous matrix and protein crystalloid regions. A number of other proteins present in smaller amounts, including some enzymes, are also found inside protein bodies.

The fluorescent antibody localization of the storage proteins legumin and vicilin in *Vicia* cotyledons, as carried out by Graham and Gunning (1970), is of interest here since it tackles the problem of whether or not there is diversity between individual protein bodies. Both legumin and vicilin not only were located within protein bodies but occurred within the same protein body in most cases. A few protein bodies were found which contained vicilin but no detectable legumin. Some protein bodies were found which stained for protein histochemically but lacked either vicilin or legumin. That protein bodies within a cell or within a tissue may store different proteins is thus a real possibility which deserves considerably more research.

B. Phytin and Other Constituents

In addition to proteins it is known that significant amounts of phytic acid are stored in seed protein bodies. The phytin molecule is a cation salt of *myo*-inositol hexaphosphoric acid (Asada and Kasai, 1962; Ashton and Wil-

---

Fig. 6. Section of two protein bodies from glutaraldehyde-OsO₄ fixed *Cassia artemisioides* cotyledon tissue. In this legume the protein bodies consist of large globoid crystals (GC) surrounded by proteinaceous matrix (PM). Globoid crystals, which are hard and do not penetrate well with embedding medium, often shatter during sectioning leaving holes in the section (unlabeled arrows). Bar = 1 μm. Micrograph courtesy of M. S. Buttrose, reproduced with permission of CSIRO from Lott and Buttrose (1978a).

Fig. 7. Freeze–fracture replica of dry cotyledon tissue from *Cassia*. The fracture revealed a globoid (G) surrounded by proteinaceous matrix (PM). Bar = 1 μm. Micrograph courtesy of M. S. Buttrose, reproduced with permission of CSIRO from Lott and Buttrose (1978a).

Fig. 8. Freeze–fracture replica of a protein body from the cotyledon of an imbibed squash (*Cucurbita maxima*) seed. Proteinaceous matrix (PM), globoid crystal (GC), soft globoid (SG), and protein crystalloid (PC) regions can be seen. Bar = 1 μm. Reproduced with permission of the National Research Council of Canada from Lott and Vollmer (1973a).

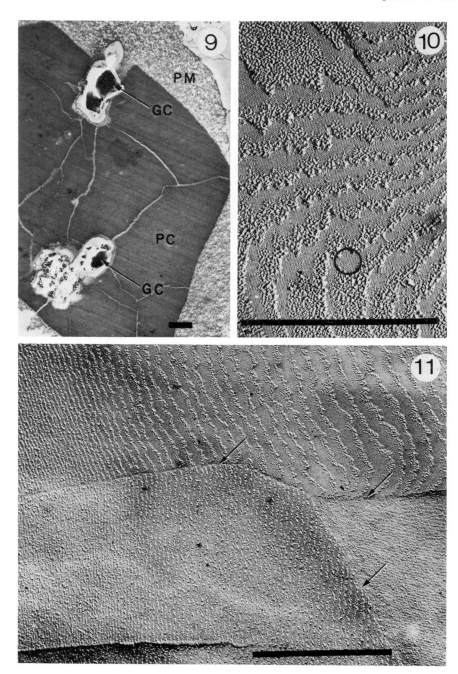

liams, 1958; Ergle and Guinn, 1959; Lui and Altschul, 1967). Phytin is a store for phosphorus, various cations, and inositol. Thus, in a seed, elements such as phosphorus are mainly stored in organic form (Ergle and Guinn, 1959; Hall and Hodges, 1966; Williams, 1970). In a variety of species 50–80% of the total seed phosphorus has been shown to be in the form of phytin (Ashton and Williams, 1958; Hall and Hodges, 1966; Lolas and Markakis, 1975; Lolas et al., 1976; Makower, 1969; Matheson and Strother, 1969; Nahapetian and Bassiri, 1976; Tluczkiewicz and Berendt, 1977).

In most species phytin is concentrated in the globoid crystals (Lui and Altschul, 1967; Ogawa et al., 1975; Sharma and Dieckert, 1975; Suvorov et al., 1971). There is also evidence for some phytin occurring in the proteinaceous parts of protein bodies, perhaps as protein–phytin complexes (Bourdillon, 1951; Fontaine et al., 1946; Lolas and Markakis, 1975) or as soluble potassium phytate (Poux, 1965). That proteinaceous regions of some species such as soybean have relatively high levels of K in readily soluble form is clear (Lott and Buttrose, 1978b). In some species, such as *Phaseolus* most of the phytic acid is in a water soluble form (Lolas and Markakis, 1975). Protein–phytin complexes could be of physiological importance because the complex can have very different solubility characteristics compared to the protein alone.

In addition to its role in seedling nutrition, phytic acid is important in considerations of nutritional qualities of seed products. Because phytic acid has the capacity to chelate minerals such as Ca, Mg, Fe, Zn, and Mo it reduces the availability of these elements in the digestive tract of animals and may cause mineral deficiencies. (See reviews in Bassiri and Nahapetian, 1977; Lolas and Markakis, 1975.)

Globoid crystals in protein bodies are naturally electron-dense regions and thus make ideal subjects for energy-dispersive x-ray (EDX) analysis. When

Fig. 9. Thin-section of part of a protein body from glutaraldehyde-OsO₄ fixed Brazil nut (*Bertholletia excelsa*) radicle tissue. Proteinaceous matrix (PM), globoid crystals (GC), and protein crystalloid (PC) can be seen. In Brazil nut there is an unusual occurrence in that the protein crystalloid formation seems to be affected by the presence of globoid crystals. The protein crystalloid in this micrograph is permeated by cracks. Some of this cracking may result from mechanical damage done when a portion of dry seed tissue was dissected out for freeze-fracturing. Alternately there may be some separation of constituent crystals during fixation. Bar = 1 μm. Micrograph courtesy of M. S. Buttrose, reproduced with permission of the National Research Council of Canada from Lott and Buttrose (1978b).

Fig. 10. Portion of a freeze–fractured protein crystalloid from a protein body in Brazil nut radicle tissue. The lattice steps and particles can be seen. The particles (circled area) are in a hexagonal system with a rhombohedral elementary cell. Bar = 1 μm. Reproduced with permission of the National Research Council of Canada from Lott and Buttrose (1978b).

Fig. 11. Portion of a freeze–fractured protein crystalloid from radicle tissue of Brazil nut. The interfaces (arrows) between regions of a crystalloid can be seen as can the lattice steps. Bar = 1 μm. Reproduced with permission of the National Research Council of Canada from Lott and Buttrose (1978b).

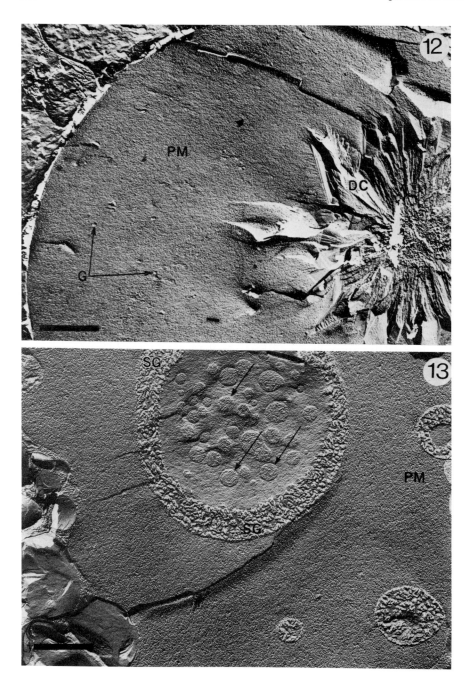

EDX analysis equipment is mounted on a transmission electron microscope it is possible to obtain specific information on the elements present in selected cell areas. The capacity to study elemental differences between regions within one cell or between regions in different cells is a particularly useful one. With most EDX analysis systems all elements with atomic number eleven and heavier can be detected with great sensitivity. The intense electron beam used for x-ray analysis destroys some of the tissue sample. Simultaneous analysis of all elements, which occurs with EDX analysis systems, is thus particularly useful for studies of biological materials. In EDX analysis studies of tissue, considerable care must be taken to ensure that elements are not removed during sample preparation. Sample preparation of seed tissue, prior to EDX analysis studies of protein body elemental composition, has been investigated (Lott *et al.*, 1978a,b). Osmium tetroxide extracts cations and often phosphorus from globoid crystals and thus should be avoided. Since the globoid crystals are naturally electron-dense it is possible to locate them in tissue that has been glutaraldehyde fixed, dehydrated, embedded, and sectioned. It is also possible to locate globoid crystals in freeze-dried tissue powders. Freeze-dried tissue powders are particularly useful since all fixation or dehydration artifacts are avoided. Where the use of tissue sections is necessary, Lott *et al.* (1978b) recommend that only glutaraldehyde be used as a fixative and that all fixation and dehydration times be kept to a minimum.

From the EDX analysis studies that have been made it is clear that P, K, and Mg are usually present in globoid crystals (see Table II). Calcium values, however, vary greatly from absent, to present in trace amounts only, to present in relatively major amounts. All the reasons for variation in calcium levels are not established but it is known that there is variation within one tissue of an organ and between organs. In almond cotyledons, large flaky globoid crystals were calcium-containing whereas small round globoid crystals appeared to contain little or no calcium (Lott and Buttrose, 1978b). In *Cucurbita maxima* cotyledons, where most globoid crystals lacked any calcium, occasional globoids with relatively high calcium levels were found (Lott *et al.*, 1978a). It is thus clear that some variation can occur within one organ. While investigations in this area are only in their infancy it is clear that differences in globoid crystal composition may also be related to the organ in question. In squash embryos, globoid crystals in the root–shoot axis

Fig. 12. Freeze-fracture replica of a portion of a *Eucalyptus erythrocorys* cotyledon protein body. The proteinaceous matrix (PM) surrounds a druse crystal (DC) plus some very small globoids (G). Bar = 1 μm. Micrograph courtesy of M. S. Buttrose, reproduced with permission of the National Research Council of Canada from Buttrose and Lott (1978b).

Fig. 13. Portion of a cotyledon protein body from a freeze–etched spongy mesophyll cell of *Cucurbita maxima* after 2 days of seedling growth. Regions of internal digestion (arrows) have appeared within the globoid crystal. Soft globoid (SG), proteinaceous matrix (PM). Bar = 1 μm.

TABLE II

Elemental Composition of Protein Body Regions as Determined by X-Ray Analysis[a]

Family and plant	Tissue	Protein body region	Method	P	K	Mg	Ca	S	Other	Reference and comments
Anacardiaceae										
Anacardium occidentale	Cotyledon	GC	E F G	+++	+++	+++	−	++?		Buttrose and Lott, 1978a
		PR	E F	++	+++	−	−	++		
Pistacia vera	Cotyledon	GC	E F G	+++	+++	++	+	+?		Fixation removes elements from PR
		PR	E F	−	−	−	−	++		
Betulaceae										
Corylus avellana	Cotyledon	GC	E F G	+++	+++	+++	−	++?		Lott and Buttrose, 1978c
		PR	E F	−	+++	−	−	+++		
Buxaceae										
Simmondsia chinensis	Cotyledon	GC	E G	+++	+++	+++	+	−		Buttrose and Lott, 1978b
Compositae										
Helianthus annuus	Cotyledon	GC and some PR	E F G	+++	+++	+++	−	++		Buttrose and Lott, 1978a
		PR	E F G	−	++	−	−	++		
Helichrysum bracteatum	Cotyledon	GC	E G	+++	+++	+++	+++	−		Buttrose and Lott, 1978a
		PR	E F	−	++	−	−	+++		
Cruciferae										
Crambe abyssinica	? Cotyledon or hypocotyl	GC	E M K	+++	−?	++	+++	++	Na ++	Hofsten, 1973; possible fixative extraction
Cucurbitaceae										
Cucurbita maxima	Cotyledon	GC	E F G	+++	+++	+++	+orR	−		Lott, 1975; Lott et al., 1978a
		PR	E G	++	−	−	−	++		
	Stem	GC	E G	+++	+++	+++	+++	−		
	Radicle	GC	E F	+++	+++	+++	+++	−		

Family / Species	Part	Region	Elements						Special	Reference
Gramineae										
Hordeum vulgare	Aleurone	W	M B	+++	+++	+++	+++	−		Liu and Pomeranz, 1975
Oryza sativa	Scutellum	GC and some PR	E O	+++	+++	+++	−	−		Tanaka et al., 1977
	Aleurone	GC and some PR	E O	+++	+++	+++	−	−		Possible fixative extraction
	Endosperm	W	E O	−	−	−	−	++		
Triticum aestivum	Aleurone	W	M U	+++	+++	+++	−	?		Tanaka et al., 1974
Juglandaceae										
Juglans regia	Cotyledon	GC	E F G	+++	+++	+++	−	++?		Lott and Buttrose, 1978c
		PR	E F	++	+++	−	++	++		
Lecythidaceae										
Bertholletia excelsa	Radicle	GC	E F G	+++	+++	+++	+++	+	Ba ++	Lott and Buttrose, 1978b
		PR	E F	−	+++	−	−	+++		Considerable variation in GC elements and levels
Leguminosae										
Acacia conferta	Cotyledon	GC	E F G	+++	+++	+++	++	+?		Lott and Buttrose, 1978a
		PR	E F	−	+++	+	+++	+++	Cl +?	
Cassia artemisioides	Cotyledon	GC	E F G	+++	+++	+++	+++	−	+++	Lott and Buttrose, 1978a
Clianthus formosus	Cotyledon	GC and PR	E F G	+++	+++	+++	−	++	Cl ++	Lott and Buttrose, 1978a
Glycine max	Cotyledon	GC	E F G	+++	+++	+++	R	−		Lott and Buttrose, 1978a; K is removed from PR by fixation
		PR	E F G	+++	+++	+++	−	+++		
Vicia faba	Cotyledon	GC and PR	E F	+++	+++	+++	++	++	Cl +	Lott and Buttrose, 1978a
Myrtaceae										
Eucalyptus erythrocorys	Cotyledon	D	E G	−	−	−	+++	−		Buttrose and Lott, 1978b
		GC	E F G	+++	+++	+++	+	+		
		PR	E F	++	++	−	−	+++	+++	

(Continued)

607

TABLE II (*Continued*)

Family and plant	Tissue	Protein body region	Method	Element						Reference and comments
				P	K	Mg	Ca	S	Other	
Proteaceae										
Macadamia integrifolia	Cotyledon	GC	E F	+++	+++	+++	−	++		Lott and Buttrose, 1978c
		or	E F	++	+++	−	+++	++		Two distinct types of GC
		PR	E F	−	+++	−	−	+++		
Protea compacta	Cotyledon	GC	E G	+++	+++	++	+++	−	Si	Van Staden and
		W	E B	+++	+++	++	+++	+++	Cl +++ ++++ ++	Comins, 1976
Rosaceae										
Prunus dulcis	Cotyledon	GC	E F G	+++	+++	+++	+++	+		Lott and Buttrose, 1978b
		or	E F	+++	+++	+++	+	−		
		PR	E F	−	+++	−	−	+++		
Santalaceae										
Santalum acuminatum	Endosperm	GC	E F G	+++	+++	+++	+++	++		Lott and Buttrose, 1978b
		PR	E F G	++	+++	−	−	+++		

contained much higher calcium levels than did globoid crystals from various cotyledon regions (Lott *et al.,* 1978a). In squash the calcium is thus stored in regions where growth occurs first. Globoid crystal composition in a given organ may also be influenced by seed size. Based upon the limited results available, Lott and Buttrose (1978c) observed that calcium content in globoid crystals is on average low or absent in large cotyledons but that calcium is usually present in the globoid crystals of small cotyledons. Research into globoid crystal composition in different embryo regions of a variety of *Cucurbita* species supports the concept of smaller embryos having more wide-spread distribution of calcium (Lott and Vollmer, 1979).

From EDX analysis results in Table II it is clear that sulfur has been reported as occurring in globoid crystals from a number of species. The situation with regard to sulfur is somewhat uncertain since sulfur is most readily detected in freeze-dried powders where some contamination of the globoid crystals with proteinaceous matrix is possible. Alternately it may be that glutaraldehyde fixation, dehydration, and embedding extracts the sulfur containing compounds. Globoid crystals isolated from peanuts contained 35.1% protein (Sharma and Dieckert, 1975). It is thus possible that globoids do contain some protein, perhaps with sulfur containing amino acids. Contamination of the isolated globoids with some protein is probable so even with isolates the situation is unclear.

Other elements beside P, K, Mg, Ca, and S have been detected in EDX analyses of globoid crystals. In Brazil nut radicle tissue the globoid crystals often contain barium (Lott and Buttrose, 1978a). Since barium can partially replace the calcium requirement of some organisms it is likely that Brazil nuts concentrate barium along with calcium. Given the fact that seeds concentrate the alkaline earth elements Mg, Ca, and Ba in globoid crystals, a hypothesis can be made that given the appropriate conditions seeds would also store strontium in globoid crystals.

Evidence indicates that chlorine, iron, manganese, silicon, and sodium may occur in protein bodies. Posternak (1905) isolated protein bodies from several species and reported the presence of Fe, Mn, and Si. Ashton and Williams (1958) reported the presence of Mn in oat phytin and Sharma and Dieckert (1975) reported the presence of some Mn in globoids isolated from peanut seeds. Using EDX analysis Buttrose (1978) has detected both Mn and Fe in globoid crystals from *Avena sativa* seeds and from seeds of several *Casuarina* species. Cl has been reported in several cases (see Table II). Some caution must be used where Cl is reported in sectioned material since epoxy resins used for embedding may contain Cl. Hofsten (1973) has reported the presence of Na in globoid crystals of *Crambe.*

While attention has concentrated on the phosphorus and cation aspects of phytin, the importance of *myo*-inositol should not be underestimated. *Myo*-inositol, which is one of the end products of phytin degradation, has been shown to be extensively utilized by germinating wheat (Matheson and Strother, 1969). More details on the biochemistry of inositol compounds can be found in Cosgrove (1966), Loewus (1971), and in Chapter 2, Vol. 3 of this series.

The composition of regions within protein bodies may well be much more complex than we generally consider. For example, the protein crystalloid region may well contain more than just protein. In barley, protein crystalloids which have a crystal lattice structure (Buttrose, 1971) have also been reported to stain for carbohydrate (Jacobsen *et al.*, 1971). Histochemical tests indicate the presence of lipid in barley globoids (Jacobsen *et al.*, 1971). Globoids of peanut have been reported to contain oxalic acid in addition to phytin and protein (Sharma and Dieckert, 1975).

Druse crystals in protein bodies, where investigated, would appear to be composed of calcium oxalate. Energy-dispersive x-ray analysis studies of druse crystals in hazelnut and *Eucalyptus* revealed the presence of calcium only (Lott and Buttrose, 1978c; Buttrose and Lott, 1978b). Acid solubility tests of druse crystals from *Eucalyptus* favor the crystal composition being calcium oxalate rather than calcium carbonate (Buttrose and Lott, 1978b).

IV. CHANGES IN PROTEIN BODIES DURING GERMINATION AND EARLY SEEDLING GROWTH

During germination and early seedling growth, the contents of seed protein bodies are mobilized to provide substrates and energy needed by the growing seedling. Since germinating seeds are convenient experimental material the structural and biochemical changes that occur during protein body loss are much better understood than are the changes occurring during seed formation. In this section some of the major events that occur in protein bodies in the early phases of seedling growth are outlined. References are not provided; readers are referred to Ashton (1976) and Matile (1975) for more detailed discussion and literature citations.

1. Protein bodies swell as the dry seed takes up water during imbibition. Subsequently proteins are degraded by a variety of proteolytic enzymes and phytin is degraded by the enzyme phytase. Proteolytic enzymes include both endopeptidases and exopeptidases. Specific dipeptidases may also be present. Evidence available suggests that proteins are generally hydrolyzed to amino acids and small peptides. The resulting amino acids and small peptides may be used in the tissue of origin or may be translocated to other regions of the growing plant.

2. Degradation of protein reserves can be observed in seedling tissue that has been chemically fixed, embedded, and sectioned. As proteinaceous

reserves are hydrolyzed the regions of protein degradation become less electron-dense. Peripheral and/or internal dissolution have been reported. In peripheral dissolution the electron-transparent regions develop first adjacent to the limiting membrane, whereas in internal degradation the cavities develop inside the proteinaceous mass.

3. As was the case with protein regions, globoid crystals have been reported to undergo peripheral and/or internal degradation (see Fig. 13).

4. As the contents of the protein bodies are used up the protein bodies become aqueous vacuoles. Fusion of the small protein body derived vacuoles results in the formation of the larger central vacuole of the cell. A number of light microscopic and electron microscopic investigations have supported the theory that during early seedling growth the protein bodies give rise to the main cell vacuole. This fusion has also been observed with Nomarski-optic studies of living cells. Protein body derived vacuoles, still containing some lumps of proteinaceous material, are frequently seen (Figs. 14, 15).

5. Proteolytic enzyme activity leading to the disappearance of protein body reserves is usually a carefully regulated process. In complex organs such as cotyledons, variation in the rate of protein degradation in different cells is often dependent upon the position relative to veins. The activity of some proteolytic enzymes has been demonstrated to be hormonally controlled.

V. DEVELOPMENT OF SEED PROTEIN BODIES

Despite the great economic importance of seeds there has been relatively little study of seed formation (Dure, 1975). With regard to the study of developing protein bodies the biochemistry of protein synthesis has received more study than has the study of ultrastructural changes occurring in cells that store protein. As pointed out by Dure (1975) the developing seed system is attractive for biochemical studies since there is a one-time and massive translation of a few genes over a short time period. The biosynthesis of seed storage proteins will be considered in detail in Chapter 11 of Vol. 6 of this series. For more information about protein body formation readers are also referred to Dieckert and Dieckert (1972, 1976b).

Protein body development has received some electron microscopic study in such genera as *Arachis* (Dieckert and Dieckert, 1972, 1976a,b); *Capsella* (Dieckert and Dieckert, 1972, 1976a); *Glycine* (Bils and Howell, 1963); *Gossypium* (Dieckert and Dieckert, 1972, 1976a,b; Engleman, 1966); *Oryza* (Harris and Juliano, 1977); *Phaseolus* (Klein and Pollock, 1968; Öpik, 1968); *Pisum* (Bain and Mercer, 1966; Savelbergh and Van Parijs, 1971); *Ricinus* (Sobolev *et al.,* 1968); *Triticum* (Barlow *et al.,* 1974; Buttrose, 1963;

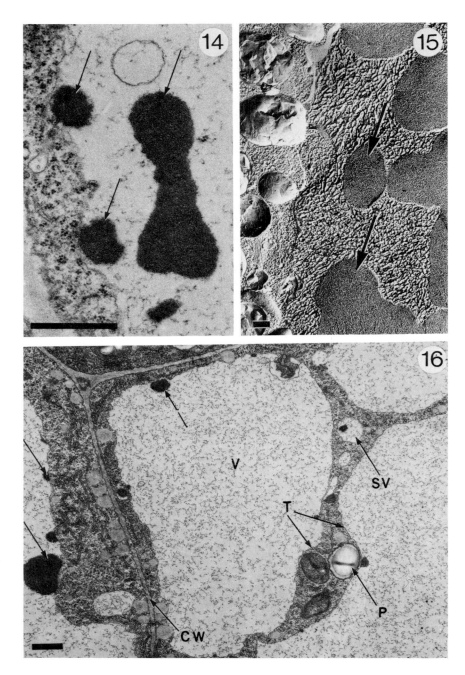

Morrison *et al.*, 1975); *Vicia* (Briarty *et al.*, 1969); *Vigna* (Harris and Boulter, 1976) and *Zea* (Khoo and Wolf, 1970; Kyle and Styles, 1977). Space does not permit a detailed survey and criticism of all the available electron microscopic evidence. A number of the published studies are technically poor by today's standards and thus readers are cautioned to judge this literature with care. Some ideas on protein body formation are reviewed as follows.

There are indications that protein bodies may be formed in different ways, even within the same cell or the same seed. Evidence from several species supports a vacuolar origin of protein bodies while evidence from other systems, including cereal endosperms, supports protein body formation from specialized regions of endoplasmic reticulum or from small cytoplasmic vesicles. In maize the protein bodies of the aleurone layer are reported to have a vacuolar origin while in the starchy endosperm cells protein bodies develop from enlargement of regions of rough endoplasmic reticulum (Kyle and Styles, 1977). In cowpea (*Vigna*) cotyledons, some protein bodies are formed by subdivision of the main vacuoles of the cotyledon cell but most protein bodies are formed by expansion of cytoplasmic vesicles (Harris and Boulter, 1976).

From several seed systems that have been studied it is clear that major changes in the vacuole system occur during protein body formation. Before protein deposition starts, cells often contain one or a few large vacuoles. During protein body formation this large vacuole is divided up into numerous smaller vacuoles (Öpik, 1968; Harris and Boulter, 1976). The enormity of this change has been most carefully documented in developing pea cotyledons by Craig *et al.* (1979). Eight days after flowering, pea cotyledon parenchyma cells were found to contain one or two large vacuoles with an average vacuole surface area of 5,500 μm^2 per cell. Proteinaceous material was beginning to appear in the vacuole at 8 days. During the next 12 days Craig *et al.* estimated that about 175,000 protein bodies per cell were formed. The

---

Fig. 14. Portion of a mesophyll cell from a squash cotyledon after 4 days of seedling growth. In this glutaraldehyde-OsO$_4$ fixed sample electron-dense pieces of proteinaceous material (arrows) can be seen inside the disintegrating protein body. Bar = 1 μm.

Fig. 15. Portion of a freeze-etched mesophyll cell from a squash cotyledon. At this stage of seedling growth (3.5 days) there are irregularly shaped proteinaceous inclusions (arrows) inside the disintegrating protein bodies. The observations obtained with chemical fixation, dehydration, embedding, and sectioning (Fig. 14) closely parallel the information obtained by freeze-etching (Fig. 15). Clumps of proteinaceous material of this sort are thus thought to be real and do not represent fixation and dehydration artifacts. Bar = 1 μm.

Fig. 16. Portion of glutaraldehyde-OsO$_4$ fixed soybean cotyledon tissue at the start of protein deposition during seed development. The cell contains numerous vacuoles (V) containing some major electron-dense deposits (arrows) plus considerable loose floculent material. Some electron-dense deposits appear in regions which may be small vesicles (SV) although serial sections would be needed to prove that this is so. Cell wall (CW), tonoplast (T), plastid with starch (P). Bar = 1 μm.

surface area of these protein bodies in each cell was estimated to be 550,000 μm^2. Thus during a 12-day time span the surface to volume ratio of the vacuole/protein body system increased 55 times. The great increase in surface area may be important in protein body formation. During this 12-day time period when pea cotyledon protein bodies are forming, the percentage of the total cell volume occupied by the vacuole/protein bodies decreased from 75% to about 20%.

With regard to storage protein synthesis there would appear to be one or more of the following possibilities: (a) synthesis on ribosomes free in the cytoplasm; (b) synthesis on ribosomes attached to endoplasmic reticulum; and (c) synthesis involving polysomes attached to the membrane surrounding the vacuole/protein body. It is possible that in some cases the last two possibilities may occur as part of an integrated system of synthesis and packaging. Structural evidence from developing maize aleurone supports the concept of free cytoplasmic ribosomes being involved in at least some synthesis of storage proteins (Kyle and Styles, 1977). Synthesis of storage proteins in rough endoplasmic reticulum is supported both by autoradiography (Bailey *et al.*, 1970) and by observations that in some systems rough endoplasmic reticulum is very prevalent at times of high storage protein synthesis (Harris and Boulter, 1976; Öpik, 1968; Savelbergh and Van Parijs, 1971). In *Vicia* cotyledons an increase in the number of membrane-bound ribosomes coincided with the onset of protein deposition (Briarty *et al.*, 1969). Protein bodies, isolated from maize endosperm at a time when storage protein synthesis was active, were found to have polysomes associated with the protein body membrane (Burr and Burr, 1976). These polysomes, when dissociated from the membrane and placed into an amino acid incorporating system, synthesized a protein believed to be the main maize storage protein called zein (Burr and Burr, 1976).

If storage proteins are synthesized on rough endoplasmic reticulum there is still the question of how the product moves from the place of synthesis to the place of storage. One or more of the following possibilities could apply: (a) direct connections from endoplasmic reticulum to the developing protein storage vacuole; (b) packaging of endoplasmic reticulum synthesized proteins into dictyosome (Golgi) vesicles which are then transported to the larger protein bodies for storage; (c) packaging of storage proteins into vesicles derived from endoplasmic reticulum membranes and the transport of these vesicles to the larger protein bodies; and (d) release of the storage protein into the cytoplasm with subsequent uptake of the protein by the developing protein body. Different investigators have supported one or more of these routes. A few investigators have not been able to come to any definite conclusions as to the route of protein movement based upon their ultrastructural studies (Craig *et al.*, 1979; Öpik, 1968). The route taken by storage proteins from the place of synthesis to the place of deposition is a particularly important topic. This difficult area of investigation deserves more detailed study.

Direct connections between the endoplasmic reticulum and developing protein vacuoles have been reported (Engleman, 1966) as has the production of vesicles by dilation of restricted areas of endoplasmic reticulum (Khoo and Wolf, 1970). Several researchers have implicated the dictyosome (Golgi) system in packaging of endoplasmic reticulum synthesized proteins for eventual deposition in protein bodies (Briarty et al., 1969; Dieckert and Dieckert, 1972, 1976a,b; Harris and Boulter, 1976). The work of Dieckert and Dieckert (1972, 1976a,b) provides the most sophisticated support for this idea in that they have used pronase digestion to prove that dictyosome vesicles present at the time of protein body formation do contain proteins. While dictyosome vesicles may contain protein there is no proof, however, that these vesicles contain the major proteins stored in a given seed. There are also no data available to suggest what proportion of the total protein stored in a protein body is deposited by dictyosome vesicles.

It is tempting to seek a single mechanism whereby storage compounds are synthesized and then channelled into developing protein bodies. Given the observed variability in protein body structure and chemical composition, however, it would seem unwise to assume that there is a single site of synthesis and a single pathway of entry for compounds into protein bodies. Just the protein component of protein bodies may well contain insoluble storage proteins, water soluble storage proteins, and enzyme proteins such as proteases. Also the various storage proteins are not synthesized at identical times during seed development. For example, in Vicia faba seeds, legumin synthesis continued long after vicilin synthesis had stopped (Manteuffel et al., 1976). Considering these factors it seems quite possible that different storage proteins and enzyme proteins might enter developing protein bodies by different routes.

One of the early structural clues that storage protein deposition has begun is the appearance of dense staining deposits along the inner surface of the tonoplast membrane. This situation is illustrated for soybean in Fig. 16 and 17 and for squash in Fig. 18. At the time deposition occurs, the cytoplasm is dense with ribosomes, endoplasmic reticulum, and other organelles (Fig. 17).

Synthesis and deposition of phytic acid reserves can occur throughout protein body development. In wheat aleurone cells some phytin globoids may appear early in protein body formation but a great deal of phytin deposition occurs later (Morrison et al., 1975). In cereals the synthesis of phytin occurs throughout development and continues until the end of maturation (Tluczkiewicz and Berendt, 1977). In Phaseolus, maximum protein synthesis and maximum phytin synthesis coincide (Walker, 1974). Figure 19, from a developing Cucurbita embryo, shows a definite globoid crystal inside a partially developed protein body.

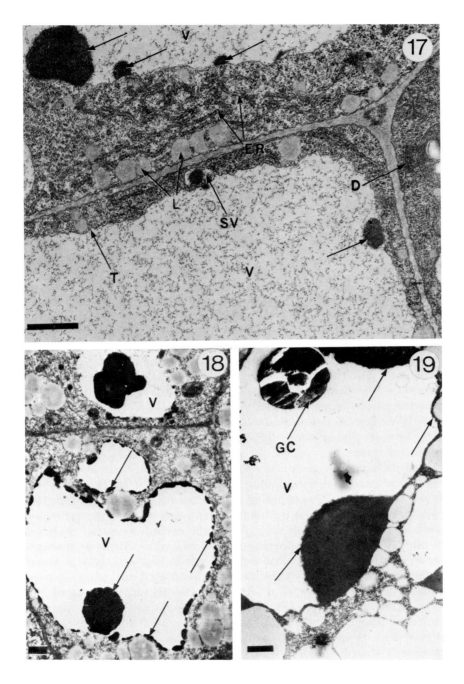

VI. PROBLEMS ENCOUNTERED IN STUDIES OF PROTEIN BODIES

An important reason for our current lack of knowledge about seed protein body structure and composition are technical problems involved in their study. Researchers interested in the biochemistry of storage proteins and in seed ultrastructure have faced major technical difficulties. Studies of seed development have been hampered by the long growing times required and by the difficulty of reliably obtaining material at the same developmental state.

A. Biochemical Studies

Many of the seed storage proteins are rather insoluble, a fact that has made them difficult to work with in biochemical studies. For example, storage proteins are generally denatured when used as substrates for protease enzyme assays. Getting high purity storage protein preparations for immunological work has proved to be very difficult. Isolation of protein bodies with intact limiting membranes has also proved to be difficult in many cases.

B. Structural Studies

Electron microscopic studies of seed tissues are generally difficult. Some of the technical problems, as reviewed by Lott and Buttrose (1978b), are outlined below. For electron microscopic studies of seed tissue, the most widely used method involves chemical fixation, dehydration, embedding in epoxy resin, and thin-sectioning. In thin-sections the cell type and location can often be specified and a large number of cells can be viewed fairly rapidly. However, due to the density of storage materials in seed storage tissues, good fixation is difficult to obtain and artifacts are a major problem. This means that the results of studies based upon chemical fixation may not give an accurate representation of the structure of the intact seed. $KMnO_4$

Fig. 17. Higher magnification portion of Fig. 16. Numerous free ribosomes can be seen in the cytoplasm in addition to those attached to the endoplasmic reticulum (ER). Vacuole (V), possible small vacuole (SV), tonoplast (T), electron-dense proteinaceous deposits (arrows), lipid droplets (L), dictyosome (D). Bar = 1 μm.

Fig. 18. Portion of cotyledon cells from a *Cucurbita maxima* seed during early stages of protein deposition in vacuoles (V). Electron-dense proteinaceous deposits (arrows) often appear next to the tonoplast but may be more centrally located in the vacuole. Glutaraldehyde-OsO$_4$ fixation. Bar = 1 μm.

Fig. 19. Portion of a squash cotyledon cell from a developing seed. This cell, which is more advanced in protein body formation than the cells seen in Fig. 18, contains a globoid crystal (GC) and proteinaceous deposits (arrows) inside the vacuole (V). Glutaraldehyde-OsO$_4$ fixation. Bar = 1 μm.

and OsO_4, which are the two fixatives most commonly used to impart electron density to portions of plant tissue, have both been reported to alter the structure and/or composition of protein body components. In *Cucurbita* the use of $KMnO_4$, even if preceded by glutaraldehyde, can result in complete extraction of globoid crystals (Lott *et al.,* 1971). While electron-dense globoid crystal material may remain after fixatives employing OsO_4 have been used, recent x-ray analysis studies on globoid crystals from a variety of seeds indicates that almost all K, Mg, and Ca and some P have been extracted (Lott *et al.,* 1978b). Glutaraldehyde fixation alone did not cause a major loss of elements from *Cucurbita* globoid crystals (Lott *et al.,* 1978a) but did cause a major extraction of K from protein bodies in certain legumes (Lott and Buttrose, 1978a). It seems pointless to argue over which chemical fixative gives the best general preservation since the degree of damage or extraction may depend upon the species involved, the types of storage tissues in a given seed, or upon the fixation conditions used.

Tears and chatter, which arise because infiltration of epoxy resins is marginal, are common problems when seed tissue is thin-sectioned. The hard globoid crystals infiltrate least and thus are often shattered during thin-sectioning. This shattering of the globoid crystals not only leaves holes in the sections but may contribute electron-dense debris to the section surface. The shattered globoid crystals are a common cause of scratches on the surface of sections.

Despite its usefulness in the interpretation of seed ultrastructure the freeze–fracture method has received little use. Almost all published freeze–fracture studies of seed storage tissues come from two laboratories, CSIRO in Adelaide and McMaster University in Ontario. Freeze–fracture studies of dry seed tissue are especially useful because freeze-fracturing of such tissue can be carried out without any chemical fixation. Since the extremely rapid freezing methods employed for sample preparation do not kill whole seeds, we have reason to believe that structural information obtained in this way very closely approximates that found in the natural state. In freeze–fracture studies of dry seed tissue the difficult step is replica cleaning. Since replicas generally break up into numerous small pieces it is unlikely that studies of major tissue regions will succeed. Difficulties in regulating sample orientation during quick freezing, uncertainty of the plane of fracture, and microtome knife smearing of portions of the sample, further reduce the chances that studies of large tissue regions will be successful. Freeze–fracture replicas do not provide any information as to the electron density of the original tissue regions, a feature that is somewhat frustrating when one is studying globoids.

The presence or absence of small, electron-translucent globoids in protein bodies is difficult to determine. If OsO_4 or $KMnO_4$ are used as the fixative any electron-translucent region could well be an artifact of extraction. Even with glutaraldehyde fixation alone the possibility of extraction of contents from small globoids cannot be ruled out. Freeze-fracturing can provide useful information on the presence of globoids but does not provide any information of the electron-density of these globoids. With freeze-dried tissue powders it is difficult to determine whether or not small electron-translucent regions are spaces incorporated during the grinding process.

VII. FUTURE RESEARCH

A chapter such as this is incomplete without some discussion of areas where future studies are needed, and my thoughts on priority areas for future work follow.

It is important that protein bodies in more species of plants be studied. From Table I it is clear that protein bodies in very few species have been studied with electron microscopy. Biochemical studies have also centered upon relatively few species. Past work has concentrated on the Gramineae and Leguminosae. The majority of species studied have been highly improved crop plants. More information on protein bodies in plants from various families, plants from various climates, and plants from various growth conditions would greatly advance our understanding of seed protein bodies. It is hoped that future studies will attempt to relate structure with composition wherever possible.

Much of the reliable information on seed protein bodies comes from studies of mature or germinating seeds while seed development has been greatly neglected. Much more information is needed from a variety of species on such questions as what are the sites of synthesis, how is synthesis regulated, what is the sequence of synthesis and deposition events, how do the various components of protein bodies get packaged into the protein bodies, do the components of complex protein bodies self-assemble into the definite structural components found in protein bodies, what structural changes occur during seed development, and are the proteins similar in all protein bodies in a cell or tissue region? Such studies should investigate possible tissue to tissue differences. For example, we should not assume that endosperm and embryo tissue are similar without investigations.

More research is needed to understand the mineral storage component of protein bodies. From evidence outlined earlier it is clear that in some plants there are variations in the elemental composition of globoid crystals in different parts of embryos. This interesting observation needs to be pursued in a range of dicot and monocot plants. Other questions to which answers are needed include what controls variations in globoid crystal composition, what effect does mineral deficiency or excess have upon the elements stored in seed protein bodies, and what trace metals can be taken up into seed protein bodies?

To date research has tended to center upon either the protein components of seed protein bodies or upon the elemental reserves in protein bodies. Most workers, if they have considered it at all, regard proteins and phytin as chance occupants of the same vacuole. Given the fact that all protein bodies carefully analyzed with x-ray analysis contain at least some mineral storage material and given observations of protein–phytin complexes it seems timely to investigate possible interrelationships between protein and phytin reserves.

More studies dealing with globoid composition are needed. As already discussed, there are technical problems involved but it may be possible to discover if there are in fact globoids that are completely electron-transparent. If such globoids exist what is their composition and function? Where there are soft globoid regions around globoid crystals we need information on the composition and function of such regions. While x-ray analysis is useful to give very specific elemental information on globoid crystals the method does not give information on the types of molecules present. There is a need for more information on the chemistry of globoids. Such work should be done with globoids from specific embryo regions.

To date crystals, thought to be calcium oxalate, have been found in protein bodies of relatively few species but future investigations will undoubtedly identify many more examples, especially among noncrop plants. Information is needed on such basic questions as whether or not such crystals are degraded during germination, when the crystals are formed during protein body formation, and what function do these crystals have?

ACKNOWLEDGMENTS

The author acknowledges with thanks the technical help, comments on the manuscript, and discussion of ideas provided by M. S. Buttrose, C. Gaston, J. Greenwood, A. Oaks, E. Spitzer, and C. M. Vollmer.

REFERENCES

Adams, C. A., and Nouvellie, L. (1975). *Plant Physiol.* **55**, 1–6.
Adams, C. A., Nouvellie, L., and Liebenberg, Nv. d. W. (1976). *Cereal Chem.* **53**, 1–12.
Altschul, A. M., Yatsu, L. Y., Ory, R. L., and Engleman, E. M. (1966). *Annu. Rev. Plant Physiol.* **17**, 113–136.
Asada, K., and Kasai, Z. (1962). *Plant Cell Physiol.* **3**, 397–406.
Ashton, F. M. (1976). *Annu. Rev. Plant Physiol.* **27**, 95–117.
Ashton, W. M., and Williams, P. C. (1958). *J. Sci. Food Agric.* **9**, 505–511.
Bailey, C. J., Cobb, A., and Boulter, D. (1970). *Planta* **95**, 103–118.
Bain, J. M., and Mercer, F. V. (1966). *Aust. J. Biol. Sci.* **19**, 49–67.

Barlow, K. K., Lee, J. W., and Vesk, M. (1974). *In* "Mechanisms of Regulation of Plant Growth" (R. L. Bieleski, A. R. Ferguson, and M. M. Cresswell, eds.), Bull. 12, pp. 793–797. Roy. Soc. Wellington, New Zealand.

Bassiri, A., and Nahapetian, A. (1977). *J. Agric. Food Chem.* **25**, 1118–1122.

Bechtel, D. B., and Pomeranz, Y. (1977). *Am. J. Bot.* **64**, 966–973.

Bils, R. F., and Howell, R. W. (1963). *Crop Sci.* **3**, 304–308.

Bourdillon, J. (1951). *J. Biol. Chem.* **189**, 65–72.

Briarty, L. G., Coult, D. A., and Boulter, D. (1969). *J. Exp. Bot.* **20**, 358–372.

Burr, B., and Burr, F. A. (1976). *Proc. Natl. Acad. Sci. U.S.A.* **73**, 515–519.

Buttrose, M. S. (1963). *Aust. J. Biol. Sci.* **16**, 768–774.

Buttrose, M. S. (1971). *Planta* **96**, 13–26.

Buttrose, M. S. (1978). *Aust. J. Plant Physiol.* **5**, 631–639.

Buttrose, M. S., and Lott, J. N. A. (1978a). *Can. J. Bot.,* **56**, 2062–2071.

Buttrose, M. S., and Lott, J. N. A. (1978b). *Can. J. Bot.,* **56**, 2083–2091.

Buttrose, M. S. and Soeffky, A. (1973). *Aust. J. Biol. Sci.* **26**, 357–364.

Cosgrove, D. J. (1966). *Rev. Pure Appl. Chem.* **16**, 209–224.

Craig, S., Goodchild, D. J., and Hardham, A. R. (1979). *Aust. J. Plant Physiol.* **6**, 81–98.

Cutter, E. G. (1978). "Plant Anatomy, Part I: Cells and Tissues," p. 45. Arnold, London.

Dengler, R. E. (1967). Ph.D. Dissertation, Univ. of California, Davis, California.

Dieckert, J. W., and Dieckert, M. C. (1972). *In* "Symposium: Seed Proteins" (G. E. Inglett, ed.), pp. 52–85. Avi, Westport, Connecticut.

Dieckert, J. W., and Dieckert, M. C. (1976a). *In* "Genetic Improvement of Seed Proteins," pp. 18–51. Natl. Acad. Sci. Washington, D. C.

Dieckert, J. W., and Dieckert, M. C. (1976b). *J. Food Sci.* **41**, 475–482.

Dure, L. S. III. (1975). *Annu. Rev. Plant Physiol.* **26**, 259–278.

Durzan, D. J., Mia, A. J., and Ramaiah, P. K. (1971). *Can. J. Bot.* **49**, 927–938.

Eggers, L. K., and Geisman, J. R. (1976). *In* "Fruit and Vegetable Processing and Food Technology," Res. Circ. 213, Ohio Agric. Res. Dev. Centre, Wooster, Ohio.

Engleman, E. M. (1966). *Am. J. Bot.* **53**, 231–237.

Ergle, D. R., and Guinn, G. (1959). *Plant Physiol.* **34**, 476–481.

Fontaine, T. D., Pons, W. A., Jr., and Irving, G. W. Jr. (1946). *J. Biol. Chem.* **164**, 487–507.

Gori, P. (1976). *J. Ultrastruct. Res.* **54**, 53–58.

Graham, T. A., and Gunning, B. E. S. (1970). *Nature (London)* **228**, 81–82.

Gruber, P. J., Trelease, R. N., Becker, W. M., and Newcomb, E. H. (1970). *Planta* **93**, 269–288.

Hall, J. R., and Hodges, T. K. (1966). *Plant Physiol.* **41**, 1459–1464.

Harris, N., and Boulter, D. (1976). *Ann. Bot. (London)* **40**, 739–744.

Harris, N., and Juliano, B. O. (1977). *Ann. Bot. (London)* **41**, 1–5.

Hensarling, T. P., Yatsu, L. Y., and Jacks, T. J. (1970). *J. Am. Oil Chem. Soc.* **47**, 224–225.

Hensarling, T. P., Jacks, T. J., and Yatsu, Y. (1974). *J. Am. Oil Chem. Soc.* **51**, 474–475.

Hofsten, A. v. (1970). *Proc. Int. Conf. Sci. Technol. Marketing Rapeseed Rapeseed Products,* pp. 70–85.

Hofsten, A. v. (1973). *Physiol. Plant.* **29**, 76–81.

Hofsten, A. v. (1974). *Sven. Bot. Tidskr.* **68**, 153–163.

Horner, H. T., and Arnott, H. J. (1965). *Am. J. Bot.* **53**, 1027–1028.

Horner, H. T., and Arnott, H. J. (1966). *Bot. Gaz. (Chicago)* **127**, 48–64.

Inglett, G. E. Ed. (1972). *Symposium: Seed Proteins.* Avi, Westport, Connecticut.

Jacobsen, J. V., Knox, R. B., and Pyliotis, N. A. (1971). *Planta* **101**, 189–209.

Khoo, V., and Wolf, M. J. (1970). *Am. J. Bot.* **57**, 1042–1050.

Kidwai, P., and Robards, A. W. (1969). *Planta* **89**, 361–368.

Kirk, J. T. O., and Pyliotis, N. A. (1976). *Aust. J. Plant Physiol.* **3**, 731–746.

Klein, S., and Pollock, B. M. (1968). *Am. J. Bot.* **55**, 658–672.

Kyle, D. J., and Styles, E. D. (1977). *Planta* **137**, 185–193.

Liu, D. J., and Pomeranz, Y. (1975). *Cereal Chem.* **52**, 620–629.

Loewus, F. (1971). *Annu. Rev. Plant Physiol.* **22**, 337–364.

Lolas, G. M., and Markakis, P. (1975). *J. Agric. Food Chem.* **23**, 13–15.

Lolas, G. M., Palamidis, N., and Markakis, P. (1976). *Cereal Chem.* **53**, 867–871.

Lott, J. N. A. (1975). *Plant Physiol.* **55**, 913–916.

Lott, J. N. A., and Buttrose, M. S. (1978a). *Aust. J. Plant Physiol.* **5**, 89–111.

Lott, J. N. A., and Buttrose, M. S. (1978b). *Can. J. Bot.* **56**, 2050–2061.

Lott, J. N. A., and Buttrose, M. S. (1978c). *Can. J. Bot.* **56**, 2072–2082.

Lott, J. N. A., and Vollmer, C. M. (1973a). *Can. J. Bot.* **51**, 687–688.

Lott, J. N. A., and Vollmer, C. M. (1973b). *Protoplasma* **78**, 255–271.

Lott, J. N. A., and Vollmer, C. M. (1979). *Plant Physiol.* **63**, 307–311.

Lott, J. N. A., Larsen, P. L., and Darley, J. J. (1971). *Can. J. Bot.* **49**, 1777–1782.

Lott, J. N. A., Greenwood, J. S., and Vollmer, C. M. (1978a). *Can. J. Bot.* **56**, 2408–2414.

Lott, J. N. A., Greenwood, J. S., Vollmer, C. M., and Buttrose, M. S. (1978b). *Plant Physiol.*
   61, 984–988.

Lui, N. S. T., and Altschul, A. M. (1967). *Archiv. Biochem. Biophys.* **121**, 678–684.

MacLeod, A. M., Johnston, C. S., and Duffus, J. H. (1964). *J. Inst. Brew. (London)* **70**,
   303–307.

Makower, R. U. (1969). *J. Sci. Food Agric.* **20**, 82–84.

Manteuffel, R., Müntz, K., Püchel, M., and Scholz, G. (1976). *Biochem. Physiol. Pflanz.* **169**,
   595–605.

Marin, L. (1975). Ph.D. Dissertation, Univ. of Toronto, Toronto, Ontario, Canada.

Matheson, N. K., and Strother, S. (1969). *Phytochemistry* **8**, 1349–1356.

Matile, Ph. (1975). "The Lytic Compartment of Plant Cells." Springer-Verlag, Berlin and New
   York.

Matile, Ph., and Wiemken, A. (1976). *In* "Encyclopedia of Plant Physiology, [N.S.], Transport
   in Plants III" (C. R. Stocking and U. Heber, eds.), Vol. 3, pp. 255–287. Springer-Verlag,
   Berlin, and New York.

Millerd, A. (1975). *Annu. Rev. Plant Physiol.* **26**, 53–72.

Mitsuda, H., Murakami, K., Kusano, T., and Yasumoto, K. (1969). *Arch. Biochem. Biophys.*
   130, 678–680.

Morrison, I. N., Kuo, J., and O'Brien, T. P. (1975). *Planta* **123**, 105–116.

Muller, L. L., Hensarling, T. P., and Jacks, T. J. (1975). *J. Am. Oil Chem. Soc.* **52**, 164–165.

Nahapetian, A., and Bassiri, A. (1976). *J. Agric. Food Chem.* **24**, 947–950.

Nartey, F., Moller, B. L., and Andersen, M. R. (1973). *Tropical Sci.* **15**, 273–275.

Nartey, F., Moller, B. L., and Andersen, M. R. (1974). *Econ. Bot.* **28**, 145–154.

Nougarède, A. (1963). *C. R. Hebd. Seances Acad. Sci.* **257**, 1335–1338.

Ogawa, M., Tanaka, K., and Kasai, Z. (1975). *Agric. Biol. Chem.* **39**, 695–700.

Ogawa, M., Tanaka, K., and Kasai, Z. (1977). *Cereal Chem.* **54**, 1029–1034.

Öpik, H. (1968). *J. Exp. Bot.* **19**, 64–76.

Paulson, R. E., and Srivastava, L. M. (1968). *Can. J. Bot.* **46**, 1437–1445.

Posternak, S. (1905). *C. R. Hebd. Seances Acad. Sci.* **140**, 322–324.

Poux, N. (1963). *J. Microsc.* **2**, 557–568.

Poux, N. (1965). *J. Microsc.* **4**, 771–782.

Rest, J. A., and Vaughan, J. G. (1972). *Planta* **105**, 245–262.

Rickson, F. R. (1968). *Am. J. Bot.* **55**, 280–290.

Rost, T. L. (1971). *Protoplasma* **73**, 475–479.

Rost, T. L. (1972). *Am. J. Bot.* **59**, 607–616.

Rothwell, N. V. (1966). *Am. J. Bot.* **53**, 7–11.

Savelbergh, R., and Van Parijs, R. (1971). *Arch. Int. Physiol. Biochim.* **79**, 1040–1041.

Sharma, C. B., and Dieckert, J. W. (1975). *Physiol. Plant.* **33**, 1–7.
Shumway, L. K., Rancour, J. M., and Ryan, C. A. (1970). *Planta* **93**, 1–14.
Shumway, L. K., Cheng, V., and Ryan, C. A. (1972). *Planta* **106**, 279–290.
Simola, L. K. (1969). *Ann. Acad. Sci. Fenn. Ser. A4* **156**, 1–18.
Simola, L. K. (1973). *Ann. Bot. Fenn.* **10**, 71–88.
Simola, L. K. (1974). *Acta Bot. Fenn.* **103**, 1–31.
Singh, A. P. (1977). *Am. J. Bot.* **64**, 1008–1022.
Smith, C. G. (1974). *Planta* **119**, 125–142.
Sobolev, A. M. (1966). *Sov. Plant Physiol.* **13**, 177–183.
Sobolev, A. M., Sveshnikova, I. N., and Ivanstov. I. A. (1968). *Dokl. Bot. Sci.* **181**, 113–115.
St. Angelo, A. J., Yatsu, L. Y., and Altschul, A. M. (1968). *Arch. Biochem. Biophys.* **124**, 199–205.
Suvorov, V. I., Buzulukova, N. P., Sobolev, A. M., and Sveshnikova, I. N. (1971). *Sov. Plant Physiol.* **18**, 1020–1027.
Swift, J. G., and Buttrose, M. S. (1972). *J. Ultrastruct. Res.* **40**, 378–390.
Swift, J. G., and O'Brien, T. P. (1972). *Aust. J. Biol. Sci.* **25**, 9–22.
Sylvester-Bradley, R., and Folkes, B. F. (1976). *Sci. Prog. London* **63**, 241–263.
Takhtajan, A. L. (1969). *In* "Flowering Plants, Origin and Dispersal" (Transl. from the Russian by C. Jeffrey), p. 221. Smithsonian Inst. Press, Washington D.C.
Tanaka, K., Yoshida, T., Asada, K., and Kasai, Z. (1973). *Arch. Biochem. Biophys.* **155**, 136–143.
Tanaka, K., Yoshida, T., and Kasai, Z. (1974). *Soil Sci. Plant Nutr. (Tokyo)* **20**, 87–91.
Tanaka, K., Ogawa, M., and Kasai, Z. (1977). *Cereal Chem.* **54**, 684–689.
Tluczkiewicz, J., and Berendt, W. (1977). *Acta Soc. Bot. Pol.* **46**, 3–14.
Tully, R. E., and Beevers, H. (1976). *Plant Physiol.* **58**, 710–716.
Van Staden, J., and Comins, N. R. (1976). *Planta* **130**, 219–221.
Van Staden, J., Gilliland, M. G., and Brown, N. A. C. (1975a). *Z. Pflanzenphysiol.* **76**, 28–35.
Van Staden, J., Gilliland, M. G., Prewes, S. E., and Davey, J. E. (1975b). *Z. Pflanzenphysiol.* **76**, 369–377.
Villiers, T. A. (1971). *New Phytol.* **70**, 751–760.
Walker, K. A. (1974). *Planta* **116**, 91–98.
Werker, E., and Vaughan, J. G. (1974). *Planta* **116**, 243–255.
Williams, S. G. (1970). *Plant Physiol.* **45**, 376–381.
Yatsu, L. Y. (1965). *J. Cell Biol.* **25**, 193–199.
Yatsu, L. Y., and Jacks, T. J. (1968). *Arch. Biochem. Biophys.* **124**, 466–471.
Youle, R. J., and Huang, A. H. C. (1976). *Plant Physiol.* **58**, 703–709.
Zamski, E. (1973). *Isr. J. Bot.* **22**, 211–230.

Plant Vacuoles | 15

FRANCIS MARTY
DANIEL BRANTON
ROGER A. LEIGH

I. INTRODUCTION

Eukaryotic cells contain a variety of membrane-bound organelles, some of which, like mitochondria or plastids, are possibly the descendants of pro-karyotic ancestors, and others, such as the endoplasmic reticulum (ER) and the Golgi apparatus, are conveniently described as components of a continuous endomembrane system. The membranes of these organelles achieve a compartmentation of cellular functions which is of functional importance and is probably a prerequisite for the differentiation of highly organized, multicel-

The Biochemistry of Plants, Vol. 1

lular organisms. In the cells of multicellular animals, lysosomes and secretion vesicles are, aside from the endoplasmic reticulum itself, among the most conspicuous components of the endomembrane system. They are generally recognized as compartments that can mediate the exchange of solutes or insoluble components between the cell and its extraprotoplasmic environment. In apparent contrast, in the cells of multicellular plants, vacuoles are the most conspicuous component of the endomembrane system. They have long been known as compartments involved in "water metabolism," solute accumulation, turgor generation, and all related functions. This contrast reflects the fact that plants use intracellular rather than extracellular compartments for transporting, storing, and depositing organic and inorganic nutrients, metabolites, and waste products. Thus, plants have developed the vacuolar portion of the endomembrane system to serve many of the functions assumed in animals by intercellular spaces. Vacuoles must therefore be recognized as compartments that mediate the exchange of components between the cytoplasm and the cell's own intracellular, but nevertheless extraprotoplasmic, spaces.

This viewpoint makes it easy to understand that vacuoles are not unique to plants but, as now evident from compelling electron microscopic evidence, a differentiated elaboration of lysosomal structures found in all eukaryotes. This elaboration of the lysosomal system must be profoundly important to plant cells which multiply by forming a cell plate rather than by cleaving themselves with a furrow; and profoundly important to plant organisms which must expand to capture energy rather than move to catch food.

II. THE LYSOSOME CONCEPT

Lysosomes were originally defined (de Duve *et al.*, 1955) by differential centrifugation and biochemical assays as distinct, saclike particles surrounded by a membrane and containing acid hydrolases in latent form (Fig. 1). Subsequent electron microscopic studies revealed the wide polymorphism of subcellular structures fitting the biochemical description of lysosomes. It became clear that the word "lysosome" was an operational term defined on the basis of certain biochemical criteria that did not actually describe a physiological body. This ambiguity was consistently pointed out by de Duve himself when he wrote, " . . . the lysosome is not really a body as its name suggests; it is part of a system . . ." (de Duve, 1969). Only when considered as part of a system involved directly or indirectly in intracellular digestion does the term lysosome describe a physiological unit.

Although the biochemical approach to lysosomes has consistently led to a quantitative description of enzyme distribution in cellular fractions from

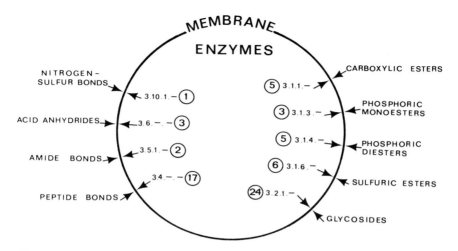

Fig. 1. The biochemical concept of lysosomes. The classification of the enzymes is that recommended by the IUPAC-IUB Commission on Biochemical Nomenclature. The classes of substrates are shown outside. The number of enzymes in each class that are judged to be lysosomal is indicated in a small circle. From Barrett and Heath (1977).

various tissues, the cellular compartmentation dynamics of these lytic functions has been unraveled by modern cytology and the use of electron microscope cytochemistry (see the reviews by Novikoff, 1973, 1976; Holtzman, 1975). The most reliable criteria for the identification of lysosomal activity are directly derived from biochemistry and require the demonstration of characteristic enzymes packaged within a membrane-bound structure accounting for the enzymatic latency. By chance, the most useful enzyme is the first enzyme to be studied in lysosomes by the biochemists, acid phosphatase.

Intracellular digestion of materials is the major physiological thread unifying the wide range of cellular processes included in the lysosome concept (Fig. 2). When materials to be digested are of extracellular origin, the digestive process is referred to as heterophagy. The extracellular materials are captured by endocytosis, which can be called phagocytosis or pinocytosis, depending upon the size of the engulfed particles. On the other hand, when components of a cell are digested inside the cell itself, the phenomenon is called autophagy, and macroautophagy and microautophagy may accordingly be distinguished.

These digestive phenomena depend upon the interaction of an endocytic or autophagic vesicle containing the material to be digested with a lysosome containing the hydrolytic enzymes. On a formal basis, the lysosomes in-

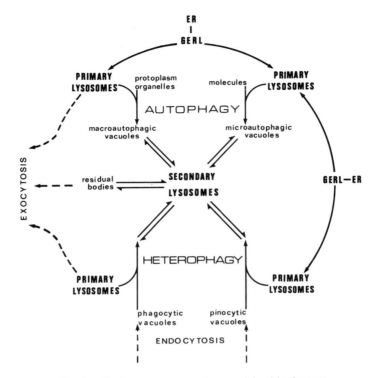

Fig. 2.   The lysosome concept, as explained in the text.

volved in the process are either primary lysosomes, whose enzymes have never been engaged in a digestive event, or secondary lysosomes, which are sites of present or past digestive activity (de Duve and Wattiaux, 1966). Thus, the intermixing of the substrates contained in an endocytic or autophagic vesicle with the digestive enzymes contained in a primary lysosome gives rise to the secondary lysosome.

De Duve (1969) has designated the potentially interconnectable locules housing these catabolic functions as the exoplasmic space of the cell. The exoplasmic space includes the locules of endocytic and exocytic vesicles, and the primary and secondary lysosomes. The anabolic functions concerned with the synthesis and packaging of macromolecules—including the hydrolytic enzymes that ineluctably move to the exoplasmic space of the lysosomes—take place in the endoplasmic space, consisting mainly of the endoplasmic reticulum. In animals and plants, the exoplasmic space mediates all interactions between the endoplasmic space and the extracellular environment. In plants, the exoplasmic space is elaborated to form vacuoles.

III. VACUOLE DEVELOPMENT

A. The Origin of Vacuole Precursors

Until recently, the origin of vacuoles was most often interpreted as swellings of ER vesicles (Buvat, 1958, 1965; Poux, 1961, 1962; Bowes, 1965; Gifford and Stewart, 1967; Matile and Moor, 1968; Mesquita, 1969; Berjak, 1972) or Golgi vesicles (Marinos, 1963; Ueda, 1966, Pickett-Heaps, 1967) and reviewed accordingly (Buvat, 1971; Matile, 1975). With improved morphological techniques and reliable data from enzyme cytochemistry, the complex biogenetic interactions between these major components and the vacuoles have been clarified.

In meristematic cells, which are the primary sites of vacuolation, the ER forms an extensive intracellular system of tubules interconnecting flattened cisternae which extend throughout the cytoplasm, from the perinuclear envelope to the cell periphery. The rough, ribosome bearing ER is frequently arranged in parallel cisternae and is much more abundant than the ribosome free, smooth ER which is mainly restricted to the immediate vicinity of the Golgi stacks. The small regions of smooth ER consist of anastomosed tubules. They form a widely fenestrated network, directly connecting the rough ER and the discoid elements of the Golgi stacks (Fig. 3, upper portion). These connections have been compared (Marty, 1973b,c, 1976, 1978) to the "boulevard périphérique" described in hepatocytes. In the past, most of the thin sections through the "boulevard périphérique" were erroneously considered to be Golgi-derived vesicles. Because of the fenestration and the thinness of the sections, the direct continuities between rough ER and Golgi apparatus were rarely seen in conventional electron microscopy. They became clearly demonstrable when thick sections were examined by high-voltage electron microscopy. At one face of the Golgi stack, this specialized region of endoplasmic reticulum is GERL, the region of smooth ER that is located at the trans face of the Golgi apparatus and which appears to produce lysosomes (Novikoff, 1973, 1976). It consists of a twisted, smooth-surfaced polygonal meshwork of anastomosing tubules with small saccular regions facing the Golgi stacks (Fig. 4a–d).

The relationship of GERL to the other elements of the Golgi stacks is probably functionally significant, but the metabolic interchanges that occur are still unknown. It has been shown that GERL is directly involved in the biogenesis of the primordial vacuole precursors or provacuoles. GERL is a specialized region of the endoplasmic reticulum where acid hydrolases are packaged into the nascent provacuoles (Fig. 4b and d). These provacuoles are by definition primary lysosomes. It has been suggested that provacuoles may also directly package a variety of newly synthesized products, such as peroxidase in common cells or polyterpenic granules in coenocytic laticifers.

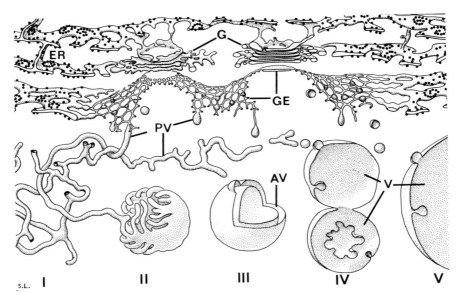

Fig. 3. Stages in the formation of vacuoles. AV, autophagic vacuole; ER, endoplasmic reticulum; G, Golgi stack; GE, GERL; M, mitochondrion; PV, provacuole; V, vacuole. ER, Golgi apparatus, and GERL are extensively drawn in the upper part of the diagram because they are present at any step of vacuole maturation. The formation of vacuoles during cell differentiation is shown from left to right in the lower portion of the drawing. Steps I–V as explained in the text. From Marty (1978).

B. Autophagy and the Formation of Vacuoles

In parenchyma cells of root meristems, where they have been best studied, nascent provacuoles elongate into branched, sinuous, 0.1-μm diameter tubes that form continuous tracts throughout the cytoplasm (Figs. 3,I and 4e,f; Marty, 1973a, 1978). This tridimensional organization was ascertained from thick sections of tissue studied by high voltage electron microscopy (Marty, 1973c, 1976; Poux *et al.*, 1974) and has been observed in a wide range of species. These tubular provacuoles were not described earlier because they are extremely fragile and readily broken into innumerable vesicles when drastic chemical fixations are used for electron microscopy or when shearing forces are used for homogenization of the tissue. On the other hand, they are reminiscent of the reticular aspect of the vacuolar system (Marty, 1973a; Buvat, 1977) seen by light microscopy in meristematic cells vitally stained with dilute solutions of neutral red (see reviews by Guilliermond *et al.*, 1933; Dangeard, 1956; Buvat, 1971). They have been characterized using modern cytochemical techniques as a compartment containing a broad set of digestive enzymes (Marty, 1973a, 1978).

In the next stage, the provacuoles undergo a program of cellular autophagy (Marty, 1972, 1973a, 1976, 1978). The lysosome tubes wrap themselves around portions of cytoplasm, like the bars of a cage (Figs. 3,II, and
5a,b). Subsequently, the tubes of each sequestering cage merge laterally to
surround the previously wrapped portion of cytoplasm with a continuous
exoplasmic space loaded with digestive enzymes (Figs. 3,III and 5c,d). Up to
this step of vacuolation, the lysosomal enzymes are kept rigorously in check
within the exoplasmic space of the enwrapping, fusing tubes as well as in the
provacuole tubes from which they derive. However, some time after the
sequestered portion of cytoplasm has been completely closed off by the
lateral fusion of wrapping tubes, the hydrolases are released from the surrounding exoplasmic space into the sequestered cytoplasm (Fig. 6a). Thus, it
is now clear that the fusion of wrapping tubes produces a fairly large, spherical, shell-like lysosome and that the volume of cytoplasm delimited by this
lysosomal shell is actually the autophagic vesicle. Unlike a heterophagic
vesicle, an autophagic vesicle shares its delimiting membrane with the surrounding lysosome.

Although the mechanism remains to be established, Marty has speculated
(1974, 1978) that the unidirectional release of enzymes into the sequestered
cytoplasm of the autophagic vesicle is the result of the physical alteration of
the inner membrane (i.e., the membrane common to the autophagic vesicle
and the surrounding shell-like lysosome) of the autophagic system. The inner
membrane of the shell is sequestered away from the cytoplasm and may thus
be deprived of interchanges required for its integrity. Conversely, the outer
membrane of the shell is in close contact with the cytoplasm and remains
unaltered and impermeable to the lysosomal hydrolases whose digestive
activities are thus confined, *intra muros,* within the forming vacuole. The
outer membrane which circumscribes the digestive activities and prevents
cellular autolysis finally becomes the vacuole membrane (or tonoplast). Because it arises from the intermixing of the contents of a lysosome and an
autophagic vesicle, the vacuole formed at the end of the autophagic process
may be regarded as a plant residual body—a kind of secondary lysosome
containing indigestible materials.

Several temporaly overlapping autophagic cycles, initiated asynchro-
nously at different foci within one cell, may be involved in vacuolation (Symillides and Marty, 1977). Direct connections between components belonging to
the different cycles are observed, and these repetitive autophagic phenomena may be compared to the successive "meals" of animal lysosomes. Once
formed, vacuoles in the central region of the cell fuse rapidly together, and
this fusion process spreads outwards reaching peripheral young vacuoles.
Thus, unlike most animal residual bodies, plant vacuoles fuse (Figs. 3,IV and
6b), behave as powerful osmotic systems, and finally give rise to the few
large vacuoles characteristic of the differentiated plant cells.

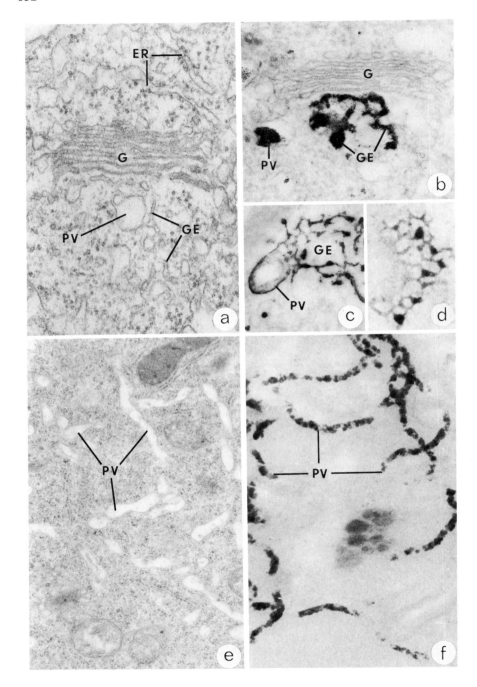

In already vacuolated, actively growing cells, studies to date indicate that provacuoles are transferred directly from GERL to the preexisting large vacuoles (Fig. 3,V). By merging with the already existing vacuoles, the GERL-derived provacuoles would account for tonoplast enlargement and vacuolar content accretion during cell growth (Marty, 1978). A complement of digestive enzymes and/or secondary metabolic products can be transported intracellularly by the provacuoles and directly accumulated in the large central vacuole (Prat *et al.*, 1977).

During their early sequential differentiation and their subsequent activity in differentiated cells, the provacuoles and vacuoles may specialize in the exclusive transport and accumulation of a compound, such as anthocyanins, glycosides, or proteins. These specializations may be restricted to a few cells in a tissue or to special vacuoles in a single cell. A high degree of specialization in vacuoles is encountered in the cells of fruits and flowers in higher plants and in the cells of algae. On the catabolic side of their activity, the large vacuoles are able to engulf bits of cytoplasm by focal invagination of the tonoplast (Figs. 3,IV and V and 6b; Matile and Moor, 1968; Villiers, 1971). These portions of cytoplasm, including even organelles, are digested within vacuoles which still house a wide variety of hydrolases (Poux, 1970; Villiers, 1971; Marty, 1972). The extent to which this autophagic activity in large vacuoles may account for the physiological turnover of cytoplasm and organelles in fully differentiated plant cells remains an open question.

C. Endocytosis and Heterophagy

Can plant cells, despite their surrounding pecto-cellulosic wall, incorporate exogenous material in a manner similar to the phagocytosis in protists, slime-molds, and a number of animal cells? Except for a few specialized cells, no definitive answers are available.

Light microscopic studies have been largely inconclusive (Bradfute *et al.*, 1964, Drew *et al.*, 1970). By electron microscopy, deep invaginations of the plasmalemma, interpreted as transient steps in the formation of endocytic vesicles, have been observed in many plant cells since the early description of Buvat (1958). However, all subsequent attempts to demonstrate that exogenous material may actually undergo intracellular digestion have had various experimental shortcomings which make them inconclusive.

Fig. 4. Portions of meristematic cells from the root tip of *Euphorbia characias* L. Abbreviations as in Fig. 3. (a) A dictyosome in a meristematic cell ($\times$ 57,600). (b) Meristematic cell incubated with β-glycerophosphate for the demonstration of acid phosphatase activity ($\times$ 64,000). (c) GERL and provacuole stained with zinc-iodide osmium ($\times$ 51,500). (d) GERL stained for the demonstration of acid phosphatase ($\times$ 46,000). (e) Elongated provacuoles in a vacuolating cell ($\times$ 31,250). (f) Provacuolar tubes stained for acid phosphatase activity ($\times$ 25,000). From Marty (1978).

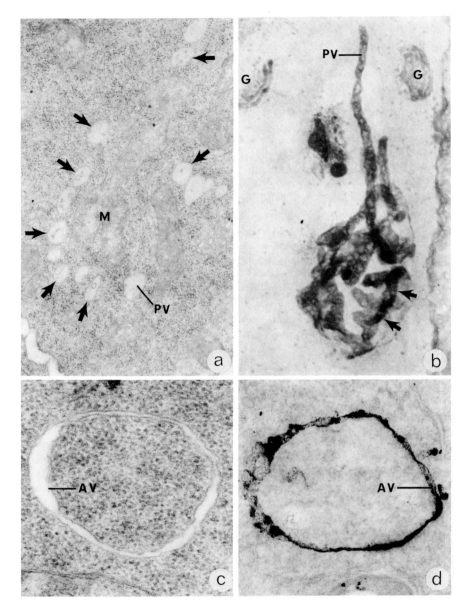

Fig. 5. Portions of meristematic cells from the root tip of *E. characias*. Abbreviations as in Fig. 3. (a) Thin equatorial section through a lysosomal cage formed by enwrapping provacuolar tubes (× 18,900). (b) Section 1 μm thick including part of a cage (stained with zinc iodide osmium, × 18,000). (c) Typical thin section through a shell-like autophagic vacuole in a vacuolating cell (× 37,100). (d) Autophagic vacuole in a root cell, incubated for the demonstration of acid esterase activity (× 41,200). From Marty (1978).

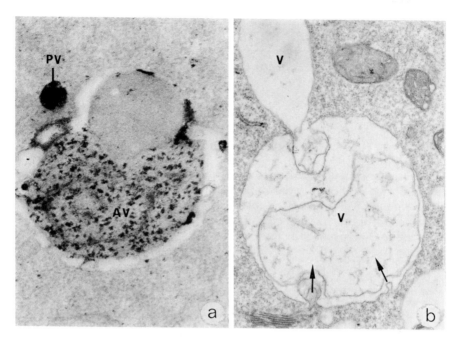

Fig. 6. Portions of meristematic cells from the root tip of *E. characias.* Abbreviations as in Fig. 3. (a) Autophagic vacuole in a vacuolating root cell, incubated for acid phosphatase activity (× 42,100). (b) Young vacuole in which the sequestered cytoplasm is almost degraded (× 20,600). From Marty (1978).

An attractive example of what might be interpreted as an evolutionary attempt at heterophagy in plant cells is shown in the symbiotic nitrogen-fixing cells of the root nodules of leguminous plants (Truchet, 1976; Robertson *et al.,* 1978; Verma *et al.,* 1978). In these cells, the prokaryotic bacteroids are taken into the eukaryotic host cells by a mechanism related to endocytosis. The bacteroid is enveloped by a membrane which is derived from the plasmalemma of the host cell through an invagination process. Although the invaginated host cell membrane retains many of its characteristics, it successively loses the ability to synthesize a cell wall and, subsequently, its complex biosynthesis is mediated by the Golgi apparatus of the host cell. Several digestive activities have been shown to occur within the endocytic vesicle housing the bacteroid. At first, during the initial endocytic process, cellulolytic enzymes destroy the engulfed cell wall material, and the bacteroid is brought in close proximity to the host endocytic membrane. Subsequently, at the terminal phase of symbiosis, hydrolases digest and kill the old bacteroid in the senescing cells. Although it is true that the endocytic vesicles containing bacteroids can maintain luminal continuity with the extracellular environment and are apparently not transferred to preexisting vacuoles, symbiosis does require internalization of an extracellular space (up to 80% of the host cell volume is involved) in which intracellular digestion does occur.

D. Exocytosis

1. General Aspects

The large osmotically sensitive vacuoles of multicellular plants have never been reported to be excreted outside the cell into the periplasmic space. They remain within the living cells, as do the residual bodies (secondary lysosomes) of most animal cells. However, exocytosis—the mechanism by which membrane-packaged materials are transported through the cytoplasm and transferred across the plasma membrane for export—is commonly found in higher and lower plant cells. Some cells, specialized in this particular function, are known as gland cells in higher plants. The most general function of exocytosis is related to the biosynthesis of the cell wall. Golgi-derived vesicles contribute, at once, to the enlargement of the plasma membrane and the secretion of the cell wall matrix, including pectins, hemicelluloses, and glycoproteins and various enzymes, including acid hydrolases (Roland, 1973). This function is well-documented in Chapter 12, this volume. A similar but hyperactive exocytic mechanism is responsible for the secretion of extracellular slimes such as those of the root cap and is beyond the scope of this discussion (Mollenhauer *et al.*, 1961; Paull and Jones, 1976; Rougier, 1976). The formation of scales (Brown and Romanovicz, 1976) and the production of extrusive organelles (''extrusomes,'' Hausmann, 1978) may be viewed as highly specialized exocytosis.

2. Digestive Gland Cells

Some plants contain gland cells that are highly specialized in the extracellular secretion of digestive enzymes. Aleurone cells of cereal grains and epidermal gland cells from the leaves of carnivorous plants excrete a number of hydrolases, some of which are presumed to be released by exocytosis. However, in both cases, the mechanisms involved are ill-defined at the subcellular level.

In the cereal grain during the early process of germination, the plant embryo secretes gibberellic acid, which stimulates the aleurone cells to secrete a number of hydrolytic enzymes finally used for the digestion of the starchy endosperm. These enzymes may be secreted by exocytic processes (discussed by Chrispeels, 1977).

In the gland cells of carnivorous plants, the subcellular mechanism of export for the digestive enzymes is poorly documented. It is currently claimed that in some instances the hydrolases diffuse directly through the plasmalemma, whereas in others, as in the sundews, the transfer occurs through local breakages of the plasma membrane, when the rapid secretion follows the trapping of the prey. Hence, the enzymes would be stored in the vacuoles and in the outer convoluted cell wall of the resting cells until they would be flushed out at the surface by an outward flow of fluid induced by the capture event (Heslop-Harrison, 1975). Biochemical data, corroborated by subcellular cytochemistry, will be required to improve the resolution of this scheme.

3. Contractile Vacuoles

Many protozoa, unicellular algae, and fungi possess contractile vacuoles, also termed water expulsion vesicles. They are osmoregulatory organelles which pulsate rythmically and expel hypotonic fluid out of the cell. The contractile vacuoles are infrequent in marine organisms and absent in terrestrial plants. Although controversial in points, the mechanism for extracellular discharge may be interpreted as a specialized exocytosis. Water and low molecular weight material are expelled, but the exocytic vesicle membrane remains associated with the plasmalemma for only a very short time. It is immediately pinched off and reused in new cycles of discharge. In many instances, as in *Chlamydomonas reinhardtii* (Weiss *et al.*, 1977), the contractile vacuole system consists of many small vesicles or tubular channels which surround and become confluent with a large main cisterna beneath the plasma membrane. In the alga *Vacuolaria virescens*, tubules have not been described, but subsidiary vacuoles, primarily associated with the Golgi apparatus, similarly empty their contents in the contractile vacuole (Heywood, 1978). The pulsation cycle is composed of a diastolic phase, during which fluid accumulates within the satellite vesicles and tubules merge with the main vacuole; and a systolic phase, during which the main vacuole contracts, and its content is discharged in the extracellular medium. In *C. reinhardtii*, the water and low molecular weight material have been reported to pass directly through hydrophilic channels which are created in a circular zone of membrane contact between the contractile vacuole and the plasma membrane (Weiss *et al.*, 1977). A different mechanism has been described in the slime-mold *Dictyostelium discoideum*, where discharge seems to occur through a large pore quickly sealed thereafter (De Chastellier *et al.*, 1978). Both models are reminiscent of Satir's concepts of membrane gate release and membrane fusion release for the discharge of secreted products (Satir, 1974a,b).

Although alkaline phosphatase has been found in the contractile vacuole of *D. discoideum* (Quivigier *et al.*, 1978), the enzymatic equipment of these vacuoles is still poorly documented.

E. Specializations

1. Maturational Changes

Differentiation in plants generally implies the specialization of the cell walls and/or of the vacuoles of the cells. Most often the functional specialization of the vacuole in the transport and accumulation of substances does not dramatically affect the morphology of the organelle. The hallmarks of some specialization are thus found in the chemical composition of the vacuolar sap and in the enzymatic activities of the surrounding membrane. These biochemical aspects are discussed later in the chapter. However, in highly specialized tissues such as the vascular tissues and the laticifers, the vacuole organization is conspicuously related to the functional specialization of the cells.

In young sieve cells, a tonoplast delimits the vacuole which can be selectively stained by neutral red, but at maturity usually no tonoplast can be identified in these fragile elements, even by electron microscopy (Esau, 1969; Cronshaw, 1974; Evert, 1977). The plasmalemma persists and is probably responsible for the osmotic behavior of the element. Therefore, the disorganized protoplasm and vacuolar sap appear to be mixed, and neutral red no longer accumulates in the mature sieve element. The vacuole behavior is different in xylem vessels where the cell dies after the end walls are perforated (Charvat and Esau, 1975; Roberts, 1976). Tonoplast is assumed to play a key function in this programmed cell death, which is somewhat reminiscent of the senescence process in ephemeral organs, such as some flower petals (Matile and Winkenbach, 1971). The tonoplast loses its integrity, and it is postulated that the hydrolases act directly on the cytoplasm and eventually on certain parts of the cell wall (Woodzicki and Brown, 1973). These are among the rare cases where vacuoles do act as "suicide bags," an activity emphasized in an early but now outdated conception of lysosome function (de Duve, 1969).

The vacuoles of laticifers are highly specialized organelles in very different ways. In the articulated laticifers which derive from contiguous cells whose intervening walls are partly or totally destroyed, the small globular vacuoles fuse together or remain in clusters in the axis of the mature cells (Esau, 1975; Nessler and Mahlberg, 1976; Giordani, 1977). They are referred to as lutoids in the laticifers of the rubber tree. Because they may be directly collected by tapping the latex vessels in the trunk and fractionated by methods conventionally used for animal lysosomes, they are probably, until now, the most extensively studied vacuoles in higher plants.

The vacuoles of the nonarticulated laticifers stand in sharp contrast to the lutoids of articulated laticifers (Marty, 1974). The nonarticulated laticifers are single gigantic multinucleate cells with a continuous central ductlike vacuole. These laticifers originate as single cells in the nodal region of the embryo and develop coenocytically with the plant body, like a fungal hypha. In the early process of embryogenesis, the initial cells of the laticifers are thought to vacuolate by an autophagic process like that which is now known to occur in meristematic cells. But, instead of terminating with the formation of the central vacuole, as in other cells, autophagy is continuous in the coenocytic articulated laticifers (Marty, 1970, 1974). The central vacuole extends tipward in a muglike shape. The rim of the mug is formed by the continuous accretion of fusing vacuole precursors which, moving continuously tipward after pinching off from GERL, circumscribe the central portion of the dense protoplasm before fusing with the rim of the mug-shaped end of the central vacuole. Tubular branched provacuoles arise from GERL and develop within the mug as well as in the parietal regions of the cytoplasm outside of the mug. In the mug, especially near its bottom, the cytoplasm becomes subdivided by these proliferating wrapping sheets of provacuoles which fuse and subsequently release their hydrolases to digest the autophagic vesicle they contain. The central vacuolar duct thus grows continuously as these autophagic vesicles are digested. The cytoplasm around the duct derives from the parietal layer unaltered by the autophagic process but extensively drained by tubular provacuoles that are connected to the central duct. Because of its autophagic origin, the duct of nonarticulated laticifers may be viewed as the largest lysosome ever described in eukaryotes. In addition to this gigantic autophagic process, small vacuolar components are involved in the removal of polyterpenic granules by more localized preferential autophagy (Marty, 1971). Provacuoles are continuously involved in packaging and transporting of the newly synthesized polyterpenic granules to the central duct. Therefore, the latex accumulating in the duct is a complex mixture of substances resulting from both the catabolic and anabolic functions of this highly specialized vacuolar system: some of the latex components are the products of autophagic digestion, and others, like polyterpenic granules, are directly products of synthesis.

2. Postdifferentiation Changes

During the cell life, the vacuoles may undergo drastic changes, encompassed by the term "vacuole instability." Sudden changes may be repeated many times during the life of a cell. They are usually related to rapid cellular movement of water. Such reversible fast changes involving the apparent breakage of a few large vacuoles into innumerable small vesicles have been described, for example, in the epidermal cells of the stalk of tentacles in carnivorous plants (Mangenot, 1929; Burian and Hofler, 1966), in algal cells, in yeast cells (Guilliermond et al., 1933), and in the cambial cells of trees (Catesson, 1974). Fast changes may also include rapid fluctuations in vacuolar pressure, such as those responsible for movements in plants. A loss of turgor by the cells causes a concomitant contraction and subsequent bending of the organ. Recovery of cell turgor presumably reverses the process, but the few histological studies carried out thus far have not clearly established the function of the vacuoles in these processes (Toriyama, 1967; Satter *et al.*, 1970).

Slow changes are mostly related to the physiological dessication of cells and organs in relation to environmental stresses. Such slow vacuole changes are directly involved in the formation of protein bodies during seed development and their fate during germination (see Chapter 14). Similar changes have been reported in many pollen grains and spores and in the cells of the vascular cambium. Besides their fast spontaneous modifications, the vacuoles of cambial cells undergo seasonal changes; active cambial cells in summer are highly vacuolate, whereas the cells of the quiescent cambium in winter contain numerous protein bodylike vacuoles (Catesson, 1974).

IV. VACUOLE FUNCTION

A. Hydrolytic Activities

Evidence showing the presence of lytic enzymes in vacuoles has been extensively reviewed by Matile (Matile, 1969, 1974, 1975, 1976; Matile and Wiemken, 1976). Acid hydrolases are readily detectable in plant tissue extracts (Matile, 1975), and recent evidence suggests that these hydrolases are localized within a separate cell compartment. Larger proportions (up to 50%) of hydrolases can be sedimented if meristematic tissues are used, and attempts have been made to fractionate the hydrolase-containing structures. They appear to be small (0.1–1.5 μm diameter), heterogeneous, membrane-bound structures which can often be resolved into several subpopulations (Matile, 1968; Balz, 1966; Hirai and Asahi, 1973; Pitt and Galpin, 1973; Parish, 1975a,b). Lysosome-like structures from maize root tips could be separated into "heavy" and "light" fractions which differed in their relative proportions of hydrolases, transaminases, and oxidoreductases (Matile, 1968). The origin of many of the structures isolated in the above studies was not determined, but it was suggested that some of the structures isolated from maize root tips were provacuoles (Matile, 1968; Parish, 1975a). In differentiated cells, hydrolases (e.g., acid phosphatase and esterase) can be readily detected in mature vacuoles by cytochemical techniques, and biochemical studies of vacuoles from tissues and cells where they are more amenable to isolation have confirmed this. Thus, hydrolases have been detected biochemically in vacuoles from giant algae (Luscher and Matile, 1974; Doi *et al.*, 1975), yeast (see Matile and Wiemken, 1976), laticifers of *Hevea brasiliensis* and *Chelidonium* (Pujarniscle, 1968; Matile *et al.*, 1970), and several different, mature, higher plant tissues (Table I). In addition, it has been suggested that the presence of cytoplasmic inclusions within vacuoles indicates that these hydrolases are actively involved in turnover of cytoplasmic components. However, it is also possible that the inclusions are remnants of the autophagic processes that lead to vacuole formation rather than the result of active ingestion. Butcher *et al.* (1977) found that of nine hydrolases detecta-

TABLE I

Acid Hydrolase Activities in Purified, Intact Vacuoles from Mature Higher Plant Tissues

Tissue used	Isolation method	Hydrolases associated	Reference
Hippeastrum petals	Protoplast lysis	RNase, DNase, acid phosphatase	Butcher *et al.* (1977)
Lycopersicon esculentum leaves	Protoplast lysis	Acid phosphatase, carboxypeptidase	Walker-Simmons and Ryan (1977)
Bryophyllum daigremontianum leaves	Protoplast lysis	Acid phosphatase, RNase	Buser and Matile (1977)
Beta vulgaris root storage organ	Mechanical slicing	Acid phosphatase, acid invertase[a]	Leigh *et al.* (1979); Goldschmidt and Branton (1977)
Ricinus communis endosperm	Protoplast lysis	Acid phosphatase, acid protease, phosphodiesterase, RNase, phytase, β-glucosidase	Nishimura and Beevers (1978)
Armoracia lapathifolia roots	Mechanical slicing	Acid phosphatase	Grob and Matile (1979)
Sorghum bicolor leaves	Protoplast lysis	Dhurrin β-glucosidase	Kojima *et al.* (1979)
Nicotiana rustica leaves	Protoplast lysis	Acid phosphatase	Saunders (1979)
Nicotiana tabacum pith-derived cells in culture	Protoplast lysis	Nuclease β-fructosidase (invertase)	Boller and Kende (1979)
Tulipa sp. petals	Protoplast lysis	α-galactosidase	Boller and Kende (1979)
Ananas comosus leaves	Protoplast lysis	Proteinase α-galactosidase	Boller and Kende (1979)
Nicotiana, Ananas, and *Tulipa*	Protoplast lysis	α-Mannosidase β-N-acetylglucosaminidase acid phosphatase phosphodiesterase	Boller and Kende (1979)

[a] Acid invertase was not detectable in vacuoles from fresh beet tissue but was present after the tissue had been aerated in distilled water for 2–3 days.

ble in *Hippeastrum* protoplasts, only acid phosphatase, RNase, and DNase were recovered in the vacuole-enriched fraction. The remainder (esterase, protease, carboxypeptidase, β-galactosidase, α-glycosidase, and β-glycosidase) were recovered in the cytosol-enriched fraction. No attempt was made to study further the localization of these hydrolases, but it was suggested that, *in vivo,* they might be associated with extravacuolar structures.

Unfortunately the results of Butcher *et al.* (1977) are not without problems of interpretation. The authors did not report a balance sheet of enzyme recoveries; thus, the possibility of enzyme activation or inhibition cannot be judged. Although ultrafiltration of fractions prior to assay made inhibition by low molecular weight compounds unlikely, inhibition by macromolecules cannot be discounted. For example, isolated vacuoles from tomato leaves contain proteins capable of inhibiting proteinases (proteinase inhibitors I and II; Walker-Simmons and Ryan, 1977). Similar proteins capable of inhibiting other hydrolases could have affected Butcher *et al.*'s results. This could have been tested by recombining cytosol and vacuole fractions and determining whether the hydrolases in the cytosol fraction were inhibited by factors present in the vacuole-enriched fraction. Also, it is feasible but less likely that the isolation procedures used by Butcher *et al.* (1977) selected for a subpopulation of vacuoles enriched only in the three hydrolases detected. As the cytosol-enriched fraction contained some hydrolases released from vacuoles which burst during isolation, the possibility that the "cytoplasmic" hydrolases were derived from broken vacuoles, other vacuoles, or vacuole precursors cannot be entirely discounted. There are certainly cases in which several different kinds of vacuoles, containing different enzymes, metabolites, or pigments coexist in one organ and even one cell. The observations of Conn and his collaborators exemplify such a compartmentation. In the leaf blades of *Sorghum,* the cyanogenic glucoside, dhurrin, was located entirely in the vacuoles of the epidermal layer, whereas the two enzymes responsible for its catabolism, i.e., dhurrin β-glucosidase and hydroxynitrile lyase, resided almost exclusively in the mesophyll tissue. (Saunders and Conn, 1978; Kojima *et al.,* 1979). An analogous situation might account for the stability of glucosinolates in the presence of myrosinase in horseradish root cells (Grob and Matile, 1979).

Nishimura and Beevers (1978) have found that six hydrolytic enzymes (acid protease, carboxypeptidase, phosphodiesterase, RNase, phytase, and β-glucosidase) were present in the isolated vacuoles from castor bean endosperm. In distinction to Butcher *et al.*'s findings, the relative amounts of the hydrolases studied by Nishimura *et al.* indicated a primarily vacuolar localization *in vivo.* More recently, Boller and Kende (1979) have shown that the intracellular activities of seven acid hydrolases, i.e., α-mannosidase, β-N-acetylglucosaminidase, β-fructosidase, nuclease, phosphatase, phosphodiesterase and proteinase were primarily localized in the vacuoles of higher plant tissues. They concluded that the central vacuole of higher plant cells has an enzyme composition analogous to that of the animal lysosome.

B. Solute Accumulation

Vacuoles accumulate solutes. This implies that specific mechanisms must exist to transport solutes into the vacuole. It is often assumed that solutes are transferred from the cytoplasm to the vacuole by the intervention of specific tonoplast-bound carriers, but this may be an oversimplification, as net transfer of solutes to the vacuole could also occur if the solutes were transported within membrane-bound vesicles which fused with preexisting vacuoles as part of the processes of vacuole formation or maintenance. MacRobbie (1976) has discussed the kinetic evidence for the involvement of such transfer processes in the accumulation of ions within the vacuoles of giant algal cells. Although transport within vesicles may be of quantitative importance in the movement of macromolecular compounds, such as enzymes or tannins, into the vacuoles of higher plants, there is no good evidence for the involvement of similar mechanisms in solute transport in mature higher cells.

1. Salts

Salts are quantitatively the most important solutes in the plant cell, and the general processes involved in their uptake and transport have been extensively reviewed (MacRobbie, 1970, 1971, 1974, 1977; Anderson, 1972; Baker and Hall, 1975; Lüttge and Pitman, 1976). The most detailed studies of ion accumulation in the vacuole have been made in giant algal cells where separation of the cytoplasmic and vacuolar phases can be achieved. In *Nitella translucens,* the vacuole contains 75 mM K$^+$, 65 mM Na$^+$, and 150–170 mM Cl$^-$ when the cell is grown in 0.1 mM K$^+$, 1.0 mM Na$^+$, and 1.3 mM Cl$^-$ (MacRobbie, 1974). The ion concentrations in the vacuole are significantly different from those in the cytoplasm immediately adjacent to the tonoplast (119 mM K$^+$, 14 mM Na$^+$, 65 mM Cl$^-$). Analysis of the electrochemical potential gradients for each of the ions indicates that both Na$^+$ and Cl$^-$ are actively transported into the vacuole, whereas K$^+$ is passively distributed (MacRobbie, 1974). In addition, giant algae maintain a lower pH in their vacuole ($<$ 5.5; Roa and Pickard, 1976) than in the cytoplasm (about pH 7.5; Walker and Smith, 1975; Spanswick and Miller, 1974) which is the result of active H$^+$ transport into the vacuole.

The small size of higher plant cells has not permitted a complete description of their ionic relations. Direct measurements of ion concentrations in the vacuole, *in situ,* can be made using ion-sensitive microelectrodes, but this technique cannot be used to measure cytoplasmic ion concentrations in the same cells. Reasonably comprehensive descriptions of the ionic relations of the vacuole can be obtained by analysis of steady-state kinetics. Using this technique, Pierce and Higinbotham (1970) were able to show that K^+, Na^+, and possibly Cl^- were actively transported into the vacuoles of *Avena* coleoptile cells. Because such studies must be made under steady-state conditions, it is difficult to assess the response of tonoplast transport processes to changing environmental conditions.

Isolated vacuoles offer a more convenient system for studying ion transport at the tonoplast. They are simple, bounded only by the membrane in question, they retain their *in vivo* orientation and their semipermeability upon isolation (Lin *et al.*, 1977a), and they lack an intrinsic energy source. Thus, they can be used to study many facets of ion transport, including compartmentation of ions, fluxes across the tonoplast, and the biochemical coupling of ion pumps. Lin *et al.* (1977a) have found that isolated *Tulipa* petal vacuoles contained 148 mM K^+, 50 mM Na^+, 33 mM Mg^{2+}, 8 mM Ca^{2+}, and 32 mM Cl^-. The balance of charge was assumed to be fulfilled by organic anions. In addition, by spectrophotometric analysis of the anthocyanin pigments in vacuoles, they were able to show that the pH gradient maintained across the tonoplast *in vivo* was abolished during isolation, possibly due to the removal of ATP. The concentrations of the ions in the vacuoles were the same as in the protoplasts from which they were isolated. This indicates that ions did not leak from vacuoles while the vacuoles were being isolated from the protoplasts, but the possibility of leakage during protoplast production and isolation was not specifically considered.

2. Organic Anions

In addition to inorganic ions, higher plants also accumulate large quantities of organic anions in their cells, particularly malate, citrate, and oxalate. There is a large body of indirect evidence which indicates that these must be compartmented in the vacuole (Osmond, 1976). Recently, Buser and Matile (1977) isolated vacuoles from the crassulacean acid metabolism plant, *Bryophyllum diagremontianum* and showed that all of the malate present in isolated *Bryophyllum* protoplasts was localized in the vacuole. Preliminary qualitative studies for oxalate, citrate, and isocitrate have been reported (Wagner, 1979). The available evidence suggests that malate accumulated in vacuoles is synthesized by a sequence involving phosphoenol pyruvate carboxylase and malate dehydrogenase (Osmond, 1976). Malate produced by this coupled reaction turns over more slowly than malate produced from acetate (Lips and Beevers, 1966), suggesting that it is rapidly transferred to the vacuole after synthesis. This may indicate that the malate-synthesizing enzymes (which are normally isolated in the soluble phase of the cell) interact loosely with the cytoplasmic surface of the tonoplast. No investigations of the possible association of these enzymes with isolated vacuoles or tonoplast have been reported.

The mechanisms by which ions are actively transported into the vacuole are not known. Although Lin *et al.* (1977b) found evidence for an ATP-dependent H$^+$ transport into isolated *Tulipa* petal vacuoles and ion-stimulated ATPases have been detected in preparations of isolated tonoplast (d'Auzac, 1977; Lin *et al.*, 1977a), direct correlations between the properties of these enzymes and the movement of specific ions across the tonoplast have not yet been established. The results of Lin *et al.* (1977a) and d'Auzac (1977) indicate that tonoplast ATPases isolated from different tissues differ in their response to salts making it difficult to draw any firm conclusions about common mechanisms of ion transport into vacuoles of different cells. In addition, it is not known to what extent all ion transport can be explained in terms of ATPase activities or whether other enzyme systems are involved in ion transport across the tonoplast.

3. Amino Acids

Plant cells contain significant quantities of free amino acids, including some not normally found in proteins. The composition of the soluble amino acid pool may differ significantly from that of the protein amino acid pool (Beevers, 1976), indicating that some of the amino acids may be sequestered in a separate cell compartment and do not participate in protein synthesis. In addition, studies with radioactive tracers indicate the presence of at least two pools of amino acids in plant tissues, only one of which is turning over rapidly. It has been suggested that the "inactive" pool is localized in the vacuole (Holleman and Key, 1967). Wiemken and Nurse (1973) were able to sequentially extract the cytoplasmic and vacuolar amino acid pools from the yeast, *Candida utilis*. In arginine-grown cells, up to 90% of the cellular α-amino nitrogen was in the vacuole with the basic amino acids, lysine, ornithine, and arginine, predominating. Similar results were obtained with *Saccharomyces cerevisiae* by Wiemken and Durr (1974). In *Hippeastrum* petals, Wagner (1977) found that 22% of the cellular serine, 25% of the glutamate, 21% of the glutamine, and 80% of the tryptophan were localized in isolated vacuoles. Wagner (1979) has also reported that glutamine, the predominant free amino acid found and γ-aminobutyric acid, a common constituent in higher plants were extravacuolar in the three tissues he investigated. However, many other amino acids showed a distribution different from tissue to tissue (Wagner, 1979), and controls that rule out accidental release of organelle bound materials are always difficult. The amino acids stored in the vacuoles of yeast and higher plant cells probably act as a reserve of soluble nitrogen which is utilized in times of nitrogen deficiency. The free amino acid content of carrot explants grown on media containing different amounts of NH_4NO_3 are consistent with this (Mott and Steward, 1972). The mechanisms by which amino acids are moved into and out of the vacuole are not known. If proteins are degraded within the vacuole, this could lead to the build up of amino acids within the vacuole. Boller *et al.* (1975) demonstrated a passive

L-arginine exchange system in isolated yeast vacuoles. The arginine is firmly retained in the isolated vacuoles despite the presence of a permease which mediates arginine diffusion through the tonoplast. The retention in the vacuoles is due to binding by polyphosphate (Durr *et al.*, 1979).

4. Sugars

Some plant tissues (e.g., sugarcane stem) can store large quantities of sugars, particularly sucrose, in a compartment which effectively removes them from metabolism (Oaks and Bidwell, 1970). By using isolated beet vacuoles, it has been shown that the vacuole is the site of sucrose storage in beet root tissue (Leigh *et al.*, 1979; Doll *et al.*, 1979). The mechanism of sucrose transport across the tonoplast of beet is not known, but experiments with whole tissue suggest that sucrose accumulation within plant cells need not involve sucrose hydrolysis (Giaquinta, 1977). In beet, stored sucrose is mobilized by the production of a vacuolar acid invertase (Goldschmidt and Branton, 1977; Leigh *et al.* 1979) as proposed by Ricardo and ap Rees (1970).

Recently, Guy *et al.* (1979) have reported studies on the uptake of monosaccharides by vacuoles isolated from the mesophyll cells of *Pisum*. The uptake of the D-glucose analog 3-O-methylglucose was very high in comparison with that of L-glucose. It is pH-dependent, stimulated by ATP, and markedly depressed by the proton ionophore SF_{6847}. These results brought evidence for the location of a selective sugar transport mechanism in the tonoplast.

5. Secondary Metabolites

The vacuole is the site of accumulation of a number of secondary products of metabolism, including anthocyanins (Swain, 1976), betacyanins (Mabry *et al.* 1972; Piatelli, 1976), cyanogenic glycosides (Saunders *et al.*, 1977; Saunders and Conn, 1978; Kojima *et al.*, 1979), alkaloids (Matile *et al.*, 1970; Saunders, 1979), and glucosinolates (Grob and Matile, 1979). A common feature of many of the secondary metabolites present in vacuoles is that they are β-glycosides (Piatelli, 1976; Swain, 1976; Conn, 1973). The most common sugar present in the aglycones is D-glucose, although other mono-, di-, and tri-saccharides may also be esterified. The presence of the sugar moiety often confers greater solubility on the aglycones (Swain, 1976) and probably also prevents their leakage out of the vacuole. For many of these compounds, there is evidence that glycosylation is the last stage in their synthesis (Conn, 1973; Swain, 1976), although for betacyanin, the evidence is conflicting (Piatelli, 1976). Transport into the vacuole presumably occurs during or immediately after glycosylation, which suggests that the enzymes responsible for glycosylation may be associated with the tonoplast.

Hradzina *et al.* (1978) investigated the activities of three enzymes involved in anthocyanin biosynthesis (flavanone synthetase, chalcone:flavanone isomerase, and UDPG:anthocyanidin glucosyltransferase) in cytoplasmic- and vacuole-enriched fractions from *Hippeastrum* and *Tulipa* petals. The activities were localized almost entirely in the soluble cytoplasmic fraction; no activities were detectable in the vacuole-enriched fractions. The authors concluded that the enzymes were cytoplasmic in their location. However, the data of Hradzina *et al.* (1978) show evidence of considerable inhibition of some of the enzymes. For instance, only 2.5% of the UDPG:anthocyanidin glucosyltransferase activity present in the soluble cytoplasmic phase of *Tulipa* petals was detectable in the original protoplast fraction from which it was prepared. This makes it difficult to draw conclusions about the location of these enzymes, as the inability of Hradzina *et al.* to detect enzyme activities in their vacuole-enriched fractions could be due to the presence of inhibitors in these fractions. Thus, firm evidence for or against the involvement of vacuoles in pigment synthesis or secondary metabolite glycosylation must await a more quantitative study.

C. Biochemical Properties of the Tonoplast

Currently very little is known about the biochemistry of the tonoplast, beyond those properties that can be inferred from the probable functions of the vacuole. For instance, as an important function of the vacuole is solute accumulation, it is assumed that the tonoplast must contain enzymes that are involved in the transport of specific solutes into the vacuole. However, only in the case of ion transport have the enzymes been tentatively identified.

In addition to ATPases (discussed previously), there is also evidence that the tonoplast contains oxidoreductase activities. Membranes isolated from the lutoid fraction of *Hevea brasiliensis* contain NADH-cytochrome c reductase and NADH-ferricyanide reductase (Moreau *et al.*, 1975). The enzymes were NADH specific and were antimycin A insensitive. A specific role for these enzymes in the maintenance of acetate levels in laticiferous cells was suggested by Moreau *et al.* (1975). Both NADH- and NADPH-cytochrome c oxidoreductases, insensitive to antimycin A, have been detected in isolated beet tonoplast (R. A. Leigh, unpublished). The tonoplast-bound activities represent only a small percentage of the total cellular activities (see Table II), and their function in this tissue is unknown.

Detailed studies of the lipid and protein components of the tonoplast have recently been initiated. The composition of yeast tonoplast was reported by Kramer *et al.* (1978) and Wilden and Matile (1978). The membrane had a lipid:protein ratio of 1.5 and contained 60–70% neutral lipid and 30–40% phospholipid. The major phospholipid components were phosphatidylcholine (33% of the total lipid phosphate), phosphatidylethanolamine (15–

TABLE II

The Effects of Added Salts on the ATPase Activities of Tulipa Leaf and Petal Vacuoles[a]

| Salt added[b] | Stimulation of Mg^{2+}-dependent ATPase[c] (%) | |
	Petal	Leaf
KCl	52	110
NaCl	22	104
NH$_4$Cl	31	122
LiCl	-23	76
Choline Cl$^-$	-12	76
KHCO$_3$	-30	22
KNO$_3$	-2	10
K$^+$ acetate	-22	50
KBr	-18	23
KI	-52	-48

[a] Adapted from Lin *et al.* (1977a).
[b] Final concentration, 50 mM.
[c] Negative values indicate inhibition of Mg^{2+}-dependent activity.

28%), phosphatidylserine and phosphatidylinositol (jointly, 33–43%). Not less than 17 protein bands were seen after SDS-polyacrylamide gel electrophoresis.

In contrast, the major phospholipid class in the vacuole membrane from the rubber tree laticifers was phosphatidic acid. It accounted for 82% of the total lipid phosphate (Dupont *et al.*, 1976) and its presence was apparently not the result of the hydrolysis of phospholipids by phospholipase D.

Tonoplasts obtained after purification of red beet vacuoles have recently been investigated (Marty, 1979). The tonoplast has a phospholipid:protein ratio of 0.7 and most of the lipids are polar. Seventeen have been tentatively identified by two-dimensional thin-layer chromatography. Five classes of glycolipids are present. The major phospholipids are phosphatidylcholine (54% of the total lipid phosphate), phosphatidylethanolamine (24%), phosphatidylglycerol (4%), and phosphatidylinositol and phosphatidylserine (4% and 1%, respectively). Phosphatidic acid was 13% and seemed not to be the result of the hydrolysis of phospholipids by phospholipase D since the search for such an activity in the vacuole fraction was negative.

The protein composition of the vacuole membrane was studied by SDS-acrylamide gel electrophoresis. The membrane polypeptides are all less than 100,000 MW, and some are easily extractable by either 0.1% Triton X-100 or NaOH (0.025 N or 0.1 N). The numerous bands in high resolution gradient gels can be conveniently grouped into nine major classes, each class consisting of one to five polypeptides. A limited number of polypeptides has a common distribution between the membrane and the sap, 15 polypeptides are considered major membrane polypeptides. Glycosidated polypeptides from the membrane and from the sap are probably of the high-mannose type characteristic of lysosomal enzymes which have undergone several stages of posttranslational modification.

V. APPENDIX: METHODS FOR ISOLATING VACUOLES AND TONOPLASTS

A. Intact Vacuoles

The large central vacuole of higher plant cells can be isolated individually by microsurgical dissection, but the need to use high shear forces to disrupt large pieces of tissue has made it difficult to isolate these fragile organelles by conventional techniques of tissue homogenization. Techniques of tissue disruption which overcome the need for high shear forces and which will permit the isolation of intact higher plant vacuoles in large numbers have therefore been developed.

Intact vacuoles can be isolated from yeast cells by lysing enzymically prepared protoplasts (Matile and Wiemken, 1967; Indge, 1968; Durr *et al.*, 1975). Higher plant vacuoles have been isolated by similar techniques (Gregory and Cocking, 1966), but Wagner and Siegelman (1975) were the first to develop a generally accepted vacuole isolation method using this procedure. Their method developed from the observation that 0.2 *M* phosphate buffer, pH 8.0, caused a gentle osmotic rupture of the protoplasts, thereby releasing the vacuole that could then be purified by centrifugation. Contamination by other organelles was slight, and the integrity of the vacuoles was demonstrated by their ability to absorb neutral red from solution and to retain pigments, salts, amino acids, and enzymes (Wagner and Siegelman, 1975; Lin *et al.*, 1977a; Butcher *et al.*, 1977; Wagner, 1977). Others have isolated vacuoles by similar techniques but have used either improved protoplast isolation methods (Saunders and Conn, 1978) or different methods of protoplast lysis (Lorz *et al.*, 1976; Buser and Matile, 1977).

Leigh and Branton (1976) developed a mechanical method for the isolation of vacuoles from the root storage tissue of the red beet *Beta vulgaris*. The method is based on earlier observations that small numbers of vacuoles can be isolated by slicing plasmolyzed tissue with a sharp razor blade. The technique depends on the use of a specially designed slicing machine that permits the large-scale application of the cutting conditions which had been successful when tissues were cut with a sharp razor blade. The vacuoles are collected in an undisturbed reservoir of collection medium and purified as shown in Fig. 7. Approximately 10^8 vacuoles containing approximately 4 mg total protein are recovered from 500 g tissue in the vacuole-enriched fraction (Fig. 7). Their ability to retain the water-soluble pigment betanin (the red pigment of beet) and to absorb neutral red from solution showed they were intact. Biochemical analyses indicated that the vacuoles were relatively free of contamination by other cell organelles (Table III). Only betanin and NADH-cytochrome c oxidoreductase were enriched in the vacuole fraction.

The method of Leigh and Branton (1976) can be used on those firmer tissues from which it is difficult to prepare protoplasts by cellulolytic digestion. The rapidity with which the vacuoles can be isolated from the intact tissue (500 g tissue sliced in less than 5 min) diminishes the possibility of redistribution of small molecules. The sucrose : betanin ratio in isolated vacuoles was the same as in intact tissue indicating that no differential leakage occurred during isolation (Leigh *et al.*, 1979). However, the method is relatively inefficient, as many vacuoles in the tissue are destroyed during the slicing operation; thus, it is not useful for tissues that are available in only small quantities. A direct comparison of vacuoles isolated by the mechanical method and by the protoplast–lysis method has not yet been made.

TABLE III

Recoveries and Activities of Markers in the Vacuole-Enriched Fraction from Beet (Fraction 2 in Fig. 7)[a]

Marker	Total particulate[b] activity recovered in fraction 2(%)	Relative specific activity[c] in fraction 2
Betanin	18.1	7.85
Protein	1.0	—
DNA	0.3	0.24
Cytochrome oxidase	1.0	0.55
NADH-cytochrome c oxidoreductase	0.5	2.73
Glucose-6-phosphate dehydrogenase	0	0
Glutamate oxaloacetate transaminase	0.7	0.46

[a] Adapted from Leigh and Branton (1976). Values greater than one indicate an increase in purity.

[b] This gives a measure of the recoveries as a percentage of the maximum possible recovery.

[c] Relative specific activity = specific activity in fraction 2/specific activity in 2000 g pellet. Values greater than one indicate an increase in purity.

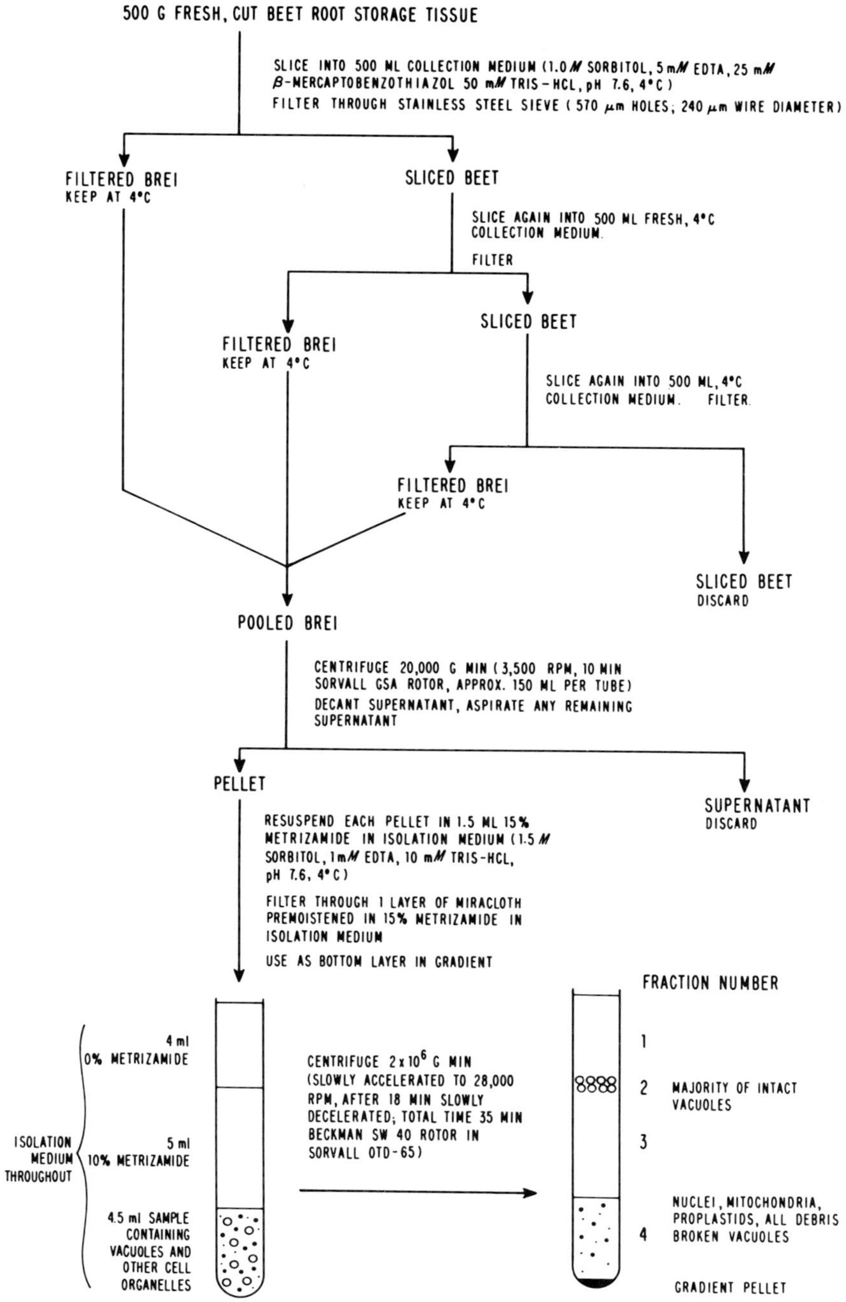

Fig. 7. Purification procedure for vacuoles following mechanical slicing of beet tissue. Modified from Leigh and Branton (1976).

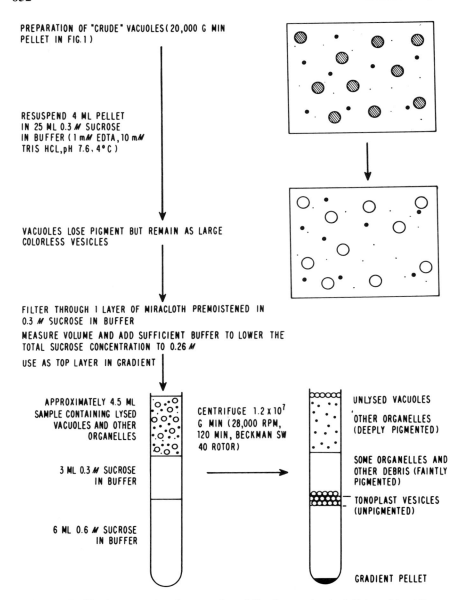

Fig. 8. Purification procedure for tonoplasts following mechanical slicing of beet tissue.

B. Tonoplast Fragments

Like other plant cell membranes (e.g., dictyosomes, endoplasmic reticulum, plasma membrane), the tonoplast is usually reduced to small vesicles during tissue homogenization. These tonoplast vesicles have no known morphological or biochemical characteristics which can be used to identify them in separated cell fractions. However, tonoplast should be the only membrane present in preparations of purified vacuoles; thus, these can be used as a source from which a tonoplast-enriched membrane fraction can be prepared and identified unambiguously. Preparations of yeast vacuoles were used by van der Wilden *et al.* (1973) as a source of yeast tonoplast, and successful attempts to isolate the membrane surrounding small vacuoles (lutoids) from laticiferous cells of *Hevea brasiliensis* have been reported (Moreau *et al.,* 1975; d'Auzac, 1975). Lin *et al.* (1977a) have centrifuged purified preparations of leaf and petal vacuoles on a 15%/35% discontinuous density gradient and recovered the tonoplast at the interface of the two sucrose layers. ATPases associated with the membranes were partially characterized, but contamination by other cell membranes was not tested for, as it was assumed that purification of the vacuoles prior to vacuole isolation made this unnecessary. Although this is a reasonable assumption, both Lorz *et al.* (1976) and Saunders and Conn (1978) have reported that membrane fragments can stick to isolated vacuoles under some conditions. A discontinuous sucrose density gradient may also be used to isolate a tonoplast-enriched fraction from beet vacuoles (R. A. Leigh, unpublished). When intact beet vacuoles are resuspended in buffered 0.3 M sucrose, they loose their pigment but remain as large (10–20 μm diameter), colorless vesicles that do not sediment through 0.6 M sucrose. This property has been used to develop a tonoplast isolation scheme which circumvents the need for extensive vacuole purification (Fig. 8). Electron microscopy and biochemical analyses indicate that the tonoplast-enriched fraction prepared by this technique is relatively free of contamination by other cell organelles and membranes, but absolute values for tonoplast enrichment cannot be determined because no unique markers for this membrane have yet been found.

VI. CONCLUSIONS

The structural polymorphism, developmental intricacies, and functional attributes of vacuoles belie their etymological designation, from the Latin *vacuum*. It is clear that vacuoles come in many forms, that their development is rooted in the dynamics of an endomembrane system fundamental to all eukaryotic cells, and that they often contain a substantial array of enzymes and metabolites. Thus, a vacuole is neither an empty space nor even necessarily a quiescent compartment, but an active participant in cellular metabolism.

The advent of more refined cytochemical techniques and detailed morphological analyses using electron microscopic methods has clarified the relation of vacuoles to the lysosomal system. Similarly, the availability of new techniques to isolate vacuoles and tonoplasts from a variety of tissues is stimulating a more thorough study of vacuolar composition, structure, and function. Real progress, however, in understanding the role of vacuoles demands that the lessons of morphology be applied continuously when physiological results are interpreted. For example, the role of the mature plant vacuole in intracellular digestion processes may be more limited than suggested when hydrolases were first localized in vacuoles. In view of the electron microscopic and cytochemical studies showing that the mature vacuole is analogous to a secondary lysosome or residual body, it would not be surprising if the spectrum of hydrolases found in some mature vacuoles is too limited to cause dissolution of large areas of cytoplasm. One might find such lytic activity in the autophagic system which is the vacuole precursor, whereas the mature vacuole may develop compartmentation functions which, albeit poorly understood, are more important than any residual lytic activities.

Although we must focus our studies on measurable attributes of the vacuolar system, it may be useful to keep in mind that, unlike animal cells, plant cells multiply in a manner that promotes a fixed spatial relation between the two daughter cells and which does not necessarily lead to the development of large intercellular spaces. As a substitute for the variable contacts and large intercellular spaces seen in many multicellular animals, the large intracellular compartments formed by the vacuoles of multicellular plants may play a more direct role in plant growth and development than heretofore recognized. It has been pointed out that the vacuole permits the cell to expand without needing to synthesize large quantities of cytoplasm (Dainty, 1968), and it is clear that by sequestering turgor-generating solutes within vacuoles a cell is not limited to using only those solute concentrations which are compatible with cytoplasmic functioning. But ascribing these passive roles to the vacuole likens the tonoplast to an inert, semipermeable membrane and should remind us that we still know little about how solutes move into and out of the vacuole and even less about how solute levels inside the vacuole are controlled, how the tonoplast responds to cytoplasmic effectors of transport activity, how the cytoplasm depends upon solutes released from the vacuole, or what function autophagy and vacuolar hydrolases play in metabolism. Isolated vacuoles provide a means for studying many aspects of these problems, but it will eventually be necessary to consider not only the mature, secondary lysosome–vacuole but also its developmental precursors in the provacuolar and autophagic stages.

REFERENCES

Anderson, W. P. (1972). *Annu. Rev. Plant Physiol.* **23**, 51–72.

Baker, D. A., and Hall, J. L., eds. (1975). "Ion Transport in Plant Cells and Tissues." North-Holland Publ., Amsterdam.

Balz, H. P. (1966). *Planta* **70**, 207–236.

Barrett, A. J., and Heath, M. F. (1977). In "Lysosomes: A Laboratory Handbook" (J. T. Dingle, ed.), 2nd ed., pp. 19–145. North-Holland Biomed. Press, Amsterdam.

Beevers, L. (1976). "Nitrogen Metabolism in Plants." Arnold, London.

Berjak, P. (1972). *Ann. Bot. (London)* **36**, 73–81.

Boller, T., and Kende, H. (1979). *Plant Physiol.* **63**, 1123–1132.

Boller, T., Durr, M., and Wiemken, A. (1975). *Eur. J. Biochem.* **54**, 81–91.

Bowes, B. G. (1965). *Cellule* **65**, 357–364.

Bradfute, O. E., Chapman-Andresen, C., and Jensen, W. A. (1964). *Exp. Cell Res.* **36**, 207–210.

Brown, R. M., Jr., and Romanovicz, D. K. (1976). *Appl. Polym. Symp.* **28**, 537–585.

Burian, K., and Hofler, K. (1966). *Protoplasma* **61**, 244–256.

Buser, Ch., and Matile, Ph. (1977). *Z. Pflanzenphysiol.* **82**, 462–466.

Butcher, H. C., Wagner, G. J., and Siegelman, H. W. (1977). *Plant Physiol.* **59**, 1098–1103.

Buvat, R. (1958). *Ann. Sci. Nat. Bot. Biol. Vég.* **19**, 121–161.

Buvat, R. (1965). In "Travaux dédiés à L. Plantefol," pp. 81–124. Masson, Paris.

Buvat, R. (1971). In "Results and Problems in Cell Differentiation" (W. Beermann, J. Reinert, and H. Ursprung, eds.), Vol. II, pp. 127–157. Springer-Verlag, Berlin and New York.

Buvat, R. (1977). *C.R. Hebd. Séances Acad. Sci.* **284**, 167–170.

Catesson, A. M. (1974). In "Dynamic Aspects of Plant Ultrastructure" (A. W. Robards, ed.), pp. 358–390. McGraw Hill, New York.

Charvat, I., and Esau, K. (1975). *J. Cell Sci.* **19**, 543–561.

Chrispeels, M. J. (1977). In "International Cell Biology 1976–1977" (B. R. Brinkley and K. R. Porter, eds.), pp. 284–292. Rockefeller Univ. Press, New York.

Chambers, R., and Hofler, K. (1931). *Protoplasma* **12**, 338–355.

Conn, E. E. (1973). *Biochem. Soc. Symp.* **38**, 277–302.

Cronshaw, J. (1974). In "Dynamic Aspects of Plant Ultrastructure" (A. W. Robarbs, ed.), pp. 391–413. McGraw-Hill, New York.

Dainty, J. (1968). In "Plant Cell Organelles" (J. B. Pridham ed.), pp. 40–46. Academic Press, New York.

Dangeard, P. (1956). *Protoplasmatologia* **3**, 1–41.

d'Auzac, J. (1975). *Phytochemistry* **14**, 671–675.

d'Auzac, J. (1977). *Phytochemistry* **16**, 1881–1885.

De Chastellier, C., Quiviger, B., and Ryter, A. (1978). *J. Ultrastruct. Res.* **62**, 220–227.

De Duve, C. (1969). In "Lysosomes in Biology and Pathology" (J. T. Dingle and H. B. Fell, eds.), Vol. 1, pp. 3–40. North-Holland Publ., Amsterdam.

De Duve, C. and Wattiaux, R. (1966). *Annu. Rev. Physiol.* **28**, 435–492.

De Duve, C., Pressman, B. C., Gianetto, R., Wattiaux, R., and Appelmans, F. (1955). *Biochem. J.* **60**, 604–617.

Doi, E., Ohtsuru, C., and Matoba, T. (1975). *Plant Cell Physiol.* **16**, 581–588.

Doll, S., Rodier, F., and Willenbrink, J. (1979). *Planta* **144**, 407–411.

Drew, M. C., Seear, J., and McLaren, A. D. (1970). *Am. J. Bot.* **57**, 837–843.

Dupont, J., Moreau, F., Lance, C., and Jacob, J.-L. (1976). *Phytochemistry* **15**, 1215–1217.

Dürr, M., Boller, T., and Wiemken, A. (1975). *Arch. Microbiol.* **105**, 319–327.

Dürr, M., Urech, K., Boller, T., Wiemken, A., Schwencke, J., and Nagy, M. (1979). *Arch. Microbiol.* **121**, 169–175.

Esau, K. (1969). *In* "Handbuch der Pflanzenanatomie" (W. Zimmermann, P. Ozenda, and H. D. Wulff, eds.), Vol. 5. Borntraeger, Berlin.

Esau, K. (1975). *Ann. Bot. (London)* **39**, 713–719.

Evert, R. F. (1977). *Annu. Rev. Plant Physiol.* **28**, 199–222.

Giaquinta, R. (1977). *Plant Physiol.* **60**, 339–343.

Gifford, E. M., and Stewart, K. D. (1967). *J. Cell Biol.* **33**, 131–142.

Giordani, R. (1977). *C.R. Hebd. Séances Acad. Sci.* **284**, 569–572.

Goldschmidt, E. E., and Branton, D. (1977). *Plant Physiol.* **59**, suppl, 104.

Gregory, D. W., and Cocking, E. C. (1966). *J. Exp. Bot.* **17**, 57–67.

Grob, K., and Matile, Ph. (1979). *Plant Sci. Lett.* **14**, 327–335.

Guilliermond, A., Mangenot, G., and Plantefol, L. (1933). "Traité de Cytologie Végétale." Le François, Jouve, Paris.

Guy, M., Reinhold, L., and Michaeli, D. (1979). *Plant Physiol.* **64**, 61–64.

Hausmann, K. (1978). *Int. Rev. Cytol.* **52**, 197–276.

Heslop-Harrison, Y. (1975). *In* "Lysosomes in Biology and Pathology" (J. T. Dingle and R. T. Dean, eds.), Vol. 4, pp. 525–578. North-Holland Publ., Amsterdam.

Heywood, P. (1978). *J. Cell Sci.* **31**, 213–224.

Hirai, M., and Asahi, T. (1973). *Plant Cell Physiol.* **14**, 1019–1029.

Holleman, J. M., and Key, J. L. (1967). *Plant Physiol.* **42**, 29–36.

Holtzman, E. (1975) "Lysosomes: A Survey," Cell Biol. Monogr., Vol. 3. Springer-Verlag, Berlin and New York.

Hradzina, G., Wagner, G. J., and Siegelman, H. W. (1978). *Phytochemistry* **17**, 153–156.

Indge, K. J. (1968). *J. Gen. Microbiol.* **51**, 441–446.

Kojima, M., Poulton, J. E., Thayer, S. S., and Conn, E. E. (1979). *Plant Physiol.* **63**, 1022–1028.

Kramer, R., Kopp, F., Neidermeyer, W., and Fuhrmann, G. F. (1978). *Biochim. Biophys. Acta* **507**, 369–380.

Leigh, R. A., and Branton, D. (1976). *Plant Physiol.* **58**, 656–662.

Leigh, R. A., ap Rees, T., Fuller, W. A., and Banfield, J. (1979). *Biochem. J.* **178**, 539–547.

Lin, W., Wagner, G. J., Siegelman, H. W., and Hind, G. (1977a). *Biocheim. Biophys. Acta* **465**, 110–117.

Lin, W., Wagner, G. J., and Hind, G. (1977b). *Plant Physiol.* **59**, Suppl., 85.

Lips, S. H., and Beevers, H. (1966). *Plant Physiol.* **41**, 709–712.

Lorz, H., Harms, C. T., and Potrykus, I. (1976). *Biochem. Physiol. Pflanz.* **169**, 617–620.

Luscher, A., and Matile, Ph. (1974). *Planta* **118**, 323–332.

Lüttge, U., and Pitman, M. G., eds. (1976). "Encyclopedia of Plant Physiology, Transport in Plants" [N.S.], Vols 2A and 2B. Springer-Verlag, Berlin and New York.

Mabry, T. J., Kimler, L., and Chang, C. (1972). *Recent Adv. Phytochem.* **5**, 105–134.

MacRobbie, E. A. C. (1970). *Q. Rev. Biophys.* **3**, 251–294.

MacRobbie, E. A. C. (1971). *Annu. Rev. Plant Physiol.* **22**, 75–96.

MacRobbie, E. A. C. (1974). *In* "Algal Physiology and Biochemistry" (W. D. P. Stewart ed.), pp. 676–713. Blackwell Scientific Publ., Oxford.

MacRobbie, E. A. C. (1976). *In* "Perspectives in Experimental Botany" (N. Sunderland, ed.), Vol. a, pp. 369–380. Pergamon, Oxford.

MacRobbie, E. A. C. (1977). *In* "International Reviews of Biochemistry" (D. H. Northcote, ed.), Vol. 13, pp. 211–247. Univ. Park Press, Baltimore, Maryland.

Mangenot, G. (1929). *Arch. Anat. Microsc. Morph. Exp.* **25**, 507–518.

Marinos, N. G. (1963). *J. Ultrastruct. Res.* **9**, 177–185.

Marty, F. (1970). *C.R. Hebd. Séances Acad. Sci.* **271**, 2301–2304.

Marty, F. (1971). *C.R. Hebd. Séances Acad. Sci.* **272**, 399–402.

Marty, F. (1972). *C.R. Hebd. Séances Acad. Sci.* **274**, 206–209.

Marty, F. (1973a). *C. R. Hebd. Séances Acad. Sci.* **276**, 1549–1552.

Marty, F. (1973b). *C.R. Hebd. Séances Acad. Sci.* **277**, 1749–1752.

Marty, F. (1973c). *C.R. Hebd. Séances Acad. Sci.* **277**, 2681–2684.

arty, F. (1974). Doctorat es-Sciences, Université Aix-Marseille France.

arty, F. (1976). *In* "Microscopie Electronique à Haute Tension 1975" (B. Jouffrey and P. Favard, eds.), pp. 373–376. Société Française de Microscopie Electronique, Paris.

ty, F. (1978). *Proc. Natl. Acad. Sci. U.S.A.* **75**, 852–856.

y, F., and Branton, D. (1979). *Plant Physiol.* **63**, Suppl., 144.

, Ph. (1968). *Planta* **79**, 181–196.

, Ph. (1969). *In* "Lysosomes in Biology and Pathology" (J. T. Dingle and H. B. Fell, ds.), Vol. 1, pp. 406–430. North-Holland Publ., Amsterdam.

Ph. (1974). *In* "Dynamic Aspects of Plant Ultrastructure" (A. W. Robards, ed.), pp. –218. McGraw-Hill, New York.

h. (1975). "The Lytic Compartment of Plant Cells," Cell Biol. Monogr., Vol. 1. ger-Verlag, Berlin and New York.

(1976). *In* "Plant Biochemistry" (J. Bonner and J. E. Varner, eds.), 3rd ed., pp. 24. Academic Press, New York.

and Moor, H. (1968). *Planta* **80**, 159–175.

and Wiemken, A. (1967). *Arch. Mikrobiol.* **56**, 148–155.

nd Wiemken, A. (1976). *In* "Encyclopedia of Plant Physiology, [N.S.], Transport " (C. R. Stocking and U. Heker, eds.), Vol. 3, pp. 255–287. Springer-Verlag, d New York.

d Winkenbach, F. (1971). *J. Exp. Bot.* **22**, 759–771.

s, B., and Rickenbacker, R. (1970). *Biochem. Physiol. Pflanzen.* **161**, 447–458.

1969). *J. Ultrastruct. Res.* **26**, 242–250.

H., Whaley, W. G., and Leech, J. H. (1961). *J. Ultrastruct. Res.* **5**, 193–200.

b, J.-L., Dupont, J., and Lance, C. (1975). *Biochim. Biophys. Acta* **396**,

teward, F. C. (1972). *Ann. Bot. (London)* **36**, 915–937.

hlberg, P. (1976). *Planta* **129**, 83–85.

Beevers, H. (1978). *Plant Physiol.* **62**, 448–48.

) *In* "Lysosomes and Storage Diseases" (G. Hers and F. van Hoof, eds.), mic Press, New York.

. *Proc. Natl. Acad. Sci. U.S.A.* **73**, 2781–2787.

G. S. (1970). *Annu. Rev. Plant. Physiol.* **21**, 43–66.

n "Encyclopedia of Plant Physiology, [N.S.] Transport in Plants" (U. Pitman, eds.), Vol. 2A, pp. 347–372. Springer-Verlag, Berlin and New

R. L. (1976). *Plant Phys.* **57**, 249–256.

nta **123**, 1–13.

ta **123**, 15–31.

mistry and Biochemistry of Plant Pigments" (T. W. Goodwin ed.), 0–596. Academic Press, New York.

J. Ultrastruct. Res.* **18**, 287–303.

am, N. (1970). *Plant Physiol.* **46**, 666–673.

). *Planta* **109**, 233–258.

Séances Acad. Sci.* **253**, 2395–2397.

, 55–66.

407–434.

sso, N. (1974). *J. Microsc.* **21**, 173–180.

Roland, J. C. (1977). *Biol. Cell.* **28**, 269–280.

g. **6**, 27–46.

and Ryter, A. (1978). *J. Ultrastruct. Res.* **62**, 228–236.

(1970). *Phytochemistry* **9**, 239–248.

Roa, R. L., and Pickard, W. F. (1976). *J. Exp. Bot.* **27**, 853–858.

Roberts, L. W. (1976). "Cytodifferentiation in Plants," Developmental and Cell Biology series (M. Abercrombie, D. R. Newth, and J. G. Torrey, eds.). Cambridge Univ. Press, London and New York.

Robertson, J. G., Lyttleton, P., Bullivant, S., and Grayston, G. F. (1978). *J. Cell Sci.* **30**, 129–149.

Roland, J. C. (1973). *Int. Rev. Cytol.* **36**, 45–92.

Rougier, M. (1976). *J. Microsc. Biol. Cell.* **26**, 161–166.

Satir, B. (1974a). *J. Supramol. Struct.* **2**, 529–537.

Satir, B. (1974b). *Symp. Soc. Exp. Biol.* **28**, 399–446.

Satter, R. L., Sabnis, D. D., and Galston, A. W. (1970). *Am. J. Bot.* **57**, 374–381.

Saunders, J. A., and Conn, E. E. (1978). *Plant Physiol.* **61**, 154–157.

Saunders, J. A., Conn, E. E., Lin, C. H., and Stocking, C. R. (1977). *Plant Physiol.* **59**, 647–652.

Spanswick, R. M., and Miller, A. G. (1974). *Proc. Natl. Acad. Sci. U.S.A.* **71**, 1413–1417.

Swain, T. (1976). *In* "Chemistry and Biochemistry of Plant Pigments" (T. W. Goodwin ed.), 2nd ed., Vol. 1, pp. 425–463. Academic Press, New York.

Symillides, Y., and Marty, F. (1977). *C.R. Hebd. Séances Acad. Sci.* **285**, 721–724.

Toriyama, H. (1967). *Proc. Jpn. Acad.* **43**, 777–782.

Truchet, G. (1976). Doctorat es-Sciences, Université Aix-Marseille, France.

Ueda, K. (1966). *Cytologia* **31**, 461–472.

van der Wilden, W., Matile, Ph., Schellenberg, M., Meyer, J., and Wiemken, A. (1973). *Naturforsch.* **28c**, 416–421.

van der Wilden, W., and Matile, Ph. (1978). *Biochem. Physiol. Pflanzen.* **173**, 285–294.

Verma, D. P. S., Kazazian, V., Zogbi, V., and Bal, A. K. (1978). *J. Cell Biol.* **78**, 919–936.

Villiers, T. A. (1971). *Nature (London) New Biol.* **233**, 57–58.

Wagner, G. J. (1977). *Plant Physiol.* **59**, Suppl., 104.

Wagner, G. J. (1979). *Plant Physiol.,* **64**, 88–93.

Wagner, G. J., and Siegelman, H. W. (1975). *Science* **190**, 1298–1299.

Walker, N. A., and Smith, F. A. (1975). *Plant Sci. Lett.* **4**, 125–132.

Walker-Simmons, M., and Ryan, C. A. (1977). *Plant Physiol.* **60**, 61–63.

Weiss, R. L., Goodenough, D. A., and Goodenough, U. W. (1977). *J. Cell Biol.* **72**, 133.

Wiemken, A., and Durr, M. (1974). *Arch. Microbiol.* **101**, 45–57.

Wiemken, A., and Nurse, P. (1973). *Planta* **109**, 293–306.

Woodzicki, T. J., and Brown, C. L. (1973). *Am. J. Bot.* **60**, 631–640.

Cyanobacteria (Blue–Green Algae)

16

C. PETER WOLK

I. INTRODUCTION

The undifferentiated blue–green algal cell is a bacterial cell that performs O_2-evolving photosynthesis, characteristic of eukaryotic plants. The structure of the wall of cyanobacteria (Fig. 1, L_I-L_{IV}) corresponds to that of gramnegative bacteria. Inside the wall and plasmalemma there is usually a region of photosynthetically active membranes, termed thylakoids (Fig. 1, T). The distribution of the thylakoids within the cell, and the identity, disposition, and function of photosynthetically active pigments within those photosynthetic lamellae, correspond closely to the properties of the chloroplasts of the eukaryotic Bangioideae, or lower red algae. Interdigitating with the thylakoids is a central region rich in nucleic acids (Fig. 1, N). A variety of inclusions is present (Fig. 1, CG, PB, L), and extracellular mucilaginous material (Fig. 1, M) is often produced.

Long appreciated as simple organisms in which to study oxygenic photosynthesis, cyanobacteria have in recent years become recognized as of theoretical and practical importance in the areas of photoreduction of dinitrogen, photoproduction of hydrogen, and development. The physiology, cytology, and biochemistry of cyanobacteria were reviewed in detail by Fogg *et al.* (1973), Carr and Whitton (1973), and Wolk (1973), and by Stanier and Cohen-Bazire in 1977 (see also Stanier, 1977, for a short, well illustrated, and insightful review). This chapter cites some of the more significant recent studies on cyanobacteria. For references to the earlier work, the reader is referred to the previously mentioned reviews.

II. CYTOLOGICAL CHEMISTRY

A. Outer Layers

It is the absence of a nuclear membrane and a mitotic apparatus, and the presence of peptidoglycan (murein) in the wall, which mark cyanobacteria as bacteria. The peptidoglycan accounts for 22–52% of the weight of isolated walls, and contains *N*-acetyl muramic acid, *N*-acetyl glucosamine, diaminopimelic acid, glutamic acid, and alanine, in molar ratio approximately $1:1:1:1:2$. The peptidoglycan, which is present in the L_{II} layer of the wall (see Fig. 1), can be degraded by lysozyme. The modified cells that result are to some extent functional analogues of isolated chloroplasts of eukaryotes (Ward and Myers, 1972; see review by Wolk, 1973).

Characteristic of gram-negative bacteria is the occurrence, at the periphery of the wall, of an outer membrane that contains proteins and lipopolysaccharide. Lipopolysaccharide with the distinctive components 2-keto-3-deoxyoctonate and/or L-acofriose has been isolated from a number

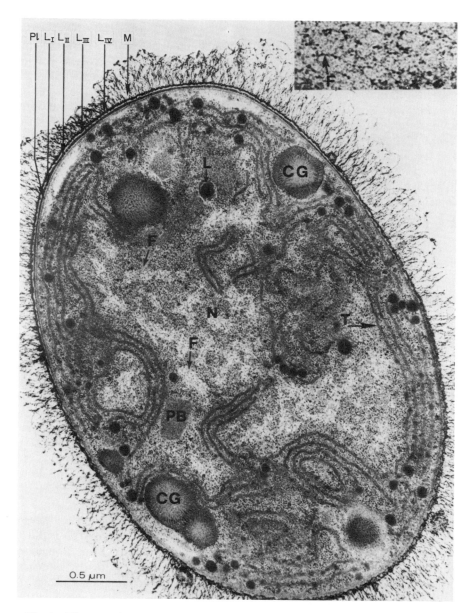

Fig. 1. Electron micrograph of a vegetative cell of *Anabaena variabilis* showing extracellular mucilage (M); layers L_I, L_{II} (peptidoglycan), L_{III}, and L_{IV} (which contains lipopolysaccharide) of the wall; plasmalemma (Pl); thylakoids (T); DNA-containing regions (N) with fine fibrils (F) which are visible in some areas (see inset); cyanophycin (structured) granules (CG); a polyhedral body (PB); and other inclusions, certain of which (L) may be lipid in nature. Courtesy of L. V. Leak (1967), with permission. × 37,000; inset, × 217,000.

of cyanobacteria (Weise *et al.*, 1970; Weckesser *et al.*, 1974; Buttke and Ingram, 1975; Katz *et al.*, 1977; Mikheyskaya *et al.*, 1977). Proteins having molecular weights similar to the molecular weights of the most abundant proteins of the outer membrane of *E. coli* are abundant in the walls of *Anacystis nidulans** (Golecki, 1977). In filamentous cyanobacteria, the outer membrane (see Fig. 1, L_{IV}) is continuous along the filament whereas the peptidoglycan-containing layer is sandwiched between consecutive cells as well as being present as part of the side walls. Formation of the end walls comes about by irislike ingrowth of the side walls. In *Plectonema boryanum*, alkaline phosphatase is localized in the L_{III} layer of the wall (Doonan and Jensen, 1977). Very little is known about the properties of the plasmalemma (Fig. 1, Pl) of cyanobacteria.

The cyanobacteria were at one time called Myxophyceae (''slime-plants'') because of the copious amounts of mucilage (Fig. 1, M) produced by many of these organisms. The mucilage is found outside of the outer membrane, and is often rich in glucose, xylose, mannose, and galactose. In some but not all cases, it contains uronic acids (see review by Wolk, 1973).

B. Nucleic Acids

The size of the genome, as determined from the kinetic complexity, has been reported to be in the range of 1.5–7.6×10^9 per cell (Herdman and Carr, 1974; Herdman, 1976; Roberts *et al.*, 1977). In *A. nidulans*, 0.6% of the DNA is complementary to ribosomal RNA and 0.062% to transfer RNA (Smith and Carr, 1977), but with the possible exception of rRNA genes, repeated sequences were not detectable (Roberts *et al.*, 1977). Plasmids have been reported (Asato and Ginoza, 1973; Roberts and Koths, 1976). Histones are absent; however, a DNA-binding protein of molecular weight 1×10^4, rich in lysine and arginine, is present to the extent of more than one dimer per 300 base pairs in unicellular and filamentous cyanobacteria (Haselkorn and Rouvière-Yaniv, 1976).

The ribosomes of cyanobacteria, like those of other bacteria, have a sedimentation coefficient of approximately 70 S, and dissociate to 50 S and 30 S subunits. In *Anacystis,* each of these ribosomal subunits forms ribosomes functional in *in vitro* protein synthesis when combined with the alternate subunit from *E. coli* (Gray and Herson, 1976). The ribosomes contain 23 S, 16 S, and 5 S RNA, all of which are derived from cleavage of larger precursors (Doolittle, 1972; Szalay *et al.*, 1972; Seitz and Seitz, 1973; Grierson and Smith, 1973; Dobson *et al.*, 1974). The half-life of messenger RNA in *A. nidulans* was determined to be approximately 12 min (Leach and Carr, 1974).

---

* This organism, frequently used for experimentation since it was shown by Kratz and Myers (1955) to grow rapidly (see Section VII), is also known as *Synechococcus* 6301 (Stanier *et al.*, 1971).

C. Inclusions

A variety of inclusions are found within the vegetative cells. (1) The principal carbon reserve is predominately α1-4 linked polyglucose ("glycogen," "cyanophycean starch"; see review by Wolk, 1973); studies of its accumulation and mobilization have shown that the cellular content of this material varies with the growth conditions. Even when present at a constant amount per cell, its turnover is taking place (Lehmann and Wöber, 1976). The biosynthesis of glycogen by ADP-glucose pyrophosphorylase is inhibited by inorganic phosphate and activated by a number of phosphorylated metabolites, notably 3-phosphoglycerate (Levi and Preiss, 1976). (2) Exposure of cells of some strains to acetate can stimulate the deposition of poly-β-hydroxybutyrate (Jensen and Sicko, 1973; Carr and Bradley, 1973). (3) The principal nitrogenous reserve ("cyanophycin granules," "structured granules"; Fig. 1, CG) consists of long polyaspartate sequences, with an arginyl residue linked by its α-amino group to the second carboxyl group of each aspartyl residue (Simon and Weathers, 1976). An enzyme capable of elongating this polymer has been purified 92-fold from *Anabaena cylindrica* (Simon, 1976). Phycobilisomes (see Section II,D) can apparently also serve as nitrogenous reserves. (4) Inorganic phosphate is also accumulated, in polymerized form, in granules (Jensen and Sicko-Goad, 1976). (5) A substantial fraction of the ribulose bisphosphate carboxylase activity in cyanobacteria, as in *Thiobacillus,* appears to be associated with inclusions termed polyhedral bodies (Fig. 1, PB). However, the enzyme may only be *stored* in these bodies because the most rapid fixation of CO_2 takes place when the frequency of these inclusions is at a minimum (Stewart, 1977). (6) Gas-filled vesicles, which aggregate into so-called gas vacuoles, form in many cyanobacteria. The vacuoles have profound effects on cellular buoyancy, so that their presence affects the light intensity and other environmental factors to which the cells are exposed (Walsby, 1975; Reynolds and Walsby, 1975). The vesicle membranes, which consist solely of protein, are permeable to a variety of gases. The highly insoluble protein has been partially sequenced (Weathers *et al.,* 1977). (7) Lipid deposits (Fig. 1, L?) are a minor feature of the cytology of cyanobacteria. (8) Other inclusions, including arrays of microtubules (Bisalputra *et al.,* 1975; Jensen and Ayala, 1976), have been reported.

D. Photosynthetic Apparatus

1. Structures

The photosynthetic apparatus of cyanobacteria consists of one or more closed discs or lamellae, termed thylakoids, bearing regularly arrayed structures, called phycobilisomes, on their outer surface. The thylakoids contain all, or almost all, of the cellular chlorophyll (Fig. 2a), and much or all of the carotenoids. The only chlorophyll found is chlorophyll a. The only exception so far reported to this structure of the photosynthetic apparatus is *Gloeobacter violacens,* which lacks thylakoids. The photosynthetic pigments in this organism appear to be associated with the plasmalemma (Rippka *et al.,* 1974).

Studies of energy transfer after absorption of light indicate that an allophycocyanin, a minor, blue, proteinaceous pigment, is present at the point of attachment of the phycobilisomes to the photosynthetic membrane. The allophycocyanin is surrounded by a layer of phycocyanin, the principal blue proteinaceous pigment of cyanobacteria. If phycoerythrin, the red proteinaceous pigment of cyanobacteria, is present, it may be located on the outside of the phycobilisome (Gray and Gantt, 1975; Gantt *et al.,* 1976; see also Glazer and Bryant, 1975). Phycobilisomes have been reported to contain, in addition, a small complement of nonpigmented proteins (Marsac and Cohen-Bazire, 1977). Information about the biliproteins has been extensively reviewed by Bogorad (1975) and Chapman (1973). The chromophores of the biliproteins are linear tetrapyrroles. Their structures, following hydrolytic cleavage from the apoproteins, are compared with those of

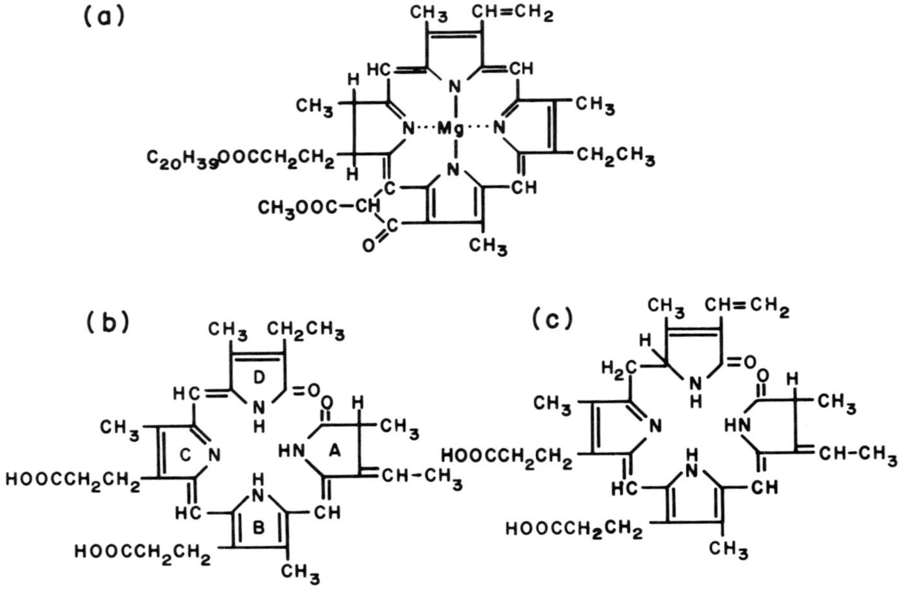

Fig. 2. Tetrapyrrole pigments of cyanobacteria: (a) chlorophyll a, (b) phycocyanobilin, and (c) phycoerythrobilin. The structures in (b) and (c) are in the form in which the chromophores appear following hydrolytic cleavage from the accessory photosynthetic pigments phycocyanin and phycoerythrin, respectively. Reproduced from Wolk (1973), with permission.

chlorophyll in Fig. 2. The β-subunit of C-phycocyanin from *Anacystis nidulans* has been sequenced, and sites of attachment of the chromophores determined (Freidenreich *et al.*, 1978; Williams and Glazer, 1978).

2. Lipid-Soluble Constituents and Related Compounds

The principal carotenoids of cyanobacteria are most frequently β-carotene, echinenone (4-keto-β-carotene), myxoxanthophyll (1',2'-dihydro-3',4'-didehydro-3,1'-dihydroxy-γ-carotene, glycosidically linked at the 2' position to rhamnose), and sometimes zeaxanthin (3,3'-dihydroxy-β-carotene), although caloxanthin (2R, 3R, 3'R-β,β-carotene-2,3,3'-triol), nostoxanthin (2R, 3R, 2'R, 3'R-β,β-carotene-2,3,2',3'-tetrol; Buchecker *et al.*, 1976), and a variety of other carotenoids have been described as accounting for 10% or more of the total carotenoid in certain species (Stransky and Hager, 1970; Hertzberg *et al.*, 1971). The four principal fatty acid-containing lipids found in chloroplasts, i.e., mono- and digalactosyl diglyceride, phosphatidyl glycerol, and sulfoquinovosyl diglyceride are also found in cyanobacteria (see reviews by Nichols, 1973; Wolk, 1973). The fatty acids found are principally C_{14}, C_{16}, and C_{18}. α-Linolenic acid (18:3 [3, 6, 9]) has been found in some but not all cyanobacteria. Hydrocarbons of chain length C_{15} to C_{19} have been found. Plastoquinones A (also called PQ 9), B, and C_{1-6}; naphthoquinones including vitamin K_1 (also called phylloquinone); tocopherols; and α-tocopherolquinone are also present.

3. Electron Transfer Carriers

Elements of electron transport chains, in addition to the quinones already mentioned, include cytochromes, low-potential carriers, and plastocyanin. Of the cytochromes, some are tightly membrane bound, whereas others can be solubilized by sonic treatment. Krogmann (1973) has presented an excellent discussion of the problems inherent in deciding whether such compounds are, *in vivo*, associated with membranes. The principal low-potential electron carriers are ferredoxin and the flavoprotein, phytoflavin, which may largely replace ferredoxin under conditions of limiting iron. Plastocyanin is a copper-containing protein which can undergo oxidation–reduction reactions.

III. METABOLISM

A. Light Reactions of Photosynthesis

To date, the available evidence is largely consistent with photosystems I and II operating in series, as normally formulated for higher plants. That is, the pathway of electron transfer is visualized as:

$$H_2O \xrightarrow{\text{Photosystem II}} Q \rightarrow \text{Cytochrome} \rightarrow P_{700} \xrightarrow{\text{Photosystem I}} X_{red} \rightarrow \text{Ferredoxin} \rightarrow \text{NADPH}.$$

Oxygen production as a function of flash number is, in *A. nidulans* as in *Chlorella* and spinach, periodic, with a period of four (Ley *et al.*, 1975). These results (see also Maxwell and Biggins, 1977) indicate that the charge–storage mechanism used to oxidize water is essentially the same in cyanobacteria as in eukaryotic plants.

Normally, only a small fraction of the energy absorbed by chlorophyll and a large fraction of the energy absorbed by biliproteins activates photosystem II, the O_2-producing photosystem, whereas the remainder activates photosystem I. The amount of chlorophyll committed to photosystem I seems, however, to be experimentally manipulable (Wang *et al.*, 1977). Energy absorbed by the biliproteins is transferred from phycoerythrin (when present) to phycocyanin to allophycocyanin(s) and then to energy-trap chlorophyll. Mohanty and Govindjee (1973) have presented evidence that light, presumably acting as an energy source, induces structural changes in the photosynthetic lamellae of *A. nidulans*. The structural changes in turn affect the structural relationship of the two photosystems and thereby regulate energy transfer between them. Energy transfer can apparently also be affected by the physical phase of the lipids in the membranes (Murata *et al.*, 1975). Light induces a flux of protons out of cells (Scholes *et al.*, 1969) as well as into thylakoids (Falkner *et al.*, 1976; Padan and Schuldiner, 1978), so that the normal cytoplasmic pH of ca. 7.0 (Falkner *et al.*, 1976) or 7.5 (Masamoto and Nishimura, 1977) is increased.

Cyclic photophosphorylation can account for 50% or more of total phosphorylation in *Anacystis* and *Anabaena* (Bedell and Govindjee, 1973; Bornefeld and Simonis, 1974; Bottomley and Stewart, 1976), but may be of much less importance at high intensities of light (Maxwell and Biggins, 1976; cf. Hiyama *et al.*, 1977). Knaff (1977) has presented evidence of a ferredoxin-catalyzed cyclic pathway of electron transport in *Nostoc muscorum*, and of a coupling site between cytochrome b_6 and cytochrome f. Membrane fragments derived from *N. muscorum* appear to have two kinetically different populations of P-700. One fraction of the P-700 responds to light as if that P-700 were in the noncyclic pathway of electron transfer from water to $NADP^+$; the other component may be related specifically to cyclic electron flow (Hiyama *et al.*, 1977).

B. Interactions between Photosynthesis and Respiration; Photorespiration

Certain data indicate that photosynthetic electron transport interacts with oxygen and with substrate-level oxidations, and may therefore have carriers in common with respiratory electron transport. For example, the presence of oxygen can apparently affect the quantum efficiency of photosynthetic production of O_2, presumably by oxidizing a carrier between photosystem II and photosystem I (Diner and Mauzerall, 1973a,b).

At low intensities of light, O_2 uptake (concurrent with photosynthetic production of O_2) decreases, and this decrease appears to be unaffected by DCMU (Hoch et al., 1963; see also Imafuku and Katoh, 1976). It was suggested that ATP generated by cyclic photophosphorylation in some way diminishes respiration. Indeed, inhibitors and uncouplers of photophosphorylation have been reported to reverse the inhibition, by light, of respiration in A. nidulans (Peschek, 1974; see also Rubin et al., 1977), although there is controversy about this point (Imafuku and Katoh, 1976). The presumably related inhibition, by light, of the activity of the oxidative pentose phosphate pathway (see Section III,C) is, however, reversed by DCMU (Rubin et al., 1977).

In Nostoc, deposits resulting from the reduction of tellurite and tetranitroblue tetrazolium are associated with the thylakoids and not with the plasmalemma, in the dark as well as in the light. It therefore appears that respiratory sites are present on the photosynthetic lamellae (Bisalputra et al., 1969).

A. nidulans also exhibits a DCMU-sensitive acceleration of uptake of O_2 by light of medium and high intensity ("photorespiration"). Unlike dark respiration, which is saturated by about 0.05 atm O_2, photorespiration increases linearly with O_2 to at least 0.2 atm O_2. Moreover, photorespiration is very sensitive to the partial pressure of CO_2, and is completely inhibited by 0.02 atm CO_2 (Lex et al., 1972).

Although the immediate substrate for photorespiration by cyanobacteria has not been identified, that photorespiration may be related to the formation or oxidation of glycolate. The results of Ingle and Colman (1976) strongly suggest that glycolate excretion by Coccochloris peniocystis is the result of ribulose bisphosphate oxygenase activity under conditions of low CO_2 and low carbonic anhydrase activity, although there is conflicting evidence about the effect of O_2 upon metabolism of CO_2 (Codd and Stewart, 1977a; Lloyd et al., 1977). O_2-Linked glycolate-oxidizing activity of Anabaena cylindrica and Oscillatoria sp. sediments with membranes (Sallal and Codd, 1975; Grodzinski and Colman, 1976). Glycolate and, with lesser activity, succinate, malate, and isocitrate can be photooxidized by membranes of Anabaena variabilis (Murai and Katoh, 1975). That photosystem I can accept electrons from some donor other than photosystem II, in vivo, is suggested by the observation that light-dependent uptake of oxygen by Anacystis continues after inactivation of photosystem II (Hammans et al., 1977).

C. Intermediate Metabolism of Carbon

1. Pathways of Carbon Dioxide Fixation and Respiration

When cyanobacteria are grown autotrophically, carbon dioxide is fixed predominately by the reductive pentose phosphate cycle (see review by

Wolk, 1973). At a low intensity of light, synthesis of aspartate (and glutamate) proceeds more rapidly than synthesis of 3-phosphoglycerate (Jansz
and Maclean, 1973; Döhler, 1976). Concordantly, several cyanobacteria
have been reported to have higher activities of phosphoenol pyruvate carboxylase than of ribulose bisphosphate carboxylase (Colman *et al.,* 1976).
The ribulose bisphosphate carboxylase from strains of *Microcystis* and
Aphanocapsa, like those from higher plants, has a molecular weight of ca.
520,000, and is composed of subunits of molecular weight ca. 50,000 and
15,000. The small subunits of the enzyme were lost if acid precipitation was
used during purification of the enzyme (Codd and Stewart, 1977a). In other
studies, which included an acid precipitation step, only large subunits were
found (Tabita *et al.,* 1976).

Although all enzymes of the Embden-Meyerhof-Parnas pathway are present (see reviews by Smith, 1973; Wolk, 1973; Sanchez *et al.,* 1975), the
principal pathway for the breakdown of carbohydrate in *Tolypothrix tenuis*
and *Anabaena variabilis* is the oxidative pentose phosphate pathway. However, energy for maintenance of viability of *Anacystis nidulans* in the dark can
be provided in mutants lacking this pathway (Doolittle and Singer, 1974).
The oxidative pentose phosphate pathway is activated in the dark (Pelroy
and Bassham, 1972; Rubin *et al.,* 1977), apparently because allosteric inhibition of glucose-6-P dehydrogenase activity by ribulose bisphosphate (Pelroy
et al., 1972; cf., however, Grossman and McGowan, 1975; Schaeffer and
Stanier, 1978), ATP, and NADPH (Grossman and McGowan, 1975; cf. also
Pelroy *et al.,* 1976a) is relieved. The complex regulatory properties of this
dehydrogenase, as isolated from a strain of *Anabaena,* involve slow, reversible transitions between aggregates of differing catalytic activity (Schaeffer
and Stanier, 1978). 6-Phospho-D-gluconate, produced by glucose-6-P dehydrogenase (and the first intermediate specific to the oxidative pentose phosphate pathway), can thereupon inhibit ribulose bisphosphate carboxylase
(Tabita and McFadden, 1972; Codd and Stewart, 1977a). In the light, fixation
of CO_2 is activated, at least in part, by activation of phosphoribulokinase
(which requires a relatively strong reducing agent, such as dithiothreitol, for
activity *in vitro*), fructose bisphosphatase, and sedoheptulose bisphosphatase (Pelroy *et al.,* 1976a,b).

2. Metabolism of Pyruvate and Acetate

In *Anabaena variabilis* and *A. cylindrica,* pyruvate is decarboxylated with
concomitant reduction of ferredoxin (Leach and Carr, 1971; Bothe *et al.,*
1974). *Anacystis nidulans,* however, contains NAD^+-linked pyruvate dehydrogenase (Bothe and Nolteernsting, 1975). [14C]Pyruvate is metabolized to
alanine; glutamate, proline, and arginine; leucine; isoleucine and valine; but
essentially not to other amino acids.

In most cyanobacteria, [14C]acetate is metabolized to glutamate, proline, arginine, and leucine, but essentially not to other amino acids (see reviews by Smith, 1973; Wolk, 1973). However, in *Chlorogloea fritschii,* a facultative heterotroph, all amino acids are labeled by acetate, and the aspartic acid family of amino acids is especially highly labeled (Miller and Allen, 1972; Lucas *et al.,* 1973). Lack of labeling of aspartate by [^{14}C]acetate in most cyanobacteria tested is evidence of an incomplete citric acid cycle in these organisms. In fact, activity of neither α-ketoglutarate dehydrogenase nor succinyl-CoA synthetase was detected (reviewed by Wolk, 1973). Formation of [^{14}C]aspartate from [^{14}C]acetate may be due to operation of the glyoxylate cycle (Lucas *et al.,* 1973). In a variety of cyanobacteria, citrate synthase, the enzyme which synthesizes citrate from acetyl-CoA and oxaloacetate, is allosterically inhibited by α-ketoglutarate and succinyl CoA (Taylor, 1973; Lucas and Weitzman, 1975, 1977).

D. Heterotrophy and Obligate Photoautotrophy

Although a majority of the cyanobacteria for which it has been attempted could not be grown heterotrophically, many can be (Kenyon *et al.,* 1972; Rippka, 1972; Khoja and Whitton, 1975; Wolk and Shaffer, 1976). The inability of many to grow in the dark with sugar was at one time attributed possibly to their lack of an intact citric acid cycle, but it is difficult then to explain why other equally deficient cyanobacteria are capable of chemoheterotrophy (Lucas *et al.,* 1973). *Aphanocapsa* 6714 has an active-transport system for uptake of glucose (Beauclerk and Smith, 1978), and can grow heterotrophically. The activities of hexokinase and of enzymes of the oxidative pentose phosphate pathway are similar in *Anacystis nidulans* and *Aphanocapsa* 6308 to the corresponding activities in *Aphanocapsa* 6714. The inability of *A. nidulans* and of strain 6308 to grow with glucose appears to be attributable to the absence of an effective mechanism for taking up glucose from solution (Pelroy *et al.,* 1972; Beauclerk and Smith, 1978). Even this explanation fails in the case of organisms which are capable of growth in the light in the presence of the photosystem-II inhibitor, DCMU, if given glucose, but which cannot grow in the dark with glucose (Rippka, 1972). The subject of the carbon metabolism of cyanobacteria in relation to heterotrophy and obligate photoautotrophy has been reviewed recently by Stanier and Cohen-Bazire (1977).

E. Respiration and Phosphorylation

Cytochrome oxidase participates in respiration, but the inability of even very high concentrations of HCN to inhibit respiration by more than 80% indicates that a significant fraction of respiratory uptake of oxygen may not involve cytochrome oxidase (see references in Wolk, 1973). Both NADH and NADPH are oxidized by membrane preparations, with concomitant phosphorylation. However, NADPH is the more rapidly oxidized, and *in vivo*, O_2 affects the steady-state oxidation level of NADPH, but not of NADH. A soluble cytochrome can serve as electron acceptor from reduced NADP in a strain of *Nostoc* (Pulich, 1977).

The relationship between the endogenous level of ATP and environmental conditions has been studied extensively in recent years (Biggins, 1969; Bornefeld and Simonis, 1974; Bornefeld, 1976a,b; Bottomley and Stewart, 1976a,b; Imafuku, 1976; Imafuku and Katoh, 1976; see also Pelroy *et al.*, 1976a,b). Upon transition from light to dark, there is a rapid decrease in the intracellular concentration of ATP, followed by a slow increase to the previous steady-state level despite a much lower rate of phosphorylation (Bottomley and Stewart, 1976b). In the dark under anaerobic conditions, the ATP concentration remains low.

F. Metabolism of Nitrogen

1. Initial Processes of Assimilation of Nitrogen

Cyanobacteria can assimilate nitrogen in a variety of forms. Many can reduce N_2 (see below), most can reduce nitrate, and probably all can utilize ammonium. At least some cyanobacteria can utilize urea (see Healey, 1977), uric acid, protein nitrogen (presumably by extracellular proteolysis), and amino acids (reviewed by Wolk, 1973; see also Pope, 1974; Diakoff and Scheibe, 1975; Kapp *et al.*, 1975; Ohki and Katoh, 1975; Nikitina and Gusev, 1976; Rowell *et al.*, 1977). Cyanobacteria also utilize a wide variety of nucleic acid precursors (Pigott and Carr, 1971; Glaser *et al.*, 1973; Charles, 1977).

Heterocyst-forming cyanobacteria fix N_2 under aerobic conditions. Many cyanobacteria without heterocysts do so under microaerobic conditions, i.e., supplied with O_2-free gas but illuminated with light which induces photosynthetic production of O_2 (Stewart and Lex, 1970; Kenyon *et al.*, 1972). Few cyanobacteria lacking heterocysts appear able to fix N_2 aerobically (Wyatt and Silvey, 1969; Rippka *et al.*, 1971; Singh, 1973; Taylor *et al.*, 1973), and in these organisms there may be a temporal (Weare and Benemann, 1974; Gallon *et al.*, 1974) or spatial (Carpenter and Price, 1976; Meeks *et al.*, 1978) separation between the processes of fixation of N_2 and photosynthetic evolution of O_2.

Nitrogenase from cyanobacteria, like the nitrogenases from other bacteria, consists of two proteins (I and II). Those two proteins from *Anabaena cylindrica* can react with certain of the complementary proteins (II and I) from other bacteria to give active nitrogenase. The cyanobacterial enzyme also resembles nitrogenase from other sources in that it requires ATP and a strong reductant for activity, is labile to cold and oxygen, and evolves hydrogen.

Membrane preparations capable of photoreducing nitrate or nitrite to ammonium with H_2O as electron source, in the light, have been derived from *Nostoc muscorum* 7119 (Candau *et al.*, 1976; Manzano *et al.*, 1976). Nitrate reductase from *Anabaena cylindrica* can be solubilized with detergent, and nitrite reductase from that same organism is a soluble enzyme. According to an action spectrum and to the consequences of addition of DCMU, photoreduction of nitrate by *Agmenellum quadruplicatum* is, in the short run, independent of photosystem II, and may therefore depend upon an unidentified organic donor of electrons to photosystem I (Stevens and Van Baalen, 1973).

2. Assimilation of Ammonium and Biosynthesis of Amino Acids

In a wide variety of N_2-grown cyanobacteria, ammonium ion, whether supplied exogenously or derived from N_2 (and presumably also from NO_3^-) is assimilated by the glutamine synthetase-glutamate synthase pathway, as shown by experiments using the radioisotope ^{13}N (Wolk *et al.*, 1976; Meeks *et al.*, 1978). This pathway is active even in ammonia-grown *Anabaena cylindrica* (Meeks *et al.*, 1977). Alanine dehydrogenase (Rowell and Stewart, 1976) also operates in the assimilation of ammonia, but is less active than the glutamine synthetase-glutamate synthase pathway (Meeks *et al.*, 1977, 1978). Glutamic acid dehydrogenase plays a very minor role (Meeks *et al.*, 1977), with the possible known exception of nitrate-grown *Anacystis nidulans* (Meeks *et al.*, 1978). Glutamine synthetase from *Anabaena* appears to be a dodecamer of molecular weight ca. 590,000 (Stacey *et al.*, 1977b). No evidence has been found that its activity is controlled by adenylylation (Dharmawardene *et al.*, 1973). Glutamate synthase from *Anabaena* and *Nostoc* is ferredoxin-dependent, but can use reduced methyl viologen as an alternative electron donor (Lea and Miflin, 1975).

Feedback inhibition has been found to be operative in a number of amino acid biosynthetic pathways in cyanobacteria: arginine, valine, leucine, isoleucine, threonine and lysine, methionine, tyrosine and phenylalanine, and tryptophan. Enzymes of the tryptophan biosynthetic pathway are repressed by tryptophan in *Agmenellum quadruplicatum* (Ingram *et al.*, 1972), but repression of other pathways of amino acid biosynthesis appears not to have been observed.

3. Nucleic Acid Metabolism and Protein Synthesis

The subunits of DNA-dependent RNA polymerase from *Anacystis nidulans* have molecular weights of 190,000 and 145,000 (β and β'), 38,000 (α), and 72,000 (σ) (Herzfeld and Rath, 1974). The enzyme can be reconstituted from its subunits only if the σ subunit is present (Herzfeld and Kiper, 1976).

Purified DNA polymerase is stabilized by DNA (Schönherr and Keir, 1972). DNAse and different RNAses active against normal RNA and against 2'-O-methyl RNA (cf. Biswas and Myers, 1960) were detected by Norton and Roth (1967). Restriction endonucleases have been isolated from *Anabaena variabilis* (Kopecka *et al.*, 1976), and their sequence specificity characterized (Murray *et al.*, 1976; see also Thibodeau and Verly, 1976). A light-activated DNA-photoreactivating enzyme has been isolated. Magic spots (ppGpp and pppGpp) are synthesized in response to shift-down of *A. nidulans* (Carr and Mann, 1975; Mann *et al.*, 1975; Smith, 1977; and see especially Adams *et al.*, 1977), but accumulation of stable RNA is apparently under some other control.

Cell-free preparations capable of polypeptide synthesis and dependent on both transfer and messenger RNA have been reported (Leach and Carr, 1974). Three methionyl tRNAs were found in *A. nidulans,* only one of which could be formylated (Ecarot and Cedergren, 1974). The same organism yielded five leucyl tRNAs and leucyl tRNA synthetase, and four seryl tRNAs and seryl tRNA synthetase (Beauchemin *et al.*, 1973; Parthier and Krauspe, 1974). The ratios of the amounts of total, transfer and ribosomal RNA to DNA appear to be largely independent of growth rate in *Anacystis* (Mann and Carr, 1974).

G. Metabolism of Sulfur

The assimilation of sulfur by cyanobacteria has been little studied. The uptake of sulfate by *A. nidulans* involves a permease, with a K_m of 0.75 μM (Utkilen *et al.*, 1976), which appears to be subject to derepression (Jeanjean and Broda, 1977). In *Anabaena cylindrica,* sulfate is assimilated by ATP sulfurylase, yielding adenosine phosphosulfate (APS), followed by APS kinase, yielding phosphoadenosine phosphosulfate (PAPS; Sawhney and Nicholas, 1976). Eleven of 21 cyanobacterial strains tested were capable of photoassimilating CO_2 in the presence of DCMU and Na_2S in 703-nm light, with $S^=$ replacing H_2O as electron donor (Garlick *et al.*, 1977). In two of these strains, hydrogen could also serve as electron donor (Belkin and Padan, 1978).

IV. GENETICS

Many types of mutants of cyanobacteria have been isolated (reviewed by Van Baalen, 1973; Wolk, 1973) including a wide variety of auxotrophic mutants of both unicellular (Herdman and Carr, 1972; Ingram *et al.*, 1972; Kaney, 1973; Herdman *et al.*, 1973; Stevens *et al.*, 1975; Singer and Doolittle,

1975; Romanova and Shestakov, 1976) and filamentous cyanobacteria (Currier *et al.*, 1977), and developmental mutants (Ingram and Aldrich, 1974; Wilcox *et al.*, 1975a; Currier *et al.*, 1977). Genetic study of oxygenic photosynthesis by cyanobacteria has been initiated by the isolation of strains defective in photoautotrophy (Stevens and Myers, 1976; Sherman and Cunningham, 1977; Shaffer *et al.*, 1978). To date, the only systematically usable means for obtaining gene transfer is by transformation with isolated DNA (Shestakov and Khyen, 1970; Orkwiszewski and Kaney, 1974; Astier and Espardellier, 1976).

V. VIRUSES

Viruses (''cyanophage'') of relatively narrow host range among the cyanobacteria have been isolated (see the reviews by Safferman, 1973, and Padan and Shilo, 1973). Thus far, all viruses found contain DNA, and all appear to adhere to the host cell by the viral tails and to inject their DNA through the wall of the host. In the DNA of cyanophage S-2L, adenine is replaced by 2-amino-adenine (Kirnos *et al.*, 1977). Temperate viruses, including a heat-inducible lysogen (Rimon and Oppenheim, 1975), have been reported in *Plectonema*. Such viruses have been found also in *Anabaena variabilis* (Hudyakov and Gromov, 1973) and in diverse unicellular cyanobacteria (Goryushin *et al.*, 1976). Nonetheless, successful transduction has not been documented as yet.

VI. MOVEMENT

The subject of the gliding movements of cyanobacteria has been reviewed by Castenholz (1973) and Wolk (1973). There is a frequent but not ubiquitous correlation between motility and the secretion of mucilage. All portions of a filament propel the organism, and ''surface waves'' have been observed. It has been suggested that motility is the consequence of torsional waves traveling in a fibrillar layer within the wall.

Movement in the dark appears to be energized by respiration. Light may or may not increase the rate of movement, depending upon the organism. From action spectra, such a ''photokinetic'' effect of light appears to be energized by photosynthetically active pigments, but by which pigment system depends again upon the organism, and also by certain pigments not active in photosynthesis (Nultsch, 1972). Movement toward or away from a source of light (''topophototaxis'') appears to make use of different steering mechanisms in the Nostocaceae (e.g., *Anabaena, Cylindrospermum*) and Oscillatoriaceae (e.g., *Oscillatoria, Phormidium*). Within the latter family, the effective pigment system again differs from organism to organism. Some species merely orient themselves perpendicular to the direction of the incident illumination. Filaments of certain species reverse direction soon after the front of the filament moves from the light into the dark, a response known as "positive photophobotaxis." In *Phormidium uncinatum*, the cyanobacterium in which this phenomenon has been most intensively studied, the two photosystems of photosynthesis appear to cooperate in controlling photophobotaxis (Nultsch and Häder, 1974).

VII. CULTURE

Conditions suitable for culture of cyanobacteria were reviewed by Wolk (1973). A neutral or slightly alkaline inorganic medium supplemented with a chelating agent to retain iron in solution suffices for growth of many cyanobacteria in the presence of light and CO_2. A small number of strains have been found to require vitamin B_{12}. *Anabaena* 6411 produces its own iron-chelating substance, the siderochrome schizokinen, first found in *Bacillus megaterium* (Simpson and Nielands, 1976). Phosphate or substituted alkyl sulfonate buffers (Good *et al.*, 1966; Smith and Foy, 1974) are normally added at concentrations sufficient for buffering.

Organic phosphates can be hydrolyzed and the phosphate utilized (Reichardt, 1971; Ihlenfeldt and Gibson, 1975; Healey and Hendzel, 1975; see also Rubin *et al.*, 1977). The rate of CO_2 assimilation, and thereby the rate of growth, can often be increased—other conditions being optimal—by increase of the partial pressure of CO_2 in the gas phase above ambient levels. The inorganic nutrients required are those typically needed for the growth of prokaryotes (including Co for the synthesis of vitamin B_{12}), photosynthetic plants (Fe, Mn, probably Cl) and, when appropriate, nitrogen-fixing organisms (Fe and Mo). The role of sodium ion remains unclear. Although sodium is required by *A. nidulans,* among other cyanobacteria, it is actively pumped out of that organism (Paschinger, 1977).

Growth rates determined for a variety of cyanobacteria have been summarized by Hoogenhout and Amesz (1965). Given sufficient light, CO_2, agitation, a proper nutrient medium, and an optimal temperature, *Anacystis nidulans* can double in about 2 h, the shortest time recorded for a cyanobacterium. The shortest doubling time reported for an organism capable of aerobic fixation of N_2, a marine *Anabaena*, is 3.6 h (4.3 h using N_2) (Stacey *et al.*, 1977a). Doubling times under heterotrophic conditions range upward from the value of 25 h observed for *Anabaena variabilis* at 34° (Wolk and Shaffer, 1976).

VIII. REGULATION

A. Metabolic and Nutritional Controls

As noted above, allosteric inhibitions affect a variety of pathways of carbohydrate and amino acid metabolism in cyanobacteria. With infrequent exceptions, such as the control by tryptophan of the concentrations of enzymes of the tryptophan biosynthetic pathway in *Agmenellum quadruplicatum* (Ingram *et al.*, 1972). exogenous organic metabolites have not been found to repress or induce the synthesis of specific cyanobacterial proteins. It remains to be determined whether such will remain the case when cyanobacteria are examined which have been isolated from sites, such as sewage oxidation ponds, which are rich in organic nutrients. As Singer and Doolittle (1975) have pointed out, it is appropriate to look for such effects in response to changes in environmental stimuli which do affect the growth of these microorganisms. In fact, as described below, dramatic alterations in protein synthesis are seen to result from changes in the provision of phosphorus, nitrogen, and light.

Phosphate deprivation results in enhanced ability of *A. variabilis* and *A. nidulans* to assimilate P_i (Bornefeld *et al.*, 1974; Healey and Hendzel, 1975) as well as in higher alkaline phosphatase activity (Healey and Hendzel, 1975). Production of extracellular phosphomonoesterase by *A. nidulans* is dependent upon the phosphorus source provided (Reichardt, 1971; see also Bone, 1971; Healey, 1973; and Ihlenfeldt and Gibson, 1975).

Nitrate reductase is induced by nitrate and repressed by ammonium (references in the review by Wolk, 1973; Stevens and Van Baalen, 1973, 1974; Camm and Stein, 1974). Nitrite reductase is also an inducible enzyme. Nitrogen deprivation leads to degradation of phycocyanin in *A. nidulans* (Lau *et al.*, 1977), and apparently similar effects are seen with other cyanobacteria. Degradation of the pigment is apparently a result of the activity of a protease with narrow substrate-specificity (Foulds and Carr, 1977; Wood and Haselkorn, 1979).

B. Effects of Light

The pigmentation of cyanobacteria can be affected somewhat by the intensity of the incident light, but much greater quantitative—and in certain instances, qualitative—changes in pigmentation result from variation in the color of the incident light. Marsac (1977) compared the pigmentation of 44 phycoerythrin-forming cyanobacteria grown in red and in green light. Twelve strains did not adapt; in seven, the amount of phycoerythrin present was modulated, but not the amount of phycocyanin; and in the remaining 25, the phycoerythrin decreased—in certain cases, to zero—while phycocyanin increased, in red light relative to green light. Because the phycobiliproteins comprise a substantial fraction of total cellular protein, and in at least certain instances can account for the majority of "soluble protein," these shifts represent major changes in the patterns of cellular protein. It has been directly demonstrated, in the case of *Fremyella diplosiphon,* that transfer to red light causes an abrupt cessation of synthesis of C-phycoerythrin, and that upon transfer from red to fluorescent light, *de novo* synthesis of C-phycoerythrin is initiated (Bennett and Bogorad, 1973). The proportions of phycocyanin and phycoerythrin that form during dark incubation of *Tolypothrix tenuis* is determined, photoreversibly, by the color of light during the final 6 min of prior illumination. A photoreversible pigment with absorption maxima close to the wavelengths of the action maxima was found in this organism (Diakoff and Scheibe, 1973), and the occurrence of similar pigments has been demonstrated in other cyanobacteria (Björn and Björn, 1976).

Light affects the synthesis of other proteins in cyanobacteria, and some phenomena other than pigmentation are under photoreversible control by light. Simply transferring cultures of *Anacystis nidulans* from light to dark results in preferential synthesis of certain proteins, including two—glycogen phosphorylase and glucose-6-P dehydrogenase—which play a role in endogenous metabolism in the dark. The rate of growth of *Fremyella diplosiphon* in the dark is determined, photoreversibly, by brief daily exposures to red or green light (Diakoff and Schiebe, 1975); and whether *Nostoc muscorum* grows in the dark as filaments or as aseriate packets of cells is similarly affected (Lazaroff, 1973).

IX. DEVELOPMENT

A. Heterocysts

1. Physiology and Biochemistry

Heterocysts, which arise by differentiation of vegetative cells in certain filamentous cyanobacteria, are foci of nitrogen fixation within the filaments during aerobic growth (Fay *et al.,* 1968; Thomas *et al.,* 1977; Peterson and Wolk, 1978b). The subject of heterocysts has been reviewed repeatedly (Wolk, 1973, 1975; Fay, 1973; Haselkorn, 1978).

Heterocysts (Fig. 3) have a wall which is, as far as known, identical to that of a vegetative cell, but outside of that wall they have a bilayered envelope. The outer, homogeneous portion of the envelope consists of polysaccharide. The envelope polysaccharide from *Anabaena cylindrica* is highly branched, and has a repeating backbone subunit (-Man-Glc-Glc-Glc-)$_n$, in which the

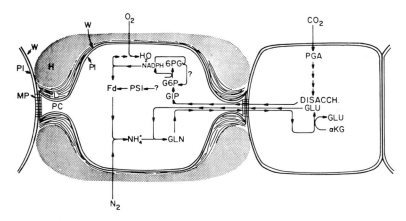

Fig. 3.   Diagram showing the principal known interactions between a heterocyst (at left) and a vegetative cell (at right). Outside of the wall (W) of the heterocyst is an envelope consisting principally of a laminated, glycolipid layer (L) and a homogeneous, polysaccharide layer (H). Microplasmodesmata (MP) join the plasma membranes (Pl) of the heterocyst and vegetative cell at the end of the pore channel (PC) of the heterocyst. Disaccharide formed by photosynthesis in the vegetative cells moves into heterocysts, where it is thought to be metabolized to glucose-6-P and oxidized by the oxidative pentose phosphate pathway. Pyridine nucleotide (NADPH) reduced by this pathway can react with O_2 via an electron transport chain to maintain reducing conditions within the heterocysts, and can reduce ferredoxin (Fd). Ferredoxin can also be reduced by photosystem I. Reduced ferredoxin can donate electrons to nitrogenase, which reduces N_2 to NH_4^+ with concomitant production of hydrogen. Glutamate produced by vegetative cells reacts in heterocysts with NH_4^+ to form glutamine. The glutamine moves into the vegetative cells, where it reacts with α-ketoglutarate (αKG) to form two molecules of glutamate.

monosaccharides are β1-3 linked (Cardemil and Wolk, 1979). Fibrous material present at the periphery of the homogeneous layer in electron micrographs may be such polymers of polysaccharide, which have not been integrated into that layer.

The inner, laminated layer of the envelope consists of a group of structurally related glycolipids. The principal glycolipid in the heterocysts of *A. cylindrica* is 1-(*O*-α-D-glucopyranosyl)-3,25-hexacosanediol. This same lipid was observed in the heterocysts of other cyanobacteria investigated, but was sometimes a minor component. The lipid layer surrounds the heterocyst except at the junction(s) with adjacent vegetative cells, and may therefore constrain heterocysts to interchange lipid-insoluble materials only with vegetative cells, rather than directly with the medium. At the junctions to vegetative cells, the walls of heterocysts are traversed by structures which may correspond to plasmodesmata. The thylakoids of heterocysts have the normal complement of acyl lipids.

Heterocysts lack a reductive pentose phosphate pathway, have very low glycolytic activity (Winkenbach and Wolk, 1973; Lex and Carr, 1974; Codd and Stewart, 1977b), and apparently lack photosystem II (Donze *et al.*, 1972; Tel-Or and Stewart, 1977), and so must rely on vegetative cells for a supply of carbon, especially carbohydrate, and reductant. Reducing equivalents may move from cell to cell principally in the form of a disaccharide (Jüttner and Carr, 1976). The disaccharide is presumed to be metabolized to glucose-6-P, because the specific activities of glucose-6-P dehydrogenase and 6-phosphogluconate dehydrogenase are far higher in heterocysts than in vegetative cells (Winkenbach and Wolk, 1973; Lex and Carr, 1974; Rowell and Stewart, 1976), and because isolated heterocysts can utilize glucose 6-P as electron donor both to oxygen (Peterson and Burris, 1976) and to substrates of nitrogenase (Peterson and Burris, 1978; Lockau *et al.*, 1978). Oxygen is also rapidly reduced by hydrogen, a byproduct of fixation of nitrogen, within heterocysts (Peterson and Burris, 1978). The uptake hydrogenase may be restricted to the heterocysts in certain aerobically grown and assayed species of *Anabaena* (Wolk and Peterson, 1978a).

Heterocysts have photosystem I by means of which they can photoreduce acetylene, a substrate of nitrogenase, although the presumed donor of electrons to photosystem I *in vivo* is unidentified. Heterocysts also appear to carry out photophosphorylation as well as oxidative phosphorylation (Tel-Or and Stewart, 1976). Experiments involving autoradiography of ^{35}S-labeled proteins showed that the biliproteins, the auxiliary pigments of photosystem II, are missing from heterocysts (Fleming and Haselkorn, 1974; cf. also Wood and Haselkorn, 1979).

Heterocysts may be anaerobic cells within oxygen-producing filaments, and for that reason, sites hospitable to activity of nitrogenase (Fay *et al.*, 1968). Lacking photosystem II, they do not generate oxygen by photosynthesis, but as noted above, they do have the capacity to reduce oxygen which enters them. In addition, their thick envelope may restrict the penetration of oxygen (Stewart, 1973).

The specific activity of glutamine synthetase is higher in heterocysts than in vegetative cells (Dharmawardene *et al.*, 1973; Thomas *et al.*, 1977). As noted in Section III,F, fixed nitrogen derived from N_2 is metabolized first by that enzyme (Stewart and Rowell, 1975; Wolk *et al.*, 1976) and then by glutamate synthase (Wolk *et al.*, 1976). The latter enzyme, however, is restricted to vegetative cells (Thomas *et al.*, 1977). Nitrogen fixed by isolated heterocysts is secreted largely as glutamine, the synthesis of which is dependent upon provision of glutamate (Thomas *et al.*, 1977), which is supplied by the vegetative cells (N. Schilling and C. P. Wolk, unpublished; see also Jüttner and Carr, 1976) (see Fig. 3).

2. Patterns of Heterocyst Formation

It had been shown earlier that the spacing of heterocysts within filaments of cyanobacteria is determined by an inhibition, by heterocysts, of heterocyst formation (Wolk, 1967; Wolk and Quine, 1975; cf. also Wilcox *et al.*, 1975b). Ammonium ion inhibits the differentiation of heterocysts, and methionine sulfoximine, an inhibitor of glutamine synthetase, prevents the inhibitory effect of ammonium (Stewart and Rowell, 1975; Ownby, 1977). It thus appears that glutamine or a derivative of glutamine is inhibitory to the differentiation of heterocysts, and therefore that glutamine produced by heterocysts may mediate the intercellular inhibition of heterocyst formation (Thomas *et al.*, 1977). Even very immature heterocysts, which are not yet capable of fixing nitrogen, inhibit the formation of heterocysts (Wolk, 1967; Wolk and Quine, 1975), perhaps by means of products of the proteolysis which takes place in the immature heterocysts (Fleming and Haselkorn, 1974). Heterocyst formation is stimulated moderately by azatryptophan and by very low concentrations of rifampicin (Mitchison and Wilcox, 1973; Wilcox *et al.*, 1975; Wolk and Quine, 1975).

B. Spores

Certain cyanobacteria also form spores (akinetes), the function of which is perennation. In *Anabaena cylindrica,* heterocysts induce adjacent cells to differentiate into spores. *Cylindrospermum licheniforme* secretes a substance which stimulates sporulation of fresh inoculum (Fisher and Wolk, 1976). Whether this substance is involved in intercellular induction of sporulation is unknown. Spores also form adjacent to heterocysts in *Cylindrospermum,* but mutants of that organism have been isolated which lack heterocysts, yet in which spores are formed nonetheless (Singh, 1976). Sporulation of *Anabaena doliolum* begins in the middle between two heterocysts and spreads towards the heterocysts. Because heterocysts are a sink for carbohydrate and a source of nitrogen, whereas glucose stimulates sporulation and sources of fixed nitrogen inhibit that process, it was suggested that in *A. doliolum,* heterocysts may influence the location of sporulation by affecting the carbohydrate and nitrogen status of the cells in the filament (Tyagi, 1974).

Spores are similar cytologically to enlarged vegetative cells with a thick envelope. The polysaccharide which accounts for much of the mass of the envelope of the spore of *A. cylindrica* is identical to the envelope polysaccharide of heterocysts of the same organism (Cardemil and Wolk, 1979). Direct measurement of spores isolated from *A. cylindrica* shows no more DNA per cell than in vegetative cells despite a 3- to 4-fold increase in RNA and protein (Simon, 1977).

X. EVOLUTION

Although cyanobacteria may have been the first organisms on earth which were capable of oxygenic photosynthesis, they appear to have postdated the first appearance of oxygen in the atmosphere and of aerobic, heterotrophic bacteria (Schwartz and Dayhoff, 1978). The application of techniques for the sequencing of proteins and of RNAs has now provided direct evidence in favor of Mereschkowsky's (1905) proposal that the photosynthetic plastids of eukaryotic plant cells are the evolutionary descendants of endosymbiotic prokaryotes (Schwartz and Dayhoff, 1978).

1. The ferredoxins of *Aphanotheca sacrum, Spirulina platensis, Spirulina maxima,* and *Nostoc muscorum* have been sequenced, and thereby shown to be much more closely related to chloroplast ferredoxins from eukaryotic algae and higher plants than to the ferredoxins from aerobic and anaerobic heterotrophic bacteria with which they were compared. Similarly, the cytochromes C_6 of *Spirulina* and *Plectonema* are much more closely related in structure to the chloroplast cytochromes of eukaryotic algae of four divisions—especially, of the red alga *Porphyra tenera*—than to the cytochromes of other photosynthetic bacteria, aerobic bacteria, anaerobic bacteria, and mitochondria with which they were compared. So, too, is the plastocyanin of *Anabaena* much more closely related to the plastocyanin of *Chlorella* and of higher plants than to a structurally related, copper-containing protein of aerobic, heterotrophic bacteria (Schwartz and Dayhoff, 1978). Moreover, catalogs of the products of digestion of 16 S ribosomal RNA from certain unicellular cyanobacteria show greater similarity to a corresponding catalog derived from the 16 S RNA of *Porphyridium* than of representatives of other bacterial groups (Bonen and Doolittle, 1975, 1976).

2. Ribosomal 5 S RNA from *A. nidulans* has been sequenced; it is much more closely related to the 5 S ribosomal RNAs of heterotrophic bacteria than to the cytoplasmic 5 S RNAs of *Chlorella,* rye, and other eukaryotes (Schwartz and Dayhoff, 1978). The sequence of formylmethionyl transfer RNA from *A. nidulans* is much more closely related to the sequence of initiator tRNAs from heterotrophic prokaryotes than from heterotrophic eukaryotes (Ecarot-Charrier and Cedergren, 1976).

The chloroplasts of the lower red algae (Bangioideae) differ from cyanobacteria in that they possess neither a peptidoglycan-containing circumferential layer, nor the keto-carotenoids and glycosylated carotenoids characteristic of cyanobacteria. In many other ways, the structure, composition and (photosynthetic) physiology of the two is extremely similar (see review by Wolk, 1973). This similarity has now been extended to the N-terminal amino acid sequences (Glazer and Apell, 1977; cf. also Williams *et al.,* 1974, and Glazer *et al.,* 1976) and to chromophore-containing oligopeptide sequences (Bryant *et al.,* 1978) of subunits of their biliproteins. The photosynthetic organelles of *Cyanophora paradoxa,* a taxonomically anomalous, blue–green pigmented, eukaryotic flagellate, have a genetic complexity less than 10% as great as the complexity of the smallest known cyanobacterial genomes

(Herdman and Stanier, 1977). However, these organelles have a lysozyme-sensitive envelope that contains diaminopimelic acid and muramic acid, constituents characteristic of peptidoglycan (Stanier and Cohen-Bazire, 1977). Of great interest relative to the origin of the chloroplasts of the green algae and higher plants is the recent discovery of prokaryotes that have chlorophylls a and b, and lack biliproteins (Lewin and Withers, 1975).

REFERENCES

Adams, D. G., Phillips, D. O., Nichols, J. M., and Carr, N. G. (1977). *FEBS Lett.* **81**, 48–52.
Asato, Y., and Ginoza, H. S. (1973). *Nature (London) New Biol.* **244**, 132–133.
Astier, C., and Espardellier, F. (1976). *C. R. Hebd. Seances Acad. Sci.* **282**, 795–797.
Beauchemin, N., LaRue, B., and Cedergren, R. J. (1973). *Arch. Biochem. Biophys.* **156**, 17–25.
Beauclerk, A. A. D., and Smith, A. J. (1978). *Eur. J. Biochem.* **82**, 187–197.
Bedell, G. W., II, and Govindjee (1973). *Plant Cell Physiol.* **14**, 1081–1097.
Belkin, S., and Padan, E. (1978). *Arch. Microbiol.* **116**, 109–111.
Bennett, A., and Bogorad, L. (1973). *J. Cell Biol.* **58**, 419–435.
Biggins, J. (1969). *J. Bacteriol.* **99**, 570–575.
Bisalputra, T., Brown, D. L., and Weier, T. E. (1969). *J. Ultrastruct. Res.* **27**, 182–197.
Bisalputra, T., Oakley, B. R. Walker, D. C., and Shields, C. M. (1975). *Protoplasma* **86**, 19–28.
Biswas, B. B., and Myers, J. (1960). *Nature (London)* **186**, 238–239.
Björn, G. S., and Björn, L. O. (1976). *Physiol. Plant.* **36**, 297–304.
Bogorad, L. (1975). *Annu. Rev. Plant Physiol.* **26**, 369–401.
Bone, D. H. (1971). *Arch. Mikrobiol.* **80**, 147–153.
Bonen, L., and Doolittle, W. F. (1975). *Proc. Natl. Acad. Sci. U.S.A.* **72**, 2310–2314.
Bonen, L., and Doolittle, W. F. (1976). *Nature (London)* **261**, 669–673.
Bornefeld, T. (1976a). *Biochem. Physiol. Pflanz.* **170**, 333–344.
Bornefeld, T. (1976b). *Biochem. Physiol. Pflanz.* **170**, 345–353.
Bornefeld, T., and Simonis, W. (1974). *Planta* **115**, 309–318.
Bornefeld, T., Lee-Kaden, J., and Simonis, W. (1974). *Ber. Deutsch. Bot. Ges.* **87**, 493–500.
Bothe, H., and Nolteernsting, U. (1975). *Arch. Microbiol.* **102**, 53–57.
Bothe, H., Falkenberg, B., and Nolteernsting, U. (1974). *Arch. Microbiol.* **96**, 291–304.
Bottomley, P. J., and Stewart, W. D. P. (1976a). *Brit. Phycol. J.* **11**, 69–82.
Bottomley, P. J., and Stewart, W. D. P. (1976b). *Arch. Microbiol.* **108**, 249–258.
Bryant, D. A., Hixson, C. S., and Glazer, A. N. (1978). *J. Biol. Chem.* **253**, 220–225.
Buchecker, R., Liaaen-Jensen, S., Borch, G., and Siegelman, H. W. (1976). *Phytochemistry* **15**, 1015–1018.
Buttke, T. M., and Ingram, L. O. (1975). *J. Bacteriol.* **124**, 1566–1573.
Camm, E. L., and Stein, J. R. (1974). *Can. J. Bot.* **52**, 719–726.
Candau, P., Manzano, C., and Losada, M. (1976). *Nature (London)* **262**, 715–717.
Cardemil, L., and Wolk, C. P. (1979). *J. Biol. Chem.* **254**, 736–741.
Carpenter, E. J., and Price, C. C., IV (1976). *Science* **191**, 1278–1280.
Carr, N. G., and Bradley, S. (1973). *Symp. Soc. Gen. Microbiol.* **23**, 161–188.
Carr, N. G., and Mann, N. (1975). *Biochem. Soc. Trans.* **3**, 368–373.
Carr, N. G., and Whitton, B. A., eds. (1973). "The Biology of Blue-Green Algae." Blackwell, Oxford.
Castenholz, R. W. (1973). *In* "The Biology of Blue-Green Algae" (N. G. Carr and B. A. Whitton, eds.), pp. 320–339. Blackwell, Oxford.

Chapman, D. J. (1973). *In* ''The Biology of Blue-Green Algae'' (N. G. Carr and B. A. Whitton, eds.), pp. 162–185. Blackwell, Oxford.

Charles, D. (1977). *Plant Sci. Lett.* **8**, 35–44.

Codd, G. A., and Stewart, W. D. P. (1977a). *Arch. Microbiol.* **113**, 105–110.

Codd, G. A., and Stewart, W. D. P. (1977b). *FEMS Microbiol. Lett.* **2**, 247–249.

Colman, B., Cheng, K. H., and Ingle, R. K. (1976). *Plant Sci. Lett.* **6**, 123–127.

Currier, T. C., Haury, J. F., and Wolk, C. P. (1977). *J. Bacteriol.* **129**, 1556–1562.

Dharmawardene, M. W. N., Haystead, A., and Stewart, W. D. P. (1973). *Arch. Mikrobiol.* **90**, 281–295.

Diakoff, S., and Scheibe, J. (1973). *Plant Physiol.* **51**, 382–385.

Diakoff, S., and Scheibe, J. (1975). *Physiol. Plant.* **34**, 125–128.

Diner, B., and Mauzerall, D. (1973a). *Biochim. Biophys. Acta* **305**, 329–352.

Diner, B., and Mauzerall, D. (1973b). *Biochim. Biophys. Acta* **305**, 353–363.

Dobson, P. R., Doolittle, W. F., and Sogin, M. L. (1974). *J. Bacteriol.* **117**, 660–666.

Döhler, G. (1976). *Planta* **131**, 129–133.

Donze, M., Haveman, J., and Schiereck, P. (1972). *Biochim. Biophys. Acta* **256**, 157–161.

Doolittle, W. F. (1972). *J. Bacteriol.* **111**, 316–324.

Doolittle, W. F., and Singer, R. A. (1974). *J. Bacteriol.* **119**, 677–683.

Doonan, B. B., and Jensen, T. E. (1977). *J. Bacteriol.* **132**, 967–973.

Ecarot, B., and Cedergren, R. J. (1974). *Biochim. Biophys. Acta* **340**, 130–139.

Ecarot-Charrier, B., and Cedergren, R. J. (1976). *FEBS Lett.* **63**, 287–290.

Falkner, G., Horner, F., Werdan, K., and Heldt, H. W. (1976). *Plant Physiol.* **58**, 717–718.

Fay, P. (1973). *In* ''The Biology of Blue-Green Algae'' (N. G. Carr and B. A. Whitton, eds.), pp. 238–259. Blackwell, Oxford.

Fay, P., Stewart, W. D. P., Walsby, A. E., and Fogg, G. E. (1968). *Nature (London)* **209**, 94–95.

Fisher, R. W., and Wolk, C. P. (1976). *Nature (London)* **259**, 394–395.

Fleming, H., and Haselkorn, R. (1974). *Cell* **3**, 159–170.

Fogg, G. E., Stewart, W. D. P., Fay, P., and Walsby, A. E. (1973). ''The Blue-Green Algae.'' Academic Press, New York.

Foulds, I., and Carr, N. G. (1977). *FEMS Microbiol. Lett.* **2**, 117–119.

Freidenreich, P., Apell, G. S., and Glazer, A. N. (1978). *J. Biol. Chem.* **253**, 212–219.

Gallon, J. R., LaRue, T. A., and Kurz, W. G. W. (1974). *Can. J. Microbiol.* **20**, 1633–1637.

Gantt, E., Lipschultz, C. A., and Zilinskas, B. (1976). *Biochim. Biophys. Acta* **430**, 375–388.

Garlick, S., Oren, A., and Padan, E. (1977). *J. Bacteriol.* **129**, 623–629.

Glaser, V. M., Al-Nuri, M. A., Groshev, V. V., and Shestakov, S. V. (1973). *Arch. Mikrobiol.* **92**, 217–226.

Glazer, A. N., and Apell, G. S. (1977). *FEMS Lett.* **1**, 113–116.

Glazer, A. N., and Bryant, D. A. (1975). *Arch. Microbiol.* **104**, 15–22.

Glazer, A. N., Apell, G. S., Hixson, C. S., Bryant, D. A., Rimon, S., and Brown, D. M. (1976). *Proc. Natl. Acad. Sci. U.S.A.* **73**, 428–431.

Golecki, J. R. (1977). *Arch. Microbiol.* **114**, 35–41.

Good, N. E., Winget, G. D., Winter, W., Connolly, T. N., Izawa, S., and Singh, R. M. M. (1966). *Biochemistry* **5**, 467–477.

Goryushin, V. A., Shatokhina, E. S., Grigor'eva, G. A., and Shestakov, S. V. (1976). *Vestn. Mosk. Univ., Biol. Pochvoved.* **31**, 82–84.

Gray, B. H., and Gantt, E. (1975). *Photochem. Photobiol.* **21**, 121–128.

Gray, J. E., and Herson, D. S. (1976). *Arch. Microbiol.* **109**, 95–99.

Grierson, D., and Smith, H. (1973). *Eur. J. Biochem.* **36**, 280–285.

Grodzinski, B., and Colman, B. (1976). *Plant Physiol.* **58**, 199–202.

Grossman, A., and McGowan, R. E. (1975). *Plant Physiol.* **55**, 658–662.

Hammans, J. W. K., Hendricks, G. M., and Teerlink, T. (1977). *Biochem. Biophys. Res. Commun.* **74**, 1560–1565.

Haselkorn, R. (1978). *Annu. Rev. Plant Physiol.* **29**, 319–344.
Haselkorn, R., and Rouvière-Yaniv, J. (1976). *Proc. Natl. Acad. Sci. U.S.A.* **73**, 1917–1920.
Healey, F. P. (1973). *J. Phycol.* **9**, 383–394.
Healey, F. P. (1977). *Can. J. Bot.* **55**, 61–69.
Healey, F. P., and Hendzel, L. L. (1975). *J. Phycol.* **11**, 303–309.
Herdman, M. (1976). *Proc. Int. Symp. Photosynthetic Prokaryotes 2nd,* pp. 229–231.
Herdman, M., and Carr, N. G. (1972). *J. Gen. Microbiol.* **70**, 213–220.
Herdman, M., and Carr, N. G. (1974). *Arch. Microbiol.* **99**, 251–254.
Herdman, M., and Stanier, R. Y. (1977). *FEMS Lett.* **1**, 7–12.
Herdman, M., Delaney, S. F., and Carr, N. G. (1973). *J. Gen. Microbiol.* **79**, 233–237.
Hertzberg, S., Liaaen-Jensen, S., and Siegelman, H. W. (1971). *Phytochemistry* **10**, 3121–3127.
Herzfeld, F., and Kiper, M. (1976). *Eur. J. Biochem.* **62**, 189–192.
Herzfeld, F., and Rath, N. (1974). *Biochim. Biophys. Acta* **374**, 431–437.
Hiyama, T., McSwain, B. D., and Arnon, D. I. (1977). *Biochim. Biophys. Acta* **460**, 76–84.
Hoch, G., Owens, O. v. H., and Kok, B. *Arch. Biochem. Biophys.* **101**, 171–180.
Hoogenhout, H., and Amesz, J. (1965). *Arch. Mikrobiol.* **50**, 10–25.
Hudyakov, I. Ya., and Gromov, B. V. (1973). Mikrobiologiya **42**, 904–907.
Ihlenfeldt, M. J. A., and Gibson, J. (1975). *Arch. Mikrobiol.* **102**, 23–28.
Imafuku, H. (1976). *Physiol. Plant.* **38**, 191–195.
Imafuku, I., and Katoh, T. (1976). *Plant Cell Physiol.* **17**, 515–524.
Ingle, R. K., and Colman, B. (1976). *Planta* **128**, 217–223.
Ingram, L. O., and Aldrich, H. C. (1974). *J. Bacteriol.* **118**, 708–716.
Ingram, L. O., Pierson, D., Kane, J. F., Van Baalen, C., and Jensen, R. A. (1972). *J. Bacteriol.* **111**, 112–118.
Jansz, E. R., and MacLean, F. I. (1973). *Can. J. Microbiol.* **19**, 497–504.
Jeanjean, R., and Broda, E. (1977). *Arch. Microbiol.* **114**, 19–23.
Jensen, T. E., and Ayala, R. P. (1976). *Arch. Microbiol.* **111**, 1–6.
Jensen, T. E., and Sicko, L. M. (1973). *Cytologia* **38**, 381–391.
Jensen, T. E., and Sicko-Goad, L. (1976). Aspects of phosphate utilization by blue-green algae. U.S. Environ. Protect. Agency, Corvallis, Oregon.
Jüttner, F., and Carr, N. G. (1976). *Proc. Int. Symp. Photosynthetic Prokaryotes 2nd,* pp. 121–123.
Kaney, A. R. (1973). *Arch. Mikrobiol.* **92**, 139–142.
Kapp, R., Stevens, S. E., Jr., and Fox, J. L. (1975). *Arch. Microbiol.* **104**, 135–138.
Katz, A., Weckesser, J., Drews, G., and Meyer, H. (1977). *Arch. Microbiol.* **113**, 247–256.
Kenyon, C. N., Rippka, R., and Stanier, R. Y. (1972). *Arch. Mikrobiol.* **83**, 216–236.
Khoja, T. M., and Whitton, B. A. (1975). *Brit. Phycol. J.* **10**, 139–148.
Kirnos, M. D., Khudyakov, I. Y., Alexandrushkina, N. I., and Vanyushin, B. F. (1977). *Nature (London)* **270**, 369–370.
Knaff, D. B. (1977). *Arch. Biochem. Biophys.* **182**, 540–545.
Kopecka, H., Hillova, J., and Hill, M. (1976). *Nature (London)* **262**, 72–74.
Kratz, W. A., and Myers, J. (1955). *Amer. J. Bot.* **42**, 282–287.
Krogmann, D. W. (1973). *In* "The Biology of Blue-Green Algae" (N. G. Carr and B. A. Whitton, eds.), pp. 80–98. Blackwell, Oxford.
Lau, R. H., MacKenzie, M. M., and Doolittle, W. F. (1977). *J. Bacteriol.* **132**, 771–778.
Lazaroff, N. (1973). *In* "The Biology of Blue-Green Algae" (N. G. Carr and B. A. Whitton, eds.), pp. 279–319. Blackwell, Oxford.
Lea, P. J., and Miflin, B. J. (1975). *Biochem. Soc. Trans.* **3**, 381–384.
Leach, C. K., and Carr, N. G. (1971). *Biochim. Biophys. Acta* **245**, 165–174.
Leach, C. K., and Carr, N. G. (1974). *J. Gen. Microbiol.* **81**, 47–58.
Leak, L. V. (1967). *J. Ultrastruct. Res.* **20**, 190–205.
Lehmann, M., and Wöber, G. (1976). *Arch. Microbiol.* **111**, 93–97.

Levi, C., and Preiss, J. (1976). *Plant Physiol.* **58**, 753–756.

Lewin, R. A., and Withers, N. W. (1975). *Nature (London)* **256**, 735–737.

Lex, M., and Carr, N. G. (1974). *Arch. Microbiol.* **101**, 161–167.

Lex, M., Silvester, W. B., and Stewart, W. D. P. (1972). *Proc. R. Soc. London Ser. B* **180**, 87–102.

Ley, A. C., Babcock, G. T., and Sauer, K. (1975). *Biochim. Biophys. Acta* **387**, 379–387.

Lloyd, N. D. H., Canvin, D. T., and Culver, D. A. (1977). *Plant Physiol.* **59**, 936–940.

Lockau, W., Peterson, R. B., Wolk, C. P., and Burris, R. H. (1978). *Biochim. Biophys. Acta* **502**, 298–308.

Lucas, C., and Weitzman, P. D. J. (1975). *Biochem. Soc. Trans.* **3**, 379–381.

Lucas, C., and Weitzman, P. D. J. (1977). *Arch. Microbiol.* **114**, 55–60.

Lucas, C., Smith, A. J., and London, M. (1973). *Biochem. Soc. Trans.* **1**, 710–713.

Mann, N., and Carr, N. G. (1974). *J. Gen. Microbiol.* **83**, 399–405.

Mann, N., Carr, N. G., and Midgley, J. E. M. (1975). *Biochim. Biophys. Acta* **402**, 41–50.

Manzano, C., Candau, P., Gomez-Moreno, C., Relimpio, A. M., and Losada, M. (1976). *Mol. Cell Biochem.* **10**, 161–169.

Marsac, N. T. de (1977). *J. Bacteriol.* **130**, 82–91.

Marsac, N. T. de, and Cohen-Bazire, G. (1977). *Proc. Natl. Acad. Sci. U.S.A.* **74**, 1635–1639.

Masamoto, K., and Nishimura, M. (1977). *J. Biochem.* **82**, 483–487.

Maxwell, P. C., and Biggins, J. (1976). *Biochemistry* **15**, 3975–3981.

Maxwell, P. C., and Biggins, J. (1977). *Biochim. Biophys. Acta* **459**, 442–450.

Meeks, J. C., Wolk, C. P., Thomas, J., Lockau, W., Shaffer, P. W., Austin, S. M., Chien, W.-S., and Galonsky, A. (1977). *J. Biol. Chem.* **252**, 7894–7900.

Meeks, J. C., Wolk, C. P., Lockau, W., Schilling, N., Shaffer, P. W., and Chien, W.-S. (1978). *J. Bacteriol.* **134**, 125–130.

Mereschkowsky, C. (1905). *Biol. Centralbl.* **25**, 593–604.

Mikheyskaya, L. V., Ovodova, R. G., and Ovodov, Yu. S. (1977). *J. Bacteriol.* **130**, 1–3.

Miller, J. S., and Allen, M. M. (1972). *Arch. Mikrobiol.* **86**, 1–12.

Mitchison, G. J., and Wilcox, M. (1973). *Nature (London) New Biol.* **246**, 229–233.

Mohanty, P., and Govindjee (1973). *Biochim. Biophys. Acta* **305**, 95–104.

Murai, T., and Katoh, T. (1975). *Plant Cell Physiol.* **16**, 789–797.

Murata, N., Troughton, J. H., and Fork, D. C. (1975). *Plant Physiol.* **56**, 508–517.

Murray, K., Hughes, S. G., Brown, J. S., and Bruce, S. A. (1976). *Biochem. J.* **159**, 317–322.

Nichols, B. W. (1973). *In* "The Biology of Blue-Green Algae" (N. G. Carr and B. A. Whitton, eds.), pp. 144–161. Blackwell, Oxford.

Nikitina, K. A., and Gusev, M. V. (1976). *Fiziol. Rast.* **23**, 1219–1224.

Norton, J., and Roth, J. S. (1967). *Comp. Biochem. Physiol.* **23**, 361–371.

Nultsch, W. (1972). "Der Einfluss des Lichtes auf die Bewegung phototropher Mikroorganismen. I. Photokinesis." Abhandl. Marburg Gelehrt. Gesellsch., no. 2, Wilhelm Fink Verlag, München.

Nultsch, W., and Häder, D.-P. (1974). *Ber. Deutsch Bot. Ges.* **87**, 83–92.

Ohki, K., and Katoh, T. (1975). *Plant Cell Physiol.* **16**, 53–64.

Orkwiszewski, K. G., and Kaney, A. R. (1974). *Arch. Microbiol.* **98**, 31–37.

Ownby, J. D. (1977). *Planta* **136**, 277–279.

Padan, E., and Schuldiner, S. (1978). *J. Biol. Chem.* **253**, 3281–3286.

Padan, E., and Shilo, M. (1973). *Bacteriol. Rev.* **37**, 343–370.

Parthier, B., and Krauspe, R. (1974). *Biochem. Physiol. Pflanz.* **165**, 1–17.

Paschinger, H. (1977). *Arch. Microbiol.* **113**, 285–291.

Pelroy, R. A., and Bassham, J. A. (1972). *Arch. Mikrobiol.* **86**, 25–38.

Pelroy, R. A., Rippka, R., and Stanier, R. Y. (1972). *Arch. Mikrobiol.* **87**, 303–322.

Pelroy, R. A., Kirk, M. R., and Bassham, J. A. (1976a). *J. Bacteriol.* **128**, 623–632.
Pelroy, R. A., Levine, G. A., and Bassham, J. A. (1976b). *J. Bacteriol.* **128**, 633–643.
Peschek, G. A. (1974). *Proc. Int. Cong. Photosynthesis, 3rd,* pp. 921–928.
Peterson, R. B., and Burris, R. H. (1976). *Arch. Microbiol.* **108**, 35–40.
Peterson, R. B., and Burris, R. H. (1978). *Arch. Microbiol.* **116**, 125–132.
Peterson, R. B., and Wolk, C. P. (1978a). *Plant Physiol.* **61**, 688–691.
Peterson, R. B., and Wolk, C. P. (1978b). *Proc. Natl. Acad. Sci. U.S.A.* **75**, 6271–6275.
Pigott, G. H., and Carr, N. G. (1971). *Arch. Mikrobiol.* **79**, 1–6.
Pope, D. H. (1974). *Can. J. Bot.* **52**, 2369–2374.
Pulich, W., Jr. (1977). *J. Phycol.* **13**, 40–45.
Reichardt, W. (1971). *Z. Allg. Mikrobiol.* **11**, 501–524.
Reynolds, C. S., and Walsby, A. E. (1975). *Biol. Rev.* **50**, 437–481.
Rimon, A., and Oppenheim, A. B. (1975). *Virology* **64**, 454–463.
Rippka, R. (1972). *Arch. Mikrobiol.* **87**, 93–98.
Rippka, R., Neilson, A., Kunizawa, R., and Cohen-Bazire, G. (1971). *Arch. Mikrobiol.* **76**, 341–348.
Rippka, R., Waterbury, J., and Cohen-Bazire, G. (1974). *Arch. Microbiol.* **100**, 419–436.
Roberts, T. M., and Koths, K. E. (1976). *Cell* **9**, 551–557.
Roberts, T. M., Klotz, L. C., and Loeblich, A. R., III (1977). *J. Mol. Biol.* **110**, 341–361.
Romanova, N. I., and Shestakov, S. V. (1976). *Dokl. Akad. Nauk SSSR* **226**, 692–694.
Rowell, P., and Stewart, W. D. P. (1976). *Arch. Microbiol.* **107**, 115–124.
Rowell, P., Enticott, S., and Stewart, W. D. P. (1977). *New Phytol.* **79**, 41–54.
Rubin, P. M., Zetooney, E., and McGowan, R. E. (1977). *Plant Physiol.* **60**, 407–411.
Safferman, R. S. (1973). *In* "The Biology of Blue-Green Algae" (N. G. Carr and B. A. Whitton, eds.), pp. 214–237. Blackwell, Oxford.
Sallal, A.-K. J., and Codd, G. A. (1975). *FEBS Lett.* **56**, 230–234.
Sanchez, J. J., Palleroni, N. J., and Doudoroff, M. (1975). *Arch. Microbiol.* **104**, 57–65.
Sawhney, S. K., and Nicholas, D. J. D. (1976). *Planta* **132**, 189–195.
Schaeffer, F., and Stanier, R. Y. (1978). *Arch. Microbiol.* **116**, 9–19.
Scholes, P., Mitchell, P., and Moyle, J. (1969). *Eur. J. Biochem.* **8**, 450–454.
Schönherr, O. Th., and Keir, H. M. (1972). *Biochem. J.* **129**, 285–290.
Schwartz, R. M., and Dayhoff, M. O. (1978). *Science* **199**, 395–403.
Seitz, U., and Seitz, U. (1973). *Arch. Mikrobiol.* **90**, 213–222.
Shaffer, P. W., Lockau, W., and Wolk, C. P. (1978). *Arch. Microbiol.* **117**, 215–219.
Sherman, L. A., and Cunningham, J. (1977). *Plant Sci. Lett.* **8**, 319–326.
Shestakov, S. V., and Khyen, N. T. (1970). *Mol. Gen. Genet.* **107**, 372–375.
Simon, R. D. (1976). *Biochim. Biophys. Acta* **422**, 407–418.
Simon, R. D. (1977). *J. Bacteriol.* **129**, 1154–1155.
Simon, R. D., and Weathers, P. (1976). *Biochim. Biophys. Acta* **420**, 165–176.
Simpson, F. B., and Nielands, J. B. (1976). *J. Phycol.* **12**, 44–48.
Singer, R. A., and Doolittle, W. F. (1975). *Nature (London)* **253**, 650–651.
Singh, P. K. (1973). *Arch. Mikrobiol.* **92**, 59–62.
Singh, P. K. (1976). *Z. Allg. Mikrobiol.* **16**, 453–463.
Smith, A. J. (1973). *In* "The Biology of Blue-Green Algae" (N. G. Carr and B. A. Whitton, eds.), pp. 1–38. Blackwell, Oxford.
Smith, R. J. (1977). *FEMS Microbiol. Lett.* **1**, 129–132.
Smith, R. J., and Carr, N. G. (1977). *J. Gen. Microbiol.* **98**, 559–567.
Smith, R. V., and Foy, R. H. (1974). *Br. Phycol. J.* **9**, 239–245.
Stacey, G., Van Baalen, C., and Tabita, F. R. (1977a). *Arch. Microbiol.* **114**, 197–201.
Stacey, G., Tabita, F. R., and Van Baalen, C. (1977b). *J. Bacteriol.* **132**, 596–603.
Stanier, R. Y. (1977). *Carlsberg Res. Commun.* **42**, 77–98.

Stanier, R. Y., and Cohen-Bazire, G. (1977). *Annu. Rev. Microbiol.* **31**, 225–274.

Stanier, R. Y., Kunisawa, R., Mandel, M., and Cohen-Bazire, G. (1971). *Bacteriol. Rev.* **35**, 171–205.

Stevens, C. L. R., and Myers, J. (1976). *J. Phycol.* **12**, 99–105.

Stevens, C. L. R., Stevens, S. E., Jr., and Myers, J. (1975). *J. Bacteriol.* **124**, 247–251.

Stevens, S. E., Jr., and Van Baalen, C. (1973). *Plant Physiol.* **51**, 350–356.

Stevens, S. E., Jr., and Van Baalen, C. (1974). *Arch. Biochem. Biophys.* **161**, 146–152.

Stewart, W. D. P. (1973). *Annu. Rev. Microbiol.* **27**, 283–316.

Stewart, W. D. P. (1977). *Br. Phycol. J.* **12**, 89–115.

Stewart, W. D. P., and Lex, M. (1970). *Arch. Mikrobiol.* **73**, 250–260.

Stewart, W. D. P., and Rowell, P. (1975). *Biochem. Biophys. Res. Commun.* **65**, 846–856.

Stransky, H., and Hager, A. (1970). *Arch. Mikrobiol.* **72**, 84–96.

Szalay, A., Munsche, D., Wollgiehn, R., and Parthier, B. (1972). *Biochem. J.* **129**, 135–140.

Tabita, F. R., and McFadden, B. A. (1972). *Biochem. Biophys. Res. Commun.* **48**, 1153–1159.

Tabita, F. R., Stevens, S. E., Jr., and Gibson, J. L. (1976). *J. Bacteriol.* **125**, 531–539.

Taylor, B. F. (1973). *Arch. Mikrobiol.* **92**, 245–249.

Taylor, B. F., Lee, C. C., and Bunt, J. S. (1973). *Arch. Mikrobiol.* **88**, 205–212.

Tel-Or, E., and Stewart, W. D. P. (1976). *Biochim. Biophys. Acta* **423**, 189–195.

Tel-Or, E., and Stewart, W. D. P. (1977). *Proc. R. Soc. London Ser. B* **198**, 61–86.

Thibodeau, L., and Verly, W. G. (1976). *FEBS Lett.* **69**, 183–185.

Thomas, J., Meeks, J. C., Wolk, C. P., Shaffer, P. W., Austin, S. M., and Chien, W.-S. (1977). *J. Bacteriol.* **129**, 1545–1555.

Tyagi, V. V. S. (1974). *Ann. Bot. (London)* **38**, 1107–1111.

Utkilen, H. C., Heldal, M., and Knutsen, G. (1976). *Physiol. Plant.* **38**, 217–220.

Van Baalen, C. (1973). *In* "The Biology of Blue-Green Algae" (N. G. Carr and B. A. Whitton, eds.), pp. 201–213. Blackwell, Oxford.

Walsby, A. E. (1975). *Annu. Rev. Plant Physiol.* **26**, 427–439.

Wang, R. T., Stevens, C. L. R., and Myers, J. (1977). *Photochem. Photobiol.* **25**, 103–108.

Ward, B., and Myers, J. (1972). *Plant Physiol.* **50**, 547–550.

Weare, N. M., and Benemann, J. R. (1974). *J. Bacteriol.* **119**, 258–265.

Weathers, P. J., Jost, M., and Lamport, D. T. A. (1977). *Arch. Biochem. Biophys.* **178**, 226–244.

Weckesser, J., Katz, A., Drews, G., Mayer, H., and Fromme, I. (1974). *J. Bacteriol.* **120**, 672–678.

Weise, G., Drews, G., Jann, B., and Jann, K. (1970). *Arch. Mikrobiol.* **71**, 89–98.

Wilcox, M., Mitchison, G. J., and Smith, R. J. (1975a). *Arch. Microbiol.* **103**, 219–223.

Wilcox, M., Mitchison, G. J., and Smith, R. J. (1975b). *In* "Microbiology-1975" (D. Schlessinger, ed.), pp. 453–463. Am. Soc. Microbiol., Washington, D.C.

Williams, V. P., and Glazer, A. N. (1978). *J. Biol. Chem.* **253**, 202–211.

Williams, V. P., Freidenreich, P., and Glazer, A. N. (1974). *Biochem. Biophys. Res. Commun.* **59**, 462–466.

Winkenbach, F., and Wolk, C. P. (1973). *Plant Physiol.* **52**, 480–483.

Wolk, C. P. (1967). *Proc. Natl. Acad. Sci. U.S.A.* **57**, 1246–1251.

Wolk, C. P. (1973). *Bacteriol. Rev.* **37**, 32–101.

Wolk, C. P. (1975). *In* "Spores VI" (P. Gerhardt, R. N. Costilow, and H. L. Sadoff, eds.), pp. 85–96. Am. Soc. Microbiol., Washington, D.C.

Wolk, C. P., and Quine, M. P. (1975). *Dev. Biol.* **46**, 370–382.

Wolk, C. P., and Shaffer, P. W. (1976). *Arch. Microbiol.* **110**, 145–147.

Wolk, C. P., Thomas, J., Shaffer, P. W., Austin, S. M., and Galonsky, A. (1976). *J. Biol. Chem.* **251**, 5027–5034.

Wood, N. B., and Haselkorn, R. (1979). *In* "Limited Proteolysis in Microorganisms" (G. N. Cohen and H. Holzer, eds.), pp. 159–166. U.S. DHEW Publication No. (NIH) 79-1591, Bethesda, Maryland.

Wyatt, J. T., and Silvey, J. K. G. (1969). *Science* **165**, 908–909.

Index

Contents of Other Volumes

VOLUME 5—AMINO ACIDS AND DERIVATIVES

VOLUME 6—PROTEINS AND NUCLEIC ACIDS

VOLUME 7—SECONDARY PLANT PRODUCTS

VOLUME 8—PHOTOSYNTHESIS